T0222536

Communications in Computer and Information Science **831**

Commenced Publication in 2007
Founding and Former Series Editors:
Alfredo Cuzzocrea, Xiaoyong Du, Orhun Kara, Ting Liu, Dominik Ślęzak,
and Xiaokang Yang

Editorial Board

Simone Diniz Junqueira Barbosa
  *Pontifical Catholic University of Rio de Janeiro (PUC-Rio),
  Rio de Janeiro, Brazil*
Phoebe Chen
  La Trobe University, Melbourne, Australia
Joaquim Filipe
  Polytechnic Institute of Setúbal, Setúbal, Portugal
Igor Kotenko
  *St. Petersburg Institute for Informatics and Automation of the Russian
  Academy of Sciences, St. Petersburg, Russia*
Krishna M. Sivalingam
  Indian Institute of Technology Madras, Chennai, India
Takashi Washio
  Osaka University, Osaka, Japan
Junsong Yuan
  University at Buffalo, The State University of New York, Buffalo, USA
Lizhu Zhou
  Tsinghua University, Beijing, China

More information about this series at http://www.springer.com/series/7899

Guilherme A. Barreto · Ricardo Coelho (Eds.)

Fuzzy Information Processing

37th Conference of the North American Fuzzy
Information Processing Society, NAFIPS 2018
Fortaleza, Brazil, July 4–6, 2018
Proceedings

 Springer

Editors
Guilherme A. Barreto (iD)
Department of Teleinformatics Engineering
Federal University of Ceará
Fortaleza, Ceará
Brazil

Ricardo Coelho (iD)
Department of Statistics & Applied
  Mathematics
Federal University of Ceará
Fortaleza, Ceará
Brazil

ISSN 1865-0929   ISSN 1865-0937   (electronic)
Communications in Computer and Information Science
ISBN 978-3-319-95311-3   ISBN 978-3-319-95312-0   (eBook)
https://doi.org/10.1007/978-3-319-95312-0

Library of Congress Control Number: 2018947460

© Springer International Publishing AG, part of Springer Nature 2018
This work is subject to copyright. All rights are reserved by the Publisher, whether the whole or part of the material is concerned, specifically the rights of translation, reprinting, reuse of illustrations, recitation, broadcasting, reproduction on microfilms or in any other physical way, and transmission or information storage and retrieval, electronic adaptation, computer software, or by similar or dissimilar methodology now known or hereafter developed.
The use of general descriptive names, registered names, trademarks, service marks, etc. in this publication does not imply, even in the absence of a specific statement, that such names are exempt from the relevant protective laws and regulations and therefore free for general use.
The publisher, the authors and the editors are safe to assume that the advice and information in this book are believed to be true and accurate at the date of publication. Neither the publisher nor the authors or the editors give a warranty, express or implied, with respect to the material contained herein or for any errors or omissions that may have been made. The publisher remains neutral with regard to jurisdictional claims in published maps and institutional affiliations.

Printed on acid-free paper

This Springer imprint is published by the registered company Springer International Publishing AG
part of Springer Nature
The registered company address is: Gewerbestrasse 11, 6330 Cham, Switzerland

Preface

In 1965, Lofti Asker Zadeh published the seminal paper "Fuzzy Sets" (Information and Control, 8, 338–353), which describe the first ideas about a formal mathematical modelling intended to bridge the gap between classic binary modelling and the subjective way that humans relate to day-to-day situations. Despite these ideas being ambitious, this preliminary work inspired many researchers around the world and today his ideas are found in almost all branches of science. According to the website Google Scholar, this seminal paper has been cited in more than 100,000 scholarly works, and many consumer products and software have been built based on its mathematical concepts. Unfortunately, Professor Zadeh died in September 2017, and this book is a modest tribute to the generous, gentle continuous and always friendly support that the authors received over the years from Professor Lofti A. Zadeh.

It can be noted that the research field of fuzzy sets and systems has undergone tremendous growth since 1965. This growth is in no small measure the result of the emergence of some important scientific societies in North America (North American Fuzzy Information Processing Society – NAFIPS – and IEEE Computational Intelligence Society – IEEE CIS), Europe (European Society for Fuzzy Logic and Technology – EUSFLAT), Asia (Japan Society for Fuzzy Theory and Intelligent Informatics – JSFTII), and South America, especially in Brazil, (Brazilian Society of Automatics – SBA – and Brazilian Society of Computational and Applied Mathematics – SBMAC). There is also a transnational scientific society (International Fuzzy Systems Association – IFSA). These societies promote scientific events in order to spread the state of the art, its applications, and technological advances. A quick search in the Scopus database gives an idea of the number of published articles about fuzzy sets and systems. By dividing time from 1965 until today into four periods, we obtain the following: (a) 4,754 published papers until 1990; (b) 27,773 published papers from 1991 to 2000; (c) 93,012 from 2001 to 2010; and (d) 105,604 from 2011 to the current date (May 2018). This search was made by using the words "Fuzzy Sets" or "Fuzzy Systems" or "Fuzzy Logic" as title, abstract, or keywords.

Among the societies mentioned, NAFIPS is the premier fuzzy society in North America, which was founded in 1981. The purpose of NAFIPS is "the promotion of the scientific study of, the development of an educational institution for the instruction in, and the dissemination of educational materials in the public interest including, but not limited to, theories and applications of fuzzy sets through publications, lectures, scientific meetings, or otherwise." In this role, we understand the importance and necessity of developing a strong intellectual base and encouraging new and innovative applications. In addition, we recognize our leading role in promoting interactions and technology transfer to other national and international organizations so as to bring the benefits of this technology to North America and the world. The scientific event organized by the NAFIPS has been contributing for more than 30 editions to the growth of the number of articles published in the fuzzy sets and systems field. The first edition took place in the city of Logan, Utah, USA, in 1982, and it is held annually.

One of the objectives of NAFIPS is to expand the network of collaborators and enthusiasts of fuzzy thinking beyond the borders of North American countries. The 37th North American Fuzzy Information Processing Society Annual Conference (NAFIPS 2018) was held during July 4–6, 2018, in the beautiful city of Fortaleza, capital of the state of Ceará, located on the sunny northeast coast of Brazil. This event was held simultaneously with the 5th Brazilian Congress on Fuzzy Systems (CBSF 2018), bringing together researchers, engineers, and practitioners to share and present the latest achievements and innovations in the area of fuzzy information processing, to discuss thought-provoking developments and challenges, and to consider potential future directions. Bearing this in mind, the NAFIPS 2018 meeting was the first edition of the meeting to be organized outside the USA, Canada, and Mexico. NAFIPS 2018 had an international Program Committee including researchers from industry and academia worldwide.

The organization of NAFIPS 2018 and CBSF 2018 was the result of a joint action of the Brazilian Computational Intelligence Society (SBIC), the Brazilian Society of Computational and Applied Mathematics (SBMAC), the Federal University of Ceará (UFC), and the Brazilian funding agencies CAPES, process 88887.155510/2017-00, and CNPq, project 407666/2017-6, in addition to the executive boards of NAFIPS and CBSF.

This book is a collection of high-quality papers ranging over a large spectrum of topics, including theory and applications of fuzzy numbers and sets, fuzzy logic, fuzzy inference systems, fuzzy clustering, fuzzy pattern classification, neuro-fuzzy systems, fuzzy control systems, fuzzy modeling, fuzzy mathematical morphology, fuzzy dynamical systems, time series forecasting, and making decision under uncertainty.

We received 73 submissions from 11 countries, from which 54 papers were accepted. The authors were from Brazil, Chile, Colombia, Czech Republic, India, Iran, Mexico, Romania, Spain, Turkey, and the USA. Each submitted paper was reviewed by at least three independent referees. The acceptance/rejection decision used the following criteria: every paper with two positive reviews was accepted, and those with two negative reviews were rejected. Borderline papers, those with one positive and one negative review, were analyzed carefully by the conference chairs in order to evaluate the reasons given for acceptance or rejection. Our final decision on these submissions took into account mainly the potential of each paper to foster fruitful discussions and the future development of the research on the theory and applications of fuzzy sets and systems in Brazil and, for extension, in the whole of Latin America.

We are enormously grateful to all reviewers for their goodwill in cooperating for the success of the aforementioned events. We very much appreciate their willingness for hard work and prompt feedback, which certainly guaranteed the high quality of the technical program.

We wish NAFIPS a long life. And we wish a long life for the Brazilian community, who organizes CBSF, with which we share this mutual congress.

June 2018   Guilherme A. Barreto
                                                                    Ricardo Coelho

Organization

General Co-chairs

Guilherme Barreto   Federal University of Ceará, Brazil
Ricardo Coelho   Federal University of Ceará, Brazil

Organizing Committee

Fernando Gomide   University of Campinas, Brazil
Guilherme Barreto   Federal University of Ceará, Brazil
Laecio Carvalho de Barros University of Campinas, Brazil
Patricia Melin   Tijuana Institute of Technology, Mexico
Ricardo Coelho   Federal University of Ceará, Brazil
Weldon Lodwick   University of Colorado Denver, USA
Centro Acadêmico de   CAMI
  Matemática Industrial

Web Masters

Felipe Albuquerque   Federal University of Ceará, Brazil
Francisco Yuri Martins   Federal University of Ceará, Brazil

NAFIPS Officers

Patricia Melin (President)
Martine Ceberio (President-Elect)
Christian Servin (Treasurer)
Valerie Cross (Secretary)

NAFIPS Board of Directors

Ildar Batyrshin
Ricardo Coelho
Martine De Cock
Scott Dick
Juan Carlos Figueroa Garcia
Weldon A. Lodwick
Marek Reformat
Shahnaz Shahbazova
Mark Wierman
Dongrui Wu

Program Committee

Giovanni Acampora	University of Naples Federico II, Italy
Plamen Angelov	Lancaster University, UK
Krassimir Atanassov	Bulgarian Academy of Science, Bulgaria
Adrian I. Ban	University of Oradea, Romania
Guilherme Barreto	Universidade Federal do Ceará, Brazil
Laecio Carvalho de Barros	University of Campinas, Brazil
Ildar Batyrshin	Instituto Politécnico Nacional, Mexico
Fernando Bobillo	University of Zaragoza, Spain
Giovanni Bortolan	Institute of Neuroscience, IN-CNR, Italy
Tadeusz Burczynski	Institute of Fundamental Technological Research, Poland
João Paulo Carvalho	Instituto Superior Tecnico/INESC-ID, Portugal
Oscar Castillo	Tijuana Institute of Technology, Mexico
Martine Ceberio	University of Texas at El Paso, USA
Wojciech Cholewa	Silesian University of Technology, Poland
Ricardo Coelho	Universidade Federal do Ceará, Brazil
Lucian Coroianu	University of Oradea, Romania
Valerie Cross	Miami University, USA
Bernard de Baets	Ghent University, Belgium
Didier Dubois	Université Paul Sabatier, France
Robert Fuller	Óbuda University, Hungary
Takeshi Furuhashi	Nagoya University, Japan
Fernando Gomide	University of Campinas, Brazil
Wladyslaw Homenda	Warsaw University of Technology, Poland
Sungshin Kim	Pusan National University, South Korea
Peter Klement	Johannes Kepler University, Austria
Vladik Kreinovich	University of Texas at El Paso, USA
Jonathan Lee	National Central University, Taiwan
Weldon A. Lodwick	University of Colorado Denver, USA
Francesco Marcelloni	University of Pisa, Italy
Radko Mesiar	Slovak University of Technology, Slovakia
Vesa Niskanen	University of Helsinki, Finland
Fabrício Nogueira	Universidade Federal do Ceará, Brazil
Vilem Novak	University of Ostrava, Czech Republic
Reinaldo Martinez Palhares	Federal University of Minas Gerais, Brazil
Irina Perfilieva	University of Ostrava, Czech Republic
Henri Prade	Université Paul Sabatier, France
Radu Emil Precup	Politehnica University of Timisoara, Romania
Alireza Sadeghian	Ryerson University, Canada
Yabin Shao	Northwest University for Nationalities, China
Andrzej Skowron	University of Warsaw, Poland
Umberto Straccia	Consiglio Nazionale delle Ricerche, ISTI-CNR, Italy
Ricardo Tanscheit	Pontifícia Universidade Católica do Rio de Janeiro, Brazil

Jose Luis Verdegay	University of Granada, Spain
Yiyu Yao	University of Regina, Canada
Hao Ying	Wayne State University, USA
Fusheng Yu	Beijing Normal University, China
Slawomir Zadrozny	Polish Academy of Sciences, Poland
M. H. Fazel Zarandi	Amirkabir University of Technology, Iran
Guangquan Zhang	University of Technology Sydney, Australia
Hans J. Zimmermann	RWTH Aachen University, Germany

Additional Reviewers

João Fernando Alcântara

Aluizio Araújo

Rodrigo Araújo

Romis Attux

Iury Bessa

Arthur Braga

Luiz Cordovil

Pedro Coutinho

Alexandre Evsukoff

Carmelo J.A. Bastos Filho

Heriberto Román Flores

João Paulo Pordeus Gomes

Jose Manuel Soto Hidalgo

Daniel Leite

Adi Lin

José Everardo Bessa Maia

Sebastia Massanet

Ajalmar Rêgo da Rocha Neto

Rudini Sampaio

Peter Sussner

George Thé

Marcos Eduardo Valle

Bin Wang

Zhen Zhang

Contents

XIV   Contents

Formal Verification of a Fuzzy Rule-Based Classifier Using the Prototype Verification System

Solomon Gebreyohannes, Ali Karimoddini$^{(\boxtimes)}$, Abdollah Homaifar, and Albert Esterline

Department of Electrical and Computer Engineering,
North Carolina A&T State University, 1601 East Market Street,
Greensboro, NC 27411, USA
{shgebrey,akarimod,homaifar,esterlin}@ncat.edu

Abstract. This paper presents the formal specification and verification of a Type-1 (T1) Fuzzy Logic Rule-Based Classifier (FLRBC) using the Prototype Verification System (PVS). A rule-based system models a system as a set of rules, which are either collected from subject matter experts or extracted from data. Unlike many machine learning techniques, rule-based systems provide an insight into the decision making process. In this paper, we focus on a T1 FLRBC. We present the formal definition and verification of the T1 FLRBC procedure using PVS. This helps mathematically verify that the design intent is maintained in its implementation. A highly expressive language such as PVS, which is based on a strongly-typed higher-order logic, allows one to formally describe and mathematically prove that there is no contradiction or false assumption in the procedure. We show this by (1) providing the formal definition of the T1 FLRBC in PVS and then (2) formally proving or deducing rudimentary properties of the T1 FLRBC from the formal specification.

Keywords: Formal verification · Fuzzy rule-based classifier
Prototype verification system

1 Introduction

Unlike many machine learning techniques, which treat systems as "black boxes" and model them using input-output relationships, Rule-Based Systems (RBSs) model systems using rules that can provide an insight into their decision making processes. RBCs are effective tools for encoding a human expert's knowledge into an automated system [1]. They can be used for different applications such as prediction and control applications. Recently, they have been used for a classification purpose. A fuzzy logic based RBS that is used for classification is called a Fuzzy Logic Rule-Based Classifier (FLRBC). It translates the expert's heuristic knowledge into fuzzy "IF-THEN" statements (i.e., rules), which, along with an

© Springer International Publishing AG, part of Springer Nature 2018
G. A. Barreto and R. Coelho (Eds.): NAFIPS 2018, CCIS 831, pp. 1–12, 2018.
https://doi.org/10.1007/978-3-319-95312-0_1

appropriate inference engine, are used for system classification [2,3]. FLRBCs have been used to classify various kinds of items such as text, image [4], gesture [5], and video [2,3] and have been applied to several areas such as battlefield ground vehicles [6], sentiment analysis [7], and so on. FLRBCs are also combined with genetic algorithms [8], deep learning [9], decision trees, and other techniques to optimize their performance.

Despite the great capabilities of FLRBCs and their wide range of applications, a major challenge is that how can we be sure (i.e., mathematically verify) that the design intent of an FLRBC is maintained in its implementation? Regarding to design procedures, FLRBC is no exception to the general point that natural language lacks the formality needed for requirement verification, i.e., with natural language, one cannot show (and hence cannot ensure) consistency and completeness of the procedure. Hence, a formal way (from a mathematical perspective) of representing the FLRBC design procedure and verifying its properties is necessary.

Formal verification can be defined as a "systematic process based on mathematical reasoning in order to verify that the design intent (specifications) is maintained during implementation" [10]. There are different tools developed for conducting formal verification such as Prototype Verification System (PVS) [11], B [12], HOL [13], and Coq [14]. Using a higher-order theorem-proving system, such as PVS, it is possible to reach a much higher level of confidence compared to lighter formal methods. Formalizing specifications using PVS has been used in a range of applications. In [15], a formal specification and verification of the requirements for an airline reservation system using PVS is presented. As a more sophisticated use of PVS in industrial applications, [16] uses PVS for verification of two hardware examples, the pipelined microprocessor and n-bit ripple-carry adder. Butler [17] also uses PVS for formal capturing of requirements of an autopilot (related to an early Boeing-737 autopilot). In [18], PVS is used for analysis of a space shuttle software requirements. Despite the progress made on employing PVS for different verification applications, we are not aware of any PVS formalization of the FLRBC.

This paper, therefore, proposes to use formal verification techniques to systemically and formally verify a fuzzy RBC whether it is consistent with required specifications. We present FLRBC specifications and verify them using the PVS framework. Our focus in this paper is on Type-1 (T1) FLRBC; however, the approach can be extended to other fuzzy RBCs. We encode PVS theories for the T1 FLRBC main components (viz., fuzzifier, inference, and defuzzifier), keeping their generality. This means that the developed PVS theories are not dependent on any particular application. Therefore, one can call and instantiate them to be used for any other application that utilizes the FLRBC technique.

The rest of the paper is organized as follows. Section 2 discusses a fuzzy logic rule-based classification. Section 3 presents a brief introduction to PVS. Section 4 presents the formal definition and verification of the T1 FLRBC. PVS theories are developed and formal proofs are shown. Section 5 concludes this paper.

2 Fuzzy Logic Rule-Based Classification: An Informal Description

In this section, we present fuzzy sets and systems preliminaries, and will discuss fuzzy rule-based classification, mainly borrowed from [2,3,19].

2.1 Fuzzy Sets and Systems Preliminaries

This section introduces fuzzy sets and fuzzy logic systems. A *fuzzy set* is characterized by a membership function (MF), mapping the elements of a domain space or universe of discourse to the interval [0, 1]. A type-1 fuzzy set can be defined as follows:

Definition 1. *A type-1 fuzzy set A is a set function on universe X into $[0,1]$, i.e., $\mu_A : X \rightarrow [0,1]$.*

$$A = \{(x, \mu_A(x)) | x \in X, 0 \le \mu_A(x) \le 1\} \tag{1}$$

where the MF of A is denoted $\mu_A(x)$ and is called a type-1 MF.

A fuzzy system that operates on type-1 fuzzy sets (and crisp sets) is called a *type-1 fuzzy system.*

Definition 2. *A Type-1 fuzzy system contains four components* - rules, fuzzifier, inference engine, *and* defuzzifier - *that are interconnected as shown in Fig. 1. Once the rules have been established, the fuzzy system can be viewed as a mapping from p inputs $\boldsymbol{x} = \{x_1, \ldots, x_p\}$ to an output y, and the mapping can be expressed quantitatively as $y = f(\boldsymbol{x})$.*

Figure 1 shows a type-1 fuzzy logic system. Note that $\mathbf{x}'$ is a specific value of $\mathbf{x}$. The components of this fuzzy system are described below.

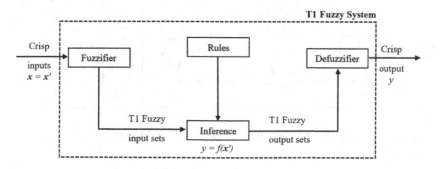

Fig. 1. A type-1 fuzzy system [2,3]

Rules are sets of IF-THEN statements that model the system and can have two commonly different structures.

1. Zadeh's l^{th} rule, $l = 1, \ldots, M$, has a form:

$$R_Z^l : \text{IF } x_1 \text{ is } F_1^l \text{ and } \ldots, x_p \text{ is } F_p^l, \text{THEN } y \text{ is } G^l \tag{2}$$

where F_i^l is the ith antecedent MF and G^l is the consequent MF of the lth rule.

2. Takagi, Sugeno, and Kang (TSK) l^{th} rule, $l = 1, \ldots, M$, has a form:

$$R_{TSK}^l : \text{IF } x_1 \text{ is } F_1^l \text{ and } \ldots, x_p \text{ is } F_p^l, \text{THEN } y \text{ is } g^l(x_1, \ldots, x_p) \tag{3}$$

where F_i^l is the ith antecedent MF and g^l is the function for the lth rule.

The *fuzzifier* maps a crisp input $\mathbf{x} = \{x_1, \ldots, x_p\}$ into a fuzzy set in X. There are two kinds of fuzzifiers: singleton and non-singleton. A singleton fuzzification maps a specific value x_i' into $\mu_{F_i^l}(x_i') \in [0, 1]$ while a non-singleton maps into a type-1 fuzzy number.

In the fuzzy *inference engine*, fuzzy logic principles are used to map fuzzy input sets in $X_1 \times \ldots \times X_p$, that flow through an IF-THEN rule (or a set of rules), into fuzzy output sets in Y. Each rule is interpreted as a fuzzy implication.

A *defuzzifier* maps a fuzzy output of the inference engine to a crisp output y.

2.2   Type-1 Fuzzy Logic Rule-Based Classifier

This section discusses a singleton T1 FLRBC. The classifier consists of five components, as shown in Fig. 2, by adding a comparator to a fuzzy system. The T1 FLRBC is used for a binary classification, i.e., it classifies its inputs as either Class 1 or Class 2.

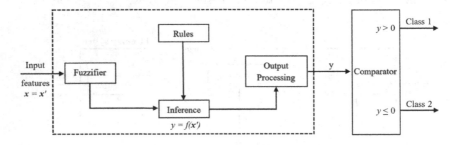

Fig. 2. A type-1 fuzzy logic rule-based classifier

Recall that $\mathbf{x}$ is a set of p input features, i.e., $\mathbf{x} = \{x_1, x_2, \ldots, x_p\}$, and y is an output of the fuzzy system.

The rules of an FLRBC are a special case of Zadeh's rule, in which the consequent is a singleton or TSK rule with a constant function [3]. They are characterized by MFs. For the lth Mamdani rule [20], $l = 1, \ldots, M$, we have R^l : $A^l \to G^l$, which can be represented by $\mu_{A^l \to G^l}(\mathbf{x}, y)$, where $A^l = F_1^l \times \ldots \times F_p^l$.

For the consequent, a crisp value $+1$ is used for Class 1 and -1 is used for Class 2.

$$y^l = \begin{cases} 1 & \text{Class 1} \\ -1 & \text{Class 2} \end{cases} \tag{4}$$

Correspondingly, for the consequent sets, G^l, the MFs can be defined as

$$\mu_{G^l}(y) = \begin{cases} 1 & y = y^l \\ 0 & \text{otherwise} \end{cases} \tag{5}$$

where y^l could be either $+1$ for Class 1 or -1 for Class 2.

The fuzzy inference engine for this rule-base classifier is shown in Fig. 3.

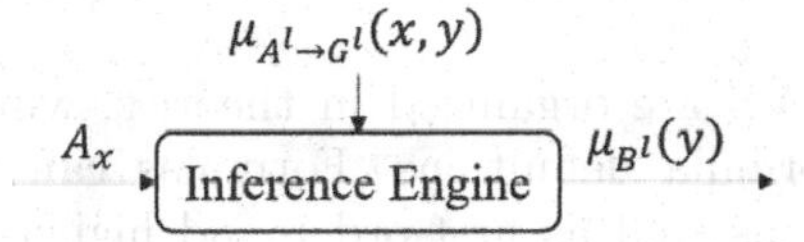

$$\mu_{A^l \to G^l}(x, y)$$

$$A_x \longrightarrow \boxed{\text{Inference Engine}} \xrightarrow{\mu_{B^l}(y)}$$

Fig. 3. Fuzzy inference engine ([3], p. 108)

The membership function of each fired rule for singleton input can be calculated using a t-norm as:

$$\mu_{B^l}(y) = \begin{cases} T_{i=1}^p \mu_{F_i^l}(x_i) = f^l(\mathbf{x}), & y = y^l \\ 0, & y \neq y^l \end{cases} \tag{6}$$

where $\mu_{F_i^l}(x_i)$, $i = 1, \ldots, p$, represents a singleton fuzzification and T is a t-norm operation.

Using height defuzzification, the output can be calculated as:

$$y_{RBC}(\mathbf{x}) = \frac{\sum_{l=1}^M f^l(\mathbf{x}) y^l}{\sum_{i=1}^M f^l(\mathbf{x})}, \quad y^l = \pm 1 \tag{7}$$

A classification then can be performed as:

$$\begin{aligned} &\text{If} \quad y_{RBC}(\mathbf{x}) > 0, \quad \text{classify } \mathbf{x} \text{ as Class 1} \\ &\text{If} \quad y_{RBC}(\mathbf{x}) \leq 0, \quad \text{classify } \mathbf{x} \text{ as Class 2} \end{aligned} \tag{8}$$

3 Prototype Verification System

PVS [11] is a formal verification environment which provides a highly expressive specification language based on a strongly-typed higher-order logic.

The type system of PVS supports the use of both interpreted and uninterpreted types. Uninterpreted types support abstraction (using *type*) with a minimum of assumptions on the type; e.g., `var1: TYPE`. Interpreted types, on the other hand, detail the type. For example, `var2: TYPE = nat` declaration introduces the type name `var2` (interpreted) as a natural number. The PVS prelude [21] provides definitions of a comprehensive collection of basic interpreted types, such as booleans, natural numbers, integers, reals, etc. The type system can be easily extended by defining new types using well-known type operators, such as functional, tuple and record combinators. It is also possible to define enumerated types and predicate subtypes as well as abstract data types, such as lists, stacks, binary trees, etc. Details on the PVS language may be found in the *PVS Language Reference* [22].

Since the PVS language is so expressive, the type checking process is not decidable. For that reason, the type checker usually generates additional proof obligations, which must be proved by the user in order to verify the type consistency of the specification. Such obligations are called Type-Correctness Conditions (TCC).

Formalizations in PVS are organized in theories, which include type, constant, variable, and formula definitions. Formulas can be constructed using propositional operators as well as first-order and higher-order quantifications. Every formula definition can be stated as an axiom or a theorem. Axioms are assumed to be valid in the specification, but formal proofs must be provided by the user for theorems. Specifications for many foundational and standard theories are preloaded into PVS as *prelude* theories.

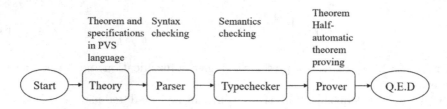

Fig. 4. Theorem proving in PVS

The theorem prover implemented in PVS is based on a formalism called *sequent calculus*. A proof in such a calculus can be seen as a tree, where every node is a *sequent* composed of two collections of formulas, called *antecedents* and *consequents*, respectively.

$$\left| \frac{\text{antecedents}}{\text{consequents}} \right. \tag{9}$$

The intuition behind the notion of a sequent is that it represents the logical consequence between the conjunction of its antecedents and the disjunction of its consequents. In order to prove that a formula, for example, $\alpha \rightarrow \beta$, is valid, the user must start with the sequent $\alpha \vdash \beta$ (the operator '$\vdash$' denotes a sequent) and try to reach trivially valid sequents by applying predefined proof rules to the leaves of the proof tree. Although proofs in PVS are constructed interactively, it does provide a considerable degree of automatization for a wide spectrum of cases. Details about proofs can be found in the PVS prover guide [23]. The steps in theorem proving using PVS as are shown in Fig. 4. There are also PVS tutorials and applications developed in [15, 16].

4   Formal Description and Verification of Fuzzy Rule-Based Classification

This section presents the formal representation and verification of a singleton T1 FLRBC using PVS. Table 1 summarizes the semantic mapping of T1 FLRBC components to PVS.

Table 1. Fuzzy RBC constructs to PVS mapping

Fuzzy RBC	PVS
Input feature	Uninterpreted Type
Output	Enumerated type
Input, antecedent	Finite sequence
μ, consequent	Subtypes
Membership function, rule	Record type
Inference, defuzzification	RECURSIVE function

To formally describe a fuzzy RBC, we start by defining basic objects in the following sections.

4.1   Basic TYPE Definition

The first step in the design of a T1 FLRBC system is selecting features that act as antecedents. They are usually represented using their MFs. The antecedent features are encoded as PVS (uninterpreted) TYPEs and their MFs as RECORD TYPEs defined in the `fbasic_defs` PVS theory. An MF consists of a range of values and a function that maps a value (within the range) into $[0, 1]$.

```
fbasic_defs: THEORY
BEGIN
   Feature:     TYPE
   m:           TYPE = nat
   mu:          TYPE = {u:real| 0 <= u AND u <= 1}
   MF:          TYPE = [# range: [real,real],
                          f:   [{x:real|range'1<=x
```

```
                                             AND x<=range'2}-> mu]  #]
Feat_MF:        TYPE = [# feat: Feature,
                         mf:    MF #]
```

A rule (R) maps a set of antecedents $(A = F_1 \times F_2 \ldots \times F_p)$ to a consequent (G), i.e., $R : A \to G$; it is usually described by the MF $\mu_R(X, y) = \mu_{A \to G}(X, y)$. In an RBC, the consequent is a singleton where 'Class 1' is represented by $+1$ and 'Class 2' is represented by -1. Hence, a consequence is represented as a subtype in PVS. Each rule has a specifier $l <= M$, where M is the total number of rules. An input X has p features, $x_1, x_2, x_3, \ldots, x_p$. We represent the input as a finite sequence of real values. A fuzzy rule is encoded using PVS RECORD with accessors rule number, antecedents, and consequents.

```
Input:       TYPE = finseq[real]
Output:      TYPE = {class1, class2}
Antecedents: TYPE = finseq[Feat_MF]
Consequent:  TYPE = {n: nat| n = -1 OR n = 1}
Rule:        TYPE = [# rulen: l,
                        antcs: Antecedents,
                        consq: Consequent #]
Rules:       TYPE = finseq[Rule]
END fbasic_defs
```

4.2   Describing Fuzzy RBC Using PVS

This section formally defines the T1 FLRBC using PVS. We put the definition in a `fls` theory. We start by importing `fbasic_defs` theory defined in Sect. 4.1 and declaring input and rule variables. Let `X`,`r`, and `rs` be input, a rule, and a set of rules, respectively. There are p inputs and M rules.

```
fls: THEORY
BEGIN
  IMPORTING fbasic_defs
  p,M: VAR nat
  X:    VAR Input
  r:    VAR Rule
  rs:   VAR Rules
```

The number of antecedents in a rule r should be the same as the length of the input vector x. This is encoded as a PVS AXIOM.

```
inpt_length: AXIOM FORALL (X,r): X'length = r'antcs'length
```

Now, we can represent the inference engine. Based on Eq. 6, the inference engine can be represented as a PVS RECURSIVE function as follows:

```
fl(X,r,p): RECURSIVE mu =
    IF p=0 THEN 0
    ELSE r'antcs'seq(p)'mf'f(X'seq(p))*fl(X,r,p-1)
    ENDIF
Measure p
```

The defuzzified output of the RBC can be obtained using Eq. 7. The numerator of Eq. 7 is encoded using a PVS RECURSIVE function as:

```
sumfy(X,rs,M): RECURSIVE mu =
    IF M=0 THEN 0
    ELSE
        fl(X,rs'seq(M),X'length)*rs'seq(M)'consq + sumfy(X,rs,M-1)
    ENDIF
Measure M
```

Similarly, the denominator of Eq. 7 is encoded using a PVS as:

```
sumf(X,rs,M): RECURSIVE mu =
    IF M=0 THEN 0
    ELSE fl(X,rs'seq(M),X'length) + sumf(X,rs,M-1)
    ENDIF
Measure M
```

Therefore, Eq. 7 can now be represented using PVS as:

```
yRBC(X,rs): real = sumfy(X,rs,rs'length)/sumf(X,rs,rs'length)
```

The final decision of the classifier is based on the sign of the defuzzified output as shown in Eq. 8. This is encoded using a PVS IF-ELSE statement.

```
decision(X,rs): Output =
    IF (yRBC(X,rs) > 0) THEN class 1
    ELSE class 2
    ENDIF
END fls
```

This completes the formal definition of the T1 FLRBC process.

4.3   Formal Verification

This section presents the formal verification - well-formedness (to verify our specification is correct, i.e., free of contradiction or false assumption) and requirement verification (to verify a given requirement or property can be deduced from the specification).

Well-Formedness. PVS requires theorem proving in order to guarantee that the specification is type correct [15]. Type checking of the `fbasic_defs` theory generates no TCCs, but the `fls` theory does generate TCCs. For example, one TCC generated is

```
yRBC_TCC1: OBLIGATION
    FORALL (X:Input, rs: Rules): sumf(X, rs, rs'length) /= 0
```

The denominator of Eq. 7 needs to be non-zero to avoid division by zero. This was not contained in our definition. Therefore, we add a PVS AXIOM for this.

```
nz_tnorm: AXIOM FORALL (X,rs): NOT (sumf(X,rs,rs'length) = 0)
```

Any TCC should be discharged if we are to guarantee the correctness (no contradiction nor false assumptions) of the specification. We revise the theory for no TCC. We successfully discharged some of the TCCs using the axioms. The rest of the TCCs are proved automatically by a PVS standard strategy (`tcc`).

Once all the TCCs are discharged, the theories are well-formed; i.e., there is no contradiction (nor false assumption) in the declaration. However, we do not know yet whether it satisfies any given properties (or requirements). We show this in Sect. 4.3.

Verification of Properties. Now, since the T1 FLRBC is formally represented, one can mathematically verify properties of the T1 FLRBC by encoding them as PVS THEOREMs and proving them interactively using appropriate prover commands. In this section, we prove some rudimentary properties.

Lemma 1. *The output of the T1 FLRBC should be either Class 1 or Class 2. We define two PVS predicates: first we check independently that if a given output is Class 1 or Class 2.*

```
class1?(X,rs): bool = decision(X,rs) = class1
class2?(X,rs): bool = decision(X,rs) = class2
```

Then, the theorem will be the XOR of the two predicates.

```
bin_class: THEOREM FORALL(X,rs):xor(class1?(X,rs),class2?(X,rs))
```

This theorem is discharged automatically by a prover command *grind*. Q.E.D.

Let us check the dependence of the output of the classifier on the rules.

Lemma 2. *(a) If the output of the classifier is Class 1, then there exists a rule with a consequent of 1.*

```
class1_det: THEOREM
    decision(X,rs) = class1 IMPLIES
        EXISTS r:member(rs,r) AND r'consq = 1
```

(b) If the output of the classifier is Class 2, then there exists a rule with a consequent of −1.

```
class2_det: THEOREM
  decision(X,rs) = class2 IMPLIES
    EXISTS r:member(rs,r) AND r'consq = -1
```

This is proved again using the prover command *grind*. Q.E.D.

Finally, let us examine the defuzzified output of the fuzzy system from Eq. 7.

Lemma 3. *The numerator in Eq. 7 is less than or equal to the denominator. The two values are equal only if $y^l = 1$, for $l = 1, \ldots, M$. Therefore, the following theorem needs to be proved from our formal definition.*

```
yrbc_def: THEOREM
  FORALL (X,rs): sumfy(X,rs,rs'length) <= sumf(X,rs,rs'length)
```

It too is proved using the prover command *grind*. Q.E.D.

Similarly, more properties can be formally represented and then deduced from the theories. Also, if a particular example uses the FLRBC technique (and so its formalization will be represented by instantiating the PVS theories developed in this paper), then application-dependent properties can be proved.

5 Conclusion

This paper presented the formalization of a T1 FLRBC using PVS. We developed PVS theories to formally represent the FLRBC procedure. We also verified some rudimentary properties of the FLRBC by discharging PVS type-checking requirements and proving theorems. For the future, we aim to formalize a specific application that utilizes the FLRBC technique by instantiating the developed PVS theories.

Acknowledgment. This research is supported by Air Force Research Laboratory and Office of the Secretary of Defense under agreement number FA8750-15-2-0116 as well as US ARMY Research Office under agreement number W911NF-16-1-0489.

References

1. Grosan, C., Abraham, A.: Rule-based expert systems. In: Grosan, C., Abraham, A. (eds.) Intelligent Systems. ISRL, vol. 17, pp. 149–185. Springer, Heidelberg (2011). https://doi.org/10.1007/978-3-642-21004-4_7
2. Mendel, J.M.: Uncertain Rule-Based Fuzzy Logic Systems: Introduction and New Directions. Prentice Hall PTR, Upper Saddle River (2001)
3. Mendel, J.M.: Uncertain Rule-Based Fuzzy Logic Systems: Introduction and New Directions. Springer, Cham (2017). https://doi.org/10.1007/978-3-319-51370-6
4. Jaworska, T.: Application of fuzzy rule-based classifier to CBIR in comparison with other classifiers. In: 2014 11th International Conference on Fuzzy Systems and Knowledge Discovery (FSKD), pp. 119–124. IEEE (2014)
5. Beke, A., Yuceler, A.A., Kumbasar, T.: A rule based fuzzy gesture recognition system to interact with Sphero 2.0 using a smart phone. In: 2017 International Artificial Intelligence and Data Processing Symposium (IDAP), pp. 1–4. IEEE (2017)

6. Wu, H., Mendel, J.M.: Classification of battlefield ground vehicles using acoustic features and fuzzy logic rule-based classifiers. IEEE Trans. Fuzzy Syst. **15**(1), 56–72 (2007)
7. Jefferson, C., Liu, H., Cocea, M.: Fuzzy approach for sentiment analysis. In: 2017 IEEE International Conference on Fuzzy Systems (FUZZ-IEEE), pp. 1–6. IEEE (2017)
8. Stavrakoudis, D.G., Theocharis, J.B.: A genetic fuzzy rule-based classifier for land cover image classification. In: 2009 IEEE International Conference on Fuzzy Systems, FUZZ-IEEE 2009, pp. 1677–1682. IEEE (2009)
9. Angelov, P.P., Gu, X.: A cascade of deep learning fuzzy rule-based image classifier and SVM. IEEE (2017)
10. Nawaz, M.S., IkramUllah Lali, M., Pasha, M.A.: Formal verification of crossover operator in genetic algorithms using prototype verification system (PVS). In: 2013 IEEE 9th International Conference on Emerging Technologies (ICET), pp. 1–6. IEEE (2013)
11. Owre, S., Rushby, J.M., Shankar, N.: PVS: a prototype verification system. In: Kapur, D. (ed.) CADE 1992. LNCS, vol. 607, pp. 748–752. Springer, Heidelberg (1992). https://doi.org/10.1007/3-540-55602-8_217
12. The B method. http://www.methode-b.com/en/
13. Gordon, M.J.C., Melham, T.F.: Introduction to HOL a theorem proving environment for higher order logic (1993)
14. Coq Development Team: The Coq proof assistant (2012). https://coq.inria.fr/distrib/current/refman/
15. Butler, R.: An elementary tutorial on formal specification and verification using PVS. NASA Technical Memorandum 108991, September 1993
16. Crow, J., Owre, S., Rushby, J., Shankar, N., Srivas, M.: A tutorial introduction to PVS. In: Workshop on Industrial-Strength Formal Specification Techniques, Boca Raton, Florida, April, Updated June 1995
17. Butler, R.: An introduction to requirements capture using PVS: specification of a simple autopilot. IEEE Trans. Softw. Eng. **24** (1996). NASA Technical Memorandum 110255
18. Crow, J., Di Vito, B.L.: Formalizing space shuttle software requirements. To be Presented at the ACM SIGSOFT Workshop on Formal Methods in Software Practice, San Diego, CA, January 1996
19. Enyinna, N., Karimoddini, A., Opoku, D., Homaifar, A., Arnold, S.: Developing an interval type-2 TSK fuzzy logic controller. In: 2015 Annual Conference of the North American Fuzzy Information Processing Society (NAFIPS) Held Jointly with 2015 5th World Conference on Soft Computing (WConSC), pp. 1–6, August 2015
20. Mamdani, E.H., Assilian, S.: An experiment in linguistic synthesis with a fuzzy logic controller. Int. J. Man Mach. Stud. **7**(1), 1–13 (1975)
21. Owre, S., Shankar, N.: PVS prelude library. CSL Technical Report SRI-CSL-03-01, SRI International, Computer Science Laboratory, March 2003
22. Owre, S., Shankar, N., Rushby, J., Stringer-Calvert, D.: PVS language reference. SRI International, Computer Science Laboratory, version 2.4 (2001)
23. Owre, S., Shankar, N., Rushby, J., Stringer-Calvert, D.: PVS prover guide. SRI International, Computer Science Laboratory, version 2.4 (2001)

Regularized Fuzzy Neural Network Based on Or Neuron for Time Series Forecasting

Paulo Vitor de Campos Souza[1,2](✉) [iD]
and Luiz Carlos Bambirra Torres[3](✉) [iD]

[1] Secretariat of Information Governance-CEFET-MG, Avenue Amazonas 5253,
Belo Horizonte, Minas Gerais 30.421-169, Brazil
goldenpaul@informatica.esp.ufmg.br
pauloc@prof.unibh.br
[2] Institute of Communication and Design-UNI-BH, Avenue Prof. Mario
Werneck 1685, Belo Horizonte, Minas Gerais 30.575-180, Brazil
[3] Post Graduate Program in Electrical Engineering- UFMG, Av. Antônio Carlos
6627, Belo Horizonte, Minas Gerais 31270-901, Brazil
luizlitc@gmail.com

Abstract. This paper presents a training algorithm for regularized fuzzy neural networks which is able to generate consistent and accurate models while adding some level of interpretation to applied problems to act in the prediction of time series. Learning is achieved through extreme learning machines to estimate the parameters and a technique of selection of characteristics using regularization concept and resampling, which is able to perform the definition of the network topology through the selection of subsets of fuzzy neuron more significant to the problem. Numerical experiments are presented for time series problems using benchmark bases on machine learning problems. The results obtained are compared to other techniques of prediction of reference series in the literature. The model made rough estimates of the responses obtained by the models of fuzzy neural networks for time series forecasting with fewer fuzzy rules.

Keywords: Fuzzy neural networks · Regularization · Bolasso
Fuzzy logic neurons · Time series forecasting

1 Introduction

Intelligent models that are composed of artificial neural networks and concepts of fuzzy systems have great utility for classification, regression or prediction of time series. The fuzzy neural networks use the structure of an artificial neural network, where classical artificial neurons are replaced by fuzzy neurons [16]. It has as its relevance factor its transparency, allowing the use of information a priori to define the initial structure of the network and the extraction of relevant information from the resulting topology. Thus, the neural network is seen as a linguistic system with level of interpretation, preserving the learning capacity of RNA. Its fuzzy neurons are composed of triangular norms, which generalize the union and intersection operations of classical clusters to the theory of fuzzy sets.

© Springer International Publishing AG, part of Springer Nature 2018
G. A. Barreto and R. Coelho (Eds.): NAFIPS 2018, CCIS 831, pp. 13–23, 2018.
https://doi.org/10.1007/978-3-319-95312-0_2

Examples of fuzzy neurons include neurons, *and, or, nullneurons* and *unineuron* where their differences are based on the way in which inputs and weights are aggregated [8]. The use of extreme learning machine theory (ELM) [9] has been used for the training of fuzzy neural networks as in [3, 14, 18, 20]. This motivation is mainly due to the performance and low computational cost presented by algorithms based on ELMs, however, most of these cases were used in pattern classification techniques. Already [12, 17, 21, 22] uses fuzzy neural networks to work with time series prediction.

This paper proposes to use the learning methodology for fuzzy neural networks of the feed forward type with a layer of fuzzification, a hidden layer composed of fuzzy neurons, and a linear output layer. The algorithm is able to generate consistent and accurate models by aggregating interpretation to the resulting structure. The learning of the model is carried out through extreme learning machine concepts, but a regularization term is added in the cost function which, together with a resampling technique, is able to perform the selection of the best neurons in internal layers, generating parsimonious models. Initially, fuzzy sets with equally spaced membership functions are defined for each input variable in the fuzzification layer. Subsequently an initial set of fuzzy candidate neurons is generated. From this initial set of neurons, the bootstrap lasso algorithm [2], is used to define the network topology, selecting a subset of the significant fuzzy neurons. Finally, the least squares algorithm is used to estimate the weights of the network output layer. This technique was used to classify binary patterns using *unineuron* [18] and *andneuron* [20], but due to the characteristics present in the ELM, the model's ability to predict time series was verified. Through all the steps the fuzzy neural network is able to act effectively in the forecast of time series. In this paper, we performed tests of time series prediction in the Box and Jenkins gas furnace [5] and the proposed model was compared to other algorithms of fuzzy neural networks for the same purpose widely used in the literature.

The remainder of the paper is organized as follows. Section 2 presents the theoretical concepts related to fuzzy neural networks and neural logic neurons. Section 3 describes the methodology used to train fuzzy neural networks. In Sect. 4 results of numerical experiments are presented. Finally, Sect. 5 presents the conclusions.

2 Fuzzy Neural Networks

2.1 Artificial Neural Networks and Fuzzy Systems

In [23] the authors defines *an* artificial neural network composed of an input layer, one or more hidden layers and an output layer. The network can be completely connected where each neuron is connected to all the neurons of the next layer, partially connected where each neuron is, or locally connected where there is a partial connection oriented to each type of functionality. To perform the training of a neural network, a set of data is required that contains patterns for training and desired outputs. In this way, the problem of neural network training is summarized in an optimization problem in which we want to find the best set of weights that minimizes the mean square error calculated between the network outputs and the desired outputs. The fuzzy systems are based on fuzzy logic, developed by [24]. His work was motivated because of the wide variety of vague and uncertain information in making human decisions. Some problems can be not solved with classical Boolean logic. In some situations only two values are insufficient to solve a problem.

2.2 Neural Logic Neurons

Among the several studies performed to simulate the behavior of the human neuron, we highlight those who sought to add fuzzy nature to the artificial neuron model, adding the ability to treat inaccurate information. This neuron is called fuzzy neuron [8]. This paper deals with a class of neurons called fuzzy logic neurons [16]. These neurons perform a mapping in the space formed by the Cartesian product between the input space and the space of the weights in the unit interval, i.e. $\mathbf{X} \times \mathbf{W} \to [0,\ 1]$ [8]. Examples of such neurons are neurons *and* and *or* [16] and unineuron [16, 14].

The logical *or* neuron uses a *t-norm* in the weighting operation and an *s-norm* in the final aggregation [8]. Given an input vector $x = [x_1, x_2, \ldots x_n]$ and a vector of weights of neuron $\mathbf{w} = [w_1, w_2, \ldots w_n]$ for $a_i \in [0, 1]$ and $w_i \in [0, 1]$ for i of $1, \ldots, n$. The output of logical neuron *or* is described as [16]:

$$z = OR(x, w) = S_{i=1}^n (x_i t w_i) \tag{1}$$

where t is *t-norms* and S is s-norms. Figure 1 presents the structure of an OR-type neuron

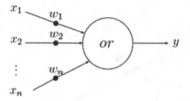

Fig. 1. *Orneuron* architecture

2.3 Fuzzy Neural Networks

The fuzzy neural networks have several characteristics from which we can distinguish the models through their basic properties, such as the way the network is connected, the type of fuzzy neurons used, the type of learning (training) and the way the inputs are handled in the first layers of the models. In this models, each layer is responsible for a specific function or task. Usually the first layer is responsible for handling the inputs and the last to bring the network response. Between these two layers there are other intermediate, which can be hidden or not. Depending on the model and what it is proposed, each layer has a specific function. Evaluating the type of training for fuzzy neural networks we can highlight that these algorithms are a set of well-defined rules for solving learning problems. These training methodologies seek to simulate human learning by learning or updating their new concepts, mainly by updating network factors, such as synaptic weights.

As the use of the extreme learning machine can be verified as a faster and more efficient alternative to adjust the parameters of a fuzzy neural network, it was proposed ways of modifying methodologies that act on these models, either the way of updating the parameters or the way of granulize the input in the model. In [4] a new methodologies to train fuzzy neural networks based also on extreme learning machine concepts, creating a model which they called XUninet. To train their fuzzy network, [17] used the extreme learning machine where weights of the hidden layer of a neural network are randomly chosen. To find the weights of the output layer we use the technique of recursive weighted least squares. For their algorithm, they defined the weights in the hidden layer and the identity elements of the uninorm [19] between zero and one. These values are updated recursively in training. Finally, [20] uses the ELM and to train the parameters of its network after the regularization method select the most representative neurons to the problem. In this context pattern classification techniques are used.

3   Fuzzy Neural Networks for Time Series Forecasting

3.1   Fuzzy Neural Networks Architecture

The fuzzy neural networks discussed below use the two types of neural logic neurons described in the previous subsection. The logical neurons that make up the network are described in (1). The structure of the network is illustrated in Fig. 2 in which the z-neurons are orneurons.

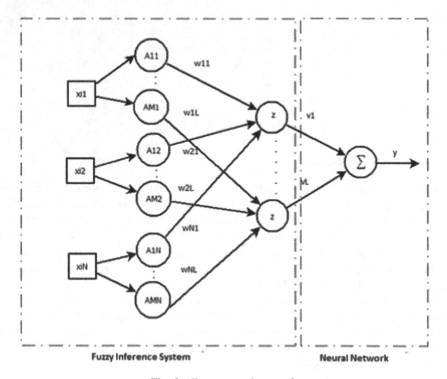

Fig. 2. Fuzzy neural network.

The architecture of the model used in this article presents its structure in [20], however changes are necessary so that the network can act as a model capable of predicting a time series. The first layer is defined as a fuzzification layer and is composed of neurons whose activation functions are membership functions of the respective fuzzy sets used in the partition of the input variables. For each input variable x_{ij} are defined M fuzzy sets A_j^m, for m varying from $1 \dots M$. The outputs of the first layer are the degrees of membership associated with the input values, that is, $a_{jm} = \mu_{A_j^m}$ for $j = 1, .., N$ at $= 1, \dots, M$, where N is the number of inputs and M is the number of fuzzy sets for each input variable. The second layer is composed of L fuzzy neurons of the *orneuron* type. Each neuron performs a weighted aggregation of some outputs of the first layer. The fuzzy logic neurons perform the aggregation using the w_{il} weights (for $i = 1, \dots, N$, and $l = 1, \dots, L$). The strategy of creation of the fuzzy neurons uses the grid partition defined by the ANFIS model [10]. Finally, the output layer is composed of a single linear neuron. In [15, 20] used the output of the neuron adapted for pattern recognition, transforming their final responses into -1 or 1. For time series problems, we consider the following neuron:

$$y = \left(\sum_{j=0}^{L} z_j v_j \right) \tag{2}$$

where $z_0 = 1$, v_0 is bias, and z_j and $v_j, j = 1, \dots, l$ are the output of each fuzzy neuron of the second layer and their corresponding weight, respectively. Fuzzy rules can be extracted from the network topology. To see how the fuzzy rules are generated, see [15, 20].

The ELM [9] is a learning algorithm developed for hidden layer feedforward neural networks (SLNFs) where random values are assigned to the weights of the first layer and the weights of the output layer are estimated analytically. In [15, 20] defined a training model for the fuzzy neural network where the parameters of the neurons are randomly assigned and the output parameters are calculated through least squares. The difference between them is the approach in creating the fuzzy rules of the first layer performed directly the amount of input data [20] and using the grid to divide the input space [15], in addition to [20] the neuron used to be the andneuron and in [20] there is also the use of unineuron [15].

This paper will use the same partitioning technique proposed in [15] where it will use equally spaced membership functions for each input variable to define the fuzzification layer neurons and the use of a smoothing technique to define the topology of the hidden layer. The model is able to generate parsimonious models, selecting more relevant neurons within the context of the problem. From the resulting model it is possible to extract a set of fuzzy rules.

The learning algorithm initially defines the neurons of the first layer through the partition of each interval of each input variable into M fuzzy sets with equally spaced Gaussian membership functions with its center at 0.5. Then, a strategy of partitioning the input space by a grid [10] is used to define an initial set of candidate neurons. The initial number of neurons in the hidden layer is defined as M^N, that is, for each possible combination of the membership functions of each input, a neuron is generated and its inputs are defined. The weights associated to the neuron inputs are randomly defined in the interval [0, 1], similarly to the ELMs. This approach to defining the network topology facilitates the interpretability of the extracted rules. For example, if three fuzzy sets are used per input variable ($M = 3$), each fuzzy set can be interpreted as "Small", "Medium" and "Large". Figure 3 presents the Gaussian relevance functions proposed for problem solving.

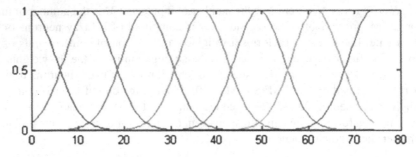

Fig. 3. Gaussian membership functions.

The final architecture of the network is defined by a feature extraction technique based on $L1$ regularization and resampling, called Bolasso [2]. The learning algorithm assumes that the output hidden layer composed of the candidate neurons can be written as:

$$f(x_i) = \sum_{l=0}^{L} v_l z_l(x_i) = z(x_i)v \tag{3}$$

where $v = [v_0, v_1, v_2, \ldots, v_L]$ is the weight vector of the output layer and $\mathbf{z}(x_i) = [z_0, z_1(x_i), z_2(x_i)]$ the output vector of the second layer, for $z_0 = 1$. In this context, $\mathbf{z}(x_i)$ is considered as the non-linear mapping of the input space for a space of fuzzy characteristics of dimension L. Since the weights connecting the first two layers are randomly assigned, the only parameters to be estimated are the weights of the output layer. Thus, the problem of network parameter estimation can be seen as a simple linear regression problem, allowing the use of regression techniques [7] for estimating parameters and selecting candidate neurons. The regression algorithm used by the model proposed by [15] for high-dimensional data estimating the regression coefficients and the subset of candidate regressors to be included in the final model is the LARS [6]. When we evaluate a set of K distinct samples (x_i, y_i), where $x_i = [x_{i1}, x_{i2}, \ldots, x_{iN}] \varepsilon \mathbb{R}$ and $y_i \varepsilon \mathbb{R}$ for $i = 1, \ldots, K$, the cost function of this regression algorithm can be defined as:

$$\sum_{i=1}^{K} \|z(x_i)v - y_i\|_2 + \lambda\|v\|_1 \tag{4}$$

where λ is a regularization parameter of L_1 norm, commonly estimated via cross-validation [15].

The LARS algorithm is used to perform the model selection, since, for a given value of λ only a fraction (or none) of the regressors have corresponding weights other than zero. For the problem considered in this work, the regressors z_{ls} are the outputs of the significant neurons. Thus, the LARS algorithm can be used to select an optimal subset of the significant neurons (L_s) that minimize (4) for a given value of λ. The approach used to increase the stability of the model selection algorithm is the use of resampling. This procedure developed by [2] is defined as a bootstrap-enhanced least absolute shrinkage operator where LARS algorithm runs on several bootstrap replications of the training data set. For each repetition, a distinct subset of the regressors is selected. The regressors to be included in the final model are defined according to the frequency with which each of them is selected through different tests. A consensus threshold is defined, say $\gamma = 60\%$, and a regressor is included, if selected in at least 60% of the assays. Finally, after the definition of the network topology, the calculations of the estimation of the vector of weights of the output layer are performed. In this paper, this vector is estimated by the Moore-Penrose pseudo Inverse:

$$v = Z^+ y \tag{5}$$

where Z^+ is pseudo-inverse of Moore-Penrose of Z which is the minimum norm of the solution of the least squares for the weights of the exit. The learning process can be synthesized as demonstrated in Algorithm 1. It has three parameters:

- the number of fuzzy sets that will partition the input space, M;
- the number of bootstrap replications, b;
- the consensus threshold, γ.

Algorithm 1- Learning Algorithm for Fuzzy Neural Networks Proposed in time series forecasting

Define M equally spaced fuzzy sets for each input variable.

Define candidate neurons (L)

For all K entries do

Calculate the mapping $\mathbf{z}(x_i)$

end for

Select Ls significant neurons using bootstrap lasso.

Estimate the weights of the output layer (5)

---

4 Tests and Experiments

The learning model of normalized fuzzy neural networks was evaluated through numerical experiments of time series prediction. A time series, $\mathbf{x}(t)$, can be defined as a function of an independent time t variable, tied to a process in which a mathematical description is considered to be unknown. Its most relevant feature is that its future behavior can not be predicted exactly, as can be predicted from a deterministic function, known at t. However, the behavior of a time series can sometimes be anticipated through stochastic procedures. The database used is the time series of Box-Jenkings (Gas-Furnace). The gas furnace of Box and Jenkins [5] consists of a furnace where u_k is the feed rate of methane gas (cubic feet per minute) and the output y_k is the concentration of carbon dioxide (% CO_2) in a gas mixture. A set of 296 samples (pairs of input and output data) is available for identification. The normalized data set represents the concentration of CO_2, and k, from the values y^{k-1} and u^{k-4}. See more in [12]. The studies in [5] state that a suitable model to act on this data set is in the form of:

$$y'^k = f\left(y^{k-1}, u^{k-4}\right) \tag{6}$$

Figure 4 shows the input data of the experiment and the output data of the base used in the experiments.

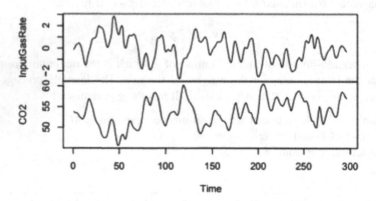

Fig. 4. Sample of the gas furnace input and output data.

The experiment was set up to use 200 samples for the neural network training phase and 96 samples for the validation phase of the model for this time series. All samples were normalized with mean zero and unit variance. In all experiments, the test assumptions defined in [15] were considered, as well as Gaussian activation functions. The performance of the proposed model was evaluated using the Root Mean Square Error (RMSE). The RMSE was calculated in the same way as in [13].

$$RMSE = \left(\frac{1}{N}\sum\nolimits_{k=1}^{n}\left(y^k - y'^k\right)\right)^{\frac{1}{2}} \tag{7}$$

In the tests carried out using Matlab, we tried to verify the ability of the learning model proposed in this paper to improve the structure of the network through the method of definition of the proposed structure, in addition to verifying that the method has the capacity to work in solving problems of time series. Table 1 shows the rules used (best value in the test), the best value of RMSE obtained in the test and the average value of RMSE for the model used in this work, besides presenting a comparison with the results of fuzzy neural networks commonly employed in time series problems. For a version where it is desired to evaluate the best network, RMSE can be considered as the value to be considered for analysis, but the mean value of RMSE can help in the definition of a more stable model. To avoid that the choice of the parameters of the models used interfere in the accuracy of the final training of the models, each algorithm was executed 30 times and the RMSE mean values were the indices used for the comparison.

The algorithm proposed in this paper was compared with other efficient methodologies to solve time series problems that are widely used in the literature. R-ORNEURON is considered the network formed by logical neurons composed by neurons or. The other fuzzy neural network models used were the DENFIS, proposed by [11], the FbeM, proposed by [12], the XUninet, developed by [4], eRFH, proposed by [17], the model of [13] based on uninorms, called in this paper of UN-RNN and also proposed by [14] FL-RNN which deals with a rapid learning approach, the model eTS, developed by [1]. Table 1 summarizes the results obtained.

Table 1. Performance evaluation of the algorithms.

Models	Rules	RMSE	Standard deviation	RMSE average
DENFIS	12	0.021	0.005	0.021
FEeM	3	0.052	0.012	0.052
XUninet	13	0.038	0.432	0.048
eRFH	9	0.027	0.004	0.031
eTS	7	0.066	0.065	0.066
NU-RNN	6	0.052	0.008	0.052
FL-RNN	6	0.047	0.014	0.046
OrNeuron	**2**	**0.019**	**0.007**	**0.020**

The results of Table 1 allow an analysis that the proposed model uses a smaller number of rules to solve the problem and presents better RMSE to the models that are traditionally used in the literature to solve time series problems. Figure 5 shows the result obtained by the *OrNeuron* model in the final validation of the results.

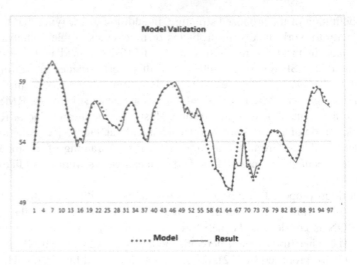

Fig. 5. Validation of the model.

5   Conclusion

This paper presents a new way to use regularized fuzzy neural networks based on the concepts of extreme learning machine to act in the forecast of time series. The method presented highlighted numerical results compared to the models of fuzzy neural networks to act in time series prediction, in addition to using a smaller number of fuzzy rules to solve the problem. The experiments performed and the results suggest that the network is able to act as a model capable of solving time series, presenting consistent results and close to the results obtained by models commonly used for this purpose in the literature. Future actions can be taken so that the model is submitted to other types of time series models and to regression problems and their results with statistical tests.

References

1. Angelov, P.P., Filev, D.P.: An approach to online identification of Takagi-Sugeno fuzzy models. IEEE Trans. Syst. Man Cybern. Part B (Cybern.) **34**(1), 484–498 (2004)
2. Bach, F.R.: Bolasso: model consistent lasso estimation through the bootstrap. In: Proceedings of the 25th International Conference on Machine learning, pp. 33–40. ACM, July 2008
3. Bordignon, F.L.: Aprendizado extremo para redes neurais fuzzy baseadas em uninormas (2013)
4. Bordignon, F., Gomide, F.: Uninorm based evolving neural networks and approximation capabilities. Neurocomputing **127**, 13–20 (2014)
5. Box, G.E., Box, G.M.J., Gregory, C.R.: Time series analysis: forecasting and control, No. 04, QA280, B6 1994 (1994)
6. Efron, B., Hastie, T., Johnstone, I., Tibshirani, R.: Least angle regression. Ann. Stat. **32**(2), 407–499 (2004)

7. Hastie, T., Tibshirani, R., Friedman, J.: The Elements of Statistical Learning. vol. 2, no. 1. Springer, New York (2009). https://doi.org/10.1007/978-0-387-84858-7
8. Hell, M.B.: Abordagem neurofuzzy para modelagem de sistemas dinamicos não lineares. Doctoral dissertation, Faculdade de Engenharia Elétrica e de Computação, Universidade Estadual de Campinas (UNICAMP) (2008)
9. Huang, G.-B., Zhu, Q.-Y., Siew, C.-K.: Extreme learning machine: theory and applications. Neurocomputing 70(1), 489–501 (2006)
10. Jang, J.S.: ANFIS: adaptive-network-based fuzzy inference system. IEEE Trans. Syst. Man Cybern. 23(3), 665–685 (1993)
11. Kasabov, N.K., Song, Q.: DENFIS: dynamic evolving neural-fuzzy inference system and its application for time-series prediction. IEEE Trans. Fuzzy Syst. 10(2), 144–154 (2002)
12. Leite, D., Gomide, F., Ballini, R., Costa, P.: Fuzzy granular evolving modeling for time series prediction. In: 2011 IEEE International Conference on Fuzzy Systems (FUZZ), pp. 2794–2801. IEEE, June 2011
13. Lemos, A., Caminhas, W., Gomide, F.: New uninorm-based neuron model and fuzzy neural networks. In: 2010 Annual Meeting of the North American Fuzzy Information Processing Society (NAFIPS), pp. 1–6. IEEE, July 2010
14. Lemos, A.P., Caminhas, W., Gomide, F.: A fast learning algorithm for uninorm-based fuzzy neural networks. In: 2012 Annual Meeting of the North American Fuzzy Information Processing Society (NAFIPS), pp. 1–6. IEEE, August 2012
15. Murphy, K.P.: Machine learning: a probabilistic perspective. MIT Press, Cambridge (2012)
16. Pedrycz, W.: Fuzzy neural networks and neurocomputations. Fuzzy Sets Syst. 56(1), 1–28 (1993)
17. Rosa, R., Gomide, F., Ballini, R.: Evolving hybrid neural fuzzy network for system modeling and time series forecasting. In: 2013 12th International Conference on Machine Learning and Applications (ICMLA), vol. 2, pp. 378–383. IEEE, December 2013
18. Souza, P.V.C., Lemos, A.P.: Redes Neurais Nebulosas para problemas de Classificação. In: Simpósio Brasileiro de Automação Inteligente, Natal-RN. XII SBAI-Simpósio Brasileiro de Automação Inteligente (2015)
19. Yager, R.R., Rybalov, A.: Uninorm aggregation operators. Fuzzy Sets Syst. 80(1), 111–120 (1996)
20. Souza, P.V.C.: Regularized fuzzy neural networks for pattern classification problems. Int. J. Appl. Eng. Res. 13(5), 2985–2991 (2018)
21. Dash, P.K., Ramakrishna, G., Liew, A.C., Rahman, S.: Fuzzy neural networks for time-series forecasting of electric load. IEE Proc.-Gener. Transm. Distrib. 142(5), 535–544 (1995)
22. Maguire, L.P., Roche, B., McGinnity, T.M., McDaid, L.J.: Predicting a chaotic time series using a fuzzy neural network. Inf. Sci. 112(1–4), 125–136 (1998)
23. Braga, A.D.P., Carvalho, A.P.L.F., Ludermir, T.B.: Redes neurais artificiais: teoria e aplicações, pp. 5–55. Livros Técnicos e Científicos (2000)
24. Zadeh, L.A.: Fuzzy sets. Inf. Control 8(3), 338–353 (1965)

Design and Implementation of Fuzzy Expert System Based on Evolutionary Algorithms for Diagnosing the Intensity Rate of Hepatitis C

Mehrnaz Behrooz$^{(\boxtimes)}$ and Mohammad Hossein Fazel Zarandi

Department of Industrial Engineering, Amirkabir University of Technology,
Tehran, Iran
Mehrnaz.behrooz@gmail.com, zarandi@aut.ac.ir

Abstract. Hepatitis, is one of the most common and dangerous diseases which affects liver. If hepatitis does not detect early, some side effects such as cirrhosis, hepatocellular carcinoma, liver failure and mature death will be occurred. Among different types of this disease, hepatitis C arises from HCV viruses, is the leading cause of liver disease. Although hepatitis C can be easily diagnosed by a simple test, the intensity rate of this disease is a qualitative and controversial issue. This paper attempts to design a fuzzy expert system for diagnosing the intensity rate of hepatitis C with FibroScan results. The proposed system includes three steps: pre-processing, create the primary fuzzy system and optimize the membership functions' parameters. KNN method is used for filling missing data; moreover, feature selection is done by decision tree and genetic algorithm. The primary fuzzy system is established and in the third step, three different evolutionary algorithms are implemented to optimize the parameters of primary system. Results portray that Differential Evolution algorithm presents better performance in learning the pattern of data and decreases the error around 30%.

Keywords: Hepatitis C · Diagnosing intensity rate · Fuzzy expert system
Evolutionary algorithms

1 Introduction

1.1 Hepatitis C

About 170 million people in the world are infected with hepatitis C virus. This is a major cause of chronic liver disease. It has been recognized as a global health problem because of the progression to cirrhosis and hepatocellular cancer. The major routes of transmission are injection drug use, blood transfusion, hemodialysis, organ transplantation and less frequently sexual intercourse. Hepatitis C virus infection can present as acute or chronic hepatitis. Acute hepatitis usually is asymptomatic and rarely leads to hepatic failure. About 60–80% people with acute infection develop chronic infection which is usually asymptomatic but can cause considerable liver damage before its recognition [1].

© Springer International Publishing AG, part of Springer Nature 2018
G. A. Barreto and R. Coelho (Eds.): NAFIPS 2018, CCIS 831, pp. 24–36, 2018.
https://doi.org/10.1007/978-3-319-95312-0_3

1.2 Fuzzy Expert System

An expert system that uses fuzzy logic instead of Boolean logic is known as fuzzy expert system. A fuzzy expert system is a collection of fuzzy rules and membership functions that are used to reason about data. Using fuzzy expert system expert knowledge can be represented that use vague and ambiguous terms in computer [2]. Fuzzy logic is a set of mathematical principles for knowledge representation based on degree of membership rather than the crisp membership of classical binary logic. Unlike two-valued Boolean logic, fuzzy logic is multi valued. Fuzzy logic is a logic that describes fuzziness. As fuzzy logic attempts to model human's sense of words, decision making and common sense of words, decision making and common sense, it is leading to more human intelligent machines [3].

2 Literature Review

Up to now, many studies have been performed in the diagnosis of hepatitis literature. In some cases, articles attempted to increase the classification accuracy, others focused on fuzzy-rules and the value membership function's parameters to design a fuzzy system. Although there are lots of works on designing a system for diagnosing hepatitis, just one fuzzy system was defined for diagnosing the intensity rate of hepatitis which utilized hepatitis B database and generated fuzzy-rules according to the specialist's experience [4]. However, our system uses the pattern of hepatitis C data for generating rules; moreover, we optimize the primary fuzzy system with evolutionary algorithms. Table 1 presents the classification of previous hepatitis diagnosis methods [5].

Table 1. Literature methods for classification

Author	Method	Author	Method
Grudzinski et al. [6]	Weighted 9NN	Ster and Dobnikar [13]	LDA
Duch and Grudziński [7]	18NN, stand, Manhattan	Ster and Dobnikar [13]	QDA
Duch and Grudziński [7]	15NN, stand, Euclidean	Ster and Dobnikar [13]	1NN
Duch and Adamczak [8]	FSM with rotations	Ster and Dobnikar [13]	ASR
Duch and Adamczak [8]	FSM without rotations	Ster and Dobnikar [13]	FDA
Duch and Adamczak [8]	RBF (Tooldiag)	Ster and Dobnikar [13]	LVQ
Duch et al. [9]	MLP + BP (Tooldiag)	Ster and Dobnikar [13]	CART
Ozyildirim et al. [10]	MLP	Ster and Dobnikar [13]	MLP with BP
Ozyildirim et al. [11]	RBF	Ster and Dobnikar [13]	ASI
Ozyildirim et al. [11]	GRNN	Ster and Dobnikar [13]	LFC
Polat and Gunes [12]	FS-AIRS with fuzzy res.	Jankowski [14]	Inc Net

3 Hepatitis C Dataset

Most of the studies on creating system of diagnosing system were utilized UCI repository of machine learning database. This database includes 155 records with 19 features. The outputs of UCI's database has 2 classes of 'live' and 'die' [15]. In addition to the low number of records, the output wasn't useful for determining intensity rate of hepatitis. Therefore, we required to gather enough sample of data with suitable outputs. Consequently, we use the database of hepatitis C patients from Liver Institute of Shariati Hospital. This Database includes 480 instances of hepatitis C patients with the FibroScan test results as an output [16].

A FibroScan is noninvasive test that measures the amount of fibrosis (thickening or scarring of tissues) in liver. It can be used alone or with other tests (such as biopsy, blood tests, ultrasounds) to see how much scarring there is on liver. Results are measured using kilopascal and range from 2 to 75. The results will vary based on the type of liver disease. Thus, they can be divided into 4 classes and illustrate the amount of scarring with 1, 2, 3 and 4 values. The stages of fibrosis and their relations with FibroScan results are presented in Table 2 [17].

Table 2. Classification of FibroScan results

Approximate cutoff value	Fibrosis stage	Qualitative parameter
[2, 7]	Mild Fibrosis	F0 to F1
[8, 9]	Moderate Fibrosis	F2
[10, 14]	Sever Fibrosis	F3
[15,75]	Advanced Fibrosis	F4

After consulting with liver specialists in Shariati hospital, we selected some of blood tests results and individual informations from patients as input features of our expert system. The 10 proposed attributes are given in Table 3 [5]. The selected attributes, except Age and Sex, are blood tests which consist liver enzymes and other blood factors affecting liver's function.

Table 3. The attributes of hepatitis C database

Number	Labels	Values
1	Age	[23, 98]
2	Sex	Male, Female
3	Platelet	[10000, 780000]
4	Creatinine	[0.4, 8.98]
5	Sodium	[130, 148]
6	Bilirubin total	[0.2, 19]
7	ALP	[36, 965]
8	ALT	[10, 525]
9	AST	[13, 316]
10	INR	[0.22, 4.06]

4 Pre-processing

4.1 Missing Values

Among the 480 sample of our database, 102 patients had 1 or 2 missing values in different features. This was an impressive number of missing data; thus, we should use an appropriate method with minimum error for filling missing values. It was decided that the best method to fill missing data was KNN algorithms. In this method we separate complete database from others with missing data. By removing the column of features with missing values, the algorithm compares second database with the complete one and tries to find nearest instance for each case with missing data. The missing value fills with the amount of nearest neighborhood sample in complete database, which has the lowest Euclidean distance from it. Therefore, all of the missing values in each sample can be filled with this method.

4.2 Feature Selection

The number of features in the raw dataset can be enormously large. This enormity may cause serious problems to many data mining systems. Feature selection is one of the oldest existing methods that deals with these problems. Its objective is to select a minimal subset of features according to some reasonable criteria so that the original task can be equally achieved well, if it was not better. By choosing a minimal subset of features, irrelevant and redundant features are removed according to the criterion [12]. In this essay, two different methods including decision tree and genetic algorithm are considered for feature selection.

Decision Tree. Decision trees classify instances by sorting them down the tree from the root to some leaf node, which provides the classification of the instance. Each node in the tree specifies a test of some attribute of the instance, and each branch descending from that node corresponds to one of the possible values for this attribute [18]. This process, which is done by learning the pattern of data, helped us to determine the low important attributes and remove them from our database.

Genetic Algorithm. As we conclude from previous algorithm, 2 attributes were removed; however, we use another algorithm to ensure. Genetic algorithm is a kind of search algorithm based on the mechanics of natural selection and natural genetics. They combine survival of the fittest among string structures with a structured yet randomized information exchange to form a search algorithm with some of the innovative flair of human search [19]. We define MSE^1 as an objective function for GA and try to calculate the error of our system with and without 2 proposed features. Hence, the population size of GA is set to 10 and we repeat the algorithm 300 times. A small sample was chosen, since comparing the presence of features in reducing error was just important for us. It means that the point of using GA was comparison and the performance of algorithm was not our case. Table 4, clearly shows that removing sodium

---

[1] Mean Squared Error.

and sex from our database, not only make any problem, but also reduce the amount of system error. However, by removing other attributes the error of the system increases comparing to total database. For example, if we delete AST, Platelet or Age the error will reach to 0.859, 0.854, and 0.81 respectively; this illustrates that the existence of other attributes have a fundamental role in our system.

Table 4. Examining the system's error with genetic algorithm

Database	MSE
Total database	0.757
Without the feature of sex	0.755
Without the feature of sodium	0.745
Without sex and sodium	0.75

5 Design the Primary Fuzzy System

5.1 Train and Test Data

We divide our database to 2 subset, including train and test database for creating the first diagnosing system. 70% of the database was randomly selected as train data. Therefore, 336 random instances were selected as training data and remaining were labeled as testing data.

5.2 Determining the Number of Rules

In a fuzzy clustering algorithm, we should use a cluster validity index to determine the most suitable number of clusters. In this paper, the validity index proposed by Fazel Zarandi et al. is used. The validity index V_{ECAS} (an exponential compactness and separation index) can find the number of clusters as the maximum of its function with respect to c. This index is defined as:

$$V_{ECAS} = ECAS(c) = \frac{EC_{comp}(c)}{max_c(EC_{comp}(c))} - \frac{ES_{sep}(c)}{max_c(ES_{sep}(c))} \tag{1}$$

Where $EC_{comp}(c)$ and $ES_{sep}(c)$ are exponential compactness and exponential separation measures respectively, and are defined as [20]:

$$EC_{comp}(c) = \sum_{i=1}^{c} \sum_{j=1}^{n} u_{ij}^m \exp(-(\frac{\| x_i - v_j \|^2}{\beta_{comp}} + \frac{1}{c+1})) \tag{2}$$

$$ES_{sep}(c) = \sum_{i=1}^{c} \exp(-min_{i \neq k}\{\frac{(c-1) \| v_i - v_k \|^2}{\beta_{sep}}\}) \tag{3}$$

In which,

$$\beta_{comp} = \left(\sum\nolimits_{k=1}^{n} \| x_i - \bar{v} \|^2 / n(i)\right) \tag{4}$$

$$\beta_{sep} = \left(\sum\nolimits_{l=1}^{c} \| v_l - \bar{v} \|^2 / c(i)\right) \text{ with } \bar{v} = \left(\sum\nolimits_{j=1}^{n} x_j / n\right) \tag{5}$$

The most suitable number of clusters based on this cluster validity is obtained in 4 clusters. Thus, the system contains 4 rules.

5.3 Implementation of the Primary System

In this step, we try to design our primary diagnosing system with 4 rules, 8 input features and FibroScan results as the output for the training database. Thus, the construction of system is based on the pattern of data. Inputs are consisting age, platelet, creatinine, bilirubin total, ALP, ALT, AST and INR. We use Sugeno-style and wtaver defuzzification method. Moreover, the 4 rules of the proposed system are as follows:

- If (AGE is in1cluster1) and (PLATELET is in2cluster1) and (CREATININE is in3cluster1) and (BILIRUBIN TOTAL is in4cluster1) and (ALP is in5cluster1) and (ALT is in6cluster1) and (AST is in7cluster1) and (INR is in8cluster1) then (output is out1cluster1)
- If (AGE is in1cluster2) and (PLATELET is in2cluster2) and (CREATININE is in3cluster2) and (BILIRUBIN TOTAL is in4cluster2) and (ALP is in5cluster2) and (ALT is in6cluster2) and (AST is in7cluster2) and (INR is in8cluster2) then (output is out1cluster2)
- If (AGE is in1cluster3) and (PLATELET is in2cluster3) and (CREATININE is in3cluster3) and (BILIRUBIN TOTAL is in4cluster3) and (ALP is in5cluster3) and (ALT is in6cluster3) and (AST is in7cluster3) and (INR is in8cluster3) then (output is out1cluster3)
- If (AGE is in1cluster4) and (PLATELET is in2cluster4) and (CREATININE is in3cluster4) and (BILIRUBIN TOTAL is in4cluster4) and (ALP is in5cluster4) and (ALT is in6cluster4) and (AST is in7cluster4) and (INR is in8cluster4) then (output is out1cluster4)

Designing the primary fuzzy system, we compare the results of system's outputs with the real output values in training database (targets). As we consider to Fig. 1, it is obvious that there is a lot of difference between outputs and targets. Therefore, we need to improve the parameters of this system.

5.4 Membership Function

As it was discussed in previous section, we require to improve our primary system which consists different rules and various membership functions for attributes. Difficulties of changing rules made us to improve the parameters of MFs. While the type of fuzzy system is Sugeno, inputs and output have Gaussian and linear MFs, respectively. We can introduce each feature with 4 MFs which includes two parameters of mean and variance. Moreover, 9 values are determined for output in each cluster. Overall 64 parameters, defined as inputs and 36 values, stored as MFs of output.

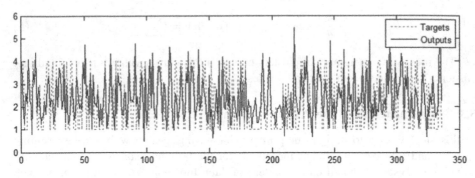

Fig. 1. Primary system for training data

6 Optimize the Parameters of Primary Fuzzy System

6.1 Objective Function

In this approach, we use RMSE[2] for comparing system's output and real values in database. It seems to be an appropriate criterion for evaluating the efficiency of diagnosing system. The RMSE in primary system is equal to 0.92, hence, we attempt to utilize suitable methods to minimize RMSE for reducing error.

$$\text{RMSE} = \sqrt{\frac{1}{N} \sum_{n=1}^{N} (y_n - \widetilde{y}_n)^2} \tag{6}$$

6.2 Optimization Approach

Due to the extraction of MFs from primary system, the values were saved in the vector of P_i^0. This section aims to find the improved vector of MFs defined as P_i^*. We assume the following linear equation, which illustrates the linear relationship between the MFs of primary and optimized system. Another hypothesis is that $x_i \in [10^{-\infty}, 10^{\infty}]$. Normalizing the variables is the reason of assuming x_i in one interval. ∞ controls the scale factor and is considered as $\infty = 0.1$ to obtain better performance on system. Figure 2 presents the exact approach for improving the primary system.

$$P_i^* = x_i \times P_i^0 \tag{7}$$

---

[2] Root Mean Square Error.

6.3 Optimization

The value of x_i would be determined with three different evolutionary algorithms consist of Adaptive-Neuro Fuzzy Inference System, Particle Swarm Optimization and Differential Evolution which are defined below. Those algorithm proved their efficient performances in training the pattern of data in literature. We aim to find the best method with minimum RMSE as a result.

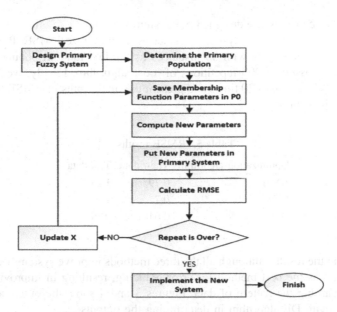

Fig. 2. Flowchart of optimization approach

ANFIS.[3] An adaptive network, as its name implies, is a network structure consisting of nodes and directional links through which the nodes are connected. Moreover, part or all of the nodes are adaptive, which means their outputs depend on the parameters pertaining to these nodes, and the learning rule specifies how these parameters should be changed to minimize a prescribe error measure. Thus, ANFIS can serve as a basis for constructing a set of fuzzy if-then rules with appropriate membership functions to generate the stipulated input-output pairs [21].

PSO. Particle Swarm Optimization is a population-based search algorithm inspired by the behavior of biological communities that exhibit both individual and social behavior, like the communities of birds, bees and fishes. PSO has become an established optimization algorithm to control and engineering design. PSO is appropriate for Problems with immense search spaces that present mane local minima [22].

---

[3] Adaptive Neuro Fuzzy Inference System.

DE. Differential Evolution reserves population-based global search strategy and uses a simple mutation operation of the differential and one-on-one competition, so it can reduce the genetic complexity of the operation. At the same time, the specific memory capacity of DE enables it to dynamically track the current search to adjust their search strategy with strong global convergence and robustness [23].

7   Results

Given the above concepts, we designed our system with three evolutionary algorithms to optimize the primary fuzzy system considering to minimize RMSE. PSO and DE start their optimization cycle with 30 instances as the first random population. Also, stop condition is assumed 900 repetitions of those algorithm. Finally, we run all the three algorithms in a loop of 30 times and present the final results of RMSE for test and train database in Table 5.

Table 5. RMSE results

Optimization algorithm	Train data	Test data
ANFIS	0.79	1.24
PSO	0.7	1.02
DE	0.64	0.99

Comparing the results, although all the three methods improve system, we conclude DE's algorithm as the best method for error reducing, resulting in improving system behavior in learning the pattern of data. Figures 3 and 4 show the outcomes of optimized system with DE algorithm in determining the outputs.

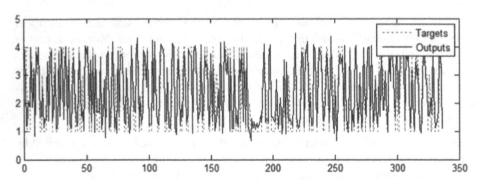

Fig. 3. Optimized system with DE algorithm for training data

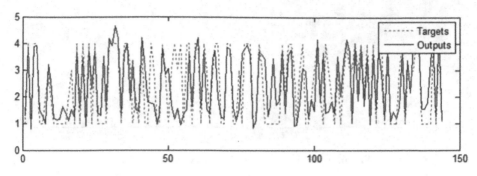

Fig. 4. Optimized system with DE algorithm for testing data

It is obvious that comparing to primary system, errors reduce and the system becomes more efficient in diagnosing the intensity rate of hepatitis c with specific inputs. Thus, compared with the RMSE of primary system, we summarized the amount of optimization with each algorithm in Table 6.

Table 6. Percentage of improvement

Optimization algorithm	Percentage of decreasing error
ANFIS	14.13
PSO	23.91
DE	30.4

Below, in Figs. 5 and 6 we present the membership functions of 2 attributes optimized with DE algorithm. Moreover, the rule's surface of improved fuzzy system with DE's algorithm is presented in Fig. 7.

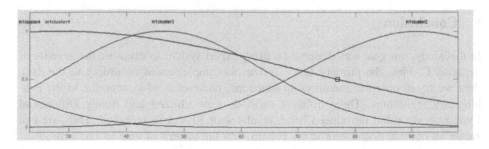

Fig. 5. Optimized membership function of age

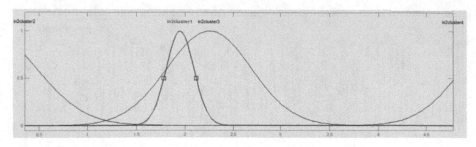

Fig. 6. Optimized membership function of platelet

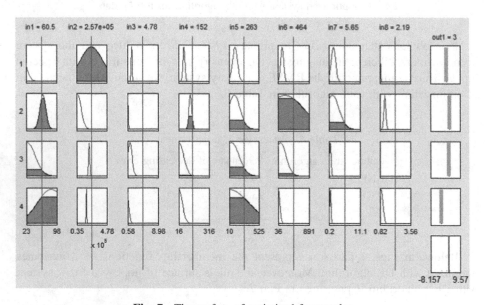

Fig. 7. The surface of optimized fuzzy rules

8 Conclusion

In this study, our goal was designing a fuzzy expert system to diagnose the severity of hepatitis C. First, the primary fuzzy system was implemented according to the data; then we try to optimize membership functions' parameters of system due to the evolutionary algorithms. Three different methods were utilized and finally Differential Evolution algorithm introduced better results with 30% error-reducing. Thus, we can create an optimal diagnosing fuzzy system with the pattern of data as our aim.

9 Future Study

Although by changing membership functions' parameters we reached to optimal fuzzy system, the other part of fuzzy systems did not consider in this approach. Defining appropriate rules has an important role in achieving optimal fuzzy system. Therefore, changing the period of rules, we will be able to introduce better systems. Moreover, we assume Sugeno-type system with Gaussian and linear MFs for inputs and outputs, respectively. However, there are various types of MFs which can be tested to achieve better results in optimization.

References

1. Modi, A.A., Liang, T.J.: Hepatitis C: a clinical review. Oral Dis. **14**(1), 10–14 (2008)
2. Patterson, D.: Introduction to Artificial Intelligence and Expert Systems. Prentice-Hall Inc., Upper Saddle River (1990)
3. Hasan, M.A., Chowdhury, A.R.: Human disease diagnosis using a fuzzy expert system. arXiv preprint arXiv:1006.4544 (2010)
4. Neshat, M., Yaghobi, M.: Designing a fuzzy expert system of diagnosing the hepatitis B intensity rate and comparing it with adaptive neural network fuzzy system. In: Proceedings of the World Congress on Engineering and Computer Science, vol. 2, pp. 797–802, October 2009
5. Sotudian, S., Zarandi, M.F., Turksen, I.B.: From Type-I to Type-II fuzzy system modeling for diagnosis of hepatitis. World Acad. Sci. Eng. Technol. Int. J. Comput. Electr. Autom. Control Inf. Eng. **10**(7), 1280–1288 (2016)
6. Duch, W., Grudzi, N.K., Diercksen, G.: Neural minimal distance methods. In: Proceedings of the 3-rd Conference on Neural Networks and Their Applications, Kule, Poland, pp. 183–188, October 1997
7. Duch, W., Grudziński, K.: Ensembles of similarity-based models. In: Kłopotek, M.A., Michalewicz, M., Wierzchoń, S.T. (eds.) Intelligent Information Systems 2001, pp. 75–85. Physica, Heidelberg (2001)
8. Duch, W., Adamczak, R.: Statistical methods for construction of neural networks. In: ICONIP, pp. 639–642 (1998)
9. Duch, W., Adamczak, R., Diercksen, G.H.: Neural networks from similarity based perspective. In: Mohammadian, M. (ed.) New Frontiers in Computational Intelligence and its Applications, pp. 93–108. IOS Press, Amsterdam (2000)
10. Ozyilmaz, L., Yildirim, T.: Artificial neural networks for diagnosis of hepatitis disease. In: Proceedings of the International Joint Conference on Neural Networks, 2003, vol. 1, pp. 586–589. IEEE, July 2003
11. Vural, R.A., Özyilmaz, L., Yıldırım, T.: A Comparative Study on Computerised Diagnostic Performance of Hepatitis Disease Using ANNs. In: Huang, D.S., Li, K., Irwin, G.W. (eds.) ICIC 2006. LNCS, vol. 4114, pp. 1177–1182. Springer, Heidelberg (2006). https://doi.org/10.1007/978-3-540-37275-2_145
12. Polat, K., Güneş, S.: Hepatitis disease diagnosis using a new hybrid system based on feature selection (FS) and artificial immune recognition system with fuzzy resource allocation. Digit. Signal Proc. **16**(6), 889–901 (2006)

13. Šter, B., Dobnikar, A.: Neural networks in medical diagnosis: comparison with other methods. In: International Conference on Engineering Applications of Neural Networks, pp. 427–430 (1996)
14. Jankowski, N.: Approximation and classification in medicine with IncNet neural networks. In: Machine Learning and Applications. Workshop on Machine Learning in Medical Applications, pp. 53–58, July 1999
15. Gong, G., Cestnik, B.: UCI Repository of Machine Learning Databases, November 1988. https://archive.ics.uci.edu/ml/machine-learning-databases/hepatitis/
16. Merat, S.: Interviewee, Shariati Hospital, Hepatitis C Records. [Interview], September 2017
17. Yim, C.: FibroScan and Liver Disease. University of Health Network, Toronto (2016)
18. Department of Computer Science, Princeton University. http://www.cs.princeton.edu/courses
19. Fang, Z., Na, L., Jinhui, L.: Application research of the genetic algorithm on the intelligent test paper composition of examination database. In: Zhang, J. (ed.) ICAIC 2011. CCIS, vol. 227, pp. 443–448. Springer, Heidelberg (2011). https://doi.org/10.1007/978-3-642-23226-8_58
20. Zarandi, M.F., Faraji, M.R., Karbasian, M.: An exponential cluster validity index for fuzzy clustering with crisp and fuzzy data. Sci. Iranica. Trans. E Ind. Eng. **17**(2), 95 (2010)
21. Jang, J.S.: ANFIS: adaptive-network-based fuzzy inference system. IEEE Trans. Syst. Man Cybern. **23**(3), 665–685 (1993)
22. Escalante, H.J., Montes, M., Sucar, L.E.: Particle swarm model selection. J. Mach. Learn. Res. **10**(Feb), 405–440 (2009)
23. Huang, Z., Chen, Y.: An improved differential evolution algorithm based on adaptive parameter. J. Control Sci. Eng. **2013**, 3 (2013)

Evolving Granular Fuzzy Min-Max Modeling

Alisson Porto$^{(\boxtimes)}$ and Fernando Gomide

University of Campinas School of Electrical and Computer Engineering,
Campinas, SP 13083-852, Brazil
{alisport,gomide}@dca.fee.unicamp.br

Abstract. This paper develops an evolving fuzzy min-max algorithm
for fuzzy rule-based systems modeling. Starting with an initially empty
rule base, the algorithm may add, modify, or delete fuzzy rules of the
rule base while processing input stream data. The data space is granulated using hyperboxes. Membership functions and affine functions are
assigned to the hyperboxes, and each hyperbox defines a corresponding
functional fuzzy rule. The model output is found combining the affine
rule consequents weighted by the normalized activation degrees of the
rules. The parameters of the consequent affine functions are updated
with the recursive least squares with forgetting factor. The algorithm is
intrinsically incremental, learns in one-pass, and allows gradual model
changes in an online like manner. Computational experiments suggest
that evolving granular fuzzy min-max modeling procedure is competitive with state of the art approaches.

Keywords: Evolving fuzzy systems · System modeling
Incremental learning · Regression

1 Introduction

Real-world system dynamics are highly nonlinear and non-stationary. Learning in these environments requires fast and efficient processing of stream data.
Because of limited computational resources, traditional system modeling may
be unpractical if all data need to be stored.

Evolving systems are an important alternative in complex system modeling
because they can learn the model structure and its parameters simultaneously
with input data processing. Evolving modeling is intrinsically incremental, can
capture shifts in data caused by abrupt changes (data shift), and track gradual
changes in data (data drifts).

Fuzzy min-max systems were originally developed as neural fuzzy networks
for classification [1] and clustering [2]. Numerous papers have improved the pio-
neering classification and clustering algorithms [3,4], but few address system
modeling, especially in the realm of fuzzy rule-based models. Exceptions include
a min-max regression technique with a gradient descent algorithm to tune rule

© Springer International Publishing AG, part of Springer Nature 2018
G. A. Barreto and R. Coelho (Eds.): NAFIPS 2018, CCIS 831, pp. 37–48, 2018.
https://doi.org/10.1007/978-3-319-95312-0_4

consequent parameters [5], and a min-max network to cluster input data with an ANFIS network [6]. However, these approaches are not suited to process stream data, neither for recursive learning.

This paper improves the evolving fuzzy min-max algorithm (eFMM) for fuzzy rule-based system modeling from stream data recently introduced in [7]. The eFMM processes stream data and simultaneously learns the rule base structure, and the parameters of the local rule consequent models in a single pass basis. The learning algorithm needs simple operations such as maximum, minimum, and comparison, an important feature for complex data processing. The efficiency of the algorithm is shown using real world data to forecast daily electrical energy load. The results show that eFMM is highly efficient and competitive with state of the art evolving algorithms.

The paper proceeds as follows: Sect. 2 details the eFMM algorithm, Sect. 3 addresses the computational experiments to evaluate the performance of eFMM against evolving, neural, and neural fuzzy evolving modeling. Section 4 concludes the paper and lists issues to be pursued in the future.

2   Evolving Fuzzy Min-Max Modeling

This section explains the evolving granular fuzzy min-max algorithm. The algorithm has three main steps. The first step develops the structure of the model using a hyperbox-based input data space granulation procedure to find the number of fuzzy rules. The idea is to associate to each hyperbox a functional fuzzy rule. The second step estimates the coefficients of affine rule consequents assigned to each hyperbox using the recursive least squares with forgetting factor. The third and final step assesses the rule base quality to identify either redundant or outdated rules. The purpose is to keep the rule base concise and representative of the current data. This step is a major improvement upon the algorithm reported in [7].

2.1   eFMM Modeling

The first step of evolving fuzzy min-max modeling is to granulate the input data space using hyperboxes. Here we assume that the domain of the data space is the n-dimensional unit cube I^n. A hyperbox in I^n is a n-dimensional rectangle defined by a maximum (W) and a minimum (V) points, as shown in Fig. 1a. Figure 1b illustrates how a collection of hyperboxes granulates the data space, and how membership functions and affine functions are assigned to the hyperboxes. The maximum and minimum points, regardless of the data space dimension, uniquely define a hyperbox. Formally, a hyperbox B_i is defined as follows:

$$B_i = \{X, V_i, W_i, b_i(x, V_i, W_i, c_i)\} \tag{1}$$

where $X \subseteq I^n$ denotes the input data space, b_i is the membership function associated with the i-th hyperbox, $x \in X$ is an input data point, and V_i, W_i, $c_i \in X$ are the minimum, maximum, and the centroid points, respectively.

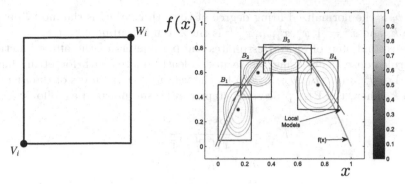

(a) Hyperbox in I^2.   (b) Data Space granulation and local models.

Fig. 1. Hyperbox and granulation

More precisely, eFMM can be viewed as a fuzzy rule-based modeling approach whose rules are endowed with local models forming their consequents, referred to as fuzzy functional models. We adopt Takagi-Sugeno [8] type of fuzzy model with affine functions in the rule consequents. These are models composed by a set of R fuzzy rules of the following form:

$$R_i : \text{If } x \text{ is } B_i \text{ then } \bar{y}_i = \theta_{0i} + \sum_{j=1}^{n} \theta_{ji} x_j \qquad (2)$$

where R_i is the i-th fuzzy rule, $\bar{y}_i$ is the i-th rule output, θ_{ji}, $i = 1, ..., R$, $j = 1, ..., n$ are the consequent parameters, and R is the total number of fuzzy rules in the rule base. The collection of the R fuzzy rules assembles the overall model as a weighted combination of the local affine models. The contribution of each local model to the overall output is proportional to the normalized firing degree of each rule. eFMM uses antecedent fuzzy sets that are an aggregation of elementwise Gaussian membership functions:

$$\sigma_{ji} = min(w_{ji} - c_{ji}, c_{ji} - v_{ji}), \ b_i = \prod_{j=1}^{n} b_{ji}, \ b_{ji} = exp\left(\frac{x_j - c_{ji}}{2\sigma_{ji}^2}\right) \qquad (3)$$

where b_{ji} is the j-th component of the i-th rule membership function, σ_{ji} is width, x_j is the j-th component of the n-dimensional input data x, and w_{ji}, v_{ji}, c_{ji} are the components of the i-th rule maximum point, minimum point, and centroid, respectively.

The output of the eFMM model at step k is computed as the weighted average of the individual rule contributions, that is

$$\hat{y} = \sum_{i=1}^{R} \psi_i \theta_i^T \bar{x}^k, \ \psi_i = \frac{b_i}{\sum_{l=1}^{R} b_l} \qquad (4)$$

where ψ_i is the normalized firing degree of the i-th rule, y_k is the model output at step k, and $\bar{x}^k = [1, x_1^k, x_2^k, ..., x_n^k]^T$ is the extended input vector.

The second step of eFMM estimates the parameters of the affine functions of the rule consequents using the recursive least squares with forgetting factor (RLS). Let $\theta_i^k = (\theta_{0i}, \theta_{1i}, \theta_{2i}, ..., \theta_{ni})$ be the vector of parameters of the i-th rule at step k. Then, the RLS processing steps can be summarized as follows:

$$K = \frac{P_i^{k-1}}{\gamma + (\bar{x})^T P_i^{k-1} \bar{x}} \bar{x} \tag{5}$$

$$\theta_i^k = \theta_i^{k-1} + K(y^k - \bar{y}_i^k) \tag{6}$$

$$P_i^k = \frac{1}{\gamma}(I - K(\bar{x}^k)^T)P_i^{k-1} \tag{7}$$

where γ is the forgetting factor, $\bar{y}_i^k$ is the i-th rule output at k, I is the identity matrix, and P is initially set as $P^0 = \omega I$, $\omega = [100, 10000]$.

The third and final step monitors the rule base to identify obsolete or redundant rules. This is done using the utility measure [9]:

$$U_i^k = \frac{\sum_{p=1}^{k} \psi_i^p}{k - I_i^*} \tag{8}$$

where I_i^* is the step at which the i-th rule was created, and ψ_i^p is the normalized firing degree of the i-th rule at step p. The utility measures how often a rule is activated. This quality measure avoids unused clusters to be kept in the rule base because outdated fuzzy rules no longer are representative of the current data. The criterion to delete rules is:

$$\textbf{IF } U_i^k < \epsilon \bar{U}^k, \textbf{ THEN } \text{delete}\{B_i\} \tag{9}$$

where $\epsilon = [0.03, 0.5]$ is a user defined parameter, and $\bar{U}^k$ is the mean utility of the R rules at k. It is important to mention that, even though ϵ is defined by the user, the threshold to delete an old rule also depends on the mean utility at k, which reduces the user responsibility. Setting a threshold by a proportion of the mean utility value also helps the user to understand if a given ϵ value is a strict, or a tolerant threshold for deleting outdated rules from the rule base.

The eFMM also has a procedure to find redundant rules. This is an automatic mechanism that uses the maximum, minimum, and center points to classify a rule pair as either redundant or not. Merging similar rules is important due to the dynamic behavior of the algorithm, which displaces hyperboxes positions as new data arrive. If two rules are sufficiently close to each other, and their consequents are similar, then they can be replaced by a unique rule with no model performance degeneration. This is desirable to avoid unnecessary rule base complexity.

The eFMM decides if rules i and l are redundant in the following way:

condition 1: $v_{ji} \leq v_{jl}$ AND $w_{ji} \geq w_{jl}$, for $j = 1, 2, ..., n$

condition 2: $v_{jl} \leq v_{ji}$ AND $w_{jl} \geq w_{ji}$, for $j = 1, 2, ..., n$

condition 3: $v_{ji} \leq c_{jl} \leq w_{ji}$ AND $v_{jl} \leq c_{ji} \leq w_{jl}$ AND $vol_{il} < vol_i + vol_l$

$$vol_l = \prod_{j=1}^{n} (w_{jl} - v_{jl}), \ vol_i = \prod_{j=1}^{n} (w_{ji} - v_{ji}) \tag{10}$$

$$vol_{il} = \prod_{j=1}^{n} (max(w_{ji}, w_{jl}) - min(v_{ji}, v_{jl})) \tag{11}$$

$$\textbf{IF} \text{ condition 1 OR condition 2 OR condition 3} \tag{12}$$
$$\text{THEN merge}\{B_i, B_l\}$$

where vol_i, vol_l and vol_{il} are the volumes of the hyperboxes i, l, and the hyperbox il formed by the union of hyperboxes i and l.

Rules i and l are merged if any of the three conditions hold. Conditions 1 and 2 say that if hyperbox l is located inside hyperbox i, or hyperbox i is located inside hyperbox l, then they should be merged. Figure 2a illustrates the situation in which hyperbox i includes hyperbox l. Even though partial superposition is allowable, eventually desirable, the complete superposition may produce contradictions if the two centers are close, but rule consequents are not alike. Condition 3 evaluates the situation in which none of the hyperboxes include each another, but there is reasonable superposition between two of them. This test requires the values of three hyperbox volumes, expressions (10) and (11). The volume vol_i is the product of the i-th hyperbox lengths in all n dimensions. The value of vol_{il} is the volume of the hyperbox formed by the union of hyperboxes i and l. Figure 2b shows hyperboxes i, l and il. Condition 3 holds in Fig. 2b (top) and the hyperboxes should be merged. At the bottom of Fig. 2b merging condition 3 does not hold. Figure 2b exhibits the union hyperbox il delimited by dashed lines. The reason to use hyperbox il volume as a merging criteria is that the volume increases if hyperboxes i and l do not have similar dispersions. This property is verified in Fig. 2b: as the example at the top shows, the volume of hyperbox il is smaller than the sum of the volumes of covered hyperboxes i and l, whereas for the example at the bottom this is not true.

If one of the three merging conditions is met, then the following merging operation is performed on boxes i and l:

$$p_i = \frac{vol_i}{vol_i + vol_l} \tag{13}$$

$$w_{ji} = p_i w_{ji} + (1 - p_i) w_{jl}, \ v_{ji} = p_i v_{ji} + (1 - p_i) v_{jl} \tag{14}$$

$$c_{ji} = p_i c_{ji} + (1 - p_i) c_{jl}, \ \theta_{ji} = p_i \theta_{ji} + (1 - p_i) \theta_{jl} \tag{15}$$

$$R = R - 1, \ \text{delete}\{B_l\} \tag{16}$$

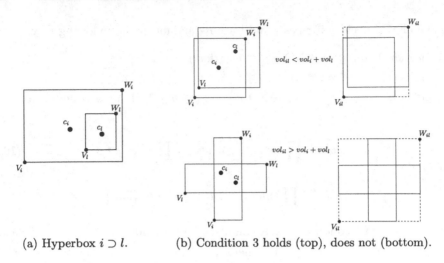

(a) Hyperbox $i \supset l$.   (b) Condition 3 holds (top), does not (bottom).

Fig. 2. Hyperbox merging

When two rules are merged, the resulting rule is a convex combination of the two original ones. The combination weights, p_i and p_l, are computed as the normalized rule volumes, where $p_l = 1 - p_i$. Behind this approach is the assumption that the rule volume is an indicative of consistency because new rules have small volumes, and volumes tend to increase as new data samples are encompassed. This procedure avoids the merging operation to discard the knowledge stored in the structure of a well established rule when merging it with smaller ones.

2.2 Automatic Parameter Selection

The eFMM includes an automatic selection mechanism of the learning parameters α and δ. The first controls the update rate of the maximum size of the most influential rule. This parameter is commonly referred to as the learning ratio. Setting the magnitude of the learning ratio is not a trivial task because the same value can speed up or delay the model updating process depending on data distribution. The eFMM estimates the value of α using:

$$\alpha^k = \left(1 - \frac{|\bar{y}^k - y^k|}{max_{err}}\right)^t max_\alpha \tag{17}$$

The coefficient max_{err} is the maximum magnitude of the difference $|\bar{y}^k - y^k|$ which produces $\alpha > 0$. The value of max_α is the maximum allowed value for the α parameter, which occurs when $\bar{y}^k = y^k$, meaning that the local affine model is fully compatible with the local functional relationship between input and output data. Parameter t sets the function decay rate. The values of these parameters are $max_{err} = 0.3$, $max_\alpha = 0.03$ and $t = 10$. These values were kept constant during the computational experiments.

The parameter δ determines the maximum size a hyperbox may achieve along any dimension. It is commonly referred to as θ in the fuzzy min-max literature, and its value critically affects the algorithm performance. Given the importance of this parameter, it is not reasonable to let to the user the responsibility to accurately choose a suitable value. The eFMM algorithm uses an automatic approach to adjust δ in an online fashion using the data local dispersion, and the α parameter. The adjustment proceeds as follows:

$$d_{ji}^k = \sqrt{\frac{M_i - 1}{M_i} \left(d_{ji}^{k-1}\right)^2 + \frac{1}{M_i} \left(x_j - c_{ji}\right)^2} \tag{18}$$

$$\delta_{ji}^k = (1 - \alpha)\,\delta_{ji}^{k-1} + 2\alpha F_d d_{ji}^k \tag{19}$$

where d_{ji}^k is the local data dispersion in the j-th dimension, and M_i is the i-th rule counter, which stores the number of data points incorporated by this rule. The chosen value of F_d is 2, so the i-th rule will cover the 2σ zone [10], since, in practice, the values of the membership functions outside this zone are negligibly small [9]. The 2 factor on the left size of α in Eq. 19 is applied since the maximum size parameter should allow the hyperbox to expand in the 2σ zone in both senses of a given direction (i.e. right and left from the center, up and down from the center, and so forth). Figure 3a illustrates hyperbox i along with its data dispersion (dashed line ellipse, with horizontal d_i length) and the maximum size $2F_d d_i$ (dashed line box). The maximum value initially is a value $\delta_{ji} = \delta_0$ given by the user. As new data are input, the value of δ_{ji} gradually converges to $2F_d d_{ji}^k$ through expression 19.

2.3   Learning Algorithm

The eFMM is a recursive algorithm that can learn on-the-fly using a stream of input data without any retraining, or storing past data. Initially, there are no fuzzy rules. As the algorithm progress processing input data, rules are created, or existing ones modified. Rule modification means to displace the maximum and/or minimum points until input data is accommodated within the boundaries of a hyperbox. Every hyperbox has a centroid whose value depends on the arrangement of all data points lying within its boundary. Each hyperbox has a counter M_i to store the number of the data samples it encompasses.

Parameters ϵ and M_{min} are user defined. ϵ controls the threshold to delete outdated rules. M_{min} is the minimum number of data samples required to accept a hyperbox as valid granule of information. A valid hyperbox is assumed to have a consistent local model. Even though it is not strictly guaranteed that the model is locally consistent, setting a minimum number of samples could prevent initial condition issues regarding the recursive least squares during parameter estimation of the affine rule consequents in the second learning step. A similar procedure is used in [11].

The first input data x^1 is the first rule centroid, minimum, and maximum points, that is, $V_1 = W_1 = c_1 = x^1$. The counter of the first hyperbox B_1 is set as $M_1 = 1$. Whenever a new data sample is input, its membership degree to all existing hyperboxes is computed using (3). Next, the hyperbox with the highest membership value undergoes the following test:

$$max(w_{ji}, x_j^k) - min(v_{ji}, x_j^k) \leq \delta_{ji}, \quad j = 1, 2, ..., n \tag{20}$$

Condition (20) is the requirement needed by the most active hyperbox to include current input data x^k. Thus, if (20) holds, then the hyperbox B_i is expanded to include sample x^k, that is, the values of W_i and V_i become:

$$W_i = max(x^k, W_i), \quad V_i = min(x^k, V_i) \tag{21}$$

Figure 3b illustrates the expansion of hyperbox B_i.

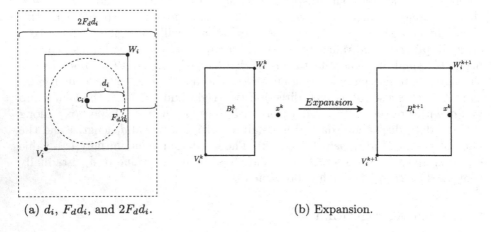

(a) d_i, $F_d d_i$, and $2F_d d_i$.   (b) Expansion.

Fig. 3. Maximum size and expansion

Whenever a hyperbox B_i includes input data x^k, the corresponding hyperbox counter and centroid are updated as follows:

$$M_i^k = M^{k-1} + 1$$
$$c_i^k = \frac{M_i^k - 1}{M_i^k} c_i^{k-1} + \frac{1}{M_i^k} x^k \tag{22}$$

Next, data dispersion d_i and maximum size δ_i are adjusted with (18) and (19).

After updating the antecedent part, the affine function of the corresponding hyperbox has its parameters adjusted using the recursive least squares with forgetting factor, using (5), (6) and (7). Only the winning hyperbox B_i is updated at each processing step k.

If (20) does not hold, then the hyperbox with the second highest membership value undergoes the same test. If (20) is satisfied, then the second hyperbox is expanded to include the data, otherwise the next hyperbox with the highest membership value is evaluated until all current existing hyperboxes are checked.

If all hyperboxes for which $b_i > 0$ are tested, and none of them satisfied (20), then it is checked if there are rules with zero firing degrees. Recently created rules have zero initial influence, and therefore, their firing degrees are zero until they expand to acquire new samples. If there are such rules, the elementwise distance between their centers and x^k are calculated:

$$dist_i = \sum_{j=1}^{n} |c_{ji} - x_j^k| \qquad (23)$$

The values $dist_i$ are then sorted in ascending order, and the corresponding rules are checked, one at a time, for Condition (20). If one of these rules satisfy (20), then the corresponding rule is expanded, and the test stops.

If there is no hyperbox for which (20) holds, then a new one is created:

$$R = R + 1 \qquad (24)$$

$$V_R = W_R = c_R = x^k \qquad (25)$$

$$\delta_{jR} = \delta_0 \text{ for } j = 1, 2, ..., n, \quad M_R = 1 \qquad (26)$$

After updating the model antecedents and consequents, the algorithm assesses the rule base quality. Outdated rules are identified and deleted using (9), whereas redundant rule pairs are merged applying (13) to (16).

The model output is produced at every at step k using (4). The algorithm generates the output for the current processing step, and then uses the actual output to update the rule consequent parameters with RLS.

3   Computational Experiments

This section evaluates the performance of the eFMM using a short-term load forecasting example. Comparisons with alternative evolving and batch modeling approaches are reported considering root mean squared error and the non-dimensional error indexes as performance measures. The expressions for these two indexes are:

$$RMSE = \left(\frac{1}{N} \sum_{k=1}^{N} \left(\hat{y}^k - y^k \right)^2 \right)^{\frac{1}{2}}, \ NDEI = \frac{RMSE}{std(y)} \qquad (27)$$

where N is the size of the test dataset, y^k and $\hat{y}^k$ are the target and the model output, respectively, and $std()$ is the standard deviation function.

The number of rules, for fuzzy rule-based methods, or the number of neurons, for neural-based approaches, gives an idea of the model complexity.

3.1 Short-Term Load Forecast

Forecast of load demand is very important to operate electric energy systems because load considerably impacts system operation planning, security analysis, and market decisions. Significant load forecast errors may result in economic losses, security constraints violations, and system operation drawbacks. Load forecast can be long, medium, or short term. Long-term forecasting is important for capacity expansion of the electric system, and medium term to organize fuel supply, maintain operations, and interchange scheduling. Short-term forecasting is needed for daily planning and operation of the electrical system, energy transfer, and demand management [12,13].

In this section, the effectiveness of the eFMM is evaluated using load data of a major electrical utility located in the southeast region of Brazil. Load data is expressed in kilowatts per hour (kW/h), and correspond to 31 days of August, 2000. Data were normalized between 0 and 1 to preserve privacy.

The eFMM model was used as a one-step ahead forecaster: the purpose is to predict the current load value using lagged load values. Previous study [13] has analyzed the sample partial autocorrelation function for the first 36 observations of the series, and suggested the use of the last two previous load values as model inputs, i.e., the forecast model is of the form:

$$\hat{y}^k = f(y^{k-1}, y^{k-2}) \tag{28}$$

The experiment was conducted as follows. The hourly load for first 30 days were input to the eFMM algorithm (718 observations). The performance of the evolved model is evaluated using data of the last day (24 observations), keeping the model structure and parameters fixed at the values found after evolving during the 30-days period. Table 1 shows how eFMM performs against evolving and fixed structure modeling methods using RMSE and NDEI error measures. The values of Table 1 for the algorithms other than eFMM were taken from [13]. The multilayer perceptron (MLP) has one hidden layer with five neurons trained with backpropagation algorithm, and the adaptive-network-based fuzzy inference system (ANFIS) has three fuzzy sets for each input variable and seven fuzzy rules. The MLP adopted the following scheme for initialization phase: small weights values randomly assigned, $\alpha = 0.9$ as momentum parameter, 1000 as the maximum number of epochs, and a adaptive learning rate starting from $\eta = 0.01$ as the initial step size. The ANFIS has 1000 as maximum number of epochs, $\eta_i = 0.01$ as initial step size, $s_d = 0.9$ as step size decrease rate, and $s_i = 1.1$ as step size increase rate. The parameters of the eTS model were set to $r = 0.4$ and $\Omega = 750$. The extended TakagiSugeno (xTS) has a $\Omega = 750$. The parameters of the ePL model were $\tau = 0.3$, $r = 0.25$, and $\lambda = 0.35$. The eMG has parameters $\lambda = 0.05$, $\Sigma_{init} = 10^{-2}I$, where I is the identity matrix, and $\alpha = 0.01$. Table 1 has two results for the eMG algorithms, one for the window size $\omega = 20$ and another for $\omega = 25$. These two results were included to compare the eMG and the eFMM performances with different rule base complexities.

The parameters of eFMM were $\gamma = 0.95$, $M_{min} = 3.5(n+1)$ and $\epsilon = 0.05$. Three forecasts are presented, for $\delta_0 = 0.1, 0.2$ and 0.5. The result for $\delta_0 = 0.1$ is considerably superior than the remaining ones, even though the model complexity is higher. Different values for δ_0 were tested to verify eFMM performance with simpler rule base structures. Notice that, with the same number of rules, eFMM outperforms eMG and xTS . Figure 4 shows the forecast results for $\delta_0 = 0.1$.

Table 1. Short-term load prediction

Model	Number of rules	RMSE	NDEI
xTS	4	0.0574	0.2135
MLP	5	0.0552	0.2053
ANFIS	9	0.0541	0.2012
eTS	6	0.0527	0.1960
ePL	5	0.0508	0.1889
eMG ($\omega = 20$)	5	0.0459	0.1707
eMG ($\omega = 25$)	4	0.0524	0.1948
eFMM ($\delta_0 = 0.1$)	39	0.0279	0.1038
eFMM ($\delta_0 = 0.2$)	13	0.0434	0.1614
eFMM ($\delta_0 = 0.5$)	4	0.0487	0.1810

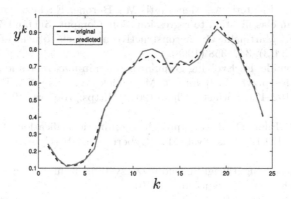

Fig. 4. Short-term load forecast.

4   Conclusion

This paper has introduced a generalized evolving fuzzy min-max regression algorithm. The algorithm uses hyperboxes to granulate the data space, and associates a functional fuzzy rule with each hyperbox to develop models from stream data.

The current version continuously monitors the rule base, excludes unnecessary rules, or merges similar ones to keep parsimonious models. Furthermore, some of the user defined parameters are automatically tuned. The algorithm uses computationally fast operations such as maximum, minimum, and comparison. Because of its recursive, one-pass learning nature, it is also memory efficient. Computational experiments show the efficiency of the algorithm when compared against more complex approaches. Future work shall address the issue of how to either automatically, or help the user to select all parameters the algorithm requires.

Acknowledgment. The authors thank the Brazilian National Council for Scientific and Technological Development (CNPq) for a fellowship, and grant 305906/2014-3, respectively.

References

1. Simpson, P.K.: Fuzzy min-max neural networks. I. classification. IEEE Trans. Neural Netw. **3**(5), 776–786 (1992)
2. Simpson, P.K.: Fuzzy min-max neural networks - part 2: clustering. IEEE Trans. Fuzzy Syst. **1**(1), 32–45 (1993)
3. Mohammed, M.F., Lim, C.P.: Improving the fuzzy min-max neural network with a k-nearest hyperbox expansion rule for pattern classification. Appl. Soft Comput. **52**, 135–145 (2017)
4. Gabrys, B., Bargiela, A.: General fuzzy min-max neural network for clustering and classification. IEEE Trans. Neural Netw. **11**(3), 769–783 (2000)
5. Tagliaferri, R., Eleuteri, A., Meneganti, M., Barone, F.: Fuzzy min-max neural networks: from classification to regression. Soft Comput. **5**(1), 69–76 (2001)
6. Mascioli, F.F., Martinelli, G.: A constructive approach to neuro-fuzzy networks. Sig. Process. **64**(3), 347–358 (1998)
7. Porto, A., Gomide, F.: Evolving granular fuzzy min-max regression. In: Melin, P., Castillo, O., Kacprzyk, J., Reformat, M., Melek, W. (eds.) NAFIPS 2017. AISC, vol. 648, pp. 162–171. Springer, Cham (2018). https://doi.org/10.1007/978-3-319-67137-6_18
8. Angelov, P.P., Filev, D.P.: An approach to online identification of Takagi-Sugeno fuzzy models. IEEE Trans. Syst. Man Cybern. Part B (Cybern.) **34**(1), 484–498 (2004)
9. Angelov, P.: Evolving Takagi-Sugeno fuzzy systems from streaming data (eTS+), pp. 21–50. John Wiley & Sons Inc. (2010)
10. Hastie, T., Tibshirani, R., Friedman, J.: The Elements of Statistical Learning. Springer, New York (2009). https://doi.org/10.1007/978-0-387-84858-7
11. Filev, D., Georgieva, O.: An extended version of the Gustafson-Kessel algorithm for evolving data stream clustering. John Wiley & Sons, Inc. pp. 273–299 (2010)
12. Rahman, S., Hazim, O.: A generalized knowledge-based short-term load-forecasting technique. IEEE Trans. Power Syst. **8**(2), 508–514 (1993)
13. Lemos, A., Caminhas, W., Gomide, F.: Multivariable gaussian evolving fuzzy modeling system. IEEE Trans. Fuzzy Syst. **19**(1), 91–104 (2011)

Combination of Spatial Clustering Methods Using Weighted Average Voting for Spatial Epidemiology

Laisa Ribeiro de Sá$^{(\boxtimes)}$ (iD), José Carlos da Silva Melo,
Jordana de Almeida Nogueira (iD), and Ronei Marcos de Moraes (iD)

Federal University of Paraíba, João Pessoa, Brazil
laisa8910@gmail.com

Abstract. The methods of spatial clustering analyze the phenomenon under study, identifying the significant and not significant clusters, which when used individually do not exactly reflect the reality of the phenomenon studied. However, with the combination of the methods it becomes possible to obtain better results. The objective of this work was to perform a combination of methods of spatial clustering, by using weighted average voting rule, for identification of municipalities in the state of Paraiba more vulnerable to the dengue fever. For methodology application, dengue fever cases in the state of Paraiba-Brazil in the year of 2011 were used. The spatial Scan statistic, Getis-Ord, Besag-Newell methods combined by the weighted average voting rule were used in order to obtain a final map with the classification of each municipality according to "priority municipalities", "transition municipalities" (which can become priority or not) and "non-priority". This method allowed the visualization of the spatial distribution of the dengue fever in all municipalities of Paraiba, allowing to identify vulnerable municipalities to the dengue fever. The levels of priority can help managers for decisions concerning the specific characteristics of each municipality.

Keywords: Weighted average voting · Spatial clustering methods
Dengue fever · Combining classifiers

1 Introduction

The methods of Spatial clustering analyze the phenomenon under study identifying the significant and not significant clusters. The methods Getis-Ord, Spatial Scan Statistics, Besag-Newell, Tango, Geary and M Statistics are the most used in the area of spatial statistics. These methods, when used individually, do not exactly reflect the reality of the phenomenon studied. With the combination of the methods, though, it is possible to obtain satisfactory results, indicating an analysis closer to the phenomenon reality. The purpose of the combination of classifiers of spatial clustering is to improve the efficiency of decision making, adopting rules of combination [1].

When implementing different spatial clustering methods, they have different decision thresholds and generalizations. To use the combined information from multiple classifiers, the output of each classifier can be combined with the others, allowing

© Springer International Publishing AG, part of Springer Nature 2018
G. A. Barreto and R. Coelho (Eds.): NAFIPS 2018, CCIS 831, pp. 49–60, 2018.
https://doi.org/10.1007/978-3-319-95312-0_5

a capability of generalization and stability of classification in the final decision [2]. The results of combining multiple classifiers bring a general improvement in accuracy when compared to individual classifiers, as it combines the independent decisions of each classifier [3].

The only one paper found in the scientific literature was Holmes [1], which suggests the combination for spatial clustering methods and provides interesting results for dengue fever epidemiology in Brazil. As limitation of that study, they used binary majority voting on significant and not significant p-values resulting from spatial clustering methods.

Latif proposed a new algorithm for the weighted average voting based on a soft threshold. It has some benefits over the previously introduced weighted average voters for computing weights. It uses two tuneable parameters, each with a ready interpretation, to provide a flexible voting performance when using the voter in different applications. The weight assignment technique is transparent to the user because the impact of the degree of agreement between any voter input and the other inputs is directly reflected in the weight value assigned to that input. The voter can be tuned to behave either as the well known inexact majority voting that is generally used in safety-critical control systems at different voting planes or as a simple average voting used in many sensory voting planes, giving better performance in terms of safety and availability [4].

This paper proposes the combination of spatial grouping methods through a new approach based on the weighted average voting in p values resulting from the spatial grouping methods directly. As advantage, all information available can be used in the voting process and it brings an innovation. Beyond that, it strives to optimize results, allowing provide more than two classes for municipalities as for instance: "priority", "transition" (which can become priority or not) and "non-priority", which is important to the knowledge of the health-disease process. In order to present the new approach characteristics, a study of case was performed with goal of identification classes of priority of Paraiba state municipalities for dengue fever in Brazil.

2   Materials and Methods

2.1   Study Characterization

The state of Paraiba-Brazil was selected as an area of research. It occupies a land area of 56,585 km^2, with a population of around 3.943.885 inhabitants, being constituted by 223 municipalities [5]. This study is characterized by being an ecological, retrospective study with a quantitative approach, which used data of the secondary type of cases reported in the Information System on Diseases of Compulsory Declaration/Dengue (SINAN/Dengue) of the Department of Health of Paraiba, in 2011. Data from resident population in each municipality were obtained from the Brazilian Institute of Geography and Statistics [5].

Dengue fever is a systemic and dynamic viral disease and has a wide clinical spectrum that includes both severe and non-severe manifestations, transmitted by the Aedes mosquito and its incidence has increased about 30 times over the last 50 years.

Estimates from the World Health Organization (WHO) indicate that in 2014, about 2.3 million individuals were diagnosed with dengue fever in the world. The registered incidence rate was around 455.4 per 100,000 inhabitants (inhab.) in the world [6].

In Brazil, 1.452.489 new cases were reported in 2013 in the SINAN. Regarding the incidence rate (IR), the year 2011 showed a rate of 400/100,000 inhab., with a decrease in 2012, with a value of 303.9/100,000 in hab. and significant growth in 2013, with a rate of 722.4/100,000 inhab. In the state of Paraiba, 7,366 suspected dengue fever cases were reported in 2014, of which 3,442 were confirmed cases. All regions of the country showed a reduction of reported cases, and the Southeast had the most significant decrease (67%), followed by the South (64%), the Midwest (58%), Northeast (42%) and the North (12%) [7].

The factors contributing to this large quantity of cases are impoverished urban areas, suburbs and rural areas, but it also affects the richest regions in tropical and subtropical countries [6]. In this context, with the high incidence of dengue fever cases in the state of Paraiba, an interesting question to raise is how this morbidity is distributed spatially.

A combination of spatial clustering methods using majority voting rule was proposed by study [1]. A general kind of combining spatial clustering methods is presented in the Fig. 1. A predefined N (N > 1) spatial clustering methods are applied on epidemiological database and their results are combined in order to provide a decision map, in which subareas are identified with different degree of priority.

A refinement of the database was performed to identify the variables of interest: year of the notification and municipality of residence of the reported case. Normality distribution for data was tested by the Lilliefors test [8]. The test identifies data that do not follow a normal distribution, and for this reason, nonparametric spatial clustering methods were used in this study.

In study [1] the authors used the Getis-Ord, Spatial Scan Statistics and Besag-Newell spatial clustering methods, aggregated by majority voting. As a result, municipalities are classified in two classes of priority: priority and non-priority. In otherwise, in this paper, combining classifiers was performed using the same three methods (Fig. 1), but preserving probability values from spatial clustering methods, which are used as input for weighted average voting rule, in order to obtain a final map with the classification of each location.

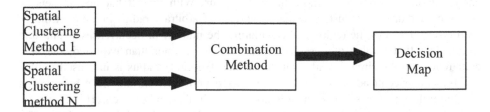

Fig. 1. The combination model of spatial clustering methods.

2.2   Relative Risk

The relative risk (RR) is a measure of frequency which measures the probability of an individual being affected by a disease. It is a measure that represents the occurrence of a phenomenon in an area with respect to the entire study area [9]. Its calculation is given by the ratio of the incidence rate in a sub-region and the incidence rate in the region in its entirety. Next, the result obtained is divided in classes, and thereby generates a choropleth map (colored) associating a color to each preset interval. Such maps allow the comparison of information from different areas, because it standardizes the data removing, thus, the effect of different populations. This study follows the interpretation of Relative Risk through classes displayed in Table 1.

Table 1. Interpretation of relative risk through classes.

Relative risk	Interpretation of relative risk
$0 \leq x < 0.5$	Very low risk
$0.5 \leq x < 1$	Low risk
$1 \leq x < 1.5$	Medium risk
$1.5 \leq x < 2$	High risk
$x \geq 2$	Very high risk

The RR map was used in a comparison with the outputs of spatial clustering methods for choosing the most appropriate map, i.e., the one which showed significant clusters of regions that coincided with the highest risk regions in the RR map.

2.3   Spatial Scan Statistic

The Scan statistic, proposed by Kulldorff and Nagarwalla [10], aims to identify clusters. To this end, the association of the information of the area with a single point within a polygon is done, be it a circle [11], an ellipse [12], a rectangle [13] or other geometrical shape [14]. This point is called centroid, representing the center of mass of each area of the region. For this study a circular shape was used, since it facilitates the observation of the functionality of the method.

The method, in turn, makes a search through the entire region to find areas where a phenomenon is significantly more likely to occur. With this, it handles all sets of possible candidates for clusters, centering circles of arbitrary radius in the centroid of each covered area of the region and calculating the number of occurrences within that circle [15]. If the value of observed occurrences is greater than expected, the area demarcated by the circle is called cluster, if not, the circle radius is increased until it involves other centroids. The process is repeated until all centroids have been tested and the null hypothesis is that there is no clusters in the study region [16].

There are two models for the Scan statistic, Binomial and Poisson [10]. The Poisson model is adopted for this study, as the number of events in each area is considered to be distributed according to the known population at risk. Therefore, the Poisson model consists of computing the centroids' radius whose $p(z)$ values, which represents cases in the circle, and $q(z)$, which are the cases out of the circle, maximize the likelihood function conditioned to the total of cases observed [16]. For the z circle, there is the following statistic Eq. 1 [17]:

$$\lambda = max_{z \in Z} \frac{L(z, \hat{p}(z), \hat{q}(z))}{L_0},$$ (1)

where $\hat{p}(z)$ and $\hat{q}(z)$ are estimators for $p(z)$ and $q(z)$ and $Z = \{z_1, \ldots, z_k\}$ is the set of all possible cluster candidates. The calculation of L_0 is given by Eq. 2:

$$L_0 = \frac{C^C (M - C)^{M-C}}{M^M},$$ (2)

where C is the total of observed cases in the region of study and M is the total population. $L(z, \hat{p}(z), \hat{q}(z))$ can be defined as Eq. 3:

$$L(z, \hat{p}(z), \hat{q}(z)) = \frac{exp[-p(z)n_z - q(z)(M - n_z)]}{C!} p(z)^{C_z} q(z)^{C - C_z} \prod_i C_i,$$ (3)

where n_z is the number of individuals at risk in circle z, in this case, at risk of contracting dengue fever; exp represents the exponential function and c_z and $c_i(i, z = 1, 2, \ldots, k)$ are, respectively, the number of cases in circle z and in circle i.

Based on the above, the circle starts in a single centroid, being calculated the λ value for each new centroid involved. Next, the λ that has greater value and the significance of the test via Monte Carlo simulation are recorded. The process is repeated for each of the centroids of the study region [18]. There is a restriction on the percentage of the population at risk, in which the search radius increases until it covers at maximum $x\%$ of the population. However, there is no standardization of this percentage, it is only recommended that it does not exceed 50% of the population, being necessary the comparison of the percentage with the relative risk maps.

Such method is non-parametric, i.e., it does not depend on the statistical distribution. Because of this assumption, tests were conducted in order to verify the normality of the data, coming to the conclusion that they do not have a distribution that approximates the normal. Therefore, it is possible to make use of statistical Scan.

2.4   Getis-Ords

The index of Getis-Ord can be applied to data with a non-normal distribution, with the purpose of measuring a nonparametric spatial autocorrelation. The indexes of Getis-Ord are estimated from neighbor groups of critical distance d in each area i. The critical distance is formed from a proximity matrix W, where each element is defined in function of the critical distance, $w_{ij}(d)$ [19].

Getis-Ord proposed two statistical functions: the global index $G(d)$, which is similar to traditional measures of spatial autocorrelation, and local indexes G_i and G_i^A, which are spatial association measures for each area i. From a level of significance, which can be defined as the probability of rejecting the null hypothesis (existence of spatial autocorrelation), the p-value is compared to the generated index. Its evaluation is made from the positive value and significance: the positive and significant value of $G(d)$ indicates spatial clustering of high values. In contrast, negative and significant values of $G(d)$ indicate spatial clustering of low values [19].

A second type of statistic suggested by Getis-Ord is a measure of association for each individual spatial unit for each observation i, where G_i and G_i^A indicate the extent to which this position is surrounded by higher values or lower values for the variable. This index is simply a ratio of the sum of the values in the surrounding positions to the sum of the values in the data series as a whole (excluding the considered position) [19].

The local index is interpreted as follows: the positive and significant standardized values (p-value lower than 5%) report a spatial clustering of high values. The negative and significant standadized values of statistics (p-value lower than 5%) indicate a spatial clustering of low values (Table 2).

Table 2. Interpretation of local index.

Negatives***	Negative Index with p-value lower than 0.005
Negatives**	Negative Index with p-value between 0.005 e 0.025
Negatives*	Negative Index with p-value between 0.025 e 0.05
Negatives	Negative Index with p-value higher than 0.05
Positives	Positive Index with p-value higher than 0.05
Positives*	Positive Index with p-value between 0.025 e 0.05
Positives**	Positive Index with p-value between 0.005 e 0.025
Positives***	Positive Index with p-value lower than 0.005

2.5 Besag-Newell

Besag-Newell proposed a method with a visual output of multiple overlapping circles of varying sizes and covering the whole of the study area, whose centroids of each subregion are their centers [20]. Denoted by Besag-Newell, the method determines the required radius the circle must have to contain at least k cases inside. The procedure starts with a circle of radius equal to zero. If it contains k or more cases, the process is interrupted; otherwise, the radius is increased until the circle includes the nearest centroid. This way, their respective cases and exposed population are added. Thus, the radii are defined in such a way to include new centroids when necessary, executing this process until there are at least k cases inserted in the circle [21].

The interpretation is done in the following way: if the p-value obtained is less than the adopted significance the cluster is said significant. Thus, after obtaining all the circles that encompass a number k of cases through the Besag-Newell method, only significant circles are drawn on the map (p-value $< \alpha$). Generally the choice of α is made to allow many simultaneous tests and values lower than the usual significance 0.05 or 0.01 are considered [21].

2.6   Combining Spatial Clustering Methods

In the scientific literature, several cases can be found, in which combining multiple classifiers provided an improvement of the results with respect to each individual classifier performance. So, that combination makes them more efficient [22–25]. Combining classifiers can be done using three architectures:in sequential (or linear) way, in parallel or hierarchically [25]. In order to provide the final decision, an architecture should be chosen, as well as a combination scheme of classifiers, which is called combiner [1].

In this study, combining spatial clustering methods was performed using Getis-Ord, Spatial Scan Statistics and Besag-Newell methods. They were aggregated by weighted average voting in order to provide a map where municipalities of a state were labeled as "priority" (colored brown on the decision map), "in transition" (municipalities can become priority or non-priority, and they are colored yellow on the decision map) and "non-priority" (colored beige on the decision map).

Given a set of inputs (redundant module results) x_1; x_2 and x_3 for a particular voting cycle, a distance measure based weighted average voter determines the numerical distance of input pairs, $d_{12} = |x_1 - x_2|$; $d_{13} = |x_1 - x_3|$ and $d_{32} = |x_3 - x_2|$ from which weighting values of individual inputs, w_1; w_2 and w_3 are computed. A module result which is far away from other module results is assigned a smaller weight value compared with a result that is close to any of the other module results. The weight values are then used to calculate a single value as the voter output [4].

Latif-Shabgahi [4] proposes a new voting that uses the concept of light threshold for determining the degree of proximity of all entries in the voting pairs. For all input voting pairs i and j; the degree of proximity S_{ij} is defined as follows Eq. 4:

$$S_{ij} = \begin{cases} 1 \text{ if } d_{ij} \leq a \\ \left(\frac{n}{n-1}\right)\left(1 - \frac{d_{ij}}{n.a}\right) \text{if } a < d_{ij} < n.a \\ 0 \text{ if } d_{ij} \geq n.a \end{cases} \tag{4}$$

where d_{ij} is the numerical distance of the input pairs i and j; one is the fixed voting threshold value (used in conventional voters of inexact majority), and n is a tunable proportionality constant, parameter to be used to control the voting. We call S_{ij} the indicator according to the input pairs x_i and x_j: for input pairs with a distance less than a threshold specified in an application; $S_{ij} = 1$; and for those with a distance greater than $n.a$ (where n is a constant of proportionality) $S\_(ij) = 0$. For input pairs with a distance between a and $n.a$, a value from the range [0, 1] is given to S_{ij}.

According to the values of the computed indicators for all voter input pairs, the weighted value of each voter i is defined based on the following Eq. 5 (m - assumed input).

$$w_i = \frac{\sum_{j=1, j\neq 1}^{m} S_{ij}}{m - 1} \Big| i, j = 1, \ldots m, \text{ where } i = j : s_{ij=0} \tag{5}$$

The voter output is given by Eq. 6:

$$y = \frac{\sum_{i=1}^{m} x_i.w_i}{\sum_{i=1}^{m} w_i} \tag{6}$$

3   Results and Discussion

In the year 2011, 11,490 cases of dengue fever were reported in the state of Paraiba, Brazil, SINAN, 2014. From that data, it was used the Poisson probability model for the spatial Scan method in order to identify spatial clusters. Population percentages of 0.1%, 0.5%, 0.7% and 1% were tested, to know which presents better results, when compared with the map of relative risk. The appropriatest percentage was the one with restriction of 0.1% of the population.

Figure 2 shows the RR map for 2011, which illustrates dengue fever behavior in the municipalities of Paraiba, where the RR presented values between 0 (zero) and 6.07 per 100,000 inhabitants. Spatial analysis of cases through the Scan statistic identified clusters of high and low risk. Forty-eight (48) municipalities belong to spatial clusters for that year, with a percentage of the population of 0.1%. It is possible to verify regions coincidence when comparing the Scan map with the relative risk map, with respect to regions with the greatest RR indexes and those ones with significant spatial clusters.

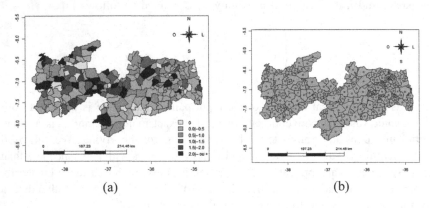

(a)   (b)

Fig. 2. (a) Relative Risk Map and (b) spatial *Scan* of dengue fever cases according to municipality of residence. Paraíba, Brazil, 2011.

In Fig. 3(a), the resultant map of applying the Getis-Ord method, it was observed that only some municipalities present spatial clusters of high values. These were located mostly in the regions of Central and West in Paraíba. The clusters of low values were distributed in an heterogeneous way in the state of Paraíba. In the map resulting from the Besag-Newell method for dengue fever data in 2011 (Fig. 3b), several spatial clusters were observed in all regions. There was a concentration of clusters in the West region in the state of Paraíba. The Lest region in the state of Paraíba had the lowest number of spatial clusters.

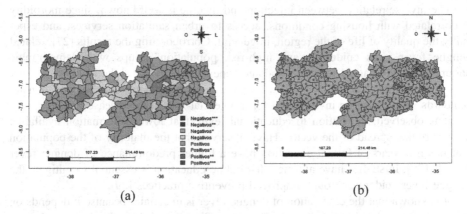

(a)   (b)

Fig. 3. (a) Getis-Ord map and (b) Besag-Newell map of dengue fever cases according to municipality of residence. Paraíba, Brazil, 2011

The combination of the three spatial clustering methods (spatial Scan statistic, Besag-Newell method and Getis-Ord statistic) by weighted average voting generated the decision map, which is shown in Fig. 4. In that Figure, it is possible to note 37 priority municipalities, 17 transition municipalities and 169 non-priority municipalities. When it is compared the maps obtaining from the each one of three methods (Figs. 2 and 3) with the decision map (Fig. 4), the coincidence in the concentration of municipalities in the region Central and West as well as in municipalities with large populations are remarkable.

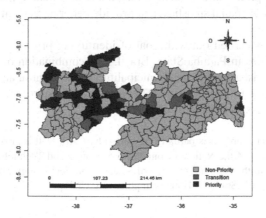

Fig. 4. Decision map for the combination of spatial clustering methods through weighted average voting for dengue fever cases. Paraíba, Brazil, 2011.

This concentration in less developed municipalities or the most densely populated municipalities can be justified by an association of demographic, ecological and socioeconomic factors to the spread of dengue fever. In literature have shown that areas with the highest human population densities are associated with the higher risk categories for dengue fever [26].

The inverse relation between infection and income is well known, since morbidity is correlated with housing conditions, access to urban sanitation services, and especially the quality of life of the region, in general. Corroborating the results [27], related dengue fever virus epidemics and high-risk population groups with demographic parameters such as age distribution and age-dependent behavior at the local level. They have affirmed that variations in socioeconomic status, housing and housing density standards should also be investigated for a potential role in the disease group.

It is observed, in relation to educational level, that lower information available, implies higher spread of the vector. Have observed that the majority of the population that has few years of formal education have a higher predominance of dengue fever records [28]. In studies have affirmed that public education is essential to mitigate the dengue fever epidemic through improved preventive practices [29].

It is known that the eradication of dengue fever is uncertain, because it depends on several factors related to the disease. The most efficient and cost-effective measure, so far, is to combat the vector (Aedes aegypti). The joint efforts made by society and government are key to an organized urban environment, whether it is sanitary supervision and access to health services, or even educational measures and population awareness so that all are active in combating the disease, particularly in the most affected districts.

4   Conclusion

In this paper was proposed a new combining of spatial clustering methods. It is based on weighted average voting approach. The previous paper used binarizing the information. This new approach allows using all available information available. It opens a new path in the area of spatial statistics, and provide better agreement with relative risk map.

A case study was carried out with goal of identifying priority classes for dengue fever for municipalities in Paraíba State, Brazil. The combination of spatial clustering methods allowed to visualize that the municipalities of the Central and West regions of Paraíba, are vulnerable to dengue fever. This result can support managers for decision making regarding the specificities of each locality.

Acknowledgments. This project is partially supported by CAPES. It is also partially supported by grants 308250/2015-0 of the National Council for Scientific and Technological Development (CNPq) and is related to the National Institute of Science and Technology Medicine Assisted by Scientific Computing (465586/2014-4) also supported by CNPq.

References

1. Holmes, D.C.S.C., Moraes, R.M., Vianna, R.P.T.: A rule for combination of spatial clustering methods. In: Patterns 2015 (2015)
2. Ponti Jr., M.P.: Combinação de múltiplos classificadores para identificação de materiais em imagens ruidosas. 2004. Dissertação (Mestrado), UFSC, São Carlos (2004)
3. Duin, R.P.W., Tax, D.M.J.: Experiments with classifier combining rules. In: Kittler, J., Roli, F. (eds.) MCS 2000. LNCS, vol. 1857, pp. 16–29. Springer, Heidelberg (2000). https://doi.org/10.1007/3-540-45014-9_2
4. Latif-Shabgahi, G.R.: A novel algorithm for weighted average voting used in fault tolerantcomputing systems. Microprocess. Microsyst. **28**(7), 357–361 (2004)
5. IBGE: Síntese de indicadores sociais 2010 uma análise das condições de vida da população brasileira. IBGE, Rio de Janeiro (2010)
6. WHO: Dengue Fever. Epidemiologic (2015)
7. BRASIL: Ministério da saúde. Situação Epidemiológica da Dengue no Brasil (2014)
8. Siegel, S., Castellan Jr., N.J.: Estatística não-paramétrica para ciências do comportamento. 2. ed. Porto Alegre: Artmed, 448 p (2006)
9. Lucena, S.E.d.F., Moraes, R.M.: Detection of space-time cluster of homicides by stabbing in the João Pessoa city, Paraíba state, Brazil. Boletim de Ciências Geodésicas. **18**(4), 605–623 (2012)
10. Kulldorff, M., Nagarwalla, N.: Spatial disease cluster: detection and inference. Stat. Mcd. **14**, 799–810 (1995)
11. Kulldorff, M., et al.: An elliptic spatial scan statistic. Stat. Med. **25**, 3929–3943 (2006)
12. Neil, D.B., Moore, A.W., Sabhnani, M., Daniel, K.: Detection of emerging space-time clusters. In: Proceedings of the Eleventh ACM SIGKDD International Conference on Knowledge Discovery (KDD '05), Chicago (USA), pp. 218–227 (2005)
13. Assunção, R.M., Duczmal, A.R.: A simulated annealing strategy for the detection of arbitrary shaped spatial clusters. Comput. Stat. Data Anal. **45**, 269–286 (2004)
14. Tango, T., Takahashi, K.: A flexibly shaped spatial scan statistic for detecting clusters. Int. J. Health Geogr. **4**(1), 11 (2005)
15. Coulston, J.W., Ritters, K.H.: Geographic analysis of forest health indicators using spatial scan statistics. Environ. Manag. **31**(6), 764–773 (2003)
16. Lucena, S.E.F., Moraes, R.M.: Análise do desempenho dos métodos Scan e Besag e Newell para identificação de conglomerados espaciais do dengue no municípios de João pessoa jan 2004 a dez 2005. Boletim de Ciências Geodésicas, **15**(3), 544–561 (2009)
17. Gómez-Rubio, V., Ferrándiz-Ferragud, J., López, A.: Detecting clusters of disease with R. J. Geogr. Syst. **7**(2), 189–206 (2005)
18. Moura, F.R.: Detecção de clusters espaciais via algoritmo scan multi-objetivo. Dissertação (Mestrado em Estatística) - Universidade Federal de Minas Gerais, Belo Horizonte (2006)
19. Anselin, L.: Spatial data analysis with GIS: an introduction to application in the social sciences. National Center for Geographic Information end Anlisis. University of California - Santa Barbara, August 1992
20. Openshaw, S., Craft, A.W., Charlton, M., Birch, J.M.: Investigation of leukaemia clusters by use of a geographical machine. Lancet **331**, 272–273 (1988)
21. Costa, M.A., Assunção, R.M.: A fair comparison between the spatial scan and the Besag-Newell disease clustering tests. Environ. Ecol. Stat. **12**(3), 301–319 (2005)
22. Li, C.S., Wang, Y., Yang, H.: Combining fuzzy c-means classifiers using fuzzy majority vote. In: Fifth International Conference on Fuzzy Systems and Knowledge Discovery (2008)

23. Li, C.S., Wang, Y., Yang, H.: Combining fuzzy partitions using fuzzy majority vote and KNN. J. Comput. **5**(5), 791–798 (2010)
24. Xu, L., Krzyzah, A., Suen, C.Y.: Methods of combining multiple classifiers and their applications to handwriting recognition. IEEE Trans. Syst. Man Cybern. **22**(3), 418–435 (1992)
25. Kittler, J., Hatef, M., Duin, R.P.W., Matas, J.: On combining classifiers. IEEE Trans. Pattern Anal. Mach. Intell. **20**(3), 226–239 (1998)
26. Rogers, D.J., et al.: The global distribution of yellow fever and dengue. Adv. Parasitol. **62**, 181–220 (2006). PMC Web
27. Lin, C.-H., et al.: Dengue outbreaks in high-income area, Kaohsiung City, Taiwan, 2003–2009. Emerg. Infect. Dis. **18**(10), 1603–1611 (2012). PMC Web, 7 Apr 2016
28. Fantinati, A.M.M.: Demographic and epidemiological profile of dengue cases in central Goiania- Goias: 2008 to March 2013. Rev Tempus Actas Saúde Col (2013)
29. Al-Dubai, S.A., et al.: Factors affecting dengue fever knowledge, attitudes and practices among selected urban, semi-urban and rural communities in Malaysia. Southeast Asian J. Trop. Med. Public Health **44**(1), 37–49 (2013)

Fuzzy Logic Applied to eHealth Supported by a Multi-Agent System

Afonso B. L. Neto[(✉)], João P. B. Andrade, Tibério C. J. Loureiro,
Gustavo A. L. de Campos, and Marcial P. Fernandez

Universidade Estadual do Ceará (UECE), Fortaleza, Ceara, Brazil
afonsoblneto@gmail.com, joaopbern7@gmail.com,
tiberiocj@gmail.com, {gustavo,marcial}@larces.uece.br

Abstract. Living better, with health and stay prepared for the challenges on getting older is becoming one of the most concerns for people. In USA, there are studies that have shown that the amount of people living alone or, at most, with just one person, is increasing over the years. Technology products applied for health are receiving prominence because they help those people to achieve their goals. Considering this, the article proposes a multi-agent system architecture that uses IoT devices to monitor patients' heart signals and, using fuzzy logic process, estimates the level of hypertension, considering systolic pressure, diastolic pressure, age and body mass index. Information of 768 patients were obtained from "Pima Indians Diabetes Data Set" public database and used to evaluate the performance of the presented fuzzy logic model. The results of such fuzzy logic were compared with an evaluation made by accredited nurses, reaching a 94.40% of positive predictiviness in the diagnosis.

Keywords: Fuzzy logic · Multi-agent system · Health · IoT

1 Introduction

In developed countries, such as USA, there are benefits from sweeping advances in nutrition, sanitation, and medicine that transformed public health practice and increased average lifespan during the first half of the twentieth century [1]. Figure 1 depicts the evolution of American's expectance of life, where in 2012 there were 43 million Americans with 65 years or more, and in 2050 the expectance is an increase of approximately 200%.

Blood pressure (BP) measurements are usually used to diagnose Hypertension. Casual BP measurements are of limited value because they do not reflect the circadian variation in BP, the "white-coat effect," regression to the mean, observer bias, and other factors [16]. There are other factors that contributes to the variability of blood pressure, such as sex, age and BMI.

In the last decade, wearable devices have attracted much attention from the academic community and industry and have recently become very popular. The most relevant definition of wearable electronics is the following: "devices that can be worn or mated with human skin to continuously and closely monitor an individual's activities, without interrupting or limiting the user's motions" [2]. Currently, mobile apps

© Springer International Publishing AG, part of Springer Nature 2018
G. A. Barreto and R. Coelho (Eds.): NAFIPS 2018, CCIS 831, pp. 61–71, 2018.
https://doi.org/10.1007/978-3-319-95312-0_6

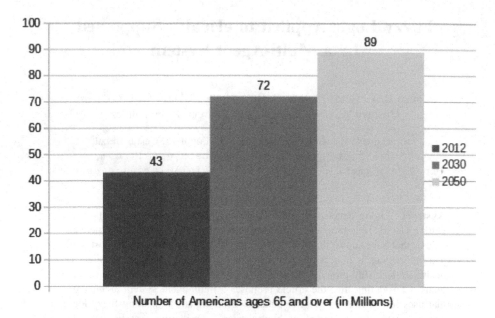

Fig. 1. Evolution of Americans' expectance of life

and wearable devices have been integrated with medicine purposes, structuring the medical internet of things. IoT has areas residing on medical and health care [3]. Interconnection of sensors and medical devices has been the goal of many industry players, considering that this is part of the core of IoT. This research has been conducted in order to provide an environment on which technology can improve people's health and welfare.

In this article, a Multi-Agent solution using IoT that monitors heart signals, processes remotely the information gathered from the device, and infers diagnosis related to high blood pressure is proposed. The solution proposed applies fuzzy logic, a multi-agent system and a publish/subscribe pattern. It is organized with theoretical reference on Sect. 2, related works on Sect. 3, the architecture of the MAS on Sect. 4, Sect. 5 presents the application and results of the solution proposed, and Sect. 6 concludes the article and brings future works.

2 Theoretical Reference

2.1 IoT Healthcare Networks

One of the most important elements of the IoT in healthcare is the IoT Healthcare Network (IoThNet) [19]. It provides access to the IoT backbone, facilitates medical data transmission and reception, and enables the use of healthcare-tailored communications. The way on which sign data are gathered using sensors, such as ECG, Blood Pressure, etc., and how IoThNet topology distributes, processes and visualizes those data are depicted over Fig. 2.

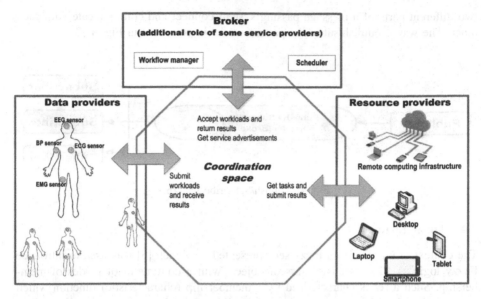

Fig. 2. A conceptual diagram of IoT-based ubiquitous healthcare solutions [19].

2.2 Sensors and Controllers

In order to get the electrical activity from the heart over a period of time, it is necessary using electrodes placed on the skin, and a shield connected to those electrodes over wires. There are several types of shields available on the market, such as the AD8232 Heart Rate Monitor Hookup [4], to provide ECG. That hardware works with an Arduino or NodeMCU board.

Arduino consists of both a physical programmable circuit board (often referred to as a microcontroller) and a piece of software, or IDE (Integrated Development Environment) which runs on a computer and is used to write and upload computer code to the physical board. NodeMCU is an open source IoT platform, that includes firmware which runs on a Wi-Fi environment. These devices are shown on Fig. 3.

(A)   **(B)**   **(C)**

Fig. 3. (A) AD8232, (B) Arduino, (C) NodeMCU.

2.3 Publish/Subscribe Pattern

Considering software architecture, there are patterns which are used as a solution that can be reused in order to solve problems related to software architecture within a given context. The publish/subscribe pattern belongs to message pattern which describes how two different parts of a message passing system connect and communicate with each other. The way a publish/subscribe pattern works is presented in Fig. 4.

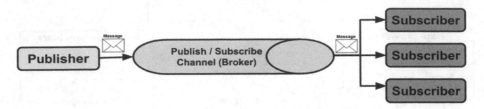

Fig. 4. Publish/Subscribe pattern.

2.4 Fuzzy Logic

The context of the theory of fuzzy sets, presented by Zadeh [5] was used to introduce Fuzzy logic. A fuzzy set is a class of objects with a continuum of grades of membership. Such a set is characterized by a membership (characteristic) function which assigns to each object a grade of membership ranging between zero and one. The notions of inclusion, union, intersection, complement, relation, convexity, etc., are extended to such sets, and various properties of these notions in the context of fuzzy sets are established.

While variables usually take numerical values in mathematics, in fuzzy logic applications non-numeric values are often used to facilitate the expression of rules and facts. A linguistic variable such as height may accept values such as tall and its antonym short. Because natural languages do not always contain enough value terms to express a fuzzy value scale, it is common practice to modify linguistic values with adjectives or adverbs [6]. As shown in Fig. 5, fuzzy logic process consists in four main steps:

- Fuzzifier Module - It transforms the system inputs, which are crisp numbers, into fuzzy sets.
- Rule Base - It stores IF-THEN rules provided by experts.
- Inference Engine - It simulates the human reasoning process by making fuzzy inference on the inputs and IF-THEN rules.
- Defuzzifier Module - It transforms the fuzzy set obtained by the inference engine into a crisp value.

Membership functions are used to quantify linguistic term and represent a fuzzy set graphically. A membership function for a fuzzy set A on the universe of discourse X is defined as

$$\mu A : X \rightarrow [0, 1] \tag{1}$$

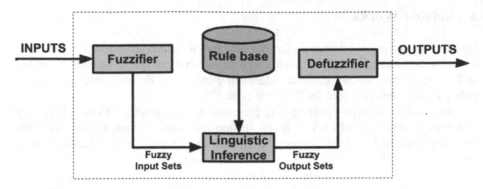

Fig. 5. Fuzzy logic overview.

2.5 Multi-Agent Systems (MAS)

In artificial intelligence research, the agent-based approach is now extremely popular in software engineering [18]. It has also infiltrated the area of operating systems, where autonomic computing refers to computer systems and networks that monitor and control themselves with a perceive–act loop and machine learning methods. According to Russell and Norving [7], an agent is "anything that can be viewed as perceiving its environment through sensors and acting upon that environment through actuators." Figure 6 shows the elements of an agent.

Increasingly, multiple agents that can work together are required by applications. A multi-agent system (MAS) is a loosely coupled network of software agents that interact to solve problems that are beyond the individual capacities or knowledge of each problem solver [8].

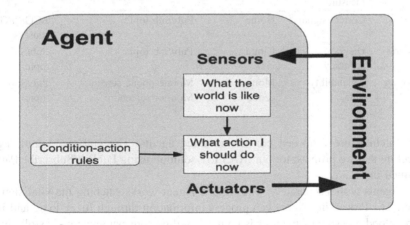

Fig. 6. Elements of an agent [7].

3 Related Works

Considering that data exchange performs a vital role in healthcare, a Publish/Subscribe based Architecture moderated through a web service that can enable an early exchange of healthcare data among different interested parties e.g. doctors, researchers, and policy makers was proposed by Wadhwa et al. [9].

Studies about lifestyle profiling were presented by Días-Rodríguez et al. [17]. They proposed an ontology which is a fuzzy version and augmentation of the wearables ontology presented in [18]. Fuzzy Description Logics was used, because it is a well-known family of logics for knowledge representation. They are considered the main formalism to represent fuzzy ontologies. The fuzzy datatypes in that fuzzy ontology were automatically learnt from data records using clustering algorithms.

The interest in keeping a good health status through the use of monitoring devices has been growing during last decade. Considering this, a two-stage fuzzy logic approach in which the device tries to learn and fit customer habits to discover outlier warning signals was proposed by Santamaria et al. [10].

4 MAS Solution Architecture Using IoT

It is important to know and understand where the agents will work. As stated by Russell [7], the task environment involves the PEAS (Performance, Environment, Actuators, Sensors) that should be described when designing an agent, as described on Table 1.

Table 1. PEAS description of the task environment

Agent type	Performance measure	Environment	Actuators	Sensors
Mobile	Correct signals	Home	Pub/Sub topic	ECG/PPG sensors
Processing	Diagnosis accuracy	Cloud	Pub/Sub topic	Pub/Sub topic
Monitoring	Availability time	Monitor centre	Mobile phone screen/ Monitor screen	Pub/Sub topic

The architecture proposed consists in three agents. The embedded multi-agent architecture that we propose for our MAS/IoT solution using Publish/Subscribe Pattern is resumed in Fig. 7.

All Agents works separately. The Mobile Agent works catching the vital signs of the user. The Processing Agent can process information through fuzzy logic and infer how the blood pressure of the user is going, based on four parameters - systolic blood pressure, diastolic blood pressure, age and body mass index, and indicates some abnormality. The Monitoring Agent gets the process result which is shown on an output device. Publish/Subscribe Pattern allows the connectivity of this environment.

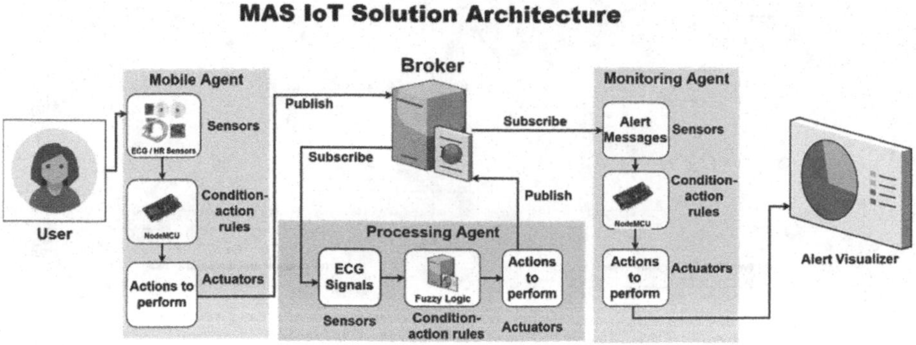

MAS IoT Solution Architecture

Fig. 7. Architecture proposed - MAS (3 agents).

In this article we proposed a MAS/IoT solution, that consists in:

1. Publish/Subscribe environment using MQTT broker;
2. Mobile Agent: consists in a mobile MQTT client encapsulated on a physical device containing ECG sensors and NodeMCU board, eliminate some noise, and publish on a topic on MQTT broker. Each mobile agent should have unique Id of the user and information of age and BMI.
3. Processing Agent: There is a subscriber service developed in python that gets the published data. As proposed in [13], we developed a fuzzy expert system using R [16], for the management of hypertension (High Blood Pressure), that classifies the hypertension risk as Mild, Moderate or Severe. The results are stored on the database and published on a topic named "alerts" on MQTT broker;
4. Monitoring Agent: The results can be shown in two ways: NodeMCU with a Led, that is a subscriber of a specific patient pre-configured topic, and a website that access the database and show the results according to available filters.

5   Application and Results

In order to implement the proposed MAS/IoT solution, we used AD8232 and MAX30100 shields and nodeMCU board considering the mobile agent. The MQQT [12] connected the environment. Regarding estimation of blood pressure using ECG and PPG signals, we followed Kumar and Ayub [4]. In [13] it is proposed a web-based fuzzy expert system for the management of hypertension (High Blood Pressure), whose classifiers can detect the hypertension risk as Mild, Moderate or Severe.

The solution has been used in two ways: first, using ECG and EPP sensors, allowing that all three agents could be used. Second, we imported the "Pima Indians Diabetes Data Set" from UCI Machine Learning Repository Database [14] into MySQL database, allowing the processing agent to be used in its fullness, using real data from 768 distinct patients.

Some adjustments were made on classifiers to keep them aligned to 7th Brazilian Director of Blood Hypertension [10]. The input membership function for input parameters is shown in Fig. 8.

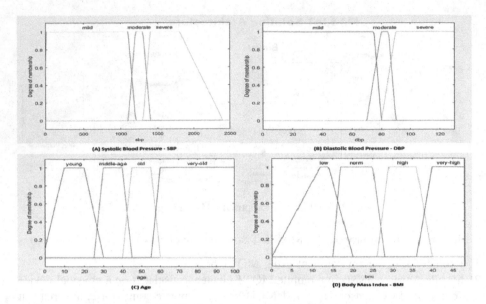

Fig. 8. Input membership functions.

The input membership functions considered the following:

- SBP (Trapezoidal): Mild, Moderate, Severe
- Diastolic (Trapezoidal): Mild, Moderate, Severe
- Age (Trapezoidal): Young, Middle-Aged, Old, Very Old
- Body Mass Index (Trapezoidal): Low, Normal, High, Very High

The output membership function considered the following, as shown in Fig. 9:
Blood Pressure (Trapezoidal): Mild, Moderate, Severe

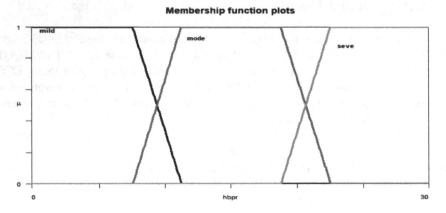

Fig. 9. Output membership function.

Using the first way, the data was validated and published on MQTT broker. After that, the published values are taken by Processing Agent and the information is processed using fuzzy logic. When some abnormality was detected, the inferred diagnostic was published on MQTT broker into an alert topic. Finally, the Monitoring Agent has a feature as subscriber of that alert topic and when any alert is read, the NodeMCU turn its LED on, meaning that some ab-normality was detected. During the tests using this first method some inconsistences where detected, indicating improvements in the code.

Considering the second way of use the application, to validate the applied fuzzy logic, the data stored into MySQL database was processed by the Processing Agent, and the results were stored into another table of the database. Figure 10 presents the achieved results by diagnosis.

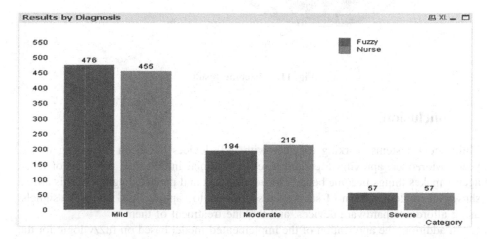

Fig. 10. Achieved results by diagnosis.

Using either fuzzy or nurse evaluation, the severe diagnosis where the same – 57 patients. The moderate diagnosis had a little difference applying those evaluation – 215 for nurse and 194 for fuzzy (difference on 21 results). The mild diagnosis was the most expressive representation, but the difference between the evaluation methods – 476 for fuzzy and 455 for nurse (difference on 21 results).

Analyzing the universe set of 768 patients on the database, 41 patients data were considered invalid due to some constraints, such as diastolic blood pressure value equals to zero. As shown in Fig. 11, the remaining universe of 727 patients, we had an amount of 41 where the diagnosis given by the nurses and fuzzy don't match, and 686 that match.

The monitoring agent has taken the results of processing stored in the database and made available through a web page with search filters. Moreover, that agent turned the led of the shield on, taking into account that some alert topic was generated on MQTT broker.

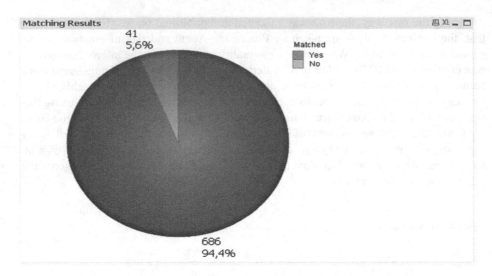

Fig. 11. Matching results.

6　Conclusion

Multi-agent systems working together with IoT devices has been increasing in last years. Moreover, applying a certain level of artificial intelligence on some of those agents makes things become better, considering fast and precise diagnosis. This article shows an implementation of such agents in order to capture ECG and PPG signals, using affordable hardware devices, and the due treatment of them.

In addition, the application of the implemented model based on fuzzy logic for the diagnosis of blood pressure level, with data obtained from public databases, was interesting, achieving a positive predictivity of 94.40% in High Blood Pressure Diagnosis.

As a future work, we intend finish the integration of this model that uses fuzzy logic within an agent as proposed in the architecture, turning all elements into a single solution and improve the code to calculate systolic and diastolic blood pressure from ECG and PPG data.

References

1. National Institute on Aging: Aging Well in the 21st Century: Strategic Directions for Research on Aging. Bethesda (2016)
2. Haghi, M., Thurow, K., Stoll, R.: Wearable devices in medical internet of things: scientific research and commercially available devices. US National Library of Medicine National Institutes of Health, January 2017
3. Pang, Z.: Technologies and architectures of the Internet-of-Things (IoT) for health and well-being. KTH Royal Institute of Technology, Kista, Sweden, xiv, 75 p (2013)

4. Kumar, S., Ayub, S.: Estimation of blood pressure by using Electrocardiogram (ECG) and Photoplethysmogram (PPG). In: Fifth International Conference on Communication Systems and Network Technologies, pp. 521–524 (2015)
5. Zadeh, L.A.: Fuzzy Sets. Inf. Control **8**, 338–358 (1965)
6. Tamir, D.E., et al.: Fifty Years off Fuzzy Logic and its Applications. Studies in Fuzziness and Soft Computing, vol. 326. Springer, Heidelberg (2015). https://doi.org/10.1007/978-3-319-19683-1
7. Russell, S.J., Norvig, P.: Artificial Intelligence A Modern Approach, 3rd edn. Prentice Hall, Upper Saddle River (2010)
8. Shoham, Y., Leyton-Brown, K.: Multiagent Systems. "Algorithmic, Game-Theoretic, and Logical Foundations", rev. 1.1 (2010)
9. Wadhwa, R., Mehra, A., Singh, P., Singh, M.: A Pub/Sub based architecture to support public healthcare data exchange. In: Net Health Workshop, COMSNETS 2015 (2015). http://ieeexplore.ieee.org/document/7098706/
10. Santamaria, A.F., Raimondo, P., Rango, F., Serianni, A.: A two stages fuzzy logic approach for internet of things (IoT) wearable devices. In: 2016 IEEE 27th Annual International Symposium on Personal, Indoor, and Mobile Radio Communications (PIMRC): Workshop: Internet of Things for Ambient Assisted Living (IoTAAL) (2016). http://ieeexplore.ieee.org/document/7794563/
11. Toader, C.G.: Multi-agent based e-health system. In: 2017 21st International Conference on Control Systems and Computer Science (2017). http://ieeexplore.ieee.org/document/7968635/
12. The MQTT home page (2017). http://mqtt.org/
13. Chandra, V., Singh, P.: Fuzzy based high blood pressure diagnosis. Int. J. Adv. Res. Comput. Sci. Amp. Technol. IJARCST, **2**(2) (2014). Ver. 1
14. Sigillito, V.: Pima indians diabetes data set. In: Lichman, M. (ed.) The Johns Hopkins University UCI Machine Learning Repository https://archive.ics.uci.edu/ml/ School of Information and Computer Science, University of California, Irvine, CA (2013)
15. Imai, Y., Aihara, A., Ohkubo, T., Nagai, K., Tsuji, I., Minami, N., Satoh, H., Hisamichi, S.: Factors that affect blood pressure variability. Am. J. Hypertens. **10**(11), 1281–1289 (1997). https://doi.org/10.1016/S0895-7061(97)00277-X
16. Díaz-Rodríguez, N., Härmä, A., Helaoui, R., Huitzil, I., Bobillo, F., Straccia, U.: Couch potato or gym addict? Semantic lifestyle profiling with wearables and knowledge graphs. In: Proceedings of the 6th Workshop on Automated Knowledge Base Construction (AKBC 2017), Long Beach (USA), December 2017
17. Díaz Rodríguez, N., et al.: An ontology for wearables data interoperability and ambient assisted living application development. In: Zadeh, L., Yager, R., Shahbazova, S., Reformat, M., Kreinovich, V. (eds.) Recent Developments and the New Direction in Soft-Computing Foundations and Applications, World Conference on Soft Computing. Studies in Fuzziness and Soft Computing, vol. 361, pp. 559–568. Springer, Heidelberg (2018). https://doi.org/10.1007/978-3-319-75408-6_43
18. Ciancarini, P., Wooldridge, M.: Agent-Oriented Software Engineering. Springer, Heidelberg (2001). https://doi.org/10.1007/3-540-44564-1
19. Islam, S.M.R., Kwak, D., Kabir, M.H., Hossain, M., Kwak, K.S.: The internet of things for health care: a comprehensive survey. IEEE Access **3**, 678–708 (2015). https://doi.org/10.1109/access.2015.2437951

Some Examples of Relations Between F-Transforms and Powerset Theories

Jiří Močkoř[(⊠)]

Institute for Research and Applications of Fuzzy Modeling,
University of Ostrava, 30. dubna 22, 701 03 Ostrava 1, Czech Republic
Jiri.Mockor@osu.cz
http://irafm.osu.cz/

Abstract. Six examples of powerset theories based on lattice-valued fuzzy sets are presented and relations between these powerset theories and F-transforms are investigated. It is proved that powerset extensions corresponding to these powerset theories are identical to, or restrictions of the F-transforms with respect to spaces with fuzzy partitions, consisting of objects of corresponding powerset theories.

1 Introduction

In fuzzy set theory there are two important methods which are frequently used both in theoretical research and applications. These methods are the *powerset theory and the F-transform*. Both these methods were in full details and theoretical backgrounds introduced relatively recently and, in the present, both methods represent very strong tools in the theory and applications.

The powerset structures are widely used in algebra, logic, topology and also in computer science. The standard example of a powerset structure $P(X) = \{A : A \subseteq X\}$ and the corresponding extension of a mapping $f : X \to Y$ to the map $f_P^{\to} : P(X) \to P(Y)$ is widely used in almost all branches of mathematics and it applications, including computer science. For illustrative examples of possible applications see, e.g., the introductory part of the paper of [22]. Because the classical set theory can be considered to be a special part of the fuzzy set theory, introduced by [25], it is natural that powerset objects associated with fuzzy sets were soon investigated as generalizations of classical powerset objects. The first approach was done again by Zadeh [25], who defined $[0, 1]^X$ to be a new powerset object $Z(X)$ instead of $P(X)$ and introduced the new powerset operator $f_Z^{\to} : Z(X) \to Z(Y)$, such that for $s \in Z(X), y \in Y$,

$$f_Z^{\to}(s)(y) = \bigvee_{x, f(x)=y} s(x).$$

A lot of papers were published about Zadeh's extension and its generalizations, see, e.g., [5,9,11,13,19–22]. Zadeh's extension was for the first time intensively

This research was partially supported by the project 18-06915S provided by the Grant Agency of the Czech Republic.

© Springer International Publishing AG, part of Springer Nature 2018

G. A. Barreto and R. Coelho (Eds.): NAFIPS 2018, CCIS 831, pp. 72–83, 2018.
https://doi.org/10.1007/978-3-319-95312-0_7

studied by Rodabaugh in [19]. This paper was, in fact, the first real attempt to uniquely derive the powerset operator $f_Z^{\rightarrow}$ from $f_P^{\rightarrow}$ and not only explicitly stipulate them. The works of Rodabaugh gave very serious basis for further research of powerset objects and operators. That new approach to the powerset theory was based on application of the *theory of monads in clone form*, introduced by Manes [8]. A special example of monads in clone form was introduced by Rodabaugh [21] as a special structure describing powerset objects. In the papers [9] and [11] we presented some examples of powerset theories based on fuzzy sets which are generated by monads in clone form.

Another important method which was recently introduced in the fuzzy set theory is the F-transform. This theory was in lattice-valued form introduced by Perfilieva [17] and elaborated in many other papers (see, e.g., [14–16,18]). Analogically as the powerset operator $f_P^{\rightarrow} : P(X) \rightarrow P(Y)$, F-transform is a special transformation map $F : \mathcal{L}^X \rightarrow \mathcal{L}^Y$, that transforms $\mathcal{L}$-valued fuzzy sets defined in the set X to $\mathcal{L}$-valued fuzzy sets defined in another set Y.

Fuzzy transforms represent new methods that have been successfully used in signal and image processing [1,2,5], signal compressions [14], numerical solutions of ordinary and partial differential equations [7,23,24], data analysis [3,4,16] and many other applications.

The aim of the paper is to show that there exists a very close relation between these two theories. Namely, we show that many of powerset theories are, in fact, identical to F-transforms, or are restrictions of the F-transforms. Especially, it holds for powerset theories $\mathbf{T} = (T, \rightarrow, V, \eta)$ in a category $\mathcal{K}$, such that for any object $X \in \mathcal{K}$ there exists a natural embedding $i_X : T(X) \hookrightarrow \mathcal{L}^X$.

In the paper we introduce six examples of powerset theories $\mathbf{T}$ in the categories **Set** of sets, the category **Set**($\mathcal{L}$) of sets with similarity relations and in the category **SpaceFP** of spaces with fuzzy partitions, respectively, and we prove that all these powerset theories are, in fact, restrictions of F-transforms. Hence, in those examples, both theories coincide. This relation between powerset theories and F-transforms enables us to apply together all methods and tools used in both theories.

2 Preliminary Notions

In this section we present some preliminary notions and definitions which could be helpful for better understanding of results concerning sets with similarity relations and other tools. A principal structure used in the paper is the *complete residuated lattice* (see e.g. [23]), i.e., a structure $\mathcal{L} = (L, \wedge, \vee, \otimes, \rightarrow, 0, 1)$ such that $(L, \wedge, \vee)$ is a complete lattice, $(L, \otimes, 1)$ is a commutative monoid with operation $\otimes$ isotone in both arguments and $\rightarrow$ is a binary operation which is residuated with respect to $\otimes$, i.e.

$$\alpha \otimes \beta \leq \gamma \quad \text{iff} \quad \alpha \leq \beta \rightarrow \gamma.$$

Recall that a set with a similarity relation (or $\mathcal{L}$-set) is a couple (X, δ), where $\delta : X \times X \rightarrow \mathcal{L}$ is a reflexive, symmetric and $\otimes$-transitive map. In the paper we use the category $\mathbf{Set}(\mathcal{L})$ of $\mathcal{L}$-sets as objects and with maps $f : X \to Y$ such that $\gamma(f(x), f(y)) \geq \delta(x, y)$, for all $x, y \in X$, as morphisms $f : (X, \delta) \to (Y, \gamma)$. If $\leftrightarrow$ is the bi-residuum operation in $\mathcal{L}$, then $\leftrightarrow$ is a similarity relation in L and we can consider the $\mathcal{L}$-set $(L, \leftrightarrow)$. A morphism $f : (X, \delta) \to (L, \leftrightarrow)$ in the category $\mathbf{Set}(\mathcal{L})$ is called *an extensional map* with respect to δ. The set of all extensional maps with respect to (X, δ) is denoted by $F(X, \delta)$.

Recall that a *cut* in a set X is a system $(C_\alpha)_{\alpha \in L}$ of subsets of X such that $C_\alpha \subseteq C_\beta$ if $\alpha \geq \beta$, and the set $\{\alpha \in L : a \in C_\alpha\}$ has the greatest element for any $a \in X$. By $D(X)$ we denote the set of all cuts in X.

Analogically as in classical sets, in $\mathcal{L}$-sets we can define the so called f-cuts.

Definition 1 ([10]). *Let (X, δ) be an $\mathcal{L}$-set. Then a system $\mathbf{C} = (C_\alpha)_\alpha$ of subsets in A is called an **f-cut** in (X, δ) in the category $\mathbf{Set}(\mathcal{L})$ if*

1. *$\forall a, b \in X, \quad a \in C_\alpha \Rightarrow b \in C_{\alpha \otimes \delta(a,b)}$,*
2. *$\forall a \in X, \forall \alpha \in \mathcal{L}, \quad \bigvee_{\{\beta : a \in C_\beta\}} \beta \geq \alpha \Rightarrow a \in C_\alpha$.*

The set of all f-cuts in (X, δ) is denoted by $C(X, \delta)$.

Any system of subsets $(C_\alpha)_\alpha$ in a set X can be extended to the f-cut $(\overline{C_\alpha})_\alpha$, defined by

$$\overline{C_\alpha} = \{a \in X : \bigvee_{\{(x,\beta) : x \in C_\beta\}} \beta \otimes \delta(a, x) \geq \alpha\}.$$

The set $C(X, \delta)$ of all f-cuts in a $\mathcal{L}$-set (X, δ) can be ordered by $(C_\alpha)_\alpha \leq (D_\alpha)_\alpha$ iff $C_\alpha \subseteq D_\alpha$, for each $\alpha \in \mathcal{L}$. Then $C(X, \delta)$ is a complete $\vee$-semilattice, such that $\bigvee_{i \in I}(C_\alpha^i)_\alpha = (\overline{\bigcup_{i \in I} C_\alpha^i})_\alpha$.

We recall some basic facts about the F-transform method. Recall that a *core* of a ($\mathcal{L}$-valued) fuzzy set $f : X \to L$ is defined by $core(f) = \{x \in X : f(x) = 1\}$. A normal ($\mathcal{L}$-)valued fuzzy set f in a set X is such that $core(f) \neq \emptyset$.

The F-transform in a form introduced by Perfilieva [18] is based on the so called fuzzy partitions on the crisp set. It should be noted that the definition introduced in [18] differs from the standard definition of a fuzzy partition, which is mostly defined by a $\mathcal{L}$-valued similarity relation.

Definition 2 ([18]). *Let X be a set. A system $\mathcal{A} = \{A_\lambda : \lambda \in \Lambda\}$ of normal $\mathcal{L}$-valued fuzzy sets in X is a fuzzy partition of X, if $\{core(A_\lambda) : \lambda \in \Lambda\}$ is a partition of X. A pair $(X, \mathcal{A})$ is called a space with a fuzzy partition. The index set of $\mathcal{A}$ will be denoted by $|\mathcal{A}|$.*

In the paper [12] we introduced the category $\mathbf{SpaceFP}$ of spaces with fuzzy partitions.

Definition 3. *The category $\mathbf{SpaceFP}$ is defined by*

1. *Fuzzy partitions $(X, \mathcal{A})$, as objects,*
2. *Morphisms $(g, \sigma) : (X, \{A_\lambda : \lambda \in \Lambda\}) \to (Y, \{B_\omega : \omega \in \Omega\})$, such that*
   (a) $g : X \to Y$ and is $\sigma : \Lambda \to \Omega$ are mappings,
   (b) $\forall \lambda \in \Lambda, A_\lambda(x) \leq B_{\sigma(\lambda)}(g(x))$, for each $x \in X$.

3. *The composition of morphisms in* **SpaceFP** *is defined by* $(h, \tau) \circ (g, \sigma) = (h \circ g, \tau \circ \sigma)$.

Objects of the category **SpaceFP** represent ground structures for a fuzzy transform, firstly proposed by Perfilieva [17] and, in the case where it is applied to $\mathcal{L}$-valued functions with $\mathcal{L}$-valued partitions, in [18]. Lattice-valued fuzzy transforms are defined in two variants - lower and upper F-transforms. In the paper we deal with the upper F-transform only.

Definition 4 ([18]). *Let* $(X, \mathcal{A})$ *be a space with a fuzzy partition* $\mathcal{A} = \{A_\lambda : \lambda \in |\mathcal{A}|\}$. *The (upper) F-transform of functions from* $\mathcal{L}^X$ *with respect to the space* $(X, \mathcal{A})$ *is a function* $F_{X,\mathcal{A}} : \mathcal{L}^X \to \mathcal{L}^{|\mathcal{A}|}$, *defined by*

$$f \in \mathcal{L}^X, \lambda \in |\mathcal{A}|, \quad F_{X,\mathcal{A}}(f)(\lambda) = \bigvee_{x \in X} (f(x) \otimes A_\lambda(x)).$$

In the residuated lattice $\mathcal{L}$ we can define a fuzzy partition $\mathbf{L} = \{L_\alpha : \alpha \in L\}$, such that $L_\alpha(\beta) = \alpha \leftrightarrow \beta$, $\beta \in L$.

The following notion of the extensional fuzzy set in the category **SpaceFP** extends the notion of the extensional mapping in the category **Set**($\mathcal{L}$).

Definition 5. *A mapping* $f : X \to L$ *is called the extensional* $\mathcal{L}$-*fuzzy set in a space with a fuzzy partition* $(X, \mathcal{A})$ *in the category* **SpaceFP**, *if there there exists a map* $\sigma : |\mathcal{A}| \to L$, *such that* (f, σ) *is a morphism* $(X, \mathcal{A}) \to (L, \mathbf{L})$ *in the category* **SpaceFP**. *By* $F(X, \mathcal{A})$ *we denote the set of all extensional fuzzy sets in* $(X, \mathcal{A})$.

In [12] we proved that for any space with a fuzzy partition $(X, \mathcal{A})$ it is possible to construct the $\mathcal{L}$-set $(X, \delta_{X,\mathcal{A}})$ with the similarity relation called *characteristic similarity relations* of $(X, \mathcal{A})$. The similarity relation $\delta_{X,\mathcal{A}}$ is the minimal similarity relation defined in X, such that for arbitrary map $f : X \to L$, f is extensional in $(X, \mathcal{A})$ iff f is extensional with respect to $\delta_{X,\mathcal{A}}$. Hence, we have

$$F(X, \mathcal{A}) = E(X, \delta_{X,\mathcal{A}}).$$

3   Examples of Powerset Theories

In this section we repeat the basic definition of the powerset theory and we introduce six examples of typical powerset theories which are used in fuzzy sets and are based on standard fuzzy objects. Our claim is to show, that all these powerset theories are, in fact, identical to, or restrictions of the F-transform for appropriate fuzzy partition.

In what follows, by $CSLAT(\vee)$ we denote the category of complete $\vee$-semilattices as objects and with $\vee$-preserving maps as morphisms. By **Set** we denote the classical category of sets with mappings. We introduce the definition of a $CSLAT(\vee)$-powerset theory, which we use in the sequel.

Definition 6 (Rodabaugh [21]). *Let* **K** *be a ground category. Then* $\mathbf{T} = (T, \rightarrow, V, \eta)$ *is called* $CSLAT(\vee)$-**powerset theory in K,** *if*

1. $T : |\mathbf{K}| \rightarrow |CSLAT(\vee)|$ *is an object-mapping,*
2. *for each* $f : A \rightarrow B$ *in* **K**, *there exists* $f_{\vec{T}} : T(A) \rightarrow T(B)$ *in* $CSLAT(\vee)$,
3. *There exists a concrete functor* $V : \mathbf{K} \rightarrow \mathbf{Set}$, *such that* η *determines in* **Set** *for each* $A \in \mathbf{K}$ *a mapping* $\eta_A : V(A) \rightarrow T(A)$,
4. *For each* $f : A \rightarrow B$ *in* **K**, $f_{\vec{T}} \circ \eta_A = \eta_B \circ V(f)$.

In the paper we will deal with several examples of $CSLAT(\vee)$-powerset theories. Some of these examples were derived by previous authors, e.g., Rodabaugh [21], Höhle [6], Solovyov [22], other examples was presented in Močkoř [9,11]. It should be observed that in all these examples the object function $T : |\mathcal{K}| \rightarrow |CSLAT(\vee)|$ is, in fact, the object function of a functor $T : \mathcal{K} \rightarrow CSLAT(\vee)$, with $T(f) = f_{\vec{T}}$, for any morphism f in $\mathcal{K}$.

Example 1. $CSLAT(\vee)$-Powerset theory $\mathcal{P} = (P, \rightarrow, id, \eta)$ in the category **Set**, where

1. $P : |\mathbf{Set}| \rightarrow |CSLAT(\vee)|$ is defined by $P(X) = (2^X, \subseteq)$, and any element S of $P(X)$ is identified with the characteristic function χ_S^X of a subset $S \subseteq X$ in X.
2. for each $f : X \rightarrow Y$ in **Set**, $f_{\vec{P}} : P(X) \rightarrow P(Y)$ is defined by $f_{\vec{P}}(\chi_S^X) = \chi_{f(S)}^Y$,
3. for each $X \in \mathbf{Set}$, $\eta_X : X \rightarrow P(X)$ is the characteristic function $\chi_{\{x\}}^X$ of a subset $\{x\}$ defined in X.

Example 2. $CSLAT(\vee)$-Powerset theory $\mathcal{Z} = (Z, \rightarrow, id, \chi)$ in the category **Set**, where

1. $Z : |\mathbf{Set}| \rightarrow |CSLAT(\vee)|$ is defined by $Z(X) = \mathcal{L}^X$,
2. for each $f : X \rightarrow Y$ in **Set**, $f_{\vec{Z}} : \mathcal{L}^X \rightarrow \mathcal{L}^Y$ is defined by $f_{\vec{Z}}(s)(y) = \bigvee_{x \in X, f(x) = y} s(x)$,
3. for each $X \in \mathbf{Set}$, $\chi^X : X \rightarrow \mathcal{L}^X$ is defined by $\chi^X(a) = \chi_{\{a\}}^X$, for $a \in X$.

Example 3. $CSLAT(\vee)$-Powerset theory $\mathcal{D} = (D, \rightarrow, id, \rho)$ in the category **Set**, where

1. $D : |\mathbf{Set}| \rightarrow |CSLAT(\vee)|$ is defined by $D(X) =$ the set of all cuts $(C_\alpha)_{\alpha \in L}$ in a set X, naturally ordered by inclusion,
2. for each $f : X \rightarrow Y$ in **Set**, $f_{\vec{D}} : D(X) \rightarrow D(Y)$ is defined by $f_{\vec{D}}((C_\alpha)_\alpha) = \overline{(f(C_\alpha))}_\alpha \in D(Y)$, where the closure $(\overline{S_\alpha})_\alpha$ in a set Y is defined by $\overline{S_\alpha} = \{a \in Y : \bigvee_{\beta : a \in C_\beta} \beta \geq \alpha\}$,
3. for each $X \in \mathbf{Set}$, $\rho_X : X \rightarrow D(X)$ and $\rho_X(x)$ is defined as the constant cut $(\{x\})_\alpha$.

Example 4. $CSLAT(\vee)$-Powerset theory $\mathcal{E} = (E, \rightarrow, V, \widehat{\chi})$ in the category $\mathbf{Set}(\mathcal{L})$, where

1. $E : |\mathbf{Set}(\mathcal{L})| \to |CSLAT(\vee)|$, where $E(X, \delta)$ is the set of all functions $f \in \mathcal{L}^X$ extensional with respect to the similarity relation δ, ordered point-wise,
2. for each morphism $f : (X, \delta) \to (Y, \gamma)$ in $\mathbf{Set}(\mathcal{L})$, $f_E^{\to} : E(X, \delta) \to E(Y, \gamma)$ is defined by $f_E^{\to}(s)(y) = \bigvee_{x \in X} s(x) \otimes \gamma(f(x), y)$,
3. $V : \mathbf{Set}(\mathcal{L}) \to \mathbf{Set}$ is the forgetful functor,
4. for each $(X, \delta) \in \mathbf{Set}(\mathcal{L})$, $\widehat{\chi}_{(X,\delta)} : X \to E(X, \delta)$ is defined by $\widehat{\chi}_{(X,\delta)}(a)(x) = \delta(a, x)$, for $a, x \in X$.

Example 5. $CSLAT(\vee)$-Powerset theory $\mathcal{C} = (C, \to, V, \overline{\chi})$ in the category $\mathbf{Set}(\mathcal{L})$, where

1. $C : |\mathbf{Set}(\mathcal{L})| \to |CSLAT(\vee)|$ is defined by $C(X, \delta) = $ set of all f-cuts in (A, δ) in the category $\mathbf{Set}(\mathcal{L})$, naturally ordered by inclusion,
2. for each morphism $f : (X, \delta) \to (Y, \gamma)$ in $\mathbf{Set}(\mathcal{L})$, $f_C^{\to} : C(X, \delta) \to C(Y, \gamma)$ is defined by

$$f_C^{\to}((E_\alpha)_\alpha) = (\overline{f(E_\alpha)})_\alpha, \quad \overline{f(E_\alpha)} = \{b \in Y : \bigvee_{(y,\beta):y \in f(E_\beta)} \beta \otimes \gamma(b, y) \geq \alpha\},$$

3. $V : \mathbf{Set}(\mathcal{L}) \to \mathbf{Set}$ is the forgetfull functor,
4. for each $(X, \delta) \in \mathbf{Set}(\mathcal{L})$, $\overline{\chi}_{(X,\delta)} : X \to C(X, \delta)$ is defined by $\overline{\chi}_{(X,\delta)}(a) = (\overline{\{a\}})_\alpha$, where $\overline{\{a\}}_\alpha = \{b \in X : \delta(a, b) \geq \alpha\}$.

Example 6. A $CSLAT(\vee)$-powerset theory $\mathcal{F} = (F, \to, W, \vartheta)$ in the category $\mathbf{SpaceFP}$ is defined by

(1) $F : |\mathbf{SpaceFP}| \to |CSLAT(\vee)|$, defined by

$$F(X, \mathcal{A}) = \{f | f : X \to L \text{ is extensional in } (X, \mathcal{A})\},$$

ordered pointwise.
(2) For each $(f, u) : (X, \mathcal{A}) \to (Y, \mathcal{B})$ in $\mathbf{SpaceFP}$, $(f, u)_F^{\to} : F(X, \mathcal{A}) \to F(Y, \mathcal{B})$ is defined by

$$g \in F(X, \mathcal{A}), y \in Y, \quad (f, u)_F^{\to}(g)(y) = \bigvee_{x \in X} g(x) \otimes \delta_{Y,\mathcal{B}}(f(x), y),$$

where $\delta_{Y,\mathcal{B}}$ is the characteristic similarity relation in Y in a space with a fuzzy partition $(Y, \mathcal{B})$.
(3) $V : \mathbf{SpaceFP} \to \mathbf{Set}$ is the forgetfull functor, $V(X, \mathcal{A}) = X$,
(4) For each $(X, \mathcal{A})$ in $\mathbf{SpaceFP}$, $\vartheta_{(X,\mathcal{A})} : V(X, \mathcal{A}) \to F(X, \mathcal{A})$, $\vartheta_{(X,\mathcal{A})}(a)(x) = \delta_{X,\mathcal{A}}(a, x)$, for each $a, x \in X$.

It is clear that the powerset extension u_P from the Example 1 is the reduction of the powerset extension u_Z from the Example 2 to the set of all characteristic functions χ_S^X of subsets from the set X. On the other hand, the powerset extension u_Z is the reduction of the powerset extension u_E from the Example 4 to the subcategory $\mathbf{Set}$ of the category $\mathbf{Set}(\mathcal{L})$.

4   Powerset Functors and F-Transform

In this section we investigate relationships between powerset theories **T** from Examples 1–6 on one hand, and F-transforms $F : T(X) \to T(Y)$, defined for appropriate fuzzy partitions. We show that all powerset theories from these examples are, in fact, identical to F-transforms or restrictions of F-transform. In what follows, for $\alpha \in L$, by $\underline{\alpha}_X$ we denote the constant function from $\mathcal{L}^X$ with the value α.

Proposition 1. *Let $\mathcal{P}$ be the powerset theory in the category* **Set** *from the Example 1 and let $u : X \twoheadrightarrow Y$ be a surjective map. Then there exists a fuzzy partition $\mathcal{A} \subseteq P(X)$, such that the powerset operator $u_P^{\rightarrow}$ is the restriction of the F-transform $F_{X,\mathcal{A}}$, i.e., the following diagram commutes:*

$$
\begin{array}{ccc}
P(X) & \xrightarrow{u_P^{\rightarrow}} & P(Y) \\
\downarrow & & \downarrow \\
\mathcal{L}^X & \xrightarrow{F_{X,\mathcal{A}}} & \mathcal{L}^Y.
\end{array}
$$

Proof. Let $\alpha \in \mathcal{L}$, $\alpha = 0$, or $\alpha = 1$. Then, for any $x \in X, S \subseteq X$,

$$u_P^{\rightarrow}(\underline{\alpha}_X \otimes \chi_S^X) = \underline{\alpha}_Y \otimes \chi_{u(S)}^Y. \tag{1}$$

holds. The equality (1) is correct, because $\underline{\alpha}_X \otimes \chi_S^X \in P(X)$. As it can be proved simply, for arbitrary $S \subseteq X$, the following equality holds:

$$\chi_S^X = \bigvee_{x \in X} \chi_S^X(x)_X \otimes \chi_{\{x\}}^X.$$

Let $\mathcal{A} = \{A_y : y \in Y\} \subseteq P(X)$ be defined by $A_y(x) = u_P^{\rightarrow}(\chi_{\{x\}}^X)(y)$, for arbitrary $x \in X, y \in Y$. Then, $\mathcal{A}$ is a fuzzy partition in $P(X)$, as follows from $A_{u(x)}(x) = u_P^{\rightarrow}(\chi_{\{x\}})(u(x)) = \chi_{\{u(x)\}}(u(x)) = 1$. Since $u_P^{\rightarrow}$ is a morphism in $CSLAT(\vee)$, from the relation (1) it follows for arbitrary $y \in Y$

$$u_P^{\rightarrow}(\chi_S^X)(y) = u_T^{\rightarrow}(\bigvee_{x \in X} \chi_S^X(x)_X \otimes \chi_{\{x\}}^X)(y) =$$

$$\bigvee_{x \in X} u_P^{\rightarrow}(\chi_S^X(x)_X \otimes \chi_{\{x\}}^X)(y) = \bigvee_{x \in X} \chi_S^X(x)_Y \otimes u_P^{\rightarrow}(\chi_{\{x\}}^X)(y) =$$

$$\bigvee_{x \in X} \chi_S^X(x) \otimes A_y(x) = F_{X,\mathcal{A}}(\chi_S^X)(y),$$

and the diagram commutes. In that case, for arbitrary $x \in X, y \in Y$, we have

$$A_y(x) = \begin{cases} 1, & \text{iff } u(x) = y, \\ 0, & \text{otherwise.} \end{cases}$$

$\square$

Proposition 2. *Let $\mathcal{Z}$ be the powerset theory in the category* **Set** *from the Example 2 and let $u : X \twoheadrightarrow Y$ be a surjective map. Then there exists a fuzzy partition $\mathcal{A} \subseteq Z(X)$, such that the powerset operator $u_P^{\rightarrow}$ equals to the F-transform $F_{X,\mathcal{A}}$,*

$$u_{\mathcal{Z}}^{\rightarrow} = F_{X,\mathcal{A}}.$$

Proof. Let $\alpha \in \mathcal{L}, f \in \mathcal{L}^X$. Then, we have

$$u_{\mathcal{Z}}^{\rightarrow}(\underline{\alpha}_X \otimes f) = \underline{\alpha}_Y \otimes u_{\mathcal{Z}}^{\rightarrow}(f), \tag{2}$$

as follows by a simple computation. The following equality was proven by Rodabaugh [20] for arbitrary $f \in \mathcal{L}^X$:

$$f = \bigvee_{x \in X} \underline{f(x)}_X \otimes \chi_{\{x\}}^X. \tag{3}$$

Let $\mathcal{A} = \{A_y : y \in Y\} \subseteq Z(X)$, where $A_y(x) = u_{\mathcal{Z}}^{\rightarrow}(\chi_{\{u(x)\}})(y)$, for arbitrary $x \in X, y \in Y$. Since $A_{u(x)}(x) = 1$, $\mathcal{A}$ is a fuzzy partition of $Z(X)$. Since $u_{\mathcal{Z}}^{\rightarrow}$ is a morphism in the category $CSLAT(\vee)$, from the equalities (2) and (3), for arbitrary $f \in \mathcal{L}^X$, it follows

$$u_{\mathcal{Z}}^{\rightarrow}(f)(y) = u_{\mathcal{Z}}^{\rightarrow}(\bigvee_{x \in X} \underline{f(x)}_X \otimes \chi_{\{x\}}^X)(y) = \bigvee_{x \in X} u_{\mathcal{Z}}^{\rightarrow}(\underline{f(x)} \otimes \chi_{\{x\}}^X)(y) =$$

$$\bigvee_{x \in X} \underline{f(x)}_Y(y) \otimes u_{\mathcal{Z}}^{\rightarrow}(\chi_{\{x\}}^X)(y) = \bigvee_{x \in X} \underline{f(x)}_Y(y) \otimes \chi_{\{u(x)\}}^Y(y) =$$

$$\bigvee_{x \in X} f(x) \otimes A_y(x) = F_{X,\mathcal{A}}(f)(y).$$

In that case, fuzzy sets A_y are the same as in the previous proposition.

$\square$

Proposition 3. *Let $\mathcal{E}$ be the powerset theory in the category* $\mathbf{Set}(\mathcal{L})$ *from the Example 4 and let $u : (X, \delta) \twoheadrightarrow (Y, \gamma)$ be an morphism in* $\mathbf{Set}(\mathcal{L})$, *such that u is surjective. Then there exists a fuzzy partition $\mathcal{A} \subseteq E(X, \delta)$, such that the powerset operator $u_E^{\rightarrow}$ is the restriction of the F-transform $F_{X,\mathcal{A}}$, i.e., the following diagram commutes:*

$$
\begin{array}{ccc}
E(X, \delta) & \xrightarrow{u_E^{\rightarrow}} & E(Y, \gamma) \\
\uparrow & & \uparrow \\
\mathcal{L}^X & \xrightarrow{F_{X,\mathcal{A}}} & \mathcal{L}^Y.
\end{array}
$$

Proof. Let $\alpha \in \mathcal{L}, f \in \mathcal{L}^X$. Then, we have

$$u_E^{\rightarrow}(\underline{\alpha}_X \otimes f) = \underline{\alpha}_Y \otimes u_E^{\rightarrow}(f), \tag{4}$$

as follows by a simple computation. The following equality was proven in [11]; Lemma 3.4, for arbitrary $f \in E(X, \delta)$, where $\widehat{\chi}_{(X,\delta)} : X \to E(X, \delta)$ is the embedding from the powerset theory $\mathcal{E}$:

$$f = \bigvee_{x \in X} \underline{f(x)}_X \otimes \widehat{\chi}_{(X,\delta)}(x). \tag{5}$$

Let $\mathcal{A} = \{A_y : y \in Y\} \subseteq E(X, \delta)$, where

$$A_y(x) = u_{\vec{E}}(\widehat{\chi}_{(X,\delta)}(x))(y) = \bigvee_{z \in X} \delta(z, x) \otimes \gamma(u(z), y) = \gamma(u(x), y),$$

for arbitrary $x \in X, y \in Y$. Since $A_{u(x)}(x) = 1$, $\mathcal{A}$ is a fuzzy partition of $E(X, \delta)$. Since $u_{\vec{E}}$ is a morphism in the category $CSLAT(\vee)$, for arbitrary $f \in E(X, \delta)$, from the equalities (4) and (5), it follows,

$$u_{\vec{E}}(f)(y) = u_{\vec{E}}(\bigvee_{x \in X} \underline{f(x)}_X \otimes \widehat{\chi}_{(X,\delta)}(x))(y) = \bigvee_{x \in X} u_{\vec{E}}(\underline{f(x)} \otimes \widehat{\chi}_{(X,\delta)}(x))(y) =$$

$$\bigvee_{x \in X} \underline{f(x)}_Y(y) \otimes u_{\vec{E}}(\widehat{\chi}_{(X,\delta)}(x))(y) = \bigvee_{x \in X} f(x) \otimes \widehat{\chi}_{(X,\delta)}(x)(y) = F_{X,\mathcal{A}}(f)(y).$$

$\square$

Proposition 4. *Let $\mathcal{C} = (C, \to, V, \overline{\chi})$ be the powerset theory in the category $\mathbf{Set}(\mathcal{L})$ from the Example 5 and let $u : (X, \delta) \twoheadrightarrow (Y, \gamma)$ be a morphism in $\mathbf{Set}(\mathcal{L})$, such that u is surjective. Then there exists a fuzzy partition $\mathcal{A} \subseteq C(X, \delta)$ and injective maps i_X, i_Y, such that the powerset operator $u_{\vec{C}}$ is the restriction of the F-transform $F_{X, i_X(\mathcal{A})}$, i.e., the following diagram commutes:*

$$
\begin{array}{ccc}
C(X, \delta) & \xrightarrow{\ u_{\vec{C}}\ } & C(Y, \gamma) \\
{\scriptstyle i_X}\big\uparrow & & \big\downarrow{\scriptstyle i_Y} \\
\mathcal{L}^X & \xrightarrow{\ F_{X, i_X(\mathcal{A})}\ } & \mathcal{L}^Y.
\end{array}
$$

Proof. Let $\mathbf{D} = (D_\alpha)_\alpha \in C(X, \delta)$. Then, we set

$$x \in X, \quad i_X(\mathbf{D})(x) = \bigvee_{\beta, x \in D_\beta} \beta.$$

In [10]; Theorem 4.2, it was proven that i_X is an injective maps. Let $\lambda \in \mathcal{L}, (C_\alpha)_\alpha \in C(X, \delta)$, then we set

$$\lambda \otimes (C_\alpha)_\alpha := (G_\alpha)_\alpha \in C(X, \delta), \quad G_\alpha = \{x \in X : \lambda \otimes \bigvee_{x \in C_\gamma} \gamma \geq \alpha\}.$$

It can be proven that $\overline{\chi}_{(X,\delta)}$ generates $C(X, \delta)$, i.e., for arbitrary $\mathbf{C} = (C_\alpha)_\alpha \in C(X, \delta)$, the following equality holds:

$$\mathbf{C} = \bigvee_{x \in X} i_X(\mathbf{C})(x) \otimes \overline{\chi}_{(X,\delta)}(x). \tag{6}$$

Further, for arbitrary $\mathbf{D} \in C(X, \delta), \alpha \in \mathcal{L}$, the following equalities can be proven simply:

$$u_{\vec{C}}(\alpha \otimes \mathbf{D}) = \alpha \otimes u_{\vec{C}}(\mathbf{D}),$$
$$i_X(\alpha \otimes \mathbf{D}) = \alpha \otimes i_X(\mathbf{D}).$$

We set

$$\mathcal{A} = \{A_y : y \in Y\} \subseteq C(X, \delta),$$
$$\forall y \in Y, y = u(x), \quad A_y = (\{z \in Y : i_Y(u_{\vec{C}}\overline{X}_{(X, \delta)}(x))(z) \geq \alpha\})_\alpha.$$

It can be proven that this definition is correct, i.e., $A_y \in C(X, \delta)$. Then, for arbitrary $\mathbf{D} \in C(X, \delta), y \in Y$, using the relation (6), we obtain

$$i_Y(u_{\vec{C}}(\mathbf{D}))(y) = i_Y u_{\vec{C}}(\bigvee_{x \in X} i_X(\mathbf{D})(x) \otimes \overline{X}_{(X, \delta)}(x))(y) =$$

$$\bigvee_{x \in X} i_X(\mathbf{D}) \otimes i_Y(u_{\vec{C}}\overline{X}_{(X, \delta)}(x))(y) =$$

$$\bigvee_{x \in X} i_X(\mathbf{D}) \otimes i_X(A_y)(x) = F_{X, i_X(\mathcal{A})}(i_X(\mathbf{D}))(y).$$

$\square$

Proposition 5. *Let $\mathcal{D} = (D, \rightarrow, id, \rho)$ be the powerset theory in the category* **Set** *from the Example 3 and let $u : X \twoheadrightarrow Y$ be a surjective mapping. Then there exists a fuzzy partition $\mathcal{A} \subseteq D(X)$ and injective maps i_X, i_Y, such that the powerset operator $u_{\vec{D}}$ is the restriction of the F-transform $F_{X, i_X(\mathcal{A})}$, i.e., the following diagram commutes:*

$$
\begin{array}{ccc}
D(X) & \xrightarrow{u_{\vec{D}}} & D(Y) \\
{\scriptstyle i_X}\downarrow & & \downarrow{\scriptstyle i_Y} \\
\mathcal{L}^X & \xrightarrow{F_{X, i_X(\mathcal{A})}} & \mathcal{L}^Y.
\end{array}
$$

Proof. Since $=$ is the similarity relation, the proof follows directly from the equation $D(X) = C(X, =)$ and from the proof of Proposition 4.

$\square$

Proposition 6. *Let $\mathcal{F} = (F, \rightarrow, W, \vartheta)$ be the powerset theory in the category* **SpaceFP** *from the Example 6 and let $(u, \sigma) : (X, \mathcal{A}) \rightarrow (Y, \mathcal{B})$ be a morphism in* **SpaceFP**, *such that u is a surjective map. Then there exists a fuzzy partition $\mathcal{A} \subseteq F(X, \mathcal{A})$, such that the powerset operator $(u, \sigma)_{\vec{F}}$ is the restriction of the F-transform $F_{X, \mathcal{A}}$, i.e., the following diagram commutes:*

$$
\begin{array}{ccc}
F(X, \mathcal{A}) & \xrightarrow{(u, \sigma)_{\vec{F}}} & F(Y, \mathcal{B}) \\
\downarrow & & \downarrow \\
\mathcal{L}^X & \xrightarrow{F_{X, \mathcal{A}}} & \mathcal{L}^Y.
\end{array}
$$

Proof: The proof follows directly from the equality $F(X, \mathcal{A}) = E(X, \delta_{X, \mathcal{A}})$ and from the proof of Proposition 3.

$\square$

5   Conclusions

We introduced six examples of powerset theories $\mathbf{T} = (T, \rightarrow, V, \eta)$ based on $\mathcal{L}$-valued fuzzy objects in the category **Set** of crisp sets, the category **Set**($\mathcal{L}$) of sets with similarity relations and in the category **SpaceFP** of spaces with fuzzy partitions, respectively. For these powerset theories we constructed special fuzzy partitions of powerset objects and we proved that for any epimorphism $f : X \rightarrow Y$ in these categories, the powerset extension $f_{\overrightarrow{T}}$ is identical, or a restriction, of the F-transform based on these fuzzy partitions. These results show that for some powerset theories $\mathbf{T}$ based on functors T from a category to the category of complete $\bigvee$-semilattices $CSLAT(\vee)$, where $T(X)$ can be embedded into the set $\mathcal{L}^{V(X)}$, the powerset extension $f_{\overrightarrow{T}} : T(X) \rightarrow T(Y)$ of a morphism $f : X \rightarrow Y$ can be substituted by the F-transform $F_{V(X), \mathcal{A}} : \mathcal{L}^{V(X)} \rightarrow \mathcal{L}^{V(Y)}$. Hence, in that case, the powerset theory and the F-transform theory coincide.

References

1. Di Martino, F., et al.: An image coding/decoding method based on direct and inverse fuzzy tranforms. Int. J. Approx. Reason. **48**, 110–131 (2008)
2. Di Martino, F., et al.: A segmentation method for images compressed by fuzzy transforms. Fuzzy Sets Syst. **161**(1), 56–74 (2010)
3. Di Martino, F., et al.: Fuzzy transforms method and attribute dependency in data analysis. Inf. Sci. **180**(4), 493–505 (2010)
4. Di Martino, F., et al.: Fuzzy transforms method in prediction data analysis. Fuzzy Sets Syst. **180**(1), 146–163 (2011)
5. Gerla, G., Scarpati, L.: Extension principles for fuzzy set theory. J. Inf. Sci. **106**, 49–69 (1998)
6. Höhle, U.: Many Valued Topology and Its Applications. Kluwer Academic Publishers, Boston (2001)
7. Khastan, A., Perfilieva, I., Alijani, Z.: A new fuzzy approximation method to Cauchy problem by fuzzy transform. Fuzzy Sets Syst. **288**, 75–95 (2016)
8. Manes, E.G.: Algebraic Theories. Springer, Berlin (1976). https://doi.org/10.1007/978-1-4612-9860-1
9. Močkoř, J.: Closure theories of powerset theories. Tatra Mt. Math. Publ. **64**, 101–126 (2015)
10. Močkoř, J.: Cut systems in sets with similarity relations. Fuzzy Sets Syst. **161**, 3127–3140 (2010)
11. Močkoř, J.: Powerset operators of extensional fuzzy sets. Iran. J. Fuzzy Syst. **15**, 143–163 (2017). https://doi.org/10.22111/IJFS.2017.3318
12. Močkoř, J., Holčapek, M.: Fuzzy objects in spaces with fuzzy partitions. Soft Comput. **21**, 7269–7284 (2017). https://doi.org/10.1007/s00500-016-2431-4
13. Nguyen, H.T.: A note on the extension principle for fuzzy sets. J. Math. Anal. Appl. **64**, 369–380 (1978)

14. Perfilieva, I.: Fuzzy transforms and their applications to image compression. In: Bloch, I., Petrosino, A., Tettamanzi, A.G.B. (eds.) WILF 2005. LNCS (LNAI), vol. 3849, pp. 19–31. Springer, Heidelberg (2006). https://doi.org/10.1007/11676935_3
15. Perfilieva, I.: Fuzzy transforms: a challange to conventional transform. In: Hawkes, P.W. (ed.) Advances in Image and Electron Physics, vol. 147, pp. 137–196. Elsevier Academic Press, San Diego (2007)
16. Perfilieva, I., Novak, V., Dvořak, A.: Fuzzy transforms in the analysis of data. Int. J. Approx. Reason. **48**, 36–46 (2008)
17. Perfilieva, I.: Fuzzy transforms: theory and applications. Fuzzy Sets Syst. **157**, 993–1023 (2006)
18. Perfilieva, I., Singh, A.P., Tiwari, S.P.: On the relationship among F-transform, fuzzy rough set and fuzzy topology. In: Proceedings of IFSA-EUSFLAT. Atlantis Press, Amsterdam, pp. 1324–1330 (2015)
19. Rodabaugh, S.E.: Powerset operator foundation for poslat fuzzy SST theories and topologies. In: Höhle, U., Rodabaugh, S.E. (eds.) Mathematics of Fuzzy Sets: Logic, Topology and Measure Theory. The Handbook of Fuzzy Sets Series, vol. 3, pp. 91–116. Kluwer Academic Publishers, Boston (1999)
20. Rodabaugh, S.E.: Powerset operator based foundation for point-set lattice theoretic (poslat) fuzzy set theories and topologies. Quaestiones Mathematicae **20**(3), 463–530 (1997)
21. Rodabaugh, S.E.: Relationship of algebraic theories to powerset theories and fuzzy topological theories for lattice-valued mathematics. Int. J. Math. Math. Sci. **2007**, 1–71 (2007)
22. Solovyov, S.A.: Powerset operator foundations for catalg fuzzy set theories. Iran. J. Fuzzy Syst. **8**(2), 1–46 (2001)
23. Štěpnička, M., Valašek, R.: Numerical solution of partial differential equations with the help of fuzzy transform. In: Proceedings of the FUZZ-IEEE 2005, Reno, Nevada, pp. 1104–1009 (2005)
24. Tomasiello, S.: An alternative use of fuzzy transform with application to a class of delay differential equations. Int. J. Comput. Math. **94**(9), 1719–1726 (2017)
25. Zadeh, L.A.: Fuzzy sets. Inf. Control **8**, 338–353 (1965)

Numerical Solutions for Bidimensional Initial Value Problem with Interactive Fuzzy Numbers

Vinícius F. Wasques$^{(\boxtimes)}$, Estevão Esmi, Laécio C. Barros, and Peter Sussner

Institute of Mathematics, Statistics and Scientific Computing Campinas,
University of Campinas, São Paulo 13081-970, Brazil
vwasques@outlook.com, eelaureano@gmail.com,
{laeciocb,sussner}@ime.unicamp.br

Abstract. We present a comparison between two approaches of numerical solutions for bidimensional initial value problem with interactive fuzzy numbers. Specifically, we focus on SI epidemiological model considering that initial conditions are given by interactive fuzzy numbers. The interactivity is based on the concept of joint possibility distribution and for this model, it is possible to observe two types of interactivities for fuzzy numbers. The first one is based on the completely correlated concept, while the other one is given by a family of joint possibility distributions. The numerical solutions are given using Euler's method adapted for the arithmetic operations of interactive fuzzy numbers via sup-J extension principle, which generalizes the Zadeh's extension principle.

Keywords: Interactive fuzzy numbers · Joint possibility distribution
Interactive arithmetic · Epidemiology

1 Introduction

This paper proposes numerical solutions for bidimensional initial value problems with interactive fuzzy numbers. The proposed method can be applied to any fuzzy initial value problems. In particular, we focus on the well-known SI epidemiological model. The mathematical model SI [1] is used to describe the dynamic of a diseases that affect the population whose is divided in two subpopulations, susceptible (S) and infected (I). In this model, a susceptible individual is a member of a population who is at risk of becoming infected by a disease and when it becomes infected there is no cure for it. AIDS is an example of disease that behaves this way.

V. F. Wasques—Grantee CNPq 142414/2017-4.

E. Esmi—Grantee FAPESP 2016/26040-7.

L. C. Barros—Grantee CNPq 306546/2017-5.

© Springer International Publishing AG, part of Springer Nature 2018
G. A. Barreto and R. Coelho (Eds.): NAFIPS 2018, CCIS 831, pp. 84–95, 2018.
https://doi.org/10.1007/978-3-319-95312-0_8

The initial values of susceptible and infected populations in the SI model may be uncertain, as well as the parameters involved (the infection rate, for example). Classical models do not consider this fact, instead of that, the fuzzy numbers are being used especially in applications involving parameters containing uncertainties and inaccuracies.

Consequently, arithmetic on fuzzy numbers is necessary. The most common is the standard arithmetic whose is based on Zadeh's extension principle and that does not consider interactivity, that is, each quantity involved in the arithmetic operation does not influences the other. However, in several areas such as in epidemiology, one can observe that, in phenomena, different uncertain compound influences on one another to a certain unknown degree, influencing the final result. For example, Grenfell *et al.* [2] show that pathogen epidemic dynamics and genetics can each potentially influence the other, depending on the biology of the host-parasite interaction. When we express these compounds through fuzzy numbers, this interaction is described using the concept of interactive fuzzy numbers. The interactive notion of fuzzy numbers can be defined in terms of a joint possibility distribution [3], which Carlsson *et al.* [4], employed to propose a generalization of Zadeh's extension principle [5].

This manuscript considers the SI model with initial conditions given by interactive fuzzy numbers. Moreover, we suppose that in each instant of time t the following equation is satisfied $S(t) + I(t) = p$, where $p > 0$. In other words, we are supposing that the population is constant, with respect to time. Under these conditions, we are not able to use the standard sum, even if the initial conditions were not interactive, since S and I are fuzzy numbers and k is a real number (cf. [6]). From the definition of sup-J extension principle [7], it is possible to define an arithmetic for interactive fuzzy numbers. In this paper, we use this concept to provide a numerical solution for the SI model using the Euler's method with the arithmetic operations adapted for fuzzy numbers. Depending on the shape of the fuzzy numbers involved, different types of interactivity can be used in the numerical solution.

One type of interactivity between fuzzy numbers is called completely correlated [4,8]. This approach assumes that two given fuzzy numbers have a special kind of relationship. In this manuscript, we provide another type of interactivity that is more general. Moreover, we present a discussion about when to use one or the other interactivity, in the SI model.

This manuscript is divided as follows. In Sect. 2, we recall some mathematical background used in subsequent sections. In Subsect. 2.1 we present the interactivity based on the concept of completely correlated. In Subsect. 2.2 we present the join possibility distribution that we propose. In Sect. 3, we present the fuzzy numerical solution for the SI model, as well as a discussion about the interactivity that can be used in different cases.

2   Mathematical Background

Let us review some pertinent concepts and results for our study.

A fuzzy set A of an universe X is associated with a function $\mu_A : X \to [0,1]$ called membership function, where $\mu_A(x)$ represents the membership degree of x in A for all $x \in X$ [9]. For notational convenience, we may simply use the symbol $A(x)$ instead of $\mu_A(x)$. The class of fuzzy subsets of X is denoted by $\mathcal{F}(X)$. Note that each subset of X can be uniquely identified with the fuzzy set whose membership function is given by its characteristic function.

The $\alpha-$cut of a fuzzy set $A \subseteq X$, denoted by $[A]^\alpha$, is defined as $[A]^\alpha = \{x \in X : A(x) \geq \alpha\}$, $\forall \alpha \in (0,1]$ [10]. If X is also a topological space, then $[A]^0 = cl\{x \in X : A(x) > 0\}$, where cl Y, $Y \subseteq X$, denotes the closure of Y.

A fuzzy set A of $\mathbb{R}$ is called a fuzzy number if all $\alpha-$cuts are bounded, closed and non empty nested intervals for all $\alpha \in [0,1]$. Thus the $\alpha-$cuts of the fuzzy number A are denoted by $[A]^\alpha = [a_\alpha^-, a_\alpha^+]$ [10]. The class of fuzzy numbers, denoted by $\mathbb{R}_\mathcal{F}$, represents a special class of fuzzy sets of $\mathbb{R}$ that includes the sets of the real numbers as well as the set of the bounded closed intervals of $\mathbb{R}$. Note that every trapezoidal fuzzy number is an example contained in $\mathbb{R}_\mathcal{F}$. Recall that a trapezoidal fuzzy number A is denoted by the quadruple $(a; b; c; d)$ for some $a \leq b \leq c \leq d$. By means of $\alpha-$cuts we have $[A]^\alpha = [a + \alpha(b - a), d - \alpha(d - c)]$, $\forall \alpha \in [0,1]$ [10]. If $b = c$, then we speak of a triangular fuzzy number, denoted $(a; b; d)$ instead of $(a; b; b; d)$.

Each fuzzy set A of X can be associated with an uncountable family of crisp sets of X namely $\alpha-$cuts of A. In particular, Negoita-Ralescu's representation Theorem established a formula to obtain the membership function of A from its $\alpha-$cuts [11].

Theorem 1 [11]. *Let $\{M_\alpha : \alpha \in [0,1]\}$ be a family of subsets that satisfies the following conditions:*

(a) M_α is a non-empty closed interval for any $\alpha \in [0,1]$;
(b) If $0 \leq r_1 \leq r_2 \leq 1$, then we have $M_{\alpha_2} \subseteq M_{\alpha_1}$;
(c) For any sequence α_n which converges from below to $\alpha \in (0,1]$ we have

$$\bigcap_{n=1}^{\infty} M_{\alpha_n} = M_\alpha;$$

(d) For any sequence α_n which converges from above to 0 we have

$$M_0 = cl\left(\bigcup_{n=1}^{\infty} M_{\alpha_n}\right).$$

Then there exists a unique $M \in \mathbb{R}_\mathcal{F}$, such that $[M]^\alpha = M_\alpha$, for any $\alpha \in [0,1]$.

Theorem 1 clarifies when a family of subsets can be uniquely associated with a fuzzy number.

Definition 1 [9] (Zadeh's extension principle). *Let $f : X \to Y$. The Zadeh's extension of f at $A \in \mathcal{F}(X)$ is the fuzzy set $\hat{f}(A) \in \mathcal{F}(Y)$ whose membership function is given by*

$$\mu_{\hat{f}(A)}(y) = \bigvee_{x \in f^{-1}(y)} \mu_A(x), \ \forall \, y \in V, \tag{2.1}$$

where $f^{-1}(y) = \{x \in X \mid f(x) = y\}$ is the inverse image of the function f at y and, by definition, $\bigvee \emptyset = 0$.

Zadeh's extension principle [9] can be viewed as a mathematical method to extend a function $f : U \to V$ to a function $\hat{f} : \mathcal{F}(U) \to \mathcal{F}(V)$. A fuzzy relation R between two universes X and Y is given by the mapping $R : X \times Y \to [0,1]$, where $R(x,y) \in [0,1]$ is the degree of relationship between $x \in X$ and $y \in Y$. An n-ary relation on $X = X_1 \times ... \times X_n$ is nothing else than a fuzzy (sub)set of X.

The projection of a fuzzy relation $R \in \mathcal{F}(X_1 \times ... \times X_n)$ onto X_i, where $1 \le i \le n$, is the fuzzy set Π_R^i of X_i given by

$$\Pi_R^i(y) = \bigvee_{x : x_i = y} R(x), \ \forall y \in X_i.$$

A fuzzy relation $J \in \mathcal{F}(\mathbb{R}^n)$ is said to be a joint possibility distribution among the fuzzy numbers $A_1, ..., A_n \in \mathbb{R}_{\mathcal{F}}$ if

$$A_i(y) = \Pi_J^i(y) = \bigvee_{x : x_i = y} J(x), \ \forall y \in \mathbb{R}, \tag{2.2}$$

for all $i = 1, ..., n$.

Let t be a t-norm, that is, an associative, commutative and increasing operator $t : [0,1]^2 \to [0,1]$ that satisfies $t(x, 1) = x$ for all $x \in [0,1]$. The fuzzy relation J_t given by

$$J_t(x_1, ..., x_n) = A_1(x_1) \ t \ ... \ t \ A_n(x_n) \tag{2.3}$$

is called the t-norm-based joint possibility distribution of $A_1, ..., A_n \in \mathbb{R}_{\mathcal{F}}$. In the special case where $t = \wedge$, the fuzzy numbers $A_1, ..., A_n$ are said to be non-interactive. Otherwise, $A_1, ..., A_n$ are called interactive. Thus, the interactivity of the fuzzy numbers $A_1, ..., A_n$ arises from a given joint possibility distribution.

Remark 1. Note that Eq. (2.2) guarantees that $J(x_1, ..., x_n) \le A_i(x_i)$, for all $i = 1, ..., n$. Thus $J(x_1, ..., x_n) \le J_\wedge(x_1, ..., x_n)$. This implies that every joint possibility distribution is contained in the joint possibility distribution based on minimum t-norm among the fuzzy numbers $A_1, ..., A_n$.

It is possible, from Definition 2 below, to obtain an interactive arithmetic operations for fuzzy numbers by taking $f(x_1, ..., x_n) = x_1 \otimes ... \otimes x_n$, where $\otimes$ represents an arithmetic operation. This is done as follows.

Definition 2 [7]. *Let $J \in \mathcal{F}(\mathbb{R}^n)$ be a joint possibility distribution of $(A_1, ..., A_n) \in \mathbb{R}_{\mathcal{F}}^n$ and $f : \mathbb{R}^n \to \mathbb{R}$. The sup-J extension of f at $(A_1, ..., A_n) \in \mathbb{R}_{\mathcal{F}}^n$, denoted $f_J(A_1, ..., A_n)$, is the fuzzy set defined by:*

$$f_J(A_1, ..., A_n)(y) = \bigvee_{(x_1, ..., x_n) \in f^{-1}(y)} J(x_1, ..., x_n), \tag{2.4}$$

where $f^{-1}(y) = \{(x_1, ..., x_n) \in \mathbb{R}^n : f(x_1, ..., x_n) = y\}$ is the inverse image of the function f at y.

Note that, if in Definition 2 the joint possibility distribution is given by (2.3) with $t = \wedge$ then the Sup-J extension principle boils down to the Zadeh's extension principle.

2.1 Completely Correlated Fuzzy Numbers

We can see from (2.2) that it is possible to obtain an interactivity relation among fuzzy numbers that is not based on t-norm. Specifically, Carlsson et al. [4] introduced a correlation concept. They stated that two fuzzy numbers are said to be completely correlated if there is a *linearly* connection between their membership functions. Mathematically, two fuzzy numbers A and B are completely correlated if there exists $q, r \in \mathbb{R}$ with $q \neq 0$ such that the corresponding joint possibility distribution $J = J_{\{q,r\}}$ is given by

$$J_{\{q,r\}}(x_1, x_2) = A(x_1)\chi_{\{qu+r=v\}}(x_1, x_2) \qquad (2.5)$$
$$= B(x_2)\chi_{\{qu+r=v\}}(x_1, x_2),$$

for all $x_1, x_2 \in \mathbb{R}$, where $\chi_{\{qu+r=v\}}$ stands for the characteristic function of the set $\{(u, qu + r = v) : \forall u\}$, that is

$$\chi_{\{qu+r=v\}}(x_1, x_2) = \begin{cases} 1, & \text{if } \quad qx_1 + r = x_2 \\ 0, & \text{otherwise} \end{cases}.$$

Remark 2. Recall that if $q < 0$ then A and B are said to be completely negatively correlated. In this case, we have that $A +_{\{q,r\}} B = A + B$, where $+$ is the standart sum. Moreover, if $q > 0$ then A and B are said to be completely positively correlated and, in this case, we have $A -_{\{q,r\}} B = A - B$, where $-$ is the standart difference [4,8].

We denoted $A +_L B$ instead of $A +_{\{q,r\}} B$ for simplicity of notation. From Definition 2, the joint possibility distribution $J_{\{q,r\}}$, given by (2.5), produces the interactive arithmetic that satisfies the following properties [12]

(a) $[A +_L B]^\alpha = (q + 1)[A]^\alpha + r$, for all $\alpha \in [0, 1]$;
(b) $[A -_L B]^\alpha = (1 - q)[A]^\alpha + r$, for all $\alpha \in [0, 1]$;
(c) $[A *_L B]^\alpha = \{qx_1^2 + rx_1 \in \mathbb{R} : A(x_1) \geq \alpha\}$, for all $\alpha \in [0, 1]$;
(d) $[A \div_L B]^\alpha = \{\frac{x_1}{qx_1+r} \in \mathbb{R} : A(x_1) \geq \alpha\}$, for all $\alpha \in [0, 1]$.

2.2 Join Possibility Distribution J_0

The joint possibility distribution (2.5) is restrictive because it can not be applied to a pair of fuzzy numbers that do not have a colinear relation, for example, it can not be applied when A is a triangular fuzzy number and B is a trapezoidal fuzzy number. Our purpose is defined a joint possibility distribution that has no restrictions, as we see in [13,14]. Thus, let A_1 and A_2 be two fuzzy numbers and the functions $g_{1,2} : \mathbb{R} \times [0, 1] \to \mathbb{R}$ given by

$$g_1(z, \alpha) = \bigwedge_{w \in [A_2]^\alpha} |w + z| \quad \text{and} \quad g_2(z, \alpha) = \bigwedge_{w \in [A_1]^\alpha} |w + z|. \qquad (2.6)$$

Also, let us consider the classical sets

$$R_\alpha^i = \begin{cases} \{a_{i_\alpha}^-, a_{i_\alpha}^+\} & \text{if} \quad \alpha \in [0,1) \\ [A_i]^1 & \text{if} \quad \alpha = 1 \end{cases} \tag{2.7}$$

and

$$L^i(z, \alpha) = [A_{3-i}]^\alpha \cap [-g_i(z, \alpha) - z, g_i(z, \alpha) - z], \tag{2.8}$$

for all $z \in \mathbb{R}, \alpha \in [0,1]$ and $i = 1, 2$.

Let J_0 be the fuzzy relation given by

$$J_0(x_1, x_2) = \begin{cases} A_1(x_1) \wedge A_2(x_2) & \text{, if} \quad (x_1, x_2) \in P \\ 0 & \text{, otherwise} \end{cases}, \tag{2.9}$$

where

$$P = \bigcup_{i=1}^{2} \bigcup_{\alpha \in [0,1]} P^i(\alpha)$$

and $P^i(\alpha)$ is given as follows

$$P^i(\alpha) = \{(x_1, x_2) : x_i \in R_\alpha^i \quad \text{and} \quad x_{3-i} \in L^i(x_i, \alpha)\}, \quad \forall \alpha \in [0,1],$$

for $i \in \{1, 2\}$.

The joint possibility distribution J_0, from Definition 2 given by (2.9), provides the arithmetic for interactive fuzzy numbers as follows

$$(A_1 \otimes_{J_0} A_2)(y) = \bigvee_{x_1 \otimes x_2 = y} J_0(x_1, x_2), \tag{2.10}$$

where J_0 is the joint possibility distribution of A_1 and A_2 given by (2.9) and $\otimes$ represents the arithmetic operations $+, -, *$ and $\div$.

This new approach towards joint possibility distribution can be applied to every pair of fuzzy numbers. Note that the norm and the width of the sum are not equivalent, that is, $\| A \|_{\mathcal{F}} \leq \| B \|_{\mathcal{F}}$ does not imply that $width(A) \leq width(B)$. For example, for $A = (-2; 0; 2)$ and $B = (1; 2; 3)$ we have that $\| A \|_{\mathcal{F}} = 2 \leq 3 = \| B \|_{\mathcal{F}}$ but $width(A) = 4 > 2 = width(B)$. As we will recall below, translations of fuzzy numbers can be used in order to control the width of the sum based on sup-J extension principle.

Definition 3. Let $A \in \mathbb{R}_{\mathcal{F}}$. The translation of A by $k \in \mathbb{R}$ is defined as the following fuzzy number $\tilde{A}$:

$$\tilde{A}(x) = A(x + k), \quad \forall x \in \mathbb{R}. \tag{2.11}$$

Next, we provide a joint possibility distribution using the concept of Definition (3)

Theorem 2 [14]. *Given $A_1, A_2 \in \mathbb{R}_{\mathcal{F}}$ and $c = (c_1, c_2) \in \mathbb{R}^2$. Let $\tilde{A}_1, \tilde{A}_2 \in \mathbb{R}_{\mathcal{F}}$ be such that $\tilde{A}_1(x) = A_1(x + c_1)$ and $\tilde{A}_2(x) = A_2(x + c_2), \forall x \in \mathbb{R}$.*

Let $\tilde{J}_0$ be the joint possibility distribution of fuzzy numbers $\tilde{A}_1, \tilde{A}_2 \in \mathbb{R}_{\mathcal{F}}$ defined as Eq. (2.9). The fuzzy relation J_0^c given by

$$J_0^c(x_1, x_2) = \tilde{J}_0(x_1 - c_1, x_2 - c_2), \ \forall (x_1, x_2) \in \mathbb{R}^2, \tag{2.12}$$

is a joint possibility distribution of A_1 and A_2.

We denote the sum between two fuzzy numbers A_1 and A_2 based on sup-J extension principle, with $J = J_0^c$, by $A_1 +_0 A_2$. In addition, we have that $A_1 -_0 A_2 = A_1 +_0 (-A_2), \forall A_1, A_2 \in \mathbb{R}_{\mathcal{F}}$ [14] for simplicity of notation.

3   Fuzzy Numerical Solution for SI Model

This Section presents a fuzzy numerical solution for the SI model with interactive fuzzy conditions. The numerical solution is given by the Euler's method with the appropriate arithmetic operations for fuzzy numbers. The arithmetic is based on sup-J extension principle and we use the joint possibility distributions given by (2.5) and (2.9).

First, we present the Euler's method. Let $y_i : \mathbb{R} \to \mathbb{R}^n$, with $i = 1, ..., n$, functions that depend on time t. Consider the following initial value problem (IVP) composed of ordinary differential equations (ODE) and initial condition

$$\begin{cases} \frac{dy_i}{dt} = f_i(t, y_1, y_2, ..., y_n) \\ y(t_0) = y_0 \in \mathbb{R}^n \end{cases}, \tag{3.13}$$

where f_i is a function that depends on $y_1, y_2, ..., y_n$ and t, for each $i = 1, ..., n$.

Euler's method consists in determining numerical solutions for (3.13). The algorithm is given by

$$y_i^{k+1} = y_i^k + h f_i(t^k, y_1^k, ..., y_n^k), \tag{3.14}$$

with $0 \le k \le N - 1$, where N is the number of partitions of the interval time divided in equally spaced intervals $[t^k, t^{k+1}]$ with size h and initial condition (t_0, y_0).

The SI model is given by [1]

$$\begin{cases} \frac{dS}{dt} = -\beta SI, \quad S(0) = S_0 \\ \frac{dI}{dt} = \beta SI, \quad\ \ I(0) = I_0 \end{cases}, \tag{3.15}$$

where β is the rate of the infection of disease and S_0 and I_0 are given by interactive fuzzy numbers.

Thus, the fuzzy numerical solution for (3.15) that we propose is given by

$$\begin{cases} S^{k+1} = S^k +_0 h(-\beta S^k *_0 I^k) \\ I^{k+1} = I^k +_0 h(\beta S^k *_0 I^k) \end{cases} . \qquad (3.16)$$

where the arithmetic operations are based on sup-J extension principle with $J = J_0$.

Now, suppose that the initial values S_0 and I_0 are given by interactive fuzzy numbers with respect the joint possibility distribution $J_{\{q,r\}}$, given by (2.5). Even if S and I are interactive, we do not have that $S *_L I$ and S are interactive. The same observation can be made for $S *_L I$ and I. Thus, we can not apply the arithmetic operations in (3.14) by sup-J extension principle with $J = J_{\{q,r\}}$, we need first to make some appropriated changes in Eq. (3.14). Since $S(t) + I(t) = p$, $\forall t \in [0, \infty)$ the following equations are equivalent

$$S^{k+1} = S^k + h(-\beta S^k I^k) \Leftrightarrow S^{k+1} = S^k(1 - ph\beta + h\beta S^k). \qquad (3.17)$$

In this case, S^k and $(1 - ph\beta + h\beta S^k)$ are also completely correlated and then we can use the sup-J extension principle for $J = J_{\{q,r\}}$, where $q = h\beta$ and $r = 1 - ph\beta$. The same holds for I^k. Thus, the fuzzy numerical solution for (3.15), with respect $J_{\{q,r\}}$, is given by

$$\begin{cases} S^{k+1} = S^k *_L (1 - ph\beta + h\beta S^k) \\ I^{k+1} = I^k *_L (1 + ph\beta - h\beta I^k) \end{cases} . \qquad (3.18)$$

3.1   Example 1

Let $S_0 = (4; 5; 6)$ and $I_0 = (0; 1; 2)$. Since S_0 and I_0 are completely correlated with respect to $J_{\{1,4\}}$ and $J_{\{-1,6\}}$, we can use the fuzzy solution provided by (3.18) whose is depicted in Fig. 1.

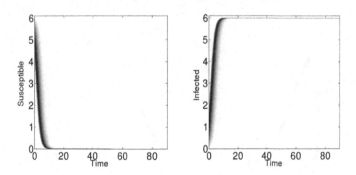

Fig. 1. The fuzzy solution provided by (3.18). The left and right Figures present the susceptible and infected populations, respectively. The parameters used were $p = 6$, $h = 0.125$ and $\beta = 0.01$. The gray lines represent the α-cuts of the fuzzy solutions, where their endpoints for α varying from 0 to 1 are represented respectively from the gray-scale lines varying from white to black.

The fuzzy solution provided by (3.16) is depicted in Fig. 2.

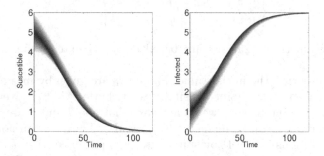

Fig. 2. The fuzzy solution provided by (3.16). The left and right Figures present the susceptible and infected populations, respectively. The parameters used were $h = 0.125$ and $\beta = 0.01$. The gray lines represent the α-cuts of the fuzzy solutions, where their endpoints for α varying from 0 to 1 are represented respectively from the gray-scale lines varying from white to black.

Note that the susceptible population given in Fig. 1 decreases faster than susceptible population given in Fig. 2. Moreover, the infected population given in Fig. 1 increases faster than infected population given in Fig. 2.

3.2   Example 2

Let $S_0 = (9; 10; 12)$ and $I_0 = (0; 1; 2)$. Even though S_0 and I_0 are triangular fuzzy numbers, there is no linear correlation between the membership functions of S_0 and I_0. Therefore the fuzzy solution provided by (3.18) can not be applied in this case. However, the fuzzy solution provided by (3.16) has no restriction. Figure 3 presents the fuzzy solution given by (3.16).

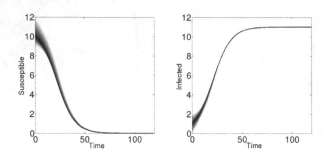

Fig. 3. The fuzzy solution provided by (3.16). The left and right Figures present the susceptible and infected populations, respectively. The parameters used were $h = 0.125$ and $\beta = 0.01$. The gray lines represent the α-cuts of the fuzzy solutions, where their endpoints for α varying from 0 to 1 are represented respectively from the gray-scale lines varying from white to black.

The behavior of the susceptible and infected populations are the same as in Example 3.1.

3.3 Example 3

Now suppose that the initial number of susceptible population is "around" 11 and the initial number of infected population is "around" the interval $[1,3]$. Thus, we model the population S_0 and I_0 by the interactive fuzzy numbers $S_0 = (10; 11; 12)$ and $I_0 = (0; 1; 3; 4)$. Since, S_0 is a triangular fuzzy number and I_0 is a trapezoidal fuzzy number, obviously S_0 and I_0 are not completely correlated. Therefore, between the two approaches only the joint possibility distribution J_0 can be used in this case. The fuzzy solution is depicted in Fig. 4.

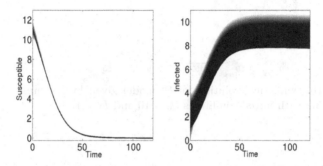

Fig. 4. The fuzzy solution provided by (3.16). The left and right Figures present the susceptible and infected populations, respectively. The parameters used were $h = 0.125$ and $\beta = 0.01$. The gray lines represent the α-cuts of the fuzzy solutions, where their endpoints for α varying from 0 to 1 are represented respectively from the gray-scale lines varying from white to black.

Note that, in Example 3.1 both joint possibility distributions produces a numerical solution such that its width decreases with respect to time. This means that the uncertainty decreases over time.

It was only possible to consider the interactivity based on joint possibility distribution J_0 in Example 3.2. Note that, the same observation that we made before with respect to width holds here.

Finally, in Example 3.3 the numerical solution for the susceptible population presents the same property with respect to the width as in the Examples 3.1 and 3.2. However, the numerical solution for the infected population present a slight growth with respect to the width.

Recall that, if the initial conditions S_0 and I_0, given in (3.16), are real numbers (crisp) then the obtained solution coincides with the classical solution as it is depicted in Fig. 5.

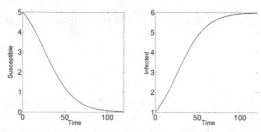

(a) Numerical solution for SI model given by Example 3.1 with initial conditions $S_0 = 5$ and $I_0 = 1$.

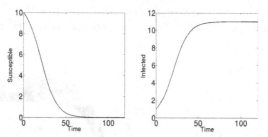

(b) Numerical solution for SI model given by Example 3.2 with initial conditions $S_0 = 10$ and $I_0 = 1$.

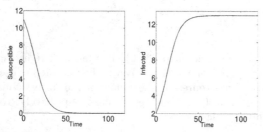

(c) Numerical solution for SI model given by Example 3.3 with initial conditions $S_0 = 11$ and $I_0 = 2$.

Fig. 5. Numerical solution for SI model given by Examples 3.1 (Subfigure 5(a)), 3.2 (Subfigure 5(b)) and 3.3 (Subfigure 5(c)). The used parameters were $h = 0.125$ and $\beta = 0.01$.

4   Conclusion

This manuscript presented a numerical method for bidimensional IVPs with interactivity on state variables. We focused on SI epidemiological model, where the initial conditions are interactive fuzzy numbers. We considered this interactivity by two approaches, completely correlated (cf. (2.5)) and via joint possibility distribution J_0 (cf. (2.9)). The fuzzy numerical solution for the SI model is given by the Euler's method, whose arithmetic operations are adopted for interactive fuzzy numbers based on sup-J extension principle.

We also presented a comparison between these two approaches for interactive fuzzy numbers. On the one hand, the numerical solution provided by (3.18) is restrictive, since can only be applied when the membership functions of the fuzzy numbers involved have a linear correlation. It was only possible to use this concept in Example 3.1. On the other hand, the numerical solution provided by (3.16) has no restrictions and it was possible to use the joint possibility distribution J_0 in Examples 3.1, 3.2 and 3.3.

Finally, differently of a fuzzy initial value problem (FIVP) where the derivative is obtained by interactive difference [8], here we only use numerical methods for the IVP, considering that initial conditions are fuzzy numbers.

References

1. Edelstein-Keshet, L.: Mathematical Models in Biology. SIAM, Philadelphia (1988)
2. Grenfell, B.T., Pybus, O.G., Gog, J.R., Wood, J.L.N., Daly, J.M., Mumford, J.A., Holmes, E.C.: Unifying the epidemiological and evolutionary dynamics of pathogens. Am. Assoc. Adv. Sci. **303**, 327–332 (2004)
3. Dubois, D., Prade, H.: Possibility Theory: An Approach to the Computerized Processing of Information. Springer, US (1988). https://doi.org/10.1007/978-1-4684-5287-7
4. Carlsson, C., Fullér, R., Majlender, P.: Additions of completely correlated fuzzy numbers. In: IEEE International Conference on Fuzzy Systems. vol. 1, pp. 535–539 (2004)
5. Zadeh, L.A.: Concept of a linguistic variable and its application to approximate reasoning, i, ii, iii. Inform. Sci. 8, 199–249, 301–357 (1975)
6. Cabral, V.M., Barros, L.C.: Fuzzy differential equation with completely correlated parameters. Fuzzy Sets Syst. **265**, 86–98 (2015)
7. Fullér, R., Majlender, P.: On interactive fuzzy numbers. Fuzzy Sets Syst. **143**, 355–369 (2004)
8. Barros, L.C., Pedro, F.S.: Fuzzy differential equations with interactive derivative. Fuzzy Sets Syst. **309**, 64–80 (2017)
9. Zadeh, L.A.: Fuzzy sets. Inf. Control **8**, 338–353 (1965)
10. Barros, L.C., Bassanezi, R.C., Lodwick, W.A.: A First Course in Fuzzy Logic, Fuzzy Dynamical Systems, and Biomathematics. Springer, Heidelberg (2017). https://doi.org/10.1007/978-3-662-53324-6
11. Negoita, C., Ralescu, D.: Application of Fuzzy Sets to Systems Analysis. Wiley, New York (1975)
12. Pedro, F.S., Barros, L.C., Esmi, E.: Population growth model via interactive fuzzy differential equation (2017, Submitted)
13. Esmi, E., Sussner, P., Ignácio, G.B.D., Barros, L.C.: A parametrized sum of fuzzy numbers with applications to fuzzy initial value problems. Fuzzy Sets Syst. **331**, 85–104 (2018)
14. Sussner, P., Esmi, E., Barros, L.C.: Controling the width of the sum of interactive fuzzy numbers with applications to fuzzy initial value problems. In: IEEE International Conference on Fuzzy Systems, pp. 1453–1460 (2016)

Relevance of Classes in a Fuzzy Partition. A Study from a Group of Aggregation Operators

Fabián Castiblanco[1]([⊠]) [iD], Camilo Franco[2] [iD], Javier Montero[3] [iD], and J. Tinguaro Rodríguez[3] [iD]

[1] Faculty of Economic, Administrative and Accounting Sciences, Gran Colombia University, Bogotá, Colombia
fabianalberto.castiblanco@ugc.edu.co
[2] Department of Industrial Engineering, Andes University, Bogotá, Colombia
[3] Department of Statistics, Complutense University, Madrid, Spain

Abstract. This paper presents a study of the relevance property in a fuzzy partition from a fuzzy classification system. This study allows establishing a stopping criterion for the inclusion of a class in a fuzzy partition based on relevance. Such a criterion is constructed from a stable relationship on the commutative group formed by two new mappings (and the aggregation operators conjunctive and disjunctive) of the fuzzy classification system. The criterion is illustrated through an example on image analysis by the fuzzy c-means algorithm.

Keywords: Relevance · Covering · Overlap · Classification · Fuzzy partition

1 Introduction

Since fuzzy logic was proposed by Zadeh [1], classification systems have presented great advances in their theoretical and practical development. Clear examples can be found in [2–4] dealing with fuzzy classification.

The fuzzy approach in this area has allowed, on the one hand, the reduction of the amount of information handled by users in complex environments, and on the other hand, the representation of information by means of gradable predicates, modeled by fuzzy set membership functions. In particular, the development of fuzzy classification system has received great attention from remote sensing or image segmentation applications [5, 6]. The aim of this paper is to study fuzzy classification systems and their application in image segmentation. An example of such an application will be presented at the end of the paper.

The fuzzy classification system developed by Del Amo et al. [5] is based on a generalization of the fuzzy partition concept proposed by Ruspini [7], and defines this classification system by means of families of aggregation functions. Based on the theoretical framework developed by Dombi [8, 9] about aggregation operators, an alternative approach for non-associative connectives is proposed in [5] through the use

© Springer International Publishing AG, part of Springer Nature 2018
G. A. Barreto and R. Coelho (Eds.): NAFIPS 2018, CCIS 831, pp. 96–107, 2018.
https://doi.org/10.1007/978-3-319-95312-0_9

of the concept of *recursiveness*. That is, from the classical approach of aggregation of information by a single index (associative conjunctive rules), the notion of associativity is extended through the establishment of recursive rules.

Such an approach proposes indexes that allow measuring the redundancy, relevance and coverage of the classes obtained in a fuzzy partition; that is to say, a system is constructed that in turn allows, evaluating the obtained classification. However, the evaluation of the quality of a partition is a problem that still needs to be addressed in greater depth (see [5, 10]) with special emphasis on the proposed indexes.

In this line of research, some first approaches led to the development of works such as those presented in [11–13], and more recently in [14, 15] about a more in-depth study of aggregation functions, which allow evaluating the redundancy and coverage of a particular classification by means of overlapping and grouping functions, respectively. Even studies on redundancy, based on other fuzzy partition concepts, have been developed (see e.g. [16]).

An issue that is still open and with a broad field of development, is the study of the relevance property as initially explored in [2]. In [6] an alternative approach is proposed from a more statistical perspective. Here we propose some first steps towards the study of relevance, its characterization, and with it, a general study of a global quality index for a fuzzy classification system.

2   The Relevance Property

In the study of the intrinsic properties of a fuzzy partition, we can highlight the covering and redundancy properties, respectively graded by the degree in which a family of classes allows explaining the object's main attributes and the degree of overlapping between that family of classes. (see [5]).

Relevance, in general terms, is a fuzzy concept and from a more general and intuitive perspective, people may be able to distinguish irrelevant information or, in some cases, more relevant information from less relevant information. The fact that there is a linguistic notion of relevance with a vague and variable meaning exposes the complexity of the problem and reveals different ways of approaching it. Moreover, intuitions of relevance are relative to contexts, and there is no way of controlling exactly which context someone will have in mind at a given moment or how to understand such a context [17].

By its nature, the concept of *relevance* requires a treatment beyond its etymological meaning; the fundamental thing is to characterize when an object is relevant with respect to a given context. Therefore, as a technical concept which can be suitable of being measured by computational methods, *relevance* requires a characterization that allows its formal understanding for computational use. Keeping this in mind, here we propose a new approach over relevance (following [5] but also [6]), and the means for evaluating and measuring it regarding a given fuzzy partition.

In particular, establishing that a proposition is relevant necessarily requires considering a space or context of reference, in such a way that the element or proposition generates changes or modifications when it is removed or added from the context or space. Therefore, relevance implies a comparison process, understanding relevance as a local property and not a global property. Hence, we evaluate the relevance of an object in a context and not the relevance of the context or space in which the object is framed.

According to the above, in a first stage, the relevance of an object in a context can be established through the comparison of diverse information provided by the conformation of three sets: the information of the context with the object, the information of the context without the object and the information provided by object in itself, without any context.

The comparison process necessary to establish relevance may or may not reveal changes in the context, however, in case of changes, these changes may be more or less intense in the context. It is possible that the changes are significant or not. Therefore, establishing the relevance of an object in a context requires identifying the kind of changes that occur in that context, and their degree of intensity. In this sense, it is desirable to establish a threshold or admissible parameter of relevance, or in general, of modification in the context and kind of changes.

According to the above, the relevance of an object depends on two fundamental aspects: on the one hand, a process of comparison between the object and the context that allows determining changes by the inclusion or elimination of the object and, a measure of the intensity of the changes.

In the framework of a fuzzy partition, let us assume a finite set of objects X. A fuzzy classification system is a finite family C of fuzzy sets or classes (each $c \in C$ with its associated membership function $\mu_c(x) : X \rightarrow [0, 1]$), together with a recursive triplet[1] (φ, ϕ, N), where:

1. ϕ is a standard recursive rule such that $\phi_2(0, 1) = \phi_2(1, 0) = 0$
2. $N : [0, 1] \rightarrow [0, 1]$ is a strict negation function[2], i.e., a bijective strictly decreasing function such that $N \circ N^{-1}(\mu(x)) = \mu(x)$ for all $\mu(x) \in [0, 1]$
3. φ is a standard recursive rule such that $\varphi_n(\mu(x_1), \ldots, \mu(x_n)) = N^{-1}[\phi_n(N(\mu(x_1)), \ldots, N(\mu(x_n)))] \forall n > 1$.

Notice that, φ_n is a disjunctive recursive rule, in the sense that $\varphi_n(\mu(x_1), \ldots, \mu(x_2)) = 1$ whenever there is j such that $\mu(x_j) = 1$, while ϕ_n is a conjunctive recursive rule in the sense that $\phi_n(\mu(x_1), \ldots, \mu(x_2)) = 0$ whenever there is j such that $\mu(x_j) = 0$.

About the relevance property in [5] it is proposed to compare the behavior of each family of non-empty classes $A \subset C$ with the behavior of the remaining classes $C - A$, taking into account the values obtained through the following expressions for each object $x \in X$.

---

[1] Recursiveness is a property of a sequence of operators $\{\phi_n\}_{n > 2}$ allowing the aggregation of any number of items: ϕ_2 tells us how to aggregate two items, ϕ_3 tells how to aggregate three items and so on. A recursive rule ϕ is a family of aggregation functions $\{\phi_n : [0, 1]^n \rightarrow [0, 1]\}_{n > 1}$ allowing a sequential reckoning by means of a successive application of binary operators, once data have been properly ordered: the ordering rule assures that new data do not introduce modifications in the relative position of items already ordered. For more details see [5, 18].

[2] Here we refer to strict negations of the type $N(x) = f^{-1}(f(1) - f(x))$ with $f : [0, 1] \rightarrow [0, 1]$ increasing, bijective, $f(0) = 0$, and $0 < f(1) \leq 1$. In particular, if $N(x) = 1 - x$, then $f(x) = x$.

1. $\varphi_n\{\mu_c(x)/c \in C\}$
2. $\varphi_n\{\mu_k(x)/k \in A\}$
3. $\varphi_n\{\mu_d(x)/d \in C - A\}$

The following criterion is established: when the value obtained through expression 1 above is significantly greater than that obtained through expression 3, then A is a family of relevant classes, as long as the value obtained through expression 2 is not high. When expression 1 produces a value not significantly different from that obtained through expression 3, then A is a family of non-relevant classes, as long as expression 2 does not produce a low value.

The above criterion requires additional developments since,

1. There may be $x \in X$ such that some of the above situations do not appear clearly, for instance, when the values of $\mu_1(x), \ldots, \mu_c(x)$, for $x \in X$, are in a highly uniform.
2. It is opportune to establish a *global index* for each of the properties studied, that is, the aggregation of the degrees of coverage, relevance and overlap for all $x \in X$ for each class c, in the perspective posed by [19].

According to the above, it would be desirable to establish one or several criteria for evaluation of the relevance property. In general, it is sought to establish a set of criteria that allows the evaluation of a fuzzy classification system.

3   The Group of Aggregation Operators

From the fuzzy classification system (C, φ, ϕ, N), with $c, d \in C$ and in particular working on $\phi_2 : [0, 1]^2 \rightarrow [0, 1]$ with

$$\varphi_2(\mu_c(x), \mu_d(x)) = N^{-1}[\phi_2[N(\mu_c(x)), N(\mu_d(x))]],$$

two new mappings are built for all aggregation operators ϕ_2 and φ_2 such that the standard strict negation is $N(x) = 1 - x$ (in this particular case, we have that $\varphi_2(\mu_c(x), \mu_d(x)) = 1 - \phi_2(1 - \mu_c(x), 1 - \mu_d(x))$. These mappings are:

1. $\sigma_2 : [0, 1]^2 \rightarrow [0, 1]$, defined as:

$$\sigma_2(\mu_c(x), \mu_d(x)) = \mu_c(x) + \mu_d(x) - \varphi_2(\mu_c(x), \mu_d(x)), \text{ and}$$

2. $\delta_2 : [0, 1]^2 \rightarrow [0, 1]$, defined as:

$$\delta_2(\mu_c(x), \mu_d(x)) = \mu_c(x) + \mu_d(x) - \phi_2(\mu_c(x), \mu_d(x))$$

The proposed mappings can be generalized by (conjunctive, disjunctive or average) aggregation operators, leaving its formal specification for future research.

When we use the strict negation $N(\mu(x))$ on $\delta_2(\mu_c(x), \mu_d(x))$ or $\sigma_2(\mu_c(x), \mu_d(x))$, this can be interpreted as the complement of the set of aggregated classes $\{\mu_c(x), \mu_d(x)\}$. In particular, if $\varphi_2(\mu_c(x), \mu_d(x))$ represents the degree of coverage of the classes, then $N(\varphi_2)$ represents the degree of non-coverage of the classes, under-standing φ_2 as a proposition and $N(\varphi_2)$ as the negation of such a proposition.

Notice that the mapping σ_2 can be understood as the degree of partial non-coverage of the aggregated classes and δ_2 as the degree up to which the aggregated classes do not partially overlap. In general, both mappings can be understood as partial complements of the aggregated classes.

The idea that motivates this construction is based on the possibility of establishing a relationship between the conjunctive, disjunctive operators and their partial comple-ments, in such a way that a fuzzy partition can be evaluated taking into account a global vision of the corresponding fuzzy classification system. It seeks to compare the degrees of coverage, overlap and partial complements of pairs of classes in fuzzy partitions with different number of classes, and determine the partition with highest quality.

A close relationship is established between the set of mappings $\phi_2, \varphi_2, \sigma_2$ and δ_2. Let $A = \{\phi_2, \varphi_2, \sigma_2, \delta_2\}$, the composition of the mappings (denoted by $\circ$) is defined as presented in Table 1.

Table 1. Composition

$\circ$	ϕ_2	σ_2	δ_2	φ_2
ϕ_2	ϕ_2	σ_2	δ_2	φ_2
σ_2	σ_2	ϕ_2	φ_2	δ_2
δ_2	δ_2	φ_2	ϕ_2	σ_2
φ_2	φ_2	δ_2	σ_2	ϕ_2

For instance, we have that: $(\sigma_2 \circ \varphi_2)(\mu_c(x), \mu_d(x)) = \mu_c(x) + \mu_d(x) - 1 + \varphi_2 (1 - \mu_c(x), 1 - \mu_d(x)) = \mu_c(x) + \mu_d(x) - \phi_2(\mu_c(x), \mu_d(x)) = \delta_2(\mu_c(x), \mu_d(x))$

Clearly, $(A, \circ)$ is a commutative group, where ϕ_2 is the neutral element and each element is its own inverse. As mentioned above, σ_2 and δ_2 mappings can be formulated under a general framework, considering the strict negation function N for each pair ϕ_2 and φ_2 and an adequate aggregation of the classes that maintains the group structure.

The composition of the defined mappings obtains a particular structure for the algebraic group and therefore, allows proposing a relation of similarity between the functions of aggregation and their partial complements in the perspective presented by [20]. Therefore, such a similarity relation allows a first comparison process between the information obtained from the aggregated classes. Based on [20], for each pair $(\mu_c(x), \mu_d(x)) \in C$, the mapping $v_0 : A \times A \rightarrow [0, 1]$ is defined in the following way:

$$v_0(\theta_2(\mu_c(x), \mu_d(x)), \lambda_2(\mu_c(x), \mu_d(x))) = \frac{\left|\sum_{i=1}^{m}[\theta_2(\mu_c(x_i), \mu_d(x_i)) - \lambda_2(\mu_c(x_i), \mu_d(x_i))]\right|}{m}$$

With $\theta_2, \lambda_2 \in A$, $m = |X|$, $\mu_c(x_i)$ is the membership degree of the element x_i in class c. For simplicity, let us consider:

$$v_0(\theta_2(\mu_c(x), \mu_d(x)), \lambda_2(\mu_c(x), \mu_d(x))) = v_0(\theta_2, \lambda_2)(c, d).$$

Let us also denote:

1. $v_0(\phi_2, \sigma_2)(c, d) = v_0(\sigma_2, \phi_2)(c, d) = \dfrac{\left|\sum_{i=1}^{m}[\sigma_2(\mu_c(x_i), \mu_d(x_i)) - \phi_2(\mu_c(x_i), \mu_d(x_i))]\right|}{m} = \beta_0$

2. $v_0(\phi_2, \varphi_2)(c, d) = v_0(\varphi_2, \phi_2)(c, d) = \dfrac{\left|\sum_{i=1}^{m}[\varphi_2(\mu_c(x_i), \mu_d(x_i)) - \phi_2(\mu_c(x_i), \mu_d(x_i))]\right|}{m} = \alpha_0$

Then, the relation for all the element of $A \times A$ is represented in Table 2.

Table 2. Mapping v_0

v_0	ϕ_2	σ_2	δ_2	φ_2
ϕ_2	0	β_0	$\beta_0 + \alpha_0$	α_0
σ_2	β_0	0	α_0	$|\beta_0 - \alpha_0|$
δ_2	$\beta_0 + \alpha_0$	α_0	0	β_0
φ_2	α_0	$|\beta_0 - \alpha_0|$	β_0	0

For instance, if $v_0(\phi_2, \sigma_2)(c, d) = v_0(\sigma_2, \phi_2)(c, d) = \beta_0$, then we have that:

$$v_0(\phi_2, \sigma_2)(c, d) = \frac{\left|\sum_{i=1}^{m}[\sigma_2(\mu_c(x_i), \mu_d(x_i)) - \phi_2(\mu_c(x_i), \mu_d(x_i))]\right|}{m}$$

$$= \frac{\left|\sum_{i=1}^{m}[\mu_c(x_i) + \mu_d(x_i) - \varphi_2(\mu_c(x_i), \mu_d(x_i)) - \phi_2(\mu_c(x_i), \mu_d(x_i))]\right|}{m}$$

$$= \frac{\left|\sum_{i=1}^{m}[\delta_2(\mu_c(x_i), \mu_d(x_i)) - \varphi_2(\mu_c(x_i), \mu_d(x_i))]\right|}{m} = v_0(\delta_2, \varphi_2)(c, d)$$

Consider the complements of the images of v_0, i.e., the mapping $v : [0, 1] \to [0, 1]$, such that:

$$v(\beta_0) = 1 - \beta_0 = \beta_{c,d}$$

$$v(\alpha_0) = 1 - \alpha_0 = \alpha_{c,d}$$

$$v(\beta_0 + \alpha_0) = 1 - (\beta_0 + \alpha_0) = \pi_{c,d}$$

$$v(|\beta_0 - \alpha_0|) = 1 - |\beta_0 - \alpha_0| = \gamma_{c,d}$$

Therefore, for this mapping v it is possible to establish the relationships shown in Table 3.

Table 3. Mapping v

v	ϕ_2	σ_2	δ_2	φ_2
ϕ_2	1	β	π	α
σ_2	β	1	α	γ
δ_2	π	α	1	β
φ_2	α	γ	β	1

Proposition 1. *Let* (C, φ, ϕ, N) *be a fuzzy classification system. If* $\phi_2(\mu_c(x), \mu_d(x)) = \sigma_2(\mu_c(x), \mu_d(x))$ *for all* $x \in X$ *then,* $\alpha_{c,d} = \pi_{c,d} = \gamma_{c,d} < \beta_{c,d} = 1$.

Proof. If $\phi_2(\mu_c(x), \mu_d(x)) = \sigma_2(\mu_c(x), \mu_d(x))$ then $\beta_0 = 0$ wherewith $\beta_{c,d} = 1$ and $\pi_{c,d} = 1 - \alpha_0 = \alpha_{c,d}$. Similarly, $\gamma_{c,d} = 1 - |-\alpha_0| = \gamma_{c,d}$.

As established in [8], a fuzzy partition is of a higher quality, if the aggregated classes have high degrees of coverage and low degrees of overlap. Therefore, if classes are considered in pairs for comparison, a similar result would be expected. That is, if the pairs of classes (c, d), are analyzed, it is expected that:

- The degree of coverage and non-overlap of the classes studied are high, so that their difference must be small, and, $\beta_{c,d}$ must be high.
- The degree of overlap is less than its partial complement and the degree of coverage is greater than its partial complement, expecting that $\pi_{c,d}$ and $\gamma_{c,d}$ are low.

Definition 1. *Given a fuzzy classification system* (C, φ, ϕ, N) *it is said that the pair of classes {c, d} are relevant in C if* $\gamma_{c,d} < \beta_{c,d}$. *The following cases are established:*

1. *If* $\phi_2(\mu_c(x), \mu_d(x)) = \sigma_2(\mu_c(x), \mu_d(x))$, *the relevance of* {c, d} *is established from a parameter* t $\in [0, 1]$, *such that,* $t = \alpha_{c,d} = \pi_{c,d} = \gamma_{c,d}$. *The lower, the better.*
2. *If* $\phi_2(\mu_c(x), \mu_d(x)) \neq \sigma_2(\mu_c(x), \mu_d(x))$, *the relevance of* {c, d} *is established from a parameter t, such that,* $t = \gamma_{c,d} < \beta_{c,d}$. *The lower, the better.*
3. *If* $\phi_2(\mu_c(x), \mu_d(x)) \neq \sigma_2(\mu_c(x), \mu_d(x))$ *and* $= \gamma_{c,d} > \beta_{c,d}$ *then, the classes {c, d} are not relevant.*

In this sense, the coverage and overlapping of the classes analyzed by pairs, allow estimating the degree of relevance of such pair of classes (comparing the information obtained from the degree of grouping, the degree of partial non-coverage, the degree of overlap and the degree of partial overlap). Therefore, in a first stage the relevance of any pair of classes offers information on their usefulness, or relative meaning regarding their significance with respect to the already considered set of classes i.e., there is a significative lose if the classes are deleted.

4 Application

In order to apply the defined criterion, the image presented in Fig. 1 has been selected, and the unsupervised classification problem of obtaining a set of classes such that similar pixels are assigned to the same class is considered. The conjunctive operator $\phi_n(\mu_1(x),\ldots,\mu_n(x)) = \dfrac{3\prod_{k=1}^{n}\mu_k(x)}{1+2\prod_{k=1}^{n}\mu_k(x)}$ and the disjunctive operator $\varphi_n(\mu_1(x),\ldots,$ $\mu_c(x)) = \dfrac{1-\prod_{k=1}^{n}(1-\mu_k(x))}{1+2\prod_{k=1}^{n}(1-\mu_k(x))}$ whose negation is $N(\mu(x)) = 1 - \mu(x)$ have been selected. The fuzzy c-means algorithm has been applied for $c = 3$.

Fig. 1. Aurora Borealis

The values $\alpha_{c,d}, \pi_{c,d}, \gamma_{c,d}$ and $\beta_{c,d}$ are presented for each pair of classes (class 1, 2 and 3) in Tables 4, 5 and 6, and the classes are presented in Fig. 2.

Table 4. $v(v_0(\theta_2,\lambda_2)(1,2))$

v	ϕ_2	σ_2	δ_2	φ_2
ϕ_2	1	0,93	0,6	0,67
σ_2	0,93	1	0,67	0,74
δ_2	0,6	0,67	1	0,93
φ_2	0,67	0,74	0,93	1
	$\alpha = 0,67$	$\gamma = 0,74$	$\pi = 0,6$	$\beta = 0,93$

Table 5. $v(v_0(\theta_2,\lambda_2)(1,3))$

v	ϕ_2	σ_2	δ_2	φ_2
ϕ_2	1	0,8	0,4	0,5
σ_2	0,8	1	0,5	0,7
δ_2	0,4	0,5	1	0,8
φ_2	0,5	0,7	0,8	1
	$\alpha = 0,5$	$\gamma = 0,7$	$\pi = 0,4$	$\beta = 0,8$

Table 6. $v(v_0(\theta_2, \lambda_2)(2,3))$

v	ϕ_2	σ_2	δ_2	ϕ_2
ϕ_2	1	0,93	0,63	0.69
σ_2	0,93	1	0,69	0,76
δ_2	0.63	0,69	1	0,93
ϕ_2	0,69	0,76	0,93	1
	$\alpha = 0,69$	$\gamma = 0,76$	$\pi = 0,63$	$\beta = 0,93$

Fig. 2. Classes applying fuzzy 3-means algorithm (top left: class 1, top right: class 2, bottom: class 3). The gray scale represents the membership degree of each pixel to each class, where black = 0 and white = 1

In the case of the fuzzy c-means algorithm for $c = 4$, the results are presented in Tables 7, 8, 9, 10, 11 and 12, and the classes are presented in Fig. 3

Table 7. $v(v_0(\theta_2, \lambda_2)(1,2))$

v	ϕ_2	σ_2	δ_2	ϕ_2
ϕ_2	1	0,92	0,71	0,79
σ_2	0,92	1	0,79	0,86
δ_2	0,71	0,79	1	0,92
ϕ_2	0,79	0,86	0,92	1
	$\alpha = 0,79$	$\gamma = 0,8$	$\pi = 0,71$	$\beta = 0,9$

Table 8. $v(v_0(\theta_2, \lambda_2)(1,3))$

v	ϕ_2	σ_2	δ_2	ϕ_2
ϕ_2	1	0,87	0,57	0,7
σ_2	0,87	1	0,7	0,82
δ_2	0,57	0,7	1	0,87
ϕ_2	0,7	0,82	0,87	1
	$\alpha = 0,7$	$\gamma = 0,82$	$\pi = 0,57$	$\beta = 0,87$

Table 9. $v(v_0(\theta_2, \lambda_2)(1, 4))$

v	ϕ_2	σ_2	δ_2	φ_2
ϕ_2	1	0,861	0,58	0,72
σ_2	0,861	1	0,72	0,864
δ_2	0,58	0,72	1	0,861
φ_2	0,72	0,864	0,861	1
	$\alpha = 0,7$	$\gamma = 0,864$	$\pi = 0,5$	$\beta = 0,861$

Table 10. $v(v_0(\theta_2, \lambda_2)(2, 3))$

v	ϕ_2	σ_2	δ_2	φ_2
ϕ_2	1	0,86	0,6	0,74
σ_2	0,86	1	0,74	0,88
δ_2	0,6	0,74	1	0,86
φ_2	0,74	0,88	0,86	1
	$\alpha = 0,74$	$\gamma = 0,88$	$\pi = 0,6$	$\beta = 0,86$

Table 11. $v(v_0(\theta_2, \lambda_2)(2, 4))$

v	ϕ_2	σ_2	δ_2	φ_2
ϕ_2	1	0,9	0,72	0,82
σ_2	0,9	1	0,82	0,91
δ_2	0,72	0,82	1	0,9
φ_2	0,82	0,91	0,9	1
	$\alpha = 0,82$	$\gamma = 0,91$	$\pi = 0,72$	$\beta = 0,9$

Table 12. $v(v_0(\theta_2, \lambda_2)(3, 4))$

v	ϕ_2	σ_2	δ_2	φ_2
ϕ_2	1	0,93	0,75	0,81
σ_2	0,93	1	0,81	0,88
δ_2	0,75	0,81	1	0,93
φ_2	0,81	0,88	0,93	1
	$\alpha = 0,81$	$\gamma = 0,88$	$\pi = 0,75$	$\beta = 0,93$

Fig. 3. Classes applying fuzzy 4-means algorithm (top left: class 1, top right: class 2, bottom left: class 3, bottom right: class 4) The gray scale represents the membership degree of each pixel to each class, where black = 0 and white = 1

From these tables it is observed that for the case of fuzzy 3-means, for all pairs of classes c, d holds that $\gamma_{c,d} < \beta_{c,d}$. By contrast, in fuzzy 4-means the inequality is not met, i.e., $\gamma_{1,4} > \beta_{1,4}$, $\gamma_{2,3} > \beta_{2,3}$ and $\gamma_{2,4} > \beta_{2,4}$. Therefore, it is established that there are pairs of classes for which their degree of non-coverage is greater than their degree of coverage without considering their overlap. In principle, class 2 and class 4 are those that most affect the relevance of the classes analyzed by pairs. Therefore, the fuzzy 3-means algorithm application obtains greater relevance, illustrating the proposed criterion for identifying the partition with highest quality.

5   Final Comments

Through this work, some of the fundamental elements that allow characterizing the property of relevance in the framework of the evaluation of a fuzzy classification system are given. Three determining aspects are considered for the study of relevance: (1) a process of comparison between classes and the way they cover the objects under consideration, (2) degrees of intensity in the changes generated by the elements in the space and (3) a stopping criterion for inclusion of classes in a fuzzy partition.

The complement of two ratios β_0 and γ_0 for each pair of classes $\{c, d\}$ have been established as elements for comparison. The ratio β_0 expresses the global degree (aggregation of the degree for all items $x \in X$) in which the overlap of the two classes covers the objects under consideration, while γ_0 expresses the global degree in which the coverage of the two classes differs in relation to their partial complement. Comparing the complements of these two ratios it is expected that $\beta = 1 - \beta_0$ is greater than $\gamma = 1 - \gamma_0$.

According to the above, the stopping criterion corresponds to a comparison process in which a pair of classes is relevant up to a degree t, provided that the degree of coverage of two classes without considering their overlap is greater than the degree of non-coverage of the classes in relation to the objects under consideration. In this sense, the class pair $\{c, d\}$ will be non-relevant when $\gamma > \beta$.

As future work, it is proposed to build a model that allows generalizing the mappings together with the stopping criterion, while maintaining the group structure. Such a model should be general enough to include cases that do not meet the Ruspini's partition.

The characterization of the relevance of classes in a fuzzy partition still requires further developments and as future research, we propose to study the kinds of changes that the inclusion or elimination of a class can generate in a partition. Although the changes are measured in degrees of intensity, such changes can also be of a different nature, for example, affecting both the grouping and overlapping of each element, as there may be changes that affect only one of the properties. Likewise, a more in-depth study is necessary to relate the degrees of coverage and overlap of the partition, with the degree of relevance for every pair of classes.

Acknowledgements. This research has been partially supported by the Government of Spain (grant TIN2015-66471-P), the Government of Madrid (grant S2013/ICE-2845), and Complutense University (UCM Research Group 910149).

References

1. Zadeh, L.: Fuzzy sets. Inf. Control **8**, 338–353 (1965)
2. Bezdek, J., Harris, J.: Fuzzy partitions and relations: an axiomatic basis for clustering. Fuzzy Sets Syst. **1**, 111–127 (1978)
3. Bellman, R., Kalaba, R., Zadeh, L.: Abstraction and pattern classification. J. Math. Anal. Appl. **13**, 1–7 (1966)
4. Pedrycz, W.: Fuzzy sets in pattern recognition: methodology and methods. Pattern Recogn. **23**, 121–146 (1990)
5. Del Amo, A., Montero, J., Biging, G., Cutello, V.: Fuzzy classification systems. Eur. J. Oper. Res. **156**, 459–507 (2004)
6. Del Amo, A., Gómez, D., Montero, J., Biging, G.: Relevance and redundancy in fuzzy classification systems. Mathw. Soft Comput. **8**, 203–216 (2001)
7. Ruspini, E.: A new approach to clustering. Inf. Control **15**, 22–32 (1969)
8. Dombi, J.: Basic concepts for a theory of evaluation: the aggregative operator. Eur. J. Oper. Res. **10**, 282–293 (1982)
9. Dombi, J.: A general class of fuzzy operators, the DeMorgan class of fuzzy operators and fuzziness measures induced by fuzzy operators. Fuzzy Sets Syst. **8**, 149–163 (1982)
10. Matsakis, P., Andrefouet, P., Capolsini, P.: Evaluation of fuzzy partition. Remote Sens. Environ. **74**, 516–533 (2000)
11. Bustince, H., Fernández, J., Mesiar, R., Montero, J., Orduna, R.: Overlap functions. Nonlinear Anal. Theory Methods Appl. **72**, 1488–1499 (2010)
12. Bustince, H., Barrenechea, E., Pagola, M., Fernández, J.: The notions of overlap and grouping functions. In: Saminger-Platz, S., Mesiar, R. (eds.) On Logical, Algebraic, and Probabilistic Aspects of Fuzzy Set Theory, Studies in Fuzziness and Soft Computing, vol. 336, pp. 137–156. Springer, Switzerland (2016). https://doi.org/10.1007/978-3-319-28808-6_8
13. Gómez, D., Rodríguez, J., Bustince, H., Barrenechea, E., Montero, J.: n-dimensional overlap functions. Fuzzy Sets Syst. **287**, 57–75 (2016)
14. Qiao, J., Hu, B.Q.: On interval additive generators of interval overlap functions and interval grouping functions. Fuzzy Sets Syst. **323**, 19–55 (2017)
15. Qiao, J., Hu, B.Q.: On the migrativity of uninorms and nullnorms over overlap and grouping functions. Fuzzy Sets Syst. (2017). http://doi.org/10.1016/j.fss.2017.11.012
16. Klement, E., Moser, B.: On the redundancy of fuzzy partitions. Fuzzy Sets Syst. **85**, 195–201 (1997)
17. Sperber, D., Wilson, D.: Relevance: Communication and Cognition, 2nd edn. Blackwell Publishers Inc., Cambridge (1995)
18. Cutello, V., Montero, J.: Recursive connective rules. Int. J. Intell. Syst. **14**, 3–20 (1999)
19. Castiblanco, F., Gómez, D., Montero, J., Rodríguez, J.: Aggregation tools for the evaluation of classifications. In: 2017 Joint 17th World Congress of International Fuzzy Systems Association and 9th International Conference on Soft Computing and Intelligent Systems (IFSA-SCIS). IEEE, Otsu, pp. 1–5 (2017)
20. Mordeson, J., Bhutani, K., Rosenfeld, A.: Fuzzy Group Theory. Springer, Heidelberg (2005). https://doi.org/10.1007/b12359

Interactive Fuzzy Process:
An Epidemiological Model

Francielle Santo Pedro[1]($\boxtimes$), Laécio Carvalho de Barros[2], and Estevão Esmi[2]

[1] Department of Physics, Chemistry and Mathematics,
Federal University of São Carlos, Sorocaba, SP 18052-780, Brazil
`fran.stopedro@gmail.com`
[2] Institute of Mathematics, Statistics and Scientific Computation,
State University of Campinas, Campinas, SP 13083-970, Brazil
`{laeciocb,eelaureano}@ime.unicamp.br`

Abstract. In this study we analyze an two-dimensional epidemiological model via fuzzy differential equation considering that the solution is an interactive fuzzy process. More particularly, we will consider the case where this process is linearly correlated.

Keywords: Epidemiological model · Fuzzy interactive derivative
Autocorrelated fuzzy process

1 Introduction

Epidemiological models of direct transmission are of great importance for epidemic prediction. Normally, they are modeled by ordinary differential equations and partial differential equations [9]. In this manuscript, we study the model with only two compartments, namely susceptible-infectious. In this type of system, the infected individual is never considered to be susceptible again (e.g., HIV).

It is difficult to accurately determine the infected and susceptible population in epidemiology. Therefore, fuzzy sets can be a good tool to model the population over time. In addition, there seems to be a relationship between these populations, so that considering interactivity in the process is indispensable. For instance, the susceptible-infected (SI) model without vital dynamics, that is, the mortality rate is equal to the birth rate, it is possible to consider that the sum of the susceptible and infected populations is constant for all $t > 0$ [4].

We solve a system of two-dimensional fuzzy differential equations by the interactive fuzzy derivative theory. The system in question is the epidemiological model of direct transmission without vital dynamics.

The manuscript is organized as follows. Section 2 presents the mathematical background. Section 3 presents the interactive fuzzy derivative theory. Section 4 presents the epidemiological fuzzy model. Lastly, our final remarks are presented in Sect. 5.

L. C. de Barros—CNPq processo 306546/2017 − 5.

E. Esmi—FAPESP processo 2016/26040 − 7.

© Springer International Publishing AG, part of Springer Nature 2018
G. A. Barreto and R. Coelho (Eds.): NAFIPS 2018, CCIS 831, pp. 108–118, 2018.
https://doi.org/10.1007/978-3-319-95312-0_10

2 Mathematical Background

A fuzzy number A is a fuzzy subset of $\mathbb{R}$ with a normal, fuzzy convex and continuous membership function $\mu_A : \mathbb{R} \to [0,1]$, and with compact support. We denote the family of fuzzy numbers by $\mathbb{R}_{\mathcal{F}}$. The α-levels of A are given by

$$[A]_\alpha = \{x \in \mathbb{R} : \mu_A(x) \geq \alpha\},$$

if $\alpha > 0$,

$$[A]_0 = cl\{x \in \mathbb{R} : \mu_A(x) > 0\} = \overline{suppA}$$

and when $\alpha = 1$ we say $[A]_1$ is a core of A [1,2].

A joint possibility distribution J, n-dimensional, is a fuzzy subset of $\mathbb{R}^n$ with a normal membership function and a compact support. We denote by $\mathcal{F}_J(\mathbb{R}^n)$ the family of joint possibility distribution of $\mathbb{R}^n$.

Let $A_1, A_2, \ldots, A_n$ be fuzzy numbers and $J \in \mathcal{F}_J(\mathbb{R}^n)$, then μ_J is a joint possibility distribution of $A_1, A_2, \ldots, A_n$ if

$$\max_{x_j \in \mathbb{R}, j \neq i} \mu_J(x_1, \ldots, x_n) = \mu_{A_i}(x_i). \tag{1}$$

Besides that, μ_{A_i} is called the i-th marginal distribution marginal of J [3]. The interactivity between fuzzy numbers is determined from a possibility distribution [3]. If J is a possibility distribution of fuzzy numbers $A_1, A_2, \ldots, A_n$ then the following relationship is satisfied

$$\mu_J(x_1, \ldots, x_n) \leq \min\{\mu_{A_1}(x_1), \ldots, \mu_{A_n}(x_n)\},$$

and

$$[J]_\alpha \subseteq [A_1]_\alpha \times \ldots \times [A_n]_\alpha,$$

for all $\alpha \in [0,1]$.

We say that the fuzzy numbers $A_1, A_2, \ldots, A_n$ are non-interactive when

$$\mu_J(x_1, \ldots, x_n) = \min\{\mu_{A_1}(x_1), \ldots, \mu_{A_n}(x_n)\},$$

or equivalently,

$$[J]_\alpha = [A_1]_\alpha \times \ldots \times [A_n]_\alpha,$$

for all $\alpha \in [0,1]$. Otherwise they are interactive. The metric used in this study is the Pompieu-Hausdorff distance $d_\infty : \mathbb{R} \times \mathbb{R} \to [0, \infty)$, and is defined [1] by equation

$$d_\infty(A, B) = \sup_{0 \leq \alpha \leq 1} \max \{|a_\alpha^- - b_\alpha^-|, |a_\alpha^+ - b_\alpha^+|\},$$

where $A, B \in \mathbb{R}_{\mathcal{F}}$, $[A]_\alpha = [a_\alpha^-, a_\alpha^+]$ and $[B]_\alpha = [b_\alpha^-, b_\alpha^+]$.

Let J be a joint possibility distribution of $A_1, \ldots, A_n \in \mathbb{R}_{\mathcal{F}}$ and $f : \mathbb{R}^n \to \mathbb{R}$ a continuous function. The function f_J is said to be the extension principle of f via J [3] and its membership function is defined by

$$\mu_{f_J(A_1,\ldots,A_n)}(y) = \sup_{y = f(x_1,\ldots,x_n)} \mu_J(x_1, \ldots, x_n). \tag{2}$$

Notice that $f_J(A_1, \ldots, A_n) \in \mathbb{R}_{\mathcal{F}}$.

The next result is a generalization of Nguyen's theorem.

Theorem 1 [3]. *Let $A_1, \ldots, A_n$ be completely correlated fuzzy numbers, J their joint possibility distribution and $f : \mathbb{R}^n \to \mathbb{R}$ a continuous function, then*

$$[f_J(A_1, \ldots, A_n)]_\alpha = f([J]_\alpha),$$

for all $\alpha \in [0, 1]$.

Carlsson et al. [3] introduced the concept of completely correlated fuzzy numbers using the concept of possibility distribution. Let A_1 and A_2 be fuzzy numbers. Then, we say that A_1 and A_2 are completely correlated fuzzy numbers if exist $q \neq 0$ and r real numbers, such that their joint possibility distribution is given by

$$\begin{aligned} \mu_C(x_1, x_2) &= \mu_{A_1}(x_1)\mathcal{X}_{\{qx_1+r=x_2\}}(x_1, x_2) \\ &= \mu_{A_2}(x_2)\mathcal{X}_{\{qx_1+r=x_2\}}(x_1, x_2) \end{aligned} \tag{3}$$

where $\mathcal{X}_{\{qx_1+r=x_2\}}(x_1, x_2)$ is the characteristic function of line $\{(x_1, x_2) \in \mathbb{R}^2 : qx_1 + r = x_2\}$. Barros and Pedro [6, 7], introduced the concept of linearly correlated fuzzy numbers to model correlations where knowledge of the joint distributions are not known.

Two fuzzy numbers A and B are called linearly correlated if there exist $q, r \in \mathbb{R}$ such that

$$\mu_B(y) = \begin{cases} \sup\limits_{y=qx+r} \mu_A(x) & \text{if } q \neq 0 \text{ or } y = r \\ 0 & \text{if } q = 0 \text{ and } y \neq r \end{cases} . \tag{4}$$

Notice that the fuzzy number B is given by the Zadeh's extension principle of the fuzzy number A by $f(x) = qx + r$, thereby $\mu_B(x) = \mu_A(\frac{x-r}{q})$. Thus, according to [1, Theorem 2.1], the α-levels is given by

$$[B]_\alpha = q[A]_\alpha + r.$$

Let A and B be linearly correlated fuzzy numbers. We define:

- The addition between two linearly correlated fuzzy numbers $B +_L A$ is given by the following membership function

$$\mu_{B+_L A}(z) = \begin{cases} \sup\limits_{x \in \Phi^{-1}(z)} \mu_A(x) & \text{if } \Phi^{-1}(z) \neq \emptyset \\ 0 & \text{if } \Phi^{-1}(z) = \emptyset \end{cases},$$

  where $\Phi^{-1}(z) = \{x | z = (q+1)x + r\}$.
- The subtraction between two linearly correlated fuzzy numbers $B -_L A$ is given by the following membership function

$$\mu_{B-_L A}(z) = \begin{cases} \sup\limits_{x \in \Phi^{-1}(z)} \mu_A(x) & \text{if } \Phi^{-1}(z) \neq \emptyset \\ 0 & \text{if } \Phi^{-1}(z) = \emptyset \end{cases},$$

  where $\Phi^{-1}(z) = \{x | z = (q-1)x + r\}$.

- The product between two linearly correlated fuzzy numbers $B \cdot_L A$ is given by the following membership function

$$\mu_{B \cdot_L A}(z) = \begin{cases} \sup_{x \in \Phi^{-1}(z)} \mu_A(x) \ if \ \Phi^{-1}(z) \neq \emptyset \\ \qquad 0 \qquad \quad if \ \Phi^{-1}(z) = \emptyset \end{cases},$$

where $\Phi^{-1}(z) = \{x | z = qx^2 + rx\}$.
- The division between two linearly correlated fuzzy numbers $B \div_L A$ is given by the following membership function

$$\mu_{B \div_L A}(z) = \begin{cases} \sup_{x \in \Phi^{-1}(z)} \mu_A(x) \ if \ \Phi^{-1}(z) \neq \emptyset \\ \qquad 0 \qquad \quad if \ \Phi^{-1}(z) = \emptyset \end{cases},$$

where $\Phi^{-1}(z) = \{x | z = q + \frac{r}{x}\}$ and $0 \notin \text{supp} A$.

We reiterate once again the fact that for linearly correlated fuzzy numbers we do not need to know the joint possibility distribution involved.

Notice that the arithmetic operations between two linearly correlated fuzzy numbers are given as restrictions of the traditional operators to the curve $(x, qx + r)$. Moreover, the operations are given by the extension principle of the functions: $\Phi(x) = (q+1)x + r$, $\Phi(x) = (q-1)x + r$, $\Phi(x) = qx^2 + rx$ and $\Phi(x) = q + \frac{r}{x}$, respectively.

So, according to [1, Theorem 2.1], in terms of α-levels, the four operations of linearly correlated fuzzy numbers are given, respectively, by

- $[B +_L A]_\alpha = (1+q)[A]_\alpha + r, \ \forall \alpha \in [0,1]$.
- $[B -_L A]_\alpha = (q-1)[A]_\alpha + r, \ \forall \alpha \in [0,1]$.
- $[B \cdot_L A]_\alpha = \{qx_1^2 + rx_1 \in \mathbb{R} | x_1 \in [A]_\alpha\}, \ \forall \alpha \in [0,1]$.
- $[B \div_L A]_\alpha = \{q + \frac{r}{x_1} \in \mathbb{R} | x_1 \in [A]_\alpha\}, \ \forall \alpha \in [0,1]$.

There is another set difference operator for subtraction that appears in the literature. Let A and B be fuzzy numbers. The generalized Hukuhara difference (gH-difference) $A -_{gH} B = C$ is the fuzzy number C (if it exists) such that $A = B + C$ or $B = A - C$ [2,5,10].

3   Fuzzy Interactive Differential Equations

The modeling of dynamical systems from an initial condition, takes into account the "past" moment. Then, we have a process memory and the derivative operator must therefore incorporate these past relations between their states. The interactive fuzzy derivative, in particular the L-derivative, expresses these relations through the interactivity between their states. For h with absolute value sufficiently small, this interaction is given, in levels, by

$$[F(t+h)]_\alpha = q(h)[F(t)]_\alpha + r(h).$$

It means that the future value $F(t + h)$ is linearly correlated with the present value $F(t)$, for each h with absolute value sufficiently small. Thus F is an auto-correlated fuzzy processes.

Let $F : [a, b] \to \mathbb{R}_{\mathcal{F}}$ be a fuzzy-number-valued function and for each h with absolute value sufficiently small, let $F(t_0 + h)$ and $F(t_0)$ with $t_0 \in [a, b]$ be linearly correlated fuzzy numbers. According to [6], F is called L-differentiable at t_0 if there exists a fuzzy number $F_L'(t_0)$ such that the limit

$$\lim_{h \to 0} \frac{F(t_0 + h) -_L F(t_0)}{h}$$

exists and is equal to $F_L'(t_0)$, using the metric d_∞. Additionally, $F_L'(t_0)$ is called linearly correlated fuzzy derivative of F at t_0. At the endpoints of $[a, b]$, we consider only one-sided derivative.

The next theorem gives us a practical way to calculate the L-derivative.

Theorem 2 [6]. *Let $F : [a, b] \to \mathbb{R}_{\mathcal{F}}$ be L-differentiable at t_0 and $[F(t)]_\alpha = [f_\alpha^-(t), f_\alpha^+(t)]$, for $\alpha \in [0, 1]$, then f_α^- and f_α^+ are differentiable at t_0 and for each h with absolute value sufficiently small, we have*

$$[F_L'(t_0)]_\alpha = \begin{cases} i. & [(f_\alpha^-)'(t_0), (f_\alpha^+)'(t_0)] & \text{if } q(h) \geq 1 \\ ii. & [(f_\alpha^+)'(t_0), (f_\alpha^-)'(t_0)] & \text{if } 0 < q(h) < 1 \\ iii. & \{(f_\alpha)'(t_0)\} & \text{if } q(h) \leq 0 \end{cases}.$$

Let $F : [a, b] \to \mathbb{R}_{\mathcal{F}}$ be a fuzzy-number-valued function. According to [2, 5, 10], F is called gH-differentiable at t_0 if there exists a fuzzy number $F_{gH}'(t_0)$ such that the limit

$$\lim_{h \to 0} \frac{F(t_0 + h) -_{gH} F(t_0)}{h}$$

exists and is equal to $F_{gH}'(t_0)$, using the metric d_∞. Additionally, $F_{gH}'(t_0)$ is the generalized Hukuhara derivative of F at t_0. At the endpoints of $[a, b]$, we consider only one-sided derivative.

A strongly measurable and integrably bounded fuzzy-valued function is called integrable according to Kaleva [8]. A fuzzy Aumann integral of $F : [a, b] \to \mathbb{R}_{\mathcal{F}}$ is defined level-wise by

$$\left[(FA) \int_a^b F(x)dx \right]_\alpha = \int_a^b [F(x)]_\alpha dx$$

$$= \left\{ \int_a^b f(x)dx / f : [a, b] \to \mathbb{R} \right.$$

$$\left. \text{is a measurable selection for } F_\alpha \right\},$$

for all $\alpha \in [0, 1]$.

Theorem 3 [6]. *Let $F : [a,b] \to \mathbb{R}_\mathcal{F}$ be an autocorrelated fuzzy processes and L-differentiable, with $[F(t)]_\alpha = [f_\alpha^-(t), f_\alpha^+(t)]$. If F_L' is Aumann integrable then, for $t \in [a,b]$, we have*

$$F(t) = F(a) +_L \int_a^t F_L'(s)ds, \tag{5}$$

where $+_L$ is the addition between fuzzy numbers linearly correlated.

The fuzzy initial value problem (FIVP)

$$x_L'(t) = F(t, x(t)), \quad x(0) = x_0, \tag{6}$$

where $F : [a,b] \times \mathbb{R}_\mathcal{F} \to \mathbb{R}_\mathcal{F}$ is a continuous function and x_0, is a fuzzy number. Let $[x(t)]_\alpha = [x_\alpha^-(t), x_\alpha^+(t)]$ and $[F(t, x(t))]_\alpha = [f_\alpha^-(t, x_\alpha^-(t), x_\alpha^+(t)), f_\alpha^+(t, x_\alpha^-(t), x_\alpha^+(t))]$.

Lemma 1 [6]. *Let F be an autocorrelated fuzzy processes. The function $x : [a,b] \to \mathbb{R}_\mathcal{F}$ is a solution for (6) if only if F is continuous and satisfies*

$$x(t) = x_0 +_L \int_a^t F_L'(s, x(s))ds, \ t \in [a,b]. \tag{7}$$

4 Biological Problem: Fuzzy Epidemiological Model of Direct Transmission with Two Compartments

Infectious diseases can be classified into two categories, namely, microparasitic and macroparasitic. The first one is related to virus and bacteria, and the second one is related to worms. The difference between them is not just the size of the infectious agent. Microparasites reproduce within their host and are transmitted from one host to another, whereas macroparasites have a somewhat more complicated life cycle normally involving more than one host. When the disease spreads within a community without migration and immigration due to the contact of healthy and infected people we have an epidemic.

Let X be the healthy population, that is, susceptible to disease, and Y be the infectious population, that is, which transmits the disease. Consider the epidemiological model of direct transmission with two compartments without vital dynamics, given by the diagram

Note that the increase of Y is proportional to the encounter between healthy people (X) and infectious people (Y). The diagram supports the following system of differential equations

$$\begin{cases} X'_L(t) = -\beta X(t) \cdot_L Y(t) \\ Y'_L(t) = \beta X(t) \cdot_L Y(t) \\ X(0) = X_0 \in \mathbb{R}_{\mathcal{F}} \\ Y(0) = Y_0 \in \mathbb{R}_{\mathcal{F}} \end{cases}, \tag{8}$$

where the parameter $\beta \in (0, \infty)$ is the rate of transmission of the disease, $X, Y \in \mathbb{R}_{\mathcal{F}}$, and

$$X(t) +_L Y(t) = 1. \tag{9}$$

From condition (9), we have

$$X(t) = 1 -_L Y(t). \tag{10}$$

Notice that there is a clear functional relation between X and Y. Therefore, one says they are fuzzy numbers linearly correlated.

Substituting (10) into the system (8), we have

$$\begin{cases} Y'_L(t) = \beta(1 -_L Y(t)) \cdot_L Y(t) \\ Y(0) = Y_0 \end{cases}. \tag{11}$$

According to the operations between fuzzy numbers linearly correlated, we have $[\beta(1 -_L Y(t)) \cdot_L Y(t)]_\alpha =$

$$\begin{cases} [\beta(1 - y_\alpha^-)y_\alpha^-, \beta(1 - y_\alpha^+)y_\alpha^+], & y_\alpha^- < y_\alpha^+ < \frac{1}{2} \\ [\min\{c_\alpha^-, c_\alpha^+\}, \frac{\beta}{4}], & y_\alpha^- \le \frac{1}{2} \le y_\alpha^+, \\ [\beta(1 - y_\alpha^+)y_\alpha^+, \beta(1 - y_\alpha^-)y_\alpha^-], & y_\alpha^+ > y_\alpha^- > \frac{1}{2} \end{cases} \tag{12}$$

where $c_\alpha^- = \beta(1 - y_\alpha^-)y_\alpha^-$ and $c_\alpha^+ = \beta(1 - y_\alpha^+)y_\alpha^+$.

Consider that the function $Y(t)$ is an autocorrelated fuzzy processes with $0 < q < 1$, since when the time increases the susceptible population tends to decreases. Hence, we expect that the uncertainty will vanish over time. Thus, we use the case *ii.* of Theorem 2.

Therefore, system (11), $\forall \alpha \in [0, 1]$, becomes

$$\begin{cases} [(y_\alpha^+)', (y_\alpha^-)'] = [f_\alpha^-(t, y_\alpha^-, y_\alpha^+), f_\alpha^+(t, y_\alpha^-, y_\alpha^+)] \\ [Y(0)]_\alpha = [y_{0\alpha}^-, y_{0\alpha}^+] \end{cases}, \tag{13}$$

where $[F(t, y)]_\alpha = [\beta(1 -_L Y(t)) \cdot_L Y(t)]_\alpha$. That is,

- For $y_\alpha^- < y_\alpha^+ < \frac{1}{2}$

$$\begin{cases} (y_\alpha^-(t))' = \beta(1 - y_\alpha^+)y_\alpha^+ \\ (y_\alpha^+(t))' = \beta(1 - y_\alpha^-)y_\alpha^- \\ y_\alpha^-(0) = y_{0\alpha}^- \\ y_\alpha^+(0) = y_{0\alpha}^+ \end{cases} \tag{14}$$

- For $y_\alpha^- \le \frac{1}{2} \le y_\alpha^+$

$$\begin{cases} (y_\alpha^-(t))' = \frac{\beta}{4} \\ (y_\alpha^+(t))' = \min\{c_\alpha^-, c_\alpha^+\} \\ y_\alpha^-(0) = y_{0\alpha}^- \\ y_\alpha^+(0) = y_{0\alpha}^+ \end{cases} \tag{15}$$

– For $y_\alpha^+ > y_\alpha^- > \frac{1}{2}$

$$\begin{cases} (y_\alpha^-(t))' = \beta(1 - y_\alpha^-)y_\alpha^- \\ (y_\alpha^+(t))' = \beta(1 - y_\alpha^+)y_\alpha^+ \\ y_\alpha^-(0) = y_{0\alpha}^- \\ y_\alpha^+(0) = y_{0\alpha}^+ \end{cases} \qquad (16)$$

Notice that (14), (15) and (16) are real systems of differential equations. Thus, for each $\alpha \in [0,1]$, we solve the system numerically using the fourth order Runge-Kutta method or any IVP numerical method. In Fig. 1, we show the curves of 0-level and 1-level and in Fig. 2, we can see the graphical representation of fuzzy solution $X(t)$ and $Y(t)$ for a contractive fuzzy process.

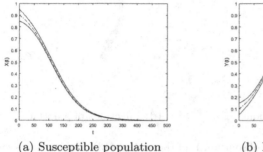

(a) Susceptible population

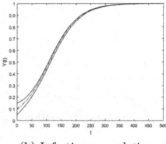

(b) Infectious population

Fig. 1. The 0-level (continuous line) and the core (dashed-dotted line) of solution Y of the system (13), and its correspondent X with $\beta = 0.02$ and initials conditions $X_0 = (0.85; 0.9; 0.95)$ and $Y_0 = (0.05; 0.1; 0.15)$.

Now if one considers that the function $Y(t)$ is an autocorrelated fuzzy processes with $q > 1$ one expects that the uncertainty will increase over time. Thus, we use the case i. of Theorem 2, and system (11), $\forall \alpha \in [0,1]$, becomes

$$\begin{cases} [(y_\alpha^-)', (y_\alpha^+)'] = [f_\alpha^-(t, y_\alpha^-, y_\alpha^+), f_\alpha^+(t, y_\alpha^-, y_\alpha^+)] \\ [Y(0)]_\alpha = [y_{0\alpha}^-, y_{0\alpha}^+] \end{cases}, \qquad (17)$$

where $[F(t,x)]_\alpha = [\beta(1 -_L Y(t)) \cdot_L Y(t)]$.

Proceeding in an equivalent way to the previous case we obtain Figs. 3 and 4. In Fig. 3, we show the curves of 0-level and 1-level and in Fig. 4, we can see the graphical representation of fuzzy solution $X(t)$ and $Y(t)$ for expansive fuzzy process.

Let us consider the gH-derivative in the system (11), that is,

$$\begin{cases} Y'_{gH}(t) = \beta(1 -_{gH} Y(t)) \cdot Y(t) \\ Y(0) = Y_0 \end{cases}. \qquad (18)$$

Thus, $\forall \alpha \in [0,1]$, we have

$$\begin{cases} [(y_\alpha^+)', (y_\alpha^-)'] = [\beta(1 - y_\alpha^+)y_\alpha^-, \beta(1 - y_\alpha^-)y_\alpha^+] \\ [Y(0)]_\alpha = [y_{0\alpha}^-, y_{0\alpha}^+] \end{cases}. \qquad (19)$$

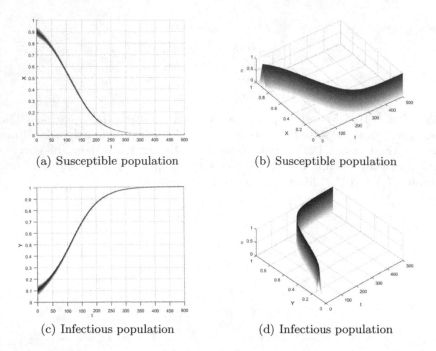

(a) Susceptible population   (b) Susceptible population

(c) Infectious population   (d) Infectious population

Fig. 2. Graphical representation of solution Y of the system (13), and its correspondent X with $\beta = 0.02$ and initials conditions $X_0 = (0.85; 0.9; 0.95)$ and $Y_0 = (0.05; 0.1; 0.15)$. The darkest region represents 1-level of $X(t)$ and $Y(t)$.

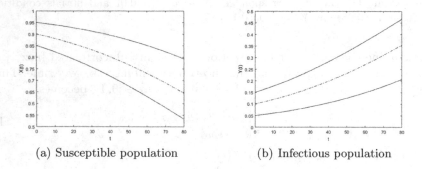

(a) Susceptible population   (b) Infectious population

Fig. 3. The 0-level (continuous line) and the core (dashed-dotted line) of solution Y of the system (13), and its correspondent X with $\beta = 0.02$ and initials conditions $X_0 = (0.85; 0.9; 0.95)$ and $Y_0 = (0.05; 0.1; 0.15)$.

We show the curves of 0-level and 1-level n Fig. 5 of the solution X and Y with L-derivative and gH-derivative. It is worth noticing that in the case with gH-derivative one considers that the difference is interactive with gH but the product operation is non-interactive. In this way, there is an inconsistency with the choice of the product operation in system (18). Comparing the solutions in

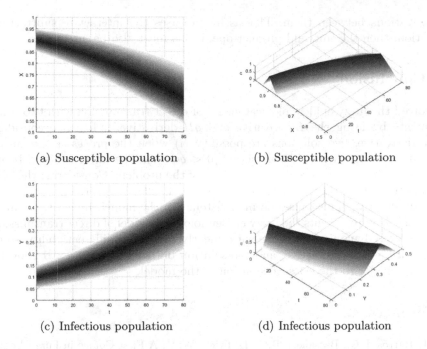

Fig. 4. Graphical representation of solution Y of the system (13), and its correspondent X with $\beta = 0.02$ and initials conditions $X_0 = (0.85; 0.9; 0.95)$ and $Y_0 = (0.05; 0.1; 0.15)$. The darkest region represents 1-level of $X(t)$ and $Y(t)$.

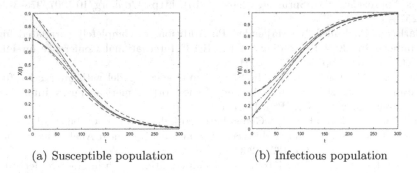

Fig. 5. The 0-level (dashed line) and the core (dashed-dotted line) of solution Y of the system (8), and its correspondent Y with L-derivative, and the 0-level (continuous line) and the core (dashed-dotted line) of solution Y of the system (8), and its correspondent X with gH-derivative. Both with $\beta = 0.02$ and initials conditions $X_0 = (0.85; 0.9; 0.95)$ and $Y_0 = (0.05; 0.1; 0.15)$.

Fig. 5, one can see the difference between the solution via gH-derivative and the solution via L-derivative.

We stress the importance of using the fuzzy interactive arithmetic. Since X and Y are interactive fuzzy numbers, one needs to maintain the coherence over all operations between them. That is, if one uses the interactive subtraction operation, then the sum and product operations must also be interactive.

5 Conclusion

We solved the fuzzy epidemiological model of direct transmission with two compartments by using the L-derivative and gH-derivative. When one considers the L-derivative, two solutions are possible: (i) when the process is expansive ($q > 1$), and (ii) when it is contractive ($0 < q < 1$). The latter solution being consistent with the biological interpretation of the problem. Considering the gH-derivative, we note that although the difference is interactive, the product used is non-interactive. This causes an inconsistency in the model. In short, the approach that best represents the fuzzy epidemiological model of direct transmission without vital dynamics is the one with linearly correlated differentiability, since all operations are interactive, and the solution obtained by $0 < q < 1$ shows a similar behavior to the classical solution of the model.

References

1. de Barros, L.C., Bassanezi, R.C., Lodwick, W.A.: A First Course in Fuzzy Logic, Fuzzy Dynamical Systems, and Biomathematics: Theory and Applications. SFSC, vol. 347. Springer, Heidelberg (2017). https://doi.org/10.1007/978-3-662-53324-6
2. Gomes, L.T., de Barros, L.C., Bede, B.: Fuzzy Differential Equations in Various Approaches. SM. Springer, Cham (2015). https://doi.org/10.1007/978-3-319-22575-3
3. Carlsson, C., Fúller, R., Majlender, P.: Additions of completely correlated fuzzy numbers. In: 2004 Proceedings of the IEEE International Conference on Fuzzy Systems, vol. 1, pp. 535–539 (2004)
4. Cabral, V.M., Barros, L.C.: The SI epidemiological model with interactive fuzzy parameters. In: 2012 Annual Meeting of the North American Fuzzy Information Processing Society (NAFIPS). IEEE (2012)
5. Stefanini, L., Barnabas, B.: Generalized Hukuhara differentiability of interval-valued functions and interval differential equations. Nonlinear Anal.: Theory Methods Appl. **71**(3–4), 1311–1328 (2009)
6. Barros, L.C., Pedro, F.S.: Fuzzy differential equations with interactive derivative. Fuzzy Sets Syst. **309**, 64–80 (2017)
7. Pedro, F.S., Barros, L.C., Esmi, E.: Population Growth Model via Interactive Fuzzy Differential Equation (2017, Submitted for publication)
8. Kaleva, O.: Fuzzy differential equations. Fuzzy Sets Syst. **24**(3), 301–317 (1987)
9. Edelstein-Keshet, L.: Mathematical Models in Biology. SIAM, Philadelphia (1988)
10. Bede, B.: Mathematics of Fuzzy Sets and Fuzzy Logic. STUDFUZZ. Springer, Berlin (2013). https://doi.org/10.1007/978-3-642-35221-8

IoT Resources Ranking: Decision Making Under Uncertainty Combining Machine Learning and Fuzzy Logic

Renato Dilli, Amanda Argou, Renata Reiser$^{(\boxtimes)}$, and Adenauer Yamin

Laboratory of Ubiquitous and Parallel Systems, Federal University of Pelotas,
Pelotas, RS, Brazil
{renato.dilli,aacardozo,reiser,adenauer}@inf.ufpel.edu.br

Abstract. The Internet of Things (IoT) is characterized by a broad
range of resources connected to the Internet, requesting and providing services simultaneously. Given this scenario, suitably selecting the
resources that best meet users' demands has been a relevant and current
research challenge. Based on the non-functional parameters of Quality
of Service (QoS), IoT plays an important role in the ranking of these
resources according to the offered services. This paper presents a proposal to classify and select the most appropriate resource for the client's
request, applying fuzzy logic to address uncertainties in the definition of
ideal weights for QoS attributes, and aggregating machine learning to the
pre-classification of EXEHDA middleware resources, in order to reduce
the computational cost generated by MCDA algorithms. As the main
contribution, the pre-classification of new resources of the EXEHDA-RR
is presented. The experimental results show the efficiency of the proposed
model.

Keywords: IoT · Resources ranking · MCDA · Machine learning
Fuzzy logic

1 Introduction

The current scenario accounts for more than six billion things connected to
the Internet, in a dynamic composition environment, by providing services to
customers and forecasting more than 100 billion services by 2025 [1].

A challenge to be overcome after resource discovery is to classify services
to select what best suits the user's request [2], which is a difficult and time-
consuming task. Classification processes have focused on user preferences, which
often establish an order based on Quality of Service (QoS) or Non-Functional
Properties (NFPs) [3].

Another challenge after discovering resources is to classify the services to
select which one best suits the user's request [2]. Classification processes focus
on user preferences that often establish an order based on Quality of Service

© Springer International Publishing AG, part of Springer Nature 2018
G. A. Barreto and R. Coelho (Eds.): NAFIPS 2018, CCIS 831, pp. 119–131, 2018.
https://doi.org/10.1007/978-3-319-95312-0_11

(QoS) [4] or the non-functional properties (NFPs). Using these preferences, the set of discovered services can be classified so that the best service can be chosen.

Considering the typical IoT infrastructure, the resource discovery process must consider both functional and non-functional requirements in order to meet the demands of the applications. Functionalities explicitly describe the features of the resources, that is, what they can offer, and the non-functional requirements define additional information about the services, such as performance and safety [5].

The resources, which are usually made available through services, have QoS attributes with values presenting units of different measurements and broad ranges in minimum and maximum values. The entity responsible for defining these values should be aware of all the characteristics of the quality attributes. Considering that in each institution the definition of these values can be done by specialists with different perceptions, the uncertainty arises when defining the values to be adopted as attributes of QoS. This paper presents a new proposal for resource ranking, considering EXEHDA middleware [6], called EXEHDA-RR (Resource Ranking). The proposal models the treatment of uncertainty when defining the importance of the different attributes of QoS. Furthermore, this model considers customer preferences and service quality attributes and aggregates Semantic Web technologies to the specification and query of resources.

The original contribution of this proposal lies in the combined use of fuzzy logic and machine learning in recognition of standards for resource classification. The resources are initially classified by Multiple-Criteria Decision Analysis (MCDA) algorithms. At each rating, a new machine learning algorithm training is performed to pre-classify new resources as they enter the computational infrastructure. This process reduces the need to process all MCDA algorithm computations at each client request. The evaluation scenario of the EXEHDA-RR showed satisfactory results for the accuracy obtained by combining fuzzy logic with machine learning. Considering the literature and the demands of a research group, the results achieved are timely for use in the EXEHDA middleware.

This paper is organized as follows: Sect. 2 presents preliminaries of fuzzy logic. Section 3 presents preliminaries in multiple-criteria decision analysis, which describe the process of the resource classification used. The model and evaluation of EXEHDA-RR are described in Sect. 4. The related works are discussed in Sect. 5. Finally, conclusion is given in Sect. 6.

2   Preliminaries in Fuzzy Logic

Fuzzy set theory (FST), which has been widely applied to model the ambiguities of human thinking, also adequately addresses the uncertainties in the information available for decision-making based on multiple criteria. The adequacy of substitutions versus criteria and the significant weight of criteria are evaluated regarding linguistic values represented by diffuse numbers. In FST, linguistic variables are used to describe fuzzy terms mapping linguistic variables to numerical ones. The truth values of Boolean logic are replaced by fuzzy values in the unit interval in the decision-making process [7].

Thus, mathematically, a set is defined as a finite, infinite, or infinite countable collection of elements. In each case, each element is either a member of a the set or not. However, in fuzzy systems, the element may be partially in or out of the set. So the answer to the question: "is X a member of a set A" does not have a definite answer, either true or false.

Fuzzy Sets

A fuzzy set A defined in the discourse universe $U\neg\emptyset$ is characterized by a membership function $\mu_A : U \to [0,1]$, given by

$$A = \{(x, \mu_A(x)) \mid x \in U \land \mu_A(x) \in [0,1]\}$$

For each $x \in X$, $\mu_A(x)$ represents the degree of relevance of x in A.

$$x \in (A, \mu) \iff x \in A \land \mu(x) \neq 0$$

Additionally, by the membership function, each element $x \in U$ has a degree of relevance in each set A, expressing how much it is possible for the element x belonging to the set A. Thus, when an element of degree of relevance 0, means that it is not included in the fuzzy set, while a grade 1 element is fully included in it.

Fuzzification

Considering specifications related to applied area of the present work, triangular membership functions were adopted. A triangular fuzzy number A can be set by a triple (a, b, c) with the membership function given by Eq. (1) below:

$$\mu_A(x) = \begin{cases} 0, & se\ x \leq a; \\ \frac{x-a}{b-a}, & se\ a \leq x \leq b; \\ \frac{c-x}{c-b}, & se\ b \leq x \leq c; \\ 0, & se\ x \geq c. \end{cases} \tag{1}$$

Defuzzification

Defuzzification is the process which produces a quantifiable real-number (crisp) result in fuzzy logic, meaning that a fuzzy number is transformed into a single number based on different methods. This work considers the weighted average of the maximum, expressed as

$$Z_0 = \frac{\mu(x)_i w_i}{\mu(x)_i}, \tag{2}$$

According to Eq. (2), where Z_0 is the defuzzified output, $\mu(x)_i$ is the degree of relevance and w_i is the fuzzy output weight value.

3 Preliminaries in Multiple-Criteria Decision Analysis

The resource discovery process encompasses the classification and selection of the best resources, which are suited to the client's request. In this section, the MCDA algorithm is used for classification and selection of resources used in this work.

An MCDA refers to decision-making in the presence of multiple and often conflicting criteria. The MCDA algorithms aim to aid in the judgment of decision-making using a set of objectives and criteria, estimating their relative importance by weights, establishing the contribution of each option about of multi-criteria performance. MCDA is not only a set of theories, methodologies, and techniques but it also includes a particular perspective in order to deal with decision-making problems [8].

The Simple Additive Weighting (SAW) algorithm, using an evaluation score to rank each available option is obtained by normalized criteria values, which are multiplied by corresponding weights. The options are sorted in descending order according to the final score, which is the sum of the scores for each criterion [9].

The proposed MCDA algorithm developed for resource classification is based on the algorithms SAW and Web Service Relevancy Function (WsRF) [10], the first stage of data normalization as proposed by [11]. For the matrix normalization, two vectors are defined. In the former, $N = \{n_1, n_2, \ldots, n_m\}$, the value of n_j can be given as binary number: (i) 1, when the increase in $q_{i,j}$ benefits the customer's request; or (ii) 0, when the increase in $q_{i,j}$ does not benefit the client's request. The latter vector $C = \{c_1, c_2, \ldots, c_m\}$ contains the related constants with the maximum normalized value for each attribute.

The following steps must be performed for the calculation of resource assessment applying the MCDA algorithm:

1. Normalize the matrix $Q = (q_{ij})_{n \times m}$ according to Eq. (3) whether the criterion should be maximized or to Eq. (4) whether the criterion should be minimized. In these equations, $\frac{1}{n} \sum_{i=1}^{n} q_{i,j}$ is the mean of the quality attributes j in the Q matrix.

$$
v_{i,j} = \begin{cases} \dfrac{q_{i,j}}{\frac{1}{n}\sum_{i=1}^{n} q_{i,j}} & if \quad \frac{1}{n}\sum_{i=1}^{n} q_{i,j} \neq 0 \\ & and \quad \dfrac{q_{i,j}}{\frac{1}{n}\sum_{i=1}^{n} q_{i,j}} < c_j \\ & and \quad n_j = 1 \\ c_j & if \quad \frac{1}{n}\sum_{i=1}^{n} q_{i,j} = 0 \\ & and \quad n_j = 1 \\ & or \quad \dfrac{q_{i,j}}{\frac{1}{n}\sum_{i=1}^{n} q_{i,j}} \geq c_j \end{cases} \tag{3}
$$

$$
v_{i,j} = \begin{cases} \dfrac{\frac{1}{n}\sum_{i=1}^{n} q_{i,j}}{q_{i,j}} & if \quad q_{i,j} \neq 0 \\ & and \quad n_j = 0 \\ & and \quad \dfrac{\frac{1}{n}\sum_{i=1}^{n} q_{i,j}}{q_{i,j}} < c_j \\ c_j & if \quad q_{i,j} = 0 \\ & and \quad n_j = 0 \\ & or \quad \dfrac{\frac{1}{n}\sum_{i=1}^{n} q_{i,j}}{q_{i,j}} \geq c_j \end{cases} \tag{4}
$$

2. Calculating the vector scoring of each available option. Each score can be calculated using Eq. (5) and the operator $max(v_j)$ representing the highest normalized attribute value in column j. Therefore, we need to define an array that represents the contribution of weights to each resource, where $w = \{w1, w2, w3, \dots, wj\}$. Each weight in this matrix represents the degree of importance or weight factor associated with a specific QoS property. The values of these weights vary from 0 to 1. The Eq. (6) sums all the quality attributes for the resource R_i, where N represents the number of attributes.

$$h_{i,j} = w_j \left[\frac{v_{i,j}}{max(v_j)} \right] \tag{5}$$

$$R_i = \sum_{j=1}^{N} h_{i,j} \tag{6}$$

$$MCDA(A_i) = \left[\frac{100 * R_i}{max(R)} \right] \tag{7}$$

3. The final result of the MCDA algorithm, Eq. (7), is reclassified, as shown in Fig. 1. The leaves of the tree are represented by a box containing the classification (1 a 4).

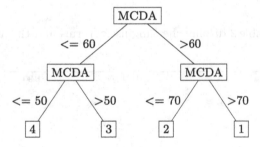

Fig. 1. Reclassification of the proposed MCDA algorithm

4   EXEHDA-RR: Model and Evaluation

The EXEHDA-RR uses ontologies for describing resources and their quality attributes (QoS) [12], MCDA algorithm, considers client preferences, treats uncertainty in defining attribute weights through fuzzy logic, and performs pre-classification of resources through of the LMT machine learning algorithm.

4.1 Exploring Treatment of Uncertainty Using Fuzzy Logic

This resource classification process, which can satisfy the user's request, the evaluation of QoS attributes, is a challenging step. The definition of the degree of importance of QoS attributes by the user and administrators of the computational infrastructure is an activity that depends on the experience and knowledge of each. The treatment of the uncertainty introduced by these divergences is one of the contributions of this work.

The evaluation process uses fuzzy logic in the specification of the ideal QoS attributes, defined by the experts of the computational environment. We use the QWS version 2.0 available from [10], with 2.505 resources and nine attributes of quality. Table 1 describes the five used attributes.

Table 1. QWS dataset attributes

Attribute	Description	Unit
RT - Response Time	Time to send a request and receive your response	ms
AV - Availability	Number of correct invocations/total invocations	%
TH - Throughput	Total Number of invocations for a given period of time	%
RE - Reliability	Ratio of the number of error messages over total messages	%
LA - Latency	Time taken for the server to process a given request	ms

Figure 2 and Table 2 display the Linguistic Terms and the assigned weights.

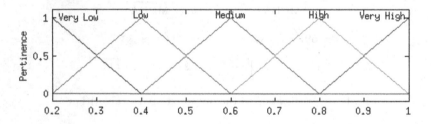

Fig. 2. Representation of fuzzy sets

Table 3 simulates the definition of the degree of importance for each quality attribute using the Linguistic Terms defined by five experts. This evaluation aims to set the weights to be assigned to each attribute providing the most adequate resources to the user's request.

Table 4 contains the conversion of the Linguistic Terms, assigned by the experts, into fuzzy triangular numbers is presented. Fuzzy mean between the specialists and the defuzzification using the weighted average method is also shown. In the end, the values of the attributes are normalized.

Table 2. Degree of importance

Linguistic Term	Value
VH - Very High	(0.8,1.0,1.0)
H - High	(0.6,0.8,1.0)
M - Medium	(0.4,0.6,0.8)
L - Low	(0.2,0.4,0.6)
VL - Very Low	(0.0,0.2,0.4)

Table 3. Expert evaluation

Attr	Exp1	Exp2	Exp3	Exp4
RT	M	H	L	H
AV	VH	VH	H	VH
TH	M	M	H	H
RE	H	H	VH	H
LA	VH	H	VL	H

Table 4. Calculation of QoS weights

Attr	Exp1	Exp2	Exp3	Exp4	Fuzzy Avg	Def	Norm
RT	(0.4,0.6,0.8)	(0.6,0.8,1.0)	(0.2,0.4,0.6)	(0.6,0.8,1.0)	(0.45,0.65,0.85)	0.65	0.70
AV	(0.9,1.0,1.0)	(0.9,1.0,1.0)	(0.6,0.8,1.0)	(0.9,1.0,1.0)	(0.83,0.95,1.00)	0.93	1.00
TH	(0.4,0.6,0.8)	(0.4,0.6,0.8)	(0.6,0.8,1.0)	(0.6,0.8,1.0)	(0.50,0.70,0.90)	0.70	0.75
RE	(0.6,0.8,1.0)	(0.6,0.8,1.0)	(0.9,1.0,1.0)	(0.6,0.8,1.0)	(0,68,0.85,1.00)	0.87	0.94
LA	(0.9,1.0,1.0)	(0.6,0.8,1.0)	(0.0,0.2,0.4)	(0.6,0.8,1.0)	(0,53,0.70,0.85)	0.70	0.75

The resulting normalized values will be applied in the process of classification and selection of resources with the application of the MCDA algorithm. In this step we evaluate the accuracy in the classification performed by the algorithm of machine learning in the process of pre-classification of new resources. In order perform the tests, the dataset QWS [10] was considered. Table 1 describes the attributes that were used. Massive Online Analysis (MOA) framework for data mining was used, prototyping decision tree algorithms. The MOA and WEKA libraries were developed in Java and Jpype was used so that Python can manipulate libraries in Java.

4.2   Exploring MCDA in the Constrution of Reference Ranking

The application of the MCDA algorithm is demonstrated in the classification of a data set containing 5 quality attributes for 10 resources. Each row represents a resource and each column has a quality attribute (Table 5). The values of the Resp.Time and Latency attributes are best if they are low, and the values of the Availability, Throughput, the higher the values, the better the Reliability attributes.

The first step of the classification is the normalization of the data. Therefore, we consider the vectors N={0,1,1,1,0}, C={6,2,3,2,50} and w={0.70, 1.00, 0.75, 0.94, 0.75}. The Resp.Time and Latency attributes qualify the resource with low values (Eq. 4) and Availability, Throughput and Reability qualify with high values (Eq. 3). All attributes are normalized with maximum value defined in "C", the result is shown in Table 6.

Table 5. Dataset example

Resp.Time	Availability	Throughput	Reliability	Latency
302.75	89	7.1	73	187.75
482	85	16	73	1
3321.4	89	1.4	73	2.6
126.17	98	12	67	22.77
107	87	1.9	73	58.33
107.57	80	1.7	67	18.21
255	98	1.3	67	40.8
136.71	76	2.8	60	11.57
102.62	91	15.3	67	0.93
200	40	13.5	67	41.66

Table 6. Normalized attributes

RT	AV	TH	RE	LA
1.70	1.07	0.97	1.06	0.21
1.07	1.02	2.19	1.06	38.56
0.15	1.07	0.19	1.06	14.83
4.07	1.18	1.64	0.98	1.69
4.80	1.04	0.26	1.06	0.66
4.78	0.96	0.23	0.98	2.12
2.02	1.18	0.18	0.98	0.95
3.76	0.91	0.38	0.87	3.33
5.01	1.09	2.10	0.98	41.46
2.57	0.48	1.85	0.98	0.93

Table 7. Classified attributes

RT	AV	TH	RE	LA	MCDA	Classif
0.24	0.91	0.33	0.94	0.00	61.18	2
0.15	0.87	0.75	0.94	0.70	85.99	1
0.02	0.91	0.07	0.94	0.27	55.67	3
0.57	1.00	0.56	0.86	0.03	76.42	1
0.67	0.89	0.09	0.94	0.01	65.68	2
0.67	0.82	0.08	0.86	0.04	62.27	2
0.28	1.00	0.06	0.86	0.02	56.14	3
0.53	0.78	0.13	0.77	0.06	57.22	3
0.70	0.93	0.72	0.86	0.75	100	1
0.36	0.41	0.63	0.86	0.02	57.59	3

Table 7 presents the values of the attributes after applying Eq. (5), that is, to divide the normalized value of the Table 6 by the highest normalized value of each column. After Eq. (6) is applied, it will add up all the attribute values in each row. The following is applied to Eq. (7) that will qualify the resource with a value ranging from 0 to 100. The value 100 will be given for the best resource of dataset. The rank of 1 to 4 is assigned through the rule shown in Fig. 1.

The MCDA algorithm presented in this step was used to classify the dataset containing the initial repository of 100, 200 or 300 resources. After classifying and grouping 1000 resources in different ways, the behavior of the machine learning algorithm is evaluated.

4.3   Improving the Reference Ranking Exploring Machine Learning

This section presents the results of 4 evaluations performed in the pre-classification of resources through the LMT machine learning algorithm. The LMT algorithm was chosen because it obtained the best accuracy among the decision tree algorithms analyzed.

The algorithms J48, LMT, RandomForest, SimpleCart, DecisionStump, RandomTree, HoeffdingTree, and REPTree were evaluated in the WEKA tool [13]. The classification of the services was done through supervised learning to classify the attribute "Service Classification" defined in the QWS dataset [10]. According to the results presented in Table 8, it is noticed that the LMT classifier obtained an accuracy of 85.99%.

Table 8. Evaluation of classifiers

Classifiers	Hits	Misses	Accuracy
LMT	313	51	85.99%
RandomForest	294	70	80.77%
J48	258	106	70.88%
RandomTree	248	116	68.13%
SimpleCart	248	116	68.13%
REPTree	239	125	65.66%
HoeffdingTree	198	166	54.40%
DecisionStump	160	204	43.95%

All resources were ranked through the MCDA algorithm presented in Sect. 3 and classified from 1 to 4. This classification was used to train the machine learning algorithm. In each evaluation, the group of resources classified through the MCDA algorithm is taken from the training dataset and used in the test dataset of the machine learning algorithm. The evaluations were carried out with the following initial repositories for training the LMT algorithm: (a) - initial repository with 100 resources; (b) - initial repository with 200 resources and (c) - initial repository with 300 resources.

Table 9 shows incorrect resource classifications in 4 evaluations. The first evaluation (E1) considered groups of 50 resources, totaling 20 groups. The second evaluation (E2) considered groups of 100 resources, totalizing 10 groups. The third evaluation (E3) considered groups of 200 resources, totaling 5 groups. The fourth evaluation (E4) considered groups of 500 resources, totaling 2 groups. For each evaluated group, the MCDA algorithm reclassifies resources and trains the machine learning algorithm. By analyzing the accuracy obtained using pre-ranking the machine learning algorithm LMT, we can observe that E1c evaluation obtained the accuracy of 82%. The initial repository was with 300 resources that were used for training and every 50 resources for new training.

Table 9. Accuracy assessment

Qty	E1a	E1b	E1c	E2a	E2b	E2c	E3a	E3b	E3c	E4a	E4b	E4c
50	14	7	8									
100	13	8	10	26	21	21						
150	8	12	8									
200	11	10	11	20	24	16	41	45	42			
250	11	10	11									
300	14	10	11	22	17	18						
350	7	6	10									
400	11	8	10	13	14	15	35	34	28			
450	5	6	6									
500	6	6	5	14	12	13				109	106	108
550	7	6	14									
600	12	14	8	16	14	27	63	61	77			
650	20	21	6									
700	17	19	9	35	38	16						
750	17	16	5									
800	16	17	6	34	31	12	73	76	24			
850	11	14	11									
900	11	9	19	22	23	45						
950	14	10	7									
1000	7	11	5	26	22	12	46	50	87	213	193	356
TOTAL	232	223	180	228	216	195	258	266	258	322	299	464
Accuracy%	76.8	77.7	82	77.2	78.4	80.5	74.2	73.4	74.2	67.8	70.1	53.6

E2b and E2c evaluations are also satisfactory, with accuracy of 78.4% and 80.5%. These evaluations were performed every 100 resources, so the data normalization and calculation processes of the MCDA algorithm was only performed every 100 resources.

Evaluation 4 proved to be inappropriate for pre-classification of resources, with accuracy below 54% (E4c). Additionally, the interval of 500 resources for recalculation of the MCDA algorithm and machine learning training showed to be very large, resulting in loss of accuracy.

5   Related Works

The Table 10 presents a comparison of the main methods, such as: (i) expressiveness in the representation of resources; (ii) use of MCDA algorithms; (iii) application of Customer Preferences; (iv) use of QoS; (v) employment of fuzzy logic and (vi) Machine Learning (ML) employment. The presence of the criterion is represented by the character " + " and the absence by " − ".

Table 10. Comparison of related work

Works	Expr.	MCDA	Pref.	QoS	Fuzzy	ML
Maheswari [14]	+	+	+	+	+	−
Salah [15]	−	+	+	+	+	−
Perera [16]	+	+	+	−	−	−
Gomes [17]	+	−	−	−	−	−
Almulla [18]	−	+	+	+	+	−
Nunes [19]	−	+	+	−	−	−
Suchithra [20]	−	+	+	+	−	−
Vaadaala [21]	−	+	+	+	−	+
EXEHDA-RR	+	+	+	+	+	+

The specification of resources through highly expressive languages is performed in the research of [14,16,17]. Only [17] does not use MCDA algorithms and does not consider client preferences because it works with context data. Quality of Service (QoS) criteria are evaluated in [14,15,18,20,21]. The fuzzy logic is discussed in [14,15,18]. None of them apply fuzzy logic to solve uncertainty among experts.

It can be seen that only [21] uses Machine Learning. But with the objective of measuring the accuracy of a single quality attribute with the J48 algorithm. The authors considered the results satisfactory and obtained an accuracy of 63%.

6   Conclusion

This paper presented a model for the IoT resource ranking, called EXEHDA-RR. As a principal contribution of the work developed so far we can highlight: (i) the use of fuzzy logic in the definition of attribute weights and its use in the MCDA algorithm to classify the adequate resources to the client's request; (ii) the use of the decision tree algorithm in the pre-classification of resources, considering different training intervals. By applying fuzzy logic in the resolution of conflicts among the experts in defining degrees of importance for each attribute of QoS, it was possible to promote adequacy of the weights used in the algorithm MCDA, thus considering the resources of better quality as the result of the clients' requests.

Moreover, using of supervised machine learning, from a dataset, ranked by the MCDA algorithm, it was possible to classify resources with considerable accuracy. The LMT algorithm was adequate, with up to 82% correct in the pre-classification. The proposed resource ranking model considers the dynamicity of the computational infrastructure provided by IoT, with a large number of consumers and resource providers.

The use of machine learning algorithms in the pre-classification of resources promotes a relevant reduction of the computational effort of the MCDA algorithms in the classification of resources to each request of the clients. This technique was modeled using semantic web technologies for the specification of the resources and their quality attributes. The objective of the EXEHDA-RR proposition was to improve the EXEHDA middleware's resource allocation process, enabling it to be highly scalable and dynamic when compiling the computational environment, a typical situation of the infrastructure provided by IoT.

References

1. BIS Research: Global Sensors in Internet of Things (IoT) Devices Market, Analysis & Forecast: 2016 to 2022. Technical report (2017)
2. García, J.M.: Improving Semantic Web Services Discovery and Ranking. Ph.D. thesis, University of Seville (2012)
3. Schröpfer, C., Schönherr, M., Offermann, P., Ahrens, M.: A flexible approach to service management-related service description in SOAs. In: CEUR Workshop Proceedings. Vol. 234 (2006)
4. Nakamura, L.H., Estrella, J.C., Santana, M.J., Santana, R.H.: Using semantic web for selection of web services with QoS. In: WebMedia'11, Proceedings of the 17th Brazilian Symposium on Multimedia and the Web, pp. 4–7 (2011)
5. Khutade, P.A., Phalnikar, R.: QoS aware web service selection and ranking framework based on ontology. Int. J. Soft Comput. Eng. (IJSCE) 4(3), 77–81 (2014)
6. Lopes, J.B.: Uma Arquitetura para Provimento de Ciência de Situação Direcionada às Aplicações Ubíquas na Infraestrutura da Internet das Coisas. Ph.D. thesis, Universidade Federal do Rio Grande do Sul (2016)
7. Alabool, H.M., Mahmood, A.K.: Trust-based service selection in public cloud computing using fuzzy modified VIKOR method. Aust. J. Basic Appl. Sci. 7(9), 211–220 (2013)
8. Figueira, J., Greco, S., Ehrgott, M.: Multiple Criteria Decision Analysis: State of the Art Surveys, vol. 78. Springer, New York (2005). https://doi.org/10.1007/b100605
9. Tzeng, G.H., Huang, J.J.: Multiple Attribute Decision Making: Methods and Applications. A Chapman & Hall book. Taylor & Francis, Boco Raton (2011)
10. Al-Masri, E., Mahmoud, Q.H.: QoS-based discovery and ranking of Web services. In: Proceedings - International Conference on Computer Communications and Networks, ICCCN, pp. 529–534 (2007)
11. Liu, Y., Ngu, A.H., Zeng, L.Z.: QoS computation and policing in dynamic web service selection. In: Proceedings of the 13th International World Wide Web Conference on Alternate Track Papers & Posters, pp. 66–73 (2004)
12. Dilli, R., Filho, H.K., Pernas, A.M., Yamin, A.: EXEHDA-RR: machine learning and MCDA with semantic web in IoT resources classification. In: Proceedings of the 23rd Brazillian Symposium on Multimedia and the Web. WebMedia 2017, pp. 293–300. ACM, New York (2017)
13. Witten, I.H., Frank, E., Hall, M.A.: Data Mining: Practical Machine Learning Tools and Techniques, 3rd edn. Morgan Kaufmann Publishers Inc., Burlington (2011)

14. Maheswari, S., Karpagam, G.R.: Comparative analysis of semantic web service selection methods (2015)
15. Salah, N.B., Saadi, I.B.: Fuzzy AHP for learning service selection in context-aware ubiquitous learning systems. In: International IEEE Conferences on Ubiquitous Intelligence & Computing, Advanced and Trusted Computing, Scalable Computing and Communications, Cloud and Big Data Computing, Internet of People, and Smart World Congress, pp. 171–179 (2016)
16. Perera, C., Zaslavsky, A., Christen, P., Compton, M., Georgakopoulos, D.: Context-aware sensor search, selection and ranking model for internet of things middleware. In: Proceedings - IEEE International Conference on Mobile Data Management. vol. 1, pp. 314–322 (2013)
17. Gomes, P., Cavalcante, E., Batista, T., Taconet, C., Chabridon, S., Conan, D., Delicato, F., Pires, P.: A QoC-aware discovery service for the internet of things. Ubiquitous Comput. Ambient Intell. **7656**, 344–355 (2016)
18. Almulla, M., Yahyaoui, H., Al-Matori, K.: A new fuzzy hybrid technique for ranking real world web services. Knowl.-Based Syst. **77**, 1–15 (2015)
19. Nunes, L.H., Estrella, J.C., Perera, C., Reiff-Marganiec, S., Delbem, A.N.: Multi-criteria IoT resource discovery: a comparative analysis. pp. 1–16. Wiley Inter-Science (2016)
20. Suchithra, M., Ramakrishnan, M.: A survey on different web service discovery techniques (2015)
21. Vaadaala, V.: Classification of web services using Jforty eight. Natl. Conf. Recent Trends Comput. Sci. Technol. Int. J. Electron. Commun. Comput. Eng. **4**(6), 181–184 (2013)

Least Squares Method with Interactive Fuzzy Coefficient: Application on Longitudinal Data

Nilmara J. B. Pinto$^{(\boxtimes)}$, Vinícius F. Wasques$^{(\boxtimes)}$,
Estevão Esmi, and Laécio C. Barros

Institute of Mathematics, Statistics and Scientific Computing,
State University of Campinas, Campinas, São Paulo 13081-970, Brazil
nilmarabiscaia@gmail.com, vwasques@outlook.com, eelaureano@gmail.com,
laeciocb@ime.unicamp.br

Abstract. This work focus on the least square method to fit a fuzzy function to longitudinal data given by fuzzy numbers. In order to consider the intrinsic correlation of longitudinal data, we assume that there exits a linear relation among the involved fuzzy numbers that arises from the concept of a joint possibility distribution. We propose a numerical method to solve a fuzzy least square problem taking into account this linear correlation. To this end, we extend the classical least square method by means of the sup-J extension principle, which consists of a generalization of Zadeh's extension principle. Finally, we use our proposal method to fit a longitudinal dataset.

Keywords: Fuzzy least square method · Interactive fuzzy numbers
Joint possibility distribution · Longitudinal data

1 Introduction

The least squares methods are used, in general, to obtain a continuous function that best fit pairs of data in a dataset [1]. The fuzzy least squares method arises when the dataset is composed by fuzzy numbers. Tanaka *et al.* proposed a fuzzy least squares method based on fuzzy regression models [2]. This method was used to find fuzzy parameters of a fuzzy linear function from a fuzzy dataset. However, this approach converts the problem to a classic linear programming problem which may lead to losing the notion of close distance between the fuzzy data and the obtained solution.

Celmins [3] proceeded with the same methodology of [2] but considered a intrinsic relation among the dataset based on conical membership functions that

---

N. J. B. Pinto—Grantee CAPES 1691227.

V. F. Wasques—Grantee CNPq 142414/2017-4.

E. Esmi—Grantee FAPESP 2016/26040-7.

L. C. Barros—Grantee CNPq 306546/2017-5.

© Springer International Publishing AG, part of Springer Nature 2018
G. A. Barreto and R. Coelho (Eds.): NAFIPS 2018, CCIS 831, pp. 132–143, 2018.
https://doi.org/10.1007/978-3-319-95312-0_12

are (geometrically) similar to joint possibility distributions. In addiction, the concept of interactive fuzzy numbers [4–6] was only considered in [7], which improved the approach presented by [3].

In contrast to these previous methods, Diamond [8] proposed a fuzzy least squares method based on distance between functions. He used projection theorems for cones in Banach spaces to find the fuzzy linear function that best fit a dataset.

It is worth noting that all these approaches were developed for data given only by triangular fuzzy numbers and for fitting only fuzzy linear functions. Nevertheless, these methods can be used to model many phenomenons, for example, in economy [9], psychology [10], medicine [11], and logistics [12].

Data correlations arise naturally in longitudinal datasets. A dataset is said to be longitudinal if it contains the same type of information on the same itens at multiple points in time. Therefore, longitudinal data is characterized by the fact that repeated observations are correlated [13]. In this work, we suppose that this correlation is given by the notion of completely correlated fuzzy numbers [5,14].

The method proposed here is based on the (sup-J) extension of classical numerical algorithm to the fuzzy context and does not take into account any distance between fuzzy numbers. Moreover, our method can be applied not only for triangular fuzzy numbers, but for any type of completely correlated fuzzy numbers, and it can approximate the dataset with higher orders functions.

In Sect. 2 we briefly recall the classical least squares method and some basic definitions and results from fuzzy set theory. In Sect. 3, we develop the extension of the classical least squares method for the case where dataset is composed by completely correlated fuzzy numbers. Finally, in Sect. 4, we apply the proposed method to fit a fuzzy function to the longitudinal dataset given in [15].

2   Mathematical Background

This section presents the least squares method [1] and some basic concepts of fuzzy set theory [16].

2.1   Least Square Method

Let $f : [c,d] \to \mathbb{R}$ be a continuous function. Given n functions $g_1, \ldots, g_n$, where $g_i : \mathbb{R} \to \mathbb{R}$ for $i = 1, \ldots, n$, we need to find n coefficients $a_1, \ldots, a_n \in \mathbb{R}$ such that the function $\varphi : \mathbb{R} \to \mathbb{R}$ given by

$$\varphi(x) = a_1 g_1(x) + \ldots + a_n g_n(x)$$

is the best approximation of the function f, $i.e.$, $\varphi \approx f$.

The function φ is obtained by minimizing the distance between f and φ. More precisely, let $||\cdot||_2$ be the $\mathcal{L}^2$-norm defined on the class of the continuous functions from $[c,d]$ to $\mathbb{R}$ (denoted by $C([c,d])$) given by $||h||_2 = \left(\int_c^d |h(s)|^2 ds \right)^{1/2}$, $\forall h \in$

$C([c, d])$. The coefficients $a_1, \ldots, a_n$ of the function φ which produces the best fit to f are obtained by solving the following minimization problem:

$$\min_{a_1,\ldots,a_n \in \mathbb{R}} {}^1\!/_2 \|\varphi - f\|_2^2.$$

In the case some values of f are known, say $D = \{f(x_1) = y_1, \ldots, f(x_m) = y_m\}$, the function φ must fit the data D, that is, $\varphi(x_i) \approx y_i$, for all $i = 1, \ldots, m$. Therefore the following minimization problem must be solved.

$$\min_{a_1,\ldots,a_n \in \mathbb{R}} {}^1\!/_2 \|(\varphi(x_1) - y_1, \ldots, \varphi(x_m) - y_m)\|_2^2. \tag{2.1}$$

The real coefficients $a_1, \ldots, a_n$ that minimize the problem (2.1), *i.e.*, that produces the best approximation φ of f, are obtained by solving the following matrix equation called normal equation:

$$Ma = b,$$

where

$$M = \begin{bmatrix} \sum_{k=1}^{m} g_1(x_k)g_1(x_k) & \cdots & \sum_{k=1}^{m} g_1(x_k)g_n(x_k) \\ \vdots & \ddots & \vdots \\ \sum_{k=1}^{m} g_n(x_k)g_1(x_k) & \cdots & \sum_{k=1}^{m} g_n(x_k)g_n(x_k), \end{bmatrix},$$

$$a = \begin{bmatrix} a_1 \\ \vdots \\ a_n \end{bmatrix} \quad \text{and} \quad b = \begin{bmatrix} \sum_{k=1}^{m} y_k g_1(x_k) \\ \vdots \\ \sum_{k=1}^{m} y_k g_n(x_k) \end{bmatrix}.$$

If the matrix M is non singular, say $P = M^{-1} = [p_{ij}]$, then the vector a is obtained by

$$a = Pb. \tag{2.2}$$

Thus, each parameter a_i is given by

$$a_i = p_{i1}b_1 + p_{i2}b_2 + \ldots + p_{in}b_n$$

$$= p_{i1}\left(\sum_{k=1}^{m} y_k g_1(x_k)\right) + \ldots + p_{in}\left(\sum_{k=1}^{m} y_k g_n(x_k)\right)$$

$$= \left(\sum_{j=1}^{n} p_{ij}g_j(x_1)\right) y_1 + \ldots + \left(\sum_{j=1}^{n} p_{ij}g_j(x_m)\right) y_m$$

$$= c_{i1}y_1 + \ldots + c_{im}y_m,$$

where $c_{ik} = \sum_{j=1}^{n} p_{ij} g_j(x_k)$, for $i = 1, \ldots, n$ and $k = 1, \ldots, m$. In general case, the matrix P stands for the pseudoinverse of M.

Since the parameters of the function φ can be obtained by computing the matrix product (2.2), we rewrite the function φ in terms of $y_1, \ldots, y_m$ as follows:

$$\begin{aligned}
\varphi(x) &= a_1 g_1(x) + \ldots + a_n g_n(x) \\
&= (c_{11} y_1 + \ldots + c_{1m} y_m) g_1(x) + \ldots + (c_{n1} y_1 + \ldots + c_{nm} y_m) g_n(x) \\
&= \left(\sum_{j=1}^{n} g_j(x) c_{j1} \right) y_1 + \ldots + \left(\sum_{j=1}^{n} g_j(x) c_{jm} \right) y_m \\
&= s_1(x) y_1 + \ldots + s_m(x) y_m,
\end{aligned} \tag{2.3}$$

where

$$s_i = \left(\sum_{j=1}^{n} g_j(x) c_{ji} \right)$$

for each $i = 1, \ldots, n$.

2.2   Fuzzy Set Theory

A fuzzy subset A of an universe X is characterized by a function $\mu_A : X \to [0, 1]$, called membership function [16], where $\mu_A(x)$, or simply $A(x)$, represents the membership degree of x in A, for all $x \in X$. The class of fuzzy sets of X is denoted by the symbol $\mathcal{F}(X)$. Each classical subset A of X is a particular fuzzy set whose membership function is given by its characteristic function $\chi_A : X \to \{0, 1\}$, i.e., $\chi_A(x) = 1$ if and only if $x \in A$.

The α-cut of a fuzzy set A of X, denoted by $[A]^\alpha$, is defined as $[A]^\alpha = \{x \in X : A(x) \geq \alpha\}$, $\forall \alpha \in (0, 1]$. If X is also a topological space, then we can define the 0-cut of A by $[A]^0 = cl\{x \in X : A(x) > 0\}$ [17], where $cl \, Y$, $Y \subseteq X$, denotes the closure of Y.

Zadeh's extension principle [18] can be viewed as mathematical method to extend a function $f : X \to Y$ to a function $\hat{f} : \mathcal{F}(X) \to \mathcal{F}(Y)$.

Definition 1 *(Zadeh's extension principle [17,18]). Let $f : X \to Y$. The Zadeh's extension of f at $A \in \mathcal{F}(X)$ is the fuzzy set $\hat{f}(A) \in \mathcal{F}(Y)$ whose membership function is given by*

$$\hat{f}(A)(y) = \bigvee_{x \in f^{-1}(y)} A(x), \ \forall \, y \in Y,$$

where $f^{-1}(y) = \{x \in X : f(x) = y\}$ is the preimage of the function f at y and, by definition, $\bigvee \emptyset = 0$.

A fuzzy set $A \in \mathcal{F}(\mathbb{R})$ is called a fuzzy number if its α-cuts are closed, bounded and non-empty intervals for all $\alpha \in [0,1]$ [17]. Since each α-cut of a fuzzy number A is an interval that satisfies the previous properties, we can write $[A]^\alpha = [a_\alpha^-, a_\alpha^+]$. We denote the class of fuzzy numbers by the symbol $\mathbb{R}_\mathcal{F}$. The next theorem indicates when a family of subsets can be uniquely associated with a fuzzy number.

Theorem 1 *(Negoita-Ralescu's characterization theorem [19,20]). Given a family of subsets $\{A_\alpha : \alpha \in [0,1]\}$ that satisfies the following conditions*

(a) A_α is a non-empty, closed, and bounded interval for any $\alpha \in [0,1]$;
(b) $A_{\alpha_2} \subseteq A_{\alpha_1}$, for all $0 \leq \alpha_1 \leq \alpha_2 \leq 1$;
(c) For any sequence α_n which converges from below to $\alpha \in (0,1]$ we have

$$\bigcap_{n=1}^{\infty} A_{\alpha_n} = A_\alpha;$$

(d) For any sequence α_n which converges from above to 0 we have

$$A_0 = cl\left(\bigcup_{n=1}^{\infty} A_{\alpha_n}\right).$$

Then there exists a unique $A \in \mathbb{R}_\mathcal{F}$, such that $[A]^\alpha = A_\alpha$, for each $\alpha \in [0,1]$.
Conversely, let $A \in \mathbb{R}_\mathcal{F}$, if $A_\alpha = [A]^\alpha$ for all $\alpha \in [0,1]$ then the family of subsets $\{A_\alpha : \alpha \in [0,1]\}$ satisfies the conditions (a)–(d).

An example of fuzzy number is a triangular fuzzy number that is denoted by the triple $(a; b; c)$, with $a \leq b \leq c$. In view of Theorem 1, the triangular fuzzy number can be defined in terms of its α-cuts as follows:

$$[A]^\alpha = [a + \alpha(b-a), c - \alpha(c-b)], \ \forall \alpha \in [0,1].$$

Note that a real number a is a particular case of triangular fuzzy number since we have $a \equiv (a; a; a)$.

A fuzzy relation R over $X = X_1 \times \ldots \times X_n$ is any fuzzy subset of $X_1 \times \ldots \times X_n$. Thus, a fuzzy relation R is associated with a membership function $R : X_1 \times \ldots \times X_n \to [0,1]$, where $R(x_1, \ldots, x_n) \in [0,1]$ represents the degree of relationship among $x_1, \ldots, x_n$ with respect to R [17].

The projection of fuzzy relation $R \in \mathcal{F}(X_1 \times \ldots \times X_n)$ onto X_i, for $i \in \{1, \ldots, n\}$, is the fuzzy set Π_R^i of X_i given by

$$\Pi_R^i(y) = \bigvee_{x \in X : x_i = y} R(x_1, \ldots, x_n).$$

A fuzzy relation $J \in \mathcal{F}(\mathbb{R}^n)$ is said to be a joint possibility distribution of $A_1, \ldots, A_n \in \mathbb{R}_\mathcal{F}$ if

$$A_i(y) = \Pi_J^i(y) = \bigvee_{x \in X : x_i = y} J(x_1, \ldots, x_n),$$

for all $y \in \mathbb{R}$ and for all $i = 1, \ldots, n$.

Given a t-norm t, that is, a commutative, associative, and increasing operator $t : [0,1]^2 \rightarrow [0,1]$ satisfying $t(x,1) = x\,t\,1 = x$ for all $x \in [0,1]$. A fuzzy relation J_t given by

$$J_t(x_1, \ldots, x_n) = A_1(x_1)\,t\,\ldots\,t\,A_n(x_n) \tag{2.4}$$

is said to be a t-norm-based joint possibility distribution of $A_1, \ldots, A_n \in \mathbb{R}_{\mathcal{F}}$ [4]. Well-known example of t-norm include the minimum t-norm "$\wedge$". In particular, when $J = J_\wedge$, that is, J is given by (2.4) with $t = \wedge$, we say that $A_1, \ldots, A_n$ are non-interactive. Otherwise, $J \neq J_\wedge$, we say that $A_1, \ldots, A_n$ are interactive [5,18,21].

Thus, the notion of interactivity between fuzzy numbers is given by means of joint possibility distributions. Carlsson $et\ al.$ [5] introduced a possible type of interactivity relation between two fuzzy numbers that is not based on t-norms. Specifically, two fuzzy numbers A and B are said to be completely correlated if there exist $q, r \in \mathbb{R}$ with $q \neq 0$ such that the corresponding joint possibility distribution $J_{\{q,r\}}$ is given by

$$\begin{aligned} J_{\{q,r\}}(x_1, x_2) &= A(x_1)\chi_{\{qu+r=v\}}(x_1, x_2) \\ &= B(x_2)\chi_{\{qu+r=v\}}(x_1, x_2), \end{aligned} \tag{2.5}$$

where $\chi_{\{qu+r=v\}}$ stands for the characteristic function of the set $\{(u,v) \in \mathbb{R}^2 : qu + v = r\} \subset \mathbb{R}^2$. In addition, if $q > 0$ ($q < 0$) then A and B are said to be completely positively (negatively) correlated. Since $q \neq 0$ in Eq. (2.5), the membership function of B can be written as $B(qx+r) = A(x)$ for all $x \in \mathbb{R}$, and consequently $[B]^\alpha = q[A]^\alpha + \{r\}$ for all $\alpha \in [0,1]$. Moreover, for each $\alpha \in [0,1]$, the α-cut of the joint possibility distribution $J_{\{q,r\}}$ is given by [5]:

$$[J_{\{q,r\}}]^\alpha = \{(x, qx + r)\ :\ x \in [A]^\alpha\}.$$

$Remark\ 1.$ Note that if the fuzzy numbers A and B are completely correlated by the line $qu + r_1 = v$, and we choose $r_2 = q(a_\alpha^- + a_\alpha^+) + r_1$, then A and B are also completely correlated if we consider $J_{\{-q,r_2\}}$, that is, A and B are also completely correlated with respect to the line $-qu + r_2 = v$. Therefore, the distribution J is not unique.

The next definition is a generalization of Zadeh's extension principle (cf. Definition 1).

Definition 2 $(Sup\text{-}J\ Extension\ Principle\ [6])$. $Let\ J \in \mathcal{F}(\mathbb{R}^n)\ be\ a\ joint\ possibility\ distribution\ of\ A_1, \ldots, A_n \in \mathbb{R}_{\mathcal{F}}\ and\ let\ f : \mathbb{R}^n \rightarrow \mathbb{R}.\ The\ \sup -J\ extension\ of\ f\ at\ (A_1, \ldots, A_n)\ is\ defined\ by$

$$f_J(A_1, \ldots, A_n)(y) = \hat{f}(J)(y) = \bigvee_{(x_1,\ldots,x_n) \in f^{-1}(y)} J(x_1, \ldots, x_n),$$

$where\ f^{-1}(y) = \{(x_1, \ldots, x_n) \in \mathbb{R}^n : f(x_1, \ldots, x_n) = y\}.$

From Definition 2, we can define arithmetic operations among n fuzzy numbers by taking the sup-J extension of the corresponding arithmetic operator. For example, let $f(x_1, \ldots, x_n) = x_1 + \ldots + x_n$ for all $x_1, \ldots, x_n \in \mathbb{R}$. If $J_\wedge$ is defined as in (2.4) with $t = \wedge$, then $f_{J_\wedge}(A_1, \ldots, A_n)$ boils down to Zadeh's extension of f at $(A_1, \ldots, A_n)$, i.e.,

$$\widehat{f}(A_1, \ldots, A_n)(y) = \bigvee_{(x_1, \ldots, x_n) \in f^{-1}(y)} A_1(x_1) \wedge \ldots \wedge A_n(x_n), \quad \forall\, y \in \mathbb{R}.$$

The next proposition ensures that the completely correlation is a transitive relation of interactivity between fuzzy numbers. Moreover, under some conditions, the sup-$J_{q,r}$ extensions of the addition operator, denoted by the symbol $+_L$, satisfies the associative property.

Proposition 1 [22]. *Let A, B, $C \in \mathbb{R}_{\mathcal{F}}$. If A and B are completely correlated with respect to $J_{\{q_1, r_1\}}$ and B and C are completely correlated with respect to $J_{\{q_2, r_2\}}$, then there are real numbers q_3 and r_3 such that A and C are completely correlated with respect to $J_{\{q_3, r_3\}}$.*

Moreover, if each A, B, $C \in \mathbb{R}_{\mathcal{F}}$ is completely correlated to $D \in \mathbb{R}_{\mathcal{F}} \backslash \mathbb{R}$, then the associative property holds true, i.e., $A +_L (B +_L C) = (A +_L B) +_L C$.

The notion of completely correlation can be extended to n fuzzy numbers as follows.

Definition 3. *The fuzzy numbers $A_1, \ldots, A_n \in \mathbb{R}_{\mathcal{F}}$ are said completely correlated if the joint possibility distribution J is given by*

$$J(x_1, \ldots, x_n) = \chi_U(x_1, \ldots, x_n) A_1(x_1) \tag{2.6}$$
$$= \chi_U(x_1, \ldots, x_n) A_2(x_2)$$
$$\vdots$$
$$= \chi_U(x_1, \ldots, x_n) A_n(x_n),$$

where $U = \{(u, q_2 u + r_2, \ldots, q_n u + r_n) : u \in \mathbb{R}\}$, $q_i, r_i \in \mathbb{R}$, with $q_i \neq 0$, $\forall i = 1, \ldots, n$.

From (2.5) and (2.6), one can see that A_1 and A_i, $i > 1$, are also completely correlated since we have $[A_i]^\alpha = q_i [A_1]^\alpha + \{r_i\}$, for all $i = 2, \ldots, n$. This implies that, for each $\alpha \in [0, 1]$, the α-cut of J is given as follows

$$[J]^\alpha = \{(x, q_2 x + r_2, \ldots, q_n x + r_n) : x \in [A_1]^\alpha\} \tag{2.7}$$

Remark 2. From Eq. (2.7), we can note that the α-cuts of the joint possibility distribution J can be expressed in terms of α-cuts of A_1 and the parameters q_i and r_i, for all $i = 2, \ldots, n$.

Theorem 2 [23, 24]. *Let $f : \mathbb{R}^n \to \mathbb{R}$ be a continuous function and $J \in \mathcal{F}(\mathbb{R}^n)$. We have that*

$$[\widehat{f}_J(A_1, \ldots, A_n)]^\alpha = f([J]^\alpha), \quad \forall \alpha \in [0, 1].$$

By Theorem 2, if the sup-J extension of f at $(A_1, \ldots, A_n)$ is a fuzzy number, then the α-cuts of $\widehat{f}_J(A_1, \ldots, A_n) = \widehat{f}(J)$ can be written as follows:

$$[\widehat{f}(J)]^\alpha = \left[\bigwedge_{(x_1, \ldots, x_n) \in [J]^\alpha} f(x_1, \ldots, x_n) \quad \bigvee_{(x_1, \ldots, x_n) \in [J]^\alpha} f(x_1, \ldots, x_n) \right]. \quad (2.8)$$

In the next section, we consider the problem given in (2.1) for the case where the known values y_i are interactive fuzzy numbers.

3   Least Squares Method for Interactive Fuzzy Data

In this paper, we deal with least squares method to fit uncertain data given by interactive fuzzy numbers. In particular, we focus on the case where these fuzzy numbers are completely correlated. A typical example of correlated data are the well-known longitudinal data, which are widely studied in the statistical area [13].

Let $D = \{(x_1, Y_1), \ldots, (x_m, Y_m)\} \subset \mathbb{R} \times \mathbb{R}_{\mathcal{F}}$ such that $Y_1, \ldots, Y_m$ are completely correlated fuzzy numbers, with respect to a joint possibility distribution J as in (2.6), and let $F : \mathbb{R} \to \mathbb{R}_{\mathcal{F}}$ be a function that satisfies $F(x_i) = Y_i$ for $i = 1, \ldots, m$. We produce a function $\Phi : \mathbb{R} \to \mathbb{R}_{\mathcal{F}}$ that approximates F given by means of the sup-J extension principle of a function $\varphi : \mathbb{R} \to \mathbb{R}$ of the form

$$\varphi(x) = a_1 g_1(x) + \ldots + a_n g_n(x),$$

where $a_1, \ldots, a_n \in \mathbb{R}$ and $g_1, \ldots, g_n$ are real-valued-functions. More precisely, we define the function Φ in terms of the sup-J extension principle of (2.3) at $(Y_1, \ldots, Y_m)$. Since Eq. (2.3) is continuous with respect to $y_1, \ldots, y_m$, from Theorem 2 and Eq. (2.7), we have that α-cuts of the fuzzy number $\Phi(x)$ is given by

$$[\Phi(x)]^\alpha = \{s_1(x)y_1 + \ldots + s_m(x)y_m : (y_1, \ldots, y_m) \in [J]^\alpha\} \quad (3.9)$$
$$= \{s_1(x)y + s_2(x)(q_2 y + r_2) + \ldots + s_m(x)(q_m y + r_m)y : y \in [Y_1]^\alpha\}.$$

Since the interval $[Y_1]^\alpha = [y_{1\alpha}^-, y_{1\alpha}^+]$ can be rewritten as the set of all convex combination of $y_{1\alpha}^-$ and $y_{1\alpha}^+$, that is, $[Y_1]^\alpha = \{(1-\lambda)y_{1\alpha}^- + \lambda y_{1\alpha}^+ : \lambda \in [0,1]\}$, the α-cut of J can also be expressed in terms of a parameter $\lambda \in [0,1]$ as follows:

$$[J]^\alpha = \{(1-\lambda)Y_\alpha^- + \lambda Y_\alpha^+ : \lambda \in [0,1]\},$$

where $Y_\alpha^- = (y_{1\alpha}^-, q_2 y_{1\alpha}^- + r_2, \ldots, q_m y_{1\alpha}^- + r_m)$ and $Y_\alpha^+ = (y_{1\alpha}^+, q_2 y_{1\alpha}^+ + r_2, \ldots, q_m y_{1\alpha}^+ + r_m)$. Thus, Eq. (3.9) can be expressed as

$$[\Phi(x)]^\alpha = \{(1-\lambda)\langle S(x), Y_\alpha^- \rangle + \lambda \langle S(x), Y_\alpha^+ \rangle : \lambda \in [0,1]\} \quad (3.10)$$

where $\langle \cdot, \cdot \rangle$ denotes the usual inner product of $\mathbb{R}^m$ and $S(x) = (s_1(x), s_2(x), \ldots, s_m(x))$, $x \in \mathbb{R}$.

In order to characterize the endpoints of each α-cut of $\Phi(x)$, we define the auxiliary function h by

$$h(x, \alpha, \lambda) = (1 - \lambda)B_1(x, \alpha) + \lambda B_2(x, \alpha), \ \forall x \in \mathbb{R} \text{ and } \forall \alpha, \lambda \in [0, 1],$$

where
$$B_1(x, \alpha) = \langle S(x), Y_\alpha^- \rangle \text{ and } B_2(x, \alpha) = \langle S(x), Y_\alpha^+ \rangle.$$

By Eqs. (3.10) and (2.8), we have that

$$[\Phi(x)]^\alpha = \{h(x, \alpha, \lambda) : \lambda \in [0, 1]\}$$

$$= \left[\bigwedge_{\lambda \in [0,1]} h(x, \alpha, \lambda), \ \bigvee_{\lambda \in [0,1]} h(x, \alpha, \lambda) \right]. \tag{3.11}$$

Note that if $B_1(x, \alpha) \leq B_2(x, \alpha)$, then the function $h(x, \alpha, \cdot)$ assumes the minimum and the maximum values at $\lambda = 0$ and $\lambda = 1$, respectively. On the other hand, if $B_1(x, \alpha) > B_2(x, \alpha)$ then the minimum and maximum values of $h(x, \alpha, \cdot)$ are achieved at $\lambda = 1$ and $\lambda = 0$, respectively. In other words, the global minimizer and maximizer of $h(x, \alpha, \lambda)$ for $\lambda \in [0, 1]$ are given at $\lambda = 0$ or $\lambda = 1$. Therefore, for each $x \in \mathbb{R}$, the α-cuts of the fuzzy solution φ is given by

$$[\Phi(x)]^\alpha = [\min\{h(x, \alpha, 0), h(x, \alpha, 1)\}, \max\{h(x, \alpha, 0), h(x, \alpha, 1)\}], \tag{3.12}$$

where
$$h(x, \alpha, 0) = B_1(x, \alpha) = \langle S(x), Y_\alpha^- \rangle$$

and
$$h(x, \alpha, 1) = B_2(x, \alpha) = \langle S(x), Y_\alpha^+ \rangle.$$

In the next section we illustrate this proposed method by means of an example.

4   Application of Least Squares Method for Completely Correlated Fuzzy Data

In this section we apply the proposed method to determine a function that fits longitudinal data obtained from [15]. The authors discussed the association between children mortality and air pollution in São Paulo, Brazil, from 1994 to 1997. In their study were collected longitudinal data of sulfur dioxide (SO_2), carbon monoxide (CO), inhalable particulate (PM_{10}) and ozone (O_3). Here, we focus on the ozone dataset.

For simplicity, suppose that the longitudinal data are given by completely correlated triangular fuzzy numbers of the form $(M - \sigma; M; M + \sigma)$, where M and σ are the mean and the standard deviation of the collected data in each year, respectively. Recall that the proposed method is not restricted to triangular fuzzy numbers, then other types of fuzzy number can be considered.

Let $D = \{(x_1, Y_1), (x_2, Y_2), (x_3, Y_3), (x_4, Y_4)\} \subset \mathbb{R} \times \mathbb{R}_{\mathcal{F}}$ be the fuzzy dataset given in Table 1. The values $x_1 = 1$, $x_2 = 2$, $x_3 = 3$, and $x_4 = 4$ represent respectively the years 1994, 1995, 1996, and 1997. The fuzzy numbers $Y_1 = (17.6; 57; 96.4)$, $Y_2 = (25.3; 60.7; 96.1)$, $Y_3 = (34.8; 76.3; 117.8)$, and $Y_4 = (29.5; 63; 96.5)$ are completely correlated with respect to joint possibility distribution J, whose membership function is given by

$$J(v_1, v_2, v_3, v_4) = \chi_U(v_1, v_2, v_3, v_4)Y_1(v_1), \ \forall \ (v_1, v_2, v_3, v_4) \in \mathbb{R}^4,$$

where

$$U = \{(u, 0.8985u + 9.4855, 1.0533u + 16.2619, 0.8502u + 14.5386) \ : \ u \in \mathbb{R}\}. \tag{4.13}$$

Table 1. Fuzzy dataset D

x:	1	2	3	4
Y:	$(17.6; 57; 96.4)$	$(25.3; 60.7; 96.1)$	$(34.8; 76.3; 117.8)$	$(29.5; 63; 96.5)$

Note that Eq. (4.13) suggests that Y_1 and Y_2 are positively completely correlated, as well as Y_1 and Y_3, Y_1 and Y_4, since $q_i > 0$, for all $i = 2, 3, 4$.

Consider the functions $g_1(x) = x^2$, $g_2(x) = x$ and $g_3(x) = 1$. From (3.12), for each $\alpha \in [0, 1]$ and $x \in [1, 4]$, the fuzzy function Φ is given by $[\Phi(x)]^\alpha = [\min\{h(x, \alpha, 0), h(x, \alpha, 1)\}, \max\{h(x, \alpha, 0), h(x, \alpha, 1)\}]$, where

$$h(x, \alpha, 0) = -3.24x^2 + 20.76x - 0.75 + \alpha(-x^2 + 3.84x + 35.34)$$

and

$$h(x, \alpha, 1) = -5.24x^2 + 28.44x + 69.93 - \alpha(-x^2 + 3.84x + 35.34).$$

Figure 1 exhibits the fuzzy function Φ produced by our proposal. One can observe in Subfigure 1(a) fits the data of Table 1 which varies from 1994 to 1997. The red triangles and the gray-scale surface depicted in Subfigure 1(b) correspond to the membership functions of fuzzy data Y_i, $i = 1, \ldots, 4$, and fuzzy solution, respectively.

Note that $Y_1, \ldots, Y_4$ are completely correlated with respect to 2^3 different joint possibility distributions. Thus, we can obtain 2^3 fuzzy functions Φ. However, in general, the choice of a joint possibility distribution is not arbitrary and depends on the context. For example, if each object is measured m times with the same n measuring devices then we can assume that the obtained values depend only on the calibration of each equipment and not on the objects. This type of assumption induces the choice of specific parameters q_i and r_i in (2.7).

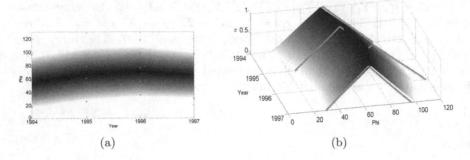

Fig. 1. Subfigures (a) and (b) exhibit respectively the top and depiction views of the fuzzy solution Φ where the greatest and smallest membership values are represented respectively by the black and white colors. In Subfigure (a), the red dots represent the endpoints of the $\alpha-$cuts of the fuzzy data Y_i for $\alpha = 0, 0.5, 1$ and $i = 1, \ldots, 4$. Each fuzzy data Y_i is represented by red lines in Subfigure (b).

5   Conclusion

In this manuscript, we considered a fuzzy least squares problem based on dataset that has some type of correlation, for example a longitudinal dataset. We assumed that the dataset is composed by completely correlated fuzzy numbers [5]. In particular, we presented a method that provides a fuzzy function that fits a given fuzzy data. This fuzzy function depends on the choice of a joint possibility distributions as in (2.6).

The α-cut of the fuzzy solution given by means of the sup-J extension principle is a non-empty, bounded, closed interval whose endpoints are obtained by solving a minimization and maximization problems given in Eq. (3.11). Investigating this problem, we concluded that the endpoints of the α-cut of the proposed solution can be evaluated by taking the minimum and maximum of two associated real functions (see Eq. (3.12)).

Finally, we applied the proposed method to determine a fuzzy function which fits a longitudinal air polution dataset [15]. The fuzzy data in this dataset was modelled using triangular fuzzy numbers, but it can be done with other types of completely correlated fuzzy numbers. The fuzzy solution was calculated considering polynomial functions g_1, g_2, and g_3. For further works, we intend to investigate fuzzy least squares method for dataset with other intrinsic type of interactivity.

References

1. Buckingham, R.A.: Numerical Methods. Sir Isaac Pitman & Sons Ltd., London (1966)
2. Tanaka, H., Uejima, S., Asai, K.: Linear regression analysis with fuzzy model. IEEE Trans. Syst. **6**, 903–907 (1982)

3. Celmins, A.: Least squares model fitting to fuzzy vector data. Fuzzy Sets Syst. **22**, 245–269 (1987)
4. Dubois, D., Prade, H.: Additions of interactive fuzzy numbers. IEEE (1981)
5. Carlsson, C., Fullér, R., Majlender, P.: Additions of completely correlated fuzzy numbers. In: IEEE International Conference on Fuzzy Systems, vol. 1, pp. 535–539 (2004)
6. Fullér, R., Majlender, P.: On interactive fuzzy numbers. Fuzzy Sets Syst. **143**, 355–369 (2004)
7. Tanaka, H., Ishibuchi, H.: Identification of possibilistic linear systems by quadratic membership functions of fuzzy parameters. Fuzzy Sets Syst. **41**, 145–160 (1991)
8. Diamond, P.: Fuzzy least squares. Inf. Sci. **46**, 141–157 (1988)
9. Wu, B., Tseng, N.-F.: A new approach to fuzzy regression models with application to business cycle analysis. Fuzzy Sets Syst. **130**, 33–42 (2002)
10. Takemura, K.: Fuzzy least squares regression analysis for social judgment study. J. Adv. Comput. Intell. Intell. Inform. **9**(5), 461–466 (2005)
11. Seng, K.-Y., Nestorov, I., Vicini, P.: Fuzzy least squares for identification of individual pharmacokinetic parameters. IEEE Trans. Biomed. Eng. **56**(12), 2796–2805 (2009)
12. Torfi, F., Farahani, R.Z., Mahdavi, I.: Fuzzy least-squares linear regression approach to ascertain stochastic demand in the vehicle routing problem. Appl. Math. **2**, 64–73 (2011)
13. Zeger, S.L., Liang, K.Y., Albert, P.S.: Models for longitudinal data: a generalized estimating equation approach. Biometrics **4**, 1049–1060 (1988)
14. Barros, L.C., Pedro, F.S.: Fuzzy differential equations with interactive derivative. Fuzzy Sets Syst. **309**, 64–80 (2017)
15. Conceição, G.M.S., Miraglia, S.G.E.K., Kishi, H.S., Saldiva, P.H.N., Singer, J.M.: Air pollution and child mortality: a time-series study in São Paulo Brazil. Environ. Health Perspect. **109**, 347–350 (2001)
16. Zadeh, L.A.: Fuzzy sets. Inf. Control **8**, 338–353 (1965)
17. Barros, L.C., Bassanezi, R.C., Lodwick, W.A.: A First Course in Fuzzy Logic, Fuzzy Dynamical Systems, and Biomathematics. Springer, Heidelberg (2017). https://doi.org/10.1007/978-3-662-53324-6
18. Zadeh, L.A.: Concept of a linguistic variable and its application to approximate reasoning, i, ii, iii. Inf. Sci. **8**, 199–249, 301–357 (1975)
19. Negoita, C., Ralescu, D.: Application of Fuzzy Sets to Systems Analysis. Wiley, New York (1975)
20. Bede, B.: Mathematics of Fuzzy Sets and Fuzzy Logic. Springer, Heidelberg (2012). https://doi.org/10.1007/978-3-642-35221-8
21. Dubois, D., Prade, H.: Possibility Theory: An Approach to the Computerized Processing of Information. Springer, New York (1988). https://doi.org/10.1007/978-1-4684-5287-7
22. Pedro, F.S.: On differential equations for linearly correlated fuzzy processes: applications in population dynamics. Ph.D. thesis, Portuguese (2017)
23. Nguyen, H.T.: A note on the extension principle for fuzzy sets. J. Math. Anal. Appl. **64**, 369–380 (1978)
24. Barros, L.C., Bassanezi, R.C., Tonelli, P.A.: On the continuity of the Zadeh's extension. Presented at Proceedings of the IFSA Congress (1997)

Using the Choquet Integral in the Pooling Layer in Deep Learning Networks

Camila Alves Dias[1(✉)], Jéssica C. S. Bueno[2], Eduardo N. Borges[1],
Silvia S. C. Botelho[1,2], Graçaliz Pereira Dimuro[1,2,4],
Giancarlo Lucca[3], Javier Fernandéz[3,4], Humberto Bustince[3,4],
and Paulo Lilles Jorge Drews Junior[1]

[1] PPGCOMP, C3, Universidade Federal do Rio Grande, Rio Grande, Brazil
cmdias@outlook.com.br
[2] PPGMC, Universidade Federal do Rio Grande, Rio Grande, Brazil
[3] Depart. of Automática y Computación, Universidad Publica de Navarra,
Pamplona, Spain
[4] Instituto of Smart Cities, Universidad Publica de Navarra, Pamplona, Spain

Abstract. This paper aims to introduce the proposal of replacing the usual
pooling functions by the Choquet integral in Deep Learning Networks. The
Choquet integral is an aggregation function studied and applied in several areas,
as, e.g., in classification problems. Its importance is related to the fact that it
considers the relationship between the data to be aggregated by means of a fuzzy
measure, unlike other aggregation functions such as the arithmetic mean and the
maximum. The idea of this paper is to use the Choquet integral to reduce the size
of an image, obtaining an abstract form of representation, that is, reducing the
perception of the network corresponding to small changes in the image. The use
of this aggregation function in the place of the max-pooling and mean-pooling
functions of Convolutional Neural Networks presented promising results. This
assertion is based on the Normalized Cross-Correlation and Structural Content
quality measures applied to the original images and resulting images. It is
important to emphasize that this preliminary study of Choquet integral as a pool
layer has not yet been implemented on Convolutional Neural Networks until the
present moment.

Keywords: Choquet integral · Deep Learning Networks
Convolutional Neural Networks · Image classification
Aggregation functions

1 Introduction

Image classification is a very common problem in Computer Vision area. The main
problems faced in this area are identifying patterns in images, distinguishing living
beings and objects, labelling collected images, among others. Most of the problems
mentioned above have complex information to be identified, using machine learning
methods such as Convolutional Neural Networks (CNN) [23] and Deep Learning
Networks (DLN) [10].

© Springer International Publishing AG, part of Springer Nature 2018
G. A. Barreto and R. Coelho (Eds.): NAFIPS 2018, CCIS 831, pp. 144–154, 2018.
https://doi.org/10.1007/978-3-319-95312-0_13

These methods use a variety of functions within the different steps that are employed in their architectures. One of them consists in the usage of aggregation functions, such as the maximum and the arithmetic mean [20], are used. Aggregation is understood as the process of combining different numeric values by returning a single value. The function that accomplishes this task is called aggregation function [3, 9], which is an increasing function satisfying appropriate boundaring conditions.

The research of different aggregation functions inside the area of DLP can be found mainly in subjects related to CNN [18]. First, we point out the importance of the role of pooling layers in CNN, which often take convolutional layer as input. The convolutional layer is a stack of feature maps where we have one feature map for each filter [1]. A complicated dataset with many different object categories requires a large number of filters, where each one is responsible for finding a pattern in the image. Then, the usage of more filters enlarges the dimensionality of the convolutional layers. Higher dimensionality means we will need to use more parameters which can lead to over-fitting [12]. Thus, we need a method for reducing this dimensionality so that we can avoid overfitting, and this is the role of pooling layers.

The goals of the application of max-pooling are the reduction of the number of parameters of the model (can observed in Fig. 1, the output is smaller than the input), which is called down-sampling or sub-sampling, in addition to a generalization of the results from a convolutional filter (making the detection of features invariant to scale or orientation). The main motivation of the application of these methods is to aggregate multiple low-level features in the neighborhood by using the Choquet integral to gain invariance mainly in object recognition.

There are several works that study different types of aggregation methods in convolutional networks and the impact of different pooling methods as [4, 19]. In [4], it is presented a detailed theoretical analysis of max-pooling and mean-pooling, and extensive empirical comparisons for object recognition tasks are presented. In [19], it is shown that a surprising fraction of the performance of certain state-of-the-art methods can be attributed to the architecture alone. In addition, the application of aggregation functions in DLN presented acceptable results with the combination of two methods presented in [10].

In this context, the objective of this work is the study of the application of state-of-the-art aggregation functions used in classification (as, e.g., in [14–16]) in Deep Learning Networks, replacing the processes of max-pooling or mean-pooling performed by the network. The purpose is to use the Choquet integral [5] to reduce the size of an image, obtaining an abstract form of representation. This preliminary study is being carried applying the aggregation functions in a group of images, obtaining the results of each image individually and comparing the results of the three functions with the original image using the Measurements of Image Quality [8]. In this way, we intend to evaluate the performance of the Choquet integral in representing, in smaller dimension and without the loss of robustness and spatial invariance, the structural information present in the images.

2 Preliminary Concepts

The Choquet integral is an extension of the integral of Lebesgue, being defined based on a fuzzy measure. The fuzzy measure has several interpretations depending on the context of the problem to be worked. In the context of aggregation functions, the fuzzy measure represents the degree of relationship between the elements to be aggregated [3]. In this way, the great use of the Choquet integral occurs due to its model to consider the importance of each attribute to be aggregated, as well as its interactions.

Definition 2.1 [3]. Consider $N = \{1, 2, \ldots, n\}$ and 2^N the power set of N. The function $\mu: 2^N \rightarrow [0, 1]$ is a fuzzy measure if, for all, $B \subseteq N$, the following conditions hold:
  (M1) Boundary conditions: $\mu(\emptyset) = 0$ and $\mu(N) = 1$;
  (M2) Increasingness: $\mu(A) \leq \mu(B)$ sempre que $A \subseteq B$.

Definition 2.2 [3]. Let μ be a fuzzy measure. The discrete Choquet integral of $\vec{x} = (x_1, x_2, \ldots, x_n) \in [0, 1]^n$ with respect to the fuzzy measure μ is the functions $C_\mu(\vec{x}) : [0, 1]^n \rightarrow [0, 1]$, defined by

$$C_\mu(\vec{x}) = \sum_{i=1}^{n} \left(x_{(i)} - x_{(i-1)} \right) \mu \left(A_{(i)} \right)$$

where $(x_{(1)}, x_{(2)}, \ldots, x_{(m)})$ is a non-decreasing permutation $\vec{x}$, that is, $0 \leq x_{(1)} \leq x_{(2)}, \leq \cdots \leq x_{(n)}$, where, by convention, $x_{(0)} = 0$ and $A_{(i)} = \{(1), \ldots (n)\}$ is the subset of indices of the $n - i + 1$ greatest components of $\vec{x}$.
  The fuzzy measure adopted in this work is the power measure [3], given by:

$$\mu_p(A) = \left(\frac{|A|}{n} \right)^q$$

where $q > 0$, $A \subseteq N$ and $|A|$ is the cardinality of the set A.
  The main characteristic of the Choquet integral, compared to other aggregation functions, lies in the fact that it considers through the fuzzy measure the interaction between the elements to be aggregated. For example, the maximum does not consider the relationship between the elements to be aggregated, discarding important information between those elements. In the context of image processing the values to be aggregated are the pixels that appear in a window. So, in this sense, it is understood that the more information of the relations between the pixels of the window, the better the output image.

3 Aggregation Functions in Deep Learning Neural Networks

The area of deep learning networks or DLN has received more attention in recent years and has stood out as a new area of research in machine learning [7]. Deep learning, according to [6], can be broadly defined as: a class of machine learning techniques that analyze many layers of non-linear data processing for extraction and alteration of supervised or unsupervised aspects and for recognition and classification of data.

The architecture of the networks is modified according to the purpose of its use. Normally, the input of a CNN takes an order 3 tensor, e.g., an image with H rows, W columns, and 3 channels (R, G, B color channels) [23]. Then the input goes through a sequence of processing.

An abstract description of the CNN structure is given by:

$$x^1 \rightarrow w^1 \rightarrow x^2 \rightarrow \ldots \rightarrow x^{L-1} \rightarrow w^{L-1} \rightarrow x^L \rightarrow w^L \rightarrow z \tag{1}$$

It can be verified how a CNN runs layer by layer in a forward pass. The input is shown as x^1 usually an image (order 3 tensor). It follows by processing the first layer, which is w^1. The output of the first layer is x^2, which also acts as the input to the second layer processing of the neural network. This process is carried out until all layers of the neural network were made, resulting in the output variable x^L. However one additional layer is added for backward error propagation, a method that learns good parameter values in the CNN. Assuming, e.g., the problem is in image classification with C classes. For resolution, a strategy used is to output x^L as a C dimensional vector, whose i-th entry encodes the prediction (posterior probability of x^1 comes from the i-th class). There is a possibility to set the processing in the $(L-1)$-th layer as a softmax transformation of x^{L-1} to make x^L a probability mass function. In other applications, the output x^L may have other forms and interpretations.

The last layer of the CNN architecture is a loss layer. The operation of this layer is given by:

$$z = \frac{1}{2}||t - x^L||^2, \tag{2}$$

where the variable t is the corresponding target (ground-truth) value for the input x^1. Equation (2) demonstrates how a cost or loss function can be used to measure the discrepancy between the CNN prediction x^L and the target t. This equation is less complex than other functions that are most commonly used.

The ground-truth in a classification problem is a categorical variable t. The categorical variable t is convert to a C dimensional vector t. Then both t and x^L are probability mass functions, and the cross entropy loss measures the distance between them. In this way it is possible to minimize cross-entropy.

Equation (1) explicitly models the loss function as a loss layer, whose processing is modeled as a box with parameters w^L. There are some layers may not have any parameters, that is, w^i may be empty for some i. The softmax layer is one such example in the structure.

The pooling layer acts by aggregating a group of data, where the input can be of type array, image, and among other data types.

Max-pooling is the application of a moving window across a 2D input space, where the maximum value within that window is the output. A visual example is show below (see Fig. 1) with some padding (it's a margin given to the data when the structure is smaller than the stride given the window).

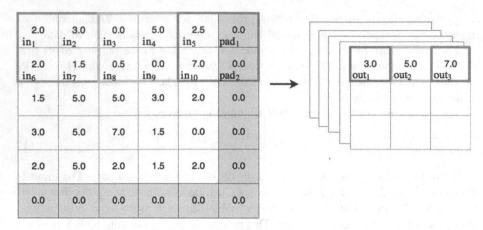

Fig. 1. Illustration of the application of the max-pooling function with stride 2 and 2 × 2 window. Each colored moving window captures the maximum value inside the 2 × 2 square and outputs it on the right hand side. (Color figure online)

The max-pooling reduces the number of parameters in the model. Observe in Fig. 1 that the output is smaller than the input. This is called down-sampling or sub-sampling and generalizes the results from a convolutional filter (making the detection of features invariant to scale or orientation changes).

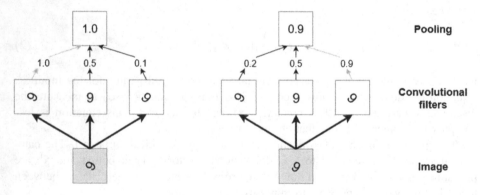

Fig. 2. A stylized representation of various convolutional feature maps that have been tuned during training to detect the digit "9" in an image, but at different orientations.

The diagram in Fig. 2 shows that the max-pooling allows a generalization which detects a "9" no matter what the orientation (or size etc.) inside the input. This is a kind of movement from low level data to higher level information.

The same procedure is done for other aggregation functions such as mean-pooling and the Choquet integral, and it is possible to vary the window size and the stride of the windows. Depending on the variations chosen, the size of the resulting matrix will vary.

4   Methodology

To improve the aggregation of meaningful information without degrading its discriminative power for image processing, we propose replace the max-pooling and mean-pooling functions by the Choquet integral in the context of pooling layers on CNN. Grouping layers reduce the size of the input by synthesizing the neurons of a small neighborhood. Figure 3 shows a general architecture of a Convolutional Neural Network similar to the models of [2, 13]. The values above the images represent the number of layers and the size in pixels of the images.

For example, in the first layer of the network represented in the figure (input layer L0: 2 @ 96 × 96), there are 2 layers of 1 image with the size of 96 × 96. After the first convolution (C1: 16 @ 92 × 92) 16 layers with the size of 92 × 92 each were generated. In the end, these layers will be fully connected (fully connected feature maps) creating a single volume F5 and F6, as shown in the figure.

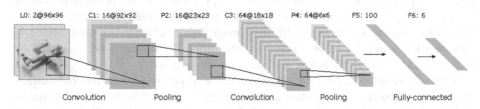

Fig. 3. Architecture of a CNN [20] for NORB (an object recognition dataset) experiments, consisting of alternating convolution and grouping layers. Grouping layers can implement subsampling or grouping operations of maxima.

Another important issue associated with the convolutional layer is the stride, which defines how many pixels of the image (or values of a numerical array) will be jumped between each neighborhood, representing the size of the next layer of the processed input.

The Choquet integral domain are the pixels values of an input image, ranging from 0 to 255, in a normalized way. In addition, to facilitate the analysis of the results, the input images are converted to grayscale. Another point for the use of grayscale images is the application of the measurements of image quality performed later. Such measurements are applied in grayscale images and not in colored images [8]. It is important to emphasize that in this work the Choquet integral is being implemented as a preliminary study in order to evaluate it as an aggregation function and dimensional reduction, able to improve the representation and summarize the information of the input data.

The stride and window size of the input images can be defined by the user. For example, if you set the window size to size 2, and stride 2, the function will be executed generating multiple 2 × 2 arrays and 2 pixels will be skewed between each window, limited to the total size of the input image. That is, a common input image of total size of 500 × 500 pixels will output as a 250 × 250 pixels image.

After setting the parameters window size and the stride, we apply the Choquet integral, in the same way as the other aggregators. However, to perform the Choquet integral it is necessary to convert the input matrix into a vector and organize it in non-decreasing form. After sort the vector, the Choquet integral is applied. Finally, the vector is transformed into an array, resulting in a smaller representation of the original image.

In order to evaluate our method automatically, without a visual inspection of each image, we have used two different quality measures, which are calculated for the image with the application of max-pooling, mean-pooling and Choquet integral, in relation to the original image in grayscale.

These image quality measures help the systematic design of coding, communication and image systems, as well as improving or optimizing image quality for a desired application quality at minimal cost [8].

The Normalized Cross-Correlation measure verifies the matching of the resulting image with the original image, which can also be quantified in terms of the correlation function, the higher the measurement result value, the more the output image is close to the input image. The Structural Content (SC) measure is used to compare original image and resulting images in several small image points, which they have in common. This measure is also based on correlation, measuring the similarity between the two images. If the value of SC is low, the image quality is good.

From the values obtained from the results of the Structural Content and Normalized Cross-Correlation measurements, a hypothesis test was applied to have a statistical analysis of the results. The non-parametric Fridman test [11] was applied, where it is desired to detect statistical differences between a group of results, that is, between the aggregation functions used. After the existence of differences between groups, a post-hoc test is applied to verify among which groups there are these differences. For this test the non-parametric Wilcoxon paired test was used [21]. The level of significance considered for the hypothesis tests was 0.05.

The purpose of this comparison is to verify if the information obtained as output of the pooling functions is close to the information of the input images.

5   Experimental Evaluation

We used 12 images containing words. The choice of images of this genre is due to the facility of visually evaluating its qualitative characteristics. The 12 images were taken from an IIIT 5K-Word dataset [17].

The experiments were performed using Matlab[1], where new images were generated from the original images applying the aggregation functions of maximum, mean and Choquet integral. The parameters presented in the methodology were configured as follows: stride s = {2 and 3}, fuzzy measure exponent q = {0, 1, 0.3, 0.5 and 0.7}, window size ws = {2×2, 3×3 and 4×4}. These values were chosen by the

---

[1]   https://www.mathworks.com/products/matlab.html.

specialists in order to obtain better performance of the function and based on a pilot experiment using the same methodology.

In Figs. 4 and 5, it is possible to do a visual comparison of two images, of the 12 used in the work, where the aggregation functions of max, mean and Choquet integral were applied. The values configured for the application of the functions were: stride 3, exponent value of fuzzy measure 0.3 and window size 3×3. It is possible to verify visually that the image resulting from the integral function of Choquet is more complete when compared to the other results. In the quantitative results of the image quality measurements, the resulting image of the Choquet Pooling also stands out: the image in Fig. 4 obtained the value 1.31 in the Normalized Cross-Correlation measure while the resulting image of the max-pooling obtained 1.10 e of mean-pooling 0.93. That is, the image in which the Choquet pooling was applied has characteristics closer to that of the original image.

The values of the other measure (Structural Content) the image of the choquet pooling also stood out: it obtained 0.52 as resulting, while the max-pooling obtained 0.71 and the mean-pooling obtained 1.06. In this case, the higher the value, the worse the quality of the image and the farther it is from the original image. The central goal of this work is to get a resulting image closer to the input image.

The results of Fig. 5 were also good, when evaluated by the measurements of image quality: the image of the choquet pooling function reached 1.23, while the result of max-pooling was 1.09 and mean-pooling was 0.92 when using the normalized cross-correlation. In the results of the structural content measure, the choquet pooling image was also highlighted with a value of 0.58, as the max-pooling image reached 0.72 and mean-pooling 1.06.

These images support this study of the application of Choquet integral in the pool layer of the deep neural network, validating the assertions that such a function can improve the interpretation of the network with the input images.

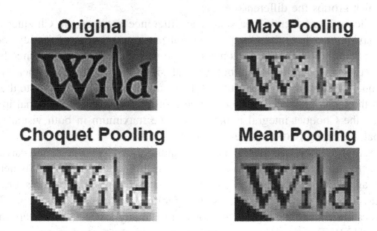

Fig. 4. Results of the aggregation functions max, mean and Choquet integral, obtained through the experiments applied to one of the images of the IIIT 5K-Word dataset called 138_4.

Fig. 5. Results of the aggregation function max, mean and Choquet integral obtained through the experiments applied to one of the images of the IIIT 5K-Word dataset called 138_6.

There were cases, such as the results of the applied images with stride 3, exponent value of the fuzzy 0.1 measurement and 3×3 windows, in which the values of the measurements of image quality for each applied arithmetic function were very different. However, the values of the images of the maximum function and Choquet integral have always been closer compared to the images of the arithmetic mean function, as can be seen previously in the shown examples.

The tests of hypotheses applied in the results of the measures of image evaluation generated the same interpretations. The Fridman test resulted in rejection of the null hypothesis, so there are statistical differences between the aggregators for both quality measures. Thus, it was necessary to apply the paired Wilcoxon post-hoc test to identify among which groups the difference exists.

The Wilcoxon test identified a statistical difference between the Choquet integral and the maximum, where the Choquet integral presented results of quality measurement better than the maximum. In contrast, there was no statistical difference between the Choquet integral and the mean, but the analysis of the results of the measurements of image quality showed that the results of the Choquet integral compared to the output images of the mean presented more concrete information about the original image.

In sum, the Choquet integral is better than the maximum in both visual and statistical analysis, using two measures of quality.

The maximum draws the most important features, such as borders, and captures the strongest activation, disregarding all other units in the pool region. By this detail, the maximum aggregation is better than the mean for extraction of extreme resources. The mean takes into account all activations in a pooling region with equal contributions [24]. This can minimize the high activations, since many low activations are included in the mean [22]. The mean can then end up not extracting good resources, since the result that will be an average value has the possibility of being or not being important.

Then, some works use the maximum for better performance and better network results. Therefore, the choice of the use of the Choquet integral in the pooling layer, substituting these functions usually employed, is valid, since the Choquet integral is better than the maximum in the analyses performed.

In this way, it can be affirmed that this study is promising within the area of Artificial Neural Networks, leading to generate new experiments not only with images but with any data that needs to pass through each pool of Neural Network, since the architectures of networks vary depending on their purpose.

6 Conclusions

In this paper, we presented a preliminary study on the implementation of the arithmetic function Choquet integral in the pool layers of a deep neural network. Based on these results, it was verified that the Choquet integral resulted in quality images and maintained the integrity of the original image, that is, at the moment the original image passes through the neural network and the pooling function is applied, at the end the network will be able to keep the input image information, which results in better performance for the deep neural network.

In general, the results presented in these initial works are promising and this is intended to refine the studies, so that the q parameter used in the Choquet integral with respect to the power fuzzy measure is learned by the neural network itself, thus generating better results than those presented.

We also intend to analyse the behaviour of generalizations of the Choquet integral introduced by Lucca et al. [14–16], which presented excellent results in classification problems.

References

1. Agarap, A.F.: An Architecture Combining Convolutional Neural Network (CNN) and Support Vector Machine (SVM) for Image Classification. Adamson University, Manila (2017)
2. Ahmed, A., Yu, K., Xu, W., Gong, Y., Xing, E.: Training hierarchical feed-forward visual recognition models using transfer learning from pseudo-tasks. In: Forsyth, D., Torr, P., Zisserman, A. (eds.) ECCV 2008. LNCS, vol. 5304, pp. 69–82. Springer, Heidelberg (2008). https://doi.org/10.1007/978-3-540-88690-7_6
3. Beliakov, G., Pradera, A., Calvo, T.: Aggregation Functions: A Guide for Practitioners, vol. 221. Springer, Berlin/Heidelberg (2007). https://doi.org/10.1007/978-3-540-73721-6
4. Boureau, Y.L., Ponce, J., Lecun, Y.: A theoretical analysis of feature pooling in visual recognition. In: Proceedings of the 27th International Conference on Machine Learning (ICML-2010), Haifa, Israel, pp. 111–118 (2010)
5. Choquet, G.: Theory of capacities. In: Annales de l'institut Fourier, Grenoble, France, vol. 5, pp. 131–295 (1954)
6. Deepika, J., Vishvanathan, S., KP, S.: Image classification using convolutional neural networks. Int. J. Sci. Eng. Res. 5(6), 1661–1668 (2014)
7. Deng, J., Dong, W., Socher, R., Li, L.J., Li, K., Fei-Fei, L.: imageNet: a large-scale hierarchical image database. In: IEEE Conference on Computer Vision and Pattern Recognition (CVPR 2009), Miami, Florida, pp. 248–255. IEEE (2009)

8. Eskicioglu, A.M., Fisher, P.S.: Image quality measures and their performance. IEEE Trans. Commun. **43**(12), 2959–2965 (1995)
9. Grabisch, M., Marichal, J.L., Mesiar, R., Pap, E.: Aggregation functions: means. Inf. Sci. **181**(1), 1–22 (2011)
10. Han, Y., Kim, J., Lee, K., Han, Y., Kim, J., Lee, K.: Deep convolutional neural networks for predominant instrument recognition in polyphonic music. IEEE/ACM Trans. Audio, Speech Lang. Process. (TASLP) **25**(1), 208–221 (2017)
11. Hodges, J.L., Lehmann, E.L.: Rank methods for combination of independent experiments in analysis of variance. Ann. Math. Stat. **33**(2), 482–497 (1962)
12. Krizhevsky, A., Sutskever, I., Hinton, G.E.: Imagenet classification with deep con-volutional neural networks. In: Advances in Neural Information Processing Systems (NIPS), Stateline, NV, pp. 1097–1105 (2012)
13. Lecun, Y., Huang, F.J., Bottou, L.: Learning Methods for generic object recognition with invariance to pose and lighting. In: Proceedings of the 2004 IEEE Computer Society Conference on Computer Vision and Pattern Recognition (CVPR 2004), Washington, D.C., USA, vol. 2, pp. II–104. IEEE (2004)
14. Lucca, G., Sanz, J.A., Dimuro, G.P., Bedregal, B., Mesiar, R., Kolesárová, A., Bustince, H.: Preaggregation functions: construction and an application. IEEE Trans. Fuzzy Syst. **24**(2), 260–272 (2016)
15. Lucca, G., Sanz, J.A., Dimuro, G.P., Bedregal, B., Asiain, M.J., Elkano, M., Bustince, H.: CC-integrals: choquet-like copula-based aggregation functions and its application in fuzzy rule-based classification systems. Knowl.-Based Syst. **119**, 32–43 (2017)
16. Lucca, G., Sanz, J.A., Dimuro, G.P., Bedregal, B., Bustince, H., Mesiar, R.: CF-integrals: a new family of pre-aggregation functions with application to fuzzy rule-based classification systems. Inf. Sci. **435**, 94–110 (2018)
17. Mishra, A., Alahari, K., Jawahar, C.V.: Scene text recognition using higher order language priors. In: 23rd British Machine Vision Conference, BMVC 2012, Paris, France (2012)
18. Pagola, M., Forcen, J.I., Barrenechea, E., Lopez-Molina, C., Bustince, H.: Use of OWA operators for feature aggregation in image classification. In: IEEE International Conference on Fuzzy Systems (FUZZ-IEEE 2017), Naples, Italy, pp. 1–6. IEEE (2017)
19. Saxe, A.M., Koh, P.W., Chen, Z., Bhand, M., Suresh, B., Ng, A.Y.: On random weights and unsupervised feature learning. In: Proceedings of the 28th International Conference on Machine Learning (ICML-2011), Bellevue, Washington, USA, pp. 1089–1096. Bellevue, Washington, USA (2011)
20. Scherer, D., Müller, A., Behnke, S.: Evaluation of pooling operations in convolutional architectures for object recognition. In: Diamantaras, K., Duch, W., Iliadis, L.S. (eds.) ICANN 2010. LNCS, vol. 6354, pp. 92–101. Springer, Heidelberg (2010). https://doi.org/10.1007/978-3-642-15825-4_10
21. Wilcoxon, F.: Individual comparisons by ranking methods. Biometrics Bull. **1**(6), 80–83 (1945)
22. Wu, H., Gu, X.: Max-pooling dropout for regularization of convolutional neural networks. In: Arik, S., Huang, T., Lai, W., Liu, Q. (eds.) ICONIP 2015. LNCS, vol. 9489, pp. 46–54. Springer, Cham (2015). https://doi.org/10.1007/978-3-319-26532-2_6
23. Wu, J.: Introduction to Convolutional Neural Networks. National Key Lab for Novel Software Technology. Nanjing University, China (2017)
24. Yu, D., Wang, H., Chen, P., Wei, Z.: Mixed pooling for convolutional neural networks. International Conference on Rough Sets and Knowledge Technology, pp. 364–375. Springer, Cham (2014). https://doi.org/10.1007/978-3-319-11740-9_34

Consensus Image Feature Extraction with Ordered Directionally Monotone Functions

Cedric Marco-Detchart[1,2]([✉]), Graçaliz Pereira Dimuro[1,3], Mikel Sesma-Sara[1,2], Aitor Castillo-Lopez[1], Javier Fernandez[1,2], and Humberto Bustince[1,2]

[1] Dpto. Estadistica, Informatica y Matematicas, Universidad Publica de Navarra, Campus Arrosadia, 31006 Pamplona, Spain
{cedric.marco,gracaliz.pereira,mikel.sesma,
fcojavier.fernandez,bustince}@unavarra.com, akastillo011@gmail.com
[2] Institute of Smart Cities, Universidad Publica de Navarra, Campus Arrosadia, 31006 Pamplona, Spain
[3] Centro de Ciências Computacionais, Universidade Federal do Rio Grande, Av. Itália km 08, Campus Carreiros, Rio Grande 96201-900, Brazil

Abstract. In this work we propose to use ordered directionally monotone functions to build an image feature extractor. Some theoretical aspects about directional monotonicity are studied to achieve our goal and a construction method for an image application is presented. Our proposal is compared to well-known methods in the literature as the gravitational method, the fuzzy morphology or the Canny method, and shows to be competitive. In order to improve the method presented, we propose a consensus feature extractor using combinations of the different methods. To this end we use ordered weighted averaging aggregation functions and obtain a new feature extractor that surpasses the results obtained by state-of-the-art methods.

Keywords: Edge detection · Feature extraction
Ordered directionally monotone functions
Ordered weighted averaging aggregation functions

1 Introduction

Whenever an object detection problem is faced in, key information has to be extracted from an image. A primary approach to obtain this information are edge detection methods. A great variety of information is involved in the definition of an edge, going from the basic concept that considers an edge as a big enough intensity jump, to the fact that texture can be considered as a jump but it should not represent an edge. For this reason, the process of extracting edges is very difficult, and many different methods have been proposed in the literature [11,15,19]. Among the most common edge detection methods there are those based on gradients, *i.e.*, vectors measuring the variation of intensity

© Springer International Publishing AG, part of Springer Nature 2018
G. A. Barreto and R. Coelho (Eds.): NAFIPS 2018, CCIS 831, pp. 155–166, 2018.
https://doi.org/10.1007/978-3-319-95312-0_14

along specific directions in the neighbourhood of each pixel. These gradients are commonly calculated applying a filter over the original image trough a convolution operation. Some examples of this type of edge detectors are Sobel and Feldman [22], Prewitt [20] and Canny [7].

Edge detection has been traditionally considered as a unique operation, but the resulting objective is mainly obtained through a concatenation of procedures. Several approaches have been presented in the literature in order to encompass this series of procedures, and explain edge detection as a multi-step technique [23]. For example, Law *et al.* [12] proposed a process consisting of three steps: *filtering*, *detection* and *tracing*. Some years later, Bezdek *et al.* [3] introduced a framework to embrace a variety of methods in the literature (primarily those based on gradient extraction), proposing a four step process: *conditioning*, where the image is enhanced in order to improve their quality, *feature extraction*, where an estimation of geometric changes at edges is computed, *blending*, which aggregate information to represent edges and *scaling*, where the original intensity scale of the image is recovered.

In our work we mainly follow the Bezdek Breakdown Structure (BSS) for the experimentation, focusing on to the *feature extraction* step, *i.e.*, the process that converts visual information from pixels into unique properties that define the image. Usually in edge detection methods these properties are based on the gradient. In this work, besides the gradient magnitude we also use the gradient direction.

The presented method consists in building a feature map of the image based on pixels neighbourhood information. The information is taken from the surrounding intensity values and fused using Ordered Directionally Monotone (ODM) functions. The importance of this type of functions resides in their ability to consider different directions of increasingness.

Finally, we use our method for the construction of a consensus feature image, combining different edge detection methods, including the one that we define and some other well known as the Canny method, the gravitational method and the fuzzy morphology.

This work is organized in the following way. In Sect. 2 we include some mathematical concepts related to image processing, as well as some notions about aggregation theory to introduce ODM functions. Section 3 is devoted to introduce our proposal. In Sect. 4 an application to edge detection is presented. Finally, in Sect. 5 we expose conclusions and future work.

2   Preliminaries

An image $\mathbb{I}_L$ is represented as a matrix of elements with positions in the set $D = X \times Y = \{1, ..., w\} \times \{1, ..., h\}$, being w and h the number of rows and columns respectively.

Each particular element, known as pixel, of the matrix can take values in a set L. This set represent the number of possible values for the intensity of the corresponding pixel. In the particular case of grey-scale images, pixels will be represented in $L = \{0, ..., 255\}$.

Let $n > 1$. We use bold letters to denote points in the hypercube $[0, 1]^n$, i.e., $\mathbf{x} = (x_1, \ldots, x_n) \in [0, 1]^n$. In particular, we write $\mathbf{0} = (0, \ldots, 0)$ and $\mathbf{1} = (1, \ldots, 1)$. Given $\mathbf{x}, \mathbf{y} \in [0, 1]^n$ we write $\mathbf{x} \leq \mathbf{y}$ if $x_i \leq y_i$ for every $i \in \{1, \ldots, n\}$. Note that this relation is a partial order which extends the usual linear order between real numbers.

We use the notation $(\vec{\cdot})$ for n-dimensional vectors in the Euclidean space $\mathbb{R}^n$, i.e., $\vec{r} = (r_1, \ldots, r_n) \in \mathbb{R}^n$.

For $n > 1$, we denote by S_n the set of permutations of $\{1, \ldots, n\}$. That is,

$$S_n = \{\sigma : \{1, \ldots, n\} \to \{1, \ldots, n\} \mid \sigma \text{ is bijective}\}.$$

Given $\sigma \in S_n$, $\mathbf{x} \in [0, 1]^n$ and $\vec{r} \in \mathbb{R}^n$, we define:

$$\mathbf{x}_\sigma = (x_{\sigma(1)}, \ldots, x_{\sigma(n)})$$

and

$$\vec{r}_\sigma = (r_{\sigma(1)}, \ldots, r_{\sigma(n)}).$$

As in this work we are dealing with aggregation and monotonicity theory we fix some concepts and notation.

Definition 1. *[2,6] A mapping $M : [0, 1]^n \to [0, 1]$ is an aggregation function if it is monotone non-decreasing in each of its components and satisfies $M(\mathbf{0}) = 0$ and $M(\mathbf{1}) = 1$.*

An aggregation function M is an averaging or mean if

$$\min(x_1, \ldots, x_n) \leq M(x_1, \ldots, x_n) \leq \max(x_1, \ldots, x_n).$$

One relevant type of aggregation functions was presented by Yager [26], and it is known as Ordered Weighted Averaging (OWA) operators.

Definition 2. *An OWA operator of dimension n is a mapping $\Phi : [0, 1]^n \to [0, 1]$ such that it exists a weighting vector $\boldsymbol{w} = (w_1, \ldots, w_n) \in [0, 1]^n$ with $\sum_{i=1}^{n} w_i = 1$, and such that*

$$\Phi(x_1, \ldots, x_n) = \sum_{i=1}^{n} w_i \cdot x_{\sigma(i)},$$

where $\mathbf{x}_\sigma = (x_{\sigma(1)}, \ldots, x_{\sigma(n)})$ is a decreasing permutation on the input $\boldsymbol{x}$.

In [25] a way to compute the weighting vector is presented:

$$w_i = Q\left(\frac{i}{n}\right) - Q\left(\frac{i-1}{n}\right),$$

where Q is a fuzzy linguistic quantifier as, for instance,

$$Q(r) = \begin{cases} 0 & \text{if } 0 \leq r < a, \\ \frac{r-a}{b-a} & \text{if } a \leq r \leq b, \\ 1 & \text{if } b < r \leq 1, \end{cases} \tag{1}$$

with $a, b, r \in [0, 1]$.

Monotonicity is a key concept, but in some applications it can be too restrictive. Many non-monotonic averaging functions, like the mode, are used in a wide variety of applications, *e.g.*, in image filtering. Due to this fact the notion of weak monotonicity was introduced by Wilkin and Beliakov [24].

This definition was later extended into the notion of directional monotonicity [5].

Definition 3. *[5] Let $\vec{r} = (r_1, \ldots, r_n)$ be a real n-dimensional vector, $\vec{r} \neq \mathbf{0}$. A function $F : [0, 1]^n \rightarrow [0, 1]$ is $\vec{r}$-increasing if for all points $(x_1, \ldots, x_n) \in [0, 1]^n$ and for all $c > 0$ such that $(x_1 + cr_1, \ldots, x_n + cr_n) \in [0, 1]^n$ it holds*

$$F(x_1 + cr_1, \ldots, x_n + cr_n) \geq F(x_1, \ldots, x_n).$$

That is, a $\vec{r}$-increasing function is a function which is increasing along the ray (direction) determined by the vector $\vec{r}$. In particular, weak monotonicity corresponds to the case $\vec{r} = (1, \ldots 1)$.

From this concept of directional monotonicity, we come to ordered directionally monotone functions, where the direction along which monotonicity is required varies depending on the relative size of the coordinates of the considered input.

Definition 4. *[4] Let $F : [0, 1]^n \rightarrow [0, 1]$ be a function and let $\vec{r} \neq \mathbf{0}$. F is said to be ordered directionally (OD) $\vec{r}$-increasing if for any $\mathbf{x} \in [0, 1]^n$, for any $c > 0$ and for any permutation $\sigma \in S_n$ with $x_{\sigma(1)} \geq \cdots \geq x_{\sigma(n)}$ and such that*

$$1 \geq x_{\sigma(1)} + cr_1 \geq \cdots \geq x_{\sigma(n)} + cr_n \geq 0,$$

it holds that

$$F(\mathbf{x} + c\vec{r}_{\sigma^{-1}}) \geq F(\mathbf{x}),$$

where $\vec{r}_{\sigma^{-1}} = (r_{\sigma^{-1}(1)}, \ldots, r_{\sigma^{-1}(n)})$.

3 Specific Example of Ordered Directionally Monotone Functions

This section is devoted to introduce a particular example to construct ODM functions. This example is obtained by considering an affine function.

Theorem 1. *Let $G : [0, 1]^n \rightarrow [0, 1]$ be defined, for $\mathbf{x} \in [0, 1]^n$ and $\sigma \in S_n$ such that $x_{\sigma(1)} \geq \ldots \geq x_{\sigma(n)}$, by*

$$G(\mathbf{x}) = a + \sum_{i=1}^{n} b_i x_{\sigma(i)},$$

for some $a \in [0,1]$ and $\vec{b} = (b_1, \ldots, b_n) \in \mathbb{R}^n$ such that $0 \le a + b_1 + \cdots + b_j \le 1$ for all $j \in \{1, \ldots, n\}$. Then G is OD $\vec{r}$-increasing for every non-null vector $\vec{r}$ such that $\vec{b} \cdot \vec{r} \ge 0$. In particular, for every non-null vector $\vec{r}$ which is orthogonal to $\vec{b}$.

Theorem 1 can be generalized taking into account the following lemma.

Lemma 1. *[4] Let $\varphi : [0,1] \to [0,1]$ be an automorphism (i.e., an increasing bijection). Then, if $G : [0,1]^n \to [0,1]$ is an ordered directionally increasing function, the function $\varphi \circ G$ is also an ordered directionally increasing function.*

Corollary 1. *Let $p > 0$. Let $G : [0,1]^n \to [0,1]$ be defined, for $\mathbf{x} \in [0,1]^n$ and $\sigma \in S_n$ such that $x_{\sigma(1)} \ge \ldots \ge x_{\sigma(n)}$, by*

$$G(\mathbf{x}) = \left(a + \sum_{i=1}^{n} b_i x_{\sigma(i)} \right)^{\frac{1}{p}}, \qquad (2)$$

for some $a \in [0,1]$ and $\vec{b} = (b_1, \ldots, b_n) \in \mathbb{R}^n$ such that $0 \le a + b_1 + \cdots + b_j \le 1$ for all $j \in \{1, \ldots, n\}$. Then G is OD $\vec{r}$-increasing for every non-null vector $\vec{r}$ such that $\vec{b} \cdot \vec{r} \ge 0$.

From Theorem 1 the following corollary is straight.

Corollary 2. *Let $A : [0,1]^n \to [0,1]$ be an OWA operator associated to the weighting vector $\mathbf{w} = (w_1, \ldots, w_n)$. Then A is OD $\vec{r}$-increasing for every non-null vector $\vec{r}$ such that $\mathbf{w} \cdot \vec{r} \ge 0$.*

4   Experimental Study

4.1   Proposed Method and Parameters

Given a grey-scale image $\mathbb{I}_g$ we normalize each pixel intensity value to the range $[0,1]$. Then, using Algorithm 1, we obtain a feature image trough ODM functions.

Algorithm 1. Algorithm to construct a feature image using ODM functions

---

Input: A normalized grey-scale image $\mathbb{I}_g$ and an ODM function G as in Corollary 1.
Output: A feature image $\mathbb{I}_f$.
 1: **for** each pixel (x, y) of $\mathbb{I}_g$ **do**
 2:     Compute the corresponding values by means of the absolute value of the difference between $\mathbb{I}_g(x, y)$ and its 8-neighbourhood;
 3:     Order the eight values of step 2 in a decreasing way;
 4:     Apply the ODM function G, with its corresponding a, p values and $\vec{r}, \vec{b}$ vectors (see Eq. (2)), to the values obtained in step 3;
 5:     Assign as intensity of the pixel (x, y) of $\mathbb{I}_f$ the value obtained in step 4.
 6: **end for**

In order to treat step 2 we consider an 8-neighbourhood for each position (x, y), taking pixels from $(x - 1, y - 1)$ to $(x + 1, y + 1)$. Let us consider pixel a_{ij} as the one to be treated, each one of the values of the neighbourhood is computed in the following way:

$$x_1 = |a_{ij} - a_{(i-1)(j-1)}|, \quad x_2 = |a_{ij} - a_{(i-1)j}|,$$
$$x_3 = |a_{ij} - a_{(i-1)(j+1)}|, \quad x_4 = |a_{ij} - a_{i(j+1)}|,$$
$$x_5 = |a_{ij} - a_{(i+1)(j+1)}|, \quad x_6 = |a_{ij} - a_{(i+1)j}|,$$
$$x_7 = |a_{ij} - a_{(i+1)(j-1)}|, \quad x_8 = |a_{ij} - a_{i(j-1)}|.$$

Note that, as we are working with a 3×3 window, the pixels from the image border are not considered in the computation of the feature image.

In step 3 these intensity differences are ordered in a decreasing way; that is,

$$x_{\sigma(1)} \geq x_{\sigma(2)} \geq \ldots \geq x_{\sigma(7)} \geq x_{\sigma(8)}.$$

Finally, in step 4 an ODM function is applied with different $a, p, \vec{r}$ and $\vec{b}$ parameters.

Using Corollary 1 and Eq. (2) we build a new expression for ODM functions for step 4 of Algorithm 1.

As indicated in [9], we propose to vary the brightness level of the feature image obtained, to better adapt our objective of finding edges. To control the brightness level we vary the parameter p in Eq. (2), taking $p > 1$ to get a brighter image or $0 < p < 1$ to get a darker one.

For the sake of the experiment we put our method to the test using the following parameters:

$$\vec{r} = (x_{\sigma(1)}, x_{\sigma(2)}, x_{\sigma(3)}, x_{\sigma(4)}, x_{\sigma(5)}, x_{\sigma(6)}, x_{\sigma(7)}, x_{\sigma(8)});$$

$$\vec{b} = \left(\frac{|x_{\sigma(1)} - x_{\sigma(8)}|}{\sum_{i=1}^{8} |x_{\sigma(i)} - x_{\sigma(8)}|}, \ldots, \frac{|x_{\sigma(7)} - x_{\sigma(8)}|}{\sum_{i=1}^{8} |x_{\sigma(i)} - x_{\sigma(8)}|}, 0 \right)$$

$$a = 0$$

$$\frac{1}{p} = 0.35$$

The specific value of a comes from the condition of Corollary 1, where $0 \leq a + b_1 + \ldots + b_8 \leq 1$, therefore $0 \leq a$ and $a + \frac{1}{2} \leq 1$, then we can use any value of a between 0 and $\frac{1}{2}$.

Note that in the case of a flat region in the image, $i.e.$, when all the pixels have the same value we would obtain a zero denominator, so we mark directly the corresponding position in the feature image as not containing an edge. That is, if $\sum_{i=1}^{8} |x_{\sigma(i)} - x_{\sigma(8)}| = 0$ we take $\vec{b} = (0, \ldots, 0)$.

Finally, $\frac{1}{p}$ has been selected as the best value from a series of tests, made with Algorithm 1 over the train images of the BDSDS dataset. That is, taking different values from 0.1 to 0.9 partitioned uniformly in steps of 0.05.

In Fig. 1 we show the results obtained by applying our proposed algorithm with the ODM construction method.

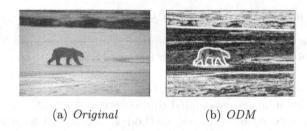

(a) *Original*   (b) *ODM*

Fig. 1. Original image from BSDS [1] (100007) along with feature image obtained after applying Algorithm 1 with ODM functions to original image.

4.2   Experimental Framework

In order to analyse the behaviour of our proposal we follow the Bezdek *et al.* [3] scheme, adding the final quantification of the results:

($S1$) Smooth the image applying a Gaussian filter (with $\sigma = 1.5$) to $\mathbb{I}_g$;
($S2$) Obtain the feature image with Algorithm 1;
($S3$) Thin the feature image using non-maxima suppression [7];
($S4$) Binarize the thinned image using the hysteresis method [18].
($S5$) Compare the binary image with ground truth images [8].

In order to measure the performance of the proposed method, we put it to the test with the following well-known edge detection approaches:

– The Canny method [7] with $\sigma_C = 2.25$ as a usual value in [14,17];
– The Gravitational Edge Detection (GED) method [13] with two configurations:
  • Probabilistic sum (G_{S_P}).
  • Maximum (G_{S_M}).
– Fuzzy morphology [10] using Schweizer and Sklar [21] t-norm and t-conorm (FM_{SS}).

4.3   Consensus Feature Image Construction with OWA Operator

As a further step in the improvement of the presented method, we consider different feature images obtained in step ($S2$), by means of different edge detection methods, and we build a consensus feature image using OWA operators.

Concretely we use the OWA operator representing the linguistic label *the majority of* constructed with Eq. (1) considering parameters $a = 0.3$ and $b = 0.8$.

To obtain the consensus feature image we order the set of intensity values of each position of the image and apply the OWA operator with the corresponding weights.

As an example of this procedure we test this construction method combining from two to five feature images obtained with different methods.

4.4   Dataset and Quantification of the Results

In order to show the behaviour of the presented method we have tested it over the images of the Berkeley Segmentation Dataset (BSDS500) [1], specifically we have selected 100 images from the test set. Each one of this images comes with a series of ground truth images that has been defined by experts. Usually there are between 4 and 9 images associated to each original image.

The performance measure of our method is considered as a classification problem, as the ground truth images are binary images indicating in each pixel if there is an edge or not. Each one of the pixels is counted in a in a confusion matrix as in the Martin *et al.* approach [16], where *True Positive, False Positive, etc.* are considered.

In order to quantify the results, we use the following well-known Precision/Recall measures:

$$Prec = \frac{TP}{TP + FP}, \ Rec = \frac{TP}{TP + FN}, \ F_\alpha = \frac{Prec \cdot Rec}{\alpha \cdot Prec + (1 - \alpha) \cdot Rec}.$$

4.5   Experimental Results

Table 1 reflects the results of each edge detection method indicating the average of *Prec*, *Rec* and $F_{0.5}$ obtained. As we can observe, in terms of *Prec* the maximum score is obtained by the Canny method, although the next best performer is one of our consensus construction method, concretely *C2*. In terms of *Rec* the highest value is obtained by the Fuzzy morphology and then we have our ODM proposal. Considering the $F_{0.5}$ measure we can see that, before going over the consensus methods, our ODM proposal, is the best performer comparing it to all the individual methods considered.

Then, in the case of the consensus methods, we show the best performers in each case. We observe that the results obtained in terms of $F_{0.5}$ is identical when using two, three and four feature images. We can see that adding feature images to the consensus image has an effect over *Prec* and *Rec*, reducing the first one and increasing the second. This behaviour makes the $F_{0.5}$ to remains unchanged. Then, when adding the fifth the results in terms of $F_{0.5}$ recovers the one obtained with the ODM function. We can extract from this results that the feature images obtained with FM_{SS}, G_{S_P} and G_{S_M} do not contribute with useful information to the consensus image, when using the OWA operator.

As complementary comparative measure we consider the number of images being the best and worst performer in terms of $F_{0.5}$. In Table 2 we show the results obtained. On one hand, we observe that with our proposed approaches,

Table 1. Comparison of ODM functions approach, along with OWA operator consensus approach from two to five combinations with respect to gravitational, fuzzy morphology and the Canny method in terms of $Prec$, Rec and $F_{0.5}$.

Edge detection methods	$Prec$	Rec	$F_{0.5}$
ODM	0.572	0.779	0.639
ODM-Canny (C2)	0.647	0.686	**0.645**
ODM-Canny-FM_{SS} (C3)	0.621	0.719	**0.645**
ODM-Canny-$FM_{SS} - G_{S_P}$ (C4)	0.612	0.735	**0.645**
ODM-Canny-$FM_{SS} - G_{S_P} - G_{S_M}$ (C5)	0.607	0.729	0.639
Canny	**0.666**	0.641	0.631
FM_{SS}	0.498	**0.807**	0.596
G_{S_P}	0.617	0.702	0.631
G_{S_M}	0.616	0.691	0.625

both when using ODM functions and with the consensus method, we obtain in all the cases the lowest results in terms of worst count. On the other hand, in terms of best count we are below the Canny method. But we are comparable when we use the consensus image combining three feature images.

Table 2. Comparison of best and worst approaches in terms of $F_{0.5}$ considering ODM functions approach and consensus approaches with two to five feature images $C2, C3, C4, C5$. ✓ and ✗ indicates the number of best images and worst images respectively.

	Edge detection methods									
	*		FM_{SS}		Canny		G_{S_P}		G_{S_M}	
	✓	✗	✓	✗	✓	✗	✓	✗	✓	✗
ODM	20	**2**	13	53	31	20	27	17	9	8
C2	18	1	19	54	27	20	**30**	18	6	7
C3	23	1	17	54	**27**	20	27	18	6	7
C4	21	1	18	53	**30**	20	25	18	6	8
C5	17	1	21	54	**30**	20	27	18	5	7

In order to see a visual example, we show in Fig. 2 the results obtained with all the approaches considered in our experiments, along with one of the ground truth images. We can clearly observe that our methods ($ODM, C2, C3, C4, C5$) detect a great majority of edges. The bear contour is detected by all the methods, but the edges separating the image in top middle and bottom are not detected in some cases, $i.e.$, G_{S_P}. Comparing the top of the ground truth we see that our methods detects many more edges than the real ones. Particularly, our methods, both the single ODM and the consensus ones detect quite well the bottom edges of the image.

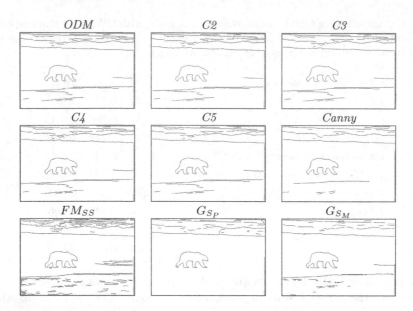

Fig. 2. Binary image obtained with ODM functions (ODM) and consensus construction using OWA ($C2$, $C3$, $C3$, $C4$, $C5$), Canny, Fuzzy Morphology (FM_{SS}) and Gravitational forces (G_{S_P}, G_{S_M}).

5   Conclusion

Our work has treated edge detection from the perspective of ODM functions. In particular, we have used this type of functions in the specific task of extracting features from gray-scale images. In addition, we have presented a method to build, using ODM functions along with other edge detection methods, consensus feature images by means of OWA operators.

Our procedure is based on a local approach, in which we extract the information at each pixel of the image using its neighbourhood. In this way we obtain intensity changes along with the direction of the variation in the intensity.

As the experimentation has shown, on one hand, ODM functions result in competitive scores with respect to classical methods, like the Canny method. On the other hand, when we combine our approach with other methods to build a consensus image, we obtain the best scores. Besides, when looking at the quantification of best and worst results, we see that we outperform all the presented methods in terms of worst count, obtaining the lowest scores.

From a preliminary study on ODM functions applied to the problem of edge detection, the results are promising and indicate that this type of function appear to be a good alternative for edge detection. In order to confirm the results obtained, more experiments should be carried out, studying different parameters, as well as using different datasets.

Acknowledgments. This work is supported by the Spanish Ministry of Science (Project TIN2016-77356-P) and the Research Services of Universidad Publica de Navarra.

References

1. Arbelaez, P., Maire, M., Fowlkes, C., Malik, J.: Contour detection and hierarchical image segmentation. IEEE Trans. Pattern Anal. Mach. Intell. **33**(5), 898–916 (2011)
2. Beliakov, G., Pradera, A., Calvo, T.: Aggregation Functions: A Guide for Practitioners, vol. 18. Springer, Heidelberg (2007). https://doi.org/10.1007/978-3-540-73721-6
3. Bezdek, J., Chandrasekhar, R., Attikouzel, Y.: A geometric approach to edge detection. IEEE Trans. Fuzzy Syst. **6**(1), 52–75 (1998)
4. Bustince, H., Barrenechea, E., Sesma-Sara, M., Lafuente, J., Dimuro, G.P., Mesiar, R., Kolesarova, A.: Ordered directionally monotone functions. Justification and application. IEEE Trans. Fuzzy Syst. **PP**(99), 1 (2017)
5. Bustince, H., Fernandez, J., Kolesárová, A., Mesiar, R.: Directional monotonicity of fusion functions. Eur. J. Oper. Res. **244**, 300–308 (2015)
6. Calvo, T., Kolesárová, A., Komorníková, M., Mesiar, R.: Aggregation operators: properties, classes and construction methods. In: Calvo, T., Mayor, G., Mesiar, R. (eds.) Aggregation Operators: New Trends and Applications. STUDFUZ, vol. 97, pp. 3–104. Springer, Heidelberg (2002). https://doi.org/10.1007/978-3-7908-1787-4_1
7. Canny, J.F.: A computational approach to edge detection. IEEE Trans. Pattern Anal. Mach. Intell. **8**(6), 679–698 (1986)
8. Estrada, F.J., Jepson, A.D.: Benchmarking image segmentation algorithms. Int. J. Comput. Vis. **85**(2), 167–181 (2009)
9. Forero-Vargas, M.G.: Fuzzy thresholding and histogram analysis. In: Nachtegael, M., Van der Weken, D., Kerre, E.E., Van De Ville, D. (eds.) Fuzzy Filters for Image Processing. STUDFUZZ, vol. 122, pp. 129–152. Springer, Heidelberg (2003). https://doi.org/10.1007/978-3-540-36420-7_6
10. Gonzalez-Hidalgo, M., Massanet, S., Mir, A., Ruiz-Aguilera, D.: On the choice of the pair conjunction-implication into the fuzzy morphological edge detector. IEEE Trans. Fuzzy Syst. **23**(4), 872–884 (2015)
11. Kass, M., Witkin, A., Terzopoulos, D.: Snakes: active contour models. Int. J. Comput. Vis. **1**(4), 321–331 (1988)
12. Law, T., Itoh, H., Seki, H.: Image filtering, edge detection, and edge tracing using fuzzy reasoning. IEEE Trans. Pattern Anal. Mach. Intell. **18**(5), 481–491 (1996)
13. Lopez-Molina, C., Bustince, H., Fernandez, J., Couto, P., De Baets, B.: A gravitational approach to edge detection based on triangular norms. Pattern Recognit. **43**(11), 3730–3741 (2010)
14. Lopez-Molina, C., De Baets, B., Bustince, H.: A framework for edge detection based on relief functions. Inf. Sci. **278**, 127–140 (2014)
15. Marr, D., Hildreth, E.: Theory of edge detection. Proc. R. Soc. Lond. B: Biol. Sci. **207**(1167), 187–217 (1980)
16. Martin, D.R., Fowlkes, C.C., Malik, J.: Learning to detect natural image boundaries using local brightness, color, and texture cues. IEEE Trans. Pattern Anal. Mach. Intell. **26**(5), 530–549 (2004)

17. Medina-Carnicer, R., Madrid-Cuevas, F.J., Carmona-Poyato, A., Muñoz-Salinas, R.: On candidates selection for hysteresis thresholds in edge detection. Pattern Recognit. **42**(7), 1284–1296 (2009)
18. Medina-Carnicer, R., Muñoz-Salinas, R., Yeguas-Bolivar, E., Diaz-Mas, L.: A novel method to look for the hysteresis thresholds for the Canny edge detector. Pattern Recognit. **44**(6), 1201–1211 (2011)
19. Perona, P., Malik, J.: Scale-space and edge detection using anisotropic diffusion. IEEE Trans. Pattern Anal. Mach. Intell. **12**(7), 629–639 (1990)
20. Prewitt, J.: Object enhancement and extraction (1970)
21. Schweiser, B., Sklar, A.: Associative functions and statistical triangle inequalities. Publ. Math. Debr. **8**, 169–186 (1961)
22. Sobel, I., Feldman, G.: A 3x3 isotropic gradient operator for image processing. In: Hart, P.E., Duda, R.O. (eds.) Pattern Classification and Scene Analysis, pp. 271–272. Wiley, Hoboken (1973)
23. Torre, V., Poggio, T.: On edge detection. IEEE Trans. Pattern Anal. Mach. Intell. **8**(2), 147–163 (1986)
24. Wilkin, T., Beliakov, G.: Weakly monotonic averaging functions. Int. J. Intell. Syst. **30**(2), 144–169 (2015)
25. Yager, R.: Quantifier guided aggregation using OWA operators. Int. J. Intell. Syst. **11**(1), 49–73 (1996)
26. Yager, R.R.: On ordered weighted averaging aggregation operators in multicriteria decisionmaking. IEEE Trans. Syst. Man Cybern. **18**(1), 183–190 (1988)

Characterization of Lattice-Valued Restricted Equivalence Functions

Eduardo Palmeira[1(✉)], Benjamín Bedregal[2], and Rogério R. Vargas[3]

[1] Programa de Pós-Graduação em Modelagem Computacional
em Ciência e Tecnologia - PPGMC, Departamento de Ciências Exatas
e Tecnológicas - DCET, Universidade Estadual de Santa Cruz - UESC,
Ilhéus, BA 45662-900, Brazil
espalmeira@uesc.br
[2] Departamento de Informtica e Matemtica Aplicada - DIMAp,
Universidade Federal do Rio Grande do Norte - UFRN, Natal,
RN 59078-970, Brazil
bedregal@dimap.ufrn.br
[3] Laboratório de Sistemas Inteligentes e Modelagem - LabSIM,
Universidade Federal do Pampa - UNIPAMPA, Itaquí, RS 97650-000, Brazil
rogvar@gmail.com

Abstract. In this paper we investigate about lattice-valued restricted equivalence functions and its characterization by means a particular class of lattice-valued implication operators.

Keywords: Restricted equivalence functions · Aggregation functions
Implication operators

1 Introduction

The study of global image and entropy comparison among fuzzy sets has been the subject of research by many professionals in recent years in different areas of knowledge due to the need for large scale data processing and classification [9, 14, 26].

There are several ways to define measures of comparison in the literature, each one trying to be the most accurate in each scenario in which it is considered. Bustince et al. in [8,9] have been defined the concept of restricted equivalence functions (REF) in the context of fuzzy sets (on [0,1]) as a particular case of the Fodor and Roubens equivalence functions [14]. Precisely speaking the REF is able to provide a local measure for comparing images by considering a pixel in one image with its corresponding pixel in the other image.

Later Julio et al. in [17] have defined the interval version of restricted equivalence functions and Palmeira and Bedregal in [24] have defined those functions on $L([0,1])$. Recently Palmeira et al. in [26] also have presented the definition of lattice-valued restricted equivalence funtions and normal $E_{e,N}$-functions.

© Springer International Publishing AG, part of Springer Nature 2018
G. A. Barreto and R. Coelho (Eds.): NAFIPS 2018, CCIS 831, pp. 167–178, 2018.
https://doi.org/10.1007/978-3-319-95312-0_15

In this paper a characterization theorem is presented for lattice-valued restricted equivalence funtions by means of a particular implication operators.

Section 2 brings some usual concepts on lattice theory and Sect. 3 presents the characterization theorem for L-REF. Main results are analyzed in Sect. 4.

2   Preliminaries

Definitions and properties presented in this section are part of the lattice theory. For a detailed review of them we strongly recommend the following references [6,11–13,15,18–20,28]. In the whole paper we write L for a lattice and M for its sublattice.

2.1   Lattices and Morphisms

This section is devoted to recall some important definitions and properties of lattices that we consider in the whole text.

Definition 1 [23]. *Let L be a nonempty set. If $\wedge_L$ and $\vee_L$ are two binary operations on L, then $\langle L, \wedge_L, \vee_L \rangle$ is an alg-lattice provided that for each $x, y, z \in L$, the following properties stand:*

1. *$x \wedge_L y = y \wedge_L x$ and $x \vee_L y = y \vee_L x$ (commutativity);*
2. *$(x \wedge_L y) \wedge_L z = x \wedge_L (y \wedge_L z)$ and $(x \vee_L y) \vee_L z = x \vee_L (y \wedge_L z)$ (associativity);*
3. *$x \wedge_L (x \vee_L y) = x$ and $x \vee_L (x \wedge_L y) = x$ (absorption law).*

If in L there are elements 0_L and 1_L such that, for all $x \in L$, $x \vee_L 0_L = x$ (bottom) and $x \wedge_L 1_L = x$ (top), then $\langle L, \wedge_L, \vee_L, 0_L, 1_L \rangle$ is a bounded lattice.

It is clear that the following equivalence

$$x \leqslant_L y \text{ if and only if } x \wedge_L y = x \tag{1}$$

provides a partial order on L and it can be proved that $\langle L, \leqslant_L \rangle$ is a lattice.

Definition 2 [6]. *A lattice L is called a complete lattice if every subset of it has a supremum and an infimum element. Notice that every complete lattice is bounded.*

Example 1. The set $[0,1]$ endowed with the operations defined by $x \wedge y = \min\{x, y\}$ and $x \vee y = \max\{x, y\}$ for all $x, y \in [0,1]$ is a complete lattice in the sense of Definitions 1 and 2 which has 0 as the bottom and 1 as the top element.

Example 2. For all $x, y \in [0,1]$ it is possible to define the interval set $L([0,1]) = \{[x,y] \; ; \; 0 \leqslant x \leqslant y \leqslant 1\}$. This set equipped with the operations

$$[x,y] \wedge_L [w,z] = [x \wedge w, y \wedge z] \text{ and } [x,y] \vee_L [w,z] = [x \vee w, y \vee z].$$

with $a \wedge b = \min(a,b)$ and $a \vee b = \max(a,b)$, is a complete lattice (in the sense of Definition 1) which has $[0,0]$ and $[1,1]$ as a bottom and a top respectively. It is easy to see that this lattice is also obtained by considered in $L([0,1])$ the partial order $[a,b] \leqslant_2 [c,d]$ if and only if $a \leqslant c$ and $b \leqslant d$.

Remark 1. When $\leqslant_L$ is a partial order on L and there are two elements x and y belonging to L such that neither $x \leqslant_L y$ nor $y \leqslant_L x$, these elements are said to be incomparable and we denote this by $x \parallel y$. Otherwise we say they are comparable (notation: $x \frown y$).

Definition 3 [13]. *Let* $(L, \leqslant_L, 0_L, 1_L)$ *and* $(M, \leqslant_M, 0_M, 1_M)$ *be bounded lattices. A mapping* $f : L \longrightarrow M$ *is said to be an order-preserving lattice homomorphism if, for all* $x, y \in L$, *it follows that*

1. *If* $x \leqslant_L y$ *then* $f(x) \leqslant_M f(y)$;
2. $f(0_L) = 0_M$ *and* $f(1_L) = 1_M$.

Definition 4 [13]. *Let* $(L, \wedge_L, \vee_L, 0_L, 1_L)$ *and* $(M, \wedge_M, \vee_M, 0_M, 1_M)$ *be bounded lattices. A mapping* $f : L \longrightarrow M$ *is said to be a lattice homomorphism if, for all* $x, y \in L$, *we have*

1. $f(x \wedge_L y) = f(x) \wedge_M f(y)$;
2. $f(x \vee_L y) = f(x) \vee_M f(y)$;
3. $f(0_L) = 0_M$ *and* $f(1_L) = 1_M$.

Definition 5 [16]. *A given lattice homomorphism* f *(in the sense of both Definitions 3 and 4) on* L *is called:*

1. *A monomorphism if it is injective;*
2. *An epimorphism if* f *is surjective;*
3. *An isomorphism when* f *is bijective. An automorphism is an isomorphism from a lattice to itself.*

Proposition 1 [6]. *Every lattice homomorphism in the sense of Definition 4 is order-preserving.*

However, in general, the reciprocal of Proposition 1 does not hold. If $f : L \longrightarrow M$ is an order homomorphism, since $x \wedge_L y \leqslant_L x$ and $x \wedge_L y \leqslant_L y$, so $f(x \wedge_L y) \leqslant_M f(x)$ and $f(x \wedge_L y) \leqslant_M f(y)$. Thus, $f(x \wedge_L y) \leqslant_M f(x) \wedge_M f(y) = \inf\{f(x), f(y)\}$, however it is possible for $f(x \wedge_L y) \neq \inf\{f(x), f(y)\}$ to occur. For example, consider the lattices L and M, as depicted in Hasse diagram shown in Fig. 1. Nevertheless, the map $f : L \longrightarrow M$ defined by $f(0_L) = 0_M$, $f(1_L) = 1_M$, $f(x) = u$ and $f(y) = v$, preserves infimum and supremum elements and, hence, is an order homomorphism, though it is not an alg-homomorphism as $\wedge$ operation is not preserved.

2.2 Negations on L

Now we are interest in study about the notion of lattice-valued negations and its properties.

Definition 6. *A function* $N : L \longrightarrow L$ *is called a **fuzzy negation** if it satisfies:*

(N1) $N(0_L) = 1_L$ *and* $N(1_L) = 0_L$;

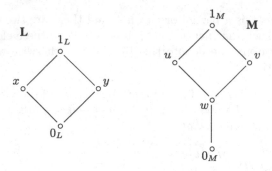

Fig. 1. Hasse diagrams of lattices L and M

(N2) If $x \leqslant_L y$ then $N(y) \leqslant_L N(x)$, for all $x, y \in L$.

Moreover, if a fuzzy negation N on L satisfies the involution property, namely

(N3) $N(N(x)) = x$, for all $x \in L$

it is called a **strong** fuzzy negation.

Negations satisfying Property (N4) are called **frontier**.

(N4) $N(x) \in \{0_L, 1_L\}$ if and only if $x = 0_L$ or $x = 1_L$

Other notions of fuzzy negation as a generalization of the classical one can be found in $[3, 5, 10, 22, 23]$.

Definition 7. *Let $N : L \to L$ a fuzzy negation on bounded lattice L. If $x \in L$ is such that $N(x) = x$ then x is called an **equilibrium point** of N.*

Example 3. Let L and M be bounded lattices as shown in the Fig. 2. The function $N_1 : M \to M$ defined by $N_1(0_M) = 1_M$, $N_1(x) = y$, $N_1(y) = x$ and $N_1(1_M) = 0_M$ is a strong M-negation. Nevertheless, N_1 has no equilibrium point.

Now, consider a function $N_2 : L \to L$ given by $N_2(0_L) = 1_L$, $N_2(a) = e$, $N_2(e) = a$, $N_2(1_L) = 0_L$ and $N_2(u) = u$ for each $u \in \{b, c, d\}$. In this case, N_2 is a strong fuzzy negation with three equilibrium points, namely b, c and d.

2.3   L-Implications

It is well-known that there are some different ways to interpret L-implications (see $[1, 5, 7, 21, 25, 29]$) but here we consider the notion considered in $[1]$.

Definition 8. *A L-implication on bounded lattice L is a function $I : L \times L \longrightarrow L$ such that for each $x, y, z \in L$ the following properties hold:*

(FPA) if $x \leqslant_L y$ then $I(y, z) \leqslant_L I(x, z)$ (First variable antitonicity);
(SPI) if $y \leqslant_L z$ then $I(x, y) \leqslant_L I(x, z)$ (Second variable isotonicity);

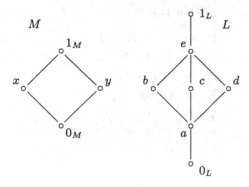

Fig. 2. Hasse diagrams of lattices M and L

(CC1) $I(0_L, 0_L) = 1_L$ *(Corner condition 1);*
(CC2) $I(1_L, 1_L) = 1_L$ *(Corner condition 2);*
(CC3) $I(1_L, 0_L) = 0_L$ *(Corner condition 3).*

Below it is presented some properties for a given function $I : L \times L \to L$:

(CC4) $I(0_L, 1_L) = 1_L$ (Corner condition 4);
(LB) $I(0_L, y) = 1_L$, for all $y \in L$;
(RB) $I(x, 1_L) = 1_L$, for all $x \in L$;
(NP) $I(1_L, y) = y$ for each $y \in L$ (left neutrality principle);
(EP) $I(x, I(y, z)) = I(y, I(x, z))$ for all $x, y, z \in L$ (exchange principle);
(IP) $I(x, x) = 1_L$ for each $x \in L$ (identity principle);
(OP) $I(x, y) = 1_L$ if and only if $x \leqslant y$ (ordering property);
(IBL) $I(x, I(x, y)) = I(x, y)$ for all $x, y, z \in L$ (iterative Boolean law);
(CP) $I(x, y) = I(N(y), N(x))$ for each $x, y \in L$ with N a fuzzy negation on L (law of contraposition);
(P) $I(x, y) = 0_L$ if and only if $x = 1_L$ and $y = 0_L$ (Positive);
(LEM) $S(N(x), x) = 1_L$ for each $x \in L$ (law of excluded middle).

Example 4. Let L be a bounded lattice (see Fig. 2) and N_2 the strong fuzzy negation on L in Example 3. The function $I : L^2 \to L$ given by

$$I(x, y) = \begin{cases} 1_L, & if \ x \leqslant_L y; \\ N_2(x), & if \ y = 0_L \ and \ x \neq 0_L; \\ y, & if \ x = 1_L; \\ e, & otherwise. \end{cases} \tag{2}$$

satisfies the properties (FPA), (OP), (CP) (with respect to N_2) and (P). It is a easy to see that it holds by Table 1.

Table 1. A function I on L

I	0_L	a	b	c	d	e	1_L
0_L	1_L	1_L	1_L	1_L	1_L	1_L	1_L
a	e	1_L	1_L	1_L	1_L	1_L	1_L
b	b	e	1_L	e	e	1_L	1_L
c	c	e	e	1_L	e	1_L	1_L
d	d	e	e	e	1_L	1_L	1_L
e	a	e	e	e	e	1_L	1_L
1_L	0_L	a	b	c	d	e	1_L

3   Lattice-Valued Restricted Equivalence Functions

Restricted equivalence functions play an important role for image processing due to it provides a similarity measure used to make a global comparison between images. Bustince et al. in [8] have been defined that operator in the context of fuzzy sets and shown how to construct similarity measures between fuzzy sets. Indeed, if $\mathcal{F}(X)$ is set of all fuzzy subsets of a given set X hence the function $SM : \mathcal{F}(X) \times \mathcal{F}(X) \to [0,1]$ given by $SM(A, B) = M_{i=1}^{n} REF(\mu_A(x_i), \mu_B(x_i))$ where M is an aggregation function and REF is a restricted equivalence function, is a similarity measure satisfying the following properties:

- $SM(A, B) = SM(B, A)$, for all $A, B \in \mathcal{F}(X)$;
- $SM(A, A_c) = 0$ if and only if is non-fuzzy;
- $SM(A, B) = 1$ if and only if $A = B$;
- $SM(A, B) = SM(A_c, B_c)$;
- if $A \le B \le C$ then $SM(A, B) \ge SM(A, C)$ and $SM(B, C) \ge SM(A, C)$.

Later, Palmeira et al. [26] have been constructed similar measures from restricted equivalence function in the context of lattice theory.

Definition 9 [26]. *Let N be a strong negation on L. A function $REF : L^2 \to L$ is called a restricted equivalence function on L with respect to N, or just an L-REF with respect to N, if it satisfies, for all $x, y, z \in L$, the following conditions:*

(L1) $REF(x, y) = REF(y, x)$;
(L2) $REF(x, y) = 1_L$ *if and only if $x = y$*;
(L3) $REF(x, y) = 0_L$ *if and only if $\{x, y\} = \{0_L, 1_L\}$*;
(L4) $REF(x, y) = REF(N(x), N(y))$;
(L5) *if $x \leqslant_L y \leqslant_L z$ then $REF(x, z) \leqslant_L REF(x, y)$*.

It is important to point out it is not easy to say when there exists a strong fuzzy negation for an arbitrary lattice L. However, negation in Definition 9 must be a strong negation in order to avoid a conflict between properties (L2) and (L4).

Notice that from $(L4)$, $(L5)$ and $(L1)$, it is also possible to conclude that $REF(x, z) \leqslant_L REF(y, z)$ whenever $x \leqslant_L y \leqslant_L z$.

Table 2. Restricted equivalence function on lattice M

R	0_L	x	y	1_L
0_L	1_L	x	y	0_L
x	x	1_L	x	y
y	y	x	1_L	x
1_L	0_L	y	x	1_L

Example 5. Let M be a bounded lattice and N_1 be the strong M-negation in Example 3. Thus, the function $R : M^2 \to M$ generated as in Table 2 is an L-REF with respect to N_1 in the sense of Definition 9.

Notice that in this case, the mapping $N_1(x) = R(0_L, x)$ defines a strong negation on the lattice M.

Below we generalize the method of constructing L-REF based on L-implication operators (see Theorem 7 of [9]). We start recalling the concept of implications on bounded lattices.

Theorem 1. *Let $N : L^2 \to L$ be a strong L-negation and $M : L^2 \to L$ be a function such that, for all $x, y \in L$, it holds:*

(M1) $M(x, y) = M(y, x)$;
(M2) $M(x, 1_L) = x$;
(M3) $M(x, y) = 1_L$ *if and only if* $x = y = 1_L$;
(M4) $M(x, y) = 0_L$ *if and only if* $x = 0_L$ *or* $y = 0_L$.

If there exist a function $I : L^2 \to L$ satisfying (FPA), (OP), (CP) for N and (P) then the function $REF : L^2 \to L$ defined by

$$REF(x, y) = M(I(x, y), I(y, x)) \tag{3}$$

is a L-REF with respect to N.

Proof. We shall prove that the conditions $(L1) - (L5)$ hold. It is clear that $(M1)$ implies $(L1)$. The proof for the other properties is given as follows:

$(L2)$
Suppose that $REF(x, y) = 1_L$. Thus, by (3) we have that $M(I(x, y), I(y, x)) = 1_L$ and hence $I(x, y) = 1_L$ and $I(y, x) = 1_L$ by $(M3)$. Therefore, by (OP) it follows that $x \leqslant_L y$ and $y \leqslant_L x$, i.e. $x = y$.
Conversely, if $x = y$, that is $x \leqslant_L y$ and $y \leqslant_L x$ then $I(x, y) = I(y, x) = 1_L$ by (OP). Thus, it is easy to see that $REF(x, y) = 1_L$.

$(L3)$
If $REF(x, y) = 0_L$ then $M(I(x, y), I(y, x)) = 0_L$ which allow us to conclude that either $I(x, y) = 0_L$ or $I(y, x) = 0_L$ since M satisfies $(M3)$. Thus, by (P) we have that either $x = 1_L$ and $y = 0_L$ or $x = 1_L$ and $y = 0_L$.
On the other hand, if $x = 1_L$ and $y = 0_L$ then $I(x, y) = 0_L$ by (P). Therefore, $REF(x, y) = M(I(x, y), I(y, x)) = M(0_L, I(y, x)) = 0_L$ by $(M3)$. Analogously it can be proved that $REF(x, y) = 0_L$ whenever $x = 0_L$ and $y = 1_L$.

(*L4*)

For all $x, y \in L$ we have that

$$
\begin{aligned}
REF(N(x), N(y)) &= M(I(N(x), N(y)), I(N(y), N(x))) \ by \ (3) \\
&= M(I(y, x), I(x, y)) && by \ (CP) \\
&= M(I(x, y), I(y, x)) && by \ (M1) \\
&= REF(x, y)
\end{aligned}
$$

(*L5*)

Given $x, y, z \in L$ such that $x \leqslant_L y \leqslant_L z$ we have that $I(x, y) = 1_L$ and $I(x, z) = 1_L$. Thus

$$REF(x, y) = M(I(x, y), I(y, x)) = M(1_L, I(y, x)) = I(y, x) \qquad (4)$$

and

$$REF(x, z) = M(I(x, z), I(z, x)) = M(1_L, I(y, x)) = I(z, x) \qquad (5)$$

Since $y \leqslant_L z$ then by (FPA) it follows that $I(z, x) \leqslant_L I(y, x)$ for all $x \in L$. Therefore, considering this fact, by Eqs. (4) and (5) it can be concluded that $REF(x, z) \leqslant_L REF(x, y)$. $\qquad \square$

Table 3. Restricted equivalence function on lattice L

REF	0_L	a	b	c	d	e	1_L
0_L	1_L	e	b	c	d	a	0
a	e	1_L	e	e	e	e	a
b	b	e	1_L	e	e	e	b
c	c	e	e	1_L	e	e	c
d	d	e	e	e	1_L	e	d
e	a	e	e	e	e	1_L	e
1_L	0_L	a	b	c	d	e	1_L

Example 6. Let L be the bounded lattice shown in Fig. 2. If $I : L^2 \to L$ is the function defined in the Example 4 and $M : L^2 \to L$ is a function given by $M(x, y) = \min\{x, y\}$ for all $x, y \in L$ then, by Theorem 1, $REF(x, y) = M(I(x, y), I(x, y))$ is a restricted equivalence function on L (see Table 3).

It is worth noting that reciprocal of Theorem 1 does not hold, in general. In other words, given an L-REF with respect to a strong L-negation N and a function $M : L^2 \to L$ satisfying (M1), (M2) and (M3) it is not always possible to define a function $I : L^2 \to L$ which satisfy (FPA), (OP), (CP), (P) and $REF(x, y) = M(I(x, y), I(y, x))$. This is due to there is not a specific way to define restricted equivalence function for pairs $(x, y) \in L^2$ such that $x \parallel y$ (incomparable elements of L). According to properties (L2) and (L3) we can just infer that $0_L <_L REF(x, y) <_L 1_L$ if $x \parallel y$.

It means that determining how to define a general condition for L-REF on incomparable elements of L^2 in order to make the reciprocal of the Theorem 1 holds constitute a very interesting open problem to be studied.

Proposition 2. *Let REF be an L-REF related to a strong L-negation N. Then the function $I_{REF} : L^2 \to L$ defined by*

$$I_{REF}(x,y) = \begin{cases} 1_L & \text{if } x \leqslant_L y \\ REF(x,y) & \text{otherwise} \end{cases} \tag{6}$$

satisfies the properties (OP), (CP) and (P).

Proof.

(OP)

Let $x, y \in L$. If $x \leqslant_L y$ then, by definition, it is trivial that $I_{REF}(x,y) = 1_L$.

Reciprocally, if $I_{REF}(x,y) = 1_L$ then either $x \leqslant_L y$ or $REF(x,y) = 1_L$ and $y <_L x$ or $x \parallel y$. But, note that if $REF(x,y) = 1_L$ then $x = y$ by (L2) which is a contradiction with both $y <_L x$ and $x \parallel y$. Therefore, it can be concluded that $x \leqslant_L y$.

(CP)

Note that if $x \leqslant_L y$ we have $N(y) \leqslant_L N(x)$ and hence $I_{REF}(x,y) = 1_L = I_{REF}(N(y), N(x))$ by definition of I_{REF}. Beyond that we can have $y <_L x$ or $x \parallel y$ which implies that $N(x) <_L N(y)$ and $N(x) \parallel N(y)$ respectively. Thus in both cases it follows that $I_{REF}(N(y), N(x)) = REF(N(y), N(x)) = REF(N(x), N(y))$ and $I_{REF}(x,y) = REF(x,y)$ what allow us to conclude that $I_{REF}(N(y), N(x)) = I_{REF}(x,y)$ since $REF(N(x), N(y)) = REF(x,y)$ by (L4).

(P)

If $I_{REF}(x,y) = 0_L$ then either $REF(x,y) = 0_L$ and $y <_L x$ or $REF(x,y) = 0_L$ and $x \parallel y$. But, the second case is a contradiction by (L3). Hence we must have $REF(x,y) = 0_L$ and $y <_L x$ and again by (L3) it follows that $x = 1_L$ and $y = 0_L$.

Reciprocally, if $x = 1_L$ and $y = 0_L$ then $I_{REF}(x,y) = REF(x,y) = REF(1_L, 0_L) = 0_L$. □

Remark 2. Notice that if *REF* and I_{REF} are functions as in Proposition 2 and $M : L^2 \to L$ is a function as in Theorem 1 (satisfying also $M(x,x) = x$ for all $x \in L$) then the Identity (3) holds.

Indeed, let $x, y \in L$ and suppose that they are comparable. In this case, if $x \leqslant_L y$ then $I_{REF}(x,y) = 1_L$ and $I_{REF}(y,x) = REF(y,x) = REF(x,y)$. Thus

$$REF(x,y) = I_{REF}(y,x) = M(1_L, I_{REF}(y,x)) = M(I_{REF}(x,y), I_{REF}(y,x))$$

Analogously, one can prove that (3) holds if $y \leqslant_L x$.
On the other hand, if $x \parallel y$ then $I_{REF}(x,y) = REF(x,y)$ and $I_{REF}(y,x) = REF(y,x) = REF(x,y)$. Therefore

$$REF(x,y) = M(REF(x,y), REF(y,x)) = M(I_{REF}(x,y), I_{REF}(y,x))$$

Table 4. The function I_{REF} on lattice M

I_{REF}	0_L	x	y	1_L
0_L	1_L	1_L	1_L	1_L
x	x	1_L	x	1_L
y	y	x	1_L	1_L
1_L	0_L	y	x	1_L

Example 7. Here we show that the reciprocal of Theorem 1 does not hold in general. Actually, we would like to highlight the fact that the property (FPA) fails for the function I_{REF} defined in (6) (by Theorem 7 in [8] it is known that I_{REF} should satisfy (FPA), (OP), (CP) and (P)). To do so, consider the bounded lattice M (see Fig. 1) and the restricted equivalence function defined in Example 5. In this case, the function I_{REF} in (6) is defined as in Table 4. Note that $y < 1_M$ but $I_{REF}(1_M, x) = y$ and $I_{REF}(y, x) = x$ which means that $I_{REF}(1_M, x) \parallel I_{REF}(y, x)$ since $x \parallel y$.

Taking into account the important fact highlighted above, we can see that the problem of proving the reciprocal of Theorem 1 becomes the problem of seeking conditions under which the function I_{REF} as in Proposition 2 satisfies (FPA). Clearly if all the elements of the bounded lattice L are comparable then the problem is sold, it means:

Theorem 2. *Let L be a bounded chain and suppose that $M : L^2 \to L$ is a function satisfying (M1), (M2), (M3) and $M(x,x) = x$ for all $x \in L$. Thus a function $REF : L^2 \to L$ is an L-REF for a strong L-negation N if and only if there exist a function $I_{REF} : L^2 \to L$ satisfying (FPA), (OP), (CP) for N and (P) in such a way that the Eq. (3) holds.*

Proof. Let REF be a *L-REF* with respect to a strong negation N. Define:

$$I_{REF}(x,y) = \begin{cases} 1_L & if\ x \leqslant_L y; \\ REF(x,y) & otherwise. \end{cases}$$

From Proposition 2 and Remark 2 we can conclude that I_{REF} satisfies (OP), (CP), (P) and $REF(x,y) = M(I_{REF}(x,y), I_{REF}(y,x))$ for all $x, y \in L$. Thus, it remains to prove that property (FPA) holds.

Take $y \in L$ and $x, z \in L$ such that $x \leq_L z$. We want to see that $I_{REF}(z,y) \leq_L I_{REF}(x,y)$. There are three possibilities.

(*i*) If $y \leq_L x \leq_L z$ we have that $I_{REF}(z,y) = REF(z,y)$ and $I_{REF}(x,y) = REF(x,y)$. But by (L5), $REF(z,y) \leq_L REF(x,y)$ and hence $I_{REF}(z,y) \preceq I_{REF}(x,y)$.

(*ii*) When $x \leq_L y \leq_L z$ it follows that $I_{REF}(z,y) = REF(z,y) \leq_L 1_L = I_{REF}(x,y)$.

(*iii*) If $x \leq_L z \leq_L y$ then $I_{REF}(z,y) = 1_L = I_{REF}(x,y)$.

The reciprocal is straightforward from Theorem 1. $\square$

4 Final Remarks

On one hand the main contribution of this is paper is related to the theoretical study on lattice-valued restricted equivalence functions and its characterization by implication operators. Results shown that REF can be generated by aggregating a suitable L-implication operator. On the other hand, it has been proven that of that result does not holds. Proposition 2 present an operator contructed from a given REF that is not an L-implication. boundcan be constructed as one can see in Theorem 2.

As future work we wish to apply the extension method via e-operator [22] for constructing similarity measures from a given one.

References

1. Baczyński, M., Jayaram, B.: Fuzzy Implications. Studies in Fuzziness and Soft Computing. Springer, Heidelberg (2008). https://doi.org/10.1007/978-3-540-69082-5
2. Bedregal, B.C., Santos, H.S., Callejas-Bedregal, R.: T-norms on bounded lattices: T-norm morphisms and operators. In: IEEE International Conference on Fuzzy Systems, pp. 22–28 (2006)
3. Bedregal, B.C.: On interval fuzzy negations. Fuzzy Sets Syst. **161**, 2290–2313 (2010)
4. Bedregal, B.C.: Xor-Implications and E-Implications: classes of fuzzy implications based on fuzzy Xor. Electron. Notes Theor. Comput. Sci. **274**, 5–18 (2009)
5. Bedregal, B., Beliakov, G., Bustince, H., Fernandez, J., Pradera, A., Reiser, R.: (S, N)-implications on bounded lattices. In: Baczyński, M., Beliakov, G., Bustince Sola, H., Pradera, A. (eds.) Advances in Fuzzy Implication Functions. Studies in Fuzziness and Soft Computing, vol. 300. Springer, Heidelberg (2013). https://doi.org/10.1007/978-3-642-35677-3_5
6. Birkhoff, G.: Lattice Theory. American Mathematical Society, Providence (1973)
7. Bustince, H., Burillo, P., Soria, F.: Automorphisms, negations and implication operators. Fuzzy Sets Syst. **134**(2), 209–229 (2003)
8. Bustince, H., Barrenechea, E., Pagola, M.: Restricted equivalence functions. Fuzzy Sets Syst. **157**(17), 2333–2346 (2006)
9. Bustince, H., Barrenechea, E., Pagola, M.: Relationship between restricted dissimilarity functions, restricted equivalence functions and normal E_N-functions: image thresholding invariant. Pattern Recogn. Lett. **29**, 525–536 (2008)
10. Calvo, T.: On mixed De Morgan triplets. Fuzzy Sets Syst. **50**, 47–50 (1992)
11. Chen, G., Pham, T.T.: Fuzzy Sets, Fuzzy Logic and Fuzzy Control Systems. CRC Press, Boca Raton (2001)
12. De Cooman, G., Kerre, E.E.: Order norms on bounded partially ordered sets. J. Fuzzy Math. **2**, 281–310 (1994)
13. Davey, B.A., Priestley, H.A.: Introduction to Lattices and Order, 2nd edn. Cambridge University Press, Cambridge (2002)
14. Fodor, J., Roubens, M.: Fuzzy Preference Modelling and Multicriteria Decision Support. Kluwer Academic Publisher, Dordrecht (1994)
15. Hajek, P.: Metamathematics of Fuzzy Logic. Kluwer Academic Publishers, Dordrecht (1998)

16. Hungerford, T.W.: Algebra. Graduate Texts in Mathematics. Springer, New York (2000)
17. Julio, A., Pagola, M., Paternain, D., Lopez-Molina, C., Melo-Pinto, P.: Interval-valued restricted equivalence functions applied on clustering techniques. In: IFSA-EUSFLAT, pp. 831–836 (2009)
18. Klement, E.P., Mesiar, R.: Logical, Algebraic, Analytic, and Probabilistic Aspects of Triangular Norms. Elsevier B.V., Amsterdam (2005)
19. Klement, E.P., Mesiar, R., Pap, E.: Triangular Norms. Kluwer Academic Publishers, Dordrecht (2000)
20. Klir, G.J., Yuan, B.: Fuzzy Sets and Fuzzy Logic, Theory and Applications. Prentice Hall PTR, Upper Saddle River (1995)
21. Mas, M., Monserrat, M., Torrens, J., Trillas, E.: A survey on fuzzy implications functions. IEEE Trans. Fuzzy Syst. **15**(6), 1107–1121 (2007)
22. Palmeira, E.S., Bedregal, B.C., Mesiar, R., Fernandez, J.: A new way to extend T-norms, T-conorms and negations. Fuzzy Sets Syst. **240**, 1–21 (2014)
23. Palmeira, E.S., Bedregal, B.C.: Extension of fuzzy logic operators defined on bounded lattices via retractions. Comput. Math. Appl. **63**, 1026–1038 (2012)
24. Palmeira, E.S., Bedregal, B.: Restricted equivalence function on $L([0, 1])$. In: Melin, P., Castillo, O., Kacprzyk, J., Reformat, M., Melek, W. (eds.) NAFIPS 2017. AISC, vol. 648, pp. 410–420. Springer, Cham (2018). https://doi.org/10.1007/978-3-319-67137-6_45
25. Palmeira, E.S., Bedregal, B.C., dos Santos, J.A.: Some results on extension of lattice-valued QL-implications. J. Braz. Comput. Soc. **22**(1), 41–49 (2016)
26. Palmeira, E.S., Bedregal, B.C., Bustince, H., Paternain, D., De Miguel, L.: Application of two different methods for extending lattice-valued restricted equivalence functions used for constructing similarity measures on L-fuzzy sets. Inf. Sci. **441**, 95–112 (2018)
27. Saminger-Platz, S., Klement, E.P., Mesiar, R.: On extensions of triangular norms on bounded lattices. Indag. Math. **19**(1), 135–150 (2008)
28. Takano, M.: Strong completeness of lattice-valued logic. Arch. Math. Logic **41**, 497–505 (2002)
29. Yager, R.R.: On the implication operator in fuzzy logic. Inf. Sci. **31**(2), 141–164 (1983)

Aggregation with T-Norms and LexiT-Orderings and Their Connections with the Leximin Principle

Henrique Viana[(✉)] and João Alcântara

Department of Computer Science, Federal University of Ceará, Fortaleza, Brazil
{henriqueviana,jnando}@lia.ufc.br

Abstract. We analyze the impact of applying families of T-norm and LexiT-ordering aggregation functions in the context of egalitarian reasoning. We compare both of them with the minimum and lexicographic minimum aggregation functions, which are well-known functions used in the aggregation approach in the decision making problem. For this task, we consider three logical properties in the Social Choice theory and Economics: Hammond Equity, Strong Pareto and Anonymity. It is known that lexicographic minimum satisfies all of these properties. We present in this paper some conditions to T-norms and LexiT-orderings satisfy these logical properties or restrictions of them.

1 Introduction

Aggregation of information are basic concerns for all kinds of knowledge based systems, from image processing to decision making, from pattern recognition to machine learning. From a general point of view we can say that aggregation has as purpose the simultaneous use of different pieces of information (provided by several sources/agents) in order to come to a conclusion or a decision [4].

The purpose of aggregation functions (or aggregation operators) is to combine inputs that are typically interpreted as degrees of membership in fuzzy sets, degrees of preference, strength of evidence, or support of a hypothesis, and so on. There exists a large number of different aggregation functions that differ on the assumptions related to data (data types) and on the properties of their results [2].

Some of most popular aggregation functions are the average, the median, the minimum and the maximum, as well as some classical generalizations like the weighted mean, lexicographic minimum, lexicographic maximum and the k-order statistics [2].

In the decision making field, an important topic, considered the heart of social choice theory [1], is the analysis of aggregation functions, which involves some aspects of their rationality. Indeed, there are two main theories approached when aggregating values: the utilitarianism and egalitarianism.

This research is supported by CNPq and CAPES.

ⓒ Springer International Publishing AG, part of Springer Nature 2018
G. A. Barreto and R. Coelho (Eds.): NAFIPS 2018, CCIS 831, pp. 179–191, 2018.
https://doi.org/10.1007/978-3-319-95312-0_16

Utilitarianism sustains the idea the best choice for a group is that maximizing the input of the group. The average function is an example of utilitarian aggregation function. Egalitarianism, on the other hand, tries to reach equality for all agents in the group. The minimum and lexicographic minimum functions are examples of egalitarian aggregation functions. Intuitively, they seek to promote equality of the group by favoring those agents in bad situation.

This work aims at exploring further egalitarian aggregation functions. The idea is to generalize the minimum function and employ fuzzy connectives. As it is known, T-norms are functions weaker than the minimum function, which can be commonly used to capture the worst case in some group decision problems. The motivation for this paper is to offer a new view about aggregation functions with T-norms, exploring their properties with respect to egalitarian conditions. We will begin our analysis with the minimum and lexicographic minimum aggregation functions and their relation with three important logical properties: Hammond Equity, Strong Pareto and Anonymity [11], which are closely related with the idea of egalitarianism. Right after, we will analyze the relation of T-norm aggregation functions with these logical properties. Furthermore, we will consider the lexicographic generalization of T-norms, called LexiT-ordering [15], and we will make a comparative between LexiT-ordering and lexicographic minimum aggregation functions.

The paper is structured as follows. In Sect. 2, we will present some basic notions about the framework of aggregation, aggregation functions, justice relations and its egalitarian logical properties. In Sect. 3, we will review the concepts of T-norms and parameterized T-norms. In Sect. 4, we will find one of the main contributions of this work, where we will consider T-norms justice relation and their connections with egalitarian logical properties, and consequently minimum and lexicographic minimum functions. In Sect. 5, we will continue this investigation with the LexiT-ordering justice relations. Finally, in Sect. 6 we will conclude the paper.

2   The Framework and Some Logical Properties

In this section, we present some fundamental notions about frameworks of aggregation and their logical properties.

We assume a fixed population of agents $A = \{1, \ldots, n\}$, and a set of outcomes $\Omega = \{\omega_1, \ldots, \omega_m\}$, where each outcome ω_i is represented by a $n-$dimensional utility vector. Each outcome $\omega_i \in \Omega$ can be viewed as a possible world or an alternative which contains the utility levels of all agents. For $\omega_i = (\omega_i^1, \ldots, \omega_i^n)$, we will refer to ω_i^j as the utility value of the agent j in the outcome i. For any ω_i^j, we will impose that $\omega_i^j \in [0, 1]$. We will use the binary relation $\leq$ to rank these utility levels. We define $<$ as follows: $\omega_i^k < \omega_j^k$ iff $\omega_i^k \leq \omega_j^k$ and $\omega_j^k \neq \omega_i^k$. Hence, the ranking of the outcomes will only depend on these utility values contained in each vector.

We assume that $\leq_f$ over Ω is a reflexive and transitive binary relation (i.e., a pre-order), where $f : [0, 1]^n \to [0, 1]$ is an aggregation function between outcomes. We will refer to $\leq_f$ as a f justice relation. A pre-order $\leq_f$ is total if

$\forall \omega_i, \omega_j \in \Omega$, $\omega_i \leq_f \omega_j$ or $\omega_j \leq_f \omega_i$. We define $<_f$ as follows: $\omega_i <_f \omega_j$ iff $\omega_i \leq_f \omega_j$ and $\omega_j \not\leq_f \omega_i$, and $\approx_f$ as $\omega_i \approx_f \omega_j$ iff $\omega_i \leq_f \omega_j$ and $\omega_j \leq_f \omega_i$. When $\omega_i <_f \omega_j$, we say ω_j is more just (or preferable) than ω_i with respect to f; when $\omega_i \leq_f \omega_j$, we say ω_j is at least as just as ω_i with respect to f; and $\omega_i \approx_f \omega_j$ denotes ω_i is as just as ω_j with respect to f.

Definition 1 (min **and** $leximin$ **justice relations).** *Let $A = \{1, \ldots, n\}$ be a set of agents, $\Omega = \{\omega_1, \ldots, \omega_m\}$ be a set of outcomes, where each $\omega_i = (\omega_i^1, \ldots, \omega_i^n)$ and $\omega_i^j \in [0,1]$, for $j \in \{1, \ldots, n\}$. Let min and $leximin$ be two aggregation functions.*

- *We define $\leq_{min}$ over Ω as $\omega_i \leq_{min} \omega_j$ iff $min(\omega_i^1, \ldots, \omega_i^n) \leq min(\omega_j^1, \ldots, \omega_j^n)$;*
- *For each ω_i we build the list $(\omega_i^{(1)}, \ldots, \omega_i^{(n)})$ of utilities sorted in ascending order. Let $\leq_{lex}$ be the lexicographical order between sequences of values in $[0,1]$, i.e., $(x_1, \ldots, x_n) \leq_{lex} (y_1, \ldots, y_n)$ if (1) for all i, $x_i \leq y_i$ or (2) there exists i such that $x_i < y_i$ and for all $j < i$, $x_j \leq y_j$. We define $\leq_{leximin}$ over Ω as $\omega_i \leq_{leximin} \omega_j$ iff $(\omega_i^{(1)}, \ldots, \omega_i^{(n)}) \leq_{lex} (\omega_j^{(1)}, \ldots, \omega_j^{(n)})$.*

When comparing outcomes, the min justice relation considers more just an outcome with a higher minimum utility value. Alternatively, we can say that we are giving absolute preference to the worst off agent in the group. The $leximin$ justice relation follows the same idea, but when the utility value of the worst off agents are equivalent, then we consider the utility value of the second worst off agents (if they are still equivalent, we continue to the third worst agents and so on). For instance, consider the following three outcomes: $\omega_1 = (0.5, 0.8, 0.2)$, $\omega_2 = (0.9, 0.5, 0.1)$ and $\omega_3 = (0.6, 0.1, 0.6)$. Thus, ω_1 is more just than ω_2 with respect to min and $leximin$, and ω_2 is as just as ω_3 only with respect to min. For $leximin$, ω_3 is more just than ω_2, since $(0.1, 0.5, 0.9) <_{lex} (0.1, 0.6, 0.6)$.

In the analysis of justice relations we will take into account the following properties of theories of social justice: Hammond Equity, Strong Pareto and Anonymity [11].

(a) Hammond Equity (HE): Let f be an aggregation function. For all $\omega_i, \omega_j \in \Omega$, there exist k, l such that: (1) $\omega_i^k < \omega_j^k$; (2) $\omega_j^l < \omega_i^l$; (3) $\omega_j^k \leq \omega_j^l$; (4) $\omega_i^m = \omega_j^m$, for any $m \neq k, l$, then $\omega_i \leq_f \omega_j$.

(HE) assigns absolute priority to the worse off agent in two-person cases, that is, cases where everyone but two agents attains the same utility value in both outcomes. For example, consider $\omega_4 = (0.3, 0.1, 0.7, 0.8)$ and $\omega_5 = (0.3, 0.1, 0.6, 0.9)$. If a justice relation $\leq_f$ satisfies **(HE)**, then ω_4 is at least as just as ω_5 with respect to f.

(b) Strong Pareto (SP): Let f be an aggregation function. For all $\omega_i, \omega_j \in \Omega$, if for all k, $\omega_i^k \leq \omega_j^k$ and there is l such that $\omega_i^l < \omega_j^l$, then $\omega_i <_f \omega_j$.

A justice relation $\leq_f$ satisfies Strong Pareto property when comparing two outcomes ω_i and ω_j, if the utility value of every agent is better or equal in ω_j than in ω_i and there is at least one agent the utility level is better in ω_j than in

ω_i, then ω_j is more just than ω_i with respect to a specific aggregation function f. For example, if $\leq_f$ satisfies **(SP)** then $\omega_6 = (0.3, 0.8, 0.5) <_f (0.3, 1, 0.6) = \omega_7$.

(c) Anonymity (A): Let f be an aggregation function. For all $\omega_i, \omega_j \in \Omega$, if ω_i is a permutation of ω_j, then $\omega_i \approx_f \omega_j$.

Anonymity is also called Symmetry or Commutativity. For instance, if $\leq_f$ satisfies **(A)**, then $\omega_6 = (0.3, 0.8, 0.5) \approx_f (0.3, 0.5, 0.8) \approx_f (0.5, 0.3, 0.8) \approx_f (0.5, 0.8, 0.3) \approx_f (0.8, 0.3, 0.5) \approx_f (0.8, 0.5, 0.3)$. With these three logical properties we can achieve the Leximin Principle.

Leximin (LM): Let f be an aggregation function. For all $\omega_i, \omega_j \in \Omega$, if there exists a position $k < n$ such that (1) $\omega_i^k < \omega_j^k$; and (2) $\omega_i^l = \omega_j^l$, for every $l < k$, then $\omega_i <_f \omega_j$. Otherwise, $\omega_i \approx_f \omega_j$.

(LM) captures the idea behind the *leximin* justice relation. Leximin Principle comes with an absolute preference to the least well off agent(s), meaning that the status of their least well off agent(s) are preferred, regardless of the utility values involved, and the rest of the agents are ignored.

Below, we have a characterization for the Leximin Principle.

Theorem 1 [11]. *A reflexive and transitive justice relation satisfies **(HE)**, **(SP)** and **(A)** if and only if it satisfies **(LM)**.*

For the justice relations presented in this section, it is known the following results.

Theorem 2. *The justice relation $\leq_{min}$ satisfies only **(A)** and $\leq_{leximin}$ satisfies **(HE)**, **(SP)** and **(A)**.*

The *min* justice relation falsifies **(HE)** and **(SP)**, which are typical properties associated to the egalitarian reasoning. A question that arises is if for other T-norms, besides *min*, they behave equivalently in terms of these logical properties. A distinction between T-norms and *leximin* (or *min*) is that not only the least well off agent(s) is/are considered, but every other agent in the population.

3   T-Norms and Parameterized T-Norms

In this section we will describe some notions about T-norms which will be employed in the remaining of this work. With regard to the *min* operation, it compares only the least value to take a decision. The minimum can also be viewed as the conjunction logic operator, i.e., $(a \wedge b) = min\{a, b\}$, and as a T-norm in the fuzzy logic literature.

Definition 2 (T-norm [8]). *A binary function $\otimes : [0,1] \times [0,1] \to [0,1]$ is a T-norm if it satisfies the following conditions: (i) $\otimes\{a, b\} = \otimes\{b, a\}$ (Commutativity); (ii) $\otimes\{a, \otimes\{b, c\}\} = \otimes\{\otimes\{a, b\}, c\}$ (Associativity); (iii) $a \leq c$ and $b \leq d \Rightarrow \otimes\{a, b\} \leq \otimes\{c, d\}$ (Monotonicity); and (iv) $\otimes\{a, 1\} = a$ (Neutral Element).*

Every T-norm has an absorbent element, also called annihilator, which is the natural number 0, i.e., $\otimes\{a, 0\} = 0$ (in this case, 0 can also be associated as an implicit veto). A T-norm is called strict if it is continuous and strictly monotone (i.e., $\forall x, y, z \otimes \{x, y\} < \otimes\{x, z\}$ whenever $x > 0$ and $y < z$). A T-norm is called nilpotent if it is continuous and if each $a \in]0, 1[$ is a nilpotent element of $\otimes$, i.e., there exists some $n \in \mathbb{N}$ such that $\otimes \underbrace{\{a, \ldots, a\}}_{n} = 0$. Besides, for all T-norm $\otimes$, we have $\otimes\{a, b\} \leq min\{a, b\}$ [5,7].

Definition 3 (Basic T-norms [6]). *The following are the four basic T-norms:*

- *Minimum T-norm:* $\otimes_M\{x, y\} = min(x, y);$
- *Product T-norm:* $\otimes_P\{x, y\} = x \cdot y;$
- *Łukasiewicz T-norm:* $\otimes_L\{x, y\} = max(x + y - 1, 0);$
- *Drastic T-norm:* $\otimes_D\{x, y\} = y$, *if* $x = 1$; x, *if* $y = 1$; 0, *otherwise.*

These four basic T-norms are remarkable for several reasons. The drastic T-norm $\otimes_D$ and the minimum $\otimes_M$ are the smallest and the largest T-norms, respectively (with respect to the pointwise order). The minimum $\otimes_M$ is the only T-norm where each $x \in [0, 1]$ is an idempotent element (recall $x \in [0, 1]$ is called an idempotent element of $\otimes$ if $\otimes\{x, x\} = x$). The product $\otimes_P$ and the Łukasiewicz T-norm $\otimes_L$ are examples of two important subclasses of T-norms, namely, the classes of strict and nilpotent T-norms, respectively (more details in [6]). Many families of T-norms can be defined by an explicit formula depending on a parameter λ. Let us give a quick overview of some of them.

Definition 4 (Schweizer-Sklar T-norms [9]). *The family of Schweizer-Sklar T-norms* $(\otimes_\lambda^{SS})_{\lambda \in [-\infty, \infty]}$ *is given by*

$$\otimes_\lambda^{SS}\{x, y\} = \begin{cases} \otimes_M\{x, y\}, & if \ \lambda = -\infty \\ \otimes_P\{x, y\}, & if \ \lambda = 0 \\ \otimes_D\{x, y\}, & if \ \lambda = \infty \\ (max((x^\lambda + y^\lambda - 1), 0))^{\frac{1}{\lambda}}, & otherwise \end{cases}$$

This family of T-norms is remarkable in the sense that it contains all four basic T-norms. When $\lambda = 1$, $\otimes_1^{SS} = \otimes_L$.

Definition 5 (Frank T-norms [3]). *The family of Frank T-norms* $(\otimes_\lambda^F)_{\lambda \in [0, \infty]}$ *is given by*

$$\otimes_\lambda^F\{x, y\} = \begin{cases} \otimes_M\{x, y\}, & if \ \lambda = 0 \\ \otimes_P\{x, y\}, & if \ \lambda = 1 \\ \otimes_L\{x, y\}, & if \ \lambda = \infty \\ log_\lambda(1 + \frac{(\lambda^x - 1)(\lambda^y - 1)}{\lambda - 1}), & otherwise \end{cases}$$

The Frank family comprehends a series of T-norms between the Łukasiewicz and the product T-norms (for $\lambda \in [2, \infty[$).

Definition 6 (Yager T-norms [14]). *The family of Yager T-norms* $(\otimes_\lambda^Y)_{\lambda \in [0,\infty]}$ *is given by*

$$\otimes_\lambda^Y\{x,y\} = \begin{cases} \otimes_D\{x,y\}, & \text{if } \lambda = 0 \\ \otimes_M\{x,y\}, & \text{if } \lambda = \infty \\ max(1 - ((1-x)^\lambda + (1-y)^\lambda)^{\frac{1}{\lambda}}, 0), & \text{otherwise} \end{cases}$$

It is one of the most popular families for modeling the intersection of fuzzy sets. The idea is to use the parameter λ as a reciprocal measure for the strength of the logical operator *"and"*. In this context, $\lambda = 1$ expresses the most demanding (i.e., smallest) *"and"*, and $\lambda = \infty$ the least demanding (i.e., largest) *"and"*. The Yager family comprehends a series of T-norms between the drastic and the minimum T-norms. When $\lambda = 1$, $\otimes_1^Y = \otimes_L$.

Definition 7 (Sugeno-Weber T-norms [13]). *The family of Sugeno-Weber T-norms* $(\otimes_\lambda^{SW})_{\lambda \in [-1,\infty]}$ *is given by*

$$\otimes_\lambda^{SW}\{x,y\} = \begin{cases} \otimes_D\{x,y\}, & \text{if } \lambda = -1 \\ \otimes_P\{x,y\}, & \text{if } \lambda = \infty \\ max(\frac{x+y-1+\lambda xy}{1+\lambda}, 0), & \text{otherwise} \end{cases}$$

Note that $(\otimes_\lambda^{SW})_{\lambda > -1}$ are increasing functions of the parameter λ.

4 T-Norms and the Leximin Principle

In section, we define a T-norm (basic or parameterized) justice relation as well as exploit its relation with the Leximin Principle.

Definition 8 ($\otimes$ justice relation). *Let $\otimes$ be a T-norm, $A = \{1,\ldots,n\}$ be a set of agents, $\Omega = \{\omega_1,\ldots,\omega_m\}$ be a set of outcomes, where each $\omega_i = (\omega_i^1,\ldots,\omega_i^n)$ and $\omega_i^j \in [0,1]$, for $j \in \{1,\ldots,n\}$. We define $\leq_\otimes$ over Ω as $\omega_i \leq_\otimes \omega_j$ iff $\otimes\{\omega_i^1,\ldots,\omega_i^n\} \leq \otimes\{\omega_j^1,\ldots,\omega_j^n\}$.*

T-norm justice relations may produce different results from *min* and *leximin* justice relations. For instance, let $\omega_1 = (0.2, 0.1, 0.6, 0.4)$ and $\omega_2 = (0.3, 0.1, 0.4, 0.3)$. It is true that $\omega_1 \approx_{min} \omega_2$, $\omega_1 <_{leximin} \omega_2$, but $\omega_2 <_{\otimes_P} \omega_1$, since $\otimes_P\{0.3, 0.1, 0.4, 0.3\} = 0.0036 < 0.0048 = \otimes_P\{0.2, 0.1, 0.6, 0.4\}$. In this context, the distinguishing aspect of a T-norm is that it gives preference for an outcome by considering all the utility values in it. On the other hand, *min* and *leximin* give preference to the worst cases of an outcome.

Observe that for any T-norm the presence of the annihilator 0 on the evaluation of ω_i works as an implicit veto for that outcome; for instance, if $\omega_8 = (0.2, 0.5, 0)$, then $\omega_8 \leq_\otimes \omega_j$, for $j \in \{1,\ldots,n\}$, since $\otimes\{0.2, 0.5, 0\} = 0$. So, if an outcome has the least utility value for an agent, that outcome will be the least preferred by the group when the T-norm is the aggregation function. It brings a principle of equality where the worst scenarios inside a group need to be avoided. In other words, the use of T-norms as a justice relation presupposes there exists a consensus among the agents stating that if a choice is the worst for an agent, then this choice has to be the worst for the group. Now, we will show which properties the justice relation with T-norms satisfy in the general case:

Theorem 3. *Let $\otimes$ be a T-norm. The justice relation $\leq_\otimes$ satisfies only (A) in general.*

This result is similar to that for the *min* justice relation $\leq_{min}$. An important concern when dealing with T-norms is the presence of the annihilator 0. The first property we need to revisit is **(SP)**. Recall that according to **(SP)**, if an outcome ω_j is strictly more preferable than an outcome ω_i for an agent l and ω_j is at least as preferable as ω_i for every other agents, then ω_j will be strictly more preferable than ω_i. Note that the presence of an annihilator is sufficient to falsify this property. Consider $\omega_1 = (0, 0.5)$ and $\omega_2 = (0, 0.6)$. We see that $\omega_1^1 = \omega_2^1 = 0$ and $\omega_1^2 < \omega_2^2$, but $\omega_1 \not<_\otimes \omega_2$. To overcome this issue, we will define a weaker version of **(SP)** which ignores the presence of the annihilator.

Definition 9 (Strong Pareto free from 0). *(SP-0) Let f be an aggregation function. For all $\omega_i, \omega_j \in \Omega$, if for all k, $\omega_i^k, \omega_j^k \neq 0$, $\omega_i^k \leq \omega_j^k$, and there is l such that $\omega_i^l < \omega_j^l$, then $\omega_i <_f \omega_j$.*

This weaker version of **(SP)** supports the idea the justice relation satisfies **(SP)** when the annihilator is absent. The second important condition we will reconsider is Hammond Equity [10].

Definition 10 (Hammond Equity Condition free from 0). *(HE-0) Let f be an aggregation function. For all $\omega_i, \omega_j \in \Omega$, there exist k, l such that: (1) $\omega_i^k < \omega_j^k$; (2) $\omega_j^l < \omega_i^l$; (3) $\omega_j^k \leq \omega_j^l$; (4) $\omega_i^m = \omega_j^m \neq 0$, for any $m \neq k, l$, then $\omega_i \leq_f \omega_j$.*

Now, let $\omega_4 = (0.3, 0, 0.7, 0.8)$ and $\omega_5 = (0.3, 0, 0.6, 0.9)$. For any T-norm $\otimes$, we have $\otimes\{0.3, 0, 0.7, 0.8\} = \otimes\{0.3, 0, 0.6, 0.9\} = 0$, which is enough to falsify **(HE)**. As in **(SP-0)**, **(HE-0)** considers **(HE)** when the annihilator is absent. We can continue with the analysis of logical properties for each specific T-norm:

Theorem 4. *$\leq_{\otimes_P}$ satisfies (SP-0), but does not satisfy (HE-0) in the general case. $\leq_{\otimes_M}$, $\leq_{\otimes_L}$ and $\leq_{\otimes_D}$ satisfy neither (SP-0) nor (HE-0) in the general case.*

One point to highlight is strict T-norms satisfies **(SP-0)** (e.g., $\otimes_P$ is a strict T-norm), whereas those nilpotent do not satisfy **(SP-0)**.

Theorem 5. *Let $\otimes$ be a strict T-norm, then $\leq_\otimes$ satisfies (SP-0).*

As the drastic T-norm is not continuous, little changes in the variables can change drastically the result, and this reflects the loss of **(SP-0)** and **(HE-0)**. The Łukasiewicz T-norm is a nilpotent T-norm; in this case, the presence of a nilpotent element leads to the loss of properties.

Theorem 6. *Let $\otimes$ be a nilpotent T-norm, then $\leq_\otimes$ satisfies neither (SP-0) nor (HE-0) in the general case.*

The nilpotent element works as a sort of implicit annihilator (that is, there exists some $n \in \mathbb{N}$ such that $\otimes \underbrace{\{a, \ldots, a\}}_{n} = 0$). Hence, it is not captured by the restrictions imposed to the annihilator in both (SP-0) and (HE-0). In the sequel, we will investigate deeper the behavior of some parameterized T-norms to show in what conditions we can achieve the properties discussed in this paper.

Theorem 7. $\leq_{\otimes_\lambda^{ss}}$ *satisfies (SP-0) when $\lambda \in \,] - \infty, 0]$. Let $\Omega = (\omega_1, \ldots, \omega_n)$. If $\omega_i^j \in \left[\frac{0}{n}, \frac{1}{n}, \ldots, \frac{n}{n}\right]$, for $j \in \{1, \ldots, n\}$, then $\leq_{\otimes_\lambda^{ss}}$ satisfies (HE-0) when $-\infty < \lambda \leq -\left\lfloor\frac{2n}{3}\right\rfloor$.*

The interval $[1, \infty]$ comprises strictly increasing T-norms from the Łukasiewicz T-norm ($\otimes_1^{SS}$) to the drastic T-norm ($\otimes_\infty^{SS}$). It is clear that all these conditions are falsified in this interval (since all T-norms in this interval are weaker than Łukasiewicz T-norm). Schweizer-Sklar T-norm is strict for the interval $\,] - \infty, 0]$, therefore its justice relation satisfies (SP-0) in this case. Considering this interval yet, we have any Schweizer-Sklar T-norm satisfies (HE-0) when $-\infty < \lambda \leq -\left\lfloor\frac{2n}{3}\right\rfloor$, where $\omega_i^j \in \left[\frac{0}{n}, \frac{1}{n}, \ldots, \frac{n}{n}\right]$, for $j \in \{1, \ldots, n\}$. For instance, if $\omega_i^j \in \left[0, \frac{1}{3}, \frac{2}{3}, 1\right]$, then $\leq_{\otimes_\lambda^{ss}}$ satisfies (HE-0), for any $\lambda \leq -2$.

Theorem 8. $\leq_{\otimes_\lambda^F}$ *satisfies (SP-0) for $\lambda \in \,]0, \infty[$. If $\omega_i^j \in \left[\frac{0}{n}, \frac{1}{n}, \ldots, \frac{n}{n}\right]$, $\leq_{\otimes_\lambda^F}$ satisfies (HE-0) when $0 < \lambda \leq 10^{-n}$.*

We showed previously that converging to the minimum T-norm tends to satisfy (HE-0). The limit of $0 < \lambda \leq 10^{-n}$ is rather loose, but it is a statement that there is an interval between minimum and product in the Frank T-norms where (HE-0) is satisfied. In the sequel, we will see the Yager and Sugeno-Weber families of T-norms.

Theorem 9. $\leq_{\otimes_\lambda^Y}$ *and $\leq_{\otimes_\lambda^{sw}}$ do not satisfy (SP-0) and (HE-0) in the general case.*

Yager T-norms comprise from drastic T-norm ($\otimes_0^Y$), passing through Łukasiewicz T-norm ($\otimes_1^Y$), to minimum T-norm ($\otimes_\infty^Y$). Unlike the previous parameterized T-norms, Yager T-norms are nilpotent for $\lambda \in]0, \infty[$. As consequence, they satisfy neither (SP-0) nor (HE-0). Sugeno-Weber T-norms are another class of nilpotent T-norms, which range from drastic T-norm ($\otimes_\infty^{SW}$) to Łukasiewicz ($\otimes_0^{SW}$) and product T-norms ($\otimes_{-1}^{SW}$). Thus, for the same reasons mentioned above, $\leq_{\otimes_\lambda^{sw}}$ does not satisfy (SP-0) and (HE-0).

Now we will use the results of [11] to characterize an egalitarian property of some parameterized T-norms with respect to the Leximin principle. It is known every T-norm $\otimes \leq min$, and min justice relation does not satisfy properties as (HE) and (SP); nonetheless some T-norms can satisfy weakened versions of them.

This analysis shows some T-norms present a similar (weaker) behavior to the *leximin* justice relation. What we want to achieve is that those T-norms can also follow some weaker versions of the Leximin principle. We introduce a restriction to the Leximin principle, named Leximin principle free from annihilator 0.

Definition 11 (Leximin free from 0). *(LM-0) Let f be an aggregation function. For all $\omega_i, \omega_j \in \Omega$, if there exists a position $k < n$ such that (1) $\omega_i^k < \omega_j^k$; and (2) $\omega_i^l = \omega_j^l \neq 0$, for every $l < k$, then $\omega_i <_f \omega_j$. Otherwise, $\omega_i \approx_f \omega_j$.*

Thus, we have the following results.

Corollary 1. *A reflexive and transitive justice relation satisfies (HE-0), (SP-0) and (A) if and only if it satisfies (LM-0).*

This characterization restricts the Leximin principle when the annihilator is excluded from the possible utility values of the agents. As a consequence, we have

Corollary 2. *If $\omega_i^j \in \left[\frac{0}{n}, \frac{1}{n}, \ldots, \frac{n}{n}\right]$, for $j \in \{1, \ldots, n\}$, then $\leq_{\otimes_\lambda^{ss}}$ satisfies (LM-0) when $-\infty < \lambda \leq -\left\lfloor\frac{2n}{3}\right\rfloor$; and $\leq_{\otimes_\lambda^F}$ satisfies (LM-0) when $0 < \lambda \leq 10^{-n}$.*

These justice relations satisfy **(HE-0)**, **(SP-0)** and **(A)** in these specific intervals and utility values. In other words, when the annihilator is not present in the aggregation, we can say that these T-norms have a behavior similar to the *leximin* justice relation. For the other T-norms considered in this paper, **(LM-0)** is not satisfied.

5   LexiT-Ordering Justice Relations

In this section we will propose a refinement for T-norm justice relations. It is the same idea behind the *leximin* refinement of *minimum* aggregation function. It is called LexiT-ordering and it was introduced in [15]. We will introduce the LexiT-ordering justice relation and we will show that for some specific LexiT-orderings and intervals, their behavior are equivalent to the *leximin* justice relation. In other words, LexiT-orderings do not have the same issues with the annihilator as T-norms have.

Definition 12 (LexiT-norm [15]). *Let $a = (a_1, a_2, \ldots, a_n) \in [0, 1]^n$ and let $\otimes$ be a T-norm. Let P_a be the power set of $\{a_1, a_2, \ldots, a_n\}$ excluding the empty set, that is, the set of all subsets of the indexed set $\{a_1, a_2, \ldots, a_n\}$ minus $\emptyset$ (the empty set). For any $A \in P_a$, we let $\otimes(A)$ indicate the T-norm of the elements of A. Let $\bar{a} = (\bar{a}_1, \bar{a}_2, \ldots, a_{2^n-1})$ be the $(2^n - 1)$-tuple of the family $\{\otimes(A) : A \in P_a\}$ put into ascending order. On $[0, 1]^{2^n - 1}$ we have the lexicographic ordering $\leq_{lex}$ which is a linear ordering. The binary relation $a \leq_{Lexi\otimes} b$ is defined as follows:*

- *For $a, b \in [0, 1]^n$, use $\otimes$ to construct $\bar{a}$ and $\bar{b} \in [0, 1]^{2^n - 1}$. Then $a \leq_{Lexi\otimes} b$ if and only if $\bar{a} \leq_{lex} \bar{b}$.*

In other words,

- $a <_{Lexi\otimes} b$ *if and only if there exists* $k \geq 1$ *such that* $\bar{a}_k < \bar{b}_k$ *and for* $1 \leq i < k$, $\bar{a}_i = \bar{b}_i$;
- $a \approx_{Lexi\otimes} b$ *if and only if* $\bar{a}_i = \bar{b}_i$ *for all* $i = 1, 2, \ldots, 2^n - 1$.

Here we show a simple example: take the product T-norm $\otimes_{\mathbf{P}}\{x, y\} = x \cdot y$. Let $a = (0.2, 0.8)$ and $b = (0.3, 0.5)$. In this case, both P_a and P_b have 3 elements: $P_a = (\{0.2\}, \{0.8\}, \{0.2, 0.8\})$ and $P_b = (\{0.3\}, \{0.5\}, \{0.3, 0.5\})$. After calculating $\otimes\{A\}$ and $\otimes\{B\}$ for each $A \in P_a$ and $B \in P_b$ we get $\bar{a} = (0.16, 0.2, 0.8)$ and $\bar{b} = (0.15, 0.3, 0.5)$. Now, comparing $\bar{a}$ and $\bar{b}$ lexicographically, we obtain $\bar{b} \leq_{lex} \bar{a}$, and consequently $b \leq_{Lexi\otimes_{\mathbf{P}}} a$.

Before proceeding with the application of LexiT-orderings as a justice relation, we emphasize the following property about LexiT-orderings:

Theorem 10 *[15]. Let* $\otimes_{\mathbf{M}}$ *be the Minimum T-norm. Then for* $a, b \in [0, 1]^n$, $a \leq_{leximin} b$ *if and only if* $a \leq_{Lexi\otimes_M} b$.

That is, if $\otimes$ is the Minimum T-norm, Leximin and Lexi$\otimes$ are the same ordering. Below, we will introduce a LexiT-ordering justice relation, based on the definition of LexiT-ordering.

Definition 13 (Lexi$\otimes$ justice relation). *Let* $\otimes$ *be a T-norm,* $A = \{1, \ldots, n\}$ *be a set of agents,* $\Omega = \{\omega_1, \ldots, \omega_m\}$ *be a set of outcomes, where each* $\omega_i = (\omega_i^1, \ldots, \omega_i^n)$ *and* $\omega_i^j \in [0, 1]$, *for* $j \in \{1, \ldots, n\}$. *Let* $\bar{\omega}_i = (\omega_i^1, \ldots, \omega_i^{2^n-1})$ *be the* $(2^n - 1)$*-tuple of the family* $\{\otimes\{S\} : S \in P_{\omega_i}\}$ *put in ascending order. Let* $\leq_{lex}$ *be the lexicographical order between sequences of numbers. We define* $\leq_{Lexi\otimes}$ *over* Ω *as* $\omega_i \leq_{Lexi\otimes} \omega_j$ *iff* $\bar{\omega}_i \leq_{lex} \bar{\omega}_j$.

For strict T-norms, computing the LexiT-ordering can be done in a simpler way.

Theorem 11 *[12]. If a T-norm* $\otimes$ *is strict, it takes at most* n *steps to determine whether or not* $(a_1, a_2, \ldots, a_n) <_{Lexi\otimes} (b_1, b_2, \ldots, b_n)$.

From this Theorem, we can calculate $(a_1, a_2, \ldots, a_n) <_{Lexi\otimes} (b_1, b_2, \ldots, b_n)$ as follows: let $\bar{a} = (\bar{a}_1, \bar{a}_2, \ldots, \bar{a}_n)$ and $\bar{b} = (\bar{b}_1, \bar{b}_2, \ldots, \bar{b}_n)$ be the n-tuples of $(a_1, a_2, \ldots, a_n)$ and $(b_1, b_2, \ldots, b_n)$ put into ascending order, and $(a_1, a_2, \ldots, a_n) <_{Lexi\otimes} (b_1, b_2, \ldots, b_n)$ iff $(\otimes\{\bar{a}_1, \bar{a}_2, \ldots, \bar{a}_n\}, \otimes\{\bar{a}_1, \bar{a}_2, \ldots, a_{n-1}\}, \ldots, \otimes\{\bar{a}_1, \bar{a}_2\}, \bar{a}_1) <_{lex} (\otimes\{\bar{b}_1, \bar{b}_2, \ldots, \bar{b}_n\}, \otimes\{\bar{b}_1, \bar{b}_2, \ldots, \bar{b}_{n-1}\}, \ldots, \otimes\{\bar{b}_1, \bar{b}_2\}, \bar{b}_1)$. It is now possible to simplify the definition of a strict Lexi$\otimes$ justice relation.

Definition 14 (Strict Lexi$\otimes$ justice relation). *Let* $\otimes$ *be a strict T-norm,* $A = \{1, \ldots, n\}$ *be a set of agents,* $\Omega = \{\omega_1, \ldots, \omega_m\}$ *be a set of outcomes, where each* $\omega_i = (\omega_i^1, \ldots, \omega_i^n)$ *and* $\omega_i^j \in [0, 1]$, *for* $j \in \{1, \ldots, n\}$. *Let* $\bar{\omega}_i = (\omega_i^1, \ldots, \omega_i^n)$ *be the* n*-tuple put into ascending order. Let* $\leq_{lex}$ *be the lexicographical order. We define* $\leq_{Lexi\otimes}$ *over* Ω *as* $\omega_i \leq_{Lexi\otimes} \omega_j$ *iff* $(\otimes\{\omega_i^1, \omega_i^2, \ldots, \omega_i^n\}, \otimes\{\omega_i^1, \omega_i^2, \ldots, \omega_i^{n-1}\}, \ldots, \otimes\{\omega_i^1, \omega_i^2\}, \omega_i^1) <_{lex} (\otimes\{\omega_j^1, \omega_j^2, \ldots, \omega_j^n\}, \otimes\{\omega_j^1, \omega_j^2, \ldots, \omega_j^{n-1}\}, \ldots, \otimes\{\omega_j^1, \omega_j^2\}, \omega_j^1)$.

For instance, consider $\omega = (0.6, 1, 0.6)$ and $\otimes_{\mathbf{P}}$, then the vector $\bar{\omega} = (0.36, 0.36,\ 0.6, 0.6, 0.6, 0.6, 1)$ can now be computed as $\bar{\bar{\omega}} = (\otimes_{\mathbf{P}}\{0.6, 0.6, 1\}, \otimes_{\mathbf{P}}\{0.6, 0.6\}, 0.6) = (0.36, 0.36, 0.6)$. In terms of results, both forms of computation are equivalent. Another example to compare the differences among the justice relations: let $\omega_1 = (0.2, 0.1, 0.6,\ 0.4, 0.6)$ and $\omega_2 = (0.3, 0.1, 0.4, 0.3, 0.8)$ be outcomes. For the *min* justice relation, $\omega_1 \approx_{min} \omega_2$. For $\otimes_{\mathbf{P}}$ justice relation, $\omega_1 \approx_{\otimes_{\mathbf{P}}} \omega_2$, because $\otimes_{\mathbf{P}}\{0.2, 0.1, 0.6, 0.4, 0.6\} = \otimes_{\mathbf{P}}\{0.3, 0.1, 0.4, 0.3, 0.8\} = 0.00288$. The scenario is different for *leximin* justice relation, i.e., $\omega_1 <_{leximin} \omega_2$, since $(0.1, 0.2, 0.4, 0.6, 0.6) <_{lex} (0.1, 0.3, 0.3, 0.4, 0.8)$. For Lexi$\otimes_{\mathbf{P}}$, the result is even different from all of them: $\omega_2 <_{leximin} \omega_1$, since $(0.00288, 0.0036, 0.009, 0.03, 0.1) <_{lex} (0.00288, 0.0048, 0.008, 0.02, 0.1)$.

As the class of strict T-norms presented previously may have a (weak) similar behavior to the *leximin* justice, we will see in the sequel that some LexiT-orderings may be equivalent to *leximin*.

Theorem 12. *Let $\otimes$ be a T-norm and $\leq_{\otimes}$ a $\otimes$ justice relation. We have the following results:*

- *If $\leq_{\otimes}$ satisfies (HE-0). then $\leq_{Lexi\otimes}$ satisfies (HE);*
- *If $\leq_{\otimes}$ satisfies (SP-0). then $\leq_{Lexi\otimes}$ satisfies (SP).*

From Theorems 1, 7, 8 and 12 we have

Corollary 3. *If $\omega_i^j \in \left[\frac{0}{n}, \frac{1}{n}, \ldots, \frac{n}{n}\right]$, for $j \in \{1, \ldots, n\}$, then $\leq_{\otimes_\lambda^{ss}}$ and $\leq_{\otimes_\lambda^F}$ satisfy (LM) when $\lambda \leq -\left\lfloor \frac{2n}{3} \right\rfloor$ and $0 < \lambda \leq 10^{-n}$, respectively.*

These results guarantee it is possible some parameterized LexiT-ordering justice relations behave as the *leximin* justice relation in some specific intervals and utility values. Finally, for the pre-orders described in this paper, we have the following results accounting their discriminating power.

Theorem 13. *Let $\otimes$ be a T-norm and $\otimes^*$ be a parameterized T-norm such that $\leq_{\otimes^*}$ satisfies (LM-0). We have the following results, for any $\omega_i, \omega_j \in \Omega$:*

1. *$\omega_i \leq_{leximin} \omega_i \Rightarrow \omega_i \leq_{min} \omega_j$;*
2. *$\omega_i \leq_{leximin} \omega_i \Rightarrow \omega_i \leq_{\otimes^*} \omega_j$;*
3. *$\omega_i \leq_{leximin} \omega_i \Leftrightarrow \omega_i \leq_{Lexi\otimes^*} \omega_j$;*
4. *$\omega_i \leq_{Lexi\otimes} \omega_i \Rightarrow \omega_i \leq_{\otimes} \omega_j$.*

This Theorem states that solutions being indifferent for an order could be distinguished by a higher discriminating power one. For instance, by item 1. if $\omega_i \approx_{min} \omega_j$, then it could be possible that $\omega_i <_{leximin} \omega_j$ or $\omega_j <_{leximin} \omega_i$. The same idea happens for T-norms that satisfy (LM-0) and *lexmin* in item 2; and for T-norms and LexiT-orderings in item 4. In item 3, it is stated that *leximin* and LexiT-orderings that satisfy (LM) have the same discriminating power.

6   Conclusions

In this paper, we proposed to use T-norms operators as aggregation functions. T-norms are a generalization of the usual two-valued logical conjunction, i.e., the minimum operator. In the aggregation context, the minimum (min) operator is equivalent to the $maximin$ rule in the decision theory: it tries to maximize the worst cases among the agents. Indeed, T-norms allow us to diversify the method of the $maximin$ rule by applying generalized versions of the min operator.

The purpose of this work is to explore the logical properties of T-norm aggregation functions in the context of egalitarian reasoning. We considered the logical properties Hammond Equity, Strong Pareto and Anonymity. Together these properties are responsible to characterize the Leximin (Lexicographic Minimum) Principle, which is a principle that gives absolute priority to worst cases in a group. As a Leximin is a generalization of min, it is natural also to see if the T-norms are compatible with these logical properties.

In terms of egalitarianism, T-norms differs from min and $leximin$ as they do not give preference only to the worst cases in a group, but everyone in group has a relevance in the aggregation. In terms of logical properties, T-norms are weaker than $leximin$, since they do not satisfy Hammond Equity (**HE**) and Strong Pareto (**SP**). The reason for this, it is the presence of an absorbent element in T-norms, also called annihilator. However, we proved that if we weaken Hammond Equity (resulting in (**HE-0**)) and Strong Pareto (resulting in (**SP-0**)), without the presence of the annihilator, some families of T-norms are compatible with it.

We chose in this paper some of the most representative classes of T-norms. First, we analyzed the four basic T-norms: drastic, Lukasiewicz, product and minimum T-norms. The highest T-norm minimum satisfies neither (**HE-0**) nor (**SP-0**). The product falsifies (**HE-0**), but satisfies (**SP-0**). Łukasiewicz and drastic falsify all of them.

When analyzing parameterized T-norms, which are basically generalizations of some of the four basic T-norms, we observed strict T-norms converging to minimum tend to satisfy (**HE-0**), as it is the case of the Schweizer-Sklar and Frank T-norms. The same idea does not follow from nilpotent T-norms, since they do not satisfy (**HE-0**) in the general case. For the logical property (**SP-0**), we proved that it is satisfied by the family of strict T-norms, while it is not the case for nilpotent T-norms.

Finally, we generalized further the T-norms into LexiT-orderings. With this generalization it is possible to solve the problem of the annihilator for those T-norms which satisfy (**HE-0**) and (**SP-0**). Their corresponding Lexicographic version satisfy (**HE**) and (**SP**), respectively. With this result, we achieved that in specific situations, some LexiT-ordering aggregation functions are equivalent to $leximin$ aggregation function. Lastly, we presented a comparison of discriminating power of the justice relations investigated in this work.

References

1. Arrow, K., Sen, A., Suzumura, K. (eds.): Handbook of Social Choice and Welfare, 1st edn. Elsevier, Amsterdam (2002)
2. Beliakov, G., Pradera, A., Calvo, T.: Aggregation Functions: A Guide for Practitioners. Studies in Fuzziness and Soft Computing. Springer, Heidelberg (2009). https://doi.org/10.1007/978-3-540-73721-6. https://books.google.com.br/books?id=ztIAvgAACAAJ
3. Butnariu, D., Klement, E.P.: Triangular Norm-Based Measures and Games with Fuzzy Coalitions, vol. 10. Springer, Dordrecht (1993). https://doi.org/10.1007/978-94-017-3602-2
4. Detyniecki, M.: Fundamentals on aggregation operators. This manuscript is based on Detynieckis doctoral thesis (2001). http://www.cs.berkeley.edu/~marcin/agop.pdf
5. Detyniecki, M., Yager, R.R., Bouchon-Meunier, B.: Reducing t-norms and augmenting t-conorms. Int. J. Gen Syst **31**(3), 265–276 (2002)
6. Klement, E.P., Mesiar, R.: Logical, Algebraic, Analytic and Probabilistic Aspects of Triangular Norms. Elsevier Science B.V, Amsterdam (2005)
7. Klement, E.P., Mesiar, R., Pap, E.: On the order of triangular norms: comments on "A triangular norm hierarchy" by E. Cretu. Fuzzy Sets Syst. **131**(3), 409–413 (2002)
8. Klement, E.P., Pap, E., Mesiar, R.: Triangular norms. Trends in logic. Kluwer Academic Publ. cop., Dordrecht, Boston, London (2000). http://opac.inria.fr/record=b1104736
9. Schweizer, B., Sklar, A.: Associative functions and statistical triangle inequalities. Publ. Math. **8**, 169–186 (1961)
10. Sen, A.K.: Choice, Welfare and Measurement. Harvard University Press, Cambridge (1997)
11. Tungodden, B.: Egalitarianism: Is leximin the only option? Working papers, Norwegian School of Economics and Business Administration- (1999). http://EconPapers.repec.org/RePEc:fth:norgee:4/99
12. Walker, C., Walker, E., Yager, R.: Some comments on lexit orderings for strict t-norms. In: The 14th IEEE International Conference on Fuzzy Systems, FUZZ 2005, pp. 669–671, May 2005
13. Weber, S.: A general concept of fuzzy connectives, negations and implications based on t-norms and t-conorms. Fuzzy Sets Syst. **11**(1–3), 103–113 (1983)
14. Yager, R.R.: On a general class of fuzzy connectives. Fuzzy Sets Syst. **4**(3), 235–242 (1980)
15. Yager, R.R., Walker, C.L., Walker, E.A.: Generalizing Leximin to t-norms and t-conorms: the LexiT and LexiS orderings. Fuzzy Sets Syst. **151**(2), 327–340 (2005). http://www.sciencedirect.com/science/article/pii/S0165011404001824

Fuzzy Formal Concept Analysis

Abner Brito[1,2,3]($\boxtimes$), Laécio Barros[1], Estevão Laureano[1], Fábio Bertato[2],
and Marcelo Coniglio[2,3]

[1] Department of Applied Mathematics, State University of Campinas,
Campinas, Brazil
abner.demattosbrito@gmail.com
[2] Centre for Logic, Epistemology and the History of Science,
State University of Campinas, Campinas, Brazil
[3] Department of Philosophy, State University of Campinas, Campinas, Brazil

Abstract. Formal Context Analysis is a mathematical theory that
enables us to find concepts from a given set of objects, a set of attributes
and a relation on them. There is a hierarchy of such concepts, from
which a complete lattice can be made. In this paper we present a generalization of these ideas using fuzzy subsets and fuzzy implications defined
from lower semicontinuous t-norms which, under suitable conditions, also
results in a complete lattice.

Keywords: Formal Concept Analysis · FCA
Fuzzy Formal Concept Analysis · Fuzzy attributes · Concept lattice
Fuzzy concept lattice

1 Introduction

Formal Concept Analysis (FCA) constitutes a powerful tool for acquisition and
representation of knowledge as well as for conceptual data analysis, based on
notions from general lattice theory. In FCA, data is represented as a conceptual
hierarchy, organized as a concept lattice that relates objects and their properties
(see Ganter and Wille 1999). FCA has applications in several fields.

According to Hardy-Vallée, a concept is "a general knowledge that [...] represents a category of objects, events or situations."[1] For example, the concept "Library" represents each individual library. One such "general knowledge"
("Library") abstracts attributes (e.g. having a *catalogue* of its *books*) common
to all objects (libraries). A concept in FCA is defined by a set of objects and a
corresponding set of attributes.

Nevertheless, real life knowledge is seldom precise. For instance, an automo-
bile manufacturer may construct "concepts" that relate car features (objects)
and consumer profiles (attributes). These concepts would be useful if they are

---

[1] "une connaissance générale qui [...] représente une catégorie d'objets, d'événements
ou de situations." See https://www.researchgate.net/profile/Benoit_Hardy-Vallee/
publication/228799196.

© Springer International Publishing AG, part of Springer Nature 2018
G. A. Barreto and R. Coelho (Eds.): NAFIPS 2018, CCIS 831, pp. 192–205, 2018.
https://doi.org/10.1007/978-3-319-95312-0_17

interested for example in selling more cars to young women. Because "young-ness" is a linguistic imprecise idea, a good strategy would be to use fuzzy FCA.

Fuzzy FCA has been previously proposed in the literature and there exists a vast body of literature on the field, a fair portion of which has been surveyed by Poelmans et al. (2014). Most authors, such as Belohlávek and Vychodil (2007), consider left continuous t-norms in order to define fuzzy implications underlying the fuzzy FCA notions. In this paper we propose the use of fuzzy implications defined from lower semicontinuous t-norms. It should be observed that, due to monotonicity (and commutativity), both notions of continuity are equivalent for t-norms. However, the techniques used in the proofs are different.

2   Formal Concept Analysis

The definitions and theorems presented in this section follow those presented by Ganter and Wille (1999) with a different notation and slightly different proofs.

Definition 1. *A formal context is an ordered triple* $\mathbb{C} := \langle \mathcal{O}, \mathcal{A}, I \rangle$, *in which* $\mathcal{O}$ *and* $\mathcal{A}$ *are non-empty sets, and* $I \subseteq \mathcal{O} \times \mathcal{A}$ *is a binary relation.*

The elements of $\mathcal{O}$ are called *objects*, and the elements of $\mathcal{A}$ *attributes*. We say that, in the context $\mathbb{C}$, object o *has attribute* a iff oIa. A finite context can be represented as a table, indexing rows by objects and columns by attributes, and marking cell (o, a) iff oIa, as in Table 1.

Table 1. A formal context of animals

$\mathcal{O}$	$\mathcal{A}$						
	Vertebrate	Lay eggs	Carnivorous	Has wings	Flies	Quadruped	Crawls
Eagle	X	X	X	X	X		
Snake	X	X	X				X
Goose	X	X		X	X		
Swan	X	X		X	X		
Lion	X		X			X	

Definition 2. *Let* $\mathbb{C} = \langle \mathcal{O}, \mathcal{A}, I \rangle$ *be a formal context. We define two maps,* $* : 2^{\mathcal{O}} \to 2^{\mathcal{A}}$ *and* $^\wedge : 2^{\mathcal{A}} \to 2^{\mathcal{O}}$ *(and write O^* and $A^\wedge$) as follows:*

$$O^* := \{a \in \mathcal{A} : oIa \ for \ all \ o \in O\} \tag{1}$$

$$A^\wedge := \{o \in \mathcal{O} : oIa \ for \ all \ a \in A\}. \tag{2}$$

Finally, we define the central object of FCA:

Definition 3. *Let* $\mathbb{C} = \langle \mathcal{O}, \mathcal{A}, I \rangle$ *be a formal context. A formal concept (or simply* concept*) of* $\mathbb{C}$ *is an ordered pair* $C = \langle O, A \rangle$ *such that* $O \subseteq \mathcal{O}$, $A \subseteq \mathcal{A}$, $O^* = A$ *and* $A^\wedge = O$. *The sets* O *and* A *are called the* extent *and* intent *of the concept* C *respectively.*

Thus, $C = \langle O, A \rangle$ is a concept when O is precisely the subset of objects that has all the attributes of A in common.

Example 4. The following are formal concepts of the context presented in Table 1:

$$C_{\text{Lion}} = \langle \{\text{Lion}\}, \{\text{Quadruped, Carnivorous, Vertebrate}\}\rangle, \qquad (3)$$

$$C_{\text{Carnivorous}} = \langle \{\text{Eagle, Snake, Lion}\}, \{\text{Carnivorous, Vertebrate}\}\rangle. \qquad (4)$$

Notice that there is an inverse relation between the numbers of elements in the intent and the extent of a concept. If we increase the number of elements in the extent (we added "Eagle" and "Snake" to it), the number of elements in the intent is reduced ("Quadrupede" is not in the intent of $C_{\text{Carnivorous}}$). In fact, the following useful properties hold:

Theorem 5. *Let* $O, O_1, O_2 \subseteq \mathcal{O}$ *and* $A, A_1, A_2 \subseteq \mathcal{A}$. *Then*

1. *If* $O_1 \subseteq O_2$ *then* $O_2^* \subseteq O_1^*$ 1'. *If* $A_1 \subseteq A_2$ *then* $A_2^\wedge \subseteq A_1^\wedge$
2. $O \subseteq O^{*\wedge}$ 2'. $A \subseteq A^{\wedge *}$
3. $O^* = O^{*\wedge *}$ 3'. $A^\wedge = A^{\wedge * \wedge}$
   4. $O \subseteq A^\wedge$ *iff* $A \subseteq O^*$ *iff* $O \times A \subseteq I$

Proof. We shall prove items 1., 2., 3. and 4.. Items with a prime can be proved analogously.

1. Let $a \in O_2^*$. Then oIa for all $o \in O_2$. In particular, oIa for all $o \in O_1$. Thus, $a \in O_1^*$.
2. Let $o \in O$. Then oIa for all $a \in O^*$, by definition of O^*. Thus, by definition of $O^{*\wedge}$, we have $o \in O^{*\wedge}$.
3. From 2'. with $A = O^*$ we already know that $O^* \subseteq O^{*\wedge *}$. Let $a \in O^{*\wedge *}$. Then

$$\text{(i) } oIa \text{ for all } o \in O^{*\wedge},$$

by definition of $O^{*\wedge *}$. Now let $\tilde{o} \in O^{*\wedge}$ be fixed. For all $\tilde{a} \in \mathcal{A}$,

$$\text{(ii) if } \tilde{o}I\tilde{a} \text{ then } \tilde{a} \in O^*,$$

by definition of $O^{*\wedge}$. From (i) we have $\tilde{o}Ia$. Thus, using (ii) we conclude that $a \in O^*$.
4. Suppose $O \subseteq A^\wedge$. By 1., $A^{\wedge *} \subseteq O^*$. Using 2'. and transitivity of $\subseteq$, we have $A \subseteq O^*$.
   Now suppose $A \subseteq O^*$. By 1'. and 2., we have $O \subseteq O^{*\wedge} \subseteq A^\wedge$.
   Assuming $A \subseteq O^*$, let $o \in O$ and $a \in A$. By definition of O^*, $oI\tilde{a}$ for all $\tilde{a} \in O^*$. By hypothesis, $a \in A \subseteq O^*$. Thus, oIa. Since $o \in O$ and $a \in A$ are arbitrary, $O \times A \subseteq I$. Hence, $A \subseteq O^* \Rightarrow O \times A \subseteq I$.

Finally, suppose $O \times A \subseteq I$. Let $o \in O$. By hypothesis, for all $a \in A$ we have oIa. By definition of O^*, if oIa then $a \in O^*$. Thus if $a \in A$ then $a \in O^*$. This completes the proof.

From properties 3. and 3'. we see that, given $O \subseteq \mathcal{O}$ and $A \subseteq \mathcal{A}$, $\langle O^{*\wedge}, O^* \rangle$ and $\langle A^{\wedge}, A^{\wedge *} \rangle$ are concepts. On the other hand if $C = \langle O, A \rangle$ is a concept then by definition

$$C = \langle A^{\wedge}, A \rangle = \langle (O^*)^{\wedge}, O^* \rangle.$$

Consequently every concept has either form $\langle O^{*\wedge}, O^* \rangle$ or $\langle A^{\wedge}, A^{\wedge *} \rangle$, where $O \subseteq \mathcal{O}$ and $A \subseteq \mathcal{A}$ are arbitrary.

This gives us a procedure for finding concepts. Choose a subset of objects (or attributes), apply $*$ to get an intent, and then apply $\wedge$ to get an extent (in fact, the smallest extent containing the original object subset). To find all the concepts of a given context, simply list all the subsets of objects (or attributes), and then apply the maps $*$ and $\wedge$ respectively (or $\wedge$ and $*$).

Example 6. Notice that

$$\{\text{Eagle}\}^{*\wedge} = \{\text{Vertebrate, Lay eggs, Carnivorous, Has wings, Flies}\}^{\wedge} = \{\text{Eagle}\}.$$

Thus, the following is a concept:

$$C_{\text{Eagle}} = \langle \{\text{Eagle}\}, \{\text{Flies, Has Wings, Lay eggs, Carnivorous, Vertebrate}\} \rangle.$$
$$(5)$$

There are also interesting properties regarding intersections of extents and intents. For instance, the intersection of the intents of (3) and (5) is the intent of (4). This is a particular case of a more general fact, presented in the following proposition.

Proposition 7. *Let J be an index set and, for each $\alpha \in J$, let $O_\alpha \subseteq \mathcal{O}$ and let $A_\alpha \subseteq \mathcal{A}$. Then*

$$\bigcap_{\alpha \in J} O_\alpha^* = \left(\bigcup_{\alpha \in J} O_\alpha \right)^*, \tag{6}$$

$$\bigcap_{\alpha \in J} A_\alpha^{\wedge} = \left(\bigcup_{\alpha \in J} A_\alpha \right)^{\wedge}. \tag{7}$$

Proof. We shall prove only (6). The proof of (7) is analogous.

$$a \in \bigcap_{\alpha \in J} O_\alpha^* \text{ iff } a \in O_\alpha^* \text{ for all } \alpha \in J$$

$$\text{iff } oIa \text{ for all } o \in O_\alpha, \text{ for all } \alpha \in J$$

$$\text{iff } oIa \text{ for all } o \in \bigcup_{\alpha \in J} O_\alpha$$

$$\text{iff } a \in \left(\bigcup_{\alpha \in J} O_\alpha \right)^*.$$

Given two sets $Y \subseteq X$ and a partial order $\leq$ on X, the *infimum* of Y (if it exists) is the element i_Y such that $i_y \leq y$ for any $y \in Y$ and if $x \leq y$ for all $y \in Y$ then $x \leq i_y$. Replacing $\leq$ by $\geq$ and i_Y by s_Y we get the definition of the supremum s_Y of Y. $\mathcal{L} = \langle X, \leq \rangle$ is called a *lattice* if every finite subset of X has both an infimum and a supremum. If every subset of X has an infimum and a supremum, then $\mathcal{L}$ is said to be *complete*.

From Proposition 7 we can show that, with an order induced by set inclusion, the set of all formal concepts of any context constitutes a complete lattice.

Theorem 8. *Let $\mathbb{C}$ be a formal context. Let $\mathfrak{B}(\mathbb{C})$ be the set of all concepts of $\mathbb{C}$. Define the relation $\leq$ on $\mathfrak{B}(\mathbb{C})^2$ by $\langle O_1, A_1 \rangle \leq \langle O_2, A_2 \rangle$ iff $O_1 \subseteq O_2$. Then $\leq$ is an order on $\mathfrak{B}(\mathbb{C})$. If $C_1 \leq C_2$ we say that C_1 is a subconcept of C_2. Correspondingly, C_2 is a* superconcept *of C_1. Furthermore, $\mathcal{L}_{\mathbb{C}} := \langle \mathfrak{B}(\mathbb{C}), \leq \rangle$ is a complete lattice, called the* concept lattice *of $\mathbb{C}$.*

If J is an index set and $C_\alpha = \langle O_\alpha, A_\alpha \rangle \in \mathfrak{B}(\mathbb{C})$ for each $\alpha \in J$ then

$$\inf_{\alpha \in J} C_\alpha = \left\langle \bigcap_{\alpha \in J} O_\alpha, \left(\bigcap_{\alpha \in J} O_\alpha \right)^* \right\rangle \tag{8}$$

$$\sup_{\alpha \in J} C_\alpha = \left\langle \left(\bigcap_{\alpha \in J} A_\alpha \right)^{\wedge}, \bigcap_{\alpha \in J} A_\alpha \right\rangle. \tag{9}$$

Proof. That $\leq$ is an order is clear from the fact that $\subseteq$ is an order. Now we prove (9). For each $\alpha \in J$ we have $A_\alpha = O_\alpha^*$. Thus, by (6),

$$\bigcap_{\alpha \in J} A_\alpha = \bigcap_{\alpha \in J} O_\alpha^* = \left(\bigcup_{\alpha \in J} O_\alpha \right)^*$$

is the intent of a concept (applying $^\wedge$ gives the concept's extent). Hence, by properties of set intersection, $\bigcap_{\alpha \in J} A_\alpha$ is the greatest intent smaller than all the C_α. Using 1'. of Theorem 5, $(\bigcap_{\alpha \in J} A_\alpha)^{\wedge}$ is the smallest extent greater than all the O_α, and so the supremum of $\leq$ is as stated.

The proof of (8) is similar to that of (9), only working with the extents of the C_α rather than their intents, and applying (7) instead of (6).

Theorem 8 allows us to use lattice theory for finding out many properties that come from a formal context. In particular, a finite concept lattice has an easy visual representation (see Example 9 below). In order to interpret the concept lattice from the diagram, one may write, for each concept on the diagram, the elements of its intent and extent. However, from the order $\leq$ of the concept lattice a tidier manner of presenting the diagram can be devised: for a given concept, instead of writing every element of its extent (or intent), we write only those objects (attributes) that did not appear below (above) in the concept lattice.

Example 9. The concept lattice of the context presented in Table 1 is shown in Fig. 1. Each concept is represented by a circle. Here animals are represented by numbers 1–5 in the order they appear in Table 1. Attributes are represented by letters a-g, also in the order they appear in the table. The extent (intent) of a given concept C has an object (attribute) iff that object (attribute) appears near a concept $\tilde{C}$ such that there is a descending (ascending) path from C to $\tilde{C}$.

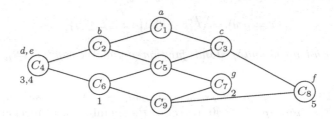

Fig. 1. Concept lattice of animals

Take, for instance, the concept C_3. It has ascending paths to concepts C_1 and C_3, and descending paths to concepts C_8, C_5, C_6, C_7 and C_9. On the other hand, C_3 has no (strictly ascending or descending) paths to C_2 or C_4. Thus, $C_3 = \langle \{1,2,5\}, \{a,c\} \rangle$.

Concept lattices are visual tools that allow us to find out relations on attributes and objects (for example, any animal with property e also has property b). However, limitations may arise on the theory, as, for example, the definition of formal context allows us only to work only with precise relations. For example, a chicken can fly for short distances, but this information could not be expressed on a formal context as defined earlier. In the next section we generalize these ideas to allow fuzzy objects, attributes and relations.

3   Fuzzy Formal Concept Analysis

We first state some basic definitions of fuzzy logic and then proceed to a fuzzy generalization of FCA. Definition of triangular norm corresponds to that presented by Klement et al. (2000), whereas definitions of fuzzy implication, fuzzy subset and fuzzy (binary) relation correspond to definitions by de Barros et al. (2017).

Definition 10. *A triangular norm (or t-norm) is a map $\triangle$: $[0,1]^2 \rightarrow [0,1]$ satisfying, for all $x, y, z \in [0,1]$:*

1. $x \triangle y = y \triangle x$	*(commutativity)*
2. $x \triangle (y \triangle z) = (x \triangle y) \triangle z$	*(associativity)*
3. If $y \leq z$ then $x \triangle y \leq x \triangle z$	*(monotonicity)*
4. $x \triangle 1 = x$	*(boundary condition)*

Definition 11. *A fuzzy implication is a map $\Rightarrow$: $[0,1]^2 \rightarrow [0,1]$ such that for all $x, y, z \in [0,1]$:*

1. $(0 \Rightarrow 0) = 1$, $(1 \Rightarrow 0) = 0$,   *(boundary conditions)*
   $(0 \Rightarrow 1) = 1$ *and* $(1 \Rightarrow 1) = 1$
2. *If* $y \leq x$ *then* $(x \Rightarrow z) \leq (y \Rightarrow z)$ *(monotonicity in the first component)*
3. *If* $x \leq y$ *then* $(z \Rightarrow x) \leq (z \Rightarrow y)$ *(monotonicity in the second component)*

Definition 12. *A* R-implication *is a fuzzy implication* $\Rightarrow_R$ *defined by*

$$(x \Rightarrow_R y) = \bigvee \{z \in [0,1] : x \bigtriangleup z \leq y\}, \tag{10}$$

where $\bigtriangleup$ *is a t-norm and* $\bigvee$ *stands for supremum. We say that* $\Rightarrow_R$ *is* induced *by* $\bigtriangleup$.

Definition 13. *Let* U *be a (classical universal) set. A* fuzzy subset F *of* U *is defined by a function* $\phi_F : U \to [0,1]$, *called the* membership function *of* F.
   Given two fuzzy subsets F_1, F_2 *of* U, *we say that* F_1 *is a (fuzzy)* subset *of* F_2 *if* $\phi_{F_1}(u) \leq \phi_{F_2}(u)$ *for all* $u \in U$, *and in this case we write* $F_1 \subseteq F_2$.
   If $\langle F_\alpha \rangle_{\alpha \in J}$ *is a family of fuzzy subsets of* U, *then their* union *and* intersection *are defined respectively by the membership functions*

$$\phi_{\cup_{\alpha \in J} F_\alpha}(u) = \bigvee_{\alpha \in J} \phi_\alpha(u) \tag{11}$$

$$\phi_{\cap_{\alpha \in J} F_\alpha}(u) = \bigwedge_{\alpha \in J} \phi_\alpha(u), \tag{12}$$

where $\bigvee$ *and* $\bigwedge$ *stand for supremum and infimum respectively.*

Definition 14. *A* fuzzy (binary) relation *on a pair of (classical) sets* U_1, U_2 *is a fuzzy subset of* $U_1 \times U_2$. *In particular, if* A_1, A_2 *are fuzzy subsets of* U_1, U_2 *respectively, and* $\bigtriangleup$ *is a t-norm, then the* (fuzzy) cartesian product *of* A_1, A_2 *induced by* $\bigtriangleup$ *is the fuzzy relation* $A_1 \times_\bigtriangleup A_2$ *on* $U_1 \times U_2$ *defined by the membership function*

$$\phi_{A_1 \times_\bigtriangleup A_2}(x_1, x_2) = \phi_1(x_1) \bigtriangleup \phi_2(x_2). \tag{13}$$

Now we can start defining the objects of Fuzzy Formal Concept Analysis (fFCA).

Definition 15. *A* fuzzy formal context *is an ordered triple* $\mathbb{C}_f := \langle \mathcal{O}, \mathcal{A}, I_f \rangle$, *in which* $\mathcal{O}$ *and* $\mathcal{A}$ *are non-empty (classical) sets, and* $I_f \subseteq \mathcal{O} \times \mathcal{A}$ *is a fuzzy binary relation.*

Notice that, in (1) and (2), the characteristic functions of O^* and $A^\wedge$ can be expressed respectively by

$$\chi_{O^*}(\tilde{a}) = \forall o \in \mathcal{O}(o \in O \longrightarrow oI\tilde{a}),$$
$$\chi_{A^\wedge}(\tilde{o}) = \forall a \in \mathcal{A}(a \in A \longrightarrow \tilde{o}Ia).$$

Since an application of the universal quantifier, $\forall$, returns the smallest truth value a predicate assumes ($\forall x P(x) = 1$ iff $P(u) = 1$ for each $u \in U$), we generalize $\forall$ as an infimum.

Definition 16. *Let* $\mathbb{C}_f = \langle \mathcal{O}, \mathcal{A}, I_f \rangle$ *be a fuzzy formal context. Let* $\Rightarrow$ *be a fuzzy implication. Then we define, for all fuzzy subsets* $O \subseteq \mathcal{O}$ *and* $A \subseteq \mathcal{A}$, *the fuzzy subsets* $O^* \subseteq \mathcal{A}$ *and* $A^\wedge \subseteq \mathcal{O}$ *by their membership functions defined as*

$$\phi_{O^*}(a) = \inf_{o \in \mathcal{O}} \left[\phi_O(o) \Rightarrow \phi_{I_f}(o,a) \right], \tag{14}$$

$$\phi_{A^\wedge}(o) = \inf_{a \in \mathcal{A}} \left[\phi_A(a) \Rightarrow \phi_{I_f}(o,a) \right]. \tag{15}$$

With these maps defined we can finally generalize the idea of formal concept.

Definition 17. *Let* $\mathbb{C}_f = \langle \mathcal{O}, \mathcal{A}, I_f \rangle$ *be a fuzzy formal context and let* $\Rightarrow$ *be a fuzzy implication. Let* O, A *be fuzzy subsets of* $\mathcal{O}$ *and* $\mathcal{A}$, *respectively. We say that* $C_f = \langle O, A \rangle$ *is a* fuzzy formal concept *of* $\mathbb{C}_f$ *if* $O^* = A$ *and* $A^\wedge = O$.

Our goal is to show that, as in the classical case, it is possible to produce a complete lattice of fuzzy formal concepts. Our definitions of semicontinuous functions are proven by Bourbaki (1966) to be equivalent with the definitions of semicontinuity he presents. The upper semicontinuous R-implication induced by a lower semicontinuous t-norm is precisely what allows us to have a complete lattice of fuzzy concepts.

In what follows, a sequence $(x_1, x_2, ...)$ is denoted by (x_n), and so x_0 is not an element of the sequence (x_n).

Definition 18. *Let* $X, Y \subseteq \mathbb{R}^2$ *be non-empty (classical) sets. Let* $f : X \to Y$ *be a function. We say that* f *is* upper semicontinuous *at* $x_0 \in X$ *if, for any sequence* (x_n) *converging to* x_0,

$$\inf_{n>0} \left(\sup_{m \geq n} f(x_m) \right) =: \limsup_{n \to \infty} f(x_n) \leq f(x_0). \tag{16}$$

If f *is upper semicontinuous at every* $x_0 \in X$ *then* f *is* upper semicontinuous.

Similarly, f *is* lower semicontinuous *at* $x_0 \in X$ *if for any sequence* (x_n) *converging to* x_0,

$$\sup_{n>0} \left(\inf_{m \geq n} f(x_m) \right) =: \liminf_{n \to \infty} f(x_n) \geq f(x_0). \tag{17}$$

If f *is lower semicontinuous at every* $x_0 \in X$ *then* f *is* lower semicontinuous.

If f *is both upper and lower semicontinuous at* x_0 *then* f *is* continuous *at* x_0, *and*

$$\limsup_{n \to \infty} f(x_n) = f(x_0) = \liminf_{n \to \infty} f(x_n). \tag{18}$$

If f *is continuous at every* $x_0 \in X$ *then it is* continuous.

We now derive a useful expression concerning lower semicontinuous t-norms in Lemma 20, and for its proof we use Proposition 19.

In the following, we shall write $x_n \nearrow x_0$ meaning that the increasing sequence (x_n) converges to x_0. Similarly, $x_n \searrow x_0$ means that the decreasing sequence (x_n) converges to x_0.

Proposition 19. *Let* $\triangle$ *be a lower semicontinuous t-norm. Let* $x, y \in [0,1]$. *Define*

$$S = \{z \in [0,1] : x \triangle z \leq y\}.$$

Then $\sup S \in S$.

Proof. Let $z_0 = \sup S$, and let (z_n) be a sequence on $[0,1]$ such that $z_n \nearrow z_0$. Then

$$x \triangle z_m \leq y \text{ for all } m > 0,$$

as $z_m \leq z_0$ for each $m > 0$. Thus, for each $n > 0$,

$$\inf_{m \geq n} (x \triangle z_m) \leq x \triangle z_n \leq y.$$

Taking the supremum for $n > 0$ on the left-hand side we have, by lower semicontinuity of $\triangle$,

$$x \triangle z_0 \leq \liminf_{n \to \infty} (x \triangle z_n) = \sup_{n > 0} \left(\inf_{m \geq n} (x \triangle z_m) \right) \leq y.$$

Therefore, $z_0 \in A$.

Lemma 20. *Let* $\triangle$ *be a lower semicontinuous t-norm and let* $\Rightarrow$ *be the R-implication induced by* $\triangle$. *Then, for all* $x, y \in [0,1]$ *we have* $x \leq [(x \Rightarrow y) \Rightarrow y]$.

Proof. Let $x, y \in [0,1]$. By Proposition 19, $x \triangle (x \Rightarrow y) \leq y$. Using commutativity of $\triangle$, it is clear that $x \in S := \{z \in [0,1] : (x \Rightarrow y) \triangle z \leq y\}$. Hence

$$x \leq \sup S = [(x \Rightarrow y) \Rightarrow y].$$

Now we have what is necessary to generalize Theorem 5, establishing dual relations for the maps * and ^ in the fuzzy case.

Theorem 21. *Let* $\mathbb{C}_f = \langle \mathcal{O}, \mathcal{A}, I_f \rangle$ *be a fuzzy context. Let* O, O_1, O_2 *be fuzzy subsets of* $\mathcal{O}$ *and* A, A_1, A_2 *be fuzzy subsets of* $\mathcal{A}$. *Let* $\triangle : [0,1]^2 \to [0,1]$ *be a lower semicontinuous t-norm, and let* $\Rightarrow$ *be the R-implication induced by* $\triangle$. *Then:*

1. *If* $O_1 \subseteq O_2$ *then* $O_2^* \subseteq O_1^*$ 1'. *If* $A_1 \subseteq A_2$ *then* $A_2^\wedge \subseteq A_1^\wedge$
2. $O \subseteq O^{*\wedge}$ 2'. $A \subseteq A^{\wedge *}$
3. $O^* = O^{*\wedge *}$ 3'. $A^\wedge = A^{\wedge * \wedge}$
4. $O \subseteq A^\wedge$ *iff* $A \subseteq O^*$ *iff* $O \times_\triangle A \subseteq I_f$

Proof. We shall prove items 1., 2., 3. and 4.. Items with a prime can be proved analogously.

1. Suppose that $O_1 \subseteq O_2$. Let $a \in \mathcal{A}$. Let $o \in \mathcal{O}$. By hypothesis, $\phi_{O_1}(o) \leq \phi_{O_2}(o)$. Since $\Rightarrow$ is decreasing in its first component we have

$$(\phi_{O_2}(o) \Rightarrow \phi_{I_f}(o, a)) \leq (\phi_{O_1}(o) \Rightarrow \phi_{I_f}(o, a)).$$

Taking the infimum over o on both sides, we see that

$$\phi_{O_2^*}(a) \leq \phi_{O_1^*}(a)$$

by definition of the map *. But $a \in \mathcal{A}$ is arbitrary. Thus $O_2^* \subseteq O_1^*$.

2. Let $o \in \mathcal{O}$. By hypothesis, $\triangle$ is lower semicontinuous and so Lemma 20 holds. Using monotonicity of $\Rightarrow$ in its first component we get

$$\phi_O(o) \leq [(\phi_O(o) \Rightarrow \phi_{I_f}(o,a)) \Rightarrow \phi_{I_f}(o,a)]$$
$$\leq \left[\inf_{\tilde{o} \in \mathcal{O}} (\phi_O(\tilde{o}) \Rightarrow \phi_{I_f}(\tilde{o},a)) \Rightarrow \phi_{I_f}(o,a) \right]$$
$$= [\phi_{O^*}(a) \Rightarrow \phi_{I_f}(o,a)].$$

Taking the infimum over a on the right-hand side, we get

$$\phi_O(o) \leq \phi_{O^{*\wedge}}(o).$$

Since $o \in \mathcal{O}$ is arbitrary, $O \subseteq O^{*\wedge}$.

3. From 2., we have $O \subseteq O^{*\wedge}$. Thus, using 1., $O^{*\wedge*} \subseteq O^*$. On the other hand, using $A = O^*$ in 2'., we have $O^* \subseteq O^{*\wedge*}$. Hence, $O^* = O^{*\wedge*}$.

4. We shall first prove that if $A \subseteq O^*$ then $O \subseteq A^\wedge$; then we show that $O \subseteq A^\wedge$ implies $O \times_\triangle A \subseteq I_f$; and finally we prove that if $O \times_\triangle A \subseteq I_f$ then $A \subseteq O^*$, concluding that the three properties are equivalent.

   (a) Suppose that $A \subseteq O^*$. By 1'., $O^{*\wedge} \subseteq A^\wedge$. Using 2. and transitivity of $\subseteq$, we have $O \subseteq A^\wedge$.

   (b) Suppose that $O \subseteq A^\wedge$. Let $o \in \mathcal{O}$ and $a \in \mathcal{A}$. Then

$$\phi_O(o) \leq \phi_{A^\wedge}(o) = \inf_{\tilde{a} \in \mathcal{A}} [\phi_A(\tilde{a}) \Rightarrow \phi_{I_f}(o,\tilde{a}))]$$
$$\leq [\phi_A(a) \Rightarrow \phi_{I_f}(o,a)]$$
$$= \sup\{z \in [0,1] : \phi_A(a) \triangle z \leq \phi_{I_f}(o,a)\}.$$

By lower semicontinuity of $\triangle$, Proposition 19 holds and so

$$\phi_O(o) \in \{z \in [0,1] : \phi_A(a) \triangle z \leq \phi_{I_f}(o,a)\}.$$

Using commutativity of $\triangle$, we have

$$\phi_{O \times_\triangle A}(o,a) = \phi_O(o) \triangle \phi_A(a) \leq \phi_{I_f}(o,a).$$

But $o \in \mathcal{O}$ and $a \in \mathcal{A}$ are arbitrary and so $O \times_\triangle A \subseteq I_f$.

   (c) Suppose that $O \times_\triangle A \subseteq I_f$. Then for all $o \in \mathcal{O}$ and for all $a \in \mathcal{A}$,

$$\phi_A(a) \in \{z \in [0,1] : \phi_O(o) \triangle z \leq \phi_{I_f}(o,a)\},$$

whence

$$\phi_A(a) \leq \sup\{z \in [0,1] : \phi_O(o) \triangle z \leq \phi_{I_f}(o,a)\} = [\phi_O(o) \Rightarrow \phi_{I_f}(o,a)].$$

Taking the infimum over o on the right-hand side, we see that

$$\phi_A(a) \leq \phi_{O^*}(a).$$

But $a \in \mathcal{A}$ is arbitrary, and so $A \subseteq O^*$.

As stated earlier, we want to prove that a lattice of fuzzy formal concepts is complete. As we shall see, the fuzzy implication we use when defining $*$ and $\wedge$ has to be upper semicontinuous. It turns out that for R-implicationsm this property follows from lower semicontinuity of the t-norm, as we show in Proposition 23. Before that we make an intermediate step.

Lemma 22. *Let $X, Y \subseteq \mathbb{R}$. Let $f : X \to Y$ be an increasing function. Let (x_n) be a sequence on X such that $x_n \searrow x \in X$. Let (y_n) be a sequence on Y such that, for all $n > 0$, $f(x_n) \leq y_n$. Then*

$$f(x) \leq \liminf_{n \to \infty} y_n. \tag{19}$$

Proof. By monotonicity of (x_n), we have $\sup_{m \geq n_0} x_m = x_{n_0}$ for all $n_0 > 0$, whence for each $n_0 > 0$ fixed, $\inf_{n_0 > 0} \left(\sup_{m \geq n} x_m \right) \leq x_{n_0}$. But (x_n) converges to x, and so for all $n_0 > 0$ we have $x = \limsup_{n \to \infty} x_n \leq x_{n_0}$. Because f is increasing we have, for each $n_0 > 0$, $f(x) \leq f(x_{n_0}) \leq y_{n_0}$. Now, taking the limit inferior over n_0 on the right-hand side and changing the variable n_0 to n, we complete the proof.

Proposition 23. *Let $\triangle$ be a lower semicontinuous t-norm, and let $\Rightarrow$ be the R-implication induced by $\triangle$. Then for all $x_0, y_0 \in [0, 1]$ fixed, the maps $x \mapsto (x \Rightarrow y_0)$ and $y \mapsto (x_0 \Rightarrow y)$ are upper semicontinuous.*

Proof. Let $x', y' \in [0, 1]$. We want to show that $x \mapsto (x \Rightarrow y_0)$ and $y \mapsto (x_0 \Rightarrow y)$ are upper semicontinuous at x' and y' respectively. Let $(x_n), (y_n)$ be sequences on $[0, 1]$ converging respectively to x' and y'. For each $n > 0$, consider the following definitions:

$$z_n^{(1)} = (x_n \Rightarrow y_0), \qquad z_n^{(2)} = (x_0 \Rightarrow y_n), \tag{20}$$

$$\tilde{x}^{(1)} = x', \qquad \tilde{x}^{(2)} = x_0, \tag{21}$$

$$\tilde{x}_n^{(1)} = \inf_{m \geq n} x_m, \qquad \tilde{x}_n^{(2)} = x_0, \tag{22}$$

$$\tilde{y}^{(1)} = y_0, \qquad \tilde{y}^{(2)} = y', \tag{23}$$

$$\tilde{y}_n^{(1)} = y_0, \qquad \tilde{y}_n^{(2)} = \sup_{m \geq n} y_m, \tag{24}$$

$$\tilde{z}_n^{(1)} = \left(\tilde{x}_n^{(1)} \Rightarrow \tilde{y}_n^{(1)} \right), \qquad \tilde{z}_n^{(2)} = \left(\tilde{x}_n^{(2)} \Rightarrow \tilde{y}_n^{(2)} \right). \tag{25}$$

Notice that, for $i = 1, 2$ the following hold:

1. $\tilde{x}_n^{(i)} \nearrow \tilde{x}^{(i)}$, as the $\tilde{x}_n^{(1)}$ are infima of decreasing sets and $\left(\tilde{x}_n^{(2)} \right)$ is constant;

2. $\tilde{y}_n^{(i)} \searrow \tilde{y}^{(i)}$, because $\left(\tilde{y}_n^{(1)} \right)$ is constant and the $\tilde{x}_n^{(2)}$ are suprema of decreasing sets;

3. For all $n > 0$ monotonicity of $\Rightarrow$ yields $z_n^{(i)} \leq \tilde{z}_n^{(i)}$, whence $\limsup_{n \to \infty} z_n^{(i)} \leq \limsup_{n \to \infty} \tilde{z}_n^{(i)}$.

4. The sequence $(\tilde{z}_n^{(i)})$ is decreasing, for $i = 1, 2$.

Now let $n_0 > 0$ be fixed, and let $n > n_0$. By Proposition 19 and (25), we have

$$\tilde{x}_{n_0}^{(i)} \vartriangle \tilde{z}_n^{(i)} \le \tilde{x}_n^{(i)} \vartriangle \tilde{z}_n^{(i)} \le \tilde{y}_n^{(i)}.$$

Thus, (19) yields

$$\tilde{x}_{n_0}^{(i)} \vartriangle \left(\limsup_{n \to \infty} \tilde{z}_n^{(i)} \right) \le \liminf_{n \to \infty} \tilde{y}_n^{(i)} = \tilde{y}^{(i)}.$$

Taking the limit inferior over n_0 on the left-hand side and using lower semicontinuity of $\vartriangle$,

$$\tilde{x}^{(i)} \vartriangle \left(\limsup_{n \to \infty} \tilde{z}_n^{(i)} \right) \le \liminf_{n_0 \to \infty} \left[\tilde{x}_{n_0}^{(i)} \vartriangle \left(\limsup_{n \to \infty} \tilde{z}_n^{(i)} \right) \right] \le \tilde{y}^{(i)}.$$

Hence by definition of $(\tilde{x}^{(i)} \Rightarrow \tilde{y}^{(i)})$ and 3. above,

$$\limsup_{n \to \infty} z_n^{(i)} \le \limsup_{n \to \infty} \tilde{z}_n^{(i)} \le \left(\tilde{x}^{(i)} \Rightarrow \tilde{y}^{(i)} \right).$$

For $i = 1$ this means $x \mapsto (x \Rightarrow y_0)$ is upper semicontinuous at x', and for $i = 2$, $y \mapsto (x_0 \Rightarrow y)$ is upper semicontinuous at y'. But $x', y' \in [0,1]$ are arbitrary. Therefore, both the maps $x \mapsto (x \Rightarrow y_0)$ and $y \mapsto (x_0 \Rightarrow y)$ are upper semicontinuous.

Proposition 24. *Let $\mathbb{C}_f = \langle \mathcal{O}, \mathcal{A}, I_f \rangle$ be a fuzzy context. Let $\vartriangle$ be a lower semicontinuous t-norm, and let $\Rightarrow$ be the implication induced by it. Let J be an index set and, for each $\alpha \in J$, let $O_\alpha \subseteq \mathcal{O}$ and $A_\alpha \subseteq \mathcal{A}$. Then*

$$\bigcap_{\alpha \in J} O_\alpha^* = \left(\bigcup_{\alpha \in J} O_\alpha \right)^*, \tag{26}$$

$$\bigcap_{\alpha \in J} A_\alpha^\wedge = \left(\bigcup_{\alpha \in J} A_\alpha \right)^\wedge. \tag{27}$$

Proof. We prove (26). The proof of (27) is analogous. Let $a \in \mathcal{A}$. For each $\alpha_0 \in J$ we have

$$\phi_{(\cup_{\alpha \in J} O_\alpha)^*}(a) = \inf_{o \in \mathcal{O}} \left[\phi_{\cup_{\alpha \in J} O_\alpha}(o) \Rightarrow \phi_{I_f}(o, a) \right]$$

$$= \inf_{o \in \mathcal{O}} \left[\left(\sup_{\alpha \in J} \phi_{O_\alpha}(o) \right) \Rightarrow \phi_{I_f}(o, a) \right]$$

$$\le \left[\left(\sup_{\alpha \in J} \phi_{O_\alpha}(o) \right) \Rightarrow \phi_{I_f}(o, a) \right]$$

$$\le \left[\phi_{O_{\alpha_0}}(o) \Rightarrow \phi_{I_f}(o, a) \right],$$

as $\Rightarrow$ is decreasing in the first component. Applying the infimum over $o \in \mathcal{O}$ and then the infimum over $\alpha \in J$ on the right-hand side yields $\phi_{(\cup_{\alpha \in J} O_\alpha)^*}(a) \le \phi_{\cap_{\alpha \in J} O_\alpha^*}(a)$. Since $a \in \mathcal{A}$ is arbitrary, $\left(\bigcup_{\alpha \in J} O_\alpha \right)^* \subseteq \bigcap_{\alpha \in J} O_\alpha^*$.

On the other hand, suppose for the sake of contradiction that for some $o \in \mathcal{O}$ and $a \in \mathcal{A}$,

$$\kappa := \inf_{\alpha \in J} \left[\inf_{\tilde{o} \in \mathcal{O}} (\phi_{O_\alpha}(\tilde{o}) \Rightarrow \phi_{I_f}(\tilde{o}, a)) \right] > \left[\left(\sup_{\alpha \in J} \phi_{O_\alpha}(o) \right) \Rightarrow \phi_{I_f}(o, a) \right].$$

Define $x_0 = \sup_{\alpha \in J} \phi_{O_\alpha}(o)$. Let (α_n) be a sequence on J such that $(\phi_{O_{\alpha_n}}(o))$ converges to x_0 and, for each $n > 0$, define $x_n = \phi_{O_{\alpha_n}}(o)$. By hypothesis, for each $n > 0$,

$$\begin{aligned}
(x_0 \Rightarrow \phi_{I_f}(o, a)) &< \inf_{\alpha \in J} \left[\inf_{\tilde{o} \in \mathcal{O}} (\phi_{O_\alpha}(\tilde{o}) \Rightarrow \phi_{I_f}(\tilde{o}, a)) \right] \quad (= \kappa) \\
&\leq \inf_{\tilde{o} \in \mathcal{O}} (\phi_{O_{\alpha_n}}(\tilde{o}) \Rightarrow \phi_{I_f}(\tilde{o}, a)) \\
&\leq (\phi_{O_{\alpha_n}}(o) \Rightarrow \phi_{I_f}(o, a)) \\
&= (x_n \Rightarrow \phi_{I_f}(o, a)).
\end{aligned}$$

Thus,

$$(x_0 \Rightarrow \phi_{I_f}(o, a)) < \kappa \leq \limsup_{n \to \infty} (x_n \Rightarrow \phi_{I_f}(o, a)),$$

and so $x \mapsto (x \Rightarrow \phi_{I_f}(o, a))$ is not upper semicontinuous, contradicting Proposition 23.

Hence, for all $o \in \mathcal{O}$ and for all $a \in \mathcal{A}$,

$$\phi_{\bigcap_{\alpha \in J} O_\alpha^*}(a) = \kappa \leq \left[\left(\phi_{\bigcup_{\alpha \in J} O_\alpha}(o) \right) \Rightarrow \phi_{I_f}(o, a) \right].$$

Taking the infimum over o on the right-hand side, and since $a \in \mathcal{A}$ is arbitrary, we conclude that $\bigcap_{\alpha \in J} O_\alpha^* \subseteq \left(\bigcup_{\alpha \in J} O_\alpha \right)^*$.

Theorem 25. *Let $\mathbb{C}_f = \langle \mathcal{O}, \mathcal{A}, I_f \rangle$ be a fuzzy formal context. Using a lower semicontinuous t-norm and the R-implication induced by it to define the maps * and $^\wedge$, define the order $\leq$ on the set $\mathfrak{B}_f(\mathbb{C}_f)$ of all the fuzzy formal concepts of $\mathbb{C}_f$ by*

$$\langle O_1, A_1 \rangle \leq \langle O_2, A_2 \rangle \quad \text{iff } O_1 \subseteq O_2 \quad (\text{iff } A_2 = O_2^* \subseteq O_1^* = A_1). \tag{28}$$

Then $\mathcal{L}_{\mathbb{C}_f} := \langle \mathfrak{B}_f(\mathbb{C}_f), \leq \rangle$ is a complete lattice, called the fuzzy concept lattice *of $\mathbb{C}_f$.*

Proof. Similar to that of Theorem 8, using Proposition 24 rather than Proposition 7.

4   Final Remarks

As mentioned in the Introduction, there exists a vast body of literature on fuzzy Formal Context Analysis. Burusco and Fuentes-González (1994) introduced fuzzy concept lattices. Belohlávek and Vychodil (2007) have shown that it is possible to define a complete lattice of fuzzy concepts, by means of fuzzy implications obtained from left continuous t-norms. These t-norms are equivalent to lower semicontinuous t-norms. In this paper we recast some basic notions and results from fuzzy FCA in terms of lower semicontinuous t-norms. Given the difference between both definitions for t-norms (left continuous and lower semicontinuous), proving the results requires somewhat different techniques. Within our framework it is shown that the set of fuzzy concepts is a complete lattice.

Acknowledgement. This research was financially supported by CAPES (Brazil) and by processes 306546/2017-5 and 308524/2014-4 from CNPq (Brazil).

References

de Barros, L.C., Bassanezi, R.C., Lodwick, W.A.: A First Course in Fuzzy Logic, Fuzzy Dynamical Systems, and Biomathematics. SFSC, vol. 347. Springer, Heidelberg (2017). https://doi.org/10.1007/978-3-662-53324-6

Belohlávek, R., Vychodil, V.: Fuzzy concept lattices constrained by hedges. JACIII **11**, 536–545 (2007)

Bourbaki, N.: Elements of Mathematics: General Topology. Volume Part 1. Addison-Wesley, Dordrecht (1966)

Burusco, A., Fuentes-González, R.: The study of the L-fuzzy concept lattice. Mathw. Soft Comput. **1**, 209–218 (1994)

Ganter, B., Wille, R.: Formal Concept Analysis. Springer, Heidelberg (1999). https://doi.org/10.1007/978-3-642-59830-2

Klement, E.P., Mesiar, R., Pap, E.: Triangular Norms. Kluwer Academic Publishers, Dordrecht (2000)

Poelmans, J., Ignatov, D.I., Kuznetsov, S.O., Dedene, G.: Fuzzy and rough formal concept analysis: a survey. Int. J. Gen. Syst. **43**(2), 105–134 (2014)

Representing Intuistionistic Fuzzy Bi-implications Using Quantum Computing

Lucas Agostini, Samuel Feitosa, Anderson Avila, Renata Reiser$^{(\boxtimes)}$,
André DuBois, and Maurício Pilla

Centro de Desenvolvimento Tecnológico, Universidade Federal de Pelotas (UFPel),
Gomes Carneiro Street, 1, Pelotas, Brazil
{lbagostini,samuel.feitosa,abdavila,reiser,dubois}@inf.ufpel.edu.br
http://www.ufpel.edu.br/

Abstract. Computer systems based on intuitionistic fuzzy logic are capable of generating a reliable output even when handling inaccurate input data by applying a rule based system, even with rules that are generated with imprecision. The main contribution of this paper is to show that quantum computing can be used to extend the class of intuitionistic fuzzy sets with respect to representing intuitionistic fuzzy bi-implications. This paper describes a multi-dimensional quantum register using aggregations operators such as t-(co)norms and implications based on quantum gates allowing the modeling and interpretation of intuitionistic fuzzy bi-implications.

Keywords: Quantum computing · Intuitionistic fuzzy sets
Intuitionistic bi-implications

1 Introduction

The similarities between Fuzzy Logic (*FL*) and Quantum Computing (*QC*) motivate researches towards a better understanding of their relationship [1–4]. Such study is relevant to understand how one can explore the phenomena of quantum mechanics to improve the efficiency of algorithms employed in the design of expert systems.

And even further, the Intuitionistic Fuzzy Logic (*IFL*) or type-2 fuzzy logic extends the concept of FL by adding the concept of imprecision when a set of rules is defined. Since both *IFL* and *QC* concern about types of uncertainties, it is important to investigate possible contributions from one area to another.

In this context, the logical structure describing the uncertainty associated with the intuitionistic fuzzy set theory can be modeled by quantum transformations (QTs) and quantum states (QSs) [5]. Thus, it is possible to model quantum

---

Work progress stage: completed.

© Springer International Publishing AG, part of Springer Nature 2018
G. A. Barreto and R. Coelho (Eds.): NAFIPS 2018, CCIS 831, pp. 206–216, 2018.
https://doi.org/10.1007/978-3-319-95312-0_18

algorithms which represent operations on intuitionistic fuzzy sets such as unions, intersections, differences, implications making use of superposition of quantum states [6].

This work introduces a methodology representing operations of Intuitionistic Fuzzy Set Theory (CFIs), as proposed by Krassemir Atanassov, making use of properties such as superposition, linearity and distributivity of the tensorial product in the state space and transformations of Quantum Computation (QC).

The information regarding each intuitionistic fuzzy connective is represented by pairs of quantum registers, guaranteeing the inherent unitarity of quantum states and transformations as well as the flexibility of the complementarity relation of membership and non-membership functions characterizing CFIs [7].

The achieved results make significant contributions:

(i) consolidating the qfuzz-Analyzer methodology for representing fuzzy sets via operators and states fo QC; and

(ii) collaborating with the development of the qfuzz2-Analyzer methodology, extending the representation of information modelled by intuitionist fuzzy sets and their operations, both of which are defined by intuitionistic fuzzy connectors via registers and quantum transformations in the Quantum Circuits model.

As the main contribution, based on the extensibility from qfuzz-Analyzer, the qfuzz2-Analyzer methodology guarantees the preservation of the representable intuitionistic fuzzy connectives. The work also considered the validation of the methodology by extending the library of intuitionistic fuzzy operators in the Visual Programming Environment for Quantum Geometric Machine Model (VPE-qGM).

The simulation of quantum intuitionistic fuzzy operators via interfaces of the VPE-qGM components (simulator, memory and process editors), contributes to the generation of the computations minimizing problems as the exponential increase in the quantum memory and the complexity of quantum measurement projection operations, facilitating the interpretation and analysis by applying quantum algorithms of qfuzz2-Analyzer.

The remainder of this paper is organized as follows: Sect. 2 presents the foundations on IFL. Section 3 brings the main concepts of QC. In Sect. 4, we present the study and modeling of fuzzy bi-implications using QC, Sect. 5 presents the results about simulating this proposal in the VPE-qGM. Finally, conclusions and further work are discussed in Sect. 6.

2   Intuitionistic Fuzzy Logic

The Atanassov's Intuitionistic Fuzzy Logic (A-IFL) is a type-2 fuzzy logic conceived as a generalization of fuzzy logic (FL) based on the intuitionistic fuzzy set theory [8], introduced to overcome the limitations related to fuzzy sets for dealing with problems where the rules applied to the system could not be defined with precision, mainly related to non-memebrship degree which cannot be defined as a complement of its membership degree.

An element $x \in \mathcal{X}$ belongs to the subset A with a membership a non-membership degrees given by $\mu_A(x)$ and $\nu_A(x)$, such that $0 \leq \mu_A(x) + \nu_A \leq 1$:

$$A = \{(x, \mu_A(x), \nu_A(x)) : x \in \mathcal{X} \text{ and } \mu_A(x) + \nu_A \leq 1\}, \tag{1}$$

denoting by $\tilde{U}$ the set of all intuitionistic fuzzy values considering the research of IFL in a narrow sense [9], extending the multi-valued logic, makes it possible to extend the usual logic connectives, as follows [10]:

- Conjunction, usually modelled by a *triangular norm (T-norm)* operator [11], which is a kind of binary operation used in the framework of multi-valued logic.
- Disjunction, similarly to the above case, are usually modelled by *t-conorms* [11] (also called *S-norms*), and represents the dual operation to *t-norms*.
- Negation, which follows the seminal work of Atanassov [8], presenting several studies on the properties of intuitionistic fuzzy negations. Although Atanassov intuitionistic fuzzy negation $\mu_A(x) = \nu_A(x)$ and $\mu_A(x) = \nu_A(x)$ is the most used in intuitionistic fuzzy systems, there are important classes of intuitionistic fuzzy negation proposed with different motivations.
- Implication, which can be represented by many different and non equivalent approaches, although some of them have been more used and well accepted [12].
- Bi-implication, which has been studied in FL under different contexts, where we can cite among others [13,17], all of them were constrained to at least one of the following restrictions [10]:
  - Satisfy the fuzzy equivalence properties.
  - Be compatible with the notion of distance on [0, 1].
  - Define the fuzzy bi-implication in terms of the conjunction and implication connectives.

Now, the connectives used to make the correlation between quantum computing and IFL will be described in function of Fuzzy Connective. Firstly, we consider the standard intuitionistic fuzzy negation [8] expressed as:

$$N_{I_S}(\tilde{x}) = N_{I_S}((x_1, x_2)) = (x_2, x_1), \forall \tilde{x} = (x_1, x_2) \tag{2}$$

Meanwhile, for all $\tilde{x} = (x_1, x_2), \tilde{y} = (y_1, y_2) \in \tilde{U}$, the intersection and union can be defined, respectively, in terms of a t-norm T and a t-conorm S, given as:

$$T_I(\tilde{x}, \tilde{y}) = T_I((x_1, x_2), (y_1, y_2)) = (T(x_1, y_1), S(x_2, y_2)); \tag{3}$$

$$S_I(\tilde{x}, \tilde{y}) = S_I((x_1, x_2), (y_1, y_2)) = (S(x_1, y_1), T(x_2, y_2)). \tag{4}$$

We consider both intuitionistic aggregations: the Product t-norm T_{I_P} and Algebraic Sum S_{I_P}, respectively described by Eqs. (5) and (6).

$$T_{I_P}(\tilde{x}, \tilde{y}) = (T_P(x_1, y_1), S_P(x_2, y_2)) = (x_1 y_1, x_2 + y_2 + x_2 y_2); \tag{5}$$

$$S_{I_P}(\tilde{x}, \tilde{y}) = (S_P(x_1, y_1), T_P(x_2, y_2)) = (x_1 + y_1 + x_1 y_1, x_2 y_2)). \tag{6}$$

According to [5], fuzzy sets can be obtained by quantum superposition of classical fuzzy states associated with a quantum register. Thus, interpretations related to the fuzzy operations as complement and intersection are obtained from the *NOT* and *AND* quantum transformations (QTs). The modeling of these operations can be found in [15], where the complement is obtained by the quantum *NOT* operator (Pauli-X gate) and the intersection is expressed using the quantum *Toffoli* transformation.

Extending this approach, other operations were introduced, such as union, difference, fuzzy (co)implication, which were derived from interpretations of *OR*, *DIV* and *(CO)IMP* quantum operators [14]. Now, the interpretation of the fuzzy bi-implication operator [10] through quantum computing is considered just making use of the *NOT* and *AND* operations:

$$A \leftrightarrow B \equiv (A \rightarrow B) \wedge (B \rightarrow A) \equiv (\neg A \vee B) \wedge (\neg B \vee A) \tag{7}$$

Based on the normal expression in Eq. (7), one can construct the Atanassov's intuitionistic fuzzy bi-implication (B_I) through the operators T_I, S_I, N_{I_S}, as shown below:

$$\begin{aligned}
B_I(\tilde{x}, \tilde{y}) &= T_I(S_I(N_{I_S}(\tilde{x}), \tilde{y}), S_I(N_{I_S}(\tilde{y}), \tilde{x})) \\
&= T_I(S_I((x_2, x_1), (y_1, y_2)), S_I((y_2, y_1), (x_1, x_2))) \\
&= T_I((S(x_2, y_1), T(x_1, y_2)), (S(y_2, x_1), T(y_1, x_2))) \\
&= (T(S(x_2, y_1), S(y_2, x_1)), S(T(x_1, y_2), T(y_1, x_2)))
\end{aligned}$$

The B_{I_P} membership and non-membership degrees are respectively, given by:

$$\mu_{B_{I_P}}(\tilde{x}, \tilde{y}) = (x_1 y_2 - x_1 - y_2)(x_2 y_1 - x_2 - y_1) \tag{8}$$

$$\nu_{B_{I_P}}(\tilde{x}, \tilde{y}) = x_1 y_2 + x_2 y_1 - x_1 x_2 y_1 y_2, \ \forall \ \tilde{x} = (x_1, x_2), \tilde{y} = (y_1, y_2) \in \tilde{U} \tag{9}$$

3   Quantum Computing

The basic unit of information in classical computing is the traditional bit, a binary classical physical system. In quantum computing the basic unit information is represented by a *quantum bit*, or qubit, a binary *quantum* physical system.

The qubit is a vector usually represented as a *superposition* of basic states, using the Dirac *braket* [1] notation [16]:

$$|\psi\rangle = \alpha|0\rangle + \beta|1\rangle.$$

The Dirac notation has the advantage that it labels the basis vectors explicitly. The basic states $|0\rangle$ and $|1\rangle$ can be explained by analogy with the classical bit,

---

[1] The name *braket* comes from the convention that a column vector is called a "ket" and is denoted by $|\ \rangle$ and a row vector is called a "bra" and is denoted by $\langle\ |$.

i.e., form a two-level system and are an orthonormal basis for the quantum vector space (usually called the standard or computational basis) [18].

The coefficients or also called *probability amplitudes*, α and β, are complex numbers, such that $|\alpha|^2 + |\beta|^2 = 1$. In other words the qubit can be formalized as a vector in a complex vector space (Hilbert space), with norm (size) equals to one.

As an example, the classical bit 0 can be represented as the basis state $|0\rangle = 1|0\rangle + 0|1\rangle$ and the classical bit 1 can be represented as the basis state $|1\rangle = 0|0\rangle + 1|1\rangle$.

Any other state with different values for α and β is said to be in a *quantum superposition* of $|0\rangle$ and $|1\rangle$, for instance the state $1/\sqrt{2}|0\rangle + 1/\sqrt{2}|1\rangle$. The interpretation of the probability amplitudes α and β can be given by the following: *when we interact or measure a quantum state such as $\alpha|0\rangle + \beta|1\rangle$ we will see/get the state $|0\rangle$ with probability $|\alpha|^2$ and the state $|1\rangle$ with probability $|\beta|^2$.*

The superposition of states gives to quantum computing a relevant characteristic called quantum parallelism. Essentially, due to superposition of states, a *qubit* can assume values of 0 and 1 at the same time. This gives an exponential power to quantum algorithms, as we can design algorithms that can verify various possibilities in parallel [18]. Table 1 shows the state space of the qubit grows with the number of qubits. We can verify that the power of a quantum computer doubles each time a qubit is added.

Table 1. State space of the qubit

# qubits	Possibilities	Power
1	0 or 1	2
2	00,01,10,11	4
3	000,001,010,011,100,101,110,111	8
⋮		⋮
N		2^N

A composite quantum state with two independent *qubits* like $|q\rangle = \alpha|0\rangle + \beta|1\rangle$ and $|p\rangle = \gamma|0\rangle + \delta|1\rangle$ are defined as $\alpha|00\rangle + \gamma|01\rangle + \delta|10\rangle + \beta|11\rangle$.

More formally, an intuitionistic fuzzy value $\tilde{x} = (x_1, x - 2)$ is represented by a composite quantum state with two independent *qubits* like $|q\rangle = \alpha|0\rangle + \beta|1\rangle$ and $|p\rangle = \gamma|0\rangle + \delta|1\rangle$, both define the following linear combination of the four classical states $|00\rangle, |01\rangle, |10\rangle, |11\rangle$, as state in Eq (10):

$$\alpha|00\rangle + \gamma|01\rangle + \delta|10\rangle + \beta|11\rangle \tag{10}$$

However, there are some combined quantum bits which are not of the form $q \otimes p$. For instance, the state: $1/\sqrt{2}|00\rangle + 1/\sqrt{2}|11\rangle$ is clearly not of the form $q \otimes p$, for any q and p. This kind of bi-dimensional quantum state which cannot be described using the tensor product operation is called *entangled*, and will be consired in further work.

4   Intuitionistic Fuzzy Bi-implications Using Quantum Computing

In this section, the description of quantum circuits for specifying the intuitionistic fuzzy bi-implication operator, extending previous works [14].

This operation was also studied in the visual programming environment VPE-qGM (Visual Programming Environment for the Quantum Geometric Machine Model), described in [19], which aims to support the modelling and simulation of quantum algorithms using a set of graphical interfaces.

The description of $IFSs$ from the QC viewpoint extends the work in [5] by modeling an element $\tilde{x}$ by a pair of quantum register:

$$(|x_1\rangle, |x_2\rangle) = \left(\sqrt{1 - x_1}|0\rangle + \sqrt{x_1}|1\rangle, \sqrt{1 - x_2}|0\rangle + \sqrt{x_2}|1\rangle\right) \qquad (11)$$

When modeling fuzzy operators in quantum computing, it is possible to represent the T_P through the *Toffoli* gate (T) and the standard negation through the *Pauli-X* gate (N). So the first step to generate the quantum representation for the B_{I_P} is to apply *De Morgan's law* to t-conorms S in Eq. 8 in order to remain only with T and N, resulting on the following fuzzy expressions for the membership and non-membership degrees:

$$\mu_{B_I}(\tilde{x}, \tilde{y}) = T(N(T(N(x_2), N(y_1))), N(T(N(y_2), N(x_1)))) \qquad (12)$$
$$\nu_{B_I}(\tilde{x}, \tilde{y}) = N(T(N(T(x_1, y_2)), N(T(y_1, x_2)))) \qquad (13)$$

Then, Eqs. (12) and (13) can be translated to the quantum representation, respectively showed in Eqs. (14) and (15). Using 10 qubits: 2 pairs for the inputs $(\tilde{x}, \tilde{y})$, 4 ancillaries qubits to store intermediate results and 1 pair for the result. The membership degree obtained is stored on qubit 9 and the non-membership degree on qubit 10.

$$|S_{\mu_{B_I}(\tilde{x}, \tilde{y})}\rangle = T_9^{5,6} \circ N_{1,4,6} \circ T_6^{1,4} \circ N_{1,4} \circ N_{2,3,5} \circ T_5^{2,3} \circ N_{2,3}$$
$$\left(|S_{\mu_{\tilde{x}}}\rangle, |S_{\nu_{\tilde{x}}}\rangle, |S_{\mu_{\tilde{y}}}\rangle, |S_{\nu_{\tilde{y}}}\rangle, |0\rangle, |0\rangle, |0\rangle, |0\rangle, |0\rangle, |0\rangle\right) \qquad (14)$$
$$|S_{\nu_{B_I}(\tilde{x}, \tilde{y})}\rangle = N_{10} \circ T_{10}^{7,8} \circ N_8 \circ T_8^{2,3} \circ N_7 \circ T_7^{1,4}$$
$$\left(|S_{\mu_{\tilde{x}}}\rangle, |S_{\nu_{\tilde{x}}}\rangle, |S_{\mu_{\tilde{y}}}\rangle, |S_{\nu_{\tilde{y}}}\rangle, |0\rangle, |0\rangle, |0\rangle, |0\rangle, |0\rangle, |0\rangle\right) \qquad (15)$$

Figure 1 shows the correspondent B_{I_P} representation as a quantum circuit given by the composition of Eq. (14) (from T1 to T7) and Eq. (15) (from T8 to T13).

The evolution of superposition quantum registers in modeling quantum circuit $B_I(\tilde{x}, \tilde{y})$ is shown in Table 2 for the most relevant points of the algorithm, presenting all non-void amplitudes related to 16 classical states of the initial multi-dimensional quantum state:

$$|\Phi\rangle = |x_1\rangle \otimes |x_2\rangle \otimes |y_1\rangle \otimes |y_2\rangle \otimes |0\rangle \otimes |0\rangle \otimes |0\rangle \otimes |0\rangle \otimes |0\rangle \otimes |0\rangle.$$

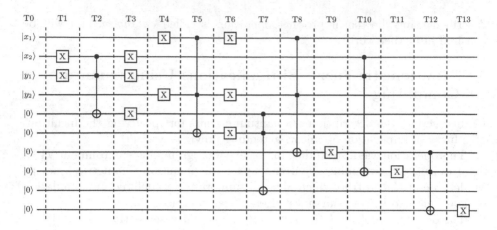

Fig. 1. Quantum circuit for fuzzy bi-implication.

One can observe that in Table 2, the evolution from $T0$ to $T13$ of the initial quantum state $|S_{\mu_{B_I}(\tilde{x},\tilde{y})}\rangle \otimes |S_{\nu_{B_I}(\tilde{x},\tilde{y})}\rangle$,. When a measure at the end of this computation is performed on 10th qubit and related to $|1\rangle$, the resulting non-normalized quantum state is given by the following expression:

$$|S_{\nu_{B_I}}(\tilde{x},\tilde{y})\rangle = \sqrt{(1-x_1)x_2y_1}|011\rangle \otimes \left(\sqrt{1-y_2}|010100\rangle + \sqrt{y_2}|111101\rangle\right) \otimes |1\rangle$$
$$+ \sqrt{x_1y_1y_2}|1\rangle \otimes \left(\sqrt{1-x_2}|01111011\rangle + \sqrt{x_2}|1111001\rangle\right) \otimes |1\rangle$$
$$+ \sqrt{x-1(1-y_1)y_2}|1\rangle \otimes \left(\sqrt{1-x_2}|00101010\rangle + \sqrt{x_2}|10111011\rangle\right) \otimes |1\rangle$$
$$+ \sqrt{x_1x_2y_1(1-y_2)}|111011101\rangle \otimes |1\rangle. \tag{16}$$

Moreover, the new amplitude is given as follows:

$$x_1y_1 + x_2y_1 - x_1x_2y_1y_2 \tag{17}$$

Therefore, according with Eq. (9), this is the expression of non-membership degree of a A-IFS related to an application of the B_I bi-implicator performed on the pair of intuitionist fuzzy value $(\tilde{x} = (\tilde{x}_1, x_2), \tilde{y} = (\tilde{y}_1, y_2)$.

Analogously, a measure at the end of this computation performed on 10th qubit and related to $|1\rangle$ provide the expression of corresponding membership degree, as given in Eq. (8).

5   Experiments and Results

For simulating the proposed operator, we build the quantum algorithm presented in Fig. 1 through VPE-qGM environment, contributing to the visualization and also the calculation of the results and the validation of the proposed algorithm.

Table 2. Evolution of superposition quantum registers in modeling quantum circuit $B_I(\tilde{x}, \tilde{y})$

Non-void amplitudes	T0	T3	T6	T7	T9	T11	T13
$(1-x_1)(1-x_2)(1-y_1)(1-y_2)$	0000000000	0000000000	0000000000	0000000000	0000001000	0000001000	0000001100
$(1-x_1)(1-x_2)(1-y_1)y_2$	0001000000	0001000000	0001010000	0001010000	0001011000	0001011000	0001011100
$(1-x_1)(1-x_2)y_1(1-y_2)$	0010000000	0010100000	0010100000	0010100000	0010101000	0010101000	0010101100
$(1-x_1)(1-x_2)y_1y_2$	0011000000	0011100000	0011110000	0011110010	0011111010	0011111010	0011111110
$(1-x_1)x_2(1-y_1)(1-y_2)$	0100000000	0100100000	0100100000	0100100000	0100101000	0100101000	0100101100
$(1-x_1)x_2(1-y_1)y_2$	0101000000	0101100000	0101110000	0101110010	0101111010	0101111010	0101111110
$(1-x_1)x_2y_1(1-y_2)$	0110000000	0110100000	0110100000	0110100000	0110101000	0110101000	0110101001
$(1-x_1)x_2y_1y_2$	0111000000	0111100000	0111110000	0111110010	0111111010	0111111010	0111111011
$x_1(1-x_2)(1-y_1)(1-y_2)$	1000000000	1000000000	1000010000	1000010000	1000011000	1000011000	1000011100
$x_1(1-x_2)(1-y_1)y_2$	1001000000	1001000000	1001010000	1001010000	1001010100	1001010100	1001010101
$x_1(1-x_2)y_1(1-y_2)$	1010000000	1010100000	1010110000	1010110010	1010111110	1010111110	1010111110
$x_1(1-x_2)y_1y_2$	1011000000	1011100000	1011110000	1011110010	1011110110	1011110110	1011110111
$x_1x_2(1-y_1)(1-y_2)$	1100000000	1100100000	1100110000	1100110010	1100111110	1100111110	1100111110
$x_1x_2(1-y_1)y_2$	1101000000	1101100000	1101110000	1101110010	1101110110	1101110110	1101110111
$x_1x_2y_1(1-y_2)$	1110000000	1110100000	1110110000	1110110010	1110111011	1110111010	1110111011
$x_1x_2y_1y_2$	1111000000	1111100000	1111110000	1111110010	1111110010	1111110010	1111110011

The VPE-qGM simulator provides interpretations of the quantum memory, quantum processes and computations related to transition quantum states obtained from the simulation of related QSs and QTs.

Values $\tilde{x} = (0.7, 0.2)$ and $\tilde{y} = (0.5, 0.3)$ were used on the simulation, given by the following quantum registers:

$$|x_1\rangle = \sqrt{0.3}|0\rangle + \sqrt{0.7}|1\rangle \ |x_2\rangle = \sqrt{0.8}|0\rangle + \sqrt{0.2}|1\rangle \qquad (18)$$

$$|y_1\rangle = \sqrt{0.5}|0\rangle + \sqrt{0.5}|1\rangle \ |y_2\rangle = \sqrt{0.7}|0\rangle + \sqrt{0.3}|1\rangle \qquad (19)$$

Figures 2 and 3 show the simulator VPE-qGM running the proposed quantum algorithm considering measurement on *qubit* 9 and 10 to obtain the result of the membership degree and non-membership degree, respectively.

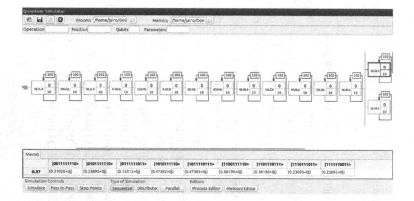

Fig. 2. Quantum process for B_I MD simulated using VPE-qGM.

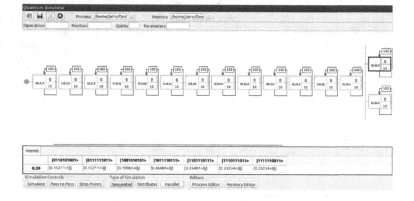

Fig. 3. Quantum process for B_I NMD simulated using VPE-qGM.

According to the results presented by the VPE-qGM simulator, *qubit* 9 will have the probability $p = 39\%$ of being one and *qubit* 10 will have the probability $p = 37\%$ of being one, which are correspondingly the membership degree and non-membership degree of the intuitionistic bi-implication for the used inputs.

6 Conclusions

This paper describes how to model bi-implications on intuitionistic fuzzy sets through concepts of QC. It was modelled using a quantum register using operations over fuzzy sets described by QTs. Therefore, this work shows another basic construction in the specification of fuzzy expert systems from QC, in order to obtain new information technologies based on intuitionistic fuzzy approach.

Computer systems based on IFL and performed over quantum computers may be able to generate an output from the manipulation of inaccurate data, by applying an imprecision rule-based system taking advantage of properties as quantum parallelism. The states are quantum registers, and the rules for the fuzzyfication process can be modeled by composition of controlled and unitary QTs.

This paper not only describes and analyses the operation of intuitionistic fuzzy bi-implication but also implements and simulates it in the VPE-qGM presenting an extension of such construction to be used in others important intuitionistic fuzzy operations.

Acknowledgements. The authors thank the partial funding of this project via 448766/2014-0 (MCTI/ CNPQ/ Universal 14/2014 - B), 310106/2016-8 (CNPq/PQ 12/2016).

References

1. Cordón, O., Herrera, F., Peregrín, A.: Applicability of the fuzzy operators in the design of fuzzy logic controllers. Fuzzy Sets Syst. **86**(1), 15–41 (1997)
2. Melnichenko, G.: Energy discriminant analysis, quantum logic, and fuzzy sets. J. Multivar. Anal. **101**(1), 68–76 (2010)
3. Nikam, S.R., Nikumbh, P.J., Kulkarni, S.P.: Fuzzy logic and neuro-fuzzy modeling. J. Artif. Intell. **3**(2), 74 (2012)
4. Rigatos, G.G., Tzafestas, S.G.: Parallelization of a fuzzy control algorithm using quantum computation. IEEE Trans. Fuzzy Syst. **10**(4), 451–60 (2002)
5. Mannucci, M.A.: Quantum fuzzy sets: Blending fuzzy set theory and quantum computation. arXiv preprint https://arxiv.org/abs/cs/0604064. 16 Apr 2006
6. Visintin, L., Maron, A., Reiser, R., Kreinovich, V.: Aggregation operations from quantum computing. In: 2013 IEEE International Conference on Fuzzy Systems (FUZZ), 7 July 2013, pp. 1–8. IEEE (2013)
7. Reiser, R., Lemke, A., Avila, A., Vieira, J., Pilla, M., Du Bois, A.: Interpretations on quantum fuzzy computing: intuitionistic fuzzy operations × quantum operators. Electron. Notes Theor. Comput. Sci. **30**(324), 135–50 (2016)

8. Atanassov, K.T.: On intuitionistic fuzzy negations. In: Reusch, B. (ed.) Computational Intelligence, Theory and Applications, pp. 159–167. Springer, Heidelberg (2006). https://doi.org/10.1007/3-540-34783-6_17
9. Hájek, P.: Metamathematics of Fuzzy Logic. Springer, Dordrecht (1998). https://doi.org/10.1007/978-94-011-5300-3
10. Olguín, C.A.: What is a Fuzzy Bi-implication? Master's thesis, Universidade Federal do Rio Grande do Norte
11. Klement, E.P., Mesiar, R., Pap, E.: Triangular Norms. Springer, Dordrecht (2013). https://doi.org/10.1007/978-94-015-9540-7
12. Baczyński, M., Jayaram, B.: An introduction to fuzzy implications. Fuzzy Implications. Studies in Fuzziness and Soft Computing, vol. 231, pp. 1–35. Springer, Heidelberg (2008). https://doi.org/10.1007/978-3-540-69082-5_1
13. Recasens, J.: Indistinguishability Operators: Modelling Fuzzy Equalities and Fuzzy Equivalence Relations, vol. 260. Springer, Heidelberg (2010). https://doi.org/10.1007/978-3-642-16222-0
14. Maron, A.K., Visintin, L., Abeijon, A., Reiser, R.H.S.: Interpreting fuzzy connectives from quantum computing- case study in reichenbach implication class. In: CBSF 2012: Congresso Brasileiro de Sistemas Fuzzy, Natal, Brasil, pp. 1–12. UFRN (2012)
15. Ávila, A., Schmalfuss, M., Reiser, R., Kreinovich, V.: Fuzzy xor classes from quantum computing. In: Rutkowski, L., Korytkowski, M., Scherer, R., Tadeusiewicz, R., Zadeh, L.A., Zurada, J.M. (eds.) ICAISC 2015. LNCS (LNAI), vol. 9120, pp. 305–317. Springer, Cham (2015). https://doi.org/10.1007/978-3-319-19369-4_28
16. Nielsen, M.A., Chuang, I.L.: Quantum Computation and Quantum Information: 10th Anniversary Edition, 10th edn. Cambridge University Press, New York (2011)
17. Fodor, J.C., Roubens, M.R.: Fuzzy Preference Modelling and Multicriteria Decision Support, vol. 14. Springer, Dordrecht (2013). https://doi.org/10.1007/978-94-017-1648-2
18. Yanofsky, N.S., Mannucci, M.A.: Quantum Computing for Computer Scientists. Cambridge University Press, Cambridge (2008)
19. Maron, A.K., Reiser, R.H.S., Pilla, M.L.: High-performance quantum computing simulation for the quantum geometric machine model. In: CCGRID 2013 IEEE/ACM International Symposium on Cluster, Cloud and Grid Computing, New York, pp. 1–8. IEEE (2013)

Interval Version of Generalized Atanassov's Intuitionistic Fuzzy Index

Lidiane Costa[1], Mônica Matzenauer[1], Adenauer Yamin[1], Renata Reiser[1(✉)],
and Benjamín Bedregal[2]

[1] Center of Tecnological Development - CDTEC,
Federal University of Pelotas - UFPel, Pelotas, RS, Brazil
`reiser@inf.ufpel.edu.br`
[2] Department of Informatics and Applied Mathematics,
Federal University of Rio Grande do Norte - UFRN, Natal, RN, Brazil

Abstract. This work extends the study of properties related to the Atanassov's interval-valued intuitionistic fuzzy entropy obtained as aggregation of generalized Atanassov's intuitionistic fuzzy index, by considering the concept of conjugate fuzzy implications and their dual constructions. Many ways to define the interval entropy were compared leading to the equation proposed in this work which is more sensitive to determine the interval entropy when using different interval-valued fuzzy sets.

Keywords: Generalized Atanassov's intuitionistic fuzzy index
Atanassov's interval-valued intuitionistic fuzzy entropy · Conjugation
Duality

1 Introduction

By allowing the expression related to the expert uncertainty in identifying a particular membership function or even to approximate the (unknown) membership degrees, the Atanassov interval-valued intuitionistic fuzzy logic (A-IvIFL) is an increasingly popular extension of fuzzy set theory.

The Atanassov's interval-valued intuitionistic fuzzy index (A-IvIFIx), called as hesitancy or indeterminacy degree of an element in an Atanassov-intuitionistic fuzzy set (A-IFS), provides either a measure of the lack of supporting information or a given incomplete/inconsistent proposition based on Atanassov-intuitionistic fuzzy logic (A-IFL). Thus, using intervals in $U = [0,1]$ that approximate the unknown data related to membership degrees, we are able to model applications in which experts do not have precise knowledge.

The concept of fuzzy entropy was introduced in order to measure how far a fuzzy set (FS) is from a crisp one [14]. Since then, this concept has been adapted to the distinct extensions of FSs and with many interpretations, describing the general measure of fuzziness through the mapping between fuzzy and real systems. Analogous interpretations lead to model data of the decision-making processes which cannot be measured precisely, taking extensions of the value of

© Springer International Publishing AG, part of Springer Nature 2018
G. A. Barreto and R. Coelho (Eds.): NAFIPS 2018, CCIS 831, pp. 217–229, 2018.
https://doi.org/10.1007/978-3-319-95312-0_19

entropy from a number to an interval value or even from an interval-valued intuitionistic value based on the definition of interval entropy [19].

This paper contributes with both approaches: (i) the new concept of generalized Atanassov's intuitionistic fuzzy index (A-GIFIx) associated with a strong intuitionistic fuzzy negation N_I [2], characterized in terms of fuzzy implication operators which is described by a construction method based on the action of automorphisms; and (ii) the Atanassov's intuitionistic fuzzy entropy (A-IFE), introduced by means of special aggregation functions of the A-GIFIx in [7].

Following the former approach, this work extends the study of the generalized Atanassov's interval-valued intuitionistic fuzzy index (A-GIvIFIx) [10], considering the concept of conjugate and dual interval-valued fuzzy implications, mainly interested in representation method [6,11] and providing impact on properties satisfied by the generated operations. Additionally, A-GIvIFIx associated with the standard negation together with the known Reichenbach interval-valued fuzzy implications are considered [13].

From the later approach, the Atanassov's interval-valued intuitionistic fuzzy entropy (A-IvIFE) is studied, describing main notions for measuring fuzziness degree or uncertain information in A-IvIFL. Such entropy is able to measure how far a set defined by actions of fuzzy connectives in A-IvIFL is from one in A-IvFL or A-IFL, and therefore, from a set in FL.

Our study mainly focuses on useful information entropy measures providing another way to explore IvIFL as a model by offering application developers as method of construction of A-IvIFE from A-GIvIFIx.

Among several papers found in the literature, see [12,19], connecting entropy measures for interval-valued intuitionistic fuzzy sets (IvIFSs) and discussing their relationships with similarity measures and inclusion measures. In [12], Jing and Min deal with the entropies of IvIFSs, proposing a λ-parametrized set of generalized entropy on IvIFSs and then it is proved that the new entropy is an increasing λ-parametrized function. In [19], a new axiomatical definition of entropy measure for A-IvIFL based on distance is proposed, which is consistent with the definition introduced in [14]. These formal studies underlying the main contribute for multi-criteria decision making problem, ranking the alternatives to study interval-valued fuzzy set models, offering application developers a method of construction of A-IvIFE from A-GIvIFIx preserving fuzziness and intuitionism based on generalized approach for an A-IvIFIx.

The preliminaries describe the basic properties of fuzzy connectives and basic concepts of A-IvIFL. The study of the A-GIvIFIx(N_I) and general results in the analysis of its properties are stated in Sect. 3. In Sect. 4, an interval version of entropy is presented based on the generalized Atanassov's intuitionistic fuzzy index. We also consider a relationship with IvIFIx and conjugate operators. Concluding, final remarks and further work are reported.

2   Preliminaries

Main results on interval-valued fuzzy connectives and IvIFSs are reported below.

2.1   Interval-Valued Fuzzy Connectives

Let $\mathbb{U} = \{[x_1, x_2] \,|\, x_1, x_2 \in U \text{ e } 0 \leq x_1 \leq x_2 \leq 1\}$. For each $x \in U$, a degenerate interval $[x, x]$ will be denoted by $\mathbf{x}$ and the subset of all degenerate interval will be denoted by $\mathbb{D}$. And, let $\leq_\mathbb{U} \subseteq \mathbb{U}^2$ be the Kulisch-Miranker (or product) order, such that for all $X, Y \in \mathbb{U}$, it is given by:

$$X \leq_\mathbb{U} Y \Leftrightarrow \underline{X} \leq \underline{Y} \text{ and } \overline{X} \leq_\mathbb{U} \overline{Y},$$

such that $\forall X, Y \in \mathbb{U}$, $\mathbf{0} = [0,0] \leq_\mathbb{U} X \leq_\mathbb{U} [1,1]$. We also consider $\preceq_\mathbb{U} \subseteq \mathbb{U}^2$ given as

$$\forall X, Y \in \mathbb{U} X \preceq_\mathbb{U} Y \Leftrightarrow \overline{X} \leq \underline{Y}.$$

By [5] an interval-valued aggregation (IvA) $\mathbb{M} : \mathbb{U}^n \to \mathbb{U}$ demands the conditions:

M1: $\mathbb{M}(\mathbf{X}) = \mathbf{0}$ and $\mathbf{X} = (\mathbf{0}, \dots, \mathbf{0})$; $\mathbb{M}(\mathbf{X}) = \mathbf{1}$ and $\mathbf{X} = (\mathbf{1}, \dots, \mathbf{1})$;
M2: If $\mathbf{X} = (X_1, \dots, X_n) \leq_{\mathbb{U}^n} \mathbf{Y} = (Y_1, \dots, Y_n)$ then $\mathbb{M}(\mathbf{X}) \leq_\mathbb{U} \mathbb{M}(\mathbf{Y})$;
M3: $\mathbb{M}(\mathbf{X}_\sigma) = \mathbb{M}(X_{\sigma_1}, \dots, X_{\sigma_n}) = \mathbb{M}(X_1, \dots, X_n) = \mathbb{M}(\mathbf{X})$.

Definition 1 *[18]. An interval function* $\mathbb{N} : \mathbb{U} \to \mathbb{U}$ *is an interval-valued fuzzy negation (IvFN) if, for all* $X, Y \in \mathbb{U}$, *it verifies the following conditions:*

N1: $\mathbb{N}([0,0]) = \mathbf{1}$; e $\mathbb{N}([1,1]) = \mathbf{0}$;
N2a: *If* $X \geq Y$ *then* $\mathbb{N}(X) \leq \mathbb{N}(Y)$.   N2b: *If* $X \subseteq Y$ *then* $\mathbb{N}(X) \supseteq \mathbb{N}(Y)$.

An IvFN $\mathbb{N}$ *is called strong IvFN [17] if* $\mathbb{N}$ *also satisfies the involutive property:*

N3: $\mathbb{N}(\mathbb{N}(X)) = X$, *for all* $X \in \mathbb{U}$,

The interval extension of the standard negation $N_S(x) = 1 - x$ is given as:

$$\mathbb{N}_S(X) = [1,1] - X = [1 - \overline{X}, 1 - \underline{X}], \forall X [\underline{X}, \overline{X}] \in \mathbb{U}. \tag{1}$$

The $\mathbb{N}$-dual operator of an interval-valued function $f : \mathbb{U}^n \to \mathbb{U}$ is given as

$$f_\mathbb{N}(X_1, \dots X_n) = \mathbb{N}(f(\mathbb{N}(X_1), \dots, \mathbb{N}(X_n)). \tag{2}$$

Definition 2 *[3]. A function* $\mathbb{I}(\mathbb{J}) : \mathbb{U}^2 \to \mathbb{U}$ *is a interval fuzzy (co)implication if for all satisfies the following boundary conditions:*
I1a: $\mathbb{I}(\mathbf{1}, \mathbf{1}) = \mathbb{I}(\mathbf{0}, \mathbf{0}) = \mathbb{I}(\mathbf{0}, \mathbf{1}) = \mathbf{1}$;   J1a: $\mathbb{J}(\mathbf{1}, \mathbf{1}) = \mathbb{J}(\mathbf{1}, \mathbf{0}) = \mathbb{J}(\mathbf{0}, \mathbf{0}) = \mathbf{0}$;
I1b: $\mathbb{I}(\mathbf{1}, \mathbf{0}) = \mathbf{0}$;   J1b: $\mathbb{J}(\mathbf{0}, \mathbf{1}) = \mathbf{1}$;
I2: *If* $X \leq Z$ *then* $\mathbb{I}(X, Y) \geq \mathbb{I}(Z, Y)$; J2: *If* $X \leq Z$ *then* $\mathbb{J}(X, Y) \geq \mathbb{J}(Z, Y)$;
I3: *If* $Y \leq Z$ *then* $\mathbb{I}(X, Y) \leq \mathbb{I}(X, Z)$; J3: *If* $Y \leq Z$ *then* $\mathbb{J}(X, Y) \leq \mathbb{J}(X, Z)$;

Additional properties can be demanded for IvFI(IvFJ):
I4: $\mathbb{I}(X, Y) = \mathbf{1} \Leftrightarrow X \leq_\mathbb{U} Y$;   J4: $\mathbb{J}(X, Y) = \mathbf{0} \Leftrightarrow X \geq_\mathbb{U} Y$;
I5: $\mathbb{I}(X, Y) = \mathbb{I}(\mathbb{N}(Y), \mathbb{N}(X))$, $\mathbb{N}$ is a SIvFN; J5: $\mathbb{J}(X, Y) = \mathbb{J}(\mathbb{N}(Y), \mathbb{N}(X))$, $\mathbb{N}$ is a SIvFN;
I6: $\mathbb{I}(X, Y) = \mathbf{0} \Leftrightarrow X = \mathbf{1}$ and $Y = \mathbf{0}$;   J6: $\mathbb{J}(X, Y) = \mathbf{1} \Leftrightarrow X = \mathbf{0}$ and $Y = \mathbf{1}$.

Analogously, these properties $\mathbb{I}_k(\mathbb{J}_k)$ can be restricted to fuzzy (co)implications by projections on $\mathbb{D}$ and will be denoted as $I_k(J_k)$.

Proposition 1. *[8, Prop 21], A fuzzy (co)implication $I(J) : U^2 \to U$ satisfies I1 (J1) and I2 (J2) iff the interval fuzzy (co)implication $\mathbb{I}(\mathbb{J})$ is given as*

$$\mathbb{I}(X,Y) = [I(\overline{X},\underline{Y}), I(\underline{X},\overline{Y})]; \quad (\mathbb{J}(X,Y) = [J(\overline{X},\underline{Y}), J(\underline{X},\overline{Y})]) . \tag{3}$$

Example 1. The interval-valued extension of the Reichenbach (co)implication $\mathbb{I}_{RH}(X,Y) = \mathbb{N}_S(X) + X \cdot Y$ ($\mathbb{J}_{RH}(X,Y) = \mathbb{N}_S(X) \cdot Y$) can be expressed as follows:

$$\mathbb{I}_{RH}(X,Y) = (N_S(\overline{X}) + \overline{X} \cdot \underline{Y}, N_S(\underline{X}) + \underline{X} \cdot \overline{Y}); \tag{4}$$

$$(\mathbb{J}_{RH}(X,Y) = [N_S(\underline{X}) \cdot \overline{Y}, N_S(\overline{X}) \cdot \underline{Y}]). \tag{5}$$

2.2 Interval-Valued Atanassov's Intuitionistic Fuzzy Sets

Based on [1] and later on [11], we briefly report main concepts and properties on interval-valued Atanassov's intuitionistic fuzzy sets (IvIFSs shortly).

An A-IvIFS $\mathbb{A}_I$ in a non-empty universe χ is expressed as

$$\mathbb{A}_I = \{(x, M_{A_I}(x), N_{A_I}(x)) : x \in \chi, M_{A_I}(x) + N_{A_I}(x) \leq_U 1\}, \tag{6}$$

and the set of all IvIFSs is denoted by $\mathcal{A}_I$. Thus, an intuitionistic fuzzy truth value of an element in $\mathbb{A}_I$ is given by a pair of intervals $(M_{A_I}(x), N_{A_I}(x))$, and

$$\tilde{U} = \{\tilde{X} = (X_1, X_2) \in U^2 : X_1 + X_2 \leq_U 1\} \tag{7}$$

denotes the set of all Atanassov's interval-valued intuitionistic fuzzy degrees[1] such that $(\tilde{U}, \leq_{\tilde{U}})$ and $(\tilde{U}, \preceq_{\tilde{U}})$ are partial ordered sets given as

$\mathbb{R}_I 1$: $\tilde{X} \leq_{\tilde{U}} \tilde{Y} \Leftrightarrow X_1 \leq_U Y_1$ and $X_2 \geq_U Y_2$;

$\mathbb{R}_I 2$: $\tilde{X} \preceq_{\tilde{U}} \tilde{Y} \Rightarrow X_1 \leq_U Y_1$ and $X_2 \leq_U Y_2$, for all $\tilde{X}, \tilde{Y} \in \tilde{U}$;

with $\tilde{0} = (0,1)$ and $\tilde{1} = (1,0)$ as the least and greatest elements on $\tilde{U}$, respectively. Additionally, an interval-valued Atanassov's intuitionistic fuzzy degree has two projections $l_{\mathbb{I}_i}, r_{\mathbb{I}_i} : \tilde{U} \to U$, defined by $l_{\mathbb{I}_i}(\tilde{X}) = X_1$ and $r_{\mathbb{I}_i}(\tilde{X}) = X_2$. When $X_1 + X_2 = 1$ then A_I is restricted to the set $\mathcal{A}$ of all interval-valued fuzzy sets. Moreover, the function $\pi_{A_I} : \chi \to U$, called an interval-valued Atanassov's intuitionistic fuzzy index (A-IvIFIx shortly), related to an IvIFS A_I, is given as

$$\pi_{A_I}(x) = \mathbb{N}_S(M_{A_I}(x) + N_{A_I}(x)). \tag{8}$$

An IvIFIx models not only the uncertainty degree but also the hesitancy/indeterminance degree of each x in A_I. The difference between A_I and B_I is given by:

$$A_I - B_I = \{\tilde{X} = (\inf(N_{A_I}(x), N_{B_I}(x)), \sup(N_{A_I}(x), M_{B_I}(x))) : \tilde{X} \in \tilde{U}, x \in \chi\}.$$

According with [18], an interval-valued Atanassov's intuitionistic fuzzy negation (IvIFN) $\mathbb{N}_I : \tilde{U} \to \tilde{U}$ satisfies, for all $\tilde{X}, \tilde{Y} \in \tilde{U}$, the following properties:

---

[1] We assume the componentwise addition on U, see [16].

$\mathbb{N}_I 1$: $\mathbb{N}_I(\tilde{\mathbf{0}}) = \mathbb{N}_I(\mathbf{0}, \mathbf{1}) = \tilde{\mathbf{1}}$ and $\mathbb{N}_I(\tilde{\mathbf{1}}) = \mathbb{N}_I(\mathbf{1}, \mathbf{0}) = \tilde{\mathbf{0}}$;

$\mathbb{N}_I 2$: If $\tilde{X} \geq_{\tilde{U}} \tilde{Y}$ then $\mathbb{N}_I(\tilde{x}) \leq_{\tilde{U}} \mathbb{N}_I(\tilde{y})$.

Moreover, $\mathbb{N}_I$ is a strong IvIFN if it also verifies the involutive property:

$\mathbb{N}_I 3$: $\mathbb{N}_I(\mathbb{N}_I(\tilde{X})) = \tilde{X}, \forall \tilde{X} \in \tilde{U}$.

Consider $\mathbb{N}_I$ as IvIFN and $\mathbb{F}_I : \tilde{U}^n \to \tilde{U}$. By [18], the $\mathbb{N}_I$-dual interval-valued Atanassov's intuitionistic function of $\tilde{f}$, denoted by $\mathbb{F}_{I\mathbb{N}_I} : \tilde{U}^n \to \tilde{U}$, is given by:

$$\mathbb{F}_{I\mathbb{N}_I}(\tilde{\mathbf{X}}) = \mathbb{N}_I(\mathbb{F}_I(\mathbb{N}_I(\tilde{X}_1), \dots, \mathbb{N}_I(\tilde{X}_n))), \forall \tilde{\mathbf{X}} = (\tilde{X}_1, \dots, \tilde{X}_n) \in \tilde{U}^n. \qquad (9)$$

When $\tilde{\mathbb{N}}_I$ is a strong IvIFN, $\tilde{f}$ is a self-dual interval-valued intuitionistic function. And, by [18], taking a strong IvFN $\mathbb{N} : U \to U$, a IvIFN $\mathbb{N}_{S_I} : \tilde{U} \to \tilde{U}$ such that

$$\mathbb{N}_{S_I}(\tilde{X}) = (\mathbb{N}(\mathbb{N}_S(X_2)), \mathbb{N}_S(\mathbb{N}(X_1))), \qquad (10)$$

is a strong IvIFN generated by the IvFNs $\mathbb{N}$ and $\mathbb{N}_S$. By [6]), a strong IvIFN is also a representable IvIFN. Additionally, if $\mathbb{N} = \mathbb{N}_S$, Eq. (10) can be reduced to $\mathbb{N}_{S_I}(\tilde{X}) = (X_2, X_1)$. Moreover, the complement of A-IvIFS A_I is defined by

$$\mathbb{A}_{I_c} = \{(x, N_{A_I}(x), M_{A_I}(x)) : x \in \chi, M_{A_I}(x) + N_{A_I}(x)) \leq_U \mathbf{1}\}, \qquad (11)$$

An interval-valued Atanassov's intuitionistic automorphism (A-IvIA) is a bijection increasing operator $\Phi : \tilde{U} \to \tilde{U}$. For all $\tilde{X}, \tilde{Y} \in \tilde{U}$, the following hold:

$\mathbb{A}_I 1$: $\Phi(\tilde{\mathbf{1}}) = \tilde{\mathbf{1}}$ and $\Phi(\tilde{\mathbf{0}}) = \tilde{\mathbf{0}}$;

$\mathbb{A}_I 2$: $\Phi \circ \Phi^{-1}(\tilde{X}) = \tilde{X}$;

$\mathbb{A}_I 3$: $\tilde{X} \leq_{\tilde{U}} \tilde{Y}$ iff $\Phi(\tilde{X}) \leq_{\tilde{U}} \Phi(\tilde{Y})$.

In the set of all A-IvIAs $(Aut(\tilde{U}))$, the conjugate function of $f_I : \tilde{U}^n \to \tilde{U}$ is a function $f_I^{\Phi} : \tilde{U}^n \to \tilde{U}$, defined as follows

$$f_I^{\Phi}(\tilde{X}_1, \dots, \tilde{X}_n) = \Phi^{-1}(f_I(\Phi(\tilde{X}_1), \dots, \Phi(\tilde{X}_n))). \qquad (12)$$

Reporting main results in [9, Theorem 17], let $\phi : U \to U$ be an interval-valued automorphism, $\phi \in Aut(U)$. Then, a ϕ-representability of Φ is given by

$$\Phi(\tilde{X}) = (\phi(l_{\tilde{U}}(\tilde{X})), \mathbf{1} - \phi(1 - r_{\tilde{U}}(\tilde{X}))), \forall \tilde{X} \in \tilde{U}; \qquad (13)$$

Moreover, if $\phi \in Aut(U)$, for all $\tilde{X} \in \tilde{U}$, a ϕ_U-representability of Φ is given by

$$\Phi(\tilde{X}) = \left([\phi_U(\underline{X_1}), \phi_U(\overline{X_1})], [1 - \phi_U(1 - \underline{X_2}), 1 - \phi_U(1 - \overline{X_2})]\right). \qquad (14)$$

Thus, if an IvIA is ϕ-representable, it is also a ϕ_U-representable automorphism [18].

3 Interval Extension of the Generalized Atanassov's Intuitionistic Fuzzy Index

Definition 3. *Let* $\mathbb{N}$ *be a strong IvFN. A function* $\Pi : \tilde{\mathbb{U}} \to \mathbb{U}$ *is called a **generalized interval-valued intuitionistic fuzzy index** $(A - GIvIFIx(\mathbb{N}))$ if, for all* $X_1, X_2, Y_1, Y_2 \in \mathbb{U}$, *it holds that:*

$\Pi 1$: $\Pi(X_1, X_2) = 1$ *iff* $X_1 = X_2 = 0$;
$\Pi 2$: $\Pi(X_1, X_2) = 0$ *iff* $X_1 + X_2 = 1$;
$\Pi 3$: *If* $(Y_1, Y_2) \preceq_{\tilde{\mathbb{U}}} (X_1, X_2)$ *then* $\Pi(X_1, X_2) \leq_{\mathbb{U}} \Pi(Y_1, Y_2)$;
$\Pi 4$: $\Pi(X_1, X_2) = \Pi(\mathbb{N}_{\mathbb{S}I}(X_1, X_2))$ *when* $\mathbb{N}_{\mathbb{S}I}$ *is given by Eq.(10).*

3.1 Relationship with Interval-Valued Fuzzy Connnectivess

In the following, Theorem 1 extends main results in [2].

Theorem 1. *Let* $\mathbb{I}(\mathbb{J}) : \mathbb{U}^2 \to \mathbb{U}$ *be a (co)implicator verifying* $\mathbb{I}1\,(\mathbb{J}2), \mathbb{I}4\,(\mathbb{J}4), \mathbb{I}5\,(\mathbb{J}5)$ *and* $\mathbb{I}6\,(\mathbb{J}6)$ *and* $\mathbb{N} : \mathbb{U} \to \mathbb{U}$ *be an involutive IvFN. A function* $\Pi_{\mathbb{N},\mathbb{I}}(\Pi_{\mathbb{N},\mathbb{J}}) : \tilde{\mathbb{U}} \to \mathbb{U}$ *is* $A\text{-}GIvIFIx(\mathbb{N})$ *iff it can be given as*

$$\Pi_{\mathbb{N},\mathbb{I}}(X) = \mathbb{N}(\mathbb{I}(\mathbb{N}_{\mathbb{S}}(X_2), X_1)) \quad (\Pi_{\mathbb{N},\mathbb{J}}(X) = \mathbb{J}(\mathbb{N}(\mathbb{N}_{\mathbb{S}}(X_2)), \mathbb{N}(X_1))). \quad (15)$$

Proof. Equation(15b) is proved below. Analogously, it can be done to Eq.(15a).

($\Rightarrow$) Consider that $\mathbb{J} : \mathbb{U}^2 \to \mathbb{U}$ verifies $\mathbb{J}2, \mathbb{J}4, \mathbb{J}5$ and $\mathbb{J}6$, it holds that:

$\Pi_1 : \Pi_{\mathbb{N},\mathbb{J}}(X_1, X_2) = 1 \Leftrightarrow \mathbb{J}(\mathbb{N}(\mathbb{N}_{\mathbb{S}}(X_2)), \mathbb{N}(X_1)) = 1$(by Eq.(15b))
$\Leftrightarrow \mathbb{N}_{\mathbb{S}}(X_2) = 1 \, and \, \mathbb{N}(X_1) = 1 \Leftrightarrow X_2 = X_1 = 0$(by $\mathbb{J}6, \mathbb{N}1$).

$\Pi_2 : \Pi_{\mathbb{N},\mathbb{J}}(X_1, X_2) = 0 \Leftrightarrow \mathbb{J}(\mathbb{N}(\mathbb{N}_{\mathbb{S}}(X_2)), \mathbb{N}(X_1)) = 0$ (by Eq.(15b))
$\Leftrightarrow \mathbb{N}(\mathbb{N}_{\mathbb{S}}(X_2)) \geq_{\mathbb{U}} \mathbb{N}(X_1)$(by $\mathbb{J}4$)
$\Leftrightarrow \mathbb{N}_{\mathbb{S}}(X_2) \leq_{\mathbb{U}} X_1$(by $\mathbb{N}3$) and $\mathbb{N}_{\mathbb{S}}(X_2) \geq_{\mathbb{U}} X_1$(by Eq.(7))
$\Leftrightarrow X_1 + X_2 = 1.$

$\Pi_3 : (Y_1, Y_2) \preceq (X_1, X_2) \Rightarrow Y_1 \leq_{\mathbb{U}} X_1 \, and \, Y_2 \leq X_2$(by $\mathbb{R}_I 2$)
$\Rightarrow \mathbb{N}(X_1) \geq_{\mathbb{U}} \mathbb{N}(Y_1)$ and $\mathbb{N}(\mathbb{N}_{\mathbb{S}}(X_2)) \leq_{\mathbb{U}} \mathbb{N}(\mathbb{N}_{\mathbb{S}}(Y_2))$(by $\mathbb{N}2$)
$\Rightarrow \mathbb{J}(\mathbb{N}(\mathbb{N}_{\mathbb{S}}(X_2)), \mathbb{N}(X_1)) \leq_{\mathbb{U}} \mathbb{J}(\mathbb{N}(\mathbb{N}_{\mathbb{S}}(Y_2)), \mathbb{N}(Y_1))$(by $\mathbb{J}1, \mathbb{J}2$)
$\Rightarrow \Pi_{\mathbb{N},\mathbb{J}}(X_1, X_2) \leq_{\mathbb{U}} \Pi_{\mathbb{N},\mathbb{J}}(Y_1, Y_2)$(by Eq.(15))

$\Pi_4 : \Pi_{\mathbb{N},\mathbb{I}}(\mathbb{N}(X_1, X_2)) = \mathbb{J}(\mathbb{N}(\mathbb{N}_{\mathbb{S}}(X_2)), \mathbb{N}_{\mathbb{S}}(\mathbb{N}(X_1))$(by Eq.(10))
$= \mathbb{J}(X_1, \mathbb{N}_{\mathbb{S}}(X_2))$(by Eq.(15))
$= \mathbb{J}(\mathbb{N}(\mathbb{N}_{\mathbb{S}}(X_2))), \mathbb{N}(X_1)) = \Pi_{\mathbb{N},\mathbb{J}}(X_1, X_2)$(by $\mathbb{J}5$ and Eq.(15))

($\Leftarrow$) Considering the function $\mathbb{J} : \mathbb{U}^2 \to \mathbb{U}$ given as $\mathbb{J}(X_1, X_2) = 1$, if $X_1 > X_2$; and $\mathbb{J}(X_1, X_2) = \Pi_{\mathbb{N},\mathbb{J}}(X_2, \mathbb{N}_{\mathbb{S}}(\mathbb{N}(X_1)))$, otherwise. The following holds:

$$\mathbb{J}2 : Y_1 \geq Y_2 \Leftrightarrow \mathbb{J}(X, Y_1) = \begin{cases} 1, \text{if } X > Y_1, \\ \Pi_{\mathbb{N}, \mathbb{J}}(Y_1, \mathbb{N}_S(\mathbb{N}(X))), \text{otherwise; (by Eq.(15);)} \end{cases}$$

$$\geq \begin{cases} 1, \text{if } X > Y_2, \\ \Pi_{\mathbb{N}, \mathbb{J}}(Y_2, \mathbb{N}_S(\mathbb{N}(X))) = \mathbb{J}(X, Y_2), \text{otherwise; (by } \Pi 3 \text{ and Eq.(15))} \end{cases}$$

$\mathbb{J}4$: Straightforward.

$$\mathbb{J}5 : \mathbb{J}(\mathbb{N}(X_2), \mathbb{N}(X_1)) = \begin{cases} 1, \text{if } \mathbb{N}(X_2) > \mathbb{N}(X_1), \\ \Pi_{\mathbb{N}, \mathbb{J}}(\mathbb{N}(\mathbb{N}_S(\mathbb{N}(X_1)), X_2), \text{otherwise; (by Eqs.(15) and (10))} \end{cases}$$

$$= \begin{cases} 1, \text{ if } X_1 \geq X_2, \\ \Pi_{\mathbb{N}, \mathbb{J}}(\mathbb{N}(\mathbb{N}_S(X_2), \mathbb{N}(X_1)), \text{otherwise (by } \Pi 4 \text{ and N3)} \end{cases}$$

$$= \mathbb{J}(X_1, X_2), (\text{Eq.(15)})$$

$\mathbb{J}6 : \mathbb{J}(X_1, X_2) = \mathbf{1} \Leftrightarrow \Pi_{\mathbb{N}, \mathbb{J}}(\mathbb{N}(X_2), \mathbb{N}_S\mathbb{N}(X_1)) = \mathbf{1}$ (by Eq.(15))

$$\Leftrightarrow \mathbb{N}(X_2) = \mathbb{N}_S(\mathbb{N}(X_1)) = \mathbf{0} \Leftrightarrow X_1 = \mathbf{0} \, and \, X_2 = \mathbf{1} \text{ (by } \Pi_1)$$

Therefore, Theorem 1 holds.

The Φ-representability and $\mathbb{N}$-dual IvIFIx constructions are discussed below.

Proposition 2. *Let* $\mathbb{I}_\mathbb{N}$ *($\mathbb{J}_\mathbb{N}$) be the $\mathbb{N}$-dual operator of a (co)implication* $\mathbb{I}(\mathbb{J})$. *The following holds:*

$$\Pi_{\mathbb{N}, \mathbb{I}_\mathbb{N}}(\tilde{\mathbb{X}}) = \Pi_{\mathbb{N}, \mathbb{I}}(\tilde{\mathbb{X}}) \quad \left(\Pi_{\mathbb{N}, \mathbb{J}_\mathbb{N}}(\tilde{\mathbb{X}}) = \Pi_{\mathbb{N}, \mathbb{J}}(\tilde{\mathbb{X}}) \right). \tag{16}$$

Proof. $\Pi_{\mathbb{N}, \mathbb{I}_\mathbb{N}}(\tilde{\mathbb{X}}) = \mathbb{I}_\mathbb{N}(\mathbb{N}(\mathbb{N}_S(X_2)), \mathbb{N}(X_1)) = \mathbb{N}(\mathbb{I}(\mathbb{N}_S(X_2), X_1)) = \Pi_{\mathbb{N}, \mathbb{I}}(\tilde{\mathbb{X}}), \forall \tilde{\mathbb{X}} \in \tilde{\mathbb{U}}$.

Corollary 1. *When* $\mathbb{N} = \mathbb{N}_S$, *Eq.(15) in Theorem 1 is given as*

$$\Pi_{\mathbb{N}_S, \mathbb{I}}(\tilde{X}) = \mathbb{N}_S(\mathbb{I}(\mathbb{N}_S(X_2), X_1)) \quad \left(\Pi_{\mathbb{N}_S, \mathbb{J}}(\tilde{X}) = \mathbb{J}(X_2, \mathbb{N}_S(X_1)) \right). \tag{17}$$

Proposition 3. *Let* $\mathbb{N}$ *be an N-representable IvFN and* $\pi_{N, I} : \tilde{U} \to U$ *be A-IFIx(N). If* $\mathbb{I}, \mathbb{J}$ *are representable (co)implications given by Eq.(3), a function* $\Pi_{\mathbb{N}, \mathbb{I}} : \tilde{U} \to U$ *given by Eq.(17) can be expressed as*

$$\Pi_{\mathbb{N}, \mathbb{I}}(\tilde{X}) = [\Pi_{N, I}(\overline{X}_2, \overline{X}_1), \Pi_{N, I}(\underline{X}_2, \underline{X}_1)] \, (\Pi_{\mathbb{N}, \mathbb{J}}(\tilde{X}) = [\Pi_{N, J}(\overline{X}_2, \overline{X}_1), \Pi_{N, J}(\underline{X}_2, \underline{X}_1)]).$$

Proof. We proof Eq.(18a), the other one can be analogously done. By taking $(X_1, X_2) \in \tilde{U}$, $X_1 = [\underline{X}_1, \overline{X}_1], X_2 = [\underline{X}_2, \overline{X}_2]$ then $X_1 + X_2 = [\underline{X}_1 + \underline{X}_2, \overline{X}_1 + \overline{X}_2] \leq \mathbf{1}$, meaning that $\overline{X}_1 + \overline{X}_2 \leq 1$ and $\underline{X}_1 + \underline{X}_2 \leq 1$. Then, we have the result $\Pi_{\mathbb{N}, \mathbb{I}}(\tilde{X}) = \mathbb{N}(\mathbb{I}([1 - \overline{X}_2, 1 - \underline{X}_2], [\underline{X}_1, \overline{X}_1])) = [N(I(1 - \overline{X}_2, \overline{X}_1)), N(I(1 - \underline{X}_2, \underline{X}_1))]$. Concluding, $\Pi_{\mathbb{N}, \mathbb{I}}(\tilde{X}) = [\Pi_{N, I}(\overline{X}_2, \overline{X}_1), \Pi_{N, I}(\underline{X}_2, \underline{X}_1)]$. So, Proposition 3 holds.

Example 2. Consider $\mathbb{I}_{RC}$ and related $\mathbb{N}_S$-dual construction $\Pi_{\mathbb{N}_S, \mathbb{J}_{RC}}$. By preserving the conditions of Proposition 3, Eq.(18) can be expressed as

$$\Pi_{\mathbb{N}_S, \mathbb{I}_{RC}}(X_1, X_2) = \begin{cases} 0, \text{ if } X_1 + X_2 = \mathbf{1}, \\ 1 - [1 - \overline{X}_2 - \overline{X}_1 + \overline{X}_2 \overline{X}_1, 1 - \underline{X}_2 - \underline{X}_1 + \underline{X}_2 \underline{X}_1], \text{otherwise.} \end{cases}$$

3.2   Relationship with Interval-Valued Automorphisms

Proposition 4. *Let* $\mathbb{N}^{\Phi} : \mathbb{U} \to \mathbb{U}$ *be the ϕ-conjugate of a strong IvFN* $\mathbb{N} : \mathbb{U} \to \mathbb{U}$ *and* $\phi : \mathbb{U} \to \mathbb{U}$ *be a ϕ-representable IvA given by Eq.(14). When* $\Phi : \tilde{\mathbb{U}} \to \tilde{\mathbb{U}}$ *is a Φ-representable IvIFa given by Eq.(13), a function* $\Pi^{\Phi} : \tilde{\mathbb{U}} \to \mathbb{U}$ *given by*

$$\Pi^{\Phi}(X_1, X_2) = (\phi^{-1}(\Pi(\phi(X_1)), 1 - \phi(1 - X_2)), \tag{18}$$

is a $A - GIvIFIx(\mathbb{N}_I)$ *whenever* $\Pi : \tilde{\mathbb{U}} \to \mathbb{U}$ *is also a* $A - GIvIFIx(\mathbb{N}_I)$.

Proof. Let $\phi : \tilde{\mathbb{U}} \to \mathbb{U}$ be a ϕ-representable A-IvA and $\Pi : \tilde{\mathbb{U}} \to \mathbb{U}$ be a $A - GIvIFIx(N_I)$. It holds that:

$\mathbf{\Pi_1} : \Pi^{\Phi}(X_1, X_2) = \mathbf{1} \Leftrightarrow \phi^{-1}(\Pi(\phi(X_1), 1 - \phi(1 - X_2))) = \mathbf{1}$ (by Eq.(18))

$\quad \Leftrightarrow \Pi(\phi(X_1), 1 - \phi(1 - X_2)) = \mathbf{1}$ (by $\mathbb{A}_I 1$)

$\quad \Leftrightarrow \phi(X_1) = \mathbf{0} \, and \, 1 - \phi(1 - X_2) = \mathbf{0}$ (by $\mathbf{\Pi_1}$)

$\quad \Leftrightarrow X_1 = \mathbf{0} \, and \, X_2 = \mathbf{0}$ (by $\mathbb{A}_I 1$)

$\mathbf{\Pi_2}$:It is analogous to $\mathbf{\Pi_1}$.

$\mathbf{\Pi_3} : (X_1, X_2) \preceq (Y_1, Y_2) \Rightarrow X_1 \leq_{\mathbb{U}} Y_1 \, and \, X_2 \leq_{\mathbb{U}} Y_2$ by $\preceq$−relation

$\quad \Rightarrow \phi(X_1) \leq_{\mathbb{U}} \phi(Y_1) \, and \, 1 - \phi(1 - X_2) \leq_{\mathbb{U}} 1 - \phi(1 - Y_2)$ by $\mathbb{A}_I 1$

$\quad \Rightarrow \Pi(\phi(X_1), 1 - \phi(1 - X_2)) \leq_{\mathbb{U}} \Pi(\phi(Y_1), 1 - \phi(1 - Y_2))$ by $\mathbf{\Pi_3}$

$\quad \Rightarrow \phi^{-1}(\Pi(\phi(X_1), 1 - \phi(1 - X_2))) \leq_{\mathbb{U}} \phi^{-1}(\Pi(\phi(Y_1), 1 - \phi(1 - Y_2)))$ by $\mathbb{A}_1$

$\quad \Rightarrow \Pi_G^{\phi}(X_1, X_2) \leq_{\mathbb{U}} \Pi_G^{\phi}(Y_1, Y_2)$ (by Eq.(13)).

Let $\mathbb{N}_I$ be a strong IvIFN given by Eq.(10) and $\mathbb{N}_I^{\Phi}$ its Φ−conjugate function.

$\mathbf{\Pi_4} : \Pi^{\Phi}\left(\mathbb{N}_I^{\Phi}(X_1, X_2)\right) = \Phi^{-1}\left(\Pi(\Phi \circ \Phi^{-1}(\mathbb{N}_I(\Phi(X_1, X_2))))\right)$ (by Eq. (18))

$\quad = \Phi^{-1}(\Pi(\mathbb{N}_I(\phi(X_1, X_2)))) = \Phi^{-1}(\Pi(\Phi(X_1, X_2))) = \Pi(X_1, X_2)$ (by $\mathbf{\Pi_4}$)

The new results follow from Proposition 4 and Theorem 1.

Corollary 2. *In conditions of Proposition 4 and also considering ϕ-representable IvA given by Eq.(14), we can express Eq.(18) as follows:*

$$\Pi^{\Phi}(X_1, X_2) = \left(\Pi^{\phi}(\underline{X_1}, \underline{X_2}), \Pi^{\phi}(\overline{X_1}, \overline{X_2})\right). \tag{19}$$

Corollary 3. *Let Φ be a ϕ-representable automorphism in $Aut(\tilde{\mathbb{U}})$ and $\mathbb{I}(\mathbb{J}) : \tilde{\mathbb{U}}^2 \to \tilde{\mathbb{U}}$ be the corresponding ϕ-conjugate operator related to a (co)implication $\mathbb{I}(\mathbb{J}) : \mathbb{U}^2 \to \mathbb{U}$, verifying the conditions of Theorem 1. And, let $\mathbb{N}^{\Phi}$ be a strong ϕ-conjugate IvFN negation. A function $\Pi_{\mathbb{N},\mathbb{I}^{\phi}}(\Pi_{\mathbb{N},\mathbb{J}^{\phi}}) : \tilde{\mathbb{U}} \to \mathbb{U}$ given by*

$$\Pi_{\mathbb{N},\mathbb{I}}^{\Phi}(X_1, X_2) = \mathbb{N}^{\Phi}(\mathbb{I}^{\Phi}(\mathbb{N}_S(X_2), X_1)) \tag{20}$$

$$\left(\Pi_{\mathbb{N},\mathbb{J}}^{\Phi}(X_1, X_2) = \mathbb{J}^{\phi}(\mathbb{N}^{\Phi}(\mathbb{N}_S(X_2), \mathbb{N}^{\Phi}(X_1)))\right). \tag{21}$$

is an $A - IvGIFIx(\mathbb{N})$ *whenever* $\Pi_{\mathbb{N},\mathbb{I}}(\Pi_{\mathbb{N},\mathbb{J}}) : \tilde{\mathbb{U}} \to \mathbb{U}$ *is also a* $A - GIvIFIx(\mathbb{N})$.

Example 3. Consider $\mathbb{I}_{\mathrm{RC}}$ and related Φ-conjugate construction $\Pi^{\Phi}_{\mathrm{N_S,J_{RC}}}$ given by Eq.(18). For a ϕ-representable IvIA, taking $\phi(X) = X^n$ and n as an integer non-negative integer, we have the following:

$$\Pi^{\Phi}_{\mathrm{N_S,I_{RC}}}(X_1, X_2) = \left[\sqrt[n]{(1 - \overline{X}_1^n)(1 - \overline{X}_2)^n}; \; \sqrt[n]{(1 - \underline{X}_1^n)(1 - \underline{X}_2)^n} \right]. \quad (22)$$

4   Interval-Valued Intuitionistic Fuzzy Entropy

This section generalizes results from [7, Definition 2] also discussing properties related to the Atanassov's interval-valued intuitionistic fuzzy entropy (A-IvIFE) which are obtained by action of an interval-valued aggregation of A-GIvIFIx.

Definition 4. *An interval-valued function* $\mathbb{E} : \mathcal{A}_I \to \mathbb{U}$ *is called an A-IvIFE if* $\mathbb{E}$ *verifies the following properties:*

$\mathbb{E}2$: $\mathbb{E}(A_I) = \mathbf{0} \Leftrightarrow A_I \in \mathcal{A}$;
$\mathbb{E}2$: $\mathbb{E}(A_I) = \mathbf{1} \Leftrightarrow M_{A_I}(x) = N_{A_I}(x) = \mathbf{0}$, $\forall x \in \chi$;
$\mathbb{E}3$: $\mathbb{E}(A_I) = \mathbb{E}(A_{I_c})$;
$\mathbb{E}4$: *If* $A_I \preceq_{\tilde{\mathrm{U}}} B_I$ *then* $\mathbb{E}(A_I) \geq_{\mathrm{U}} \mathbb{E}(B_I)$, $\forall A_I, B_I \in \mathcal{A}_I$.

Now, main properties of A-IvIFE obtained by A-GIvIFI are studied [15].

Theorem 2. *Consider* $\chi = \{x_1, \ldots, x_n\}$. *Let* $\mathbb{M} : \mathbb{U}^n \to \mathbb{U}$ *be an automorphism,* $\mathbb{N}$ *be a strong IvFN and* $\Pi \in Aut(\tilde{U})$. *A function* $\mathbb{E} : \mathcal{A}_I \to \mathbb{U}$ *given by*

$$\mathbb{E}(A_I) = \mathbb{M}_{i=1}^n \Pi(A_I(x_i)), \forall x_i \in \chi, \quad (23)$$

is an A-IvIFE in the sense of Definition 4.

Proof. Let A_{I_c} be the complement of A_I given by Eq.(11). For all $x_i \in \chi$ and $A_I, B_I \in \mathcal{A}_I$, we have that:

$\mathbb{E}1$: $\mathbb{E}(A_I) = \mathbf{0} \Leftrightarrow \mathbb{M}_{i=1}^n \Pi(A_I(x_i)) = \mathbf{0}$. By M1, $\mathbb{E}(A_I) = \mathbf{0} \Leftrightarrow M_{A_I}(x_i) + N_{A_I}(x_i) = \mathbf{1}$, $\forall x_i \in \chi$. Then, by $\Pi2$, $\mathbb{E}(A_I) = \mathbf{0} \Leftrightarrow A_I \in \mathcal{A}$.
$\mathbb{E}2$: $\mathbb{E}(A_I) = \mathbf{1} \Leftrightarrow \mathbb{M}_{i=1}^n \Pi(A_I(x_i)) = \mathbf{1}$. By M1, $\mathbb{E}(A_I) = \mathbf{1} \Leftrightarrow M_{A_I}(x_i) + N_{A_I}(x_i) = \mathbf{0}$, meaning that $M_{A_I}(x_i) = N_{A_I}(x_i) = \mathbf{0}$.
$\mathbb{E}3$: $\mathbb{E}(A_I)_c = \mathbb{M}_{i=1}^n \Pi(A_{I_c}(x_i)) = \Pi(\mathbb{N}_I(X_1, X_2))$. By $\Pi3$, the following holds $\mathbb{E}(A_I)_c = \Pi(X_1, X_2)$. Concluding, $\mathbb{E}(A_I)_c = \mathbb{E}(A_I)$.
$\mathbb{E}4$: If $A_I \preceq_{\tilde{\mathrm{U}}} B_I$ then $A_I(x_i) \preceq_{\tilde{\mathrm{U}}} B_I(x_i)$. Based on $\Pi3$, it holds that $\Pi(B_I(x_i)) \leq_{\mathrm{U}} \Pi(A_I(x_i))$. By M3, we obtain that $\mathbb{M}_{i=1}^n \Pi(B_I(x_i)) \leq_{\mathrm{U}} \mathbb{M}_{i=1}^n \Pi(B_I(x_i))$. As conclusion, $\mathbb{E}(A_I) \geq_{\mathrm{U}} \mathbb{E}(B_I)$.

Therefore, Theorem 2 is verified.

Proposition 5. *Consider* $\chi = \{x_1, \ldots, x_n\}$. *Let* $\mathbb{M} : \mathbb{U}^n \to \mathbb{U}$ *be an IvA,* $\mathbb{N}$ *be a strong IvFN and* $\Pi_{\mathrm{N,I}}(\Pi_{\mathrm{N,J}}) : \tilde{\mathbb{U}} \to \mathbb{U}$ *is A-GIvIFIx($\mathbb{N}$) given by Eq.(15). Then, for all* $x_i \in \chi$, *an A-IvIFE* $\mathbb{E} : \mathcal{A}_I \to \mathbb{U}$ *can be given by*

$$\mathbb{E}_{\Pi_{\mathrm{N,I}}}(A_I) = \mathbb{M}_{i=1}^n \Pi_{\mathrm{N,I}}(A_I(x_i)) \quad \left(\mathbb{E}_{\Pi_{\mathrm{N,I}}}(A_I) = \mathbb{M}_{i=1}^n \Pi_{\mathrm{N,J}}(A_I(x_i)) \right). \quad (24)$$

Proof. Straightforward Theorems 1 and 2.

Corollary 4. *Consider* $\mathbb{N} = \mathbb{N}_S$, *A-GIvIFIx* $(\mathbb{N}_S)$ $\varPi_{\mathbb{N},\mathbb{I}}$ *given by Eq.(15). Then, by taking* $A_I(x_i) = (M_{A_I}(x_i), N_{A_I}(x_i)) = (X_{1i}, X_{2i})$ *for all* $x_i \in \chi$, *an A-IvIFE* $\mathbb{E} : \mathcal{A}_I \rightarrow \mathbb{U}$ *which is given in Eq.(24) can be expressed as*

$$\mathbb{E}_{\varPi_{\mathbb{N},\mathbb{I}}}(A_I) = \mathbb{M}_{i=1}^n \left(\mathbb{N}_S(\mathbb{I}(\mathbb{N}_S(X_{2i}), X_{1i})) \right) \left(\mathbb{E}_{\varPi_{\mathbb{N},\mathbb{J}}}(A_I) = \mathbb{M}_{i=1}^n \mathbb{J}(X_{2i}, \mathbb{N}_S(X_{1i})) \right). \quad (25)$$

Proof. Straightforward Proposition 5 and Theorem 1.

Example 4. By taking the arithmetic mean as an aggregation operator, $\mathbb{I}_{RC}$ in Eq.(4) and related IvIFIx given in Eq.(18). Let A_I be an IvIFS defined by pairs $(X_{1i}, X_{2i}) \in \tilde{\mathbb{U}}$, for all $x_i \in \chi$, an IvIFE as $\mathbb{E}_{\varPi_{\mathbb{N}_S, \mathbb{I}_{RB}}}(X_{1i}, X_{2i}) = \frac{1}{n} \sum_{i=1}^n \mathbb{N}_S(\mathbb{I}_{RB}(\mathbb{N}_S(X_{2i}), X_{1i})$ can be given as follows:

$$\mathbb{E}_{\varPi_{\mathbb{N}_S, \mathbb{I}_{RB}}}(X_{1i}, X_{2i}) = \frac{1}{n} \sum_{i=1}^n [1 - \overline{X}_{2i} - \overline{X}_{1i} + \overline{X}_{2i} \overline{X}_{1i}, 1 - \underline{X}_{2i} - \underline{X}_{1i} + \underline{X}_{2i} \underline{X}_{1i}]. \quad (26)$$

4.1   Relationship with Intuitionistic Index and Conjugate Operators

Conjugation operator and duality properties related to generalized Atanassov's Intuitionistic Fuzzy Index are reported from [10].

Proposition 6. *Consider* $\chi = \{x_1, \ldots, x_n\}$ *and* $\varPhi \in Aut(\tilde{\mathbb{U}})$ *a* ϕ-representable *A-IvIFA given by Eq.(13). When* $\varPi$ *is* $A - GIvIFIx(\mathbb{N})$, *an A-IvIFE is a function* $\mathbb{E}^\varPhi : \mathcal{A}_I \rightarrow \mathbb{U}$ *defined by*

$$\mathbb{E}^\varPhi(A_I) = \mathbb{M}^{\phi}{}_{i=1}^n \varPi^\phi(A_I(x_i)), \forall x_i \in \chi. \quad (27)$$

Proof. Based on Eqs.(12) and (13), the following holds:

$$\mathbb{E}^\varPhi(A_I(x_i)) = \mathbb{E}^\varPhi(A_I) = \phi^{-1}(\mathbb{E}(\phi(l_{\tilde{\mathbb{U}}}(A_I(x_i))), \mathbf{1} - \phi(\mathbf{1} - r_{\tilde{\mathbb{U}}}(A_I(x_i)))$$
$$= \phi^{-1}\left(\mathbb{M}_{i=1}^n (\phi \circ \phi^{-1}) \varPi(\phi(l_{\tilde{\mathbb{U}}}(A_I(x_i))), \mathbf{1} - \phi(\mathbf{1} - r_{\tilde{\mathbb{U}}}(A_I(x_i))) \right)$$
$$= \phi^{-1}\left(\mathbb{M}_{i=1}^n (\phi(\varPi^\phi(A_I(x_i)))) = \mathbb{M}^{\phi}{}_{i=1}^n \varPi^\phi(A_I(x_i)) \right)$$

Figure 1 summarizes the main results related to the classes of A-GIvIFIx($\mathbb{N}$) and A-IvIFE denoted by $\mathcal{C}(\varPi)$ and $\mathcal{C}(\mathbb{E})$, respectively. This A-IvIFE is obtained not only from generalized IvIFIx [4] but also from dual and conjugate operators.

Proposition 7. *Let* $\varPhi$ *be a* ϕ-representable *automorphism in* $Aut(\tilde{\mathbb{U}})$ *and* $\mathbb{I}(\mathbb{J}) :$ $\tilde{\mathbb{U}}^2 \rightarrow \tilde{\mathbb{U}}$ *be the corresponding* ϕ-conjugate *operator related to a (co)implication* $\mathbb{I}(\mathbb{J}) : \mathbb{U}^2 \rightarrow \mathbb{U}$, *verifying the conditions of Theorem 1. Additionally, let* $\mathbb{N}^\varPhi$ *be a strong* ϕ-conjugate *IvFN negation and* $\mathbb{M} : \mathbb{U}^n \rightarrow \mathbb{U}$ *be an aggregation function. Then, for* $A \in \mathcal{A}$, *the functions* $\mathbb{E}_{\mathbb{N},\mathbb{I}}, \mathbb{E}_{\mathbb{N},\mathbb{I}}^\varPhi(\mathbb{E}_{\mathbb{N},\mathbb{I}}, \mathbb{E}_{\mathbb{N},\mathbb{I}}^\varPhi) : \mathbb{A} \rightarrow \mathbb{U}$ *given by*

$$\mathbb{E}_{\mathbb{N},\mathbb{I}}(A)(x_i) = \mathbb{M}_{i=1}^n \mathbb{N}(\mathbb{I}(\mathbf{1} - N_{A_I}(x_i), M_{A_I}(x_i))), \quad (28)$$

$$\mathbb{E}_{\mathbb{N},\mathbb{I}}^\varPhi(A)(x_i) = \mathbb{M}^{\phi}{}_{i=1}^n \mathbb{N}^\phi(\mathbb{I}^\phi(\mathbf{1} - N_{A_I}(x_i), M_{A_I}(x_i))); \quad (29)$$

$$\mathbb{E}_{\mathbb{N},\mathbb{J}}(A)(x_i) = \mathbb{M}_{i=1}^n \mathbb{J}(N_{A_I}(x_i), \mathbf{1} - M_{A_I}(x_i)) \quad (30)$$

$$\mathbb{E}_{\mathbb{N},\mathbb{J}}^\varPhi(A)(x_i) = \mathbb{M}^{\phi}{}_{i=1}^n \mathbb{J}^\phi(N_{A_I}(x_i), \mathbf{1} - M_{A_I}(x_i)). \quad (31)$$

express an interval-valued Atanassov's intuitionistic fuzzy entropy.

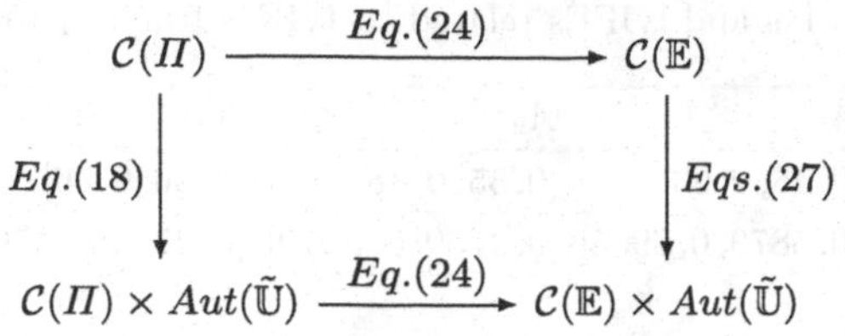

$$C(\Pi) \xrightarrow{\quad Eq.(24) \quad} C(\mathbb{E})$$

$$Eq.(18) \Big\downarrow \qquad\qquad\qquad \Big\downarrow Eqs.(27)$$

$$C(\Pi) \times Aut(\tilde{\mathbb{U}}) \xrightarrow{\quad Eq.(24) \quad} C(\mathbb{E}) \times Aut(\tilde{\mathbb{U}})$$

Fig. 1. Conjugate construction of $A - GIFIx(N)$ and $A - IFE$ on $Aut(\tilde{U})$

Proof. Straightforward from Proposition 6.

Example 5. By Eqs.(28) and (26), an IvIFE expression is obtained as follows:

$$\mathbb{E}^{\Phi}_{\mathbb{N}_S, \mathbb{I}_{RB}}(A)(x_i) = \frac{1}{n} \sum_{i=1}^{n} \left[\sqrt[n]{(1 - \overline{X}_{1i}^n)(1 - \overline{X}_{2i})^n}; \sqrt[n]{(1 - \underline{X}_{1i}^n)(1 - \underline{X}_{2i})^n} \right] \quad (32)$$

4.2 Preserving Fuzzyness and Intuitionism Based on IvIFE

Based on [12], assuming that $\chi = \{u\}$, $A_1 = \{(u, [0.1, 0.2], [0.3, 0.4])\}$ and $A_2 = \{(u, [0.2, 0.3], [0.4, 0.5])\}$ in order to calcule the entropies by equations below

$$\mathbb{E}_Y(A) = \frac{1}{n} \sum_{i=1}^{n} \left[\sqrt{2} \, cos \frac{\mu_A(x_i) + \overline{\mu_A}(x_i) - \nu_A(x_i) - \overline{\nu_A}(x_i)}{8} \pi - 1 \right] \frac{1}{\sqrt{2} - 1} \quad (33)$$

$$\mathbb{E}_G(A) = \frac{1}{n} \sum_{i=1}^{n} cos \frac{|\mu_A(x_i) - \nu_A(x_i)| + |\overline{\mu_A}(x_i) - \overline{\nu_A}(x_i)|}{8} \pi. \quad (34)$$

Thus, $\mathbb{E}(A_1)$ and $(E(A_2))$ contains the difference between the membership and non-membership degrees related to the hesitancy degree. However, despite the differences, the same value for related IvIFEs are matched, making impossible to distinguish the fuzziness and intuitionism of these two cases. Intuitively, it is easy to observe that A_1 is more fuzzy than A_2, meaning that $\pi_{A_1} \geq \pi_{A_2}$. However, this cannot be seen by using the above Eqs.(33) and (34). So, a more sensitive definition of IvIFE is introduced in order to deal with this problem.

In our proposed methodology, we calculate the related IvIFEs by using Eqs.(26) and (32) together with corresponding IvIFIx given by Eqs.(18) and (22). See these results presented in 1st and 2nd columns of Table 1 when the inputs are given as A_1 and A_2. Since χ is singleton IvIFS, the resulting hesitant degree and corresponding entropy measure coincide. Additionally, it is possible to naturally preserve properties of related interval entropy, meaning that IvIFE is an order preserving index, by including IFE. Moreover, taking $A_3 = [0.2, 0.2], [0.3, 0.3]$ and $A_4 = [0.3, 0.3], [0.4, 0.4]$ as inputs, the entropy values obtained with the degenerate intervals related to membership and non-membership degrees are included in the interval entropy obtained with non-degenerated interval-valued inputs. See these results in the 3rd and 4th columns of Table 1.

Table 1. IvIFIxs and IvIFEs related to IvIFSs from A_1 to A_4IvIFSs.

$IvIFIx$	A_1	A_2	A_3	A_4
$\Pi(A_i) = \mathbb{E}_{\Pi(A_i)}$	$[0,48;0,63]$	$[0,35;0,48]$	$[0,56;0,56]$	$[0,42;0,42]$
$\Pi^\phi(A_i) = \mathbb{E}_\Pi^\phi(A_i)$	$[0,5879;0,6965]$	$[0,4769;0,5879]$	$[0,4704;0,4704]$	$[0,3276;0,3276]$

5 Conclusion

The generalized concept of the Atanassov's interval-valued intuitionistic fuzzy index was studied by dual and conjugate construction methods. We also extend the study of Atanassov's intuitionistic fuzzy entropy based on such two constructors. Further work considers the extension of such study related to properties verified by the $A - GIvIFIx(N)$ and $A - IvIFE$ and also the use of admissible linear orders to compare the results of the interval entropy, since, in some cases, the values of interval entropy cannot be compared using the Moore's method.

Acknowledgment. Work supported by the Brazilian funding agencies *CAPES*, *MCTI/CNPQ*, Universal (448766/2014-0), *PQ* (310106/2016-8), *CNPq/PRONEX/ FAPERGS* and *PqG* 02/2017.

References

1. Atanassov, K., Gargov, G.: Elements of intuitionistic fuzzy logic. Fuzzy Sets Syst. **95**(1), 39–52 (1998)
2. Barrenechea, E., Bustince, H., Pagola, M., Fernàndez, J., Sanz, J.: Generalized Atanassov's intuitionistic fuzzy index: construction method. In: Proceedings of IFSA EUSFLAT Conference, Lisbon, Portugal, pp. 20–24 (2009)
3. Bedregal, B., Santiago, R., Dimuro, G., Reiser, R.: Interval valued R-implications and automorphisms. In: Pre-Proceedings of the 2nd Workshop on Logical and Semantic Frameworks, with Applications, pp. 82–97 (2007)
4. Bustince, H., Burillo, P., Soria, F.: Automorphisms, negations and implication operators. Fuzzy Sets Syst. **134**, 209–229 (2003)
5. Bustince, H., Montero, J., Barrenechea, E., Pagola, M.: Semiautoduality in a restricted family of aggregation operators. Fuzzy Sets Syst. **158**, 1360–1377 (2007)
6. Bustince, H., Barrenechea, E., Pagola, M.: Generation of interval-valued fuzzy and Atanassov's intuitionistic fuzzy connectives from fuzzy connectives and from $K-$alpha operators: laws for conjunctions and disjunctions, amplitude. Intell. Syst. **23**, 680–714 (2008)
7. Bustince, H., Barrenechea, E., Pagola, M., Fernandez, J., Guerra, C., Couto, P., Melo-Pinto, P.: Generalized Atanassov's intuitionistic fuzzy index: construction of Atanassov's fuzzy entropy from fuzzy implication operators. Int. J. Uncertain. Fuzziness Knowl.-Based Syst. **19**, 51–69 (2011)
8. Baczyński, M., Jayaram, B.: On the characterization of (S, N)-implications. Fuzzy Sets Syst. **158**, 1713–1727 (2007)
9. Costa, C., Bedregal, B., Dória Neto, A.: Relating De Morgan triples with Atanassov's intuitionistic De Morgan triples via automorphisms. Int. J. Approx. Reason. **52**, 473–487 (2011)

10. Costa, L., Matzenauer, M., Zanottelli, R., Nascimento, M., Finger, A., Reiser, R., Yamin, A., Pilla, M.: Analysing fuzzy entropy via generalized Atanassov's intuitionistic fuzzy indexes. Mathw. Soft Comput. **42**, 22–31 (2017)
11. Cornelis, G., Deschrijver, G., Kerre, E.: Implications in intuitionistic fuzzy and interval-valued fuzzy set theory: construction, classification and application. Int. J. Approx. Reason. **35**, 55–95 (2004)
12. Jing, L., Min, S.: Some entropy measures of interval-valued intuitionistic fuzzy sets and their applications. Adv. Model. Optim. **15**, 211–221 (2013)
13. Lin, L., Xia, Z.: Intuitionistic fuzzy implication operators: expressions and properties. J. Appl. Math. Comput. **22**, 325–338 (2006)
14. Luca, A., Termini, S.: A definition of nonprobabilistic entropy in the setting of fuzzy sets theory. Inf. Control **20**, 301–312 (1972)
15. Miguel, L., Santos, H., Sesma-Sara, M., Bedregal, B., Jurio, A., Bustince, H.: Type-2 fuzzy entropy sets. IEEE Trans. Fuzzy Syst. **25**, 993–1005 (2017)
16. Moore, E.: Interval arithmetic and automatic error analysis in digital computing. Stanford University (1962)
17. Reiser, R.H.S., Dimuro, G.P., Bedregal, B.C., Santiago, R.H.N.: Interval valued QL-implications. In: Leivant, D., de Queiroz, R. (eds.) WoLLIC 2007. LNCS, vol. 4576, pp. 307–321. Springer, Heidelberg (2007). https://doi.org/10.1007/978-3-540-73445-1_22
18. Reiser, R., Bedregal, B.: Correlation in interval-valued Atanassov's intuitionistic fuzzy sets - conjugate and negation operators. Int. J. Uncertain. Fuzziness Knowl.Based Syst. **25**, 787–820 (2017)
19. Zhang, Q., Xing, H., Liu, F., Ye, J., Tang, P.: Some new entropy measures for interval-valued intuitionistic fuzzy sets based on distances and their relationships with similarity and inclusion measures. Inf. Sci. **283**, 55–69 (2014)

Fuzzy Ontologies: State of the Art Revisited

Valerie Cross[(✉)] and Shangye Chen

Computer Science and Software Engineering,
Miami University, Oxford, OH 45056, USA
crossv@miamioh.edu

Abstract. Although ontologies have become the standard for representing knowledge on the Semantic Web, they have a primary limitation, the inability to represent vague and imprecise knowledge. Much research has been undertaken to extend ontologies with the means to overcome this and has resulted in numerous extensions from crisp ontologies to fuzzy ontologies. The original web ontology language, and tools were not designed to handle fuzzy information; therefore, additional research has focused on modifications to extend them. A review of the fuzzy extensions to allow fuzziness in ontologies, web languages, and tools as well as several very current examples of fuzzy ontologies in real-world applications is presented.

Keywords: Ontologies · Fuzzy logic · Web ontology language
OWL · Fuzzy formal concept analysis · Ontology tools · Semantic web

1 Introduction

An ontology is a shared explicit specification of a conceptualization formalizing concepts pertaining to a domain, properties of these concepts, and relationships existing between the concepts [1]. An ontology is the main knowledge representation method for describing information on the Semantic Web and promotes the inclusion of semantic content in web pages. Ontologies are both understandable to humans and expressed in a machine-readable format using a web ontology language. Being understandable to humans is important to representing knowledge, but one capability initially missing in ontologies is the ability to represent and manage imprecision and vagueness that often exists in domain knowledge as understood by humans.

This need to handle uncertainty in ontologies motivated researchers in knowledge representation to propose fuzzy ontologies. With fuzzy ontologies, the ability is provided to model real world environments that naturally include uncertainty with fuzzy set theory and mathematics and through "computing with words" [2], that is, the use of linguistic terms represented by fuzzy sets. This ability is extremely important to knowledge being extracted from human experts since they often are more comfortable with the use of inexact, fuzzy linguistic terms rather than precise numbers. Besides human imprecision in domain knowledge modeling, some concepts themselves cannot be precisely model and require a formalism to allow for a vague specification and yet still require the ability to be used in a reasoning process. Other factors also contribute to the need for modeling uncertainty in ontologies. For example, a wide variety of

© Springer International Publishing AG, part of Springer Nature 2018
G. A. Barreto and R. Coelho (Eds.): NAFIPS 2018, CCIS 831, pp. 230–242, 2018.
https://doi.org/10.1007/978-3-319-95312-0_20

knowledge sources may require the integration of diverse inexact specifications in modeling the domain knowledge. Although many different logical formalisms have been proposed to extend ontologies for handling uncertainty, this paper addresses only extensions using type-1 fuzzy sets.

The more recent research in ontological knowledge representation for the Semantic Web has had a proliferation of approaches to defining, constructing and using fuzzy ontologies. Numerous places exists where uncertainty can occur in the specification of a fuzzy ontology since there are many ways that domain knowledge can contain uncertainty and vagueness from human description. This paper outlines some of the progression of the development of fuzzy ontologies from simplest to the more complex and provides examples illustrating the fuzzy extensions. It is hoped that this paper can provide readers an introduction and an a more intuitively understanding of fuzzy ontology development so that formalisms used in other fuzzy ontology research papers which are referenced further in the paper are better understood. Section 2 examines where and why an ontology becomes a fuzzy ontology and how "fuzzy" is the ontology, that is, can only certain features of the ontology be allowed to have uncertainty. Once an ontology designer has identified places and/or levels where uncertainty needs to be specified what languages and tools are available to specify this uncertainty in the fuzzy ontology. Section 3 discusses current fuzzy languages and tools to describe fuzzy ontologies and examples of the different approaches to actually building a fuzzy ontology. Several very current uses of fuzzy ontologies in a wide variety of domains ranging from medical and transportation and for different tasks such as information retrieval and opinion-mining of social media platforms are described in Sect. 4. Section 5 summarizes and discusses some limitations of fuzzy ontologies that hinder them from becoming more widespread and presents areas for future work to increase their use.

2 What Makes an Ontology Fuzzy?

The answer to the question varies depending what the research on fuzzy ontologies is focused on. At what place and for what purpose is the uncertainty, imprecision or vagueness introduced? The simplest answer is given as "a fuzzy ontology is simply an ontology which uses fuzzy logic to provide a natural representation of imprecise and vague knowledge and eases reasoning over it" [3]. This vague definition indicates no universal standard definition of fuzzy ontologies exists since the needs of different applications require different fuzzy extensions to an ontology.

There are many places where fuzzy extensions may be made since an ontology has many components. An ontology consists of concepts C, instances of those concepts I, hierarchical or taxonomic relationships between concepts H, attributes specified in defining concepts A, properties that are nonhierarchical relationships between concepts P, and axioms that must hold for the concepts X. Some of these components may not be present; for example, simple ontologies, do not have axioms specified. All these components are similar to those found in an object-oriented database. Research on fuzzy object oriented databases previously addressed many of these issues when fuzzy extensions were proposed to object oriented databases [43]. The natural fuzzy extensions made to ontologies to produce fuzzy ontologies parallel those seen in fuzzy object oriented databases.

One of the earliest and instinctive ways to declare an ontology as fuzzy is in the definition of the concept hierarchy, that is, in the taxonomic or hierarchical relationships H such as equivalence, generalization, specialization, and part of. Here the purpose for introducing fuzziness is that a concept may not be a crisp subconcept of another concept. The concepts themselves are not fuzzy. Initial proposals for fuzzy ontologies centered on subjective judgments of membership degrees in defining the concept hierarchy. For example, suppose a concept of "vehicle" has been defined and a subconcept of "bicycle" is defined. Then a domain expert's subjective judgment might be that a "bicycle" has a 0.8 membership with respect to a "vehicle." Here what is fuzzy is the hierarchical structure of the ontology. Typically, this approach is more focused on an intensional ontology such as a vocabulary structure and there is no actual extensional ontology, i.e., no instances for the ontology.

Another approach used in terminological ontologies first assumes an initial crisp ontology hierarchy as a starting point and then uses information obtained from a data source such as a corpus to assist in determining membership degrees between concepts in the ontology hierarchy. An example of this approach can be found in [4] where an ontological terminology is created and the hierarchical relationships are interpreted as "narrower-than" and "broader-than". A fuzzy binary relation is naturally used to specify the fuzziness between two ontological concepts. Given a set of domain concepts C, a set of fuzzy binary relations R is defined for the ontology. Each fuzzy binary relation r_C in R_C is of the form $r_C: C \times C \rightarrow d_C$, where d_C in [0, 1] is the strength of the relationship between the two concepts. The relationship could be hierarchical or associative; for example, the concept 'purchase' and the concept 'bill' may have an associative relationship.

Another very instinctive place for fuzziness is simply using the notion that a fuzzy concept is specified by a fuzzy set of instances that belong to it. Each instance has a membership degree in the fuzzy concept. Given a set of domain concepts C and a set of instances I, a set of fuzzy sets F_C is defined for the ontology. Each element in F_C is a fuzzy concept that has its own membership function u_{fc} where fc is the fuzzy concept label. The membership function takes the form $u_{fc}: I \rightarrow d_{fc}$, where d_{fc} is the membership degree in [0, 1] of the instance in the fuzzy concept. Again this is nothing more than using the natural definition of a fuzzy set for defining a fuzzy concept, i.e., the instances of the fuzzy concept have a degree of membership in [0, 1]. The complicating factor is how to determine these membership degrees. As before, a domain expert might specify the membership degree subjectively. For example, the fuzzy concept 'likeable person' could be created that contains person instances but with a membership degree based on how likeable the rater considers that person.

A more formal method may be used to determine the instance's membership degree in the fuzzy concept such as specifying what attributes or properties specified for the concept are the basis for the membership function defining the fuzzy concept [44]. For example, using the attribute age, a fuzzy concept 'old person' might be defined as those instances that are crisply people but that have an age that is considered old. The membership degree in the fuzzy concept 'old person' is then determined using a comparison or fuzzy set similarity measure [5] between the fuzzy linguistic term 'old' and the person's age. Here if the value of the attribute age is precise, the membership degree of that precise value in 'old' can be used to specify the membership degree of that instance of person in the fuzzy concept 'old person'. In this example, there is no fuzziness between the concept 'person' and 'old person' since an 'old person' is crisply a 'person', i.e., no fuzziness exists in the taxonomic relationship between the two concepts. The fuzzy extension is needed to define the fuzzy concept 'old person' only.

Some researchers have described a fuzzy concept as a concept with attributes A and/or properties P that may have fuzzy set values. In the previous example for 'old person', the 'person' concept itself might be defined to have an attribute age, which in addition to precise values, permits fuzzy set values such as 'middle-age' or 'old'. Other researchers, however, might not consider 'person' a fuzzy concept just because it has attribute such as age that may have a fuzzy set or fuzzy linguistic term for its value. That fuzzy set value may still be compared to the fuzzy linguistic term 'old' to determine its similarity degree. That similarity degree then specifies the membership degree of the instance of the concept 'person' whose age is, for example, 'middle-age' for belonging to the fuzzy concept 'old person'.

A property for an instance of a concept has a value, which is a set of instances that have a non-taxonomic relationship to that instance. For example, the property 'friendOf' for an instance of 'person' has links to all instances of 'person' that are considered friends of that instance. Concept properties may allow fuzzy sets for their values. For example, the property 'friendOf' for an instance of 'person' along with the link may be associated the membership degree of this instance in the fuzzy property 'friendOf'. This membership degree most likely is a subjective membership degree indicating the strength of the friendship. A membership function, however, could be defined based on other attribute or property values contained within the instances themselves, for example, number of years knowing the person.

Fuzzy extensions become more complicated when fuzziness is introduced both in the hierarchical structure of the ontology and in defining concepts, i.e., a fuzzy hierarchy and fuzzy concepts in the same ontology. One issue becomes defining the interaction between these two different places of fuzziness. For example, consider a fuzzy concept N which is defined as a fuzzy subconcept of the fuzzy concept M with a strength of d in the hierarchical relation. A generalization principle for fuzzy extensions to the hierarchy of an ontology is that instance i in N with membership degree $u_N(i)$ requires $u_N(i) \leq u_M(i)$. The instance i cannot belong more to the subconcept than it belongs to the superconcept. The issue is the fuzzy operation used to determine the membership degree of i in the fuzzy concept M; that is, how do d, the strength of the hierarchical relation between N and M, and $u_N(i)$ interact in the calculation of this membership degree $u_M(i)$.

Fuzzy extensions to axioms are necessary to allow fuzziness in the various places needed within the ontology. In order to have fuzzy concepts, properties and individuals, the axioms must allow specifying this possibility. The relationships among these concepts, properties and individuals use the fuzzy extensions to the axioms. Description logics (DL) is a family of logics for representing structured knowledge that is concept-based. DL has a well-defined model-theoretic semantics [6] and has as its basis Attributive Language with Complement (ALC). To allow the ontology hierarchy to permit fuzziness, fuzzy extensions to the axioms defining the ontology hierarchy are necessary. The first introduction of fuzziness to DL was used only in terminological ontologies and concept knowledge [7]. This early proposal was followed up with simple fuzzy extensions to *ALC* [8, 9]. To handle concept instances, that is, concrete instances knowledge, not simply concept definitions, numerous researchers proposed fuzzy extensions to produce more sophisticated fuzzy DLs.

The first tableaux based reasoning algorithm for a fuzzy←—DL uses specifically min for conjunction, max for union, and 1 − x for negation, i.e., the standard fuzzy operators and the Kleene-Dienes implication operator [10]. As previously discussed, many places exist where fuzziness may be needed in an ontology. Researchers have focused on fuzzy extensions dependent on the specific need, for example, fuzzy concept inclusions [11], fuzzy quantifiers [12], fuzzy modifiers [13], fuzzy nominals and fuzzy comparison expressions [14]. Due to space limitations, this paper is unable to review the numerous proposals for fuzzy extensions to DL and tableau reasoning methods. A more detailed discussion is presented in [15]. An earlier survey on fuzzy DL is given in [16]. A very recent survey on fuzzy description logics in [17] presents a clear explanation of a prototypical fuzzy DL but does not provide an intuitive explanation of the natural fuzzy extensions to ontologies. In [18] fuzzy DLs are classified into four different categories that are described based on their usage or their properties, for example, their tractability. A more recent presentation on fuzzy reasoners can be found in [45].

Since numerous researchers have introduced fuzziness in a wide variety of locations within an ontology, the simplest definition of a fuzzy ontology is an ontology that depends on fuzzy set theory and logic in any needed ontology location to express uncertainty that also can take many different forms such as vagueness, imprecision, or degree of truth within an ontological representation.

3 Approaches to Constructing and Managing a Fuzzy Ontology

In order to use fuzzy ontologies, a means of constructing and representing them for real world applications is required. First, the primary representation languages for fuzzy ontologies are summarized mostly in historical order. Then examples of tools created to develop and manage fuzzy ontologies are presented. Finally various approaches to building and manipulating fuzzy ontologies are examined.

3.1 Representation Languages

The Semantic Web has had numerous proposals and iterations for an ontology representation language over the years. The earlier accepted standard languages are Resource Description Framework (RDF) and its RDF Schema (RDFS) languages. RDF and RDFS have been used mostly for creating vocabularies in hierarchically structured organization. Early attempts extended the syntax and semantics to include fuzzy information for these languages such as in [19] where the focus is representing uncertainty in the trust of RDF statements and in [20] where they introduced fuzzy extensions for RDFS. Later the semantics of fuzzy logic is more formally introduced for the purposes of querying over RDF triples [21] with the use of t-norms and r-implications. The natural extension for this feature is associating a truth degree with an RDF triple, for example, (Boise rdf:type StateCapital):0.9. Since the World Wide Web Consortium (W3C), however, later recommended the web ontology language OWL [22] as its standard, much research has focused on proposing extensions to OWL for representing a fuzzy ontology.

An easy approach for fuzzy extensions is to add to the current capabilities of OWL to specify vague information without modification of the original OWL language. The fuzzy OWL language simply adds to the same syntax of the crisp OWL language. The additions are only needed to specify, for example, extensions specifying membership degrees of instances in fuzzy concepts and for the definition of fuzzy concepts. One of the early approaches [23] FOWL began with fuzzy DL and created fuzzy constructors, axioms, and constraints by mapping fuzzy terms to fuzzy DL. It defines fuzzy classes, fuzzy relationships and based on the fuzzy DL ALC and extends some OWL axioms. This proposal, however, only presented a theoretical analysis and did not provide any algorithms for reasoning algorithm. A fuzzy relation is also specified as simply a set of triples $\{<u, v, \mu_R(u, v)> \mid u \in U, v \in V\}$ where the $\mu_R(u, v)$ is the membership degree of relation R between u and v. Another early but incomplete proposal [24] is fuzzy OWL which provided the RDF/XML syntax of several axioms in fuzzy OWL language.

A more difficult approach to introduce handling fuzzy data in OWL is to design and implement fuzzy extensions to the OWL language itself. This approach actually modifies the structure and semantics of the language. Since OWL has DL as its basis and numerous fuzzy DL versions exist based on the selected DL, diverse fuzzy extensions to OWL exist using the various fuzzy DLs. Fuzzy OWL was proposed in [25] and used the fuzzy DL *f-SHIN* which is a more expressive DL as its basis for representing fuzzy information. f-OWL [26] resulted from addressing problems related to modifying the OWL language to handle fuzzy extensions. The f-OWL language augmented OWL instance axioms by allowing degrees for fuzzy instance relations. In addition, f-OWL provided model-theoretic semantics and made syntactic modification to both the abstract and RDF/XML syntax. Then f-OWL further made fuzzy extension to allow fuzzy nominal and fuzzy subsumption. These researchers also studied other fuzzy extensions for role range axioms, disjointness axioms and functional role axioms.

With the introduction OWL 2, researchers again proposed adding fuzzy extensions to OWL 2. The focus initially was to minimize changes to OWL 2 itself and to use features of the language to incorporate fuzzy logic. In one early approach [27] instead of changing OWL 2, an ontology was created to handle fuzzy concepts and was referred to as FuzzyOwl2Ontology. The objective of FuzzyOwl2Ontology was to define fuzzy extensions but to specify them through an ontology definition and not by making modifications to OWL 2 itself. The FuzzyOwl2Ontologyhas 8 main classes representing different elements of a fuzzy ontology with each class having several subclasses. This approach has numerous advantages such as easy extensions for other fuzzy OWL 2 statements and easy development of a fuzzy ontology using standard OWL editors.

Continuing with this approach, a later proposal in [28] identifies the syntactic differences that must be managed in a fuzzy ontology and provides a specific method to represent fuzzy ontologies using OWL 2 annotation properties. Annotation properties represent the features of the fuzzy ontology that cannot be directly specified in the OWL 2 ontology. Fuzzy concepts denote fuzzy sets of individuals. Fuzzy roles denote fuzzy binary relations and five different fuzzy data types such as standard trapezoidal (Z) and the triangular membership functions are provided. Using this approach, a fuzzy ontology could also be created using OWL 2 editors and have reasoning performed on it with OWL 2 reasoners. More research followed on how best to include aggregation operators and resulted in a method handle aggregation operators in fuzzy OWL [29].

3.2   Tools for Creating Fuzzy Ontologies

As discussed, the primary means of representing fuzzy ontologies is the use of a "fuzzy" OWL or OWL 2 language. The form of this language, that is, whether the language simply uses features of the original language to specify fuzzy information or actually extends the language with new capabilities to represent fuzzy information, determines the requirements for the tools used in constructing and managing a fuzzy ontology. With the simple approaches, standard ontology editors such as Protégé can be used.

An early effort to add fuzzy logic handling into Protégé is Fuzzy Protégé [30], built as a plug-in for Protégé 3.3.1 and released over ten years ago. Fuzzy Protégé extended Protégé to permit defining fuzzy concepts and fuzzy roles or relations and their instantiations. The two top level additional metaclasses are Fuzzy-Class and Fuzzy-Relation. For each of these only parameterized trapezoidal, triangular, L-shoulder, and R-Shoulder membership function classes are defined. The functionality provided is computing of instance membership degree of a crisp attribute value in the fuzzy set value used to define a fuzzy concept. For example, 'young person' can be defined as fuzzy concept with a fuzzy set value for the attribute age in the crisp concept 'person'. The person's age in the membership function for 'young' is then used to determine that person's membership in 'young person'. In addition a query can be made to determine those persons meeting the fuzzy condition of young person to a specified membership degree, that is, threshold degree that must be met. An advantage of Protégé is its full support of OWL 2 and permits extensions easily through its plug-in architecture. Knowledge-based tools and applications can be built using a Java-based Application Programming Interface (API). To build fuzzy ontologies using OWL 2 annotation properties as proposed in [28], the researchers also developed a Protégé plug-in. First the non-fuzzy part of the ontology is created using Protégé, and then the user can transparently use the *FuzzyOWL* tab to build the fuzzy information using OWL 2 annotation properties. This tab can be used to create fuzzy datatypes, fuzzy modified concepts, weighted concepts, weighted sum concepts, fuzzy nominals, fuzzy modifiers, fuzzy modified roles, fuzzy axioms, and fuzzy modified datatypes.

3.3   Construction Methods

The construction of ontologies differ based on the data sources used, the complexity or level of details needed in the ontology, and the process or algorithm applied on the data sources. An examination of the early approaches to constructing ontologies can be found in [31]. The kinds of data sources include simple text based documents, dictionaries, XML documents, relational database models, object-oriented database models, UML models and others depending on the application.

Early on, the construction focused on creating simple terminological ontologies, a hierarchical vocabulary with fuzziness introduced as a strength of the association relations between the term concepts. As previously discussed, the work in [4] represents a fuzzy terminological ontology where the term relationships "narrower-than" and "broader-than" may be fuzzy since the degree of strength is automatically determined from information directly obtained from a corpus. This fuzzy ontology is used to refine a user's query and is incorporated in a domain-specific search engine for intelligent text information retrieval. Abstracts of research papers are manually tagged. The tagging step uses items such as the article's title, body of the abstract and keywords provided by the authors. A fuzzy ontology is constructed from the collection of keywords by creating the hierarchy from co-occurrence measures. The shortcoming of this process is relying on user judgment for determining the relevance of articles to user queries.

Since the process of creating an ontology is time-consuming much research have proposed methods of automating and/or semi-automating this process. One approach that has been much researched is the use of formal concept analysis (FCA) [32, 33]. FCA is an unsupervised learning method commonly applied for knowledge discovery that groups data into formal concepts which comprise of two components, the set of instance objects referred to as the formal concept's extent, and the set of attributes describing the objects referred to as its intent. Fuzzy formal concept analysis (FFCA) is a natural extension of FCA and it use in developing fuzzy ontologies is a predictable approach that follows. The main difference is for FFCA an attribute has a degree of membership when describing the object. For FCA, an attribute has a crisp membership when describing an object, i.e., the membership degree is either a 0 or a 1.

One of the earliest approaches to use FFCA is the Fuzzy Ontology Generation Framework (FOGA) [34]. An issue with the use of FFCA is the manual assignment of labels to the concepts, attributes and relations which necessitates domain experts to provide expressive labels for the ontology components. FOGA is also not able to produce fuzzy relational concepts from unstructured or semi-structured text documents. Others have produced fuzzy formal contexts but then use an α-cut on them to produce a crisp formal context [35, 36]. Only those entries in the fuzzy formal context with a membership degree greater than or equal to α are kept and converted to a crisp entry of 1. FCA is used on the crisp context to create the formal concepts and its lattice structure. This method, referred to as the one-sided α-cut thresholding approach, because of its simplicity, has been used as the basis of research to create fuzzy ontologies from fuzzy formal contexts.

The crisp concept lattice is modified to add fuzziness back in by using several fuzzy operators on the formal context. The methods used to re-introduce fuzziness, however, vary somewhat. One simple approach in [35] adds fuzziness by determining each object's membership in the extent as the minimum of the membership degrees of the attributes that the object possesses in the intent of the concept. Although the work in [36] uses the same method of creating the fuzzy formal context, instead of using clustering methods on the fuzzy formal concepts to create the hierarchy of the fuzzy ontology, it uses a direct transformation method. As part of constructing the hierarchy, it adds a membership degree between child and parent concepts calculated using a fuzzy set similarity measure between the fuzzy set extents of the child and parent concepts.

The research in [37] differs from the one-sided α-cut thresholding approach since it does not create a crisp concept lattice from the fuzzy formal context. The method directly uses the fuzzy formal context to create the fuzzy formal concepts from all the fix-points produced using a fuzzy closure operator over a finite chain of truth degrees. Both fuzzy extents and fuzzy intents defined a fuzzy formal concept. This aspect differs from the one-sided α-cut approach which has fuzzy extents only. The structure of the fuzzy concept lattice is determined by the partial order on the set of all fixpoints. This research only produces the fuzzy concept lattice and does not produce a resulting fuzzy ontology. Further research in [38] studies the fuzzy closure approach in order to compare the fuzzy concept lattices it produces to those produced by the much simpler α-cut approach.

Some researchers prefer to use existing techniques to first develop the crisp ontology and then follow a methodology to extend the crisp ontology into a fuzzy ontology. IKARUS-Onto [39], a detailed methodology for extending crisp ontologies to fuzzy ontologies, provides specific guidelines to correctly detect vague knowledge within a domain and not confuse different forms of uncertain knowledge such as ambiguity and inexactness with vagueness. Another objective is the explicit modeling of vague knowledge using fuzzy elements in a fuzzy ontology as accurately as possible. This research provides a methodology to take a traditional ontology and convert it into a fuzzy ontology. Others have presented how crisp OWL ontologies can be automatically enriched with fuzziness [46, 47].

4   Where are Fuzzy Ontologies Currently Being Used?

Current examples of fuzzy ontologies in the past two years and from a variety of domains and for different tasks include information retrieval with a variation in its usage for image retrieval, case-based reasoning in the medical domain, and interesting sentiment analysis in transportation monitoring for a sensor network based traffic system.

The Semantic Web uses ontologies as its primary means of knowledge representation. A key task is information retrieval. An early uses of fuzzy set theory is fuzzy information retrieval; that is, representing documents as a fuzzy set of keywords and queries as fuzzy sets of relevant terms. A natural application of fuzzy ontologies is information retrieval as can be seen by one of the earliest proposals for fuzzy ontologies [4]. Numerous examples exist where systems employ fuzzy ontologies to improve the performance of a general information retrieval system. Fuzzy ontologies are also used for specialized domains such as user profiling, medical documents, multimedia retrieval, industrial documents, and user profiling for retrieval tasks. An example of a recent use of fuzzy ontologies is for image retrieval [40]. The standard approach is to annotate images with keywords based on human judgment. The user queries for images using keywords meeting the user's requirements. To improve the efficiency and the query results an ontology is used to resolve semantic heterogeneities.

The crisp annotation and retrieval methods, however, are limited because the process uses human perceptions. Fuzzy ontologies, therefore, are used to overcome these problems to improve the performance of the image retrieval system. The semantic description of an image is modeled by dividing it into regions. The regions are then classified into concepts. A combination of concepts create a category. The concepts, categories and images are linked among themselves with fuzzy link degrees in the ontology. The retrieved results are ranked based on the relevancy between the keywords of a query and images.

Ontologies have also been used for case-based reasoning, especially in the medical domain which is also using fuzzy set theory and logic more and more. The combination of the two in a fuzzy ontology is a natural fit as a tool to improve performance of a Knowledge-Intensive Case-Based Reasoning Systems (KI-CBR) [41]. KI-CBR diagnoses diabetes with a case-based fuzzy OWL2 ontology developed using a fuzzy Extended Entity Relation (EER) data model. It contains 63 (fuzzy) classes, 54 (fuzzy) object properties, 138 (fuzzy) datatype properties, and 105 fuzzy datatypes. This fuzzy ontology contains 60 instance cases. It can be queried using SPARQL-DL. It is accurate, consistent, and covers terminologies and logic for diagnosing diabetes mellitus.

A fuzzy ontology and inference system [42] for sentiment analysis has as its domain real-time monitoring of an entire transportation system for mega-cities. Such systems must intelligently deliver emergency services and provide timely and useful information to users about city transportation and travel. The system relies on the information on social network platforms and performs sentiment analysis. Current conventional ontology-based systems cannot extract vague information from reviews and often provide inadequate results. The fuzzy ontology contains vague and imprecise information found on these platforms and aids in monitoring transportation activities such as accidents, traffic volume, street conditions and closures and city features such as bus and train stations, bridges, parks, restaurants, and airports. Reviews and tweets about the features and activities are retrieved and then extraction for feature opinions is performed. The fuzzy ontology is used to process the feature opinions to determine transportation and city-feature polarity and to provide a city-feature polarity or opinion map for travelers. The fuzzy inference layer has four components: fuzzification, inference, knowledge base and rule base, and defuzzification exist. The fuzzification specifies membership values for the opinion words. The fuzzy ontology contains the SWRL rules and linguistic values. Fuzzy inference is applied using these rules with the fuzzy interval memberships. Five linguistic values (Strongly positive, Positive, Neutral, Negative and Strongly Negative) exist for each input variable. Defuzzification converts the fuzzy output into normal terms and provides the result in the form of a value, the polarity value. The Protégé OWL language has been used to create a prototype system with the fuzzy ontology and intelligent systems software implemented using Java.

5 Conclusions

This updated review on fuzzy ontologies shows that they are becoming more widely used as a means of knowledge representation. Many researchers have contributed to both their theoretical development and practical, real world use. Recent applications of fuzzy ontologies demonstrate their advantages for domains where vague, imprecise or uncertain information must be managed for the application's success. Current progress on fuzzy ontology research to enable more advanced uses appears to have slowed down. Further research needs more tightly integrate fuzzy set theory and logic into OWL. Better tools for construction and evaluation are essential whether they result from learning methods or conversion methods from a crisp ontology. Fuzzy ontologies are proving their usefulness. With more research contributions to make their construction, reasoning with, and use more integrated for practical Semantic Web and other real world applications, fuzzy ontologies can become the go-to choice for knowledge representation.

References

1. Gruber, T.R.: A translation approach to portable ontology specifications. Knowl. Acquis. **5** (2), 199–220 (1993)
2. Zadeh, L. (ed.): Computing with Words in Information/Intelligent Systems. Springer, Heidelberg (1999). https://doi.org/10.1007/978-3-7908-1873-4
3. Bobillo, F.: Managing vagueness in ontologies. Ph.D. dissertation, University of Granada, Spain (2008)
4. Widyantoro, D.H., Yen, J.: Using fuzzy ontology for query refinement in a personalized abstract search engineer. In: Joint 9th IFSA World Congress and 20th NAFIPS International Conference, vol. 1, pp. 610–615 (2001)
5. Cross, V., Sudkamp, T.: Similarity and Compatibility in Fuzzy Set Theory: Assessment and Applications. Physica-Verlag, New York (2002). https://doi.org/10.1007/978-3-7908-1793-5
6. Baader, F., Calvanese, D., McGuinness, D.L., Nardi, D., Patel-Schneider, P.F.: The Description Logic Handbook: Theory, Implementation, Applications. Cambridge University Press, Cambridge (2003). ISBN 0-521-78176-0
7. Yen, J.: Generalizing term subsumption languages to fuzzy logic. In: Proceedings of the 12th International Joint Conference on Artificial Intelligence (IJCAI-91), pp. 472–477 (1991)
8. Straccia, U.: A fuzzy description logic. In: Proceedings of the 15th National Conference on Artificial Intelligence (AAAI 1998), pp. 594–599 (1998)
9. Tresp, C.B., Molitor, R.: A description logic for vague knowledge. In: Proceedings of the 13th European Conference on Artificial Intelligence (ECAI 1998), pp. 361–365 (1998)
10. Straccia, U.: Reasoning within fuzzy description logics. J. Artif. Intell. Res. **14**, 137–166 (2001)
11. Straccia, U.: A fuzzy description logic for the semantic web. In: Sanchez, E. (ed.) Fuzzy Logic and the Semantic Web, Capturing Intelligence, pp. 73–90. Elsevier, Amsterdam (2006)
12. Sanchez, D., Tettamanzi, V.: Fuzzy quantification in fuzzy description logics. In: Sanchez, E. (ed.) Capturing Intelligence: Fuzzy Logic and the Semantic Web. Elsevier (2006)
13. Holldobler, S., Storr, H.P., Tran, D.K.: The fuzzy description logic ALC FH with hedge algebras as concept modifiers. J. Adv. Comput. Intell. **7**(3), 294–305 (2003)
14. Bobillo, F., Delgado, M., Gómez-Romero, J.: A crisp representation for fuzzy SHOIN with fuzzy nominals and general concept inclusions. In: da Costa, P.C.G., et al. (eds.) URSW 2005–2007. LNCS (LNAI), vol. 5327, pp. 174–188. Springer, Heidelberg (2008). https://doi.org/10.1007/978-3-540-89765-1_11

15. Cross, V.V.: Fuzzy ontologies: state of the art. In: IEEE Conference on Norbert Wiener in the 21st Century (21CW) (2014)
16. Lukasiewicz, T., Straccia, U.: Managing uncertainty and vagueness in description logics for the semantic web. J. Web Semant. 6(4), 291–308 (2008)
17. Borgwardt, S., Peñaloza, R.: Fuzzy description logics – a survey. In: Moral, S., Pivert, O., Sánchez, D., Marín, N. (eds.) SUM 2017. LNCS (LNAI), vol. 10564, pp. 31–45. Springer, Cham (2017). https://doi.org/10.1007/978-3-319-67582-4_3
18. Zhang, F., Cheng, J., Ma, Z.: A survey on fuzzy ontologies for the semantic web. Knowl. Eng. Rev. 31(3), 278–321 (2016)
19. Mazzieri, M.: A fuzzy RDF semantics to represent trust metadata. In: Proceedings of the 1st Italian Semantic Web Workshop: Semantic Web Applications and Perspectives (2004)
20. Mazzieri, M., Dragoni, A.F.: A fuzzy semantics for semantic web languages. In: Proceedings of the International Workshop on Uncertainty Reasoning for the Semantic Web, pp. 12–22 (2005)
21. Straccia, U.: A minimal deductive system for general fuzzy RDF. In: Polleres, A., Swift, T. (eds.) RR 2009. LNCS, vol. 5837, pp. 166–181. Springer, Heidelberg (2009). https://doi.org/10.1007/978-3-642-05082-4_12
22. Smith, M.K., Welty, C., McGuinness, D.L.: OWL web ontology language. W3C Recommendation (2004). http://www.w3.org/TR/2004/REC-owl-guide-20040210/
23. Gao, M., Liu, C.: Extending OWL by fuzzy description logic. In: Proceedings of the 17th IEEE International Conference on Tools with Artificial Intelligence, pp. 562–567 (2005)
24. Calegari, S., Ciucci, D.: Fuzzy ontology, fuzzy description logics and fuzzy-OWL. In: Masulli, F., Mitra, S., Pasi, G. (eds.) WILF 2007. LNCS (LNAI), vol. 4578, pp. 118–126. Springer, Heidelberg (2007). https://doi.org/10.1007/978-3-540-73400-0_15
25. Stoilos, G., Stamou, G., Tzouvaras, V., Pan, J.Z., Horrocks, I.: Fuzzy OWL: uncertainty and the semantic web. In: Proceeding of the International Workshop on OWL: Experiences and Directions (2005)
26. Stoilos, G., Stamou, G., Pan, J.Z.: Fuzzy extensions of OWL: logical properties and reduction to fuzzy description logics. Int. J. Approx. Reason. 51, 656–679 (2010)
27. Bobillo, F., Straccia, U.: An OWL ontology for fuzzy OWL 2. In: Rauch, J., Raś, Z.W., Berka, P., Elomaa, T. (eds.) ISMIS 2009. LNCS (LNAI), vol. 5722, pp. 151–160. Springer, Heidelberg (2009). https://doi.org/10.1007/978-3-642-04125-9_18
28. Bobillo, F., Straccia, U.: Fuzzy ontology representation using OWL 2. Int. J. Approx. Reason. 52(7), 1073–1094 (2011)
29. Bobillo, F., Straccia, U.: Aggr academic medical center, egations operators and fuzzy OWL 2. In: Proceedings of the 2011 International Conference on Fuzzy Systems (FUZZ-IEEE 2011), pp. 1727–1734 (2011)
30. Ghorbel, H., Bahri, A., Bouaziz, R.: Fuzzy protégé for fuzzy ontology models. In: Proceedings of the 11th International Protégé Conference IPC 2009. University of Amsterdam (2009)
31. Cristani, M., Cuel, R.: A survey of ontology creation methodologies. Int. J. Semant. Web Inf. Syst. 1(2), 49–69 (2005)
32. Wang, J., He, K.: Towards representing FCA-based ontologies in semantic web rule language. In: 6th IEEE International Conference on Computer & Information Technology, p. 41 (2006)
33. Stumme, G., Maedche, A.: FCA-MERGE: bottom-up merging of ontologies. In: Proceeding of International Joint Conference on Artificial Intelligence (IJCAI), pp. 225–234 (2001)
34. Quan, T.T., Hui, S.C., Cao, T.H.: FOGA: a fuzzy ontology generation framework for scholarly semantic web. In: Proceedings of the 2004 Knowledge Discovery and Ontologies Workshop (KDO 2004), Pisa, Italy (2004)

35. Tho, Q.T., Hui, S.C., Fong, A.C.M., Cao, T.H.: Automatic fuzzy ontology generation for the semantic web. IEEE TKDE **18**(6), 842–856 (2006)
36. De Maio, C., Fenza, G., Loia, V., Senatore, S.: Towards automatic fuzzy ontology generations. In: Proceedings of the 2009 IEEE International Conference on Fuzzy Systems, Jeju Island, Korea, 20–24 August, pp. 1044–1049 (2009)
37. Belohlavek, R., De Baets, B., Outrata, J., Vychodil, V.: Computing the lattice of all fixpoints of a fuzzy closure operator. IEEE Trans. Fuzzy Syst. **18**(3), 546–557 (2010)
38. Cross, V., Kandasamy, M., Yi, W.: Comparing two approaches to creating fuzzy concept lattices. In: Proceeding of the NAFIPS, El Paso, TX, 18–19 March (2011)
39. Alexopoulos, P., Wallace, M., Kafentzis, K., Askounis, D.: IKARUS-Onto: a methodology to develop fuzzy ontologies from crisp ones. KAIS **32**(3), 667–695 (2012)
40. Liaqat, M., Khan, S., Majid, M.: Image retrieval based on fuzzy ontology. Multimed. Tools Appl. **76**, 22623–22645 (2017)
41. El-Sappagh, S., Elmogy, M.: A fuzzy ontology modeling for case base knowledge in diabetes mellitus domain. Eng. Sci. Technol. **20**, 1025–1040 (2017)
42. Ali, F., Kwak, D., Khan, P., Riazul Islam, S.M., Kim, K.H., Kwak, K.S.: Fuzzy ontology-based sentiment analysis of transportation and city feature reviews for safe traveling. Transp. Res. Part C **77**, 33–48 (2017)
43. Cross, V.: Fuzzy extensions to the object model. In: 1995 IEEE International Conference on Systems, Man and Cybernetics. Intelligent Systems for the 21st Century, Vancouver, BC, vol. 4, pp. 3630–3635 (1995)
44. Yeung, C.A., Leung, H.: Ontology with likeliness and typicality of objects in concepts. In: Embley, D.W., Olivé, A., Ram, S. (eds.) ER 2006. LNCS, vol. 4215, pp. 98–111. Springer, Heidelberg (2006). https://doi.org/10.1007/11901181_9
45. Bobillo, F., Straccia, U.: The fuzzy ontology reasoner fuzzyDL. Knowl.-Based Syst. **95**(1), 12–34 (2016)
46. Straccia, U., Mucci, M.: pFOIL-DL: learning (fuzzy) EL concept descriptions from crisp OWL data using a probabilistic ensemble estimation. In: Proceedings of the 30th Annual ACM Symposium on Applied Computing (SAC-15) (2015)
47. Lisi, F.A., Straccia, U.: Learning in description logics with fuzzy concrete domains. Fundamenta Informaticae **140**(3–4), 373–391 (2015)

Image Processing Algorithm to Detect Defects in Optical Fibers

Marcelo Mafalda[1(✉)], Daniel Welfer[1],
Marco Antônio De Souza Leite Cuadros[2],
and Daniel Fernando Tello Gamarra[1]

[1] Federal University of Santa Maria, Santa Maria, Brazil
Marcelomml03@hotmail.com
[2] Federal Institute of Espirito Santo, Vitória, Brazil

Abstract. This work proposes a system to detect visual defects in an optical fiber. Fibers of different types and with different simulated deformations were used, looking for an approximation of a real case of defect in an optical fiber. Some continuous fiber patterns were detected in images captured with a microscopic camera. The identification of these patterns was searched using different image processing techniques, such as edge detection, line detection and feature descriptors. In order to classify images of the fibers in good and defective ones, a fuzzy classifier was used. Experimental results of the algorithm are shown and is demonstrated that the proposed method helps to detect defects and classify optical fibers.

Keywords: Image processing · Optical fiber · Fuzzy logic

1 Introduction

The demand for optical fibers has increased rapidly in the last decades due to the reduction of impurities of the material in the manufacturing process and the use of better materials. The physical characteristics such as dimensions, immunity to electromagnetic interference and chemical resistance make it more attractive for industrial applications [1]. Optical fibers are also recognized as the superior medium for transmitting broadband signals over long distances. The key attribute that allows this performance is its low attenuation, that is, the signals have little loss of power as they propagate along the optical fiber [2]. To maintain this low attenuation, several factors must be taken into account, such as environmental and mechanical effects. Some of the most common effects that cause signal attenuation in fiber are macro and micro curvatures, however, there are no industry standard specifications or test methods for micro curvatures [2]. It was found that optical fiber need to undergo through an inspection process before its utilization. Due to the high cost of installation and maintenance, the industry uses a system that aids in the optical fiber inspection. However, this system still depends on the operator to judge the state of the fiber.

An image processing algorithm is described in this paper. The algorithm uses, in order to take a final decision about the quality of the fiber, an intelligent system based on fuzzy logic. It inspects visually and automatically a segment of fiber optic cable,

© Springer International Publishing AG, part of Springer Nature 2018
G. A. Barreto and R. Coelho (Eds.): NAFIPS 2018, CCIS 831, pp. 243–252, 2018.
https://doi.org/10.1007/978-3-319-95312-0_21

using a microscopic camera, in order to detect deformations and curvatures that can reduce its performance, since the lateral section length, depth, and surface roughness have great influence on the sensor sensitivity, hysteresis, and linearity [3]. The algorithm proposed in this paper is novel. As far as we know there are not similar methods to the one proposed in this paper.

The second section of this paper explains the related work, the third section explains the materials and methods applied, the forth section relates the results obtained. Finally, the conclusion is presented.

2 Related Work

After a literature review, was found that this is one of the first works that uses image processing from an external transversal point of view, in order to detect defects in fiber optics. One close related work is the Defect Detection, Classification and Quantification in Optical Fiber Connectors work, from Shahraray et al. in [4], which visually inspects the fiber from a transversal perspective. Some close references of the use of image processing to detect defects in other type of materials such as tubes or ducts could be found in the work of Sinha and Fieguth [5] on automatic inspection of underground ducts, in which a technique for inspection is developed using image recognition. We also have the application developed by Hägele et al. in [6], which consists in determining the angle of curvature of the fiber through the graphical analysis of the fiber specular pattern. The contours of the speckles in a region of interest in the specular pattern are visible by an edge detection algorithm and their quantity is adjusted in relation to the angle of bending. The work of Xiao-rong et al. in [7], shows us a Research on the Algorithm about Optical Fiber Parameters Measurement, which demonstrates a comparison between a developed method and traditional optical fiber measurement algorithm. There is also the work of Schneider [8] on a methodology to be applied in the automatic segmentation of defects in welds and pipes radiographic images, as well as to extract characteristics for the defects recognition.

3 Materials and Methods

The fiber images are obtained with an USB digital microscope thinking of the need for the fiber to remain extended during the recognition process, due to the constructive characteristics of the fiber. Images of good quality, clean and fibers with simulated deformation and micro curvatures were collected. The image processing algorithm is divided in three parts. In the first part we preprocessed the image using some filters as Canny, median, etc. The second part we extract the lines using Hough transform and more characteristics using the features descriptors, SURF and MSER. The third part is the fuzzy classification with the results of the second part. Figure 1 shows the steps followed by the image processing algorithm.

Figure 2 shows the experimental setup used for the experimental results, a micro USB camera and optical fibers of different colors are shown in the figure. The optical fibers used have been previously deformed in order to simulate the defects, because it

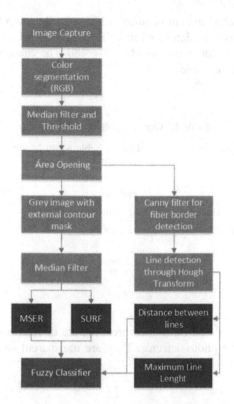

Fig. 1. Summarized steps of the proposed method.

Fig. 2. Experimental setup: USB camera and optical fibers.

would've required a large amount of fibers, to have sufficient deformed fibers to tune the algorithm. It was used a microscopical USB Lenovo EasyCamera with 2MP, video framerate of 30 fps that capture images of 640 × 480. The characteristics of the optical fiber cables are shown in Table 1.

Table 1. Optical cables used in tests.

	Optical cable Single Mode UT 09/125	Optical cable Multi Mode 62,5/125	Optical cable Single Mode Drop Cm 01f	
Color	Red/green	Yellow	Blue/light green	
Minimum bending radius (installation)	124 mm		90 mm	30 mm
Minimun bending radius (post installation)	62 mm		60 mm	15 mm
External diameter	6.2 mm		6.0 mm	5.0 mm

3.1 Optical Fiber

Optical fibers, in a simplified form, are wires that are directed to a luminous emitting power of light until the photo detector. They are transparent structures, composed of two dielectric materials (Fig. 3).

Fig. 3. Optical fiber. Font: (Basics of fiber optics [11])

Several factors can cause signals power losses in optical fibers. These factors are divided into intrinsic and extrinsic. Intrinsic attenuation is a result of fiber inherent materials. It is mainly caused by impurities in the glass during manufacturing. As precise as a manufacturing process have become, there is no way to eliminate all impurities [9]. The extrinsic attenuation is caused mainly by curvatures, which cause considerable reductions in optical power due to a stress appearing in the region of the curvature, as a result, refractions begin to happen, as well as signal attenuation. The curvatures are divided into macro (Fig. 4a) or micro curvatures (Fig. 4b).

The proposed algorithm in this paper deal with macro and micro curvatures. Macro curvatures are characterized by being curves visible to the human eye. Macro curvature attenuation occurs when the fiber is curved physically beyond the point at which the critical reflection angle is exceeded. This attenuation is reversible after the curvatures are corrected. The critical reflection angle is what determines how much a fiber can bend before there are signal losses in the fiber.

Micro curvatures are deformations caused by lateral contact of the fiber with other surfaces, by bending or twisting, and by defects in the fiber production process. They have very punctual location and are not clearly visible in inspections. Curves that cause micro deformations typically have a radius of less than 1 mm, and commonly described as a random distribution of spacing and amplitude. Micro-bending losses occur when small pressure points generate retreats on the fiber surface, even if the fiber is straight, thus changing the critical angle and refracting light for the coating. There are no industry standards or specifications for detecting micro-curvatures [2].

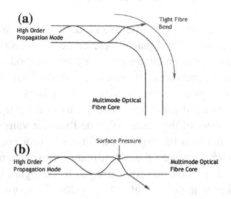

Fig. 4. Macro (a) and micro curvatures (b). Font: (Fiber optics technologies [10])

3.2   Morphology

Image processing techniques were used in this paper, one of these techniques are morphological operations, which are employed to extract data related to the shape, topology and geometry of objects in images. They interact between a structuring element implicit in the operations, which functions as a reference, with sets in the image, modifying their shape and thus obtaining information about the set. The morphological operations used in this paper are: opening, closing, and reconstruction.

3.3   Hough Transform

Another technique used was the Hough Transform, which is able to detect groups of pixels belonging to a straight line. The idea is to apply a transformation in the image such that all the points belonging to the same curve are mapped in a single point of a new parameterization space of the searched curve. The transform creates a plane with the dimensions Θ and ρ. Theta (Θ) is the angle with the origin and rho (ρ) the shortest distance to the origin of the center of coordinates. Analyzing a pixel in the image, look for the values of Θ and ρ for all the lines that pass at this point, because when the points are collinear in the space x-y, there will be an intersection of their sine-wave curves in the plane Θ - ρ. A necessary step in the process of applying the Hough Transform is edge detection and thresholding. The Canny method was used to accomplish this task.

3.4 Image Feature Descriptors

We have used two image descriptors, SURF and MSER. The Speeded Up Robust Features descriptor (SURF) is inspired by another descriptor, SIFT, and makes use of a Hessian determinant approach to detect points of interest in the image [12]. Another method to detect features would be the Maximum Stable Extreme Region (MSER). It draws regions of interest based on the connected components and the intensity of the region. It's based on the idea of joining regions that satisfy certain thresholds [13].

4 Experimental Results and Discussion

To simulate the micro curvatures in the fibers used in this work, the fibers were removed from the optical cables and placed between two sheets of paper, after vertical pressure was applied gradually. There are no actual test methods for micro-curvatures. Some solutions are to apply pressures between two sheets of paper by pressing on some cavity or pin or wrapping over a wire, however, the results are not faithfully reproducible, so there is no standard for testing micro curvatures [10].

It was noticed that fibers of the same color had similar values, so that the parameters of the filters were made in the algorithm, for each color specifically, through the color segmentation, which improved the identification in some cases of defects. Thus, sufficient data were obtained from the fiber images to configure a fuzzy classifier that decides whether the fiber is good or bad. The results of the application of the image processing algorithms will be detailed in the next subsections:

4.1 Hough Transform

The Hough Transform was introduced to detect straight lines, initially as many lines as possible, then only two lines. We also used the Otsu method for thresholding, and area opening for eliminate noise, together with the Canny filter, to detect only the outer edges of the fiber. This made the detection of Hough lines more accurate. The Hough transform algorithm detected lines with a minimum length of 30 and fill gap of 100. Figure 5 shows the step of the algorithm using the Hough Transform for a good and a bad fiber. The Hough lines are in red color.

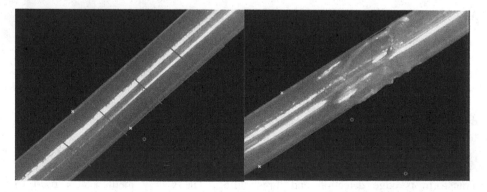

Fig. 5. Detected fiber edge using the Hough Transform. (Color figure online)

4.2 Feature Descriptors

Figures 6 and 7 show the application of the features descriptors in cases of good fiber and bad fiber. The SURF descriptors are depicted as green circles in Fig. 6 and the MSER regions in Fig. 7. In the case of SURF we can see that the points are concentrated in regions of abrupt change in uniformity in the image, which may indicate a failure in the fiber. And the MSER highlights regions of the image that have a "uniformity", which can indicate a good fiber with few and extensive regions. Both were applied on a gray image of the fiber after the application of a mask that separates the contour of the fiber and its internal part from the rest of the image, thus increasing its effectiveness.

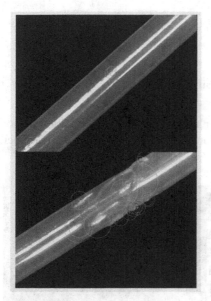
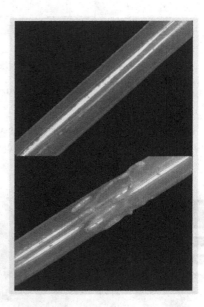

Fig. 6. Detected features using SURF. (Color figure online)

Fig. 7. Detected regions using MSER.

4.3 Fuzzy Classifier

It was used a fuzzy classifier, and the type chosen was Mamdani. The inputs of the classifier are the characteristics extracted from the images by the algorithm, there were three inputs: number of MSER regions and SURF points, the length of the lines detected by the Hough transform and the distance between them. The classifier has one output, that is to define if the quality of fiber is good or bad.

The set of instructions or fuzzy rules had its implementation facilitated by the time spent in obtaining the image data, since it is already clear which values of which variables constitute a good fiber and which mean that the fiber is damaged. In Fig. 8 we can see how the fuzzy rules were designed. obtaining the four input variables of the fuzzy classifier.

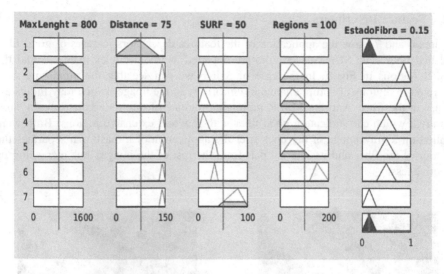

Fig. 8. Fuzzy classifier rules

The output of the classifier tells us whether the fiber is in good condition or not. Figure 9 shows us the output of the classifier for a case of good fiber and a bad fiber.

Fig. 9. Fuzzy classifier outputs

It is noticed that the fuzzy system made the correct decision in both cases. In the image of the good fiber (left), we have as a good indicator the detected lines, as well as the distance between them, a good length of axis and of detected lines, and still few SURF points and MSER regions. While in the case of the defective fiber we have the detection of straight lines, but these are not parallel because their distance cannot not be calculated, which already indicates a defect. Also, we still have many SURF points and MSER regions, even though the lines are within acceptable values, but the fuzzy system has different weights for the variables in its decision making.

After several runs of the algorithm in different fibers with different conditions, it was obtained the following table of results (Table 2):

Table 2. Results

Number of tested fiber images	Fiber condition	Fuzzy Classifier Correct Results	Fuzzy Classifier Incorrect Results	Accuracy Rate
46	GOOD	40	6	86,95%
35	BAD	31	4	88,57%

5 Conclusions

It has been proposed an algorithm to detect defects in optical fibers. As far as we know, our method is unique, because after a review in the scientific literature we have not found other related works. Fibers of different types and different simulated deformations were analyzed in order to approach a real case of defect in an optical fiber. It was then sought to identify a pattern in the fiber through image recognition techniques, such as edge detection, line detection and feature descriptors. It was also used a fuzzy classifier, tuned to the parameters that interest us of the image obtained through the algorithm, to classify the images of the fibers in good and defective. At the end of the work it was concluded that the fiber characteristics obtained by the image processing program were relevant to the classification of the fiber in good or defective, as we can observe by the results of the fibers inspected by the fuzzy classifier, we have 87% accuracy on the good fibers that are good and 89% on the bad fibers that are bad. Obtaining the features represented a certain degree of difficulty due to variations of the microscopic images, thus necessitating a more robust program, requiring many tests with several different versions of the program. However, this difficulty translates into ease in setting the fuzzy controller, since the values that defined the good and bad fiber were already clear.

References

1. Ziemann, O., Zamzow, P., Daum, W.: POF Handbook: Optical Short-Range Transmission Systems, vol. 1, pp. 1–2. Springer Science & Business, Heidelberg (2008). https://doi.org/10.1007/978-3-540-76629-2. Nar
2. Jay, J.A.: An overview of macrobending and microbending of optical fibers. White Paper WP1212, Corning (2010)
3. Leal-junior, A., Frizera, A., Pontes, M.: Sensitive zone parameters and curvature radius evaluation for polymer optical fiber curvature sensors. Opt. Laser Technol. **100**, 272–281 (2018)

4. Shahraray, B., Schmidt, A.T., Palmquist, J.M.: Defect detection, classification and quantification in optical fiber connectors. In: IAPR Workshop on Machine Vision Applications, Tokyo (1990)
5. Sinha, S.K., Fieguth, P.W.: Morphological segmentation and classification of underground pipe images. Mach. Vis. Appl. **17**, 45–56 (2006)
6. Hagele, C., Neto, A.F., Pontes, M.J.: Polymer optical fiber curvature measuring technique based on speckle pattern image processing. In: Simpósio Brasileiro de Automação Inteligente, Natal-Brazil (2015)
7. Xiao-rong, C., Yuan, C., Chuan-li, X.: Research on the Algorithm about Optical Fiber Parameters Measurement. TELKOMNIKA **11**(11), 6693–6698 (2013)
8. Schneider, G.A.: Segmentation and extraction of defects characteristics in radiographic images of ducts and soldering cords. 154 f. Master thesis, Electrical Engineering, CEFET-PR, Curitiba (2005)
9. Alwayns, V.: Optical Network Design and Implementation. Cisco Press, Indianapolis (2004)
10. The Fiber Optic Association, Inc.: Guide to Fiber Óptics & Premises Cabling: Micro-Bending (2013)
11. Curran, M., Shirk, B.: Basics of Fiber Optics, Whitepapers
12. Bay, H., Ess, A., Tuytelaars, T., Gool, L.V.: Speeded-Up robust features (SURF). Comput. Vis. Image Underst. **100**(3), 346–359 (2008)
13. Matas, J., Chum, O., Urban, M., Pajdla, T.: Robust wide-baseline stereo from maximally stable extremal regions. Image Vis. Comput. **10**(22), 761–767 (2004)

A Comparative Study Among ANFIS, ANNs, and SONFIS for Volatile Time Series

Jairo Andres Perdomo-Tovar, Eiber Arley Galindo-Arevalo, and Juan Carlos Figueroa-García[(✉)]

Universidad Distrital Francisco José de Caldas, Bogotá, Colombia
{japerdomot,eagalindoa}@correo.udistrital.edu.co,
jcfigueroag@udistrital.edu.co

Abstract. This paper presents a comparison among ANFIS, ANNs, and a Self Organized Neuro Fuzzy Inference System (SONFIS) for time series prediction. The Turkish stock index (ISE) series is analyzed using the three methods, a statistical analysis of the residuals per method is performed, and the advantages/disadvantages per method are discussed.

Keywords: Fuzzy logic systems · Self organized neural networks
Volatile time series

1 Introduction and Motivation

Volatile time series analysis has been extensively treated from a statistical perspective, but complex problems have been analyzed using neural networks in most cases. Neuro-fuzzy approaches like ANFIS (see Jang [13,17], Jang et al. [14], Ben et al. [3], Wang and Fsas [30], and Figueroa and Soriano [6]) have shown interesting results in complex problems, and ANNs also have been applied to time series forecasting (see Huang and Du [12], and Huang [11]).

Intelligent algorithms also offer efficient methods to train FLSs (see Soto et al. [26,27], Melin et al. [20], Wang and Mendel [32,33], Obayashi et al. [24], Numberger and Kruse [23], Wu and Goo [34], and Figueroa-García et al. [5,7]). We focus on the proposal of Juang and Tsao [16], and Figueroa-García et al. [2,8,9] applied to a volatile time series example.

The paper is divided into five principal sections. Section 1 shows the introduction and motivation of the work. Section 2 presents some basics on fuzzy logic systems. In Sect. 3, the Self Organized Fuzzy Inference System (SONFIS) method is described. Section 4 shows an application example, and finally some concluding remarks are presented in Sect. 5.

2 Basic Definitions of FLSs

A fuzzy set A generalizes crisp/interval sets. It is defined over an universe of discourse X using a membership function $\mu_A(x) : X \rightarrow [0,1]$. Then the set

© Springer International Publishing AG, part of Springer Nature 2018
G. A. Barreto and R. Coelho (Eds.): NAFIPS 2018, CCIS 831, pp. 253–264, 2018.
https://doi.org/10.1007/978-3-319-95312-0_22

A may be represented as a set of ordered pairs of a generic element x and its membership degree, $\mu_A(x)$, i.e.,

$$A = \{(x, \mu_A(x)) \mid x \in X\}. \tag{1}$$

A is a linguistic label that defines the sense of a fuzzy set. Now, x can be represented with $j \in \mathbb{R}_n$ sets $\{A_1, A_2, \cdots, A_n\}$, each one defined by a membership function $\{\mu_{A_1}(x), \mu_{A_2}(x); \cdots, \mu_{A_n}(x)\}$, so basically a particular $x \in X$ can have different membership degree/affinity to different labels/sets A_j.

An FLS is defined by R^j rules that relates a set of inputs to a set of outputs (consequences of operationalizing of rules). This way, each rule R^j is represented as follows (see Mendel [22], and Klir and Folger [18]):

$$R^i : if \ x_1 \ is \ A_1^i \ and \ \cdots \ and \ x_n \ is \ A_n^i, \ then \ \hat{y} \ is \ G^i; \quad i = 1, \cdots, M \tag{2}$$

where G^i represents the output of the i_{th} rule.

In this paper, we consider Mandami FLSs in which inference is made using *t-norms* and *t-conorms* to represent *and* $(\wedge) - or$ $(\vee)$ operators. Also, the output of each rule G^i is considered as a singleton, as described as follows:

$$\mu_G(x) = \begin{cases} 1 & for \ x, \\ 0 & for \ X \neq x. \end{cases} \tag{3}$$

The last step of fuzzy inference is defuzzification which is basically a function $\hat{y} : G^i \rightarrow \mathbb{R}$. As we use singletons as outputs, we use the center of sets defuzzification method (which is the average of outputs), as shown as follows:

$$\hat{y} = \frac{\sum_{i=1}^{M} G^i w_i}{\sum_{i=1}^{M} w_i} = \sum_{i=1}^{M} G^i / M. \tag{4}$$

Note that each rule in an FLS leads to a singleton output. This means that an FLS has as many outputs as rules it has, so the selection of all input fuzzy sets and rule-base is an important issue to be addressed when using FLS. The main idea is to select rules that improve the performance of the FLS only.

3   A Neuro-Fuzzy Approach for Rule Generation

As proposed by Figueroa-García et al. [9] based on Juang and Tsao [16], the goal is to generate rules of an FLS using self organized neural networks. The general structure of the FLS considered in this paper is displayed in Fig. 1.

Input data is composed by n vectors $\{X_1, \cdots, X_n\}$ of size k, and output data (a.k.a. goal or desired response) is defined as $\hat{y}$ where $\{X_j, \hat{y}\} \in \mathbb{R}$ (see Fig. 2). Now, the proposed structure is described as follows:

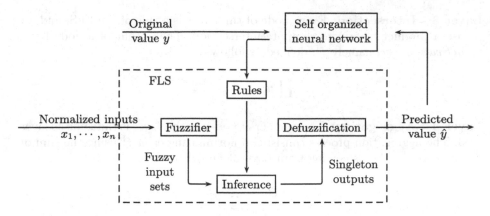

Fig. 1. Neuro-FLS design methodology

The selected method uses a five-layer neural network, as shown in Fig. 2.

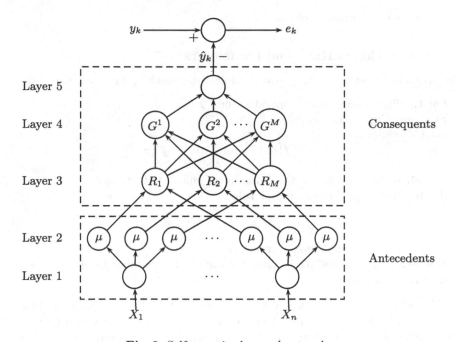

Fig. 2. Self organized neural network

Layer 1 - Normalization: In this layer, all input data $X_1, \cdots, X_n$ should be normalized to any of two choices: either the interval of $[-1, 1]$ or $[0, 1]$.

Layer 2 - Fuzzification: Every input X_j^i is characterized by a fuzzy set A_j^i whose μ_{ji} must be derivable, so we use Gaussian shapes (see Eq. (9)).

Layer 3 - Intersection: Every node of this layer is a rule of the FLS, and we use a product t-norm to compute intersection. The output of a node is its *activation level* namely f^i, defined as follows:

$$\prod_{j=1}^{n} \mu_{ji}, \quad \forall\, i \in M$$

Layer 4 - Aggregation: The number of nodes in this layer is the same as layer 3. The aggregation process consists on normalizing each f^i using the sum of all activation levels coming from Layer 3, in other words:

$$G^i = \frac{f^i}{\sum_i f^i} \tag{5}$$

Layer 5 - Defuzzification: This layer computes the defuzzified output of the system $\hat{y}$ using the average among all aggregated values of Layer 5, as follows:

$$\hat{y} = \sum_i G^i / M \tag{6}$$

where M is the amount of rules.

3.1 Generation of Rules and Parameters

Figueroa-García et al. [9] proposed the following method for generating rules:

1. For the first input data x generate a new rule.
2. For a new input data $\breve{x}$, do:
   (a) Compute:

$$f^I(\breve{x}) = \arg \max_{1 \leqslant i \leqslant M(t)} f^i(x) \tag{7}$$

   where $M(t)$ is the existant amount of rules at the time t.
   (b) If $f^I(\breve{x}) \leqslant \phi$, generate a new rule:

$$M(t+1) = M(t) + 1 \tag{8}$$

   where ϕ is a predefined parameter.
3. Define a new fuzzy set $\mu_{j,i=M(t)+1}$ for each input variable $j = 1, \cdots, n$ and a new node in Layer 3.

This algorithm has two parameters: a threshold ϕ which operates as the activation level for incoming data $\breve{x}$ to generate a new rule, and β which represents how much space an initial rule covers. If ϕ is small, then a small amount of rules centered on $X_1, \cdots, X_n$ are generated. On the other hand, if ϕ is high then the algorithm creates a large amount of rules. All new data for which $f^I(\breve{x}) \geqslant \phi$ is contained into an existent rule.

3.2 Initialization of the Algorithm

The membership functions of the antecedents/inputs are Gaussian:

$$\mu_{ji}(x_j) = \exp\left(-0.5\left((x_j - m_j^i)/\delta_j^i\right)^2\right) \tag{9}$$

where m_j^i and δ_j^i are the center and spread of μ_{ji}.

The consequents/outputs of the FLS are singletons w where we start by adding a new group $(M + 1)$ of data x_j, and the parameters of $\mu_{ji}(x_j)$ are initialized as follows:

$$m_j^{(M+1)} = x_j \tag{10}$$

$$\delta_j^{(M+1)} = -\beta \cdot \ln(f^I) \tag{11}$$

$$w^{(M+1)} = \hat{y} \cdot K; \quad K \in [0, 1] \tag{12}$$

where $\beta > 0$ is the initial cluster width.

As proposed by Juang and Lin [15] and Juang and Tsao [16], we generate K using a uniform generator and β using a chi-square generator to broaden the search space.

3.3 Learning Algorithm

The selected learning method is the fuzzy-based backpropagation algorithm (See Wang and Mendel [31]) based on ϵ_t as error function:

$$\epsilon_t = \sum_{k=1}^{N}(y_k - \hat{y}_k)^2$$

where t is the actual iteration of the algorithm, y_t is the target data, and $\hat{y}$ is the output of the FLS (see Eq. (6)).

The updating algorithm for m and δ to compute $\mu_{ji}(x_j)$ is based on a Kalman filter and a gradient descend algorithm (see Mendel [21]):

$$m_j^i(t + 1) = m_j^i(t) - \eta \cdot \frac{\hat{y} - y}{\sum_{i=1}^{M}} f_i \cdot (w^i(t) - \hat{y}) \cdot f_i \cdot \frac{2(x_j - m_j^i(t))}{\delta_j^i(t)^2} \tag{13}$$

$$\delta_j^i(t + 1) = \delta_j^i(t) - \eta \cdot \frac{\hat{y} - y}{\sum_{i=1}^{M}} f_i \cdot (w^i(t) - \hat{y}) \cdot f_i \cdot \frac{2(x_j - m_j^i(t))^2}{\delta_j^i(t)^3} \tag{14}$$

$$w^i(t + 1) = w^i(t) - \eta \cdot \frac{\hat{y} - y}{\sum_{i=1}^{M}} f_i \tag{15}$$

where $\eta \in [0, 1]$ is the learning rate which can be modified to user convenience.

Usually, higher values of η lead to a faster convergence of the algorithm with higher deviations from its objective, and smaller values of η lead to larger computing efforts with a possible overtraining of the algorithm.

3.4 Evaluation of the Performance of the Algorithm

As error measures we use the RMSE (root mean squared error) since it is derivable and positive semi-definite:

$$\text{RMSE} = \sqrt{\frac{1}{N}\sum_{k=1}^{N}(y_k - \hat{y}_k)^2}, \tag{16}$$

and the EMA (absolute mean deviation), defined as follows:

$$\text{EMA} = \sum_{k=1}^{N} |y_k - \hat{y}_k|. \tag{17}$$

Gaussian white noise residuals are highly desirable in time series as well since Gaussian residuals are less prone to be autocorrelated. Thus, we test ϵ for normality i.e. $\epsilon \approx N(0, \delta^2)$. To do so, we use the F-test (See Thode [28]), T-test (See Tukey [29]), Walf-Wolfowitz test (See Gujarati [10]), ARCH (See Engels [4]), and Ljung-Box test (See Ljung and Box [19]).

3.5 Selecting a Configuration of the Network

Two information criteria are used to select the most adequate model: *Akaike Information Criterion (AIC)* (see Akaike [1]) and *Schwarz/Bayesian Information Criterion (BIC)* (see Schwarz [25]):

$$S_e^2 = \text{RMSE}^2, \tag{18}$$

$$\text{AIC} = \ln(S_e^2) + 2m/N, \tag{19}$$

$$\text{BIC} = \ln(S_e^2) + m\ln(N)/N \tag{20}$$

where N is the sample size, and m is the number of regressors (the number of inputs of the model).

Thus, a model l_1 is better than the model l_2 when l_1 has less AIC and BIC i.e. the model l_1 is better than L_2 if $\text{AIC}_{l_1} < \text{AIC}_{l_2}$ and $\text{BIC}_{l_1} < \text{BIC}_{l_2}$.

4 Application Example

The Istanbul Stock Index (ISE) benchmark dataset taken from the UCI repository https://archive.ics.uci.edu/ml/datasets/ISTANBUL+STOCK+EXCHANGE composed by the daily closing value of the ISE index removing weekends, and seven related indexes Standard & poors 500 return index (SP), Stock market return index of Germany (DAX), Stock market return index of UK (FTSE), Stock market return index of Japan (NIKKEI), Stock market return index of Brazil (BOVESPA), MSCI European index (EU), and the MSCI emerging markets index (EM), since they have volatility i.e. heteroscedasticity. Figure 4 shows the ISE stock index series for 426 training data and 107 validation data.

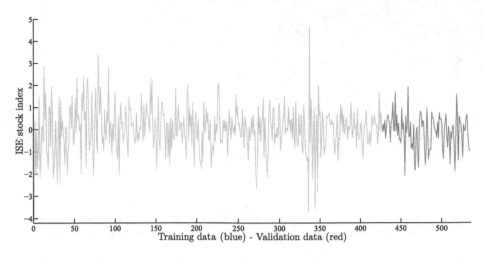

Fig. 3. Pre-processed ISE stock index (Color figure online)

Figure 4 shows differentiated data at the first lag i.e. $\Delta x_t = x_t - x_{t-1}$ and standardized using the z transformation $z_t = (\Delta x_t - \bar{\Delta x})/s_{\Delta x}$, where $\bar{\Delta x}$ is the sample mean of ΔX and $s_{\Delta x}$ is the sample variance of ΔX.

426 observations were used for training, and 107 observations for validation. Each algorithm was ran 10 times per combination, so we have performed a total of 1250 experiments changing β, ϕ (for SONFIS), activation functions (for ANN), shapes of fuzzy sets (for ANFIS), and training algorithm (for ANNs and ANFIS).

The foreign indexes SP, DAX, FTSE, NIKKEI, BOVESPA, EU, and EM were used for prediction. A brief description of the selected methods is as follows:

ANN: The selected network architecture was a feedforward backpropagation with 5 hidden layers, linear activation functions, max amount of iterations: 1000 as stoping criteria, training algorithm: Levenberg-Mardquart.

ANFIS: We used 2 Gaussian fuzzy sets per input for a total of 128 rules, max amount of iterations: 1000 as stoping criteria, training algorithm: hybrid (backpropagation for learning and least squares for training).

SONFIS: The proposed method generated a total of 73 rules, max amount of iterations: 1000 as stoping criteria, training algorithm: hybrid (backpropagation for training and threshold for creating rules).

4.1 Obtained Results

The best SONFIS experiment was reached with $\alpha = 0.05, \beta = 0.26$ and $\phi = 0.85$. The best ANN experiment was obtained with 5 hidden layers, 7 neurons per layer, and linear activation functions. The best ANFIS trial was obtained using two fuzzy sets per input, Gaussian fuzzy sets, and singleton outputs.

Some statistics and tests applied for contrasting the adequation of the model (independence and normality of the residuals) are shown in Tables 1 and 2, and the obtained residuals are shown in Fig. 4.

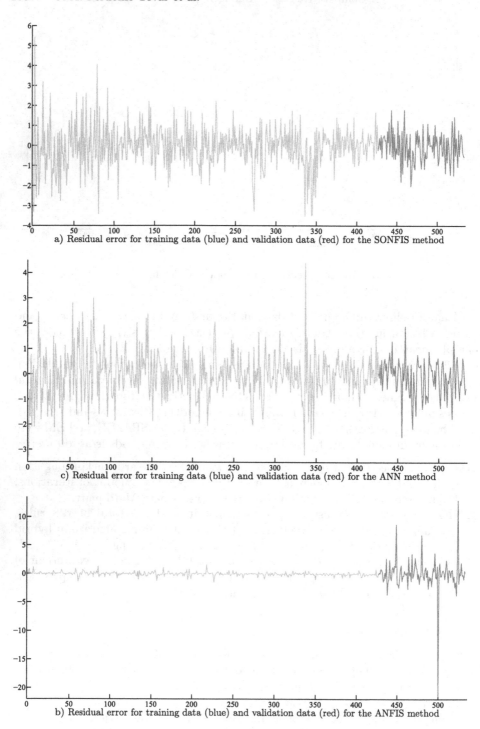

Fig. 4. Residual errors for the three selected methods (Color figure online)

Table 1. Statistical measures

Measure	SONFIS		ANN		ANFIS	
	Training	Test	Training	Test	Training	Test
MSE	1.2785	0.5906	0.8979	0.5386	0.1006	8.82
MAD	356.74	66.88	298.41	60.336	85.83	156.79
AIC	279.4	283.71	204.11	207.89	304.48	298.3
AICc	433.25	−789.39	278.38	−11564	486.58	−681.38
BIC	285.45	285.45	210.16	212.56	310.54	302.97
Runs	0.0000675	0.4316	0.7343	0.1735	0.62808	0.62513
WW	0.0000769	0.1652	0.9388	0.0374	0.8781	0.9382
SW	≈ 0	0.764	≈ 0	0.021	≈ 0	≈ 0
KS	≈ 0	0.15	≈ 0	0.135	≈ 0	≈ 0
AD	≈ 0	0.461	≈ 0	0.007	≈ 0	≈ 0

Table 1 shows the Mean Squared Error (MSE), Mean Absolute Deviation (MAD), Akaike Information Criterion (AIC), Corrected Akaike Information Criterion (AICc), Bayesian Information Criterion (BIC); Runs test and Walf-Wolfowitz (WW) tests on randomness; Shapiro-Wilks (SW), Kolmogorov-Smirnov (KS) and Anderson-Darling (AD) tests for normality.

Table 2. Statistical tests

Lag	ARCH test SONFIS		ARCH test ANN		ARCH test ANFIS	
	Training	Test	Training	Test	Training	Test
1	0.001	0.721	0.000	0.873	0.272	0.802
2	0.000	0.715	0.000	0.753	0.542	0.957
3	0.000	0.768	0.000	0.834	0.717	0.980
4	0.000	0.081	0.001	0.069	0.821	0.996
5	0.000	0.137	0.002	0.114	0.905	0.997
Lag	Ljung-Box test SONFIS		Ljung-Box test ANN		Ljung-Box test ANFIS	
	Training	Test	Training	Test	Training	Test
1	0.002	0.152	0.786	0.035	0.953	0.616
2	0.003	0.340	0.727	0.033	0.668	0.639
3	0.006	0.514	0.878	0.075	0.214	0.736
4	0.000	0.612	0.948	0.140	0.309	0.825
5	0.001	0.611	0.979	0.225	0.192	0.902

In Table 2, the ARCH (autoregressive conditional heteroscedasticity) and Ljung-Box (serial correlation) tests show mixed results, but we can conclude that SONFIS and ANFIS outperform ANNs since validation residuals have neither heteroscedasticity nor serial correlation in 1, 2, 3, 4 and 5 lags.

Roughly speaking, ANFIS has a good performance over training data, but bad performance over validation data (see its MSE, MAD, AIC, AICc, and BIC). The ANN model shows a good performance (see its MSE, MAD, AIC, AICc, and BIC) over both training/validation data but rejects normality tests. SONFIS has very similar MSE, MAD, AIC, AICc, and BIC to ANN, but SONFIS produces normally distributed residuals over validation data (it is the only method that accepted normality tests).

This way, SONFIS arises as a good method since it produces satisfactory results while outperforming ANFIS. When compared to the ANN, SONFIS produces normally distributed validation residuals without ARCH effect and serial correlation, and most important: SONFIS is a rule-based fuzzy inference system that provides knowledge about ISE while ANNs are black-box models.

Table 3. Performance of SONFIS for some β and ϕ

β	ϕ	S_e^2 training	S_e^2 test
0.2	0.2	1.322	0.795
0.2	0.4	1.335	0.624
0.2	0.6	1.348	0.640
0.22	0.6	1.361	0.697
0.24	0.6	1.299	0.643
0.24	0.7	1.290	0.737
0.22	0.75	1.294	0.686
0.22	0.8	1.294	0.621
0.26	0.85	1.279	0.591

And finally Table 3 shows the performance of SONFIS for different selections of β and ϕ. Note that higher values of β and ϕ lead to better residuals S_e^2.

5   Concluding Remarks

The application of our proposal to the ISE stock index obtained adequate results, with only 73 rules. Its results are similar to the applied ANN model and outperforms ANFIS. We experimented with different β and ϕ, and our findings are that convergence is proportionally inverse to β while the amount of rules is directly proportional to ϕ.

While ANFIS shows over training, SONFIS shows heteroscedastic residuals, and ANN rejects normality test over its residuals. This means that there is no perfect method for forecasting volatile time series, and every method shows different performance over different measures.

Other rule generation algorithms (see Jang [17], and Obayashi et al. [24]) can be combined to our proposal to improve convergence and fit of our proposal.

References

1. Akaike, H.: An information criterion (AIC). Math. Sci. **14**(153), 5–9 (1976)
2. Avellaneda, A., Ochoa, C., Figueroa-García, J.C.: A self-organizing neural fuzzy system to forecast the price of Ecopetrol shares. In: IEEE (ed.) Proceedings of 2012 IEEE Computational Intelligence for Financial Engineering and Economics Conference, pp. 1–6. IEEE (2012)
3. Ben, D., Dayan, R., Baselli, G., Invar, G.F.: An adaptive neuro-fuzzy method (ANFIS) for estimating single-trial movement-related potentials. Biol. Cybern. **91**(2), 63–75 (2004)
4. Engel, R.: Autoregressive conditional heteroscedasticity with estimates of the variance of United Kingdom inflation. Econometrica **1**(50), 987–1007 (1982)
5. Kalenatic, D., Figueroa-García, J.C., Lopez, C.A.: A neuro-evolutive interval type-2 TSK fuzzy system for volatile weather forecasting. In: Huang, D.-S., Zhao, Z., Bevilacqua, V., Figueroa, J.C. (eds.) ICIC 2010. LNCS, vol. 6215, pp. 142–149. Springer, Heidelberg (2010). https://doi.org/10.1007/978-3-642-14922-1_19
6. Figueroa-García, J.C., Soriano, J.J.: A comparison of ANFIS, ANN and DBR systems on volatile time series identification. In: IEEE (ed.) 2007 Annual Meeting of the North American Fuzzy Information Processing Society, vol. 26, pp. 321–326. IEEE (2007). https://doi.org/10.1109/NAFIPS.2007.383858
7. Figueroa-García, J.C.: An evolutive interval type-2 TSK fuzzy logic system for volatile time series identification. In: Conference on Systems, Man and Cybernetics, pp. 1–6. IEEE (2009). https://doi.org/10.1109/ICSMC.2009.5346687
8. Figueroa-García, J.C., Ochoa-Rey, C., Avellaneda-González, J.: A self-organized fuzzy neural network approach for rule generation of fuzzy logic systems. In: Huang, D.-S., Gupta, P., Wang, L., Gromiha, M. (eds.) ICIC 2013. CCIS, vol. 375, pp. 25–30. Springer, Heidelberg (2013). https://doi.org/10.1007/978-3-642-39678-6_5
9. Figueroa-García, J.C., Ochoa, C., Avellaneda, A.: Rule generation of fuzzy logic systems using a self-organized fuzzy neural network. Neurocomputing **151**(3), 955–962 (2015). https://doi.org/10.1016/j.neucom.2014.09.079
10. Gujarati, D.: Econometrics. McGraw Hill International, New York (2009)
11. Huang, D.S.: Radial basis probabilistic neural networks: model and application. Int. J. Pattern Recogn. Artif. Intell. **13**(7), 1083–1101 (1999)
12. Huang, D.S., Du, J.X.: A constructive hybrid structure optimization methodology for radial basis probabilistic neural networks. IEEE Trans. Neural Netw. **19**(12), 2099–2115 (2008)
13. Jang, J.S.R.: ANFIS: adaptive neural based fuzzy inference system. IEEE Trans. Syst. Man Cybern. **23**(3), 665–685 (1993)
14. Jang, J.S.R., Sun, C.T., Mizutani, E.: Neuro-Fuzzy and Soft Computing: A Computational Approach to Learning and Machine Intelligence. Prentice Hall, Englewood Cliffs (1997)
15. Juang, C., Lin, C.: An on-line self-constructing neural fuzzy inference network and its application. IEEE Trans. Fuzzy Syst. **6**(1), 12–32 (1998)
16. Juang, C.F., Tsao, Y.W.: A type-2 self-organizing neural fuzzy system and its FPGA implementation. IEEE Trans. Syst. Man Cybern. Part B Cybern. **38**, 1537–1548 (2008)
17. Jang, J.-S.R.: Input selection for ANFIS learning. In: IEEE (ed.) The 15th IEEE World Conference on Computational Intelligence, Reno, p. 825. IEEE (2005)
18. Klir, G.J., Folger, T.A.: Fuzzy Sets, Uncertainty and Information. Prentice Hall, Englewood Cliffs (1992)

19. Ljung, G.M., Box, G.E.P.: Information theory and the extension of the maximum likelihood principle. Biometrika **65**(1), 553–564 (1978)
20. Melin, P., Soto, J., Castillo, O., Soria, J.: A new approach for time series prediction using ensembles of ANFIS models. Expert Syst. Appl. **39**(3), 3494–3506 (2012). https://doi.org/10.1016/j.eswa.2011.09.040
21. Mendel, J.M.: Computing derivatives in interval type-2 fuzzy logic systems. IEEE Trans. Fuzzy Syst. **12**(1), 84–98 (2004)
22. Mendel, J.M.: Uncertain Rule-Based Fuzzy Logic Systems: Introduction and New Directions. Prentice Hall, Englewood Cliffs (2001)
23. Numberger, A., Kruse, R.A.: A neuro-fuzzy approach to optimize hierarchical recurrent fuzzy systems. Fuzzy Optim. Decis. Making **1**(2), 221 (2002)
24. Obayashi, M., Kuremoto, T., Kobayashi, K.: A self-organized fuzzy-neuro reinforcement learning system for continuous state space for autonomous robots. In: IEEE (ed.) Proceedings of CIMCA/IAWTIC/ISE, pp. 551–556. IEEE (2008)
25. Schwarz, G.E.: Estimating the dimension of a model. Ann. Stat. **6**(2), 461–464 (1978)
26. Soto, J., Melin, P., Castillo, O.: Time series prediction using ensembles of ANFIS models with genetic optimization of interval type-2 and type-1 fuzzy integrators. Int. J. Hybrid Intell. Syst. **11**(3), 211–226 (2014). https://doi.org/10.3233/HIS-140196
27. Soto, J., Melin, P., Castillo, O.: Particle swarm optimization of the fuzzy integrators for time series prediction using ensemble of IT2FNN architectures. In: Melin, P., Castillo, O., Kacprzyk, J. (eds.) Nature-Inspired Design of Hybrid Intelligent Systems. SCI, vol. 667, pp. 141–158. Springer, Cham (2017). https://doi.org/10.1007/978-3-319-47054-2_9
28. Thode, H.C.J.: Testing for Normality. Marcel Drecker Inc., New York (2002)
29. Tukey, J.W.: The Problem of Multiple Comparisons: Course Notes. Princeton University, Princeton (1985)
30. Wang, L., Fsas, J.: Extracting fuzzy rules. Fuzzy Sets Fuzzy Syst. **101**, 353–362 (1999)
31. Wang, L.X., Mendel, J.M.: Back-propagation fuzzy system as nonlinear dynamic system identifiers. In: Proceedings of the IEEE International Conference on Fuzzy Systems, vol. 8, pp. 1537–1548. IEEE (1992)
32. Wang, L.X., Mendel, J.M.: Generating fuzzy rules by learning from examples. In: IEEE (ed.) Proceedings of IEEE International Symposium on Control, pp. 263–268. IEEE (1991)
33. Wang, L.X., Mendel, J.M.: Generating fuzzy rules by learning from examples. IEEE Trans. Syst. Man Cybern. **22**(6), 1414–1427 (1992)
34. Wu, I.F., Goo, Y.J.: A neuro-fuzzy computing technique for modeling the time series of short-term nt/us exchange rates. J. Am. Acad. Bus. **7**(2), 176 (2005)

Equilibrium Point of Representable Moore Continuous n-Dimensional Interval Fuzzy Negations

Ivan Mezzomo[1](✉), Benjamín Bedregal[2](✉), and Thadeu Milfont[1,2](✉)

[1] Center of Exact and Natural Sciences - CCEN,
Rural Federal University of SemiArid - UFERSA,
Mossoró, Rio Grande do Norte, Brazil
{imezzomo,thadeuribeiro}@ufersa.edu.br
[2] Department of Informatics and Applied Mathematics - DIMAp,
Federal University of Rio Grande do Norte - UFRN,
Natal, Rio Grande do Norte, Brazil
bedregal@dimap.ufrn.br

Abstract. n-dimensional interval fuzzy sets are a type of fuzzy sets which consider ordered n-tuples in $[0,1]^n$ as membership degree. This paper considers the notion of representable n-dimensional interval fuzzy negations, in particular, these that are Moore continuous, proposed in a previous paper of the authors, and we study some conditions that guarantee the existence of equilibrium point in classes of representable (Moore continuous) n-dimensional interval fuzzy negations. In addition, we prove that the changing of the dimensions of representable Moore continuous n-dimensional fuzzy negations inherits their equilibrium points.

Keywords: n-dimensional interval fuzzy sets · Fuzzy negations
Moore metric · Representable · Equilibrium point

1 Introduction

The concept of the fuzzy set was introduced by Zadeh (1965) and since then, several mathematical concepts such as number, group, topology, differential equation, etc have been fuzzified. Several extensions or types of fuzzy set theory had been proposed in order to solve the problem of constructing the membership degrees functions of fuzzy sets or/and to represent the uncertainty associated to the considered problem in a way different from fuzzy set theory [11]. In particular, Shang et al. in [27] propose a new type of fuzzy sets, namely n-Dimensional fuzzy sets, where the membership values are n-truples of real numbers in the unit interval $[0,1]$ ordered in increasing order, called n-dimensional intervals. n-dimensional fuzzy sets are a special class of L-fuzzy sets introduced by Goguen in [14] and a discrete kind of Type-2 fuzzy sets introduced in [29], is a kind of (ordered) fuzzy multiset introduced by Yager in [28] and generalize some extensions of fuzzy sets, such as interval-valued fuzzy sets and interval-valued Atanassov intuitionistic fuzzy sets [11]. In addition, n-dimensional fuzzy sets are

© Springer International Publishing AG, part of Springer Nature 2018
G. A. Barreto and R. Coelho (Eds.): NAFIPS 2018, CCIS 831, pp. 265–277, 2018.
https://doi.org/10.1007/978-3-319-95312-0_23

adequated in situations where the memberships degrees are provided for a fixed numbers of experts or methods, and the information of which expert/method given a determined degree is unrelevant. The set of n-dimensional intervals is and denoted by $L_n([0,1])$.

In the context of L-fuzzy logic, the main classes of fuzzy connectives (t-norm, t-conorm, fuzzy implication and fuzzy negations) were generalized for lattice-valued fuzzy logics, as we can be seen in [2,23]. In [4] the construction of bounded lattice negations from bounded lattice t-norms is considered.

In [6], it was considered the main properties of an n-dimensional fuzzy set A on $L_n[0,1]$ defined over a set X and was introduced the notion of n-dimensional interval fuzzy negations (nDIFN) and the notion of representable functions. A deeper study of nDIFN was made in [9].

In [18], we investigate the class of nDIFN which are continuous and strictly decreasing, called n-dimensional strict fuzzy negations. In particular, we investigate the class of representable n-dimensional strict fuzzy negations, i.e., n-dimensional strict fuzzy negations which are determined by strict fuzzy negation. The main properties of strict fuzzy negations on $[0,1]$ are preserved by representable strict fuzzy negations on $L_n([0,1])$.

A metric space on a set S is a real-valued function $d : S \times S \to \mathbb{R}$ such that satisfy the axioms of positiveness, symmetry and triangular inequality [22]. In [21], Moore et al. generalized the usual metric space of real numbers for real intervals and extends the notion of continuity of real functions for interval functions based on the Moore metric space. The Moore metric has been restricted to subintervals of $[0,1]$ for the study of interval-valued fuzzy connectives [3,8]. In [20], we extent this restricted Moore metric for n-dimensional interval fuzzy sets for characterizing the notion of Moore continuous nDIFN and prove some results about them. In addition, we consider the intuitive notion of strict nDIFN and study the way of changing of the dimensions of Moore continuous nDIFN.

In this work, we consider the notion of the equilibrium point (or fixed point[1]) of the fuzzy negations, investigated by [3,15,17,25,26] for defining an equilibrium point of nDIFN and we obtain results envolving representable (Moore continuous) nDIFN and equilibrium points. The remaining parts of this paper are organized as following. In Sect. 2, we introduce some preliminary concepts for the paper as continuity, Moore metric, n-dimensional fuzzy sets and equilibrium point of n-dimensional fuzzy negations. In Sect. 3, we consider the notion of representable nDIFN these that are Moore continuous, proposed in [20], and study some conditions that guarantee the existence of equilibrium point in classes of representable (Moore continuous) nDIFN. In Sect. 4, we characterize the increase and decrease of the dimension of nDIFN and provide conditions for a change of the dimensions of representable Moore continuous nDIFN inherits their equilibrium point.

---

[1] In the literature on fuzzy negations had been widely used both terms for the same notion, namely, an element $e \in [0,1]$ such that $N(e) = e$, with N being a fuzzy negation. We choice "equilibrium point" over "fixed point" but this not means that we consider the term equilibrium point more correct or better than the fixed point.

2 Preliminaries

2.1 Fuzzy Negations

A function $N : [0,1] \rightarrow [0,1]$ is a *fuzzy negation* if

N1: $N(0) = 1$ and $N(1) = 0$;

N2: If $x \leq y$, then $N(x) \geq N(y)$, for all $x, y \in [0,1]$.

A fuzzy negation N satisfying the involutive property

N3: $N(N(x)) = x$, for all $x \in [0,1]$,

is called *strong fuzzy negation*. And, a continuous fuzzy negation N is *strict* if it satisfies

N4: $N(x) < N(y)$ when $y < x$, for all $x \in [0,1]$.

Strong fuzzy negations are also strict fuzzy negations [16]. The standard fuzzy negation is defined as $N_S(x) = 1 - x$ is strong and, therefore, strict. The fuzzy negation defined as $N_{S^2}(x) = 1 - x^2$ is an example of the fuzzy negation that is strict, but not strong.

An *equilibrium point* of a fuzzy negation N is a value $e \in [0,1]$ such that $N(e) = e$.

Proposition 1. *[3, Proposition 2.1] Let N_1 and N_2 be fuzzy negations such that $N_1 \leq N_2$. Then, if e_1 and e_2 are the equilibrium points of N_1 and N_2, respectively, then $e_1 \leq e_2$.*

See [3, Remarks 2.1 and 2.2] for additional studies related to main properties of equilibrium points.

2.2 Topology and Metric Spaces

According to Dugundji [13], a topology on a set A is a collection of subsets of A which is closed under finite intersections and arbitrary unions, including the empty set and the set A. A set A together with a topology $\mathcal{T}$ on A is a topological space denoted by $(A, \mathcal{T})$. The elements of $\mathcal{T}$ are the open sets of the space.

Note that a distance or metric on A is a function $d : A \times A \rightarrow \mathbb{R}^+$ such that, for all $x, y, z \in A$, satisfies the following properties:

1. $d(x, y) = 0 \Leftrightarrow x = y$;
2. $d(x, y) = d(y, x)$;
3. $d(x, z) \leq d(x, y) + d(y, z)$.

A metric space is a set A endowed with a metric d and denoted by (A, d).

Example 1. Let $\mathbb{I}(\mathbb{R})$ be the set of the reals intervals $X = [\underline{x}, \overline{x}]$, where $\underline{x}, \overline{x} \in \mathbb{R}$ and $\underline{x} \leq \overline{x}$. The function d_M defined on $\mathbb{I}(\mathbb{R})$ by

$$d_M(X, Y) = \max\{|\underline{x} - \underline{y}|, |\overline{x} - \overline{y}|\} \tag{1}$$

is called *Moore metric.* ∎

Let A and B be two sets. For all distance $d_A : A \times A \to \mathbb{R}$ and $d_B : B \times B \to \mathbb{R}$ we have the following notion of continuity $f : A \to B$ is (d_A, d_B)-continuous, if for each $x \in A$ and $\epsilon > 0$ there exists $\delta > 0$ such that, for all $y \in A$,

$$d_A(x, y) \leq \delta \Rightarrow d_B(f(x), f(y)) \leq \epsilon \tag{2}$$

Continuous functions with respect to d_M will be called *Moore-continuous*. For more details see: [1,7,24].

2.3 *n*-Dimensional Fuzzy Sets

Let X be a non empty set and $n \in \mathbb{N}^+ = \mathbb{N} - \{0\}$. According to [27], *an n-dimensional fuzzy set A over X* is given by

$$A = \{(x, \mu_{A_1}(x), \ldots, \mu_{A_n}(x)) : x \in X\},$$

where, for each $i = 1, \ldots, n$, $\mu_{A_i} : X \to [0, 1]$ is called i-th membership degree of A, which also satisfies the condition: $\mu_{A_1}(x) \leq \ldots \leq \mu_{A_n}(x)$, for each $x \in X$.

In [5], for $n \geq 1$, an n-dimensional upper simplex is given as

$$L_n([0, 1]) = \{(x_1, \ldots, x_n) \in [0, 1]^n : x_1 \leq \ldots \leq x_n\}, \tag{3}$$

and its elements are called *n-dimensional intervals*.

For each $i = 1, \ldots, n$, the i-th projection of $L_n([0, 1])$ is the function $\pi_i^{(n)} : L_n([0, 1]) \to [0, 1]$ defined by

$$\pi_i^{(n)}(x_1, \ldots, x_n) = x_i. \tag{4}$$

When is clear the value of (n) in $\pi_i^{(n)}$, this indice will be omitted by simplicity of notation.

Notice that $L_1([0, 1]) = [0, 1]$ and $L_2([0, 1])$ reduces to the usual lattice of all the closed subintervals of the unit interval $[0, 1]$.

A *degenerate element* $\mathbf{x} \in L_n([0, 1])$ satisfies the following condition

$$\pi_i(\mathbf{x}) = \pi_j(\mathbf{x}), \quad \forall i, j = 1, \ldots, n. \tag{5}$$

The degenerate element $(x, \ldots, x)$ of $L_n([0, 1])$, for each $x \in [0, 1]$, will be denoted by $/x/$ and the set of all degenerate elements of $L_n([0, 1])$ will be denoted by $\mathcal{D}_n$.

An m-ary function $F : L_n([0, 1])^m \to L_n([0, 1])$ is called $\mathcal{D}_n$-*preserve* function or a function preserving degenerate elements if the following condition holds

(DP) $F(\mathcal{D}_n^m) = F(/x_1/, \ldots, /x_m/) \in \mathcal{D}_n, \forall x_1, \ldots, x_m \in [0, 1].$

By considering the natural extension of the order $\leq$ on $L_2([0, 1])$ as in [3,10] to higher dimensions, for all $\mathbf{x}, \mathbf{y} \in L_n([0, 1])$, it holds that

$$\mathbf{x} \leq \mathbf{y} \text{ iff } \pi_i(\mathbf{x}) \leq \pi_i(\mathbf{y}), \quad \forall\, i = 1, \ldots, n. \tag{6}$$

Based on [5], the supremum and infimum on $L_n([0,1])$ are both given as

$$\mathbf{x} \vee \mathbf{y} = (\max(x_1, y_1), \ldots, \max(x_n, y_n)), \tag{7}$$
$$\mathbf{x} \wedge \mathbf{y} = (\min(x_1, y_1), \ldots, \min(x_n, y_n)), \ \forall \ \mathbf{x}, \mathbf{y} \in L_n([0,1]). \tag{8}$$

Definition 1. *[6] A function* $\mathcal{N} : L_n([0,1]) \to L_n([0,1])$ *is a n-dimensional interval fuzzy negation (nDIFN) if, for each* $\mathbf{x}, \mathbf{y} \in L_n([0,1])$:

N1 $\mathcal{N}(/0/) = /1/$ *and* $\mathcal{N}(/1/) = /0/$;
N2 *If* $\mathbf{x} \leq \mathbf{y}$, *then* $\mathcal{N}(\mathbf{y}) \leq \mathcal{N}(\mathbf{x})$.

Proposition 2. *[6, Proposition 3.1] Let* $N_1, \ldots, N_n$ *be fuzzy negations such that* $N_1 \leq \ldots \leq N_n$. *Then* $\widetilde{N_1 \ldots N_n} : L_n([0,1]) \to L_n([0,1])$ *defined by*

$$\widetilde{N_1 \ldots N_n}(\mathbf{x}) = (N_1(\pi_n(\mathbf{x})), \ldots, N_n(\pi_1(\mathbf{x}))) \tag{9}$$

is an n-dimensional fuzzy negation.

Definition 2. *A n-dimensional interval fuzzy negation (nDIFN)* $\mathcal{N}$ *is representable if there exists fuzzy negation* $N_1, \ldots, N_n$ *such that* $N_i \leq N_{i+1}$ *for each* $i = 1, \ldots, n-1$ *and* $\mathcal{N} = \widetilde{N_1 \ldots N_n}$.
  The tuple $(N_1, \ldots, N_n)$ *will be called the representant of* $\mathcal{N}$.

Observe that $\pi_i(\mathcal{N}(\mathbf{x})) = N_i(\pi_{n-i+1}(\mathbf{x}))$ for each $i = 1, \ldots, n$.
  In particular, when $N_i = N_j$ for each $i, j = 1, \ldots, n$, we say that (N) is the representant of $\mathcal{N}$ and denote $\widetilde{N \ldots N}$ by $\tilde{N}$.

Proposition 3. *[9, Proposition 9] Let* $\mathcal{N}$ *be an n-dimensional fuzzy negation. Then, for all* $i = 1, \ldots, n$, *the function* $N_i : [0,1] \to [0,1]$ *defined by*

$$N_i(x) = \pi_i(\mathcal{N}(/x/)) \tag{10}$$

is a fuzzy negation.

Definition 3. *Let* $\mathcal{N}$ *be a nDIFN and* $i \in \{1, \ldots, n\}$. $\mathcal{N}$ *is i-representable if* $N_i : [0,1] \to [0,1]$ *defined by Eq. (10) is a fuzzy negation such that, for all* $\mathbf{x} \in L_n([0,1])$

$$N_i(\pi_{n-i+1}(\mathbf{x})) = \pi_i(\mathcal{N}(\mathbf{x})). \tag{11}$$

Obviously, $\mathcal{N}$ is i-representable, for all $i = 1, \ldots, n$, iff $\mathcal{N}$ is representable.
  If an nDIFN $\mathcal{N}$ satisfies

N3 $\mathcal{N}(\mathcal{N}(\mathbf{x})) = \mathbf{x}, \ \forall \ \mathbf{x} \in L_n([0,1])$,

it is called *strong n-dimensional interval fuzzy negation*.

Theorem 1. *[9, Theorem 24]* $\mathcal{N}$ *is a strong n-dimensional fuzzy negation iff there exists a strong fuzzy negation* N *such that* (N) *is the representant of* $\mathcal{N}$.

3   Moore Continuous n-dimensional Interval Functions

In this section we generalize the Moore metric for n-dimensional intervals.

Proposition 4. *[20, Proposition 3.1] Let $d_M^n : L_n([0,1]) \times L_n([0,1]) \to \mathbb{R}^+$ be the function defined by*

$$d_M^n(\mathbf{x}, \mathbf{y}) = \max(|\pi_1(\mathbf{x}) - \pi_1(\mathbf{y})|, \ldots, |\pi_n(\mathbf{x}) - \pi_n(\mathbf{y})|) \tag{12}$$

Then d_M^n is a metric on $L_n([0,1])$ called n-dimensional interval Moore metric on $L_n([0,1])$.

Remark 1. Observe that d_M^1 is the usual distance on real numbers restricted to $[0,1]$ and d_M^2 is the Moore metric [12].

Definition 4. *[20, Definition 3.1] Let $F : L_n([0,1]) \to L_n([0,1])$ be a n-dimensional interval function. F is Moore continuous if F is (d_M^n, d_M^n)-continuous.*

Theorem 2. *[20, Theorem 3.1] Let $\mathcal{N}$ be a representable nDIFN with $(N_1, \ldots, N_n)$ as representant. $\mathcal{N}$ is Moore continuous iff every N_i is continuous.*

Corollary 1. *[20, Corollary 3.1] Each strong nDIFN is Moore continuous.*

Proposition 5. *If $\mathcal{N}$ is i-representable nDIFN and Moore continuous, then N_i is continuous.*

Proof. Let $\epsilon > 0$, $i \in \{1, \ldots, n\}$, $\mathcal{N}$ is Moore continuous and $x, y \in [0,1]$. Since $\mathcal{N}$ is (d_M^n, d_M^n)-continuous then there exists $\delta > 0$ satisfying the Eq. (2).

$$|x - y| \leq \delta$$
$$\Rightarrow \max(\underbrace{|x - y|, \ldots, |x - y|}_{n-times}) \leq \delta$$
$$\Rightarrow d_M^n(/x/, /y/) \leq \delta$$
$$\Rightarrow d_M^n(\mathcal{N}(/x/), \mathcal{N}(/y/)) \leq \epsilon$$
$$\Rightarrow \max(|\pi_1(\mathcal{N}(/x/)) - \pi_1(\mathcal{N}(/y/))|, \ldots, |\pi_n(\mathcal{N}(/x/)) - \pi_n(\mathcal{N}(/y/))|) \leq \epsilon$$
$$\Rightarrow |\pi_i(\mathcal{N}(/x/)) - \pi_i(\mathcal{N}(/y/))| \leq \epsilon$$
$$\Rightarrow |N_i(x) - N_i(y)| \leq \epsilon$$

Therefore, N_i is continuous.   ∎

Proposition 6. *[20, Proposition 4.2] Let $\mathcal{N}$ be a Moore continuous nDIFN such that $\mathcal{N}$ is an i and $i+1$-representable for some $1 \leq i \leq n-1$ and N be a continuous fuzzy negation satisfying $N_i \leq N \leq N_{i+1}$. Then the function $\mathcal{N}_+ : L_{n+1}([0,1]) \to L_{n+1}([0,1])$ defined by*

$$\mathcal{N}_+(\mathbf{x}) = (\pi_1^{(n)}(\mathcal{N}(\mathbf{x}_0)), \ldots, \pi_i^{(n)}(\mathcal{N}(\mathbf{x}_0)), N(x_{n-i+2}),$$
$$\pi_{i+1}^{(n)}(\mathcal{N}(\mathbf{x}_0)), \ldots, \pi_n^{(n)}(\mathcal{N}(\mathbf{x}_0)))$$

where

$$\mathbf{x}_0 = (\pi_1^{(n+1)}(\mathbf{x}), \ldots, \pi_{n-i+1}^{(n+1)}(\mathbf{x}), \pi_{n-i+3}^{(n+1)}(\mathbf{x}), \ldots, \pi_{n+1}^{(n+1)}(\mathbf{x}))$$

and $x_{\frac{n+1}{2}} = \pi_{\frac{n+1}{2}}^{(n+1)}(\mathbf{x})$, *is Moore continuous* $(n+1)DIFN$.

3.1 Equilibrium Point of Moore Continuous n-dimensional Interval Fuzzy Negations

Definition 5. *An element* $\mathbf{e} \in L_n([0,1])$ *is an equilibrium point for an nDIFN* $\mathcal{N}$ *if* $\mathcal{N}(\mathbf{e}) = \mathbf{e}$. *In addition, if* $\pi_i(\mathbf{e}) \in (0,1)$, *for each* $i = 1, \ldots, n$, *then* $\mathbf{e}$ *is called of the positive equilibrium point.*

Remark 2. Let $\mathcal{N}$ be a strict nDIFN. If $\mathbf{x} < \mathbf{e}$ then $\mathcal{N}(\mathbf{x}) > \mathbf{e}$ and if $\mathbf{e} < \mathbf{x}$ then $\mathcal{N}(\mathbf{x}) < \mathbf{e}$.

Proposition 7. *Let N be a fuzzy negation with the equilibrium point e. Then, $/e/$ is an n-dimensional equilibrium point of $\widetilde{N}$.*

Proof. Straightforward.   ∎

Proposition 8. *Let $\mathcal{N}$ be a representable nDIFN. If n is even then $\mathcal{N}$ has at least one equilibrium point.*

Proof. If $\mathcal{N}$ is a representable nDIFN then there exist $N_1, \ldots, N_n$ fuzzy negations such that $\mathcal{N} = \widetilde{N_1 \ldots N_n}$. Consider the equilibrium point $\mathbf{e} = (\underbrace{0, \ldots, 0}_{\frac{n}{2}-times}, \underbrace{1, \ldots, 1}_{\frac{n}{2}-times})$.

Since $N_i(0) = 1$ and $N_i(1) = 0$, for all $i \in \{1, \ldots, n\}$, then $\widetilde{N_1 \ldots N_n}(\mathbf{e}) = \mathbf{e}$.   ∎

Theorem 3. *All representable Moore continuous nDIFN $\mathcal{N}$ has an equilibrium point.*

Proof. If n is even the proof is similar to Proposition 8.

If n is odd, let $(N_1, \ldots, N_n)$ be the representant of $\mathcal{N}$. Then by Theorem 2, $N_{\frac{n+1}{2}}$ is continuous and so, it has an equilibrium point $e \in (0,1)$. Clearly, $(\underbrace{0, \ldots, 0}_{(\frac{n-1}{2})-times}, e, \underbrace{1, \ldots, 1}_{(\frac{n-1}{2})-times})$ is an equilibrium point of $\mathcal{N}$.   ∎

Proposition 9. *Let $\mathcal{N}$ be a representable nDIFN. If n is even and N_i is crisp, for all $i \in \{1, \ldots, n\}$, then $\mathcal{N}$ has an unique equilibrium point.*

Proof. Let $\mathbf{e} = (e_1, \ldots, e_n)$. If, for some $i \in \{1, \ldots, n\}$, $e_i \notin \{0,1\}$, then $\mathcal{N}(\mathbf{e}) = (N_1(e_n), \ldots, N_n(e_1))$. Since N_j is crisp, for all $j \in \{1, \ldots, n\}$, then there exists $0 \leq k \leq n$ such that $\mathcal{N}(\mathbf{e}) = (\underbrace{0, \ldots, 0}_{k-times}, \underbrace{1, \ldots, 1}_{(n-k)-times}) \neq \mathbf{e}$. Hence, $\mathbf{e}$ is not an equilibrium point of $\mathcal{N}$. Thus, if $\mathbf{e}$ is an equilibrium point of $\mathcal{N}$, then there exist a $0 \leq k \leq n$ such that $\mathbf{e} = (\underbrace{0,\ldots,0}_{k-times},\ \underbrace{1,\ldots,1}_{(n-k)-times}\ )$. So,

$$(\underbrace{0,\ldots,0}_{k-times},\ \underbrace{1,\ldots,1}_{(n-k)-times}\ ) = (N_1(1),\ldots,N_{n-k}(1),N_{n-k+1}(0),\ldots,N_n(0))$$

$$= (\ \underbrace{0,\ldots,0}_{(n-k)-times}\ ,\ \underbrace{1,\ldots,1}_{(k)-times}\ ).$$

Hence, $k = n - k$. Therefore, by proof of the Proposition 8, $\mathcal{N}$ has a unique equilibrium point. ∎

Lemma 1. *Let $\mathcal{N}$ be a $\frac{n+1}{2}$-representable nDIFN for n odd. If $\mathcal{N}$ has an equilibrium point, then $N_{\frac{n+1}{2}}$ has an equilibrium point.*

Proof. Let $\mathbf{e} = (e_1,\ldots,e_{\frac{n+1}{2}},\ldots,e_n)$ be an equilibrium point of $\mathcal{N}$. Then,

$$\mathcal{N}(\mathbf{e}) = (\pi_1(\mathcal{N}(\mathbf{e})),\ldots,\pi_{\frac{n-1}{2}}(\mathcal{N}(\mathbf{e})),\pi_{\frac{n+1}{2}}(\mathcal{N}(\mathbf{e})),\pi_{\frac{n+3}{2}}(\mathcal{N}(\mathbf{e})),\ldots,\pi_n(\mathcal{N}(\mathbf{e})))$$

$$= (\pi_1(\mathcal{N}(\mathbf{e})),\ldots,\pi_{\frac{n-1}{2}}(\mathcal{N}(\mathbf{e})),N_{\frac{n+1}{2}}\left(\pi_{\frac{n+1}{2}}(\mathbf{e})\right),\pi_{\frac{n+3}{2}}(\mathcal{N}(\mathbf{e})),\ldots,\pi_n(\mathcal{N}(\mathbf{e}))).$$

But,

$$\begin{aligned}
N_{\frac{n+1}{2}}\left(e_{\frac{n+1}{2}}\right) &= N_{\frac{n+1}{2}}\left(\pi_{\frac{n+1}{2}}(\mathbf{e})\right) \\
&= \pi_{\frac{n+1}{2}}(\mathcal{N}(\mathbf{e})) \quad \text{by Eq. (11)} \\
&= \pi_{\frac{n+1}{2}}(\mathbf{e}) \\
&= e_{\frac{n+1}{2}}.
\end{aligned}$$

Therefore, the proposition holds. ∎

Proposition 10. *Let $\mathcal{N}$ be a representable nDIFN and n be odd. If N_i has no equilibrium point, for all $i \in \{1,\ldots,\frac{n-1}{2},\frac{n+3}{2},\ldots,n\}$, then $\mathcal{N}$ has just one equilibrium point when $N_{\frac{n+1}{2}}$ has an equilibrium point.*

Proof. In this case, the unique equilibrium point is $(\ \underbrace{0,\ldots,0}_{(\frac{n-1}{2})-times}\ ,e,\ \underbrace{1,\ldots,1}_{(\frac{n+3}{2})-times}\ )$ where e is the equilibrium point of $N_{\frac{n+1}{2}}$. ∎

Proposition 11. *Let $\mathcal{N}$ be a representable Moore continuous n-dimensional interval fuzzy negation with $(N_1,\ldots,N_n)$ as representant. $\mathcal{N}$ has a positive equilibrium point with equilibrium point $(e_1,\ldots,e_n)$ such that $N_1(e_n) = e_n$ iff all representants have the same equilibrium point.*

Proof. Since $\mathcal{N}$ is a representable Moore continuous n-dimensional interval fuzzy negation, then by Theorem 2, all representant N_i are continuous.

($\Rightarrow$) By Eq. (9), we have that $N_1(e_n) = e_1$ and by hypothesis, $N_1(e_n) = e_n$. Because $e_1 \leq \ldots \leq e_n$, then $e_i = e_n$, for all $i = \{1, \ldots, n\}$. Therefore, for all $i = 1, \ldots, n$, by Eq. (9), $N_i(e_n) = N_i(e_{n-i+1}) = e_i = e_n$, that is, all representants have the same equilibrium point e_n.

($\Leftarrow$) Since, every N_i has the same equilibrium point $e \in (0,1)$ then, by Proposition 7, $/e/$ is a positive equilibrium point of $\mathcal{N}$.   ∎

Remark 3. [9] Note that, if $(e_1, \ldots, e_n)$ is an equilibrium point of $\widetilde{N_1 \ldots N_n}$, but it may not be the unique. For example, $(0, e_2, \ldots, e_{n-1}, 1)$ also is an equilibrium point.

4   Change of the Dimension on Representable Moore Continuous nDIFN and Equilibrium Point

In [20], we proposed a way to increasing and decreasing the dimension of Moore continuous nDIFN preserving the Moore continuity. However, these methods not preserving the equilibrium point. Now, to maintain the equilibrium points we will provide new ways of changing the dimension, considering only when the dimension n is odd.

Proposition 12. *Let n be odd and $\mathcal{N}$ be a $\frac{n+1}{2}$-representable Moore continuous nDIFN. Then the function $\mathcal{N}_- : L_{n-1}([0,1]) \to L_{n-1}([0,1])$ defined by*

$$\mathcal{N}_-(x_1, \ldots, x_{n-1}) = (\pi_1^{(n)}(\mathcal{N}(\mathbf{z})), \ldots, \pi_{\frac{n-1}{2}}^{(n)}(\mathcal{N}(\mathbf{z})), \pi_{\frac{n+3}{2}}^{(n)}(\mathcal{N}(\mathbf{z})), \ldots, \pi_n^{(n)}(\mathcal{N}(\mathbf{z}))))$$

where $\mathbf{z} = (x_1, \ldots, x_{\frac{n-1}{2}}, e, x_{\frac{n+1}{2}}, \ldots, x_{n-1})$ and e is the equilibrium point of $N_{\frac{n+1}{2}}$, is Moore continuous $(n-1)$DIFN.

Proof. Clearly $\mathcal{N}_-$ is well defined and $\mathcal{N}_-(/0/) = /1/$ and $\mathcal{N}_-(/1/) = /0/$. Let $\mathbf{x}_0, \mathbf{y}_0 \in L_{n-1}([0,1])$, then

$$\mathbf{x} = (\pi_1^{(n-1)}(\mathbf{x}_0), \ldots, \pi_{\frac{n-1}{2}}^{(n-1)}(\mathbf{x}_0), \pi_{\frac{n}{2}}(e), \pi_{\frac{n+1}{2}}^{(n-1)}(\mathbf{x}_0), \ldots, \pi_{n-1}^{(n-1)}(\mathbf{x}_0)) \in L_n([0,1])$$

and

$$\mathbf{y} = (\pi_1^{(n-1)}(\mathbf{y}_0), \ldots, \pi_{\frac{n-1}{2}}^{(n-1)}(\mathbf{y}_0), \pi_{\frac{n}{2}}(e), \pi_{\frac{n+1}{2}}^{(n-1)}(\mathbf{y}_0), \ldots, \pi_{n-1}^{(n-1)}(\mathbf{y}_0)) \in L_n([0,1]).$$

Suppose that

$$\mathbf{x}_0 \leq \mathbf{y}_0 \Rightarrow \mathbf{x} \leq \mathbf{y}$$
$$\Rightarrow \mathcal{N}(\mathbf{y}) \leq \mathcal{N}(\mathbf{x})$$
$$\Rightarrow [\pi_1^{(n)}(\mathcal{N}(\mathbf{y})), \ldots, \pi_{n-1}^{(n)}(\mathcal{N}(\mathbf{y}))] \leq [\pi_1^{(n)}(\mathcal{N}(\mathbf{x})), \ldots, \pi_{n-1}^{(n)}(\mathcal{N}(\mathbf{x}))]$$
$$\Rightarrow \mathcal{N}_-(\mathbf{y}_0) \leq \mathcal{N}_-(\mathbf{x}_0).$$

Hence, $\mathcal{N}_-$ is an $(n-1)$DIFN.

Let $\epsilon > 0$. By the continuity of $\mathcal{N}$ there exists $\delta > 0$ satisfying for each $\mathbf{x}, \mathbf{y} \in L_n([0,1])$

$$d_M^n(\mathbf{x}, \mathbf{y}) \leq \delta \Rightarrow d_M^n(\mathcal{N}(\mathbf{x}), \mathcal{N}(\mathbf{y})) \leq \epsilon.$$

Thus, if

$d_M^{n-1}(\mathbf{x}_0, \mathbf{y}_0) \leq \delta$

$\Rightarrow \max(|\pi_1^{(n-1)}(\mathbf{x}_0) - \pi_1^{(n-1)}(\mathbf{y}_0)|, \ldots, |\pi_{\frac{n-1}{2}}^{(n-1)}(\mathbf{x}_0) - \pi_{\frac{n-1}{2}}^{(n-1)}(\mathbf{y}_0)|,$

$\quad |\pi_{\frac{n+1}{2}}^{(n-1)}(\mathbf{x}_0) - \pi_{\frac{n+1}{2}}^{(n-1)}(\mathbf{y}_0)|, \ldots, |\pi_{n-1}^{(n-1)}(\mathbf{x}_0) - \pi_{n-1}^{(n-1)}(\mathbf{y}_0)|) \leq \delta$

$\Rightarrow \max(|\pi_1^{(n-1)}(\mathbf{x}_0) - \pi_1^{(n-1)}(\mathbf{y}_0)|, \ldots, |\pi_{\frac{n-1}{2}}^{(n-1)}(\mathbf{x}_0) - \pi_{\frac{n-1}{2}}^{(n-1)}(\mathbf{y}_0)|, |\pi_{\frac{n}{2}}(e) - \pi_{\frac{n}{2}}(e)|,$

$\quad |\pi_{\frac{n+1}{2}}^{(n-1)}(\mathbf{x}_0) - \pi_{\frac{n+1}{2}}^{(n-1)}(\mathbf{y}_0)|, \ldots, |\pi_{n-1}^{(n-1)}(\mathbf{x}_0) - \pi_{n-1}^{(n-1)}(\mathbf{y}_0)|) \leq \delta$

$\Rightarrow \max(|\pi_1^{(n)}(\mathbf{x}) - \pi_1^{(n)}(\mathbf{y})|, \ldots, |\pi_{\frac{n-1}{2}}^{(n)}(\mathbf{x}) - \pi_{\frac{n-1}{2}}^{(n)}(\mathbf{y})|, |\pi_{\frac{n+1}{2}}(e) - \pi_{\frac{n+1}{2}}(e)|,$

$\quad |\pi_{\frac{n+3}{2}}^{(n)}(\mathbf{x}) - \pi_{\frac{n+3}{2}}^{(n)}(\mathbf{y})|, \ldots, |\pi_n^{(n)}(\mathbf{x}) - \pi_n^{(n)}(\mathbf{y})|) \leq \delta$

$\Rightarrow d_M^n(\mathbf{x}, \mathbf{y}) \leq \delta$

$\Rightarrow d_M^n(\mathcal{N}(\mathbf{x}), \mathcal{N}(\mathbf{y})) \leq \epsilon$

$\Rightarrow \max(|\pi_1^{(n)}(\mathcal{N}(\mathbf{x})) - \pi_1^{(n)}(\mathcal{N}(\mathbf{y}))|, \ldots, |\pi_{\frac{n-1}{2}}^{(n)}(\mathcal{N}(\mathbf{x})) - \pi_{\frac{n-1}{2}}^{(n)}(\mathcal{N}(\mathbf{y}))|, |\pi_{\frac{n+1}{2}}(\mathcal{N}(e)) -$

$\quad \pi_{\frac{n+1}{2}}(\mathcal{N}(e))|, |\pi_{\frac{n+3}{2}}^{(n)}(\mathcal{N}(\mathbf{x})) - \pi_{\frac{n+3}{2}}^{(n)}(\mathcal{N}(\mathbf{y}))|, \ldots, |\pi_n^{(n)}(\mathcal{N}(\mathbf{x})) - \pi_n^{(n)}(\mathcal{N}(\mathbf{y}))|) \leq \epsilon$

$\Rightarrow \max(|\pi_1^{(n)}(\mathcal{N}(\mathbf{x})) - \pi_1^{(n)}(\mathcal{N}(\mathbf{y}))|, \ldots, |\pi_{\frac{n-1}{2}}^{(n)}(\mathcal{N}(\mathbf{x})) - \pi_{\frac{n-1}{2}}^{(n)}(\mathcal{N}(\mathbf{y}))|,$

$\quad |\pi_{\frac{n+3}{2}}^{(n)}(\mathcal{N}(\mathbf{x})) - \pi_{\frac{n+3}{2}}^{(n)}(\mathcal{N}(\mathbf{y}))|, \ldots, |\pi_n^{(n)}(\mathcal{N}(\mathbf{x})) - \pi_n^{(n)}(\mathcal{N}(\mathbf{y}))|) \leq \epsilon$

$\Rightarrow \max(|\pi_1^{(n-1)}(\mathcal{N}_-(\mathbf{x}_0)) - \pi_1^{(n-1)}(\mathcal{N}_-(\mathbf{y}_0))|, \ldots, |\pi_{\frac{n-1}{2}}^{(n-1)}(\mathcal{N}_-(\mathbf{x}_0)) - \pi_{\frac{n-1}{2}}^{(n-1)}(\mathcal{N}_-(\mathbf{y}_0))|$

$\quad |\pi_{\frac{n+1}{2}}^{(n-1)}(\mathcal{N}_-(\mathbf{x}_0)) - \pi_{\frac{n+1}{2}}^{(n-1)}(\mathcal{N}_-(\mathbf{y}_0))|, \ldots, |\pi_{n-1}^{(n-1)}(\mathcal{N}_-(\mathbf{x}_0)) - \pi_{n-1}^{(n-1)}(\mathcal{N}_-(\mathbf{y}_0))|) \leq \epsilon$

$\Rightarrow d_M^{n-1}(\mathcal{N}_-(\mathbf{x}_0), \mathcal{N}_-(\mathbf{y}_0)) \leq \epsilon.$

Therefore, $\mathcal{N}_-$ is Moore continuous $(n-1)$DIFN.

Proposition 13. *Let n be odd and $\mathcal{N}$ be a Moore continuous nDIFN such that $\mathcal{N}$ is an i and $\frac{n+1}{2}$-representable for some $1 \leq i \leq n-1$ and N be a continuous fuzzy negation satisfying $N_i \leq N \leq N_{i+1}$. Then the function $\mathcal{N}_+ : L_{n+1}([0,1]) \to L_{n+1}([0,1])$ defined by*

$$\mathcal{N}_+(\mathbf{x}) = (\pi_1^{(n)}(\mathcal{N}(\mathbf{x}_0)), \ldots, \pi_{\frac{n+1}{2}}^{(n)}(\mathcal{N}(\mathbf{x}_0)), N(x_{n-i+2}), \pi_{\frac{n+3}{2}}^{(n)}(\mathcal{N}(\mathbf{x}_0)), \ldots, \pi_n^{(n)}(\mathcal{N}(\mathbf{x}_0)))$$

where

$$\mathbf{x}_0 = (\pi_1^{(n+1)}(\mathbf{x}), \ldots, \pi_{\frac{n+1}{2}}^{(n+1)}(\mathbf{x}), \pi_{\frac{n+5}{2}}^{(n+1)}(\mathbf{x}), \ldots, \pi_{n+1}^{(n+1)}(\mathbf{x}))$$

and $x_{n-i+2} = \pi_{n-i+2}^{(n+1)}(\mathbf{x})$, is Moore continuous $(n+1)$DIFN.

Proof. Analogously from Proposition 6.

Proposition 14. *Let n be odd and $\mathcal{N}$ be a $\frac{n+1}{2}$-representable Moore continuous nDIFN. If $(e_1, \ldots, e_n)$ is an equilibrium point of $\mathcal{N}$, then $(e_1, \ldots, e_{\frac{n-1}{2}}, e_{\frac{n+3}{2}}, \ldots, e_n)$ is an equilibrium point of $\mathcal{N}_-$.*

Proof. Once $\mathcal{N}$ is $\frac{n+1}{2}$-representable Moore continuous nDIFN then by Proposition 5, $N_{\frac{n+1}{2}}$ is continuous. So, $N_{\frac{n+1}{2}}$ has a unique equilibrium point. If $\mathbf{z} = (e_1, \ldots, e_n)$ is an equilibrium point of $\mathcal{N}$, then $N_{\frac{n+1}{2}}\left(e_{\frac{n+1}{2}}\right) = e_{\frac{n+1}{2}}$. Let $\mathbf{y} = (e_1, \ldots, e_{\frac{n-1}{2}}, e_{\frac{n+1}{2}}, \ldots, e_n)$, then by Proposition 12

$$\mathcal{N}_-(\mathbf{y}) = (\pi_1^{(n)}(\mathcal{N}(\mathbf{z})), \ldots, \pi_{\frac{n-1}{2}}^{(n)}(\mathcal{N}(\mathbf{z})), \pi_{\frac{n+3}{2}}^{(n)}(\mathcal{N}(\mathbf{z})), \ldots, \pi_n^{(n)}(\mathcal{N}(\mathbf{z})))$$

$$= (\pi_1^{(n)}(\mathbf{z}), \ldots, \pi_{\frac{n-1}{2}}^{(n)}(\mathbf{z}), \pi_{\frac{n+3}{2}}^{(n)}(\mathbf{z}), \ldots, \pi_n^{(n)}(\mathbf{z}))$$

$$= (e_1, \ldots, e_{\frac{n-1}{2}}, e_{\frac{n+3}{2}}, \ldots, e_n)$$

$$= \mathbf{y}.$$

Therefore, $(e_1, \ldots, e_{\frac{n-1}{2}}, e_{\frac{n+3}{2}}, \ldots, e_n)$ is an equilibrium point of $\mathcal{N}_-$. ∎

Proposition 15. *Let n be odd and $\mathcal{N}$ be a $\frac{n+1}{2}$-representable Moore continuous nDIFN. If $(e_1, \ldots, e_n)$ is an equilibrium point of $\mathcal{N}$, then $(e_1, \ldots, e_{\frac{n-1}{2}}, e, e, e_{\frac{n+3}{2}}, \ldots, e_n)$ is an equilibrium point of $\mathcal{N}_+$, where e is the equilibrium point of $N_{\frac{n+1}{2}}$.*

Proof. Analogous from Proposition 14. ∎

Proposition 16. *Let $n \geq 2$, $\mathcal{N}$ be a Moore continuous nDIFN with representant $N_1, \ldots, N_n$. If $/e/^{(n)}$ is an n-dimensional equilibrium point of $\mathcal{N}$, then $/e/^{(n-1)}$ is an equilibrium point of $\mathcal{N}_-$.*

Proof. Straightforward. ∎

5   Conclusion

In this paper, we characterizing the notion of the equilibrium point of representable Moore continuous nDIFN and prove some results about them. Our aim was to investigate the existence of equilibrium point of this kind of nDIFN as well as the conditions for changing the dimensions of representable Moore continuous nDIFN and inherits their equilibrium point.

As further works, we intend to deepen the study in Moore continuous n-dimensional intervals fuzzy sets exploring the topological aspects.

Acknowledgment. This work is supported by Brazilian National Counsel of Technological and Scientific Development CNPq (Proc. 307781/2016-0 and 404382/2016-9).

References

1. Acióly, B.M., Bedregal, B.: A quasi-metric topology compatible with inclusion-monotonicity property on interval space. Reliab. Comput. **3**(3), 305–313 (1997)
2. Bedregal, B., Santos, H.S., Callejas-Bedregal, R.: T-norms on bounded lattices: t-norm morphisms and operators. In: IEEE International Conference on Fuzzy Systems, pp. 22–28 (2006)
3. Bedregal, B.: On interval fuzzy negations. Fuzzy Sets Syst. **161**(17), 2290–2313 (2010)
4. Bedregal, B., et al.: Negations generated by bounded lattices t-norms. In: Greco, S., Bouchon-Meunier, B., Coletti, G., Fedrizzi, M., Matarazzo, B., Yager, R.R. (eds.) IPMU 2012. CCIS, vol. 299, pp. 326–335. Springer, Heidelberg (2012). https://doi.org/10.1007/978-3-642-31718-7_34
5. Bedregal, B., et al.: A characterization theorem for t-representable n-dimensional triangular norms. In: Melo-Pinto, P., Couto, P., Serodio, C., Fodor, J., De Baets, B. (eds.) Eurofuse 2011. Advances in Intelligent and Soft Computing, vol. 107, pp. 103–112. Springer, Heidelberg (2011). https://doi.org/10.1007/978-3-642-24001-0_11
6. Bedregal, B., Beliakov, G., Bustince, H., Calvo, T., Mesiar, R., Paternain, D.: A class of fuzzy multisets with a fixed number of memberships. Inf. Sci. **189**, 1–17 (2012)
7. Bedregal, B., Santiago, R.H.N.: Some continuity notions for interval functions and representation. Comput. Appl. Math. **32**, 435–446 (2013)
8. Bedregal, B., Santiago, R.H.N.: Interval representations, Łukasiewicz implicators and Smets-Magrez axioms. Inf. Sci. **221**, 192–200 (2013)
9. Bedregal, B., Mezzomo, I., Reiser, R.H.S.: n-Dimensional Fuzzy Negations. CoRR abs/1707.08617 (2017)
10. Bustince, H., Montero, J., Pagola, M., Barrenechea, E., Gomes, D.: A survey of interval-value fuzzy sets. In: Pretrycz, W., Skowron, A., Kreinovich, V. (eds.) Handbook of Granular Computing, pp. 491–515. Wiley, West Sussex (2008). Chapter 22
11. Bustince, H., Barrenechea, E., Pagola, M., Fernandez, J., Xu, Z., Bedregal, B., Montero, J., Hagras, H., Herrera, F., De Baets, B.: A historical account of types of fuzzy sets and their relationships. IEEE Trans. Fuzzy Syst. **4**(1), 179–194 (2016)
12. Dimuro, G.P., Bedregal, B., Santiago, R.H.N., Reiser, R.H.S.: Interval additive generators of interval t-norms and interval t-conorms. Inf. Sci. **181**(18), 3898–3916 (2011)
13. Dugundji, J.: Topology, Allyn and Bacon, New York (1966)
14. Goguen, J.: L-fuzzy sets. J. Math. Anal. Appl. **18**, 145–174 (1967)
15. Higashi, M., Klir, G.J.: On measure of fuzziness and fuzzy complements. Int. J. Gen Syst **8**(3), 169–180 (1982)
16. Klement, E.P., Mesiar, R., Pap, E.: Triangular Norms. Kluwer Academic Publishers, Dordrecht (2000)
17. Klir, G.J., Yuan, B.: Fuzzy Sets and Fuzzy Logics: Theory and Applications. Prentice Halls PTR, Upper Saddle River (1995)
18. Mezzomo, I., Bedregal, B., Reiser, R., Bustince, H., Partenain, D.: On n-dimensional strict fuzzy negations. In: 2016 IEEE International Conference on Fuzzy Systems (FUZZ-IEEE), pp. 301–307 (2016). https://doi.org/10.1109/FUZZ-IEEE.2016.7737701

19. Mezzomo, I., Bedregal, B., Reiser, R.H.S.: Natural n-dimensional fuzzy negations for n-dimensional t-norms and t-conorms. In: 2017 IEEE International Conference on Fuzzy Systems (FUZZ-IEEE), pp. 1–6 (2017). https://doi.org/10.1109/FUZZ-IEEE.2017.8015506

20. Mezzomo, I., Bedregal, B.: Moore continuous n-dimensional Interval fuzzy negations. In: WCCI/Fuzz-IEEE 2018 (2018)

21. Moore, R.E., Kearfott, R.B., Cloud, M.J.: Introduction to Inverval Analysis. Society for Industrial and Applied Mathematics Philadelphia, Philadelphia (2009)

22. Mukherjee, M.N.: Elements of Space Metrics. Academic Publishers, Kolkata (2005)

23. Palmeira, E.S., Bedregal, B., Mesiar, R., Fernandez, J.: A new way to extend t-norms, t-conorms and negations. Fuzzy Sets Syst. **240**, 1–21 (2014)

24. Santiago, R.H.N., Bedregal, B., Acióly, B.M.: Formal aspects of correctness and optimality in interval computations. Form. Asp. Comput. **18**(2), 231–243 (2006)

25. Trillas, E.: Sobre funciones de negación en la teoria de conjuntos difusos. Stochastica **3**, 47–59 (1979)

26. Wagenknecht, M., Batyrshin, I.: Fixed point properties of fuzzy negations. J. Fuzzy Math. **6**, 975–981 (1998)

27. Shang, Y., Yuan, X., Lee, E.S.: The n-dimensional fuzzy sets and Zadeh fuzzy sets based on the finite valued fuzzy sets. Comput. Math. Appl. **60**, 442–463 (2010)

28. Yager, R.R.: On the theory of bags. Int. J. Gen Syst **13**, 23–37 (1986)

29. Zadeh, L.A.: Quantitative fuzzy semantics. Inf. Sci. **3**, 159–176 (1971)

Color Mathematical Morphology Using a Fuzzy Color-Based Supervised Ordering

Mateus Sangalli and Marcos Eduardo Valle[✉]

Department of Applied Mathematics, University of Campinas,
Rua Sérgio Buarque de Holanda, 651, Campinas, SP 130830-859, Brazil
mateussangalli@gmail.com, valle@ime.unicamp.br

Abstract. Mathematical morphology is a theory with applications in image processing and analysis. In a supervised approach to mathematical morphology, pixel values are ranked according to sets of foreground and background elements specified a priori by the user. In this paper, we introduce a supervised fuzzy color-based approach to color mathematical morphology that provides an elegant alternative to the support vector machine-based approach developed by Velasco-Forero and Angulo. Briefly, color elements are ranked according to the degree of truth of the proposition "the considered color is a foreground color but it is not a background color" in the new supervised color morphological approach. Furthermore, the vagueness and uncertainty inherent to the description of colors by humans can be naturally incorporated in the new approach using the concept of fuzzy colors.

Keywords: Image processing · Mathematical morphology
Complete lattice · Fuzzy color · Supervised learning

1 Introduction

Mathematical morphology (MM) is a powerful non-linear image processing framework based on geometrical and topological concepts [1,2]. Applications of MM include, edge detection, segmentation and automatic image reconstruction, pattern recognition and image decomposition [3–6].

The first morphological operators have been developed by Matheron and Serra in the 1960s for the analysis of binary images. Later, binary MM operators have been successfully generalized to deal with gray-scale images [7]. Some gray-scale morphological operators were also developed using concepts from fuzzy logic and fuzzy set theory [8–11].

Morphological operators are very well defined on complete lattices [1,12]. A complete lattice $\mathbb{L}$ is a partially ordered non-empty set in which any subset admits both a supremum and an infimum [13,14]. Since the only requirement is a partial order with well-defined extreme operations, complete lattices allowed the development of morphological operators to multivalued data, including color images [15,16]. In contrast to gray-scale approaches, however, there is no natural

© Springer International Publishing AG, part of Springer Nature 2018
G. A. Barreto and R. Coelho (Eds.): NAFIPS 2018, CCIS 831, pp. 278–289, 2018.
https://doi.org/10.1007/978-3-319-95312-0_24

ordering for colors. Hence, most researches on color MM consist on finding an appropriate ordering scheme for a given color image processing task. The interested reader can find a detailed discussion on many approaches to multivalued MM, including color MM, on [15].

Among the many partial orderings used on color MM, total orderings have been widely used because they avoid the appearance of false colors [17,18]. For example, Hanbury and Serra introduced a conditional ordering on the CIELab space to color MM [19]. Reduced orderings followed by a lexicographical cascade to remove ambiguities have also been successfully applied on color MM. Specifically, in a reduced ordering (R-ordering) color elements are ranked according to a surjective – often real-valued – mapping h [20]. For example, Louverdis and Adreadis proposed a reduced ordering in which colors are ranked using fuzzy IF-THEN rules [21]. Also, Velasco-Forero and Angulo proposed a reduced ordering scheme using statistical depth functions [22]. Another promising unsupervised ordering scheme, in which the surjective mapping is constructed from the values of an image, have been proposed by Lézoray [16].

In contrast to the unsupervised reduced ordering schemes, supervised orderings are defined using a set of color references. For example, Sartor and Weeks proposed a reduced ordering scheme based on the distance to a reference color [23]. Ordering schemes based on distance have also been investigated by many other researchers [24–29]. It turns out, however, that many distance-based approaches to color MM can be viewed as particular cases of the supervised ordering proposed by Velasco-Forero and Angulo, in which the surjective mapping is determined using support vector machines (SVMs) [30,31].

In this paper we introduce an R-ordering for color MM using concepts from fuzzy set theory. Precisely, we introduce a total ordering scheme which ranks color elements according to their membership on two families of fuzzy colors; one family corresponding to the foreground and the other representing the background fuzzy colors. Fuzzy colors, which attempt to solve the problem known as "semantic gap", address the vagueness and subjectivity in the modeling of colors [32]. Like the supervised ordering proposed by Velasco-Forero and Angulo, our R-ordering generalizes many distance-based approaches to color MM. In contrast to the SVM approach, however, the approach based on fuzzy colors does not involve the solution of an optimization problem. Furthermore, it can naturally take into account the imprecision used by humans to describe and perceive colors.

The paper is organized as follows. The next section reviews some mathematical background on color images and mathematical morphology. Some approaches to color MM are described in Sect. 3. Section 4 presents the new approach to color MM using fuzzy set theory. The paper finishes with some concluding remarks on Sect. 5.

2   Color Images and Mathematical Morphology

First of all, recall that a color image can be modeled as a mapping from a point set $\mathcal{D}$ into a color space $\mathcal{C}$. Let us denote the family of all images from $\mathcal{D}$ to $\mathcal{C}$

by $\mathcal{C}^{\mathcal{D}}$. In this paper, we assume that the point set is either $\mathcal{D} \subseteq \mathbb{R}^2$ or $\mathcal{D} \subseteq \mathbb{Z}^2$ and the color space $\mathcal{C}$ is a subset of $\bar{\mathbb{R}}^3$, where $\bar{\mathbb{R}} = \mathbb{R} \cup \{+\infty, -\infty\}$. There are many color spaces in the literature but we shall restrict our attention to the RGB space [33,34].

The RGB color space is based on the tristimulus theory of vision in which a color is decomposed into the primitives: red (R), green (G), and blue (B) [34]. Geometrically, this color space is represented by the cube $\mathcal{C}_{RGB} = [0,1] \times [0,1] \times [0,1]$ whose axes correspond to the intensities in each primitive. In the RGB space, a certain color $\mathbf{x} = (x_R, x_G, x_B)$ is a point in or inside the cube $\mathcal{C}_{RGB}$. The origin corresponds to "black" while the edge $(1,1,1)$ represents "white". The RGB color space is widely used in hardware devices including image scanners, digital cameras, and liquid-crystal display systems. In fact, most capturing devices are equipped with three sensors that are sensitive to the red, green, and blue spectrum [34]. Therefore, we may assume that a natural color image is captured using the RGB color space.

2.1 Mathematical Morphology

Briefly, morphological operators examine an image by probing it with a small pattern called structuring element [1,2]. Precisely, the structuring element is used to extract useful information about the geometrical and topological structures on an image. Such as the domain of a color image, we assume that a structuring element S corresponds to a subset of either $\mathbb{R}^2$ or $\mathbb{Z}^2$.

As pointed out in the introduction, complete lattices constitute an appropriate framework for a general theory of MM [1,12]. A partially ordered set $(\mathbb{L}, \leq)$ is a complete lattice if any subset $X \subseteq \mathbb{L}$ admits a supremum and an infimum, denoted respectively by $\bigvee X$ and $\bigwedge X$. Examples of complete lattices used in color MM are given in the following sections.

Let us assume that the color space $\mathcal{C}$, equipped with a certain partial ordering "$\leq$", is a complete lattice. The erosion of a color image $\mathbf{I} \in \mathcal{C}^{\mathcal{D}}$ by a structuring element S, denoted by $\varepsilon_S(\mathbf{I})$, is the color image defined by:

$$\varepsilon_S(\mathbf{I})(p) = \bigwedge \{\mathbf{I}(p+s) : s \in S, p+s \in \mathcal{D}\}, \quad \forall p \in \mathcal{D}. \tag{1}$$

Dually, the dilation of a color image $\mathbf{I} \in \mathcal{C}^{\mathcal{D}}$ by a structuring element S, denoted by $\delta_S(\mathbf{I})$, is the color image given by:

$$\delta_S(\mathbf{I})(p) = \bigvee \{\mathbf{I}(p+s) : s \in S, p+s \in \mathcal{D}\}, \quad \forall p \in \mathcal{D}. \tag{2}$$

Erosions and dilations are the two elementary operations of mathematical morphology [2]. Many other morphological operators are obtained by combining erosions and dilations. For example, their compositions yield the so-called opening and closing, which have interesting topological properties and are used as non-linear image filters [2].

3   Some Approaches to Color Mathematical Morphology

3.1   Marginal and Lexicographical Approaches

A straightforward extension of the gray-scale MM to color images, referred to as
the marginal or component-wise approach, is obtained by processing separately
each color component [15,26]. In mathematical terms, the marginal approach is
obtained by ordering the colors $\mathbf{x} = (x_1, x_2, x_3)$ and $\mathbf{y} = (y_1, y_2, y_3)$ as follows:

$$\mathbf{x} \leq_{\text{marg}} \mathbf{y} \iff x_1 \leq_{\mathbb{R}} y_1, \ x_2 \leq_{\mathbb{R}} y_2, \ x_3 \leq_{\mathbb{R}} y_3, \tag{3}$$

where "$\leq_{\mathbb{R}}$" denotes the usual ordering scheme of real numbers. One can easily
check that "$\leq_{\text{marg}}$" is a partial ordering on $\mathcal{C}_{RGB}$. Also, it is not hard to show that
$(\mathcal{C}_{RGB}, \leq_{\text{marg}})$ is a complete lattice. The elementary morphological operators of
the marginal approach, given by (1) and (2) with the ordering defined by (3),
are denoted respectively by ε_S^M and δ_S^M.

Although the marginal approach yielded excellent results in computational
experiments concerning the removal of Gaussian noise [15], it does not take into
account the correlations between the color components. In fact, certain features
can be removed or enhanced in one of the color components but not in the others.
As a consequence, there is the possibility of introducing false colors, changing
the color balance, or altering the edges of objects [26]. These undesired effects
can be avoided by endowing the color space with a total ordering instead of a
partial ordering.

In contrast to the marginal approach, colors are ranked sequentially in the
lexicographical approach. Formally, the lexicographical ordering, denoted by the
symbol "$\leq_{\text{lex}}$", is defined by means of the following equation for $\mathbf{x}, \mathbf{y} \in \mathcal{C}_{RGB}$:

$$\mathbf{x} \leq_{\text{lex}} \mathbf{y} \iff \begin{cases} x_1 <_{\mathbb{R}} y_1, \\ x_1 = y_1 \text{ and } x_2 <_{\mathbb{R}} y_2, \\ x_1 = y_1, x_2 = y_2 \text{ and } x_3 \leq_{\mathbb{R}} y_3. \end{cases} \tag{4}$$

One can easily show that "$\leq_{\text{lex}}$" is a total ordering and $(\mathcal{C}_{RGB}, \leq_{\text{lex}})$ is a complete
lattice. The lexicographical erosion and the lexicographical dilation of a color
image by a structuring element S, denoted by ε_S^L and δ_S^L, are given respectively
by (1) and (2) with the ordering defined by (4).

The lexicographical approach has been widely used in color MM partially
because it prevents the apparition of "false colors". It turns out, however, that
this ordering scheme prioritizes excessively the first condition in the lexicographical cascade [15]. As a consequence, many fruitful approaches to color MM are
obtained by reducing the colors to a scalar whose comparison is evaluated in the
first condition of a lexicographical cascade.

3.2   Some Approaches Based on Reduced Orderings

In a reduced ordering (R-ordering), the elements are ranked according to a map-
ping h from the color space $\mathcal{C}$ into a complete lattice $\mathbb{L}$. In this paper, we only

consider continuous real-valued mappings $h : \mathcal{C}_{RGB} \to \bar{\mathbb{R}}$. Furthermore, the colors are ranked by comparing the value of h followed by a lexicographical cascade. Precisely, we define an R-ordering, denoted by the symbol "$\leq_h$", as follows for any $\mathbf{x}, \mathbf{y} \in \mathcal{C}_{RGB}$:

$$
\mathbf{x} \leq_h \mathbf{y} \iff \begin{cases} h(\mathbf{x}) <_{\mathbb{R}} h(\mathbf{y}) \\ h(\mathbf{x}) = h(\mathbf{y}) \text{ and } \mathbf{x} \leq_{\text{lex}} \mathbf{y}. \end{cases} \tag{5}
$$

It is not hard to show that "$\leq_h$" yields a total ordering on a color space $\mathcal{C}_{RGB}$ when $h : \mathcal{C}_{RGB} \to \bar{\mathbb{R}}$ is a continuous mapping. Hence, $(\mathcal{C}_{RGB}, \leq_h)$ is a complete lattice and we can define the h-erosion ε_S^h and the h-dilation δ_S^h of an image by a structuring element S by means of Eqs. (1) and (2). Since $(\mathcal{C}_{RGB}, \leq_h)$ is a totally ordered set, the h-morphological operators prevent the apparition of false colors. Furthermore, the function h allows us to define adaptive morphological operators [31]. Examples of adaptive mappings include distance-based mappings [25] and, more generally, supervised mappings [30].

In the distance-based approach to color MM, we assume that the color space $\mathcal{C}$ is equipped with a metric d. For instance, the metric d can be the Euclidean distance or the Mahanolabis distance [25]. Given a reference color $\mathbf{r} \in \mathcal{C}$, the distance-based ordering "$\leq_{d,\mathbf{r}}$", which depends on the metric d as well as the reference $\mathbf{r}$, corresponds to the R-ordering obtained by considering in (5) the mapping $h_{\mathbf{r}}(\mathbf{x}) = \kappa(\mathbf{x}, \mathbf{r})$, where κ is the Gaussian radial basis function kernel defined by the following equation for $\sigma > 0$:

$$
\kappa(\mathbf{x}, \mathbf{r}) = \exp\left(-\frac{d^2(\mathbf{x}, \mathbf{r})}{2\sigma^2}\right). \tag{6}
$$

Note that, using the distance-based ordering "$\leq_{d,\mathbf{r}}$", a color element $\mathbf{y}$ is larger than or equal to another color $\mathbf{x}$ if $\mathbf{y}$ is closer to the color reference $\mathbf{r}$ than $\mathbf{x}$. One the one hand, the largest element of the complete lattice $(\mathcal{C}, \leq_{d,\mathbf{r}})$ is the reference $\mathbf{r}$. As a consequence, the dilation $\delta_S^{h_{\mathbf{r}}}(\mathbf{I})$, defined by (2), the R-ordering given by (5), and the h-mapping $h_{\mathbf{r}}(\mathbf{x}) = \kappa(\mathbf{x}, \mathbf{r})$, expands objects of color $\mathbf{r}$ in the color image $\mathbf{I}$. On the other hand, the least element is the color farthest from the reference. Thus, the erosion $\varepsilon_S^{h_{\mathbf{r}}}$ given by (1) with the distance-based reduced ordering does not have a simple interpretation. Concluding, the distance-based morphological operators are usually efficient if we intent to treat objects with a specific color $\mathbf{r}$.

In many practical situations, however, we are interested in objects composed of many different color elements. Alternatively, we may want to discriminate foreground and background colors. A h-supervised ordering proposed by Velasco-Forero and Angulo generalizes the distance-based approach by allowing the user to inform sets of foreground and background color references [30]. Precisely, let

$$
\mathcal{F} = \{\mathbf{f}_1, \mathbf{f}_2, \ldots, \mathbf{f}_K\} \quad \text{and} \quad \mathcal{B} = \{\mathbf{b}_1, \mathbf{b}_2, \ldots, \mathbf{b}_M\}, \tag{7}
$$

denote respectively the set of foreground and the set of background color references. In a h-supervised ordering, the mapping h is expected to satisfy the conditions

$$h(\mathbf{f}_i) = \top, \quad \forall i = 1, \ldots, K, \quad \text{and} \quad h(\mathbf{b}_j) = \bot, \quad \forall j = 1, \ldots, M, \quad (8)$$

where $\top = \bigvee h(\mathcal{C})$ and $\bot = \bigwedge h(\mathcal{C})$ denote respectively the largest and the least values of the image of the mapping h. Therefore, the supremum and the infimum on the complete lattice $(\mathcal{C}, \leq_h)$ are interpretable with respect to the color sets $\mathcal{B}$ and $\mathcal{F}$.

An effective approach is obtained by considering in (5) the supervised mapping h_{SVM} given by

$$h_{\mathsf{SVM}}(\mathbf{x}) = \sum_{i=1}^{K} \alpha_i \kappa(\mathbf{x}, \mathbf{f}_i) - \sum_{j=1}^{M} \beta_j \kappa(\mathbf{x}, \mathbf{b}_j), \quad \forall \mathbf{x} \in \mathcal{C}, \quad (9)$$

where κ is the Gaussian kernel given by (6) and, α_i's and β_j's are the Lagrange multipliers of the dual formulation of the support vector machine (SVM) trained to discriminate background and foreground color elements [30]. In mathematical terms, α_i's and β_j's are determined by solving the quadratic problem

$$\begin{cases} \text{maximize} \sum_{i=1}^{K} \alpha_i + \sum_{j=1}^{M} \beta_j - \dfrac{1}{2} \sum_{i,\ell=1}^{K} \alpha_i \alpha_\ell \kappa(\mathbf{f}_i, \mathbf{f}_\ell) - \dfrac{1}{2} \sum_{j,\ell=1}^{M} \beta_j \beta_\ell \kappa(\mathbf{b}_j, \mathbf{b}_\ell) \\ \qquad + \dfrac{1}{2} \sum_{i=1}^{K} \sum_{j=1}^{M} \alpha_i \beta_j \kappa(\mathbf{f}_i, \mathbf{b}_j), \\ \text{subject to} \sum_{i=1}^{K} \alpha_i - \sum_{j=1}^{M} \beta_j = 0 \quad \text{and} \quad 0 \leq \alpha_i, \beta_j \leq C, \end{cases}$$
$$(10)$$

where C is a user-specified positive parameter which controls the tradeoff between complexity and the number of nonseparable points [35]. Geometrically, h_{SVM} yields the signed distance to the separating surface between background and foreground color references.

In spite of its elegant formulation, the R-ordering given by (5) with h_{SVM} defined by (9) may fail do satisfy the desired conditions in (8). In terms of the elementary morphological operators, we cannot ensure that the h-supervised dilation $\delta_S^{h_{\mathsf{SVM}}}$ defined by (2), (5), and (9) will expand the foreground and shrink the background. Dually, the h-supervised erosion $\varepsilon_S^{h_{\mathsf{SVM}}}$ given by (1), (5), and (9) may fail to expand the background and shrink the foreground. The following example illustrate this remark.

Example 1. Consider the following sets of foreground and background colors

$$\mathcal{F} = \{(0,0,0), (1,1,0)\} \quad \text{and} \quad \mathcal{B} = \{(1,0,0), (0,1,0), (0.5,0.5,0)\}. \quad (11)$$

In words, black and yellow are the foreground while red, green, and olive constitute the background reference colors. Note that, purposefully, the foreground as well as the background color references belong to $\{\mathbf{x} = (x_R, x_G, x_B) : x_B = 0\}$, i.e., the red-green plane of the RGB color space. By solving the quadratic programming problem defined by (10) with $C = 10$ and the kernel κ defined by (6)

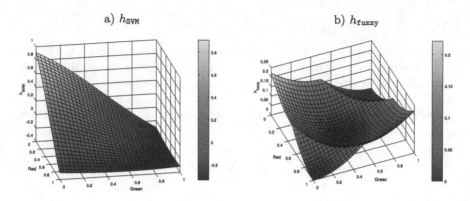

Fig. 1. Surfaces of the functions h_{SVM} and h_{fuzzy} in the red-green plane. (Color figure online)

with $\sigma = 1$, we obtain the mapping h_{SVM} whose surface[1] in the red-green plane is shown in Fig. 1(a). In particular, we obtain

$$h_{\text{SVM}}(1,1,0) = -0.37486 <_{\mathbb{R}} h_{\text{SVM}}(0,1,0) = -0.35992. \tag{12}$$

Therefore, although we admitted yellow as a foreground color and green as a background color, we conclude from the the h_{SVM}-supervised ordering that yellow is less than green. A visual interpretation of the h_{SVM}-supervised elementary morphological operators is given in Fig. 2. Precisely, Fig. 2(a) shows a synthetic color image $\mathbf{I}$ composed of the nine squares of size 3×3 with colors in the red-green plane. Figures 2(b) and (c) depict respectively the erosion $\varepsilon_S^{h_{\text{SVM}}}(\mathbf{I})$ and the dilation $\delta_S^{h_{\text{SVM}}}(\mathbf{I})$ of the image $\mathbf{I}$ shown in Fig. 2(a) by a 3×3 square structuring element S. Since yellow is a foreground color and red, green, and olive belong all to the background color set, a h-supervised dilation and a h-supervised erosion are expected to respectively expand and shrink the yellow square towards the green, red, and olive squares. Figure 2(b) and (c), however, show that the dilation $\delta_S^{h_{\text{SVM}}}$ and the erosion $\varepsilon_S^{h_{\text{SVM}}}$ have not performed as expected.

4   Fuzzy Color-Based Approach to Color Morphology

Like the h_{SVM}-ordering proposed by Velasco-Forero and Angulo, in this section we generalize the distance-based approach by considering sets $\mathcal{B}$ and $\mathcal{F}$ of background and foreground colors. However, instead of defining the mapping h in terms of the solution of a quadratic problem, we use concepts from fuzzy set theory, namely, the concept of fuzzy color [32]:

---

[1] The surface of a mapping h in the red-green plane is formally defined by the set $\{(x,y,z) : z = h(x,y,0), 0 \le x \le 1, 0 \le y \le 1\}$.

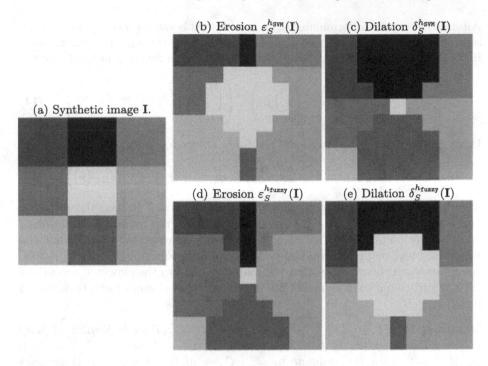

Fig. 2. Synthetic color image and its h-supervised erosions and dilations obtained by considering black and yellow as foreground and red, green, and olive as background. (Color figure online)

Definition 1. *A fuzzy color on a color space C is a linguistic label whose semantics are given by a normal fuzzy subset of C.*

Fuzzy colors attempts to model colors by considering the imprecision, subjectivity, and context dependency widely used by humans to describe them.

It turns out that the mapping $h_{\mathbf{r}} : C \to [0,1]$ defined by $h_{\mathbf{r}}(\mathbf{x}) = \kappa(\mathbf{x}, \mathbf{r})$ with the Gaussian radial basis function kernel can be interpreted as membership function of a fuzzy color. As a consequence, given sets of color elements $\mathcal{F}$ and $\mathcal{B}$, we can define the families $\tilde{\mathcal{F}} = \{F_1, \ldots, F_K\}$ and $\tilde{\mathcal{B}} = \{B_1, \ldots, B_M\}$ of fuzzy colors whose membership functions are given by the following equations:

$$\varphi_{F_i}(\mathbf{x}) = \kappa(\mathbf{x}, \mathbf{f}_i) \quad \text{and} \quad \varphi_{B_j}(\mathbf{x}) = \kappa(\mathbf{x}, \mathbf{b}_j), \quad \forall \mathbf{x} \in C. \tag{13}$$

Let us now assume that we have families $\tilde{\mathcal{F}}$ and $\tilde{\mathcal{B}}$ of foreground and background fuzzy colors. These two families can be defined from sets of color references using (13) or using a user-friendly framework such as the one developed by Soto-Hidalgo et al. [36]. Furthermore, we can define the mapping $h_{\mathtt{fuzzy}} : C \to [0,1]$ such that $h_{\mathtt{fuzzy}}(\mathbf{x})$ corresponds to the degree of truth of the proposition

"the color $\mathbf{x}$ is a foreground color but it is not a background color".

Alternatively, we can formulate h_{fuzzy} as the membership function of the fuzzy set difference between the union of foreground and the union of background fuzzy colors. In other words, h_{fuzzy} is the membership function of the fuzzy set

$$H = \left(\bigcup_{i=1}^{K} F_i \right) \setminus \left(\bigcup_{j=1}^{M} B_j \right). \tag{14}$$

Using the membership functions of the fuzzy colors F_i's and B_j's, we obtain

$$h_{\text{fuzzy}}(\mathbf{x}) \equiv \left(\varphi_{F_1}(\mathbf{x}) \triangledown \cdots \triangledown \varphi_{F_K}(\mathbf{x}) \right) \triangle \eta \left(\varphi_{B_1}(\mathbf{x}) \triangledown \cdots \triangledown \varphi_{B_M}(\mathbf{x}) \right), \tag{15}$$

where $\triangle$, $\triangledown$, and η denote respectively a triangular norm, a triangular co-norm, and a strong fuzzy negation [37]. Note that (15) does not require the solution of an optimization problem. Therefore, it can be implemented more easily than the supervised h_{SVM}-mapping proposed by Velasco-Forero and Angulo [30]. Moreover, the following theorem shows that the mapping satisfies the expected conditions (8) if a foreground color element does not belong to the support of a background fuzzy color B_j:

Theorem 1. *Let $\tilde{\mathcal{F}} = \{F_1, \ldots, F_K\}$ and $\tilde{\mathcal{B}} = \{B_1, \ldots, B_M\}$ be families of fuzzy colors whose membership functions satisfy $\varphi_{F_i}(\mathbf{f}_i) = 1$ and $\varphi_{B_j}(\mathbf{b}_j) = 1$. If $\varphi_{B_j}(\mathbf{f}_i) = 0$, then the mapping $h_{\text{fuzzy}} : \mathcal{C} \to [0,1]$ defined by (15) satisfies $h_{\text{fuzzy}}(\mathbf{f}_i) = 1$ and $h_{\text{fuzzy}}(\mathbf{b}_j) = 0$ for all $i = 1, \ldots, K$ and $j = 1, \ldots, M$.*

In view of Theorem 1, the fuzzy set-based mapping $h_{\text{fuzzy}} : \mathcal{C} \to [0,1]$ constitutes a promising alternative for the development of h-morphological operators. Let us confirm this remark with a simple illustrative example where $\delta_S^{h_{\text{fuzzy}}}$ and $\varepsilon_S^{h_{\text{fuzzy}}}$ denote respectively the h-supervised dilation and the h-supervised erosion given by (2) and (1) with the R-ordering defined by (5) and (15).

Example 2. Consider the sets of foreground and background colors references given by (11). Let us define the foreground fuzzy colors F_{black}, F_{yellow} and the background fuzzy colors B_{red}, B_{green} and B_{olive} using the Gaussian radial basis function kernel defined by (6) with $\sigma = 1$ and the Euclidean distance. For example, the membership function of fuzzy color yellow is given by the following equation for all color $\mathbf{x} = (x_R, x_G, x_B) \in \mathcal{C}_{RGB}$:

$$\varphi_{F_{\text{yellow}}}(\mathbf{x}) = \exp \left(-\frac{(x_R - 1)^2 + (x_G - 1)^2 + x_B^2}{2} \right). \tag{16}$$

Note that the foreground colors black and yellow belong to the support of any background fuzzy color. For example, yellow belongs to the support of the fuzzy color red because $\varphi_{B_{\text{red}}}(1,1,0) = 0.61$. Hence, the families $\tilde{\mathcal{F}} = \{F_{\text{black}}, F_{\text{yellow}}\}$ and $\tilde{\mathcal{B}} = \{B_{\text{red}}, B_{\text{green}}, B_{\text{olive}}\}$ do not satisfy the hypothesis in Theorem 1. Nevertheless, the mapping h_{fuzzy} defined by (15) performs as expected using the maximum ($\triangledown \equiv \vee$), the minimum ($\triangle \equiv \wedge$), and the standard fuzzy negation ($\eta(x) = 1 - x$). The surface of h_{fuzzy} in the red-green plane is shown in

Fig. 1(b). Note that $h_{\texttt{fuzzy}}$ has local maximums at the foreground colors $(0,0,0)$ and $(1,1,0)$, and global minimums at the background colors $(1,0,0)$, $(0,1,0)$, and $(0.5,0.5,0)$. Indeed, we have $h_{\texttt{fuzzy}}(\mathbf{b}_j) = 0$ for any background color $\mathbf{b}_j$, $j = 1,2,3$. Moreover, we clearly have $h_{\texttt{fuzzy}}(\mathbf{f}_i) \geq h_{\texttt{fuzzy}}(\mathbf{b}_j)$ for all $i = 1,2$ and $j = 1,2,3$. As a consequence, the $h_{\texttt{fuzzy}}$-supervised morphological operators $\delta_S^{h_{\texttt{fuzzy}}}$ and $\varepsilon_S^{h_{\texttt{fuzzy}}}$ perform as expected. For instance, consider the synthetic color image $\mathbf{I}$ depicted in Fig. 2(a). The erosion $\varepsilon_S^{h_{\texttt{SVM}}}(\mathbf{I})$ and the dilation $\delta_S^{h_{\texttt{SVM}}}(\mathbf{I})$ of $\mathbf{I}$ by a 3×3 square structuring element S are shown in Fig. 2(d) and (e), respectively. As expected, the yellow square in the middle have been expanded by the dilation $\delta_S^{h_{\texttt{fuzzy}}}$ while it have been shrank by the erosion $\varepsilon_S^{h_{\texttt{fuzzy}}}$.

5 Concluding Remarks

Mathematical morphology is a theory widely used for image processing and analysis. From the mathematical point of view, the elementary morphological operators are very well defined using complete lattices. Briefly, a complete lattice is a partially ordered set in which every subset has a supremum and an infimum. It turns out, however, that there is no natural and widely accepted partial ordering for colors. In this sense, reduced orderings (R-ordering) provides an powerful and elegant tool. In a loose sense, colors are ranked according to real-valued mapping h in an R-ordering. We speak of a supervised R-ordering if the mapping h attains its maximum and minimum respectively at the sets of foreground and background color elements specified a priori by the user. The support vector machine-based (SVM-based) supervised ordering proposed by Velasco-Forero and Angulo is an example of a supervised R-ordering that have been effectively applied for processing hyper-spectral images [30].

In this paper, we introduced a supervised R-ordering for color mathematical morphology which provides an elegant alternative to the SVM-based approach of Velasco-Forero and Angulo. The new supervised R-ordering can address the vagueness and uncertainty inherent to the description of colors by humans using the concept of fuzzy colors. Precisely, given families $\tilde{\mathcal{F}} = \{F_1, \ldots, F_K\}$ and $\tilde{\mathcal{B}} = \{B_1, \ldots, B_M\}$ of foreground and background fuzzy colors, we propose to rank colors according to the value of the $h_{\texttt{fuzzy}}$ defined by (15). In words, $h_{\texttt{fuzzy}}(\mathbf{x})$ corresponds to the degree of truth of the proposition "$\mathbf{x}$ is a foreground (fuzzy) color but it is not a background (fuzzy) color". We would like to point out that the families $\tilde{\mathcal{F}}$ and $\tilde{\mathcal{B}}$ of fuzzy colors can be defined from sets of color references using (13) or using a user-friendly framework such as the one developed by Soto-Hidalgo et al. [36]. We provided in this paper a synthetic example in which the fuzzy color-based approach performs as expected while the SVM-based approach failed to treat foreground and background colors in an appropriate manner.

In the future, we plan to investigate further the theoretical properties of the fuzzy color-based approach to color mathematical morphology. We also intent to study applications of the new approach for color image processing and analysis.

Acknowledgment. This work was supported in part by CNPq under grant no 310118/2017-4.

References

1. Heijmans, H.J.A.M.: Mathematical morphology: a modern approach in image processing based on algebra and geometry. SIAM Rev. **37**(1), 1–36 (1995)
2. Soille, P.: Morphological Image Analysis. Springer, Berlin (1999). https://doi.org/10.1007/978-3-662-03939-7
3. Braga-Neto, U., Goutsias, J.: Supremal multiscale signal analysis. SIAM J. Math. Anal. **36**(1), 94–120 (2004)
4. Gonzalez-Hidalgo, M., Massanet, S., Mir, A., Ruiz-Aguilera, D.: On the choice of the pair conjunction-implication into the fuzzy morphological edge detector. IEEE Trans. Fuzzy Syst. **23**(4), 872–884 (2015)
5. Rittner, L., Campbell, J., Freitas, P., Appenzeller, S., Pike, G.B., Lotufo, R.: Analysis of scalar maps for the segmentation of the corpus callosum in diffusion tensor fields. J. Math. Imaging Vis. **45**, 214–226 (2013)
6. Serra, J.: A lattice approach to image segmentation. J. Math. Imaging Vis. **24**, 83–130 (2006)
7. Sternberg, S.: Grayscale morphology. Comput. Vis. Graph. Image Process. **35**, 333–355 (1986)
8. Bloch, I.: Lattices of fuzzy sets and bipolar fuzzy sets, and mathematical morphology. Inf. Sci. **181**(10), 2002–2015 (2011)
9. De Baets, B.: Fuzzy morphology: a logical approach. In: Ayyub, B.M., Gupta, M.M. (eds.) Uncertainty Analysis in Engineering and Science: Fuzzy Logic, Statistics, and Neural Network Approach, pp. 53–67. Kluwer Academic Publishers, Norwell (1997)
10. Nachtegael, M., Kerre, E.E.: Connections between binary, gray-scale and fuzzy mathematical morphologies. Fuzzy Sets Syst. **124**(1), 73–85 (2001)
11. Sussner, P., Valle, M.E.: Classification of fuzzy mathematical morphologies based on concepts of inclusion measure and duality. J. Math. Imaging Vis. **32**(2), 139–159 (2008)
12. Ronse, C.: Why mathematical morphology needs complete lattices. Sig. Process. **21**(2), 129–154 (1990)
13. Birkhoff, G.: Lattice Theory, 3rd edn. American Mathematical Society, Providence (1993)
14. Grätzer, G., et al.: General Lattice Theory, 2nd edn. Birkhäuser Verlag, Basel (2003)
15. Aptoula, E., Lefèvre, S.: A comparative study on multivariate mathematical morphology. Pattern Recogn. **40**(11), 2914–2929 (2007)
16. Lézoray, O.: Complete lattice learning for multivariate mathematical morphology. J. Vis. Commun. Image Represent. **35**, 220–235 (2016)
17. Aptoula, E., Lefèvre, S.: On lexicographical ordering in multivariate mathematical morphology. Pattern Recogn. Lett. **29**(2), 109–118 (2008)
18. Serra, J.: The "false colour" problem. In: Wilkinson, M.H.F., Roerdink, J.B.T.M. (eds.) ISMM 2009. LNCS, vol. 5720, pp. 13–23. Springer, Heidelberg (2009). https://doi.org/10.1007/978-3-642-03613-2_2
19. Hanbury, A., Serra, J.: Mathematical morphology in the CIELAB space. Image Anal. Stereol. **21**, 201–206 (2002)
20. Barnett, V.: The ordering of multivariate data. J. Roy. Stat. Soc. A **3**, 318–355 (1976)
21. Louverdis, G., Andreadis, I.: Soft morphological filtering using a fuzzy model and its application to colour image processing. Formal Pattern Anal. Appl. **6**(4), 257–268 (2004)

22. Velasco-Forero, S., Angulo, J.: Random projection depth for multivariate mathematical morphology. IEEE J. Sel. Top. Sig. Process. **6**(7), 753–763 (2012)
23. Sartor, L.J., Weeks, A.R.: Morphological operations on color images. J. Electron. Imaging **10**(2), 548–559 (2001)
24. Al-Otum, H.M.: A novel set of image morphological operators using a modified vector distance measure with color pixel classification. J. Vis. Commun. Image Represent. **30**, 46–63 (2015)
25. Angulo, J.: Morphological colour operators in totally ordered lattices based on distances: application to image filtering, enhancement and analysis. Comput. Vis. Image Underst. **107**(1–2), 56–73 (2007)
26. Comer, M.L., Delp, E.J.: Morphological operations for color image processing. J. Electron. Imaging **8**(3), 279–289 (1999)
27. Deborah, H., Richard, N., Hardeberg, J.Y.: Spectral ordering assessment using spectral median filters. In: Benediktsson, J.A., Chanussot, J., Najman, L., Talbot, H. (eds.) ISMM 2015. LNCS, vol. 9082, pp. 387–397. Springer, Cham (2015). https://doi.org/10.1007/978-3-319-18720-4_33
28. Ledoux, A., Richard, N., Capelle-Laizé, A.S., Fernandez-Maloigne, C.: Perceptual color hit-or-miss transform: application to dermatological image processing. SIViP **9**(5), 1081–1091 (2015)
29. Valle, M.E., Valente, R.A.: Mathematical morphology on the spherical CIELab quantale with an application in color image boundary detection. J. Math. Imaging Vis. **57**(2), 183–201 (2017)
30. Velasco-Forero, S., Angulo, J.: Supervised ordering in $\mathbb{R}^p$: application to morphological processing of hyperspectral images. IEEE Trans. Image Process. **20**(11), 3301–3308 (2011)
31. Velasco-Forero, S., Angulo, J.: Vector ordering and multispectral morphological image processing. In: Celebi, M.E., Smolka, B. (eds.) Advances in Low-Level Color Image Processing. LNCVB, vol. 11, pp. 223–239. Springer, Dordrecht (2014). https://doi.org/10.1007/978-94-007-7584-8_7
32. Chamorro-Martínez, J., Soto-Hidalgo, J.M., Martínez-Jiménez, P.M., Sánchez, D.: Fuzzy color spaces: a conceptual approach to color vision. IEEE Trans. Fuzzy Syst. **25**(5), 1264–1280 (2017)
33. Acharya, T., Ray, A.: Image Processing: Principles and Applications. Wiley, Hoboken (2005)
34. Pratt, W.: Digital Image Processing, 4th edn. Wiley, Hoboken (2007)
35. Haykin, S.: Neural Networks and Learning Machines, 3rd edn. Prentice-Hall, Upper Saddle River (2009)
36. Soto-Hidalgo, J.M., Martinez-Jimenez, P.M., Chamorro-Martinez, J., Sanchez, D.: JFCS: a color modeling java software based on fuzzy color spaces. IEEE Comput. Intell. Mag. **11**(2), 16–28 (2016)
37. Nguyen, H.T., Walker, E.A.: A First Course in Fuzzy Logic, 2nd edn. Chapman & Hall/CRC, Boca Raton (2000)

Fuzzy Kernel Associative Memories
with Application in Classification

Aline Cristina de Souza and Marcos Eduardo Valle$^{(\boxtimes)}$

Department of Applied Mathematics, University of Campinas,
Campinas, São Paulo 13081-970, Brazil
s.alinedesouza@gmail.com, valle@ime.unicamp.br

Abstract. In this paper we introduce the class of fuzzy kernel associative memories (fuzzy KAMs). Fuzzy KAMs are derived from single-step
generalized exponential bidirectional fuzzy associative memories by interpreting the exponential of a fuzzy similarity measure as a kernel function.
The output of a fuzzy KAM is obtained by summing the desired responses
weighted by a normalized evaluation of the kernel function. Furthermore,
in this paper we propose to estimate the parameter of a fuzzy KAM by
maximizing the entropy of the model. We also present two approaches
for pattern classification using fuzzy KAMs. Computational experiments
reveal that fuzzy KAM-based classifiers are competitive with well-known
classifiers from the literature.

Keywords: Fuzzy associative memory · Similarity measure
Entropy · Pattern classification

1 Introduction

Associative memories (AMs) are mathematical models inspired by the human
brain ability to store and recall information by means of associations [1]. More
specifically, they are designed for the storage of a finite set of association pairs
$\{(\mathbf{x}^\xi, \mathbf{y}^\xi), \xi = 1, \cdots, p\}$, called *fundamental memory set*. We refer to $\mathbf{x}^1, \ldots, \mathbf{x}^p$ as
the stimuli and $\mathbf{y}^1, \ldots, \mathbf{y}^p$ as the desired responses. Furthermore, an AM should
exhibit some error correction capability, i.e, it should yield the desired response
$\mathbf{y}^\xi$ even upon the presentation of a corrupted version $\tilde{\mathbf{x}}^\xi$ of the stimulus $\mathbf{x}^\xi$. An
AM is said autoassociative if $\mathbf{x}^\xi = \mathbf{y}^\xi$ for all $\xi = 1, \cdots, p$, and heteroassociative
if there is at least one $\xi \in \{1, \cdots, p\}$ such that $\mathbf{x}^\xi \neq \mathbf{y}^\xi$. Applications of AM
models include diagnosis [2], emotional modeling [3], pattern classification [4–7],
and image processing and analysis [8,9].

The Hopfield neural network is one of the most widely known neural network
used to implement an AM [10]. Despite its many successful applications, the
Hopfield network suffers from a very low storage capacity [1]. A simple but
significant improvement in storage capacity of the Hopfield network is achieved
by the *recurrent correlation associative memory* (RCAMs) [11]. RCAMs are
closely related to the dense associative memory model introduced recently by

© Springer International Publishing AG, part of Springer Nature 2018
G. A. Barreto and R. Coelho (Eds.): NAFIPS 2018, CCIS 831, pp. 290–301, 2018.
https://doi.org/10.1007/978-3-319-95312-0_25

Krotov and Hopfield to establish the duality between AMs and deep learning [12, 13]. Furthermore, a particular RCAM, called *exponential correlation associative memory* (ECAM), is equivalent to a certain recurrent kernel associative memory proposed by Garcia and Moreno [14,15].

Like the traditional Hopfield neural network, the original ECAM is an autoassociative memory designed for the storage and recall of bipolar vectors. Many applications of AMs, however, require either an heteroassociative memory or the storage and recall of real-valued data. A heteroassociative version of the ECAM, called *exponential bidirectional associative memory* (EBAM), have been proposed by Jeng et al. [16]. As to the storage and recall of real-valued vectors, Chiueh and Tsai introduced the *multivalued exponential recurrent associative memory* (MERAM) [17].

Recently, we introduced the class of *generalized recurrent exponential fuzzy associative memories* (GRE-FAMs), which have been effectively applied for pattern classification [5,18]. Briefly, GRE-FAMs are autoassociative fuzzy memories obtained from a generalization of a fuzzy version of the MERAM. Recall that a fuzzy associative memory is a fuzzy system designed for the storage and recall of fuzzy sets [19,20]. The *generalized exponential bidirectional fuzzy associative memories* (GEB-FAMs), which generalize the GRE-FAMs for the heteroassociative case, have been applied for face recognition [21]. We would like to point out, however, that the dynamic of the GEB-FAMs are not fully understood yet. In view of this remark, we mostly considered single-step versions of these two AM models.

Summarizing, on the one hand, GEB-FAMs can be viewed as a fuzzy version of the bidirectional (heteroassociative) ECAM. On the other hand, ECAM is equivalent to a certain kernel associative memory. Put together, these two remarks suggest us to interpret the single-step GEB-FAMs using (fuzzy) kernel functions and, from now on, we shall refer to them as fuzzy kernel associative memories (fuzzy KAMs). Recall that a kernel is informally defined as a similarity measure that can be thought of as a dot product on a high-dimensional feature space [22]. Disregarding formal definitions, we interpret a fuzzy similarity measure as a fuzzy kernel. Such interpretation lead us to information theoretical learning whose goal is to capture the information in the parameters of a learning machine [23]. In this paper, we propose to fine tune the parameter of a fuzzy KAM using information theoretical learning.

The paper is structured as follows. Next section presents the fuzzy kernel associative memory and some theoretical results. In this section, we also describe how information theoretical learning can be used to determine the parameter of a fuzzy KAM. An application of autoassociative and heteroassociative fuzzy KAMs for pattern classification is given in Sect. 3. Computational experiments on pattern classification are provided in Sect. 4. The paper finishes with the concluding remarks in Sect. 5.

2   Fuzzy Kernel Associative Memory

A fuzzy similarity measure, or simply similarity measure, is a function that associates to each pair of fuzzy sets a real number that expresses the degree of equality of these sets. According to De Baets and Meyer [24], a fuzzy similarity measure is a symmetric binary fuzzy relation on the family of all fuzzy sets $\mathcal{F}(U)$. In mathematical terms, a similarity measure is a mapping $\mathcal{S} : \mathcal{F}(U) \times \mathcal{F}(U) \to [0,1]$ such that $\mathcal{S}(A, B) = \mathcal{S}(B, A)$ for all fuzzy sets $A, B \in \mathcal{F}(U)$. We speak of a strong similarity measure if $\mathcal{S}(A, B) = 1$ if, and only if, $A = B$.

Given a fuzzy similarity measure, the mapping $\kappa : \mathcal{F}(U) \times \mathcal{F}(U) \to [0,1]$ defined by the following equation for $\alpha > 0$ is also a fuzzy similarity measure

$$\kappa(A, B) = e^{\alpha(S(A,B)-1)}. \tag{2.1}$$

Disregarding formal definitions, we shall refer to κ as a fuzzy kernel.

We would like to call the reader's attention to the dependence of the fuzzy kernel κ on the parameter α when $\mathcal{S}$ is a strong similarity measure. On the one hand, κ approximates the strict equality of fuzzy sets as α increases. Precisely, we have $\kappa(A, B) = 1$ if $A = B$ and $\kappa(A, B) = 0$ otherwise as $\alpha \to \infty$. On the other hand, $\kappa(A, B) = 1$ for all $A, B \in \mathcal{F}(U)$ as $\alpha \to 0$. In other words, κ is unable to discriminate fuzzy sets for sufficiently small $\alpha > 0$. Hence, the parameter α controls the capability of the fuzzy kernel to distinguish fuzzy sets.

Let us now introduce the fuzzy kernel associative memory (fuzzy KAM):

Definition 1 (Fuzzy KAM). *Consider a fundamental memory set $\{(A^\xi, B^\xi) : \xi = 1, \ldots, p\} \subset \mathcal{F}(U) \times \mathcal{F}(V)$. Let $\alpha > 0$ be a real number, $\mathcal{S} : \mathcal{F}(U) \times \mathcal{F}(U) \to [0,1]$ a similarity measure, and H a $p \times p$ real-valued matrix. A fuzzy KAM is a mapping $\mathcal{K} : \mathcal{F}(U) \to \mathcal{F}(V)$ defined by the following equation where $X \in \mathcal{F}(U)$ is the input and $Y = \mathcal{K}(X) \in \mathcal{F}(V)$ is the output:*

$$Y(v) = \varphi \left(\frac{\sum_{\xi=1}^{p} \sum_{\mu=1}^{p} h_{\xi\mu} \kappa(A^\mu, X) B^\xi(v)}{\sum_{\eta=1}^{p} \sum_{\mu=1}^{p} h_{\eta\mu} \kappa(A^\mu, X)} \right). \tag{2.2}$$

Here, the piece-wise linear function $\varphi(x) = \max(0, \min(1, x))$ ensures $Y(v) \in [0,1]$, for all $v \in V$.

Alternatively, we can write the output of a fuzzy KAM as

$$Y(v) = \varphi \left(\sum_{\xi=1}^{p} w_\xi B^\xi(v) \right) \quad \text{where} \quad w_\xi = \frac{\sum_{\mu=1}^{p} h_{\xi\mu} \kappa(A^\mu, X)}{\sum_{\eta=1}^{p} \sum_{\mu=1}^{p} h_{\eta\mu} \kappa(A^\mu, X)}. \tag{2.3}$$

In words, the output $Y = \mathcal{K}(X)$ is given by a linear combinations of the desired responses B^ξ's. Moreover, the coefficients of the linear combinations are calculated by using the parametrized fuzzy kernel κ given by (2.1).

As pointed out in the introduction, an autoassociative fuzzy KAM is equivalent to the single-step generalized recurrent exponential fuzzy associative memories (GRE-FAM) designed for the storage of a finite family of fuzzy sets $\mathcal{A} = \{A^i, i = 1, \cdots, p\} \subset \mathcal{F}(U)$ [18]. Similarly, the heteroassociative fuzzy KAM corresponds to the single-step generalized exponential bidirectional fuzzy associative memories (GEB-FAM) in the heteroassociative case [25].

The matrix H plays a very important role in the storage capacity and noise tolerance of a fuzzy KAM. The next theorem shows how to define the matrix H so that the fundamental memories are all correctly encoded in the memory.

Theorem 1. *Let $\mathcal{A} = \{(A^\xi, B^\xi) : \xi = 1, \cdots, p\} \subset \mathcal{F}(U) \times \mathcal{F}(V)$ be the fundamental memory set, $\mathcal{S} : \mathcal{F}(U) \times \mathcal{F}(U) \to [0,1]$ a similarity measure, $\alpha > 0$ a real number, and $\kappa : \mathcal{F}(U) \times \mathcal{F}(U) \to [0,1]$ a fuzzy kernel defined by (2.1). If the matrix $K = (k_{ij}) \in \mathbb{R}^{p \times p}$, whose entries are defined by*

$$k_{ij} = \kappa(A^i, A^j), \quad \forall i, j = 1, \ldots, p, \tag{2.4}$$

is invertible, then the fuzzy KAM obtained by considering the matrix $H = K^{-1}$ satisfies the identity $\mathcal{K}(A^\xi) = B^\xi$ for all $\xi = 1, \ldots, p$.

Let us briefly address the computational effort required to synthesize a fuzzy KAM based on Theorem 1. First, the fuzzy kernel κ is evaluated $(p^2 + p)/2$ times to compute the symmetric $p \times p$ matrix K defined by (2.4). Then, instead of determining the inverse $H = K^{-1}$, we compute the LU factorization (or the Cholesky factorization if H is symmetric and positive definite) of K using $\mathcal{O}(p^3)$ operations. Then, the multiplication of H by a vector is replaced by the solution of two triangular systems during the recall phase. Summarizing, $\mathcal{O}(p^3)$ operations are performed to synthesize a fuzzy KAM.

The parameter α plays an important role on the noise tolerance of a fuzzy KAM. Briefly, the higher the parameter α, the greater the weight of the fundamental memories most similar to the input X in the calculation of the output $\mathcal{K}(X)$. In other words, increasing α emphasizes the role of the fundamental memories most similar to the input. Thus, in some sense, the parameter α controls how each fundamental memory contributes to the output of a fuzzy KAM. Formally, the next theorem states that, as α tends to infinity, the output $\mathcal{K}(X)$ converges point-wise to the arithmetic mean of the desired responses B^ξ's whose associated stimulus A^ξ's are the most similar to the input X.

Theorem 2. *Let $\mathcal{A} = \{(A^\xi, B^\xi) : \xi = 1, \cdots, p\} \subseteq \mathcal{F}(U) \times \mathcal{F}(V)$ be a family of fundamental memories and $\mathcal{S}$ a strong similarity measure. Suppose that the matrix K given by (2.4) is invertible for any $\alpha > 0$. Given a fuzzy set $X \in \mathcal{F}(U)$, define $\Gamma \subseteq \{1, \ldots, p\}$ as the set of the indexes of the stimulus which are the most similar to the input X in terms of $\mathcal{S}$, that is:*

$$\Gamma = \{\gamma : \mathcal{S}(A^\gamma, X) \geq \mathcal{S}(A^\xi, X), \forall \xi = 1, \ldots, p\}. \tag{2.5}$$

Then,

$$\lim_{\alpha \to \infty} Y(v) = \frac{1}{\text{Card}(\Gamma)} \sum_{\gamma \in \Gamma} B^\gamma(v), \quad \forall v \in V. \tag{2.6}$$

where $Y = \mathcal{K}(X)$ *is the output of a fuzzy KAM. Furthermore, the weights* w_ξ *given by* (2.3) *satisfy the following equation for all* $\xi = 1, \dots, p$:

$$\lim_{\alpha \to \infty} w_\xi = \begin{cases} \dfrac{1}{\text{Card } \Gamma}, & \xi \in \Gamma, \\ 0, & \textit{otherwise.} \end{cases} \tag{2.7}$$

2.1 Estimating the Parameter of a Fuzzy KAM

In this section, we propose to estimate the parameter α of a fuzzy KAM using information theoretical learning [23]. The basic idea is to maximize the capability of the fuzzy kernel κ to discriminate between two different stimulus. To this end, we use the information-theoretic descriptor of entropy.

The concept of entropy, introduced by Shannon in 1948 [26], represents a quantitative measure of uncertainty and information of a probabilistic system [27,28]. The entropy of a n-state system is defined by the following equation where p_i denotes the probability of occurrence of the i-th state [27]:

$$E = \sum_{i=1}^{n} p_i \log(1/p_i). \tag{2.8}$$

The entropy given by (2.8) can be used as a measure of the amount of uncertainty of a system. Given a fundamental memory set $\mathcal{A} = \{(A^\xi, B^\xi) : \xi = 1, \dots, p\} \subseteq \mathcal{F}(U) \times \mathcal{F}(V)$ and a strong similarity measure $\mathcal{S} : \mathcal{F}(U) \times \mathcal{F}(U) \to [0, 1]$, we define the entropy of a fuzzy KAM $\mathcal{K}$ by means of the equation

$$E_\mathcal{K}(\alpha) = \sum_{i=1}^{p} \sum_{j=1}^{p} \kappa(A^i, A^j) \log\left(1/\kappa(A^i, A^j)\right) \tag{2.9}$$

$$= \sum_{i=1}^{p} \sum_{j=1}^{p} - e^{\alpha(\mathcal{S}(A^i, A^j)-1)} \log\left(e^{\alpha(\mathcal{S}(A^i, A^j)-1)}\right) \tag{2.10}$$

$$= -\sum_{i=1}^{p} \sum_{j=1}^{p} \alpha(\mathcal{S}(A^i, A^j) - 1)) e^{\alpha(\mathcal{S}(A^i, A^j)-1)}. \tag{2.11}$$

Note that the entropy $E_\mathcal{K}$ of a fuzzy KAM is a function of the parameter α. Furthermore, $E_\mathcal{K}(\alpha)$ tend to zero if either $\alpha \to 0$ or $\alpha \to \infty$. Intuitively, $E_\mathcal{K}$ quantifies the capability of the fuzzy kernel κ to discriminate between A^i and A^j as a function of α. By maximizing $E_\mathcal{K}$, we expect to improve the noise tolerance of the fuzzy KAM. In view of this remark, we suggest to choose the parameter α^* that maximizes (2.11). Formally, we propose to define

$$\alpha^* = \operatorname*{argmax}_{\alpha > 0} E_\mathcal{K}(\alpha). \tag{2.12}$$

At this point, we would like to recall that Shannon derived (2.8) using classical probability theory. A fuzzy entropy that does not take into account probabilistic concepts in its definition have been provided by De Luca and Termini [29]. Although the fuzzy entropy would be more appropriate in our context, we have not observed significant improvements in our preliminary computational experiments using the fuzzy entropy compared to those obtained using the entropy of Shannon. Moreover, besides presenting lower computational cost, Shannon's entropy showed to be more robust. Therefore, we only consider the entropy of Shannon in this paper.

3   Classifiers Based on Fuzzy KAMs

A classifier is a mapping $\mathcal{C} : W \rightarrow \mathcal{L}$ that associates to each pattern $w \in W$ a label $l \in \mathcal{L}$ that represents the class which w belongs to. Classifiers are usually synthesized using a family of labeled samples, called *training set*. In this section, we present two approaches to define classifiers based on fuzzy KAMs. The first approach, which is inspired by sparse representation classifiers [30], is based on autoassociative fuzzy KAMs. The second approach contemplates the heteroassociative case.

3.1   The Autoassociative-Based Approach

Sparse representation classifiers [30] are based on the hypothesis that a sample Y from class i is approximately equal to a linear combination of the training data from class i. Formally, let $\mathcal{A_L} = \{(A^\xi, \ell_\xi), \xi = 1, \cdots, p\} \subset \mathcal{F}(U) \times \mathcal{L}$ be the training set, where A^ξ are distinct non-empty fuzzy sets on U and $\mathcal{L}$ is a finite set of labels. If Y belongs to class i, then

$$Y(u) \approx \sum_{\xi:\ell_\xi=i} \alpha_\xi A^\xi(u), \quad \forall u \in U. \tag{3.13}$$

Equivalently, Y can be written as:

$$Y(u) \approx \sum_{\xi=1}^{p} \alpha_\xi A^\xi(u), \quad \forall u \in U, \tag{3.14}$$

where $\alpha_\xi = 0$ if $\ell_\xi \neq i$. In other words, Y can be written as a sparse linear combination of the training data.

Assume we have an autoassociative fuzzy KAM $\mathcal{K} : \mathcal{F}(U) \rightarrow \mathcal{F}(U)$ designed for the storage of the fundamental memory set $\{\mathcal{A}^1, \ldots, \mathcal{A}^p\}$. Given a pattern X from class i (or a noisy version $\tilde{X}$ of X), the autoassociative fuzzy KAM is expected to produce a pattern $\mathcal{K}(X) = Y$ that also belongs to class i. From (2.3), the output of the autoassociative fuzzy KAM satisfies

$$Y(u) = \varphi \left(\sum_{\xi=1}^{p} w_\xi A^\xi(u) \right), \tag{3.15}$$

where $\varphi(x) = \max(0, \min(1, x))$. By comparing (3.14) and (3.15), except for the piece-wise linear function φ which can be disregarded if $\sum_{\xi=1}^{p} w_\xi A^\xi(u) \in [0, 1]$, we conclude that the linear combination in (3.15) should also be sparse. Therefore, the coefficients α_ξ in (3.14) can be approximated by

$$\alpha_\xi = w_\xi \chi_i(\ell_\xi), \quad \forall \xi = 1, \ldots, p, \tag{3.16}$$

where $\chi_i : \mathcal{L} \to \{0, 1\}$, for $i \in \mathcal{L}$, denotes the indicator function:

$$\chi_i(x) = \begin{cases} 1, & x = i, \\ 0, & \text{otherwise.} \end{cases} \tag{3.17}$$

Observe that (3.16) implies $\alpha_\xi = w_\xi$ if $\ell_\xi = i$ and $\alpha_\xi = 0$ otherwise. Concluding, if the input X belongs to class i, we presuppose that

$$Y(u) \approx \sum_{\xi=1}^{p} w_\xi \chi_i(\ell_\xi) A^\xi(u), \quad \forall u \in U, \tag{3.18}$$

In practice, however, we do not know a priori to which class the input X belongs. As a consequence, we assign to X the class $i \in \mathcal{L}$ that minimizes the distance between Y and the linear combination $\sum_{\xi=1}^{p} w_\xi \chi_\ell(\ell_\xi) A^\xi$. Formally, we attribute to X a class label $i \in \mathcal{L}$ such that

$$d_2\left(Y, \sum_{\xi=1}^{p} w_\xi \chi_i(\ell_\xi) A^\xi\right) \leq d_2\left(Y, \sum_{\xi=1}^{p} w_\xi \chi_j(\ell_\xi) A^\xi\right), \forall j \in \mathcal{L}, \tag{3.19}$$

where d_2 denotes the L_2-distance.

3.2 Heteroassociative-Based Approach

In the second approach, we define a classifier using the heteroassociative case. Precisely, we synthesize a heteroassociative fuzzy KAM designed for the storage of a fundamental memory set $\{(A^\xi, B^\xi), \xi = 1, \ldots, p\} \subset \mathcal{F}(U) \times \{0, 1\}^n$, where $n = \text{Card}(\mathcal{L})$ denotes the number of classes, A^ξ represents a sample from a certain class, and $B^\xi \subset \{0, 1\}^n$ indicates to which class A^ξ belongs. In mathematical terms, the (fuzzy) set B^ξ associated to the stimulus A^ξ of class i is defined by:

$$B^\xi(v) = \begin{cases} 1, & \text{if } v = i \\ 0, & \text{if } v \neq i. \end{cases} \tag{3.20}$$

Now, given an input $X \in \mathcal{F}(U)$, the fuzzy KAM yields a fuzzy set $Y = \mathcal{K}(X)$. According to Eqs. (2.2) and (3.20), $Y(i)$ is the sum of the weights w_ξ's for ξ such that A^ξ belongs to the class i. Hence, we associate the input X to the i-th class, where i is the first index such that $Y(i) \geq Y(j)$, for all $j = 1, \ldots, n$.

4   Computational Experiments and Results

In this section, we carry out computational experiments to evaluate the performance of the fuzzy KAM-based classifiers. Let us begin by clarifying the benchmark classification problems that we have used.

4.1   Classification Problems

Let us consider the following twenty two classification problems available at the *Knowledge Extraction Based on Evolutionary Learning* (KEEL) database repository as well as at the UCI Machine Learning Repository: appendicitis, cleveland, crx, ecoli, glass, heart, iris, monks, movementlibras, pima, sonar, spectfheart, vowel, wdbc, wine, satimage, texture, german, yeast, spambase, phoneme, and page-blocks [31]. We would like to point out that, due to computational limitations, we refrained to consider the classification problems: magic, pen-based, ringnorm, and twonorm. Precisely, recall that $\mathcal{O}(p^3)$ operations are performed to synthesize a fuzzy KAM and, in these four databases, we have $p \approx 10^4$.

Similar to previous experiments described on the literature, the experiments were conducted by using ten-fold cross validation technique. This method consists of dividing the data-set in ten parts and performing 10 tests, each one using one of the parts as a test set and the others nine parts as developing/training set. Afterward, we compute the mean of the ten accuracy values obtained in each one of the ten tests. In order to ensure a fair comparison, we used the same partitioning as in [31,32].

Some of the data sets considered in this experiment contain both categorical and numerical features. Therefore, a pre-processing step to convert the original data into fuzzy sets was necessary. First, in order to have only numerical values, we transformed each categorical feature $f \in \{v_1, \ldots, v_c\}$, with $c > 1$, into a c-dimensional numerical feature $\mathbf{n} = (n_1, n_2, \ldots, n_c) \in \mathbb{R}^c$ as follows for all $i = 1, \ldots, c$:

$$n_i = \begin{cases} 1, & f = v_i, \\ 0, & \text{otherwise.} \end{cases} \tag{4.21}$$

For example, the crx data set contain nine categorical features, one of them with 14 possibilities. Such categorical feature was transformed into 14 numerical features using (4.21). At the end, an instance of the transformed crx data set contain 46 numerical features instead of 9 categorical and 6 numerical features of the original classification problem.

After all categorical features were converted into numerical values, an instance from a data set can be written as a pair $(\mathbf{x}, \ell)$, where $\mathbf{x} = [x_1, \ldots, x_n]^T \in \mathbb{R}^n$ is a vector of numerical features and $\ell \in \mathcal{L}$ denotes its class label. Moreover, each feature vector $\mathbf{x} \in \mathbb{R}^n$ can be associated with a fuzzy set $A = [a_1, a_2, \ldots, a_n]^T$ by means of the equation

$$a_i = \frac{1}{1 + e^{-(x_i - \mu_i)/\sigma_i}} \in [0, 1], \quad \forall i = 1, \ldots, n, \tag{4.22}$$

where μ_i and σ_i represent respectively the mean and the standard deviation of ith component of all training instances. Concluding, any training set can be written as a labeled family of fuzzy sets $\mathcal{A}_\mathcal{L} = \{(A^\xi, \ell_\xi) : \xi = 1, \dots, p\}$.

Besides, we have removed some repeated elements from the fundamental memories sets of the spambase and page-blocks data sets.

In our computational experiments, we considered the fuzzy KAM defined by using the Gregson similarity measure and the parameter α^* that maximizes the entropy. The Gregson similarity measure $\mathcal{S}_G : \mathcal{F}(U) \times \mathcal{F}(U) \to [0, 1]$ is given by:

$$\mathcal{S}_G(A, B) = \frac{\displaystyle\sum_{i=1}^{n} \min(A(u_i), B(u_i))}{\displaystyle\sum_{i=1}^{n} \max(A(u_i), B(u_i))}. \tag{4.23}$$

Note that $\mathcal{S}_G$ given by (4.23) is a strong similarity measure which can be interpreted as the quotient between the cardinality of the intersection by the cardinality of the union of A and B. Finally, we would like to point out that we studied extensively the role of a fuzzy similarity measure in GEB-FAM models applied for face recognition and the Gregson similarity measure achieved competitive results in comparison with others models from the literature [21].

Figure 1 shows the boxplot of the average accuracy produced by the autoassociative and heteroassociative fuzzy KAM-based classifiers as well as other nine models from the literature, namely: 2SLAVE [33], FH-GBML [34], SGERD [35], CBA [36], CBA2 [37], CMAR [38], CPAR [39], C4.5 [40], and FARC-HD [32]. The accuracy of the nine first classifiers have been extract from [32]. We can observe from Fig. 1 that the fuzzy KAM-based classifiers outperformed (or are at least competitive!) with the other classifiers from the literature. Let us

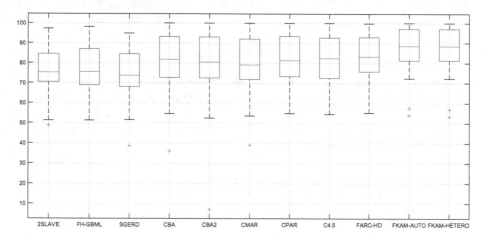

Fig. 1. Boxplot of classification accuracies of several models of the literature in twenty two problems. The accuracy of the nine first classifiers have been extract from [32].

conclude by pointing out that the outliers in the boxplot of the fuzzy KAM-based classifiers correspond to the Cleveland and Yeast classification problems.

5   Concluding Remarks

In this paper, we introduced the class of fuzzy kernel associative memories (fuzzy KAMs). Basically, a fuzzy KAM corresponds to single-step generalized exponential bidirectional fuzzy associative memories (GEB-FAMs), which have been introduced and investigated by us in the last years [5,18,21,25]. Like the single-step GEB-FAMs, fuzzy KAMs can be applied in classification problems. Indeed, in this paper we reviewed two approaches for pattern classification: one using the autoassociative case and the other based on the heteroassociative case.

The main contribution of this paper is the new interpretation of the exponential of a similarity measure as a kernel function. Although we did not elaborated rigorously on the notion of a fuzzy kernel, it allowed us to apply concepts from information theoretical learning to fine tune the parameter α of a fuzzy KAM. Precisely, in view of its simplicity, we proposed to determine the parameter α^* that maximizes the Shannon entropy of a fuzzy KAM.

Computational experiments with some well-know benchmark classification problems showed a superior performance, in terms of accuracy, of the fuzzy KAM-based classifiers over many other classifiers from the literature. In the future, we plan to formalize the notion of a fuzzy kernel and to investigate further the performance of the fuzzy KAM-based classifiers. We also intent to study other applications of the fuzzy KAM models.

Acknowledgment. This work was supported in part by FAPESP and CNPq under grants nos 2015/00745-1 and 310118/2017-4, respectively.

References

1. Hassoun, M.H.: Fundamentals of Artificial Neural Networks. MIT Press, Cambridge (1995)
2. Njafa, J.P.T., Engo, S.N.: Quantum associative memory with linear and non-linear algorithms for the diagnosis of some tropical diseases. Neural Netw. **97**, 1–10 (2018)
3. Masuyama, N., Loo, C.K., Seera, M.: Personality affected robotic emotional model with associative memory for human-robot interaction. Neurocomputing **272**, 213–225 (2018)
4. Esmi, E., Sussner, P., Sandri, S.: Tunable equivalence fuzzy associative memories. Fuzzy Sets Syst. **292**, 242–260 (2016)
5. Valle, M.E., de Souza, A.C.: Pattern classification using generalized recurrent exponential fuzzy associative memories. In: George, A., Papakostas, A.G.H., Kaburlasos, V.G. (eds.) Handbook of Fuzzy Sets Comparison Theory, Algorithms and Applications, vol. 6, pp. 79–102. Science Gate Publishing (2016)
6. Li, L., Pedrycz, W., Li, Z.: Development of associative memories with transformed data. Appl. Soft Comput. **61**, 1141–1152 (2017)

7. Ramírez-Rubio, R., Aldape-Pérez, M., Yáñez-Márquez, C., López-Yáñez, I., Camacho-Nieto, O.: Pattern classification using smallest normalized difference associative memory. Pattern Recogn. Lett. **93**, 104–112 (2017)
8. Grana, M., Chyzhyk, D.: Image understanding applications of lattice autoassociative memories. IEEE Trans. Neural Netw. Learn. Syst. **27**(9), 1920–1932 (2016)
9. Valdiviezo-N, J.C., Urcid, G., Lechuga, E.: Digital restoration of damaged color documents based on hyperspectral imaging and lattice associative memories. SIViP **11**(5), 937–944 (2017)
10. Hopfield, J.J.: Neural networks and physical systems with emergent collective computational abilities. Proc. Nat. Acad. Sci. **79**, 2554–2558 (1982)
11. Chiueh, T.D., Goodman, R.M.: Recurrent correlation associative memories. IEEE Trans. Neural Netw. **2**(2), 275–284 (1991). https://doi.org/10.1109/72.80338
12. Demircigil, M., Heusel, J., Löwe, M., Upgang, S., Vermet, F.: On a model of associative memory with huge storage capacity. J. Stat. Phys. **168**(2), 288–299 (2017)
13. Krotov, D., Hopfield, J.J.: Dense associative memory for pattern recognition (2016)
14. García, C., Moreno, J.A.: The hopfield associative memory network: improving performance with the kernel "Trick". In: Lemaître, C., Reyes, C.A., González, J.A. (eds.) IBERAMIA 2004. LNCS (LNAI), vol. 3315, pp. 871–880. Springer, Heidelberg (2004). https://doi.org/10.1007/978-3-540-30498-2_87
15. Perfetti, R., Ricci, E.: Recurrent correlation associative memories: a feature space perspective. IEEE Trans. Neural Netw. **19**(2), 333–345 (2008)
16. Jeng, Y.J., Yeh, C.C., Chiueh, T.D.: Exponential bidirectional associative memories. Eletron. Lett. **26**(11), 717–718 (1990). https://doi.org/10.1049/el:19900468
17. Chiueh, T.D., Tsai, H.K.: Multivalued associative memories based on recurrent networks. IEEE Trans. Neural Netw. **4**(2), 364–366 (1993)
18. Souza, A.C., Valle, M.E., Sussner, P.: Generalized recurrent exponential fuzzy associative memories based on similarity measures. In: Proceedings of the 16th World Congress of the International Fuzzy Systems Association (IFSA) and the 9th Conference of the European Society for Fuzzy Logic and Technology (EUSFLAT), vol. 1, pp. 455–462. Atlantis Press (2015). https://doi.org/10.2991/ifsa-eusflat-15.2015. 66
19. Kosko, B.: Neural Networks and Fuzzy Systems: A Dynamical Systems Approach to Machine Intelligence. Prentice Hall, Englewood Cliffs (1992)
20. Valle, M.E., Sussner, P.: A general framework for fuzzy morphological associative memories. Fuzzy Sets Syst. **159**(7), 747–768 (2008)
21. Souza, A.C., Valle, M.E.: Generalized exponential bidirectional fuzzy associative memory with fuzzy cardinality-based similarity measures applied to face recognition. In: Trends in Applied and Computational Mathematics (2018). Accepted for publication
22. Schölkopf, B., Smola, A.J.: Learning with Kernels: Support Vector Machines, Regularization, Optimization, and Beyond. MIT press, Cambridge (2002)
23. Principe, J.C.: Information theory, machine learning, and reproducing kernel Hilbert spaces. Information Theoretic Learning. ISS, pp. 1–45. Springer, New York (2010). https://doi.org/10.1007/978-1-4419-1570-2_1
24. Baets, B.D., Meyer, H.D.: Transitivity-preserving fuzzification schemes for cardinality-based similarity measures. Eur. J. Oper. Res. **160**(3), 726–740 (2005). https://doi.org/10.1016/j.ejor.2003.06.036

25. Souza, A.C., Valle, M.E.: Memória associativa bidirecional exponencial fuzzy generalizada aplicada ao reconhecimento de faces. In: Valle, M.E., Dimuro, G., Santiago, R., Esmi, E. (eds.) Recentes Avanços em Sistemas Fuzzy, vol. 1, pp. 503–514. Sociedade Brasileira de Matemática Aplicada e Computacional (SBMAC), São Carlos - SP (2016). ISBN 978-85-8215-079-5
26. Shannon, C.E.: A mathematical theory of communication. Bell Syst. Tech. J. **27**, 379–423 (1948)
27. Pal, N., Pal, S.: Higher order fuzzy entropy and hybrid entropy of a set. Inf. Sci. **61**(3), 211–231 (1992)
28. Klir, G.J.: Uncertainty and Information: Foundations of Generalized Information Theory. Wiley-Interscience, Hoboken (2005)
29. De Luca, A., Termini, S.: A definition of a nonprobabilistic entropy in the setting of fuzzy sets theory. Inf. Control **20**(4), 301–312 (1972)
30. Wright, J., Yang, A., Ganesh, A., Sastry, S., Ma, Y.: Robust face recognition via sparse representation. IEEE Trans. Pattern Anal. Mach. Intell. **31**(2), 210–227 (2009)
31. Alcalá-Fdez, J., Fernández, A., Luengo, J., Derrac, J., García, S., Sánchez, L., Herrera, F.: Keel data-mining software tool: data set repository, integration of algorithms and experimental analysis framework. J. Multiple-Valued Log. Soft Comput. **17**(2–3), 255–287 (2011)
32. Alcalá-Fdez, J., Alcalá, R., Herrera, F.: A fuzzy association rule-based classification model for high-dimensional problems with genetic rule selection and lateral tuning. IEEE Trans. Fuzzy Syst. **19**(5), 857–872 (2011)
33. González, A., Pérez, R.: Selection of relevant features in a fuzzy genetic learning algorithm. IEEE Trans. Syst. Man Cybern. Part B (Cybern.) **31**(3), 417–425 (2001)
34. Ishibuchi, H., Yamamoto, T., Nakashima, T.: Hybridization of fuzzy GBML approaches for pattern classification problems. IEEE Trans. Syst. Man Cybern. Part B (Cybern.) **35**(2), 359–365 (2005)
35. Mansoori, E.G., Zolghadri, M.J., Katebi, S.D.: SGERD: a steady-state genetic algorithm for extracting fuzzy classification rules from data. IEEE Trans. Fuzzy Syst. **16**(4), 1061–1071 (2008)
36. Liu, B., Hsu, W., Ma, Y.: Integrating classification and association rule mining. In: Proceedings of the Fourth International Conference on Knowledge Discovery and Data Mining (1998)
37. Liu, B., Ma, Y., Wong, C.-K.: Classification using association rules: weaknesses and enhancements. In: Grossman, R.L., Kamath, C., Kegelmeyer, P., Kumar, V., Namburu, R.R. (eds.) Data Mining for Scientific and Engineering Applications. MC, vol. 2, pp. 591–605. Springer, Boston, MA (2001). https://doi.org/10.1007/978-1-4615-1733-7_30
38. Li, W., Han, J., Pei, J.: CMAR: accurate and efficient classification based on multiple class-association rules. In: 2001 Proceedings of IEEE International Conference on Data Mining. ICDM 2001, pp. 369–376. IEEE (2001)
39. Yin, X., Han, J.: CPAR: classification based on predictive association rules. In: Proceedings of the 2003 SIAM International Conference on Data Mining, pp. 331–335. SIAM (2003)
40. Quinlan, J.: C4. 5: Programs for Machine Learning. Morgan Kaufmann Publishers Inc., San Francisco (1993). ISBN 1-55860-238-0

(T, N)-Implications and Some Functional Equations

Jocivania Pinheiro[1,2(✉)], Benjamin Bedregal[2], Regivan Santiago[2],
Helida Santos[3], and Graçaliz Pereira Dimuro[3]

[1] Group for Theory of Computation, Logic and Fuzzy Mathematics,
Center of Exact and Natural Sciences, Rural Federal University
of SemiArid (UFERSA), Mossoró, RN, Brazil
`vaniamat@ufersa.edu.br`
[2] Group for Logic, Language, Information, Theory and Applications (LoLITA),
Informatics and Applied Mathematics Department,
Federal University of Rio Grande do Norte (UFRN), Natal, RN, Brazil
`{bedregal,regivan}@dimap.ufrn.br`
[3] Centro de Ciências Computacionais, Federal University of Rio Grande (FURG),
Rio Grande, RS, Brazil
`{helida,gracaliz}@furg.br`

Abstract. Fuzzy implications has drawn attention of many authors
along the years, as their theoretical features seem to be a useful tool
in a fair amount of applications. Meanwhile, functional equations are
those in which the unknowns are functions instead of a traditional variable, and within the fuzzy logic, they can be considered generalizations
of some tautologies of the classical logic. In this paper we investigate the
validity of five functional equations for the class of (T, N)-implications,
namely, we have selected the law of importation and four distributivity
properties and have studied them in the context of the aforementioned
operator.

1 Introduction

Fuzzy implications [1,4,18] are one of the most relevant operators in fuzzy logics.
Many applications have been constructed making use of them as we can see in
[2,3,18] and they are also applied in different areas such as approximate reasoning, control and decision-making theories, expert systems, fuzzy mathematical
morphology, image processing, among others [8,9,11,15,20,21,24,27].

In [22,23], a new class of fuzzy implication named (T, N)-implication (firstly
presented by [5]) was studied. Such implications were given by the composition
of a fuzzy negation and a t-norm. The conditions under which such functions
preserved the principal properties of fuzzy implications were also investigated
and it was proved the necessary and sufficient conditions for a function $I :
[0, 1]^2 \to [0, 1]$ to be a (T, N)-implication.

It is important to recall that the classical implication is found in various tautologies in classical logic. It is clear that not all generalizations of these tautologies hold for all fuzzy operators. That is the reason why one should have a deep

© Springer International Publishing AG, part of Springer Nature 2018
G. A. Barreto and R. Coelho (Eds.): NAFIPS 2018, CCIS 831, pp. 302–313, 2018.
https://doi.org/10.1007/978-3-319-95312-0_26

and careful study on such tautologies in order to convert them into functional equations that would include fuzzy operations. In [1] it is stated that considering the generalized equivalences, some properties of implications had received more attention due to their value on the many applications, namely the properties of contrapositive symmetry, the law of importation, the distributivity properties of fuzzy implications over t-norms and t-conorms, and also the T-conditionality property. In [11], a fuzzy generalization for $I(x, I(y, z)) = I(I(x, y), I(x, z))$ law was made, providing the requirements for that Boolean-like law to be valid within some classes of fuzzy implications, among which the (T, N)-implications. In this sense, the aim of this work is to study the law of importation and four distributivity properties regarding the class of (T, N)-implications over t-norms and t-conorms.

The paper is organized as follows. Sect. 2 recalls some of the basic concepts demanded to comprehend the developments in this work, including the concept of fuzzy implication and related properties. The study of (T, N)-implications and functional equations is done in Sect. 3, including the most important results. Al last, we conclude in Sect. 4 with our final remarks and discuss some ideas for future works.

2 Preliminares

Definition 1. *A function $T : [0, 1]^2 \rightarrow [0, 1]$ is called a **triangular norm (t-norm**, for short) if it satisfies the following conditions:*

(T1) $T(x, y) = T(y, x)$ for all $x, y \in [0, 1]$;
(T2) $T(x, T(y, z)) = T(T(x, y), z)$ for all $x, y, z \in [0, 1]$;
(T3) If $x_1 \leq x_2$ and $y_1 \leq y_2$ then $T(x_1, y_1) \leq T(x_2, y_2)$, for all $x_1, x_2, y_1, y_2 \in [0, 1]$;
(T4) $T(x, 1) = x$, for all $x \in [0, 1]$.   *(boundary condition)*

Proposition 1. *Let T be a t-norm. Then $T(0, y) = 0$ for each $y \in [0, 1]$.*

In fuzzy logic, the conjunction is often represented by a t-norm. The standard fuzzy conjunction $T_M : [0, 1]^2 \rightarrow [0, 1]$ given by $T_M(x, y) = min\{x, y\}$ is the only idempotent t-norm (see [17] - Theorem 3.9).

Definition 2. *A function $S : [0, 1]^2 \rightarrow [0, 1]$ is called a **triangular conorm (t-conorm**, for short) if it satisfies the following conditions, for all $x, y, z \in [0, 1]$:*

(S1) $S(x, y) = S(y, x)$ for all $x, y \in [0, 1]$;
(S2) $S(x, S(y, z)) = S(S(x, y), z)$ for all $x, y, z \in [0, 1]$;
(S3) If $x_1 \leq x_2$ and $y_1 \leq y_2$ then $S(x_1, y_1) \leq S(x_2, y_2)$, for all $x_1, x_2, y_1, y_2 \in [0, 1]$;
(S4) $S(x, 0) = x$ for all $x \in [0, 1]$.   *(boundary condition)*

The standard fuzzy disjunction $S_M : [0, 1]^2 \rightarrow [0, 1]$ given by $S_M(x, y) = max\{x, y\}$ is the only idempotent t-conorm (see [17] - Theorem 3.14).

Definition 3. *A function* $N : [0,1] \to [0,1]$ *is a **fuzzy negation** if*

(N1) N *is antitonic, i.e.* $N(x) \le N(y)$ *whenever* $y \le x$;
(N2) $N(0) = 1$ *and* $N(1) = 0$.
     A fuzzy negation N *is said to be **strict** if*
(N3) N *is continuous and*
(N4) $N(x) < N(y)$ *whenever* $y < x$.
     A fuzzy negation N *is said to be **strong** if*
(N5) $N(N(x)) = x$, *for each* $x \in [0,1]$.
     A fuzzy negation N *is said to be **crisp** if*
(N6) $N(x) \in \{0,1\}$, *for all* $x \in [0,1]$.

By [14], a fuzzy negation $N : [0,1] \to [0,1]$ is crisp if and only if there exists $\alpha \in [0,1)$ such that $N = N_\alpha$ or there exists $\alpha \in (0,1]$ such that $N = N^\alpha$, where

$$N_\alpha(x) = \begin{cases} 0, & \text{if } x > \alpha \\ 1, & \text{if } x \le \alpha \end{cases} \tag{1}$$

and

$$N^\alpha(x) = \begin{cases} 0, & \text{if } x \ge \alpha \\ 1, & \text{if } x < \alpha \end{cases}. \tag{2}$$

Definition 4. *Let* T *be a t-norm,* S *be a t-conorm and* N *be a strict fuzzy negation. Then* S *is said to be* **N-dual to T** *if, for all* $x, y \in [0,1]$,

$$N(S(x,y)) = T(N(x), N(y)) \tag{3}$$

and T *is said to be* **N-dual to S** *if, for all* $x, y \in [0,1]$,

$$N(T(x,y)) = S(N(x), N(y)). \tag{4}$$

Definition 5. *A function* $I : [0,1]^2 \to [0,1]$ *is a **fuzzy implication** if the following properties are satisfied, for all* $x, y, z \in [0,1]$:

(I1) *If* $x \le z$ *then* $I(x,y) \ge I(z,y)$;   *(left antitonicity)*
(I2) *If* $y \le z$ *then* $I(x,y) \le I(x,z)$;   *(right isotonicity)*
(I3) $I(0,y) = 1$;   *(left boundary condition)*
(I4) $I(x,1) = 1$;   *(right boundary condition)*
(I5) $I(1,0) = 0$.   *(boundary condition)*

Definition 6. *[1] Let* I *be a fuzzy implication and* T *be a t-norm. We say that* I *satisfies the **Law of importation (LI)** with respect to a t-norm* T *if*

$$I(T(x,y), z) = I(x, I(y,z)), \tag{5}$$

for all $x, y, z \in [0,1]$.

3   Functional Equations and (T, N)-Implications

As already mentioned, functional equations are the ones in which the unknowns are functions instead of being a traditional variable. In this section we investigate the validity of some functional equations by the function I_T^N, introduced in [22]. In [1], Baczyński states that functional equations come up as generalizations of the corresponding tautologies in classical logic involving boolean implications. The results presented in the sequel consider the law of importation (LI), Eq. 5, and four basic distributive equations involving an implication, which will be discussed later.

Proposition 2. *Let T be a t-norm and let N be a fuzzy negation. Then the function $I_T^N : [0,1]^2 \rightarrow [0,1]$ defined by*

$$I_T^N(x,y) = N(T(x,N(y))) \tag{6}$$

is a fuzzy implication, for all $x, y \in [0,1]$.

Definition 7. *Let T be a t-norm and let N be a fuzzy negation. The function I_T^N defined by Eq. (6) is called **(T, N)-implication.***

The principle of exchange is one of the crucial properties of fuzzy implications. Due to the commutativity property of the t-norm T, one of the conditions for an implication to satisfy it is that (LI) is also satisfied. The well-known fuzzy implications called (S, N), R, QL and D-implications satisfy (LI) under some conditions (see [15,19]). In addition, some possible applications were pointed out in [15]. As follows, we show under which conditions (T, N)-implications satisfy (LI).

Proposition 3. *Let I_T^N be a (T, N)-implication. Then:*

(i) If N is strong then I_T^N satisfies (LI) with respect to the t-norm T;
(ii) If N is continuous and I_T^N satisfies (LI) with respect to the t-norm T, then N is strong.

Proof. (i) Indeed, for all $x, y, z \in [0,1]$

$$
\begin{aligned}
I_T^N(x, I_T^N(y,z)) &= N(T(x, N(N(T(y, N(z)))))) \\
&= N(T(x, T(y, N(z)))) \\
&\overset{(T2)}{=} N(T(T(x,y), N(z))) \\
&= I_T^N(T(x,y), z).
\end{aligned}
$$

(ii) As I_T^N satisfies (LI) with respect to the t-norm T, then, for $x = y = 1$, $I_T^N(1, I_T^N(1,z)) = I_T^N(T(1,1), z) \overset{(T4)}{\Rightarrow} N(T(1, N(N(T(1, N(z)))))) = N(T(1, N(z)))$ for all $z \in [0,1]$, still by (T4),

$$N(N(N(N(z)))) = N(N(z)). \tag{7}$$

Given that N is continuous, for all $y \in [0,1]$ there exists $x' \in [0,1]$ such that $N(x') = y$. Only for this x' there exists $x \in [0,1]$ such that $N(x) = x'$. Thus, for all $y \in [0,1]$ there exists $x \in [0,1]$ such that $N(N(x)) = y$. Therefore, by Eq. (7), $N(N(N(N(x)))) = N(N(x)) \Rightarrow N(N(y)) = y$, for all $y \in [0,1]$.

Note that if N is continuous and non-strong then I_T^N does not satisfy (LI). However, there are non-continuous negations N such that I_T^N satisfies (LI) for some t-norm T. See the following example:

Example 1. Take a crisp negation N given by $N = N_\alpha$ and the minimum t-norm T, so

$$
\begin{aligned}
I_T^{N_\alpha}(x, I_T^{N_\alpha}(y, z)) &= N_\alpha(T(x, N_\alpha(N_\alpha(T(y, N_\alpha(z)))))) \\
&= \begin{cases} N_\alpha(T(x, N_\alpha(N_\alpha(y)))), & \text{if } z \le \alpha \\ 1, & \text{if } z > \alpha \end{cases} \\
&= \begin{cases} N_\alpha(x), & \text{if } z \le \alpha \text{ and } y > \alpha \\ 1, & \text{if } z > \alpha \text{ or } y \le \alpha \end{cases} \\
&= \begin{cases} 0, & \text{if } z \le \alpha \text{ and } y > \alpha \text{ and } x > \alpha \\ 1, & \text{otherwise} \end{cases}
\end{aligned}
$$

and

$$
\begin{aligned}
I_T^{N_\alpha}(T(x, y), z) &= N_\alpha(T(T(x, y), N_\alpha(z))) \\
&= \begin{cases} N_\alpha(T(T(x, y), 1)), & \text{if } z \le \alpha \\ 1, & \text{if } z > \alpha \end{cases} \\
&= \begin{cases} N_\alpha(T(x, y)), & \text{if } z \le \alpha \\ 1, & \text{if } z > \alpha \end{cases} \\
&= \begin{cases} 0, & \text{if } z \le \alpha \text{ and } T(x, y) > \alpha \\ 1, & \text{if } z > \alpha \text{ or } T(x, y) \le \alpha \end{cases} \\
&= \begin{cases} 0, & \text{if } z \le \alpha \text{ and } x > \alpha \text{ and } y > \alpha \\ 1, & \text{otherwise} \end{cases}
\end{aligned}
$$

thus, $I_T^{N_\alpha}$ satisfies (LI).

Another example can be given by taking the crisp fuzzy negation $N = N_\alpha$ with $\alpha = 0$ and any t-norm T. In this case, by Proposition 1 we also have that $I_T^{N_\alpha}$ satisfies (LI).

In classic logic, the distributivity of binary operators over one another can somehow define the framework of the algebra imposed by these operators. In fuzzy logic, one can find a variety of studies on the distributivity of t-norms over t-conorms [6,7,10,16]. In this sense, taking into account the four basic distributive equations involving an implication, Eqs. 8, 9, 10, 11, we present in the next proposition the generalizations of them which yields to the distributivity of (T, N)-implications over t-norms and t-conorms.

$$I(T(x, y), z) = S(I(x, z), I(y, z)) \tag{8}$$
$$I(S(x, y), z) = T(I(x, z), I(y, z)) \tag{9}$$
$$I(x, S_1(y, z)) = S_2(I(x, y), I(x, z)) \tag{10}$$
$$I(x, T_1(y, z)) = T_2(I(x, y), I(x, z)) \tag{11}$$

Proposition 4. *Let I_T^N be a (T, N)-implication and S be a t-conorm. Then:*

(i) If T is N-dual of S and the range of N is a subset of the idempotent elements of T then I_T^N satisfies Eq. (8) with respect to the t-norm T and to the t-conorm S;

(ii) If I_T^N satisfies Eq. (8) with respect to the t-norm T and to the t-conorm S, then
  (1) T is N-dual of S and,
  (2) If N is strict then the range of N is a subset of the idempotent elements of T.

Proof. (i) As T is N-dual of S and the range of N is a subset of the idempotent elements of T, i.e., $T(N(x), N(x)) = N(x)$ for all $x \in [0, 1]$, then, for all $x, y, z \in [0, 1]$:

$$
\begin{aligned}
S(I_T^N(x, z), I_T^N(y, z)) &= S(N(T(x, N(z))), N(T(y, N(z)))) \\
&\overset{Eq.\ (4)}{=} N(T(T(x, N(z)), T(y, N(z)))) \\
&\overset{(T2)(T1)}{=} N(T(T(x, y), T(N(z), N(z)))) \\
&= N(T(T(x, y), N(z))) \\
&= I_T^N(T(x, y), z).
\end{aligned}
$$

(ii) (1) As I_T^N satisfies Eq. (8) with respect to the t-norm T and to the t-conorm S, then, for $z = 0$, $N(T(T(x, y), N(0))) = S(N(T(x, N(0))), N(T(y, N(0))))$, so by (T4), $N(T(x, y)) = S(N(x), N(y))$ for all $x, y \in [0, 1]$ and

(2) For $x = y = 1$, $S(I_T^N(1, z), I_T^N(1, z)) = N(T(T(1, 1), N(z)))$, so by (T4), $S(N(N(z)), N(N(z))) = N(N(z))$ for all $z \in [0, 1]$, since T is N-dual of S we have $N(T(N(z), N(z))) = N(N(z)) \overset{N\ strict}{\Rightarrow} T(N(z), N(z)) = N(z)$, for all $z \in [0, 1]$.

Corollary 1. *Let N be a strict negation and T be a t-norm. Then, I_T^N satisfies Eq. (8) if and only if $T = T_M$ and $S = S_M$.*

In the previous corollary, the continuity of N ensures that if I_T^N satisfies Eq. (8) then T is minimum. However, there are non-continuous negations such that I_T^N satisfies Eq. (8) for some t-norms. See the following example:

Example 2. Take a crisp negation N given by $N = N_\alpha$ and take T as the minimum t-norm, so

$$S(\, I_T^{N_\alpha}(x,z),\, I_T^{N_\alpha}(y,z)) =$$
$$= \quad S(N_\alpha(T(x,N_\alpha(z))), N_\alpha(T(y,N_\alpha(z))))$$
$$= \quad \begin{cases} S(N_\alpha(x), N_\alpha(y)), & \text{if } z \le \alpha \\ 1, & \text{if } z > \alpha \end{cases}$$
$$= \quad \begin{cases} 0, & \text{if } z \le \alpha \text{ and } x > \alpha \text{ and } y > \alpha \\ 1, & \text{otherwise} \end{cases}$$

and, by Example 1

$$I_T^{N_\alpha}(T(x,y),z) = \begin{cases} 0, & \text{if } z \le \alpha \text{ and } x > \alpha \text{ and } y > \alpha \\ 1, & \text{otherwise} \end{cases}$$

thus, $I_T^{N_\alpha}$ satisfies Eq. (8).

Another example can be given for any t-norm T. Just take the crisp fuzzy negation $N = N_\alpha$ with $\alpha = 0$. Then, by Proposition 1 we also have that $I_T^{N_0}$ satisfies Eq. (8).

Proposition 5. *Let I_T^N be a (T, N)-implication. Then,*

(i) I_T^N satisfies Eq. (9) for T_M and S_M, i.e., considering T_M as T and S_M as S in Eq. (9);

(ii) If I_T^N satisfies Eq. (9) with respect to the t-norm T and to the t-conorm S, then
   (1) S is N-dual of T and
   (2) If N is strict then the range of N is a subset of the idempotent elements of S.

Proof. (i) For all $x, y, z \in [0,1]$, if $x \le y$ then $S_M(x,y) = y$ and, by (T3) and (N1), $I_T^N(y,z) \le I_T^N(x,z)$, so

$$T_M(I_T^N(x,z), I_T^N(y,z)) = I_T^N(y,z) = I_T^N(S_M(x,y),z).$$

Therefore, I_T^N satisfies Eq. (9). Similarly, if $x > y$ the result follows.

(ii) (1) As I_T^N satisfies Eq. (9) with respect to the t-norm T and to the t-conorm S, then, for $z = 0$, $N(T(S(x,y), N(0))) = T(N(T(x,N(0))), N(T(y,N(0))))$, so by (T4), $N(S(x,y)) = T(N(x), N(y))$ for all $x, y \in [0,1]$ and

(2) for $x = y = 1$, $T(I_T^N(1,z), I_T^N(1,z)) = I_T^N(S(1,1),z)$, so by (T4), $T(N(N(z)), N(N(z))) = N(N(z))$ for all $z \in [0,1]$, since S is N-dual of T we have $N(S(N(z), N(z))) = N(N(z)) \overset{N \text{ strict}}{\Rightarrow} S(N(z), N(z)) = N(z)$, for all $z \in [0,1]$.

Corollary 2. *Let N be a strict negation and T be a t-norm. Then, I_T^N satisfies Eq. (9) if and only if $T = T_M$ and $S = S_M$.*

Proposition 6. *Let I_T^N be a (T, N)-implication and S_1 and S_2 be t-conorms. Then:*

(i) *If $S_1 = S_2 = S_M$ then, for any t-norm T and any negation N, I_T^N satisfies Eq. (10);*

(ii) *If I_T^N satisfies Eq. (10) with respect to t-conorms S_1 and S_2, then:*
   (1) *The range of N is a subset of the idempotent elements of S_2 and*
   (2) *If N is strict then $S_1 = S_2 = S_M$.*

Proof. (i) For all $x, y, z \in [0, 1]$, if $y \le z$ then $S_M(y, z) = z$ and, by (N1) and (T3), $I_T^N(x, y) \le I_T^N(x, z)$, so

$$S_M(I_T^N(x, y), I_T^N(x, z)) = I_T^N(x, z) = I_T^N(x, S_M(y, z)).$$

Therefore, I_T^N satisfies Eq. (10). Similarly, if $y > z$ the result follows.
(ii) (1) As I_T^N satisfies Eq. (10) then, in particular for $y = z = 0$,

$$N(T(x, N(S_1(0, 0)))) = S_2(N(T(x, N(0))), N(T(x, N(0)))),$$

so by (T4), $N(x) = S_2(N(x), N(x))$, for all $x \in [0, 1]$. (2) Since N is strict and $S_2(N(x), N(x)) = N(x)$ for all $x \in [0, 1]$, then

$$S_2(y, y) = S_2(N(N^{-1}(y)), N(N^{-1}(y))) = N(N^{-1}(y)) = y$$

for all $y \in [0, 1]$, so $S_2 = S_M$. On the other hand, for $x = 1$ and $z = y$, $N(T(1, N(S_1(y, y)))) = S_2(N(T(1, N(y))), N(T(1, N(y))))$ for all $y \in [0, 1]$, so by (T4),

$$N(N(S_1(y, y))) = S_2(N(N(y)), N(N(y))) \stackrel{S_2 = S_M}{=} N(N(y)),$$

for all $y \in [0, 1]$. Thus, $S_1(y, y) = y$ for all $y \in [0, 1]$, since N is strict. Therefore, $S_1 = S_M$.

Corollary 3. *Let N be a strict negation and T be a t-norm. Then, I_T^N satisfies Eq. (10) if and only if $S_1 = S_2 = S_M$.*

Proposition 7. *Let I_T^N be a (T, N)-implication and T_1 and T_2 be t-norms. Then:*

(i) *If $T_1 = T_2 = T_M$ then, for any t-norm T and any negation N, I_T^N satisfies Eq. (11);*

(ii) *If I_T^N satisfies Eq. (11) with respect to t-norms T_1 and T_2, then:*
   (1) *The range of N is a subset of the idempotent elements of T_2 and*
   (2) *If N is strict then $T_1 = T_2 = T_M$.*

Proof. (i) For all $x, y, z \in [0,1]$, if $y \leq z$ then $T_M(y,z) = y$ and, by (N1) and (T3), $I_T^N(x,y) \leq I_T^N(x,z)$, so

$$T_M(I_T^N(x,y), I_T^N(x,z)) = I_T^N(x,y) = I_T^N(x, T_M(y,z)).$$

Therefore, I_T^N satisfies Eq. (11). Similarly, if $y > z$ the result follows.
(ii) (1) As I_T^N satisfies Eq. (11) then, in particular for $y = z = 0$,

$$N(T(x, N(T_1(0,0)))) = T_2(N(T(x, N(0))), N(T(x, N(0)))),$$

so by (T4), $N(x) = T_2(N(x), N(x))$, for all $x \in [0,1]$. (2) Since N is strict and the range of N a subset of the idempotent elements of T_2, we have that $T_2(x,x) = T_2(N(N^{-1}(x)), N(N^{-1}(x))) = N(N^{-1}(x)) = x$. On the other hand, for $x = 1$ and $z = y$, $N(T(1, N(T_1(y,y)))) = T_2(N(T(1, N(y))), N(T(1, N(y))))$, so by (T4),

$$N(N(T_1(y,y))) = T_2(N(N(y)), N(N(y))) \overset{T_2 = T_M}{=} N(N(y)),$$

for all $y \in [0,1]$. Thus, $T_1(y,y) = y$ for all $y \in [0,1]$, since N is strict. Therefore, $T_1 = T_M$.

There are other conditions for t-norms and negations that imply that a (T, N)-implication satisfies Eq. (11). The following example ensures that if we take $T_1 = T_M$ and the crisp negation N, given by $N = N_\alpha$ with $\alpha \in [0,1)$, then, independently from t-norms T and T_2, I_T^N satisfies Eq. (11).

Example 3. Take the crisp negation N given by $N = N_\alpha$ and take T_1 as the minimum t-norm, so

$$
\begin{aligned}
T_2(\, & I_T^{N_\alpha}(x,y), I_T^{N_\alpha}(x,z)) = \\
= \quad & T_2(N_\alpha(T(x, N_\alpha(y))), N_\alpha(T(x, N_\alpha(z)))) \\
= \quad & \begin{cases} N_\alpha(T(x, N_\alpha(y))), & \text{if } z > \alpha \\ T_2(N_\alpha(T(x, N_\alpha(y))), N_\alpha(x)), & \text{if } z \leq \alpha \end{cases} \\
= \quad & \begin{cases} 1, & \text{if } z > \alpha \text{ and } y > \alpha \\ N_\alpha(x), & \text{if } z > \alpha \text{ and } y \leq \alpha \\ N_\alpha(x), & \text{if } z \leq \alpha \text{ and } y > \alpha \\ T_2(N_\alpha(x), N_\alpha(x)), & \text{if } z \leq \alpha \text{ and } y \leq \alpha \end{cases} \\
= \quad & \begin{cases} 0, & \text{if } x > \alpha \text{ and } T_1(y,z) \leq \alpha \\ 1, & \text{otherwise} \end{cases}
\end{aligned}
$$

and,

$$I_T^{N_\alpha}(x, T_1(y,z)) = N_\alpha(T(x, N_\alpha(T_1(y,z))))$$

$$= \begin{cases} 1, & \text{if } T_1(y,z) > \alpha \\ N_\alpha(x), & \text{if } T_1(y,z) \leq \alpha \end{cases}$$

$$= \begin{cases} 1, & \text{if } T_1(y,z) > \alpha \\ 0, & \text{if } T_1(y,z) \leq \alpha \text{ and } x > \alpha \\ 1, & \text{if } T_1(y,z) \leq \alpha \text{ and } x \leq \alpha \end{cases}$$

$$= \begin{cases} 0, & \text{if } T_1(y,z) \leq \alpha \text{ and } x > \alpha \\ 1, & \text{otherwise} \end{cases}$$

thus, I_T^N satisfies Eq. (11).

4   Final Remarks

In this work, we carried on the study on (T, N)-implications and presented some results considering functional equations, namely the law of importation and properties related to distributivity. It is well-known that fuzzy implications can used to construct many types of measures such as fuzzy subsethood measures, penalty functions and fuzzy entropy [9,14,25,26], which are useful for several practical applications. Thus, similarly to the works mentioned previously, we believe that (T, N)-implications can also be used to construct fuzzy subsethood measures. Besides that, we are willing to investigate other operators to define different classes of implications, for instance, functions given by the composition of overlaps and negations yielding what we call (O, N)-implication, that possibly can be related to (G, N)-implications [12], (R, O)-Implications [13] and (O, G, N)-implications [14].

Acknowledgments. This work was partially supported by the Brazilian funding agency CNPq (Conselho Nacional de Desenvolvimento Científico e Tecnológico), under the processes No. 306970/2013-9 and 307781/2016-0.

References

1. Baczyński, M., Balasubramaniam, J.: Fuzzy Implications. Springer, Heidelberg (2008). https://doi.org/10.1007/978-3-540-69082-5
2. Baczyński, M.: On the applications of fuzzy implication functions. In: Balas, V.E., Fodor, J., Várkonyiczy, A.R., Dombi, J., Jain, L.C. (eds.) Soft Computing Applications. AISC, vol. 195, pp. 9–10. Springer, Berlin (2013). https://doi.org/10.1007/978-3-642-33941-7_4
3. Baczyński, M., Beliakov, G., Bustince, H., Pradera, A. (eds.): Advances in Fuzzy Implication Functions. Studies in Fuzziness and Soft Computing, vol. 300. Springer, Heidelberg (2013). https://doi.org/10.1007/978-3-642-35677-3

4. Baczynski, M., Jayaram, B., Massanet, S., Torrens, J.: Fuzzy implications: past, present, and future. In: Kacprzyk, J., Pedrycz, W. (eds.) Springer Handbook of Computational Intelligence, pp. 183–202. Springer, Heidelberg (2015). https://doi. org/10.1007/978-3-662-43505-2_12

5. Bedregal, B.C.: A normal form which preserves tautologies and contradictions in a class of fuzzy logics. J. Algorithms **62**(3–4), 135–147 (2007)

6. Bertoluzza, C.: On the distributivity between t-norms and t-conorms. In: Proceedings of 2nd IEEE International Conference on Fuzzy Systems (FUZZ-IEEE 1993), San Francisco, USA, pp. 140–147 (1993)

7. Bertoluzza, C., Doldi, V.: On the distributivity between t-norms and t-conorms. Fuzzy Sets Syst. **142**, 85–104 (2004)

8. Bloch, I.: Duality vs. adjunction for fuzzy mathematical morphology and general form of fuzzy erosions and dilations. Fuzzy Sets Syst. **160**(13), 1858–1867 (2009)

9. Bustince, H., Fernández, J., Sanz, J., Baczyński, M., Mesiar, R.: Construction of strong equality index from implication operators. Fuzzy Sets Syst. **211**, 15–33 (2013)

10. Carbonell, M., Mas, M., Suner, J., Torrens, J.: On distributivity and modularity in De Morgan triplets. Int. J. Uncertain. Fuzziness Knowl.-Based Syst. **4**, 351–368 (1996)

11. Cruz, A., Bedregal, B.C., Santiago, R.H.N.: On the characterizations of fuzzy implications satisfying $I(x, I(y, z)) = I(I(x, y), I(x, z))$. Int. J. Approx. Reason. **93**, 261–276 (2018)

12. Dimuro, G.P., Bedregal, B., Santiago, R.H.: On (G, N)-implications derived from grouping functions. Inf. Sci. **279**, 1–17 (2014)

13. Dimuro, G.P., Bedregal, B.C.: On residual implications derived from overlap functions. Inf. Sci. **312**, 78–88 (2015)

14. Dimuro, G.P., Bedregal, B., Bustince, H., Jurio, A., Baczyński, M., Mis, K.: QL-operations and QL-implication functions constructed from tuples (O, G, N) and the generation of fuzzy subsethood and entropy measures. Int. J. Approx. Reason. **82**, 170–192 (2017)

15. Jayaram, B.: On the law of importation $(x \wedge y) \rightarrow z \equiv (x \rightarrow (y \rightarrow z))$ in fuzzy logic. IEEE Trans. Fuzzy Syst. **16**(1), 130–144 (2008)

16. Klement, E.P., Mesiar, R., Pap, E.: Triangular Norms. Kluwer Academic Publishers, Dordrecht (2000)

17. Klir, G.J., Yuan, B.: Fuzzy Sets and Fuzzy Logic: Theory and Applications. Prentice Hall, USA (1995)

18. Mas, M., Monserrat, M., Torrens, J., Trillas, E.: A survey on fuzzy implication functions. IEEE Trans. Fuzzy Syst. **15**(6), 1107–1121 (2007)

19. Mas, M., Monserrat, M., Torrens, J.: The law of importation for discrete implications. Inf. Sci. **179**(24), 4208–4218 (2009)

20. Štěpnička, M., De Baets, B.: Implication-based models of monotone fuzzy rule bases. Fuzzy Sets Syst. **232**, 134–155 (2013)

21. Pradera, A., Beliakov, G., Bustince, H., De Baets, B.: A review of the relationships between implication, negation and aggregation functions from the point of view of material implication. Inf. Sci. **329**, 357–80 (2016)

22. Pinheiro, J., Bedregal, B., Santiago, R.H., Santos, H.: (T, N)-implications. In: 2017 IEEE International Conference on Fuzzy Systems (FUZZ-IEEE), pp. 1–6 (2017)

23. Pinheiro, J., Bedregal, B., Santiago, R.H., Santos, H.: A study of (T, N)-implications and its use to construct a new class of fuzzy subsethood measure. Int. J. Approx. Reason. **97**, 1–16 (2018)

24. Reiser, R., Bedregal, B., Baczyński, M.: Aggregating fuzzy implications. Inf. Sci. **253**, 126–146 (2013)
25. Santos, H.S.: A new class of fuzzy subsethood measures. Ph.D. thesis, Universidade Federal do Rio Grande do Norte, Natal (2016)
26. Santos, H., Bedregal, B., Dimuro, G.P., Bustince, H.: Penalty functions constructed from QL subsethood measures. In: 2017 IEEE International Conference on Fuzzy Systems (FUZZ-IEEE), pp. 1–5 (2017)
27. Yager, R.R.: On some new classes of implication operators and their role in approximate reasoning. Inf. Sci. **167**(1–4), 193–216 (2004)

The Category of Semi-BCI Algebras

Jocivania Pinheiro[1,3]($\boxtimes$), Rui Paiva[2,3], and Regivan Santiago[3]

[1] Group for Theory of Computation, Logic and Fuzzy Mathematics,
Department of Natural Sciences, Mathematics and Statistics (DCME),
Center of Exact and Natural Science (CCEN),
Rural Federal University of SemiArid (UFERSA),
Mossoró, RN, Brazil
`vaniamat@ufersa.edu.br`
[2] Federal Institute of Education, Science and Technology of Ceará - IFCE,
Canindé, CE, Brazil
`rui.brasileiro@ifce.edu.br`
[3] Group for Logic, Language, Information, Theory and Applications (LoLITA),
Informatics and Applied Mathematics Department,
Federal University of Rio Grande do Norte (UFRN), Natal, RN, Brazil
`regivan@dimap.ufrn.br`

Abstract. This paper introduces briefly the notion of SBCI algebras and its role as candidate to model Fuzzy and Interval Fuzzy Logics. Its main goal is to provide how such algebraic structures behaves from the categorical theoretical standpoint.

1 Introduction

Algebraic structures are commonly used to model logical systems (Logical Algebras); e.g. MV-algebras (Łukasiewicz Logics), BL-algebras (Fuzzy Logics), Boolean Algebras (Classical Logic), etc. In the 1960's Iséki [15] introduced the notion of BCI-algebras to model Curry combinators : (B) $\lambda xyx.x(yz)$, (C) $\lambda xyz.xzy$ and (I) $\lambda x.x$. This algebra is the basic building block for many logical algebras including BL-algebras. However, when we deal with Interval Fuzzy Logic as an extension of usual Fuzzy Logic, BCI-algebras are not more suitable as a logical algebra. To overcome this situation Santiago *et al.* provided the notion of Semi-BCI algebras **(SBCI)** [17].

Every BCI algebra is a SBCI-algebra and every correct intervalization of a BCI-algebra is a SBCI-algebra. It means that such structure is a candidate to model both Fuzzy and Interval Fuzzy Logics.

SBCI-algebra is not the first extension of BCI algebras, another approach called Pseudo-BCI algebras was proposed by Dudek and Jun [9], although they have a similar signature, they are completely different from SBCIs [17]. Recently, category theory was applied to model the construction of pseudo-BCI-algebras [7]. This work goes in the same direction, here we provide an investigation of which categorical entities are available in the category of SBCIs. The paper is organized in the following way: Sect. 2 provides a brief review of SBCI-algebras,

© Springer International Publishing AG, part of Springer Nature 2018
G. A. Barreto and R. Coelho (Eds.): NAFIPS 2018, CCIS 831, pp. 314–323, 2018.
https://doi.org/10.1007/978-3-319-95312-0_27

their properties and some required concepts. Section 3 shows the category of SBCI-algebras and proves some of its properties. Finally, Sect. 4 provides some final remarks.

2   SBCI Algebras

The interval counterpart of Łukasiewicz implication, introduced by Bedregal and Santiago [3], fails to satisfy a basic property of BCI's; namely the order property **(OP)**. **(OP)** expresses the internalization (codification) of underlying partial order by implication;

$$X \to Y = 1 \Leftrightarrow X \leq Y. \tag{1}$$

The authors, however, revealed that the resulting implication satisfy, for all intervals $X = [\underline{X}, \overline{X}]$ and $Y = [\underline{Y}, \overline{Y}]$:

1. if $X \ll Y^1$, then $X \to Y = 1$;
2. if $X \to Y = 1$, then $X \leq_{KM} Y^2$.

$$\tag{$*$}$$

The relation "$\ll$" is precisely the way-below relation [4] of the usual Kulisch-Miranker order on intervals "$\leq_{KM}$". Way-below relations, "$\lll$", are auxiliary relations [4] of partial orders "$\preccurlyeq$"; they have the following properties:

1. if $x \lll y$, then $x \preccurlyeq y$.
2. if $u \preccurlyeq x \lll y \preccurlyeq z$, then $u \lll z$.
3. if a smallest element 0 exists, then $0 \lll x$.

Since partial orders are connected to auxiliary relations Bedregal *et al.* observed that (*) reveals that if Łukasiewcz implication is intervalized the internalization is spitted in terms of the partial order and its auxiliary relation [17]. In order to formalize this fact and provide an investigation about that Santiago *et al.* proposed an algebraic structure called **semi-BCI algebra**. SBCI-algebras model both: BCIs and the intervalization of BCIs, therefore they seems to model logics in which the notion of impreciseness is required. In what follows we present SBCI-algebras:

Definition 1 (Semi-BCI (SBCI) algebra). *Given a set X endowed with two binary operations: "$\to$" and "$\leadsto$", a structure $\langle X, \to, \leadsto, \top \rangle$ is a **Semi-BCI (SBCI) algebra** whenever for all $x, y, z \in X$:*

(SBCI1) $x \to (y \to z) = y \to (x \to z)$
(SBCI2) $x \leadsto (y \leadsto z) = y \leadsto (x \leadsto z)$
(SBCI3) $x \to y \preceq (z \to x) \leadsto (z \to y)$
(SBCI4) $\top \to x = x$

---

[1] $X \ll Y$ iff $\overline{X} < \underline{Y}$.
[2] $X \leq_{KM} Y$ iff $\underline{X} \leq \underline{Y}$ and $\overline{X} \leq \overline{Y}$.

(SBCI5) $x \ll y \preceq z \Longrightarrow x \ll z$
(SBCI6) $x \preceq y \ll z \Longrightarrow x \ll z$
(SBCI7) $x \preceq y$ e $y \preceq x \Longrightarrow x = y,$

where $x \ll y \Longleftrightarrow x \rightarrow y = \top$ and $x \preceq y \Longleftrightarrow x \rightsquigarrow y = \top.$

A SBCI-algebra which satisfies: $x \preceq \top$, for all $x \in X$, is called **Semi-BCK (SBCK) algebra**. An element $x \in X$ of a SBCI-algebra which satisfies: "$x \ll x$", is called **total**.

Example 1. The structure $A = \langle [0,1], \rightarrow, \rightsquigarrow, 1 \rangle$, where $x \rightarrow y = 1 - x + xy$ and $x \rightsquigarrow y = \min\{1, 1 - x + y\}$ is a SBCI-algebra. The corresponding relations are given by: (1) $x \ll y$ if and only if $x \rightarrow y = 1$ if and only if $x = 0$ or $y = 1$; (2) $x \preceq y$ if and only if $x \rightsquigarrow y = 1$ if and only if $x \leq y.$

Proposition 1. *[17] For all $x, y, z \in X$ the following properties are valid:*

(SBCI8) $x \rightsquigarrow x = \top$
(SBCI9) $x \ll y \Longrightarrow x \preceq y$
(SBCI10) $x \ll y$ and $y \ll z \Longrightarrow x \ll z,$
(SBCI11) $x \ll y$ and $y \ll x \Longrightarrow x = y,$
(SBCI12) $(y \rightarrow z) \preceq ((z \rightarrow x) \rightsquigarrow (y \rightarrow x)),$
(SBCI13) If $\top \preceq x$ then $x = \top,$
(SBCI14) $x \rightarrow y \preceq x \rightsquigarrow y,$
(SBCI15) $x \rightsquigarrow ((x \rightsquigarrow y) \rightsquigarrow y) = \top,$
(SBCI16) $x \rightarrow ((x \rightarrow y) \rightarrow y) = \top$, when $x \ll y.$

Given a SBCI-algebra $\langle X, \rightarrow, \rightsquigarrow, \top \rangle$, we can observe that the relations "$\ll$" and "$\preceq$" are not necessarily partial orders. The relation $\ll$ is transitive and antisymmetric, but it is not necessarily reflexive, whereas $\preceq$ is antisymmetric and reflexive, but is not necessarily transitive. The following proposition provides a condition to have partial orders:

Proposition 2. *[17] The relation "$\ll$" coincides with "$\preceq$" if and only if "$\ll$" is reflexive.*

Example 2. The structure $A = \langle [0,1], \rightarrow_{GD}, \rightarrow_{FD}, 1 \rangle$ is a SBCI, where

$$x \rightarrow_{GD} y = \begin{cases} 1, \text{if } x \leq y \\ y, \text{if } x > y \end{cases}$$

and

$$x \rightarrow_{FD} y = \begin{cases} 1, \text{if } x \leq y \\ max(1 - x, y), \text{if } x > y \end{cases}.$$

In this case, the relations $\ll$ and $\preceq$ coincide with the usual order $\leq.$

Definition 2. *Given a SBCI-algebra $\mathcal{X} = \langle X, \rightarrow, \rightsquigarrow, \top \rangle$ and $S \subseteq X$, we say that $\langle S, \rightarrow, \rightsquigarrow, \top \rangle$ is a **SBCI sub-algebra of** $\mathcal{X}$ if, for all $x, y \in S$, $x \rightarrow y \in S$ and $x \rightsquigarrow y \in S.$*

Definition 3. *Let $\mathcal{X}_1 = \langle A_1, \twoheadrightarrow_1, \rightsquigarrow_1, \top_1 \rangle$ and $\mathcal{X}_2 = \langle A_2, \twoheadrightarrow_2, \rightsquigarrow_2, \top_2 \rangle$ be SBCI-algebras. An application, $\varphi : A_1 \to A_2$, is called **SBCI-homomorphism**, whenever $\varphi(x \twoheadrightarrow_1 y) = \varphi(x) \twoheadrightarrow_2 \varphi(y)$ and $\varphi(x \rightsquigarrow_1 y) = \varphi(x) \rightsquigarrow_2 \varphi(y)$ for all $x, y \in A_1$.*

Note that if $\varphi : A_1 \to A_2$ is a SBCI-homomorphism, then $\varphi(\top_1) = \top_2$, since for any $x \in A_1$ we have $\varphi(\top_1) \overset{(SBCI8)_1}{=} \varphi(x \rightsquigarrow_1 x) \overset{def.hom.}{=} \varphi(x) \rightsquigarrow_2 \varphi(x) \overset{(SBCI8)_2}{=} \top_2$.

Proposition 3. *Let $\mathcal{X}_1 = \langle A_1, \twoheadrightarrow_1, \rightsquigarrow_1, \top_1 \rangle$ and $\mathcal{X}_2 = \langle A_2, \twoheadrightarrow_2, \rightsquigarrow_2, \top_2 \rangle$ be SBCI-algebras and $\varphi : A_1 \to A_2$ a SBCI-homomorphism. The relations "$\ll_i$" and "$\preceq_i$" are preserved by φ, i.e. for $x, y \in A_1$, $x \ll_1 y$ implies $\varphi(x) \ll_2 \varphi(y)$ and $x \preceq_1 y$ implies $\varphi(x) \preceq_2 \varphi(y)$.*

Proof. Indeed, given $x, y \in A_1$ such that $x \ll_1 y \Leftrightarrow x \twoheadrightarrow_1 y = \top_1$ (respectively $x \preceq_1 y \Leftrightarrow x \rightsquigarrow_1 y = \top_1$). Then applying φ to both equality members we get $\varphi(x) \twoheadrightarrow_2 \varphi(y) = \top_2$ (respectively $\varphi(x) \rightsquigarrow_2 \varphi(y) = \top_2$). But this last result is valid if and only if $\varphi(x) \ll_2 \varphi(y)$ (respectively $\varphi(x) \preceq_2 \varphi(y)$).

3 Category of Semi-BCI Algebra

In what follows we introduce the *Category of Semi-BCI algebras* and provide some results about it. We also give the required categorial notions. For more details and further examples the reader can see the following references: [1, 10, 12].

Definition 4 (Categories). *A **category** C comprises:*

1. *a collection of objects: Obj_C;*
2. *a collection of arrows (often called morphisms): Mor_C;*
3. *operations assigning to each arrow φ an object dom φ, its domain, and an object cod φ, its codomain (we write $\varphi : A \to B$ or $A \overset{\varphi}{\to} B$ to show that dom $\varphi = A$ and cod $\varphi = B$). The collection of all arrows with domain A and codomain B is written $\mathbf{C}(A, B)$;*
4. *a composition operator assigning to each pair of arrows φ and ψ, with cod $\varphi =$ dom ψ, a composite arrow $\psi \circ \varphi :$ dom $\varphi \to$ cod ψ, satisfying the following associative law: for any arrows $\varphi : A \to B, \psi : B \to C$, and $\alpha : C \to D$ (with $A, B, C,$ and D not necessarily distinct),*

$$\alpha \circ (\psi \circ \varphi) = (\alpha \circ \psi) \circ \varphi;$$

5. *for each object A, an identity arrow $id_A : A \to A$ satisfying the following identity law: for any arrow $\varphi : A \to B$,*

$$id_B \circ \varphi = \varphi \text{ and } \varphi \circ id_A = \varphi.$$

The class of all sets together with the class of all total functions, the function composition and the operations which respectively return the domain and codomain of a function forms a category called **Set**. Further examples, mainly not set-based can be found in [10]. The next proposition shows that the class of all SBCI-algebras together with SBCI-homomorphisms form a category.

Proposition 4 (Category $\mathcal{SBCI}$). *The structure*

$$\mathcal{SBCI} = \langle Ob_{\mathcal{SBCI}}, Mor_{\mathcal{SBCI}}, \partial_0, \partial_1, \iota, \circ \rangle$$

such that:

(a) $Ob_{\mathcal{SBCI}}$ *is the collection of all SBCI algebras;*
(b) $Mor_{\mathcal{SBCI}}$ *is a collection of all SBCI-homomorphisms;*
(c) $\partial_0, \partial_1 : Mor_{\mathcal{SBCI}} \to Ob_{\mathcal{SBCI}}$ *are such that, for any SBCI-homomorphism $\varphi : \mathcal{A}_1 \to \mathcal{A}_2$ for $\mathcal{A}_1 = \langle A_1, \twoheadrightarrow_1, \rightsquigarrow_1, \mathsf{T}_1 \rangle$ and $\mathcal{A}_2 = \langle A_2, \twoheadrightarrow_2, \rightsquigarrow_2, \mathsf{T}_2 \rangle$, $\partial_0(\varphi) = \mathcal{A}_1$ e $\partial_1(\varphi) = \mathcal{A}_2$;*
(d) $\circ : Mor_{\mathcal{SBCI}} \times Mor_{\mathcal{SBCI}} \to Mor_{\mathcal{SBCI}}$ *is the composition of SBCI-homomorphisms;*
(e) $\iota : Ob_{\mathcal{SBCI}} \to Mor_{\mathcal{SBCI}}$ *is such that each $\mathcal{A} = \langle A, \twoheadrightarrow, \rightsquigarrow, \mathsf{T} \rangle$ is associated with the identity homomorphism $\iota_A : A \to A$, i.e. $\iota_A(x) = x$.*

is a category.

Proof. In order to verify that $\mathcal{SBCI}$ is a category, it is sufficient to verify the associativity of composition and the property of identity, which is given, for any morphism $\varphi \in \mathrm{Hom}_{\mathcal{SBCI}(A,B)}$ by:

$$\varphi \circ \iota_A = \iota_B \circ \varphi = \varphi.$$

Indeed, given $\mathcal{A}_i = \langle A_i, \twoheadrightarrow_i, \rightsquigarrow_i, \mathsf{T}_i \rangle$, with $i = 1, 2, 3, 4$, consider the homomorphisms: $\varphi : \mathcal{A}_1 \to \mathcal{A}_2$, $\psi : \mathcal{A}_2 \to \mathcal{A}_3$, $\alpha : \mathcal{A}_3 \to \mathcal{A}_4$ and, for any $a \in A_1$,

$$((\alpha \circ \psi) \circ \varphi)(a) = (\alpha \circ \psi)(\varphi(a)) = \alpha(\psi(\varphi(a))) = \alpha(\psi \circ \varphi(a)) = (\alpha \circ (\psi \circ \varphi))(a).$$

Therefore, the composition is associative. Moreover, $\iota : Ob_{\mathcal{SBCI}} \to Mor_{\mathcal{SBCI}}$ satisfies propriety of identity. Indeed, for any morphism $\varphi : \mathcal{A} \to \mathcal{B}$ and $a \in A$,

$$(\varphi \circ \iota_A)(a) = \varphi(\iota_A(a)) = \varphi(a) = \iota_B(\varphi(a)) = (\iota_B \circ \varphi)(a).$$

So, $\mathcal{SBCI}$ is a category.

$$\textbf{Q.E.D}$$

The next two results show that in category $\mathcal{SBCI}$, injective morphisms coincide with the categorial notion of monomorphism, but surjective morphisms generally do not match with the categorial notion of epimorphism.

Definition 5. *Given a category $\mathbf{C}$ and objects A, B, C, an arrow $\varphi : B \to C$ is a **monomorphism** (or "is **monic**") if, for any pair of $\mathbf{C}$-arrows $\psi : A \to B$ and $\alpha : A \to B$, the equality $\varphi \circ \psi = \varphi \circ \alpha$ implies that $\psi = \alpha$.*

Proposition 5. *For any morphism* $\varphi : X \to Y$ *in SBCI, the following conditions are equivalent:*

  (i) φ *is injective;*
  (ii) $\varphi \circ \psi = \varphi \circ \alpha$ *implies* $\psi = \alpha$ *for any morphisms* $\psi, \alpha : Z \to X$;
  (iii) $\ker(\varphi) = \{\top_X\}$, *where* $Ker(\varphi) = \{x \in X \mid \varphi(x) = \top_Y\}$.

Proof. $(i) \Rightarrow (ii)$ Assuming that $\varphi : X \to Y$ is an injective morphism, then $\varphi \circ \psi = \varphi \circ \alpha$ implies $\psi = \alpha$ for any morphisms ψ and α. In fact, let Z be another object of category $SBCI$ and $\psi, \alpha : Z \to X$ morphisms such that $\varphi \circ \psi = \varphi \circ \alpha$. Then, for all $z \in Z$, $\varphi(\psi(z)) = \varphi(\alpha(z))$. Thus, since φ is injective, then $\psi(z) = \alpha(z)$, for all $z \in Z$. Therefore $\psi = \alpha$.

$(ii) \Rightarrow (iii)$ Assuming $\ker(\varphi) \neq \{\top_X\}$, then there is $z \in \ker(\varphi)$ such that $z \neq \top_X$. Consider the morphisms $i : \ker(\varphi) \to X$ and $j : \ker(\varphi) \to X$ such that $i(x) = x$ and $j(x) = \top_X$, for all $x \in \ker(\varphi)$. Then, $\varphi \circ i = \varphi \circ j$, so by hypothesis, $i = j$. Contradiction! Therefore, $\ker(\varphi) = \{\top_X\}$.

$(iii) \Rightarrow (i)$ Given $x_1, x_2 \in X$ such that $\varphi(x_1) = \varphi(x_2)$, then $x_1 = x_2$. Indeed, since φ is a homomorphism, $\varphi(x_1 \leadsto_X x_2) = \varphi(x_1) \leadsto_Y \varphi(x_2) \overset{(SBCI8)_Y}{=} \top_Y$ and $\varphi(x_2 \leadsto_X x_1) = \varphi(x_2) \leadsto_Y \varphi(x_1) \overset{(SBCI8)_Y}{=} \top_Y$, so $x_1 \leadsto_X x_2, x_2 \leadsto_X x_1 \in Ker(\varphi) = \{\top_X\}$. Therefore, $x_1 \leadsto_X x_2 = x_2 \leadsto_X x_1 = \top_X$, i.e., $x_1 \preceq_X x_2$ and $x_2 \preceq_X x_1$, so by $(SBCI7)_X$, $x_1 = x_2$.   **Q.E.D**

We conclude that morphisms are mono if and only if they are injective.

Definition 6. *An arrow* $\varphi : A \to B$ *in a category* **C** *is an* **epimorphism** *(or "is* **epic***") if, for any pair of* **C***-arrows* $\alpha, \psi : B \to C$, *the equality* $\psi \circ \varphi = \alpha \circ \varphi$ *implies that* $\psi = \alpha$.

Proposition 6. *For any morphism* $\varphi : X \to Y$ *in SBCI, if* φ *is surjective then it is also epic.*

Proof. Assuming that $\varphi : X \to Y$ is a surjective morphism, Z is an object and $\psi, \alpha : Y \to Z$ are morphisms such that $\psi \circ \varphi = \alpha \circ \varphi$. Since φ is surjective, for any $y \in Y$ there is $x \in X$ such that $y = \varphi(x)$. Then, for all $y \in Y$,

$$\psi(y) = \psi(\varphi(x)) = \alpha(\varphi(x)) = \alpha(y).$$

Therefore, $\psi = \alpha$.   **Q.E.D**

Proposition 7. *There are epimorphisms which are not surjective homomorphisms.*

Proof. In 2010, Busneag shows examples of epimorphisms of Hilbert algebras whose are not surjective functions (See [8, Example 4.1]). Since all Hilbert algebra is a BCK-algebra (see [2]) and in turn all BCK-algebra is a SBCI-algebra (see [17]), then not all SBCI epimorphism is a surjective function.   **Q.E.D**

Definition 7. *In a Category* **C***, an object* 0 *is called an* **initial object** *if, for every object A, there is exactly one morphism from* 0 *to A. And dually, an object* 1 *is called a* **terminal** *or* **final object** *if, for every object A, there is exactly one morphism from A to* 1.

Proposition 8. $\mathcal{SBCI}$ *has as initial and final object.*

Proof. In fact, given any SBCI algebra $\mathcal{A} = \langle X, \twoheadrightarrow_X, \rightsquigarrow_X, \top_X \rangle$ and the SBCI $\mathcal{O} = \langle \{\top\}, \twoheadrightarrow, \rightsquigarrow, \top \rangle$, there is an unique morphism $0 : \mathcal{O} \to \mathcal{A}$; namely: $0(\top) = \top_X$, so $\mathcal{O}$ is a initial object. On the other hand, there is also a unique morphism $1 : \mathcal{A} \to \mathcal{O}$; namely: $1(x) = \top, \forall x \in X$, in fact: (1) 1 is a morphism because $1(x \twoheadrightarrow_X y) \overset{def}{=} \top \overset{(SBCI4)}{=} \top \twoheadrightarrow \top \overset{def}{=} 1(x) \twoheadrightarrow 1(y)$ and $1(x \rightsquigarrow_X y) \overset{def}{=} \top \overset{(SBCI8)}{=} \top \rightsquigarrow \top \overset{def}{=} 1(x) \rightsquigarrow 1(y)$; (2) 1 is unique, since $\{\top\}$ has an unique element then any morphism $g : X \to \{\top\}$ satisfies $g(x) = \top, \forall x \in X$, so $g(x) = \top = 1(x), \forall x \in X$, therefore $g = 1$, i.e., 1 is unique.   **Q.E.D**

Definition 8. *Let C be a category. A* **product of a family of objects** $(X_i)_{i \in I}$ *of C indexed by a set I consists of an object* $P = \prod_{i \in I} X_i$ *and a family of projection morphisms* $(\pi_i : P \to X_i)_{i \in I}$ *such that for each object P′ and a family of morphisms* $(f_i : P' \to X_i)_{i \in I}$ *there is only one morphism* $u : P' \to P$ *such that* $\pi_i \circ u = f_i$, *for all* $i \in I$, *i.e., the following diagram commute*[3] *for all* $i \in I$:

A category C has **products** *if there is a product of any family of objects of C.*

Lemma 1. *Let* $\{\langle A_i, \twoheadrightarrow_i, \rightsquigarrow_i, \top_i \rangle\}_{i \in I}$ *be a family of SBCI algebras and* $P = \prod_{i \in I} A_i = \{f : I \to \bigcup_{i \in I} A_i \mid f(i) \in A_i, \forall i \in I\}$. $\langle P, \twoheadrightarrow, \rightsquigarrow, \top \rangle$ *is a SBCI, if for any* $f, g \in P$, *the binary operations "* $\twoheadrightarrow$ *" and "* $\rightsquigarrow$ *" are given by:* $(f \twoheadrightarrow g)(i) = f(i) \twoheadrightarrow_i g(i)$ *and* $(f \rightsquigarrow g)(i) = f(i) \rightsquigarrow_i g(i)$ *for all* $i \in I$, *and the function* $\top : I \to \bigcup_{i \in I} A_i$ *satisfies* $\top(i) = \top_i \in A_i, \forall i \in I$.

Proof. Indeed, given $f, g, h \in P$, (SBCI1): By definition $(f \twoheadrightarrow (g \twoheadrightarrow h))(i) = f(i) \twoheadrightarrow_i (g \twoheadrightarrow h)(i) = f(i) \twoheadrightarrow_i (g(i) \twoheadrightarrow_i h(i))$, since $\langle A_i, \twoheadrightarrow_i, \rightsquigarrow_i, \top_i \rangle$ is SBCI, $f(i) \twoheadrightarrow_i (g(i) \twoheadrightarrow_i h(i)) = g(i) \twoheadrightarrow_i (f(i) \twoheadrightarrow_i h(i))$, so $(f \twoheadrightarrow (g \twoheadrightarrow h))(i) = g(i) \twoheadrightarrow_i (f(i) \twoheadrightarrow_i h(i)) = (g \twoheadrightarrow (f \twoheadrightarrow h))(i)$ for all $i \in I$, therefore $f \twoheadrightarrow (g \twoheadrightarrow h) = g \twoheadrightarrow (f \twoheadrightarrow h)$. (SBCI2): Analogous. (SBCI4): By definition, $(f \twoheadrightarrow g)(i) = f(i) \twoheadrightarrow_i g(i)$, since $\langle A_i, \twoheadrightarrow_i, \rightsquigarrow_i, \top_i \rangle$ is SBCI then $f(i) \twoheadrightarrow_i g(i) \preceq_i (h(i) \twoheadrightarrow_i f(i)) \rightsquigarrow_i (h(i) \twoheadrightarrow_i g(i)) = ((h \twoheadrightarrow f) \rightsquigarrow (h \twoheadrightarrow g))(i), \forall i \in I$, therefore, $f \twoheadrightarrow g \preceq (h \twoheadrightarrow f) \rightsquigarrow (h \twoheadrightarrow g)$. (SBCI4): By definition, $(\top \twoheadrightarrow f)(i) = \top(i) \twoheadrightarrow_i$

---

[3] i.e. $f_i = \pi_i \circ u$.

$f(i) \overset{(SBCI4)_i}{=} f(i), \forall i \in I$, so $\top \twoheadrightarrow f = f$. (SBCI5): $f \ll g \preceq h \Rightarrow f(i) \twoheadrightarrow_i g(i) = \top(i)$ e $g(i) \leadsto_i h(i) = \top(i), \forall i \in I \Rightarrow f(i) \ll_i g(i)$ e $g(i) \preceq_i h(i), \forall i \in I \Rightarrow f(i) \ll_i g(i) \preceq_i h(i), \forall i \in I \overset{(SBCI15)_i}{\Rightarrow} f(i) \ll_i h(i), \forall i \in I \Rightarrow (f \twoheadrightarrow h)(i) = \top(i), \forall i \in I \Rightarrow f \twoheadrightarrow h = \top \Rightarrow f \ll h$. (SBCI6): Analogous. (SBCI7): $f \preceq g$ e $g \preceq f \Rightarrow (f \leadsto g)(i) = \top(i)$ e $(g \leadsto f)(i) = \top(i), \forall i \in I \Rightarrow f(i) \leadsto_i g(i) = \top(i)$ e $g(i) \leadsto_i f(i) = \top(i), \forall i \in I \Rightarrow f(i) \preceq_i g(i)$ e $g(i) \preceq_i f(i), \forall i \in I \overset{(SBCI7)_i}{\Rightarrow} f(i) = g(i), \forall i \in I \Rightarrow f = g$.   **Q.E.D**

Theorem 1. $\mathcal{SBCI}$ *has products.*

Proof. Given a family of objects $(\langle A_i, \twoheadrightarrow_i, \leadsto_i, \top_i \rangle)_{i \in I}$ of $\mathcal{SBCI}$, consider the set $P = \prod_{i \in I} A_i = \{f : I \to \bigcup_{i \in I} A_i \mid f(i) \in A_i, \forall i \in I\}$. By Lemma 1, the structure $\langle P, \twoheadrightarrow, \leadsto, \top \rangle$ is a SBCI. For each $i \in I$, there is a projection $p_i : P \to A_i$ defined by $p_i(f) = f(i), \forall f \in P$. Moreover, given any object P' and morphisms $p_i' : P' \to A_i$ for all $i \in I$, define the application $u : P' \to P$ by $(u(x))(i) = p_i'(x), \forall x \in P'$. Since $(p_i \circ u)(x) = p_i(u(x)) = (u(x))(i) = p_i'(x)$, for all $x \in P'$, then $p_i \circ u = p_i'$. The application u is a morphism between the objects $\langle P', \twoheadrightarrow', \leadsto', \top' \rangle$ and $\langle P, \twoheadrightarrow, \leadsto, \top \rangle$, since $(u(x \twoheadrightarrow' y))(i) = p_i'(x \twoheadrightarrow' y) \overset{p_i' morphism}{=} p_i'(x) \twoheadrightarrow_i p_i'(y) = (u(x))(i) \twoheadrightarrow_i (u(y))(i) = (u(x) \twoheadrightarrow u(y))(i), \forall i \in I$, therefore, $u(x \twoheadrightarrow' y) = u(x) \twoheadrightarrow u(y)$. Similarly, $u(x \leadsto' y) = u(x) \leadsto u(y)$. Suppose now, there exist a morphism $v : P' \to P$ such that $p_i \circ v = p_i', \forall i \in I$. For each $x \in P'$, $(p_i \circ v)(x) = p_i'(x) = (p_i \circ u)(x)$ for all $i \in I$, so $p_i(v(x)) = p_i(u(x)), \forall i \in I$, therefore $v(x) = u(x)$. Since this holds for every $x \in P'$, then $v = u$. Therefore, $u : P' \to P$ is the unique morphism that satisfies $p_i \circ u = p_i'$.   **Q.E.D**

Definition 9. *In a Category* **C**, *an arrow* $e : X \to A$ *is an* **equalizer** *of a pair of arrows* $f, g : A \to B$ *whenever*

1. $f \circ e = g \circ e$;
2. *if* $e' : X' \to A$ *satisfies* $f \circ e' = g \circ e'$, *then there is a unique arrow* $u : X' \to X$ *such that* $e \circ u = e'$:

$$X \xrightarrow{e} A \overset{f}{\underset{g}{\rightrightarrows}} B$$

with $u : X' \to X$ and $e' : X' \to A$ (diagram).

A category **C** *has* **equalizer** *if there is a equalizer for any pair of arrows of* **C**.

Theorem 2. *Category* $\mathcal{SBCI}$ *has equalizer.*

Proof. Given two morphisms $f, g : X_1 \to X_2$, consider the set $E_1 = \{x \in X_1 \mid f(x) = g(x)\}$. Since f and g are morphisms, then $f(\top_1) = \top_2 = g(\top_1)$, therefore $\top_1 \in E_1$, so $E_1 \neq \varnothing$. Also, given $x, y \in E_1$, $f(x \twoheadrightarrow_1 y) = f(x) \twoheadrightarrow_2 f(y) = $

$g(x) \twoheadrightarrow_2 g(y) = g(x \twoheadrightarrow_1 y)$ and analogously, $f(x \rightsquigarrow_1 y) = g(x \rightsquigarrow_1 y)$, so $x \twoheadrightarrow_1 y, x \rightsquigarrow_1 y \in E_1$, therefore $\langle E_1, \twoheadrightarrow_1, \rightsquigarrow_1, \top_1 \rangle$ is a subalgebra of $\langle X_1, \twoheadrightarrow_1, \rightsquigarrow_1, \top_1 \rangle$. Consider the embedding $e : E_1 \to X_1$, then $f \circ e = g \circ e$. Given any other object $\langle A, \twoheadrightarrow_A, \rightsquigarrow_A, \top_A \rangle$ and a morphism $e' : A \to X_1$ such that $f \circ e' = g \circ e'$, if we define $u : A \to E_1$ s.t. $u(a) = e'(a)$, then the following diagram commutes.

$$E_1 \xrightarrow{\;e\;} X_1 \xrightarrow[g]{\;f\;} X_2$$

$$u \uparrow \quad \nearrow e'$$

$$A$$

In fact, given a morphism e', we have that u is a morphism and $e \circ u = e'$. Since $(e \circ u)(a) = e(u(a)) = u(a) = e'(a), \forall a \in A$. To prove the uniqueness, suppose there is a morphism $v : A \to E_1$ such that $e \circ v = e'$, then for all $a \in A$, $(e \circ v)(a) = e'(a) = (e \circ u)(a)$, so $e \circ v = e \circ u$. Since e is a embedding, then, by Proposition 5 "e" is a monomorphism, therefore $v = u$.   **Q.E.D**

4   Final Remarks

In this paper we have shown how SBCI algebras behave from the categorical theoretical standpoint. We presented how some categorial entities are realized by such kind of algebra. From the algebraic viewpoint an important result remains to be proved; namely if this category is closed under homomomorphical images (which together with some of the results presented here establishes that there exists a set of equations which is able to describe the class of all SBCI algebras). Another open problem is the verification of co-Products and Exponenciation.

References

1. Awodey, S.: Category theory. Oxford logic guides, vol. 49. Oxford University Press (2006)
2. Iorgulescu, A.: Algebras of Logic as BCK Algebras. Academy of Economic Studies Press, Bucharest (2008)
3. Bedregal, B.C., Santiago, R.H.N.: Interval representations, Łukasiewicz implicators and Smets-Magrez axioms. Inf. Sci. **221**, 192–200 (2013)
4. Gierz, G., Hofmann, K.H., Keimel, K., Lawson, J.D., Mislove, M., Scott, D.S.: Continuous Lattices and Domains. Encyclopedia of Mathematics and its Applications. Cambridge University Press, Cambridge (2003)
5. Dymek, G.: p-Semisimple pseudo-BCI-algebras. J. Multiple-Valued Log. Soft Comput. **19**, 461–474 (2012)
6. Georgescu, G., Iorgulescu, A.: Pseudo-BCK algebras: an extension of BCK algebras. In: Calude, C.S., Dinneen, M.J., Sburlan, S. (eds.) Combinatorics, Computability and Logic. DMTCS, pp. 97–114. Springer, London (2001). https://doi.org/10.1007/978-1-4471-0717-0_9
7. Dymek, G.: On the category of pseudo-BCI-algebras. Demonstr. Math. **46**(4), 631–644 (2013)

8. Busneag, D., Ghita, M.: Some properties of epimorphisms of Hilbert algebras. Open Math. **8**, 41–52 (2010)
9. Dudek, W.A., Jun, Y.B.: Pseudo-BCI algebras. East Asian Math. J. **24**, 187–190 (2008)
10. Pierce, B.C.: Basic Category Theory for Computer Scientists. MIT Press, Cambridge (1991)
11. Dymek, G.: On two classes of pseudo-BCI-algebras. Discuss. Math. Gen. Algebra Appl. **31**, 217–229 (2011)
12. Goldblatt, R.: Topoi: The Categorial Analysis of Logic. Elsevier Science, Amsterdam (1986)
13. Shohani, J., Borzooei, R.A., Jahanshahi, M.A.: Basic BCI-algebras and Abelian groups are equivalent. Sci. Math. Jpn. **66**(2), 243–245 (2007)
14. Meng, J.: BCI-algebras and Abelian groups. Math. Jpn. **32**, 693–696 (1987)
15. Iséki, K.: An algebra related with a propositional calculus. Proc. Jpn. Acad. **42**(1), 26–29 (1966)
16. Huang, Y.S.: BCI-Algebra. Science Press, Beijing (2006)
17. Santiago, R.H.N., Bedregal, B., Marcos, J., Caleiro, C., Pinheiro, J.: Semi-BCI Algebras. arXiv:1803.04808 [cs.LO] (2018)

A Fuzzy Based Recommendation System for Stock Trading

Érico Augusto Nunes Pinto[⊠], Leizer Schnitman,
and Ricardo A. Reis

Escola Politécnica, Universidade Federal Da Bahia,
Salvador, BA 40210-630, Brazil
ericonunes@gmail.com, leizer@ufba.br,
ricardo.andre.reis@gmail.com

Abstract. Predicting prices, as well as giving buy and sells recommendations in a Stock market, is considered difficult given the very complex behavior of the price itself. It is considered dynamic, non-linear and stochastic. In similarity to other fields, stock trading firms may profit from using computational intelligence systems, reducing the analysis time and enhancing the accuracy of recommendations. This paper shows a Fuzzy trading system based in technical analysis developed in two steps: primarily, a fuzzy trading system based on common technical indicators used by technical analysts is proposed, and after, the system is supported with a price prediction methodology. The first result shows that not all the stock tickers are eligible for a simple indicator operation, which may result in a severe loss. The usage of a price prediction methodology shows real improvement on the recommendation system.

Keywords: Fuzzy · Stock market · Forecasting · Recommendation systems

1 Introduction

Although the process of buying or selling a stock may be very simple, the task of choosing which stock to buy/sell is tiresome and challenging, also involving psychological factors and lifetime savings. To avoid risks people often recur to recommendations systems. The objective of this paper is to build a recommendation system based on a fuzzy algorithm that uses basically technical analysis.

Such systems may be comprised of simply following a trusted individual indication or analyzing common criteria often used by experts, such as fundamental analysis and technical analysis. Fundamental Analysis recommendations are drawn from company's operational periodic results and indicate their economical healthiness and effectiveness in pursuing goals. Technical analysis, on the other hand, relies only on candlestick chart patterns (figures built over time that indicates a probable movement) and indicators which are built over the price time series.

The recommendations systems based on trusted individual indications may be roughly split into paid recommendation services and discussion groups- that naturally occur for the investors to discuss, trade, invest, learn and share knowledge.

© Springer International Publishing AG, part of Springer Nature 2018
G. A. Barreto and R. Coelho (Eds.): NAFIPS 2018, CCIS 831, pp. 324–335, 2018.
https://doi.org/10.1007/978-3-319-95312-0_28

According to [1], their key components are the trustworthiness and expertise. These systems, however, are vulnerable to fake news and bad intentions, and may deliberately guide the unaware to enter a loss operation. Aiming to enhance confidence these authors developed a recommendation system based on trust analysis between entities, leading to a trust-based network after verified by a trusted agency. Another paper [2] uses tweets as a measure to forecast stock movements. Using data mining and hashtags during a short period of 2016 UK local elections, a good correlation (trustable model) was noted between Tweeter mood and changes in FTSE 100 index price. Big Data methods using financial website news were also applied [3] and shows good prediction capability. Hybrid methods [4] consisting of price time series also shows good performance.

Is worth mentioning the work made in [5] where an analysis of recommendations of Brazilian capital market analysts shows that, in the period from 2000 to 2010, a bias in favorable recommendations were measurable and also, that analysts were unable to identify another opportunity of greater return, confirming the effectiveness of analysts.

There are many works using fuzzy logic and fuzzy inference systems applied to stock trading strategies. A fuzzy metagraph was used in [6] to predict an occurrence of a buy/sell signal using MACD, RSI, EMA and SMA technical indicators, aiming at the future to apply it in daily operations. Also, [7] uses a fuzzy approach to classify data based on time, sequencing, and associative rules. They described the pattern of the prices rise and fall of a company in respect of another. Reversion patterns in Japanese candlestick charts were also studied in [8] with use of fuzzy systems. Open, low, high and close prices were converted to fuzzy candlestick chart, showing the evidence of a symptom sequence before a reversal point.

At first, to clarify the approach, a fuzzy system will classify a group of technical indicators and predict buy/sell operations for a list of BOVESPA stocks. After, it will be supported by a price prediction algorithm to compose a new recommendation system. The results of both systems will be compared.

The remainder of the paper is organized as follows: Sect. 2 describes the technical indicators used, the preliminary fuzzy system built and the result of its application. Section 3 describes the model used for price prediction and shows how it was used to assist the fuzzy system as well as the result. In Sect. 4 results will be compared and Sect. 5 provides an overall conclusion.

2 Momentum Indicators in Technical Analysis and Its Application into a Fuzzy System

According to [9] indicators are calculations based on the price and the volume traded over time and measures factors such as money flow, trends, volatility, and momentum. Indicators can be used as the basis for trading as they can show buy-and-sell signals. Their two main types are leading and lagging. Lagging indicators tracks price movement and therefore are always late as predictors, such as moving averages. Leading indicators gives predictive qualities, pointing a direction to price. In this work, only leading indicators that relate to price's momentum were used, applied as inputs on the fuzzy system. They are clarified in the following subsection. After, they are applied to a BOVESPA set of tickers to verify its effectiveness.

2.1   Momentum Indicators

Momentum indicators measure the rate price changes over time. This work used RSI, Williams %R and Stochastic Oscillator as momentum indicators.

RSI – Relative Strength Index – compares the magnitude of recent gains and losses over a specified time period to measure changes in price. It is primarily used to identify overbought and oversold conditions. It is calculated by the formula:

$$RSI = 100 - 100/(1 + RS) \tag{1}$$

where RS, during the time frame desired, is the average gain of up periods divided by an average loss in down periods. Traditional usage of RSI is to sell in overbought periods (RSI > 70) and buy in oversold periods (RSI < 30).

Stochastic Oscillator (%K) measures how strong a trend is by comparing the direction of closing prices to the direction of price movement. It is measured relative to past absolute maxima and minima, according to the formula:

$$\%K = 100 \cdot (\text{Actual Closing Price} - \text{Minima})/(\text{Maxima} - \text{Minima}) \tag{2}$$

It oscillates between 0 and 100 and the traditional usage signals overbought conditions above 80 (sell opportunity) and oversold below 20 (buy opportunity).

Williams %R also determine oversold and overbought conditions by measuring how the closing price is trending over the last periods, according to the formula:

$$\%R = -100 \cdot (\text{Maxima} - \text{Actual Closing Price})/(\text{Maxima} - \text{Minima}) \tag{3}$$

oscillating between 0 and −100. It indicates buy signals from −80 to −100 and sell signals from 0 to −20.

2.2   The Momentum Based Fuzzy System

Those 3 momentum indicators shown in the last subsection where used as inputs in a Fuzzy system. The objective was to build a structure capable of evaluating how strong the indication of buy and sell are, given the inputs. A set of membership functions were built to evaluate each of the indicators. With classifications combined, the system should supply the investor with a more robust recommendation. Figure 1 depicts the form of the membership functions.

The membership functions were designed with the MATLAB Fuzzy Logic Toolbox. For the sake of simplicity, they were taken as trapezoidal functions, assuming that not all investors decide to buy and sell exactly on the same value of the indicator, thus creating a blur region for the decision to be made. However, the very extremes and middles of the indicators are respected by traditional usage.

The set of rules qualifies a buy or sell signal according to the number of indicators pointing towards the same direction, e.g. if RSI and Williams are both in oversold region, while stochastic is in the middle, a weak buy is predicted. If all indicators are in the same region then a strong operation is predicted. Even though for each indicator listed there exists three possible actions, only 15 rules are meaningful from the whole

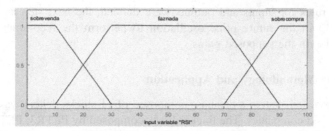

Fig. 1. Membership functions for RSI. Other indicators are also classified accordingly to traditional their traditional usage and trapezoidal membership functions.

27, since the they trend to the same directions. It is unlikely to have two indicators in oversold and other in the overbought region, so these possibilities were disconsidered. The output membership functions are depicted in Fig. 2.

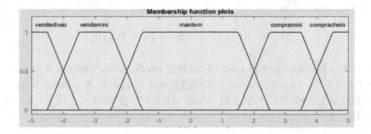

Fig. 2. Output's membership functions.

Figure 3 show the decision surface plotted pairing Stochastic with RSI. The other pairs - Stochastic with Williams and RSI with Williams - provide a similar surface, so they were omitted. A prediction of negative momentum indicates a sell signal while positive a buy signal. The absolute value of the output indicates the degree of robustness of the operation, given the regions the indicators are - strong ($M_t > 3$) and weak ($M_t < 3$), loosely. So, if the fuzzy system infers M_t as -2.47 the operation is said to be selling 2.47 standard set of stocks.

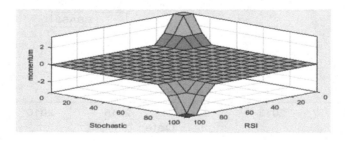

Fig. 3. Decision surface for RSI and stochastic indicators. Similar for other pairs.

After the rules for inputs and outputs were defined, the next step is to apply the multiplier (M_t) in the future price oscillation to perform the recommended trading operation and earn the supposed gains.

2.3   Trading Methodology and Application

To test the proposed system a database composed of 11 stocks tickers was taken from BOVESPA listed companies. The thicker list is composed by PETR3, PETR4, ITSA4, ITUB4, VALE3, VALE5, BBDC4, BBAS3, BVMF3, USIM5, and ABEV3. The data contains date and hour, open (P_o), close (P_c), maximum (P_{max}) and minimum (P_{min}) prices, and also volume (V) traded. The period tested is comprised of Jan 2nd 2012 to Dec 29th 2016.

To establish if the investor should buy or sell, a vector containing the information of the momentum indicators at time $t - 1$ is input into the system, which gives, as output, the operational multiplier M_t at time t. So, if one supposes an initial capital of $C_0 = $ R\$100,000.00, maximum capital allocation r = 20% and assume the investor would start the operation on open price and finishes at close price, a daily capital variation is expected to be

$$\Delta C_t = M_t \cdot C_{t-1} \cdot r \cdot (P_o - P_c)/P_o \tag{4}$$

Equation 4 is positive just in case of correct predictions. Figures 4 and 5 show the results for two of the tickers, namely BBAS3 and VALE3. A similar trend to Fig. 4 occurs for USIM5, PETR3, VALE5; while ABEV3, BVMF3, ITSA4 are similar to VALE3 in Fig. 5.

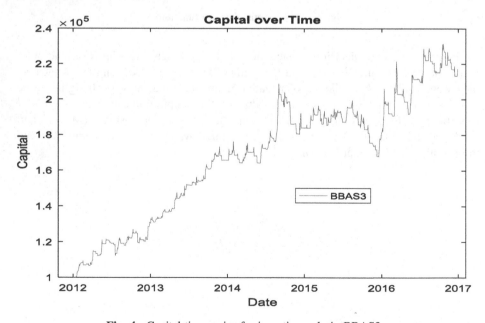

Fig. 4. Capital time series for investing only in BBAS3.

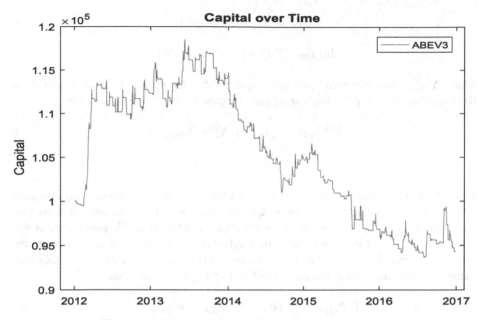

Fig. 5. Capital time series for investing only in VALE3.

The conclusion drawn from these preliminary results is that the strategy is not universal, lacking consistency, motivating the creation of a more robust fuzzy system that considers a price prediction, complementing the decisions of the first.

3 Price Prediction Methodology and the New Fuzzy System

A new parameter, derived from price and volume traded (P_{eq}), is explained in this section. It will be used as input in the new fuzzy system together with the outputs of the first and the on balance volume. The new outputs will be tested against the same market data in Sect. 2 to evaluate the new system performance.

3.1 Price Prediction

Predict the price of an asset just a time step after actual is a very complicated task. It is considered to be dynamic, highly non-linear and stochastic. Many papers [10–13] were published applying artificial neural networks and some other very powerful technics to perform the job.

In this paper, a prediction strategy considering the weighing of a moving average and an expected value is made, based on the price's Brownian movement. According to this kind of motion, the price variation of an asset Y at time t is given by Eq. 5, which solves into Eq. 6.

$$dY_t = Y_t(\mu dt + \sigma dW_t) \tag{5}$$

$$\ln(Y_{pred}/Y_t) = \mu - \sigma^2/2 + \sigma Wt \tag{6}$$

where Y_{pred} is the predicted price, μ is the drift, σ is the standard deviation and W_t a Brownian random step. In this paper, the expected value of Y_{t+1} is taken as

$$E[Y_{t+1}] = (Y_{t-1} + Y_t + Y_{pred})/3 \tag{7}$$

Because Eq. 6 overestimates the variations, a moving average is made in Eq. 7, thus smoothening the price series. Since a price is just one measure at which two investors wants to close a deal, its prediction for a trading system may become meaningless. Considering the stochastic nature of price movement, close, max and min prices are just one realization set of many possible for the process at that time. However, predict the price at which most deals will be made might give important information about the future and trend. To perform this task, a volume weighted mean of the dealt prices was made for each candlestick and named as P_{eq}. For P_{eq} Eq. 7 becomes:

$$E[P_{eq\,t+1}] = (P_{eq\,t-1} + P_{eq\,t} + P_{eq\,pred})/3 \tag{8}$$

3.2 On Balance Volume

On Balance Volume (OBV) is also a momentum indicator and measures the accumulated amount of capital traded. If after a trading period, the closing price ($P_{c,t}$) is lower than before ($P_{c,t-1}$), OBV is reduced by the amount traded. If $P_{c,t}$ is higher than $P_{c,t-1}$, OBV adds the amount. It is wise to note that its derivative direction normally coincides with price trend. Its derivative can be positive even if, at some period, the price went down. OBV's derivative is more important than the numerical value itself, given its cumulative nature. It's commonly accepted in technical trading that volume variations precedes price variation.

3.3 Trend Prediction and Application to the Fuzzy System

A simple estimate of price's second derivative at time t (d^2P_{eq}/dt^2) was made to be used as another input to the new fuzzy system. The second derivative's model was taken as

$$d^2P_{eq}/dt^2 = (P_{eq\,t-1} - 2P_{eq\,t} + P_{eq\,t+1})/2 \tag{9}$$

and it will serve as a prediction of price trends variation.

The block representation of the proposed recommendation system is given as follows in Fig. 6. The three inputs are evaluated simultaneously at each time to estimate the correct action at the beginning of next period: Momentum measured from the first system, trend taken by OBV's inclination and the second derivative of price's mean calculated by Eq. 9.

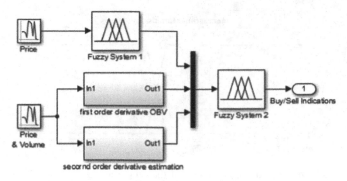

Fig. 6. Overview of the proposed system.

The input membership function for Fuzzy System 1 into Fuzzy System 2 is the same as in Fig. 2. For OBV's inclination and trend variations a relative measure is taken with respect to actual time OBV's and mean price, respectively. Their input membership functions are also trapezoidal had parameters estimated so any changes within 0.5% are considered neutral. Their membership functions can be seen in Fig. 7 while the output membership functions of system 2, which are also trapezoidal, can be seen in Fig. 8.

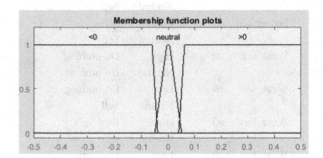

Fig. 7. Membership function of the OBV and $\left(d^2 P_{eq}/dt^2\right)$ at fuzzy system 2

For each pair of momentum measure and non-zero OBV's inclination two possible second order variations were evaluated to generate the output orders. For neutral trend (zero OBV's variation) and neutral trend variations one should go with the measured momentum. In general, the main strategy is to avoid going against a persisting trend motivated only by a strong momentum indication, what has been proved to be no better than a coin toss. Also using the momentum predictive capability together with the trend's variation $\left(d^2 P_{eq}/dt^2\right)$ allows one to make a correct operation on reversal movements. Lastly, if the two trends estimations are pointing to the same direction

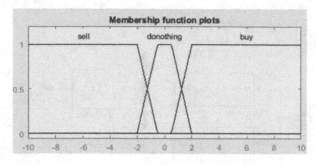

Fig. 8. Membership function of the output at fuzzy system 2

while no action is predicted by the momentum indicators, it is still possible to receive a trend following recommendation. Table 1 summarizes the relations used for input-output membership functions. Figure 9 shows the cumulative results of System 2 testing the same data for System 1.

Table 1. Laws for the second fuzzy system

Inputs			Output orders
M_t	dOBV/dt	d^2P_{eq}/dt^2	
Strong sell	>0	High	Do nothing
		Neutral	Sell
Strong sell	<0	Low	Sell
		Neutral	Sell
Weak sell	>0	High	Do nothing
		Neutral	Do nothing
Weak sell	<0	Low	Do nothing
		Neutral	Sell
Weak buy	>0	High	Buy
		Neutral	Buy
Weak buy	<0	Low	Do nothing
		Neutral	Buy
Strong buy	>0	High	Buy
		Neutral	Buy
Strong buy	<0	Low	Do nothing
		Neutral	Buy
Do nothing	>0	High	Buy
		Neutral	Buy
Do nothing	<0	Low	Sell
		Neutral	Sell
Strong sell	0	Neutral	Sell
Weak sell	0	Neutral	Sell
Weak buy	0	Neutral	Buy
Strong buy	0	Neutral	Buy

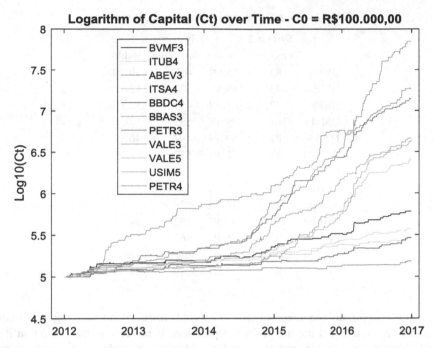

Fig. 9. Capital Time series for investing $0.2C_0$ only in one of the tickers depicted.

4 Systems Comparison

From Fig. 9 in the last section, it is possible to notice that system 2 performs well. The objective of this section is to quantify how well it performs by counting the number of recommendations made in the period and the number of positive results from them.

Since Eq. 4 can only be different from 0 when a recommendation occurs, it was used to count the number of recommendations and the positive variations (hits). The values of M_t may be summed for each set of recommendations. The total of correct recommendations greater than 50% does not guarantee positive earnings, since the absolute value of M_t in wrong recommendations may be greater. So, even if a correct sequence of small M_t operations ends in a large M_t wrong operation it is possible to lose money. This happened for tickers similar to that in Fig. 5, using system 1.

For system 2, the sum of M_t for all correct operations is much bigger, given the higher accuracy of the system. Table 2 summarizes the measurements.

For comparison reasons it is necessary to remember that both systems evaluated the same amount of possible operations. From Table 2 we conclude that not only the System 2 is more accurate but it also prevents losses being more conservative. The total number of operations possible in the time span for the tickers analyzed is equal to 1237 per ticker, or 13607 overall. System 2 indicates that only 3.5% to 15.5% operations per ticker may be profitable, while System 1 indicates 36.6% to 47.1%. In whole time spam, System 2 indicates only 10% of all operations for all tickers as profitable while System 1 indicates 42.9%.

Table 2. Summary of recommendations and statistics

	System 2		System 1	
	#recomm.	% hits	#recomm.	% hits
BVMF3	83	75.9%	453	49.2%
ITUB4	93	68.8%	532	52.8%
ABEV3	43	65.1%	457	50.1%
USIM5	181	79.0%	533	54.2%
ITSA4	81	59.3%	497	46.5%
BBDC4	85	62.4%	552	53.1%
BBAS3	172	76.7%	583	54.5%
PETR3	169	79.3%	575	54.6%
VALE3	142	76.1%	529	48.4%
VALE5	159	80.5%	563	54.5%
PETR4	192	85.4%	566	52.3%
	Sum	Mean	Sum	Mean
	1400	73.5%	5840	51.8%

So, it is also possible to infer that system 2 avoids losses by not entering in so many operations as System 1. Entering in fewer operations is not bad, in the sense that it is safer to keep money in the pocket instead of risking it at random. Of course, better would be to know for sure how and when to enter into a profitable operation, and for the data analyzed System 2 can perform with 73.5% accuracy.

5 Conclusion

In this paper, two fuzzy systems for the recommendation of financial operations were presented, both based on technical analysis. They were tested against real market data, spanning from Jan 2nd 2012 to Dec 29th 2016.

Since the strategy in System 1 was not so much better than random, it was upgraded to System 2. Both price prediction parameters added provided reliable indications that, together with a set of rules, achieved a high bias of profitable recommendations, and thus, high profitability on the long term.

References

1. Shankar, P.C., Vidyaraj, R., Kumar, K.S.: Trust-based stock recommendation system – a social network analysis approach. Procedia Comput. Sci. **46**, 299–305 (2015)
2. Nisar, T.M., Yeung, M.: Twitter as a tool for forecasting stock market movements: a short-window event study. J. Financ. Data Sci. **4**, 1–19 (2018)
3. Hayashi, A.H.: Processo para predição de preços das ações no mercado financeiro com o uso de Big Data. Dissertação de Mestrado. IPT-São Paulo (2017)
4. Enke, D., Grauer, M., Mehdiyev, N.: Stock market prediction with multiple regression, fuzzy type-2 clustering and neural networks. Procedia Comput. Sci. **6**, 201–206 (2011)

5. Antônio, R.M., Lima, F.G., Pimenta Junio, T.: Stock recommendations and investment portfolio formation: a study in the Brazilian market. Contaturia Administración **60**, 874–892 (2015)
6. Anbalagan, T., Maheswari, S.U.: Classification and prediction of stock market index based on fuzzy metagraph. Procedia Comput. Sci. **47**, 214–221 (2015)
7. Arafah, A.A., Mukhlash, I.: The application of fuzzy associative rule on co-movement analyse of indonesian stock price. Procedia Comput. Sci. **59**, 235–243 (2015)
8. Lan, Q., Zhang, D., Xiong, L.: Reversal pattern discovery in financial time series based on fuzzy candlestick lines. Syst. Eng. Procedia **2**, 182–190 (2011)
9. Investopedia Homepage. http://www.investopedia.com. Accessed 12 Dec 2018
10. Gomes, L.F.A.M., Machado, M.A.S., Caldeira, A.M., Santos, D.J., Nascimento, W.J.D.: Time series forecasting with neural networks and choquet integral. Procedia Comput. Sci. **91**, 1119–1129 (2016)
11. Rout, A.K., Dash, P.K., Dash, R., Bisoi, R.: Forecasting financial time series using a low complexity recurrent neural network and evolutionary learning approach. J. King Saud Univ. Comput. Inf. Sci. **29**, 536–552 (2017)
12. Falat, L., Pancikova, L.: Quantitative modelling in economics with advanced artificial neural networks. Procedia Econ. Financ. **34**, 194–201 (2015)
13. Lahmiri, S.: Wavelet low- and high-frequency components as features for predicting stock prices with backpropagation neural networks. J. King Saud Univ. Comput. Inf. Sci. **26**, 218–227 (2014)

Evolving Fuzzy Kalman Filter: A Black-Box Modeling Approach Applied to Rocket Trajectory Forecasting

Danúbia Soares Pires[✉] and Ginalber Luiz de Oliveira Serra

Department of Electroelectronics, Laboratory of Computational Intelligence Applied to Techonology, Federal Institute of Education, Science and Technology, São Luís, MA, Brazil
{danubiapires,ginalber}@ifma.edu.br

Abstract. A methodology to system identification based on Evolving Fuzzy Kalman Filter is proposed in this paper. The mathematical formulation using an evolving Takagi-Sugeno (TS) structure is presented: the offline Gustafson Kessel (GK) Algorithm is used to a initial data set; after that, an evolving GK algorithm estimate the antecedent parameters. A fuzzy version OKID (Observer/Kalman Filter Identification) algorithm is formulated to obtain the matrices A, B, C, D, and K (state matrix, input influence matrix, output influence matrix, direct transmission matrix, and Kalman gain matrix, respectively), recursively, composing the consequent parameters. Experimental results from black-box modeling applied to rocket trajectory forecasting show the efficiency and applicability of the proposed methodology.

Keywords: Black-box modeling · Evolving Fuzzy Kalman Filter Rocket trajectory

1 Introduction

Algorithms for state estimation aim to recover some desired state variables from a dynamic system with incomplete measurements and/or in the presence of noise. State estimation is a significant problem in the area of control and signal processing, with many research. In 1940, Wiener, founder of the modern theory of statistical estimation, established the Wiener filtering theory, which solves the problem of estimation with minimum variance for stationary stochastic processes. In the early 1960s, Kalman filtering theory was developed from a novel recursive filtering algorithm which does not require the assumption of stationarity, although it is applicable only to linear systems [1].

Several practical dynamic systems are nonlinear. Once that Kalman filter theory is applicable only to linear systems, many researchers have been motivated in extending Kalman filtering theory to non-linear systems. The Extended Kalman Filter (EKF), first proposed by Smith et al. [2] is an efficient recursive algorithm and widely used method for nonlinear systems estimation, but

© Springer International Publishing AG, part of Springer Nature 2018
G. A. Barreto and R. Coelho (Eds.): NAFIPS 2018, CCIS 831, pp. 336–347, 2018.
https://doi.org/10.1007/978-3-319-95312-0_29

it has a number of limitations: efficiency is guaranteed only in quasi-linear systems in the updates time scale; linearization can be only applied if the Jacobian matrix exists; computation of Jacobian matrices can be a very difficult process and susceptible to errors, in some applications; limitations in implementation, tuning and reliability. To overcome issues, Julier and Uhlmann proposed the UKF (Unscented Kalman Filter) [3]. It is a non-linear estimation method that propagates information about mean and covariance of parameters, recursively, through a non-linear transformation.

Althought the UKF is more accurate than EKF, it still has some limitations of the traditional KF: in some applications [4], an incompatibility between the noise characteristic of the actual process and the filter, is observed; added to this, UKF can respond to sudden disturbance slowly, decreasing filter performance. Attempting to overcome these limitations, many adaptive techniques of Kalman filtering, which guarantee robustness in the modeling of system uncertainties and perturbations, are observed in the literature: In [5], an adaptive KF approach is proposed for stochastic short-term traffic flow rate prediction; In [6], a study of satellite motion estimation algorithm is performed, with fault detection; In [7], an Asynchronous Adaptive Direct Kalman Filter (AADKF) algorithm for underwater integrated navigation system is developed.

Since 1980, fuzzy systems have been applied to dynamic systems modeling and control. Among fuzzy systems, there is a very important class called Takagi-Sugeno (TS). It has recently become a powerful tool for modeling and control. This is due to its structure based on rules as universal approximator of non-linearities and uncertainties [8,9]. It is observed the fuzzy systems in the KF literature for design of filters that require robustness in relation uncertainties and perturbations modeling. In [10], a scheme for improving the estimation accuracy and convergence speed for the agriculture industry, with an accurate estimation of the environmental changes, combining Kalman filter, fuzzy neural network with PID control algorithm, is proposed; In [11], a position estimation of AUV (Autonomous Underwater Vehicle) based on the Ensemble Kalman Filter (EnKF) and the Fuzzy Kalman Filter (FKF), is presented; In [12], the use of fuzzy based reasoning in conjunction with a Kalman filtering like approach in order to enhance the localization accuracy of mobile positioning in cellular network, is proposed.

In mid-2002, the evolving fuzzy systems emerged as a version of fuzzy systems with adaptive adjustment of parameters and structure [13,14]. Since then, this type of advanced fuzzy system has been of great interest to academy and industry. In [15], a new methodology for learning evolving fuzzy systems (EFS) from data streams in terms of online regression/system identification problems, is proposed. In [16], a new evolving fuzzy system referred to as evolving Heterogeneous Fuzzy Inference System (eHFIS), which can simultaneously perform local input selection and system identification in an evolving and integrative manner, is proposed. In [17], a Self-evolving Probabilistic Fuzzy Neural Network with Asymmetric Membership Function (SPFNN-AMF) controller for the position servo control of a Permanent Magnet Linear Synchronous Motor (PMLSM) servo drive system, is presented.

However, despite important contributions on fuzzy Kalman filters and Evolving Fuzzy Systems, the integration of both approaches is still open issue. Therefore, in this paper, a novel methodology for modeling of Evolving Fuzzy Kalman Filters from experimental data, is proposed. The mathematical formulation using an evolving Takagi-Sugeno (TS) structure is presented: the offline Gustafson Kessel (GK) algorithm is used for initial parametrization of antecedent terms of fuzzy Kalman filter inference system, considering an initial data set; and an evolving version of the GK algorithm is developed for online parametrization of antecedent of the fuzzy Kalman filter inference system. A fuzzy recursive version of OKID (Observer/Kalman Filter Identification) algorithm is proposed for parametrizing the matrices A, B, C, D and K (state matrix, input influence matrix, output influence matrix, direct transmission matrix, and Kalman gain matrix, respectively), in the consequent of the fuzzy Kalman filter inference system.

This paper is organized as follows: in Sect. 2, the formulation for evolving fuzzy Kalman filter parametric estimation, is presented, where the structure of the rule base is presented in the Sect. 2.1. The antecedent estimation by evolving GK clustering algorithm is formulated in the Sect. 2.2 and in the Sect. 2.3 the consequent estimation using OKID algorithm based on clustering, is proposed. In Sect. 3, experimental results show the efficiency and applicability of the proposed methodology, respectively. Finally, the Sect. 4 presents conclusions.

2   Evolving Fuzzy Kalman Filter Parametric Estimation: Formulation

2.1   Fuzzy Kalman Filter Model

The TS Evolving Fuzzy Kalman Filter presents the $i|^{[i=1,2,\cdots,c]}$-th rule, given by:

$R^{(i)}$: IF $\widetilde{\mathbf{Z}}_k$ IS $M^i_{j|\widetilde{\mathbf{Z}}_k}$ THEN

$$\hat{\mathbf{x}}^i_k = \mathbf{A}^i\hat{\mathbf{x}}^i_{k-1} + \mathbf{B}^i\widetilde{\mathbf{u}}^i_k + \mathbf{K}^i\epsilon_{r_k}$$

$$\widetilde{\mathbf{y}}^i_k = \mathbf{C}^i\hat{\mathbf{x}}^i_k + \mathbf{D}^i\widetilde{\mathbf{u}}^i_k + \epsilon_{r_k} \tag{1}$$

where $\mathbf{A}^i$, $\mathbf{B}^i$, $\mathbf{C}^i$, $\mathbf{D}^i$, and $\mathbf{K}^i$ (state matrix, input influence matrix, output influence matrix, direct transmission matrix, and Kalman gain matrix, respectively) are estimated by OKID algorithm based on clustering. The matrix $\widetilde{\mathbf{Z}}_k = [\widetilde{\mathbf{u}}_k \quad \widetilde{\mathbf{y}}_k]^T$ belongs to fuzzy set $M^i_{j|\widetilde{\mathbf{Z}}_k}$ with a value $\mu^i_{M_{j|\widetilde{\mathbf{Z}}_k}}$ defined by a membership function $\mu^i_{\widetilde{\mathbf{Z}}_k} : \mathbb{R} \rightarrow [0,1]$, with $\mu^i_{M_{j|\widetilde{\mathbf{Z}}_k}} \in \mu^i_{M_{1|\widetilde{\mathbf{Z}}_k}}, \mu^i_{M_{2|\widetilde{\mathbf{Z}}_k}}, \mu^i_{M_{3|\widetilde{\mathbf{Z}}_k}}, \cdots, \mu^i_{M_{p_{\widetilde{\mathbf{Z}}_k}|\widetilde{\mathbf{Z}}_k}}$, where $p_{\widetilde{\mathbf{Z}}_k}$ is the partitions number of the universe of discourse related to linguistic variable $\widetilde{\mathbf{Z}}_k$; and, a sequence of residual,

$$\epsilon_{r_k} = \mathbf{y}_k - \sum_{i=1}^{l} \gamma^i \widetilde{\mathbf{y}}_k^i, \text{ with } \sum_{i=1}^{l} \gamma^i \widetilde{\mathbf{y}}_k^i = \mathbf{C}^i \hat{\mathbf{x}}_k^i + \mathbf{D}^i \widetilde{\mathbf{u}}_k^i, \text{ where } \sum_{i=1}^{l} \gamma^i = 1.$$

2.2   Parametric Estimation of Antecedent

The antecedent estimation by evolving GK clustering algorithm is formulated in this section. Firstly, an offline approach of GK algorithm is formulated for initial estimation of antecedent parameters. After that, an evolving approach of GK algorithm is formulated for online estimation of antecedent parameters.

Initial Estimation: Offline

It is assumed that an offline clustering algorithm has been applied to identify an initial set of c clusters about previously collected data. The offline clustering algorithm used in this paper was Gustafson-Kessel [18], described as follow:

Given the data set $\mathbf{Z}$, an initial set of experimental data, the number of clusters $1 < c < N$, where c is the cluster number and N is the number of samples of data set $\mathbf{Z}$; the weighting exponent $m > 1$ and the termination tolerance $\epsilon > 0$. Randomly, the partition matrix is chosen, such that $\mathbf{U}^0 \in M_{fc}$, where M_{fc} is the set that represent the fuzzy partitioning space for $\mathbf{Z}$.

The cluster prototypes (means), $\mathbf{v}_i$, are computed as follow:

$$\mathbf{v}_i^{(l)} = \frac{\sum_{k=1}^{N} \left[\left(\mu_{ik}^{(l-1)} \right)^m \mathbf{z}_k \right]}{\sum_{k=1}^{N} \left(\mu_{ik}^{(l-1)} \right)^m}, \quad 1 \leq i \leq c \tag{2}$$

where $\mathbf{z}_k$ is the sample at instant k and μ_{ik} is its membership degree in the i-th cluster, at instant k. The cluster covariance matrices, $\mathbf{F}_i$, are computed, as follow:

$$\mathbf{F}_i = \frac{\sum_{k=1}^{N} \left[\left(\mu_{ik}^{(l-1)} \right)^m \left(\mathbf{z} - \mathbf{v}_i^{(l)} \right) \left(\mathbf{z} - \mathbf{v}_i^{(l)} \right)^T \right]}{\sum_{k=1}^{N} \left(\mu_{ik}^{(l-1)} \right)^m} \tag{3}$$

with $1 \leq i \leq c$. The GK algorithm employs an adaptive distance norm, in order to detect clusters of different geometrical shapes in a data set. Each cluster has its own norm-inducing matrix $\mathbf{O}_i$, which yields the following inner-product norm, computed by:

$$D_{ik\mathbf{O}_i} = \sqrt{\left(\mathbf{z}^{(k)} - \mathbf{v}_i^{(l)} \right)^T \mathbf{O}_i \left(\mathbf{z}^{(k)} - \mathbf{v}_i^{(l)} \right)}, \tag{4}$$

with $1 \leq i \leq c$ and $1 \leq k \leq N$. Finally, if $D_{ik\mathbf{O}_i} > 0$ for $1 \leq i \leq c, 1 \leq k \leq N$, the partition matrix is updated as follow:

$$\mu_{(i)}^{(k)} = \frac{1}{\sum_{j=1}^{c} \left(\frac{D_{(i)k\mathbf{A}_{(i)}}}{D_{(i)k\mathbf{A}_{(i)}}} \right)^{2/(m-1)}} \tag{5}$$

Evolving Estimation: Online

In the evolving Takagi-Sugeno (eTS) Fuzzy Systems, the density of the data evaluated recusively around the last data point, $D_t(z_t)$, is given by [19]:

$$D_t(z_t) = \frac{t-1}{(t-1)\left(\sum_{j=1}^{n+m} z_{tj}^2 + 1 \right) + b_t - 2 \sum_{j=1}^{n+m} z_{tj}g_{tj}} \tag{6}$$

where $D_1(z_1) = 1$; $t = 2, 3, \ldots$; $b_t = b_{t-1} + \sum_{j=1}^{n+m} z_{(t-1)j}^2$; $b_1 = 0$; $g_{tj} = g_{(t-1)j} + z_{(t-1)j}$; $g_{1j} = 0$; z_t is the data stream provided to algorithm; n is the dimension of the inputs vector; m is the dimension of the outputs vector; and, t is the number of points for which information about z is available.

The density of the focal points is uptaded recursively as follow:

$$D_t(z^{i*}) = \frac{t-1}{D^*} \tag{7}$$

where $i*$ corresponds to focal points of the ith fuzzy rule, and:

$$D^* = t - 1 + (t-2) \left(\frac{1}{D_{t-1}(z^{i*})} - 1 \right) + \sum_{j=1}^{n+m} \left(z_{tj} - z_{(t-1)j} \right) \tag{8}$$

Forming representative clusters with high generalization capability can be achieved by analyzing focal points that have high density, checking the Condition A, given by:

Condition A_1:

$$\eta D_t(z_t) > \max_{i=1}^{R} D_t(z_t^{i*});$$

$$\eta = \begin{cases} 1 & \mu_j^i(x_t) > e^{-2}, \forall i, \forall j \\ \dfrac{N_t - 3}{\log t} & otherwise \end{cases} \quad ; t = 2, 3, \ldots$$

Condition A_2:

$$D_t(z_t) > \max_{i=1}^{R} D_t(z_t^{i*}) \text{ OR } D_t(z_t) < \min_{i=1}^{R} D_t(z_t^{i*}).$$

If Condition A is attended, form a new focal point ($R \leftarrow R + 1$; $z^{i*} \leftarrow z^t$; $D(z^{i*}) \leftarrow 1$; $I^{i*} \leftarrow 1$).

To avoid redundancy and to control the level of overlap, condition B is checked. It is given by:

Condition B: IF $(\exists i, i = [1, R] : \mu_i^j(x_t) > e^{-1}, \forall j, j = [1, n], t = 2, 3, \ldots)$
THEN $(R \leftarrow R - 1)$

If Condition B is satisfied, remove the rule for which it holds, because this rule describes any of the previously existing cluster focal points.

2.3 Parametric Estimation of Consequent

The OKID (Observer/Kalman Filter Identification) method is a direct Kalman filter gain approach. It is a similar to an adaptive Kalman filter, which requires no prior statistical information and does not rely on sample correlation or covariance calculations [20, 21]. This method has been successfully applied to several real systems identification, and it can effectively identify state-space models using experimental data [22–24].

The original OKID is formulated in [25], but in this paper this algorithm is presented in the clustering context. A fuzzy formulation of the OKID (Observer/Kalman Filter Identification) algorithm, proposed in this paper, has the following steps: Given l (number of samples), p (appropriate number of observer Markov parameters from the given set of input-output data), $\mathbf{u}_{r \times l}$ (input data, r is the number of inputs), and $\mathbf{y}_{m \times l}$ (output data, m is the number of outputs).

Step 1: Compute the matrix of regressors, called V, given by:

$$\mathbf{V} = \begin{bmatrix} \mathbf{u}_0 & \mathbf{u}_1 & \ldots & \mathbf{u}_p & \ldots & \mathbf{u}_{l-1} \\ 0 & \mathbf{Z}_0 & \ldots & \mathbf{Z}_{p-1} & \ldots & \mathbf{Z}_{l-2} \\ 0 & 0 & \ldots & \mathbf{Z}_{p-2} & \ldots & \mathbf{Z}_{l-3} \\ 0 & 0 & \ddots & \vdots & \ldots & \vdots \\ 0 & 0 & \ddots & \mathbf{Z}_0 & \ldots & \mathbf{Z}_{l-p-1} \end{bmatrix} \tag{9}$$

where $\mathbf{Z}_k = [\mathbf{u}_k \quad \mathbf{y}_k]^T$ corresponds to input and output data at time k.

Step 2: Obtain from experimental data the Observer Markov Parameters $\tilde{\mathbf{Y}}$ based on fuzzy sets, as follow:

$$\tilde{\mathbf{y}} = \sum_{i=1}^{l} \tilde{\mathbf{Y}}^i \mathbf{V} \tag{10}$$

$$\tilde{\mathbf{Y}}^i = \left[\mathbf{V} \Gamma^i \mathbf{V}^T \right]^{-1} \mathbf{V}^T \Gamma^i \mathbf{y} \tag{11}$$

where $\mathbf{y} = [\mathbf{y}_0 \quad \mathbf{y}_1 \quad \ldots \quad \mathbf{y}_p \quad \ldots \quad \mathbf{y}_{l-1}]$ is the output matrix $m \times l$ of the dynamical system and

$$\Gamma^i = \begin{bmatrix} \gamma_0^i & 0 & \ldots & 0 \\ 0 & \gamma_1^i & \ldots & 0 \\ \vdots & \vdots & \ddots & \vdots \\ 0 & 0 & \ldots & \gamma_{l-1}^i \end{bmatrix} \tag{12}$$

is the diagonal weighting matrix of membership values from the i-th rule, and

$$\tilde{\mathbf{Y}}^i = [\mathbf{D}^i \left(\gamma^i \right) \quad \mathbf{C}^i \left(\gamma^i \right) \bar{\mathbf{B}}^i \left(\gamma^i \right) \quad \mathbf{C}^i \left(\gamma^i \right) \bar{\mathbf{A}}^i \left(\gamma^i \right) \bar{\mathbf{B}}^i \left(\gamma^i \right)$$

$$\dots \quad \mathbf{C}^i \left(\gamma^i \right) \bar{\mathbf{A}}^{i^{(p-1)}} \left(\gamma^i \right) \bar{\mathbf{B}}^i \left(\gamma^i \right)] \tag{13}$$

is the observer Markov Parameters of the i-th rule.

Step 3: Construct a block correlation matrix $\aleph_\tau^i$ with the elements $\mathbf{G}_{hh_{k+\tau}}^i$, which corresponds to product between $\mathbf{H}_\tau^i$ and $\mathbf{H}_{k+\tau}^i$, as follow:

$$\mathbf{G}_{hh_{k+\tau}}^i = \mathbf{H}_{k+\tau}^i \mathbf{H}_\tau^{i^T}, \text{ with } \tau = 0 \tag{14}$$

and

$$\mathbf{H}_{k+\tau}^i = \begin{bmatrix} \mathbf{Y}_{k+1}^i & \cdots & \mathbf{Y}_{k+\beta}^i \\ \mathbf{Y}_{k+2}^i & \cdots & \mathbf{Y}_{k+\beta+1}^i \\ \vdots & \ddots & \vdots \\ \mathbf{Y}_{k+\alpha}^i & \cdots & \mathbf{Y}_{k+\alpha+\beta+1}^i \end{bmatrix} \tag{15}$$

where $\mathbf{Y}_k^i$ is a matrix $m \times r$, whose columns are the Markov parameters (sampled pulse response) corresponding to m inputs. The size of $\mathbf{H}_k^i$ and $\mathbf{H}_0^i$ is $\alpha m \times \beta r$, since the size of $\mathbf{G}_{hh_k}^i$ is $\alpha m \times \alpha m$. So, it has:

$$\aleph_k^i = \begin{bmatrix} \mathbf{G}_{hh_k}^i & \cdots & \mathbf{G}_{hh_{k+\xi\tau}}^i \\ \mathbf{G}_{hh_{k+\tau}}^i & \cdots & \mathbf{G}_{hh_{k+(\xi+1)\tau}}^i \\ \vdots & \ddots & \vdots \\ \mathbf{G}_{hh_{k+\epsilon\tau}}^i & \cdots & \mathbf{G}_{hh_{k+(\epsilon+\xi)\tau}}^i \end{bmatrix} \tag{16}$$

$$= \begin{bmatrix} \mathbf{P}_\alpha^i \\ \mathbf{P}_\alpha^i \mathbf{A}^{i^\tau} \\ \vdots \\ \mathbf{P}_\alpha^i \mathbf{A}^{i^{\epsilon\tau}} \end{bmatrix} \mathbf{A}^{i^k} \left(\gamma^i \right) \begin{bmatrix} \mathbf{Q}_c^i & \mathbf{A}^{i^\tau} \mathbf{Q}_c^i & \cdots & \mathbf{A}^{i^{\xi\tau}} \mathbf{Q}_c^i \end{bmatrix} = \mathbf{P}_\epsilon^i \mathbf{A}^{i^k} \left(\gamma^i \right) \mathbf{Q}_\xi^i \tag{17}$$

Step 4: Decompose $\aleph_0^i$ using singular value decomposition, that is, $\aleph_0^i = \mathbf{R}^i \mathbf{\Sigma}^i \mathbf{S}^{i^T}$.

Step 5: Determine the order of the system by examining the singular values of Hankel matrix $\aleph_0^i$.

Step 6: Construct a minimum order realization $\left[\mathbf{A}^i, \mathbf{Q}_c^i, \mathbf{P}_\alpha^i \right]$ using a shifted block Hankel matrix $\aleph_1^i$ according to Eq. (14), as follow:

$$\mathbf{G}_{hh_1}^i = \mathbf{H}_1^i \mathbf{H}_0^{i^T} = \mathbf{P}_\alpha^i \mathbf{A}^{i^1} \left(\gamma^i \right) \mathbf{Q}_c^i \tag{18}$$

Step 7: Calculate the controllability matrix $\mathbf{Q}_\beta^i$ and determine a minimum order realization $[\mathbf{A}^i(\gamma^i), \mathbf{B}^i(\gamma^i), \mathbf{C}^i(\gamma^i)]$, as follow:

$$\mathbf{H}_0^i = \mathbf{P}_\alpha^i \mathbf{Q}_\beta^i \tag{19}$$

where:

$$\mathbf{P}_\alpha^i = \begin{bmatrix} \mathbf{C}^i(\gamma^i) \\ \mathbf{C}^i(\gamma^i)\,\mathbf{A}^i(\gamma^i) \\ \mathbf{C}^i(\gamma^i)\,\mathbf{A}^{i^2}(\gamma^i) \\ \vdots \\ \mathbf{C}^i(\gamma^i)\,\mathbf{A}^{i^{\alpha-1}}(\gamma^i) \end{bmatrix} \tag{20}$$

$$\mathbf{Q}_\beta^i = [\mathbf{B}^i(\gamma^i) \quad \mathbf{A}^i(\gamma^i)\,\mathbf{B}^i(\gamma^i) \quad \cdots \quad \mathbf{A}^{i^{\beta-2}}(\gamma^i)\,\mathbf{B}^i(\gamma^i) \quad \mathbf{A}^{i^{\beta-1}}(\gamma^i)\,\mathbf{B}^i(\gamma^i)] \tag{21}$$

and

$$\mathbf{Q}_\beta^i = \mathbf{P}_\alpha^{i^+} \mathbf{H}_0^i \tag{22}$$

$$\mathbf{A}^i = \left(\mathbf{\Sigma}^i\right)^{-1/2} \mathbf{R}_n^{i^T} \mathbf{H}_1^i \mathbf{S}_n^i \left(\mathbf{\Sigma}^i\right)^{-1/2} \tag{23}$$

$$\mathbf{D}^i = \bar{\mathbf{Y}}_{m\times 1}^i; \quad \mathbf{B}^i = \text{first } r \text{ columns of } \mathbf{Q}_\beta^i; \quad \mathbf{C}^i = \text{first } m \text{ rows of } \mathbf{P}_\alpha^i \tag{24}$$

Step 8: Compute the Kalman gain matrix:

$$\mathbf{K}^i = -\left[\psi_p^{i^T} \psi_p^i\right]^{-1} \psi_p^{i^T} \tilde{\mathbf{Y}}_k^{i^0} \tag{25}$$

where

$$\psi_p = \begin{bmatrix} \mathbf{C}^i \\ \mathbf{C}^i \mathbf{A}^i \\ \vdots \\ \mathbf{C}^i \mathbf{A}_{p-1}^i \end{bmatrix}; \quad \tilde{\mathbf{Y}}^{i^0} = \begin{bmatrix} \mathbf{C}^i \\ \mathbf{C}^i \mathbf{A}^i \\ \vdots \\ \mathbf{C}^i \mathbf{A}_{p-1}^i \end{bmatrix} \tag{26}$$

$$\tilde{\mathbf{Y}}_k^{i^0} = \tilde{\tilde{\mathbf{Y}}}^{i^{(2)}} + \sum_{i=1}^{k-1} \tilde{\mathbf{Y}}_{k-i}^{i^0} \tilde{\tilde{\mathbf{Y}}}^{i^{(2)}}, \qquad k = 1, 2, \ldots, p \tag{27}$$

$$\tilde{\tilde{\mathbf{Y}}}^{i^{(2)}} = [0 \quad -\text{third column of } \tilde{\tilde{\mathbf{Y}}}^i \quad -\text{fifth column of } \tilde{\tilde{\mathbf{Y}}}^i \quad \cdots \quad 0] \tag{28}$$

3   Experimental Results

To illustrate the applicability of the proposed methodology, consider a black-box modeling applied to rocket trajectory forecasting. The real experimental data was obtained from a rocket FTI or Fogtrein-I (*Foguetes de Treinamento - Intermediário*, in portuguese), observed in Fig. 1. It is a medium vehicle of a family of Brazilian training rockets created jointly by the Air Force Command (COMAER) and Avibrá, with a diameter of 0.3 m and a length of 5.4 m (the 490 kg rocket can achieve an apogee height of 60 km). These vehicles are used for testing, qualification, training and are designed to be launched in adverse conditions such as high salinity, winds up to 10 m/s and rain up to 10 mm/h. All models admit payloads (5 kg to 30 kg), with electric networks and telemetry equipment, along experiments of interest in the academic and scientific community. The mission of FTI is to be a model for operational training of a launch center, in isolation, without participation of the remote station for redundant monitoring vehicle, with telemetry in S-band, C-band transponders, flight termination, and height above 60 km [26].

The FTI outputs are pitch angle $\mathbf{y}_1$, yaw angle $\mathbf{y}_2$ and distance $\mathbf{y}_3$. In order to estimate the fuzzy sets, the offline GK Clustering Algorithm was implemented for 2 clusters, and tolerance 10^{-6}, using the 82 first samples. From the output data obtained by FTI, from the time step 83 to 3500 the evolving GK Clustering Algorithm, described in Sect. 2, is applied to identify the fuzzy Kalman filter, in every new sample of experimental data, taking into account the weights of fuzzy sets. The TS Evolving Fuzzy Kalman Filter presents the $i|^{[i=1,2,\cdots,c]}$-th rule, given by:

$R^{(i)}$: IF $\widetilde{\mathbf{Z}}_k$ IS $M^i_{j|\widetilde{\mathbf{z}}_k}$ THEN

$$\hat{\mathbf{x}}^i_k = \mathbf{A}^i \hat{\mathbf{x}}^i_{k-1} + \mathbf{B}^i \mathbf{u}_k + \mathbf{K}^i \epsilon_{r_k}$$

$$\mathbf{y}^i_k = \mathbf{C}^i \hat{\mathbf{x}}^i_k + \mathbf{D}^i u_k + \epsilon_{r_k} \tag{29}$$

where $\widetilde{\mathbf{Z}}_k = \begin{bmatrix} \widetilde{\mathbf{y}}_{1_{k-1}} & \widetilde{\mathbf{y}}_{2_{k-1}} & \widetilde{\mathbf{y}}_{3_{k-1}} & \widetilde{\mathbf{y}}_{1_k} & \widetilde{\mathbf{y}}_{2_k} & \widetilde{\mathbf{y}}_{3_k} \end{bmatrix}^T$ and $\epsilon_{r_k} = \mathbf{y}_k - \sum\limits_{i=1}^{l} \hat{\mathbf{y}}^i_k$ is a sequence of residual, with $\hat{\mathbf{y}}^i_k = \mathbf{C}^i \hat{\mathbf{x}}^i_k + \mathbf{D}^i \mathbf{u}_k$.

The outputs obtained by FTI, that are pitch angle $\mathbf{y}_1$, yaw angle $\mathbf{y}_2$ and distance $\mathbf{y}_3$, to the proposed, are shown in Fig. 2. It is observed that the proposed methodology follows the system outputs. The number of rules, the degrees during model evolution and the absolute error between real distance and estimated distance by proposed methodology, are observed in Fig. 3.

Fig. 1. A rocket FTI [26].

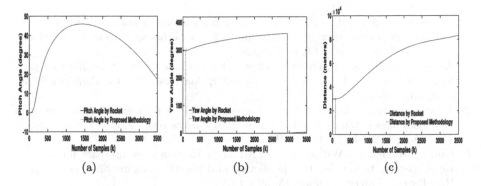

Fig. 2. (a) Pitch angle y_1; (b) yaw angle y_2; (c) distance y_3.

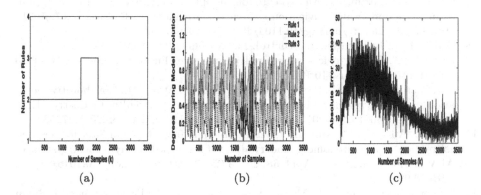

Fig. 3. (a) Number of rules; (b) degrees during model evolution; (c) absolute error between real distance and estimated distance by proposed methodology.

4   Conclusion

The methodology presented in this paper to system identification based on Evolving Fuzzy Kalman Filter applied to rocket trajectory forecasting allowed to adapt the model and Kalman Filter structure in real time. Experimental results from a black-box modeling with real experimental data of a rocket to trajectory forecasting show efficiency and applicability of the proposed methodology.

Acknowledgment. This work was supported by FAPEMA, IFMA and encouraged by Ph.D. Program in Electrical Engineering of Federal University of Maranhão (PPGEE/UFMA).

References

1. Kalman, R.E.: A new approach to linear filtering and prediction problems. Trans. ASME **82**, 35–45 (1960)
2. Smith, G.L., Schmidt, S.F., McGee, L.A.: Application of statistical filter theory to the optimal estimation of position and velocity on board a circumlunar vehicle. NASA, Technical report TR R-135 (1962)
3. Julier, S., Uhlmann, J., Durrant-Whyte, H.F.: A new approach for filtering nonlinear systems. In: Proceedings of the IEEE American Control Conference, pp. 1628–1632 (1995)
4. Teixeira, B.O.S., Aguirre, L.A., Tôrres, L.A.B.: Filtragem de Kalman com restrições para sistemas não-lineares: revisão e novos resultados. Revista Controle e Automação **21**(2), 28–146 (2010)
5. Guo, J., Huang, W., Williams, B.M.: Adaptive Kalman filter approach for stochastic short-term traffic flow rate prediction and uncertainty quantification. Transp. Res. Part C: Emerg. Technol. **43**, 50–64 (2014)
6. Hajiyev, C., Soken, H.E.: Robust adaptive unscented Kalman filter for attitude estimation of pico satellites. Int. J. Adapt. Signal Process. 107–120 (2014). https://doi.org/10.1002/acs.2393
7. Davari, N., Gholami, A.: An asynchronous adaptive direct Kalman filter algorithm to improve underwater navigation system performance. IEEE Sens. J. **17**, 1061–1068 (2016). https://doi.org/10.1109/JSEN.2016.2637402
8. Zadeh, L.A.: Information and control. Fuzzy Sets **8**, 338–353 (1965)
9. Zadeh, L.A.: Fuzzy logic - a personal perspective. Fuzzy Sets Syst. **281**, 4–20 (2015). https://doi.org/10.1016/j.fss.2015.05.009
10. Bai, X., Liu, L., Cao, M., Panneerselvam, J., Sun, Q., Wang, H.: Collaborative actuation of wireless sensor and actuator networks for the agriculture industry. IEEE Access **5**, 13286–13296 (2017). https://doi.org/10.1109/ACCESS.2017.2725342
11. Ngatini, Apriliani, E., Nurhadi, H.: Ensemble and fuzzy Kalman filter for position estimation of an autonomous underwater vehicle based on dynamical system of AUV motion. Expert Syst. Appl. **68**, 29–35 (2017). https://doi.org/10.1016/j.eswa.2016.10.003
12. Bouzera, N., Oussalah, M., Mezhoud, N., Khireddine, A.: Fuzzy extended Kalman filter for dynamic mobile localization in urban area using wireless network. Appl. Soft Comput. **57**, 452–467 (2017). https://doi.org/10.1016/j.asoc.2017.04.007

13. Angelov, P., Lughofer, E., Zhou, X.: Evolving fuzzy classifiers using different model architectures. Fuzzy Sets Syst. **159**, 3160–3182 (2008). https://doi.org/10.1016/j. fss.2008.06.019
14. Lughofer, E.: FLEXFIS: a robust incremental learning approach for evolving Takagi-Sugeno fuzzy models. IEEE Trans. Fuzzy Syst. **16**(6), 1393–1410 (2008). https://doi.org/10.1109/TFUZZ.2008.925908
15. Lughofer, E., Cernuda, C., Kindermann, S., Pratama, M.: Generalized smart evolving fuzzy systems. Evol. Syst. **6**(4), 269–292 (2015). https://doi.org/10.1007/s12530-015-9132-6
16. Alizadeh, S., Kalhor, A., Jamalabadi, H., Araabi, B.N., Ahmadabadi, M.N.: Online local input selection through evolving heterogeneous fuzzy inference system. IEEE Trans. Fuzzy Syst. **24**, 1364–1377 (2016). https://doi.org/10.1109/TFUZZ.2016.2516580
17. Chen, S., Liu, T.: Intelligent tracking control of a PMLSM using self-evolving probabilistic fuzzy neural network. IET Electr. Power Appl. **11**, 1043–1054 (2017). https://doi.org/10.1049/iet-epa.2016.0819
18. Babuska, R.: Fuzzy Modeling Control. Kluwer Academic Publishers, Dordrecht (1998)
19. Angelov, P.: An approach for fuzzy rule-base adaptation using on-line clustering. Int. J. Approx. Reason. **35**, 275–289 (2004). https://doi.org/10.1016/j.ijar.2003.08.006
20. Juang, J.N.: Applied System Identification. Prentice Hall, Englewood Cliffs (1994)
21. Juang, J.N., Pappa, R.S.: An eigensystem realization algorithm for modal parameter identification and model reduction. J. Guid. Control Dyn. **8**(5), 620–627 (1985)
22. Chien, T., Chen, Y.: An on-line tracker for a stochastic chaotic system using observer/Kalman filter identification combined with digital redesign method. Algorithms **10**, 25 (2017). https://doi.org/10.3390/a10010025
23. Wu, C.Y., Tsai, J.S.-H., Guo, S.-M., Shieh, L.-S., Canelon, J.I., Ebrahimzadeh, F., Wang, L.: A novel on-line observer/Kalman filter identification method and its application to input-constrained active fault-tolerant tracker design for unknown stochastic systems. J. Frankl. Inst. **352**, 1119–1151 (2015). https://doi.org/10.1016/j.jfranklin.2014.12.004
24. Oh, S.-K., Lee, J.M.: Stochastic iterative learning control for discrete linear time-invariant system with batch-varying reference trajectories. J. Process Control **36**, 64–78 (2015). https://doi.org/10.1016/j.jprocont.2015.09.008
25. Cheng, C.W., Huang, J.K., Phan, M., Juang, J.N.: Integrated system identification and modal state estimation for control of large flexible space structures. J. Guidance Control Dyn. **15**(1), 88–95 (1992)
26. Gunter's Space Page: FTI. http://space.skyrocket.de/doc_lau/fti.htm. Accessed 21 Mar 2018

Crisp Fuzzy Implications

Jocivania Pinheiro[1,2]($\boxtimes$), Benjamin Bedregal[2], Regivan Santiago[2],
and Helida Santos[3]

[1] Center of Exact and Natural Sciences,
Rural Federal University of SemiArid (UFERSA), Mossoró, RN, Brazil
`vaniamat@ufersa.edu.br`
[2] Department of Informatics and Applied Mathematics,
Federal University of Rio Grande do Norte (UFRN), Natal, RN, Brazil
`{bedregal,regivan}@dimap.ufrn.br`
[3] Centro de Ciências Computacionais, Federal University of Rio Grande (FURG),
Rio Grande, RS, Brazil
`helida@furg.br`

Abstract. Implications play an important role in fuzzy logics as they can be used both in practical and theoretical works. There exist many works in the literature where fuzzy implications behave in a crisp manner, i.e., implications that map to either zero or one. In this sense, we call those implications as crisp fuzzy implications and our goal is to study some their main features.

1 Introduction

A great deal of studies involving fuzzy implications can be found in the literature over the last years [1,2,4–6,12]. Fuzzy implications are interesting from the theoretical point of view to its use on a variety of applications. For instance, they can be used to perform any fuzzy "if-then" rule in fuzzy systems and inference processes, which basically combine membership functions with the control rules to derive the fuzzy output. Regarding the theoretical aspect, many works have also been done aiming to generalize the traditional implication into fuzzy logic, explaining why there exists so many classes of fuzzy implications. The existence of those classes of fuzzy implications is justified by the fact that depending on the context or/and on the rules and their behavior, different implications with different properties can be adequate.

In the literature, it is possible to find examples of fuzzy implications with a crisp behavior, i.e., fuzzy implications that always map to either 0 or 1. For instance, in [13], it was defined two crisp-valued operators, named standard sharp and standard strict, as follows:

1. Standard sharp

$$I_s(x,y) = \begin{cases} 1, & \text{if } x < 1 \text{ or } y = 1 \\ 0, & \text{otherwise} \end{cases}$$

© Springer International Publishing AG, part of Springer Nature 2018
G. A. Barreto and R. Coelho (Eds.): NAFIPS 2018, CCIS 831, pp. 348–360, 2018.
https://doi.org/10.1007/978-3-319-95312-0_30

2. Standard strict (also called Rescher-Gaines implication [8])

$$I_G(x,y) = \begin{cases} 1, & \text{if } x \leq y \\ 0, & \text{otherwise} \end{cases}$$

Those implications were used in various applications, for instance, in the domains of approximate reasoning [8], relational databases [9], fuzzy control [10,16], face recognition [17].

Thus, in this paper we intend to study the class of crisp fuzzy implications, i.e. fuzzy implications which always map to 0 or 1.

The paper is organized as follows. Section 2 summarizes some of the basic concepts demanded to understand the proposal in this work, including the concept of fuzzy implication and related properties. The study of crisp fuzzy implications is done in Sect. 3, including the most important results. At last, we finish in Sect. 4 with our final conclusions.

2   Preliminaries

Definition 1. *A function $T : [0,1]^2 \to [0,1]$ is said to be a **triangular norm** (**t-norm**, for short) if it satisfies the following conditions, for all $x, y, z \in [0,1]$:*

(T1) Symmetry: $T(x,y) = T(y,x)$;
(T2) Associativity: $T(x, T(y,z)) = T(T(x,y), z)$;
(T3) Monotonicity: If $x_1 \leq x_2$ and $y_1 \leq y_2$ then $T(x_1, y_1) \leq T(x_2, y_2)$;
(T4) 1-identity: $T(x,1) = x$.   *(boundary condition)*

In fuzzy logic, the conjunction is often represented by a t-norm. The standard fuzzy conjunction $T_M : [0,1]^2 \to [0,1]$ given by $T_M(x,y) = min\{x,y\}$ is the only idempotent t-norm (see [11] - Theorem 3.9).

Proposition 1 *[3]. Let T be a t-norm. Then $T(0,y) = 0$, for each $y \in [0,1]$.*

Definition 2. *A t-norm T is called **positive** if satisfies the following condition: $T(x,y) = 0$ if and only if $x = 0$ or $y = 0$.*

Definition 3. *A function $S : [0,1]^2 \to [0,1]$ is said to be a **triangular conorm** (**t-conorm**, for short) if it satisfies the following conditions, for all $x, y, z \in [0,1]$:*

(S1) Symmetry: $S(x,y) = S(y,x)$;
(S2) Associativity: $S(x, S(y,z)) = S(S(x,y), z)$;
(S3) Monotonicity: If $x_1 \leq x_2$ and $y_1 \leq y_2$ then $S(x_1, y_1) \leq S(x_2, y_2)$;
(S4) 0-identity: $S(x,0) = x$.   *(boundary condition)*

From an axiomatic point of view, the difference between t-norms and t-conorms is just their boundary conditions.

Definition 4. *A function* $N : [0,1] \to [0,1]$ *is called a **fuzzy negation** if*

(N1) N is antitonic, i.e. $N(x) \leq N(y)$ whenever $y \leq x$;
(N2) $N(0) = 1$ and $N(1) = 0$.
   *A fuzzy negation N is **strict** if*
(N3) N is continuous and
(N4) $N(x) < N(y)$ whenever $y < x$.
   *A fuzzy negation N is **strong** if*
(N5) $N(N(x)) = x$, for each $x \in [0,1]$.
   *A fuzzy negation N is **crisp** if*
(N6) $N(x) \in \{0,1\}$, for all $x \in [0,1]$ (see [7]).
   *A fuzzy negation N is **frontier** if it satisfies the following property:*
(N7) $N(x) \in \{0,1\}$ if and only if $x = 0$ or $x = 1$.

Remark 1. By [7], a fuzzy negation $N : [0,1] \to [0,1]$ is crisp if and only if there exists $\alpha \in [0,1)$ such that $N = N_\alpha$ or there exists $\alpha \in (0,1]$ such that $N = N^\alpha$, where

$$N_\alpha(x) = \begin{cases} 0, & \text{if } x > \alpha \\ 1, & \text{if } x \leq \alpha \end{cases} \qquad N^\alpha(x) = \begin{cases} 0, & \text{if } x \geq \alpha \\ 1, & \text{if } x < \alpha \end{cases}$$

Theorem 1 *[1]. If a function $N : [0,1] \to [0,1]$ satisfies (N1) and (N5), then it also satisfies (N2) and (N3). Moreover, N is a bijection, i.e., it satisfies (N4).*

Corollary 1 *[1]. Every strong negation is strict.*

Definition 5. *A function $I : [0,1]^2 \to [0,1]$ is a **fuzzy implication** if the following properties are satisfied, for all $x, y, z \in [0,1]$:*

(I1) If $x \leq z$ then $I(x,y) \geq I(z,y)$;
(I2) If $y \leq z$ then $I(x,y) \leq I(x,z)$;
(I3) $I(0,y) = 1$;
(I4) $I(x,1) = 1$;
(I5) $I(1,0) = 0$.

The set of all fuzzy implications will be denoted by $\mathcal{FI}$.

Definition 6. *Let $I \in \mathcal{FI}$. The function $N_I : [0,1] \to [0,1]$ defined by $N_I(x) = I(x,0)$, $x \in [0,1]$ is called the **natural negation of** I or the negation induced by I.*

Definition 7. *Let T be a t-norm, S a t-conorm and N a fuzzy negation, then:*

– A function $I : [0,1]^2 \to [0,1]$ is called a **(S, N)-implication** (denoted by $I_{S,N}$) if $I(x,y) = S(N(x),y)$.
– A function $I : [0,1]^2 \to [0,1]$ is called an **R-implication** (denoted by I_T) if $I(x,y) = \sup\{t \in [0,1] \mid T(x,t) \le y\}$.
– A function $I : [0,1]^2 \to [0,1]$ is called a **QL-implication** (denoted by $I_{S,N,T}$) if $I(x,y) = S(N(x), T(x,y))$.
– A function $I : [0,1]^2 \to [0,1]$ is called a **D-implication** (denoted by $I_{S,T,N}$) if $I(x,y) = S(T(N(x), N(y)), y)$.

Definition 8. *A fuzzy implication I is said to satisfy:*

(i) *the **exchange principle** if, for all $x,y,z \in [0,1]$*
$$I(x, I(y,z)) = I(y, I(x,z)); \tag{EP}$$
(ii) *the **left neutrality property**, if*
$$I(1,y) = y, \qquad y \in [0,1]; \tag{NP}$$
(iii) *the **identity principle**, if*
$$I(x,x) = 1, \qquad x \in [0,1]; \tag{IP}$$
(iv) *the **left-ordering property** if, for all $x,y \in [0,1]$*
$$I(x,y) = 1 \text{ whenever } x \le y; \tag{LOP}$$
(v) *the **right-ordering property** if for all $x,y \in [0,1]$*
$$I(x,y) \ne 1 \text{ whenever } x > y. \tag{ROP}$$

Definition 9. *Let $I \in \mathcal{FI}$ and let N be a fuzzy negation. I is said to satisfy the:*

(i) ***contraposition law** (or in other words, the contrapositive symmetry) with respect to N, if*
$$I(x,y) = I(N(y), N(x)), \ x,y \in [0,1]; \tag{CP}$$
(ii) ***left contraposition law** with respect to N, if*
$$I(N(x), y) = I(N(y), x), \ x,y \in [0,1]; \tag{L-CP}$$
(iii) ***right contraposition law** with respect to N, if*
$$I(x, N(y)) = I(y, N(x)), \ x,y \in [0,1]. \tag{R-CP}$$

If I satisfies the (left, right) contrapositive symmetry with respect to a specific N, then we will denote this by $L{-}CP(N)$, $R{-}CP(N)$ and $CP(N)$, respectively.

In [14,15], Pinheiro et al. introduced a new class of implication, named (T,N)-implications which was defined by means of fuzzy negations and a t-norm.

Definition 10 *[15]. Let N and N' be fuzzy negations and T be a t-norm. The function $I_{T,N}^{N'}$ defined by $I_{T,N}^{N'}(x,y) = N'(T(x, N(y)))$ is called a **($\mathbf{N'}, \mathbf{T}, \mathbf{N}$)-implication.***

Actually, in [15] for (T,N)-implications we had $N' = N$ which is different from the previous definition where we have different negations. In order to avoid misunderstanding between definitions, from here forth, implications defined according to Definition 10 we be called (N', T, N)-implications.

3   Crisp Fuzzy Implications

In classical logic there is only one bivalent implication, however in the fuzzy setting the notion of bivalence gives rise to an uncountable family of such implications. They are called here Crisp fuzzy implication.

Definition 11. *Let $I : [0,1]^2 \to [0,1]$ be a fuzzy implication. We say that I is a* **crisp fuzzy implication** *if $I(x,y) \in \{0,1\}$ for all $x,y \in [0,1]$.*

Proposition 2. *Let $I : [0,1]^2 \to [0,1]$ be a fuzzy implication. Then I is crisp if and only if one of the following conditions are satisfied, for all $x,y \in [0,1]$:*

(C1) If there exists $\alpha \in (0,1]$ and $\beta \in [0,1)$ such that $I(x,y) = I_{\alpha,\beta}(x,y)$, where

$$I_{\alpha,\beta}(x,y) = \begin{cases} 0, & \text{if } x \geq \alpha \text{ and } y \leq \beta \\ 1, & \text{otherwise} \end{cases};$$

(C2) If there exists $\alpha \in [0,1)$ and $\beta \in (0,1]$ such that $I(x,y) = I^{\alpha,\beta}(x,y)$, where

$$I^{\alpha,\beta}(x,y) = \begin{cases} 0, & \text{if } x > \alpha \text{ and } y < \beta \\ 1, & \text{otherwise} \end{cases};$$

(C3) If there exists $\alpha, \beta \in (0,1]$ such that $I(x,y) = I_{\alpha}{}^{\beta}(x,y)$, where

$$I_{\alpha}{}^{\beta}(x,y) = \begin{cases} 0, & \text{if } x \geq \alpha \text{ and } y < \beta \\ 1, & \text{otherwise} \end{cases};$$

(C4) If there exists $\alpha, \beta \in [0,1)$ such that $I(x,y) = I^{\alpha}{}_{\beta}(x,y)$, where

$$I^{\alpha}{}_{\beta}(x,y) = \begin{cases} 0, & \text{if } x > \alpha \text{ and } y \leq \beta \\ 1, & \text{otherwise} \end{cases}.$$

Proof. First, suppose I is crisp, then as $I(0,0) = 1$ and $I(1,0) = 0$, we have by (I1) that there exists: (1) $\alpha \in (0,1]$ such that $I(x,0) = \begin{cases} 0, & \text{if } x \geq \alpha \\ 1, & \text{otherwise} \end{cases}$ or (2) $\alpha \in [0,1)$ such that $I(x,0) = \begin{cases} 0, & \text{if } x > \alpha \\ 1, & \text{otherwise} \end{cases}$. By (I2), we have for case (1) that there exist: $(i)_1$ $\beta \in [0,1)$ such that $I(x,y) = \begin{cases} 0, & \text{if } x \geq \alpha \text{ and } y \leq \beta \\ 1, & \text{otherwise} \end{cases}$ or $(ii)_1$ $\beta \in (0,1]$ such that $I(x,y) = \begin{cases} 0, & \text{if } x \geq \alpha \text{ and } y < \beta \\ 1, & \text{otherwise} \end{cases}$. Hence, $I(x,y) = I_{\alpha,\beta}(x,y)$ or $I(x,y) = I_{\alpha}{}^{\beta}(x,y)$, for all $x,y \in [0,1]$, respectively.

Similarly, still by (I2), for case (2) there exist: $(i)_2$ $\beta \in [0,1)$ such that
$$I(x,y) = \begin{cases} 0, & \text{if } x > \alpha \text{ and } y \leq \beta \\ 1, & \text{otherwise} \end{cases} \text{ or } (ii)_2 \ \beta \in (0,1] \text{ such that } I(x,y) =$$
$\begin{cases} 0, & \text{if } x > \alpha \text{ and } y < \beta \\ 1, & \text{otherwise} \end{cases}$. Hence, $I(x,y) = I^{\alpha}{}_{\beta}(x,y)$ or $I(x,y) = I^{\alpha,\beta}(x,y)$,
for all $x,y \in [0,1]$, respectively.

The reciprocal case follows straightforward.

Definition 12. *Let I be a crisp fuzzy implication. Independently from I being of type C1, C2, C3 or C4, the pair (α,β) is called the **threshold pair of** I.*

In the following proposition, we can observe that we can obtain a crisp fuzzy implications from any fuzzy implication I.

Proposition 3. *Let $I \in \mathcal{FI}$. Then, for any $\gamma \in (0,1]$, $I_\gamma(x,y) = \begin{cases} 1, & \text{if } I(x,y) \geq \gamma \\ 0, & \text{if } I(x,y) < \gamma \end{cases}$ is a crisp fuzzy implication.*

Proof. We will first prove that I_γ satisfies the conditions demanded in Definition 5. Indeed,

(I_γ1) For all $x,y,z \in [0,1]$, such that $x \leq y$, we have by (I1) that $I(y,z) \leq I(x,z)$. We will analyze the following cases: (1) If $I(x,z) \geq I(y,z) \geq \gamma$, then $I_\gamma(x,z) = I_\gamma(y,z) = 1$; (2) If $I(x,z) \geq \gamma > I(y,z)$, then $I_\gamma(x,z) = 1 > 0 = I_\gamma(y,z)$ and (3) if $\gamma > I(x,z) \geq I(y,z)$, then $I_\gamma(x,z) = I_\gamma(y,z) = 0$. Therefore, I_γ satisfies (I1).

(I_γ2) For all $x,y,z \in [0,1]$, such that $y \leq z$, we have by (I2) that $I(x,y) \leq I(x,z)$. We will analyze the following cases: (1) If $\gamma \leq I(x,y) \leq I(x,z)$, then $I_\gamma(x,y) = I_\gamma(x,z) = 1$; (2) If $I(x,y) < \gamma \leq I(x,z)$, then $I_\gamma(x,y) = 0 < 1 = I_\gamma(x,z)$ and (3) if $I(x,y) \leq I(x,z) < \gamma$, then $I_\gamma(x,y) = I_\gamma(x,z) = 0$. Therefore, I_γ satisfies (I2).

(I_γ3) For all $y \in [0,1]$, we have by (I3) that $I(0,y) = 1$. So, $I(0,y) \geq \gamma$ and thereby $I_\gamma(0,y) = 1$. Therefore, I_γ satisfies (I3).

(I_γ4) For all $x \in [0,1]$, we have by (I4) that $I(x,1) = 1$. So, $I(x,1) \geq \gamma$ and thereby $I_\gamma(x,1) = 1$. Therefore, I_γ satisfies (I4).

(I_γ5) By (I5), $I(1,0) = 0$. So, $I(1,0) < \gamma$ and thereby $I_\gamma(1,0) = 0$. Therefore, I_γ satisfies (I5).

We conclude that I_γ is a fuzzy implication. As, $I_\gamma(x,y) \in \{0,1\}$ for all $x,y \in [0,1]$, then I_γ is a crisp fuzzy implication.

Notice that if we take $\gamma \in [0,1)$, and $I^\gamma(x,y) = \begin{cases} 1, & \text{if } I(x,y) > \gamma \\ 0, & \text{if } I(x,y) \leq \gamma \end{cases}$, I^γ is also a crisp fuzzy implication.

Proposition 4. *Let $I^{N'}_{T,N}$ be a (N',T,N)-implication. Then $I^{N'}_{T,N}$ is crisp if and only if N' is a crisp fuzzy negation.*

Proof. Suppose N' is not crisp, then there is $z \in (0,1)$ such that $N'(z) \notin \{0,1\}$. So, for any t-norm T and any fuzzy negation N we have that $T(z, N(0)) = T(z,1) = z$, thus $I^{N'}_{T,N}(z,0) = N'(T(z, N(0))) = N'(z) \notin \{0,1\}$. Therefore $I^{N'}_{T,N}$ is not crisp. Conversely, if N' is crisp then, for any t-norm T and any fuzzy negation N, $I^{N'}_{T,N}(x,y) = N'(T(x, N(y))) \in \{0,1\}$.

Corollary 2. *Let I^N_T be a (T,N)-implication. Then I^N_T is crisp if and only if N is a crisp fuzzy negation.*

Proposition 5. *Let I be a (T,N)-implication for any crisp negation N and any t-norm T. Then:*

  (i) *I satisfies (LOP);*
  (ii) *I does not satisfy (ROP).*

Proof. Since N is crisp we have, by Corollary 2, that $I = I^N_T$ is crisp. Note that:

$$I(x,y) = N(T(x, N(y))) = \begin{cases} 1, & \text{if } N(y) = 0 \\ N(x), & \text{if } N(y) = 1 \end{cases}.$$

Then,

  (i) For all $x,y \in [0,1]$ such that $x \leq y$, since N is crisp, we will analyze two cases:
  (1) if there exists $\alpha \in [0,1)$ such that $N = N_\alpha$, then

$$I(x,y) = \begin{cases} 1, & \text{if } y > \alpha \\ N(x), & \text{if } y \leq \alpha \end{cases}. \tag{1}$$

  For $y \leq \alpha$, as $x \leq y$, then $x \leq \alpha$. So $N_\alpha(x) = 1$ and hence $I(x,y) = 1$.
  (2) If there exists $\alpha \in (0,1]$ such that $N = N^\alpha$, then

$$I(x,y) = \begin{cases} 1, & \text{if } y \geq \alpha \\ N(x), & \text{if } y < \alpha \end{cases}. \tag{2}$$

  For $y < \alpha$, as $x \leq y$, then $x < \alpha$. So $N_\alpha(x) = 1$ and therefore, $I(x,y) = 1$. Thus, I satisfies (LOP).
  (ii) We will analyze two cases again:
  (1) if $N = N_\alpha$ for some $\alpha \in [0,1)$, then there exists $x,y \in [0,1]$ such that $x > y > \alpha$. So, by Eq. (1), $I(x,y) = 1$.
  (2) If $N = N^\alpha$ for some $\alpha \in (0,1]$, then there exists $x,y \in [0,1]$ such that $y < x < \alpha$. So, by Eq. (2) $I(x,y) = N^\alpha(x) = 1$. In any case, there exists $x > y$, but $I(x,y) = 1$, therefore I does not satisfy (ROP).

Proposition 6. *Let I be a crisp fuzzy implication. Then:*

  (i) *I satisfies (EP);*
  (ii) *I satisfies $R{-}CP(N_I)$, where N_I is the natural negation of I;*
  (iii) *I does not satisfy (NP);*

Proof.   (i) Since I is crisp, then by Proposition 2, I satisfies one of the conditions (C1), (C2), (C3) or (C4). If I satisfies (C1), then there exists $\alpha \in (0,1]$ and $\beta \in [0,1)$ such that $I(x,y) = I_{\alpha,\beta}(x,y)$. So, (1)

for $z \leq \beta$, we have $I(y,z) = \begin{cases} 0, & \text{if } y \geq \alpha \\ 1, & \text{if } y < \alpha \end{cases}$. Thus,

$$I(x, I(y,z)) = \begin{cases} I(x,0), & \text{if } y \geq \alpha \\ I(x,1), & \text{if } y < \alpha \end{cases} = \begin{cases} 0, & \text{if } x \geq \alpha \text{ and } y \geq \alpha \\ 1, & \text{otherwise} \end{cases},$$

and we also have $I(x,z) = \begin{cases} 0, & \text{if } x \geq \alpha \\ 1, & \text{if } x < \alpha \end{cases}$. Thus,

$$I(y, I(x,z)) = \begin{cases} I(y,0), & \text{if } x \geq \alpha \\ I(y,1), & \text{if } x < \alpha \end{cases} = \begin{cases} 0, & \text{if } x \geq \alpha \text{ and } y \geq \alpha \\ 1, & \text{otherwise} \end{cases}.$$

Then, for $z \leq \beta$, (EP) is satisfied. Now, (2) for $z > \beta$, we have $I(y,z) = I(x,z) = 1$. So $I(x, I(y,z)) = I(x,1) = 1 = I(y,1) = I(y, I(x,z))$.

Cases (C2), (C3) and (C4) are similar to the previous one. Therefore, I satisfy (EP).

(ii) Indeed, for all $x, y \in [0,1]$,
$$I(x, N_I(y)) = I(x, I(y,0)) \stackrel{(EP)}{=} I(y, I(x,0)) = I(y, N_I(x)).$$

(iii) As $I(x,y) \in \{0,1\}$, since I is crisp then, for all $y \in (0,1)$, $I(x,y) \neq y$.

Proposition 7. *Let I be a crisp fuzzy implication with (α, β) as its threshold. If $\beta < \alpha$ then I satisfies (IP).*

Proof. Indeed, if I is of type (C1) then there is no $x \in [0,1]$ such that $x \geq \alpha$ and $x \leq \beta$ simultaneously, since $\beta < \alpha$. So $I(x,x) = 1$, for all $x \in [0,1]$. Similarly, if I is of type (C2), (C3) or (C4) we prove that $I(x,x) = 1$. Therefore, in any case, I satisfies (IP).

Proposition 8. *Let I be a crisp fuzzy implication with (α, β) as its threshold. If $\alpha < \beta$ then I does not satisfy (IP).*

Proof. Indeed, because there exists x between α and β such that $I(x,x) = 0$, therefore I does not satisfy (IP).

Proposition 9. *Let I be a crisp fuzzy implication with (α, β) as its threshold. If $\alpha = \beta$ then:*

(i) *If I is of type (C1) then I does not satisfy (IP);*
(ii) *If I is of type (C2), (C3) or (C4) then I satisfies (IP).*

Proof. In fact,

(i) There is $x = \alpha$ such that $I(x,x) = 0 \neq 1$, so I does not satisfy (IP).
(ii) If I is of type (C2) then there is no $x \in [0,1]$ such that $x > \alpha$ and $x < \beta$ simultaneously, since $\beta = \alpha$. So $I(x,x) = 1$, for all $x \in [0,1]$. Similarly, if I is of type (C3) or (C4) we prove that $I(x,x) = 1$. Therefore, in any case, I satisfies (IP).

In [4], the conditions under which the Boolean-like law holds for some classes of fuzzy implications were given. Here, we prove that it is valid for a crisp fuzzy implication whenever (IP) is satisfied.

Proposition 10. *Let I be a crisp fuzzy implication. Then:*

(i) If I satisfies (IP) then $I(x, I(y,x)) = 1$, for all $x, y \in [0,1]$;
(ii) If I does not satisfy (IP) then there are $x, y \in [0,1]$ such that $I(x, I(y,x)) \neq 1$.

Proof. Indeed,

(i) If I satisfies (IP) then $I(x,x) = 1$, for all $x \in [0,1]$, so by Proposition 6:
$$I(x, I(y,x)) \overset{(EP)}{=} I(y, I(x,x)) = I(y,1) = 1.$$
(ii) If I does not satisfy (IP) then there exist $x \in (0,1)$ such that $I(x,x) = 0$. So, for $y = 1$, $I(x, I(1,x)) = I(1, I(x,x)) = I(1,0) = 0 \neq 1$.

Corollary 3. *Let I be a crisp fuzzy implication with (α, β) as its threshold.*

(i) If $\beta < \alpha$ then $I(x, I(y,x)) = 1$, for all $x, y \in [0,1]$;
(ii) If $\alpha < \beta$ then there exists $x, y \in [0,1]$ such that $I(x, I(y,x)) \neq 1$;
(iii) If $\alpha = \beta$ then there exists $x, y \in [0,1]$ such that $I(x, I(y,x)) \neq 1$ whenever I is of type (C1). And $I(x, I(y,x)) = 1$, for all $x, y \in [0,1]$ whenever I is of type (C2), (C3) or (C4).

Proof. It follows straight from Propositions 7, 8 and 9.

Definition 13. *Let I be a crisp fuzzy implication with (α, β) as its threshold and let N be a fuzzy negation. We say that IN is a **dual crisp fuzzy implication of I with respect to** N, or dual NCrisp, if it satisfies one of the following types, for all $x, y \in [0,1]$:*

(NC1) $IN(x,y) = I_{N(\beta), N(\alpha)}(x,y)$, *whenever I satisfies (C1);*
(NC2) $IN(x,y) = I^{N(\beta), N(\alpha)}(x,y)$, *whenever I satisfies (C2);*
(NC3) $IN(x,y) = I^{N(\beta)}{}_{N(\alpha)}(x,y)$, *whenever I satisfies (C3);*
(NC4) $IN(x,y) = I_{N(\beta)}{}^{N(\alpha)}(x,y)$, *whenever I satisfies (C4).*

Definition 14. *Let I be a crisp fuzzy implication and N be a fuzzy negation. I is said to be a*

*(i) **Crisp-CP** with respect to N, if*
$$I(x,y) = IN(N(y), N(x)), \qquad \text{(C-CP)}$$

(ii) **Crisp Left-CP** *with respect to N, if*
$$I(N(x), y) = IN(N(y), x),$$
(C-LCP)

(iii) **Crisp Right-CP** *with respect to N, if*
$$I(x, N(y)) = IN(y, N(x)),$$
(C-RCP)

where IN is its dual NCrisp.

Proposition 11. *Let I be a crisp fuzzy implication and N be a fuzzy negation. If N is strict, then I is C-CP with respect to N.*

Proof. (1) If I satisfies $(C1)$, then, by Proposition 2, there exist $\alpha \in (0, 1]$ and $\beta \in [0, 1)$ such that $I(x, y) = I_{\alpha,\beta}(x, y)$. So, as N is strict, $x \geq \alpha$ if and only if $N(x) \leq N(\alpha)$ and $y \leq \beta$ if and only if $N(y) \geq N(\beta)$. Therefore, by (NC1)

$$IN(N(y), N(x)) = I_{N(\beta), N(\alpha)}(N(y), N(x))$$
$$= \begin{cases} 0, & \text{if } N(y) \geq N(\beta) \text{ and } N(x) \leq N(\alpha) \\ 1, & \text{otherwise} \end{cases}$$
$$= \begin{cases} 0, & \text{if } x \geq \alpha \text{ and } y \leq \beta \\ 1, & \text{otherwise} \end{cases} = I_{\alpha,\beta}(x, y) = I(x, y).$$

(2) If I satisfies $(C2)$, then, by Proposition 2, there exist $\alpha \in [0, 1)$ and $\beta \in (0, 1]$ such that $I(x, y) = I^{\alpha,\beta}(x, y)$. So, as N is strict, $x > \alpha$ if and only if $N(x) < N(\alpha)$ and $y < \beta$ if and only if $N(y) > N(\beta)$. Therefore, by (NC2)

$$IN(N(y), N(x)) = I^{N(\beta), N(\alpha)}(N(y), N(x))$$
$$= \begin{cases} 0, & \text{if } N(y) > N(\beta) \text{ and } N(x) < N(\alpha) \\ 1, & \text{otherwise} \end{cases}$$
$$= \begin{cases} 0, & \text{if } x > \alpha \text{ and } y < \beta \\ 1, & \text{otherwise} \end{cases} = I^{\alpha,\beta}(x, y) = I(x, y).$$

(3) If I satisfies $(C3)$, then, by Proposition 2, there exist $\alpha, \beta \in (0, 1]$ such that $I(x, y) = I_\alpha{}^\beta(x, y)$. So, as N is strict, $x \geq \alpha$ if and only if $N(x) \leq N(\alpha)$ and $y < \beta$ if and only if $N(y) > N(\beta)$. Therefore, by (NC3)

$$IN(N(y), N(x)) = I^{N(\beta)}{}_{N(\alpha)}(N(y), N(x))$$
$$= \begin{cases} 0, & \text{if } N(y) > N(\beta) \text{ and } N(x) \leq N(\alpha) \\ 1, & \text{otherwise} \end{cases}$$
$$= \begin{cases} 0, & \text{if } x \geq \alpha \text{ and } y < \beta \\ 1, & \text{otherwise} \end{cases} = I_\alpha{}^\beta(x, y) = I(x, y).$$

(4) If I satisfies $(C4)$, then, by Proposition 2, there exist $\alpha, \beta \in [0, 1)$ such that $I(x, y) = I^\alpha{}_\beta(x, y)$. So as N is strict, $x > \alpha$ if and only if $N(x) < N(\alpha)$ and

$y \leq \beta$ if and only if $N(y) \geq N(\beta)$. Therefore, by (NC4)

$$IN(N(y), N(x)) = I_{N(\beta)}{}^{N(\alpha)}(N(y), N(x))$$

$$= \begin{cases} 0, & \text{if } N(y) \geq N(\beta) \text{ and } N(x) < N(\alpha) \\ 1, & \text{otherwise} \end{cases}$$

$$= \begin{cases} 0, & \text{if } x > \alpha \text{ and } y \leq \beta \\ 1, & \text{otherwise} \end{cases} = I^{\alpha}{}_{\beta}(x, y) = I(x, y).$$

Therefore, in any case, I is C-CP.

Notice that for C-LCP and C-RCP, the requirements are different from Proposition 11. The proof for both is analogous as we can see in the following proposition.

Proposition 12. *Let I be a crisp fuzzy implication and N be a fuzzy negation. If N is strong, then I is C-LCP and C-RCP with respect to N.*

Proof. Straightforward.

Finally, we also studied that is impossible for some implications to be crisp fuzzy implications.

Proposition 13. *None of the following classes of fuzzy implications (S, N)-, R-, QL- and D-implication is a crisp fuzzy implication.*

Proof. In fact,

(1) if I is an (S, N)-implication, then there exist a t-conorm S and a fuzzy negation N such that $I(x, y) = S(N(x), y)$ for all $x, y \in [0, 1]$. In particular, for $x = 1$, $I(1, y) = S(N(1), y) = S(0, y) \stackrel{(S4)}{=} y$. So, for all $y \in (0, 1)$, $I(1, y) \notin \{0, 1\}$. Therefore, I is not crisp.

(2) If I is an R-implication, then there exists a t-norm T such that $I(x, y) = sup\{t \in [0, 1] \mid T(x, t) \leq y\}$ for all $x, y \in [0, 1]$. In particular, for $x = 1$, $I(1, y) = sup\{t \in [0, 1] \mid T(1, t) \leq y\} \stackrel{(T4)}{=} sup\{t \in [0, 1] \mid t \leq y\} = y$. So, for all $y \in (0, 1)$, $I(1, y) \notin \{0, 1\}$. Therefore, I is not crisp.

(3) If I is a QL-implication, then there exist a t-norm T, a t-conorm S and a fuzzy negation N such that $I(x, y) = S(N(x), T(x, y))$ for all $x, y \in [0, 1]$. In particular, for $x = 1$, $I(1, y) = S(N(1), T(1, y)) = S(0, T(1, y)) \stackrel{(S4)}{=} T(1, y) \stackrel{(T4)}{=} y$. So, for all $y \in (0, 1)$, $I(1, y) \notin \{0, 1\}$. Therefore, I is not crisp.

(4) If I is a D-implication, then there exist a t-norm T, a t-conorm S and a fuzzy negation N such that $I(x, y) = S(T(N(x), N(y)), y)$ for all $x, y \in [0, 1]$. In particular, for $x = 1$, $I(1, y) = S(T(N(1), N(y)), y) = S(T(0, N(y)), y) = S(0, y) \stackrel{(S4)}{=} y$. So, for all $y \in (0, 1)$, $I(1, y) \notin \{0, 1\}$. Therefore, I is not crisp.

4   Final Remarks

One can find many examples of studies which use fuzzy implications with crisp behavior such as [8–10,17]. Our purpose in this work was to study those fuzzy implications which always map to 0 or 1, therefore named crisp fuzzy implications. We provided a characterization for those implications, presenting four possible classes (Proposition 2) and studied some properties and conditions under which fuzzy implications need in order to be considered crisp.

Acknowledgments. This work was partially supported by the Brazilian funding agency CNPq (Conselho Nacional de Desenvolvimento Científico e Tecnológico), under the process No. 307781/2016-0.

References

1. Baczyński, M., Balasubramaniam, J.: Fuzzy Implications. Springer, Heidelberg (2008). https://doi.org/10.1007/978-3-540-69082-5
2. Baczyński, M., Beliakov, G., Bustince, H., Pradera, A. (eds.): Advances in Fuzzy Implication Functions, Studies in Fuzziness and Soft Computing, vol. 300. Springer, Heidelberg (2013). https://doi.org/10.1007/978-3-642-35677-3
3. Bedregal, B.C.: A normal form which preserves tautologies and contradictions in a class of fuzzy logics. J. Algorithms **62**(3–4), 135–147 (2007)
4. Cruz, A., Bedregal, B.C., Santiago, R.H.N.: On the characterizations of fuzzy implications satisfying I(x, I(y, z)) = I(I(x, y), I(x, z)). Int. J. Approx. Reason. **93**, 261–276 (2018)
5. Dimuro, G.P., Bedregal, B., Santiago, R.H.: On (G, N)-implications derived from grouping functions. Inf. Sci. **279**, 1–17 (2014)
6. Dimuro, G.P., Bedregal, B.C.: On residual implications derived from overlap functions. Inf. Sci. **312**, 78–88 (2015)
7. Dimuro, G.P., Bedregal, B., Bustince, H., Jurio, A., Baczyński, M., Mis, K.: QL-operations and QL-implication functions constructed from tuples (O, G, N) and the generation of fuzzy subsethood and entropy measures. Int. J. Approx. Reason. **82**, 170–192 (2017)
8. Dubois, D., Prade, H.M., Bezdek, J.C.: Fuzzy Sets in Approximate Reasoning and Information Systems. Kluwer Academic Publishers, Norwell (1999)
9. Dubois, D., Nakata, M., Prade, H.: Extended divisions for flexible queries in relational databases. In: Pons, O., Vila, M.A., Kacprzyk, J. (eds.) Knowledge Management in Fuzzy Databases. Studies in Fuzziness and Soft Computing, vol. 39, pp. 105–121. Physica, Heidelberg (2000). https://doi.org/10.1007/978-3-7908-1865-9_6
10. Godo, L., Sandri, S.: Dealing with imprecise inputs in a fuzzy rule-based system using an implication-based rule model. In: Bouchon-Meunier, B., Gutiérrez-Ríos, J., Magdalena, L., Yager, R.R. (eds.) Technologies for Constructing Intelligent Systems 1. Studies in Fuzziness and Soft Computing, vol. 89. Physica, Heidelberg (2002). https://doi.org/10.1007/978-3-7908-1797-3_4
11. Klir, G.J., Yuan, B.: Fuzzy Sets and Fuzzy Logic: Theory and Applications. Prentice Hall, Upper Saddle River (1995)
12. Mas, M., Monserrat, M., Torrens, J., Trillas, E.: A survey on fuzzy implication functions. IEEE Trans. Fuzzy Syst. **15**(6), 1107–1121 (2007)

13. Oh, K.W., Bandler, W.: Properties of fuzzy implication operators. Int. J. Approx. Reason. **1**(3), 273–285 (1987)
14. Pinheiro, J., Bedregal, B., Santiago, R.H., Santos, H.: (T, N)-implications. In: 2017 IEEE International Conference on Fuzzy Systems (FUZZ-IEEE), pp. 1–6 (2017)
15. Pinheiro, J., Bedregal, B., Santiago, R.H., Santos, H.: A study of (T, N)-implications and its use to construct a new class of fuzzy subsethood measure. Int. J. Approx. Reason. **97**, 1–16 (2018)
16. Sakly, A., Benrejeb, M.: On the choice of the adequate fuzzy implication operator with the center of gravity defuzzification method based on precision criterion in fuzzy control. In: Proceedings of the 2002 IEEE International Conference on Fuzzy Systems (FUZZ-IEEE 2002), Honolulu, HI, pp. 482–487 (2002)
17. Santos, A.S., Valle, M.E.: A fast and robust max-c projection fuzzy autoassociative memory with application for face recognition. In: IEEE World Congress on Computational Intelligence (2018, submitted)

Stock Market Price Forecasting Using a Kernel Participatory Learning Fuzzy Model

R. Vieira[1(✉)], L. Maciel[2], R. Ballini[3], and Fernando Gomide[1]

[1] School of Electrical and Computer Engineering, University of Campinas,
Campinas, São Paulo, Brazil
{giordano,gomide}@dca.fee.unicamp.br
[2] São Paulo School of Politics, Economics and Business,
Federal University of São Paulo, Osasco, São Paulo, Brazil
maciel.leandro@unifesp.br
[3] Institute of Economics, University of Campinas, Campinas, São Paulo, Brazil
ballini@unicamp.br

Abstract. This paper suggests an enhanced fuzzy rule-based evolving participatory learning with kernel recursive least squares algorithm for stock market index forecasting. The algorithm combines an incremental clustering algorithm to learn the antecedent part of functional fuzzy rules, and a kernel recursive least squares method to compute the parameters of the consequents of the rules. The algorithm uses a small number of user-defined parameters to enhance its autonomy. Computational experiments concerning one-step-ahead forecasts of the S&P 500 stock market index from January 2010 to December 2017 is conducted to compare the algorithm with traditional forecasting and state-of-the-art evolving fuzzy algorithms. Accuracy and computational effort evaluation indicate the high potential of the kernel recursive participatory learning algorithm for stock market index time series forecasting.

Keywords: Evolving fuzzy systems · Adaptive modeling
Time series forecasting

1 Introduction

The financial economics literature has a long research history on the predictability of stock markets [5,6]. Asset price forecasting is a challenging task because of the high degree of noise and the semi-strong form of market efficiency [8]. Despite the divergence of opinions on the efficiency of markets, several works show that financial time series are to some extent predictable [16,17]. Econometric and statistical methods such as autoregressive models have been the widely adopted in finance and economics forecasting [2]. Artificial neural networks (ANN) is a nonlinear approach that also became popular in financial forecasting [1,9] to overcome restrictive assumptions on data distribution made by traditional

© Springer International Publishing AG, part of Springer Nature 2018
G. A. Barreto and R. Coelho (Eds.): NAFIPS 2018, CCIS 831, pp. 361–373, 2018.
https://doi.org/10.1007/978-3-319-95312-0_31

forecasting algorithms. Financial market data are affected by press news, expectations and the psychological state of investors, and ANN does not account for the imprecision and uncertainty induced in data.

This paper addresses stock price forecasting using evolving fuzzy systems to account for financial data uncertainty and to process stream data, both essential characteristics in decision making with market data in intraday frequency. Evolving fuzzy systems are incremental procedures to process stream data, and to simultaneously adapt the structure and parameters of fuzzy models. In particular, functional evolving fuzzy rule-based models are adaptive, incremental fuzzy models in which the number of rules and the rules parameters are continuously evaluated whenever data are input.

A wide range of evolving fuzzy models can be found in the literature. For instance, an autonomous user-free control parameters modeling scheme called eTS+ is given in [4]. The eTS+ uses criteria such as age, utility, local density, and zone of influence to update the model structure. In [11], the concept of participatory learning (PL) [23] was joined with the evolving fuzzy modeling idea, resulting in the ePL model. Later, ePL+ was developed in the realm of participatory learning clustering and extends the ePL by using the updating strategy of eTS+ [14]. An evolving Takagi-Sugeno model that uses least squares support vector machine (eTS-LS-SVM) to update the consequent parameters of each rule was conceived in [10]. In [20] it is also proposed an evolving Takagi-Sugeno fuzzy model with non-linear parameter estimation (eTS-KRLS). The authors have used a modified version of the recursive least squares method, called KRLS [7], for this purpose. A comprehensive review of evolving approaches is found in [13].

Recently, the authors conceived an evolving fuzzy rule-based participatory learning model called ePL-KRLS, for time series forecasting [22]. The ePL-KRLS brings a novel recursive, incremental approach to update the structure and parameters of a fuzzy rule-based model. Specifically, it combines the concepts of evolving systems with participatory learning and kernel methods to achieve higher forecast accuracy, robustness, stability, with acceptable computational cost. The participatory learning concept appears as an unsupervised clustering algorithm, whereas the KRLS [7] updates the parameters of the affine functions of fuzzy rules consequents.

This paper suggests an enhanced fuzzy rule-based evolving Participatory Learning with Kernel Recursive Least Squares (ePL-KRLS+) for stock market prices forecasting. Differently from ePL-KRLS developed in [22], the ePL-KRLS+ uses an updating mechanism similar to that of ePL+. Kernel-based methods are more sensitive to variations in the input data and are able to approximate nonlinear systems efficiently. Because financial data are affected by noise and outliers (press news, reversion practitioners expectations, etc.) the enhanced algorithm appears as a potential tool for asset price prediction, as experiments with the S&P 500 index suggest. Performance of the ePL-KRLS+ is compared against traditional econometric benchmarks, and with state-of-the-art evolving fuzzy approaches in terms of accuracy and computational costs.

The paper proceeds as follows. After this brief introduction, Sect. 2 details the ePL-KRLS+. Section 3 shows the results achieved by the computational experiments with the S&P 500 data. Finally, Sect. 4 concludes the paper and suggests issues for further investigation.

2   The ePL-KRLS+ Algorithm

An overview of the ePL-KRLS+ algorithm developed in this paper is shown in Fig. 1. It uses a fuzzy rule base composed by functional fuzzy rules of Takagi-Sugeno (TS) type. The antecedent of the functional rules contain linguistic variables associated with the input data, and consequents in the form of functions of the input variables.

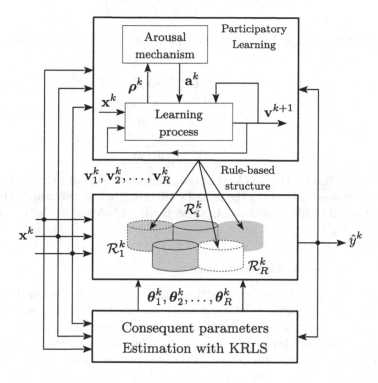

Fig. 1. ePL-KRLS+ modeling.

The fuzzy rule adopted by ePL-KRLS+ is of the following form:

$$\mathcal{R}_i : \textbf{IF} \ \underbrace{\textbf{x} \ \text{is} \ \mathcal{A}_i}_{\text{Antecedent}} \ \textbf{THEN} \ \underbrace{\hat{y}_i = f_i(\textbf{x}, \boldsymbol{\theta}_i)}_{\text{Consequent}} \tag{1}$$

where $\mathcal{R}_i$ corresponds to the i-th fuzzy rule, $i = 1, 2, \ldots, R$, R is the number of fuzzy rules, $\textbf{x} = [x_1, \ldots, x_m]^T \in \mathbb{R}^m$ is the input, $\mathcal{A}_i$ is the fuzzy set of the antecedent of the i-th fuzzy rule whose membership function is $\mathcal{A}_i(\mathbf{x}) : \mathbb{R}^m \rightarrow [0,1]$ and

$$\hat{y}_i = f_i(\mathbf{x}, \boldsymbol{\theta}_i) = \sum_{j=1}^{n_i} \theta_{ij}\, \kappa(\mathbf{d}_{ij}, \mathbf{x}) \tag{2}$$

is the output of the i-th rule, as a function of the input data and the consequent parameters $\boldsymbol{\theta}_i = [\theta_{i1}, \ldots, \theta_{in_i}]^T \in \mathbb{R}^{n_i}$ of the i-th rule. In (2), $\mathcal{D}_i = [\mathbf{d}_{i1}, \ldots, \mathbf{d}_{in_i}] \in \mathbb{R}^{n_i \times m}$ is the dictionary of the i-th rule, $\mathbf{d}_{ij} \in \mathbb{R}^m$ is the j-th element of $\mathcal{D}_i$, n_i is the number of elements in $\mathcal{D}_i$, $\theta_{ij} \in \mathbb{R}$ is the j-th consequent parameter of the i-th rule and $\kappa(\cdot, \cdot)$ is the Gaussian kernel function [19] of sample vectors $\mathbf{x}^i$ and $\mathbf{x}^j$:

$$\kappa(\mathbf{x}^i, \mathbf{x}^j) = \exp\left(-\frac{||\mathbf{x}^i - \mathbf{x}^j||^2}{2\nu_{ij}^2}\right) \tag{3}$$

where $\nu_{ij} > 0$ is the kernel size. The model output, $\hat{y} \in \mathbb{R}$, is computed using:

$$\hat{y} = \sum_{i=1}^{R} \hat{y}_i\, \Gamma_i^N(\mathbf{x}) \tag{4}$$

with

$$\Gamma_i^N(\mathbf{x}) = \frac{\mathcal{A}_i(\mathbf{x})}{\sum_{j=1}^{R} \mathcal{A}_j(\mathbf{x})} \tag{5}$$

being the normalized activation level of the i-th rule. We recall that TS modeling requires two tasks: (i) learning the model structure, that is, the number of rules R and the membership functions $\mathcal{A}_i(\mathbf{x})$ and; (ii) estimation of the parameters $\boldsymbol{\theta}_i = [\theta_{i1}, \ldots, \theta_{in_i}]^T$.

2.1 Learning the Model Structure

Learning the rule-based structure of ePL-KRLS+ is done using the participatory learning clustering algorithm (PL) [21] which assigns a rule to a unique cluster. Participatory learning [23] assumes that the learning process depends on what the system has already learned from the data. The current knowledge is part of the learning process itself and influences the way in which new data are used for self-organization. A characteristic property of PL is the impact of new data cause in self-organization or model revision. The impact depends on the compatibility of the current knowledge with input data.

Let $\mathbf{v}^k = \{\mathbf{v}_i^k\}_{i=1}^{R} \subset \mathbb{R}^{R \times m}$ as a set of $i = 1, \ldots, R$ rule centers at step k. The goal of PL clustering algorithm is to learn each vector $\mathbf{v}_i^k \in [0,1]^m$ using the input data stream $\mathbf{x}^k \in \mathbb{R}^m$. In participatory learning, a rule center is updated using a compatibility measure $\rho_i^k \in [0,1]$ and an arousal index, $a_i^k \in [0,1]$. While ρ_i^k measures how much a data point is compatible with the rule-base structure, the arousal index a_i^k acts as a critic to remind when the rule base should be revised in front of new information.

Due to its unsupervised nature, PL may create or delete a new cluster (rule, merge, or modify the existing ones at each learning step k. If a_i^k is greater than a threshold $\tau \in [0,1]$, then a new rule is created. The arousal index a_i^k is updated as follows:

$$a_i^k = a_i^{k-1} + \beta(1 - \rho_i^k - a_i^{k-1}), \tag{6}$$

where $\beta \in [0,1]$ controls the rate of change of arousal, the closer β is to one, the faster the system is to sense compatibility variations, $a_i^0 = 0$, and

$$\rho_i^k = 1 - \frac{||\mathbf{x}^k - \mathbf{v}_i^k||}{m} \tag{7}$$

with m as the size of the input space, and $\mathbf{v}_i^k$ is the i-th rule center at step k. Otherwise, the most compatible rule center, $\hat{\mathbf{v}}_i^k | i = \mathrm{argmax}_j \{\rho_j^k\}$, is updated using:

$$\hat{\mathbf{v}}_i^{k+1} = \hat{\mathbf{v}}_i^k + \alpha(\rho_i^k)^{1-a_i^k}(\mathbf{x}^k - \hat{\mathbf{v}}_i^k) \tag{8}$$

where $\alpha \in [0,1]$ is the learning rate. If a_i^k increases, the similarity measure has a reduced effect. Hence, the arousal index can be interpreted as the complement of the confidence we have in the truth of the current rule-based structure. Likewise, PL assumes that a rule must be deleted if its center $\mathbf{v}_i^k$ has compatibility greater than a threshold $\gamma \in [0,1]$ in relation to the remaining centers $\mathbf{v}_j^k$ as:

$$\rho_{ij}^k = \rho_{ij}^k(\mathbf{v}_i^k, \mathbf{v}_j^k) = 1 - \frac{1}{m}\sum_{l=1}^{m}|v_{il}^k - v_{jl}^k| \tag{9}$$

Two additional mechanisms were added in ePL-KRLS+ to improve its autonomy and adaptability in complex systems modeling: a utility measure to shrink the rule base, and the use of distinc radius for each rule. The quality of each rule is monitored by the utility measure introduced in [4]. The utility measure is an indicator of the accumulated relative firing level of rule i at step k:

$$\mathcal{U}_i^k = \frac{\sum_{j=1}^{k} \Gamma_i(\mathbf{x}^j)}{k - I^i}, \tag{10}$$

where I^i is the step in which the fuzzy rule i is created. Once a rule is created, the utility indicates how often the rule has been fired. Therefore it is a way to avoiding unused rules kept as part of the model. Therefore, low-quality rules can be deleted:

$$\mathbf{IF} \ \mathcal{U}_i^k < \epsilon \ \mathbf{THEN} \ R \leftarrow R - 1, \tag{11}$$

where $\epsilon \in [0.03, 0.1]$ is a threshold that controls the utility of each rule [4]. The second mechanism updates the radius of the most compatible rule center recursively. In general, the size of the radius has considerable influence in the activation level of the fuzzy rule, and ultimately in the algorithm output. As in [4], the radius is updated according to:

$$\sigma_i^k = \zeta\sigma_i^{k-1} + (1 - \zeta)\varrho_i^k \tag{12}$$

where $\zeta \in [0.3, 0.5]$ is a constant that regulates the compatibility of the new information with the old one [4] and ϱ_i^k is the local density over the input data space, found as:

$$\varrho_i^k = \sqrt{\frac{1}{S_i^k - 1} \sum_{j=1}^{S_i^k - 1} ||\mathbf{v}_i^k - \mathbf{x}^j||^2} \tag{13}$$

having $S_i^k = S_i^{k-1} + 1$ as the support of the i-th rule, i.e., the number of data points that are in the zone of influence of that rule.

2.2   Estimation of the Consequent Parameters

After PL clustering, the ePL-KRLS+ proceeds to estimate the consequent parameters of the rules using the kernel recursive least squares (KRLS) [7]. The KRLS method is a modified version of the traditional RLS method, where operations with the input data are viewed in a high dimensional Hilbert space $\mathbb{H}$. In this case, a transformation must be applied in $\mathbf{x}^k$ using a nonlinear function $\phi : \mathbb{R}^m \rightarrow \mathbb{H}$, so that $\mathbf{x}^k \rightarrow \phi(\mathbf{x}^k)$. It should be noted that thanks to the kernel trick [12], neither the ϕ function, nor the $\mathbb{H}$ space dimension need to be explicitly found.

Recall that the KRLS in Hilbert spaces considers a dataset $\{(\phi^1, y^1), \ldots, (\phi^k, y^k)\}$ with k pairs of values, where $\phi^k \in \mathbb{H}$ is the k-th input data in a high dimensional space, and y^k is the k-th model output. At each step, the weight vector $\boldsymbol{\omega}$ is determined solving

$$\min_{\boldsymbol{\omega}} L(\boldsymbol{\omega}) = \sum_{j=1}^{k} |y^j - \boldsymbol{\omega}^T \phi^j|^2 + \lambda ||\boldsymbol{\omega}||^2 \tag{14}$$

where $\lambda \in [10^{-5}, 10^{-2}]$ is a regularization parameter. The solution of (14) is:

$$\boldsymbol{\omega}^k = \left[\lambda \mathbf{I} + \boldsymbol{\Phi}^k (\boldsymbol{\Phi}^k)^T\right]^{-1} \boldsymbol{\Phi}^k \mathbf{Y}^k \tag{15}$$

where $\mathbf{I} \in \mathbb{R}^{k \times k}$ is the identity matrix, $\boldsymbol{\Phi}^k = [\phi^1, \ldots, \phi^k]$ and $\mathbf{Y}^k = [y^1, \ldots, y^k]^T$. Expression (15) cannot be used directly because of the dimension of $\boldsymbol{\Phi}^k$. To addres this issue, it is necessary to describe $\boldsymbol{\omega}^k$ in terms of inner products using the matrix inversion lemma. It is easy to verify that:

$$\left[\lambda \mathbf{I} + \boldsymbol{\Phi}^k (\boldsymbol{\Phi}^k)^T\right]^{-1} \boldsymbol{\Phi}^k = \boldsymbol{\Phi}^k \left[\lambda \mathbf{I} + (\boldsymbol{\Phi}^k)^T \boldsymbol{\Phi}^k\right]^{-1} \tag{16}$$

Replacing (16) for (15), one obtains

$$\boldsymbol{\omega}^k = \boldsymbol{\Phi}^k \left[\lambda \mathbf{I} + (\boldsymbol{\Phi}^k)^T \boldsymbol{\Phi}^k\right]^{-1} \mathbf{Y}^k \tag{17}$$

From this transformation, one can apply the kernel trick to reduce the dimension of $(\boldsymbol{\Phi}^k)^T \boldsymbol{\Phi}^k$ and turn feasible the calculation of $\boldsymbol{\omega}^k$, as follows:

$$\mathbf{K}^k = (\boldsymbol{\Phi}^k)^T \boldsymbol{\Phi}^k = \begin{bmatrix} \kappa(\mathbf{x}^1, \mathbf{x}^1) & \cdots & \kappa(\mathbf{x}^1, \mathbf{x}^k) \\ \vdots & \ddots & \vdots \\ \kappa(\mathbf{x}^k, \mathbf{x}^1) & \cdots & \kappa(\mathbf{x}^k, \mathbf{x}^k) \end{bmatrix} \tag{18}$$

where $\mathbf{K}^k$ is the kernel matrix. Updating of $\mathbf{K}^k$ occurs incrementally to avoid redundancy in the calculation of its elements. This means that, for each new data,

$$\mathbf{K}^k = \begin{bmatrix} \mathbf{K}^{k-1} & \mathbf{g}^k \\ (\mathbf{g}^k)^T & 1 \end{bmatrix} \tag{19}$$

where $\mathbf{g}^k = [\boldsymbol{\Phi}^{k-1}]^T \phi^k = [\kappa(\mathbf{x}^1, \mathbf{x}^k), \ldots, \kappa(\mathbf{x}^{k-1}, \mathbf{x}^k)]^T$. Now, one can show that the weight vector $\boldsymbol{\omega}^k$ can be expressed as a linear combination of the input data in $\mathbb{H}$:

$$\boldsymbol{\omega}^k = \boldsymbol{\Phi}^k \boldsymbol{\theta}^k \tag{20}$$

$$\boldsymbol{\theta}^k = \left[\lambda \mathbf{I} + \mathbf{K}^k\right]^{-1} \mathbf{Y}^k \tag{21}$$

The updated solution of $\boldsymbol{\theta}^k$ requires the calculation of the $k \times k$ inverse matrix $\left[\lambda \mathbf{I} + \mathbf{K}^k\right]$, which is computationaly expensive [12]. Hence, recursive calculation of $\left[\lambda \mathbf{I} + \mathbf{K}^k\right]$ is indispensable. Defining $\mathbf{Q}^k = \left[\lambda \mathbf{I} + \mathbf{K}^k\right]^{-1}$, the inverse of $\mathbf{Q}^k$ can be calculated recursively as follows:

$$\mathbf{Q}^k = (r^k)^{-1} \begin{bmatrix} \mathbf{Q}^{k-1} r^k + \mathbf{z}^k (\mathbf{z}^k)^T & -\mathbf{z}^k \\ -(\mathbf{z}^k)^T & 1 \end{bmatrix} \tag{22}$$

where $\mathbf{z}^k = \mathbf{Q}^{k-1} \mathbf{g}^k \in \mathbb{R}^n$ and $r^k = \lambda + (\boldsymbol{\Phi}^k)^T \phi^k - (\mathbf{z}^k)^T \mathbf{g}^k \in \mathbb{R}$. The updated solution $\boldsymbol{\theta}^k$ can now be calculated:

$$\begin{aligned} \boldsymbol{\theta}^k &= \mathbf{Q}^k \mathbf{Y}^k \\ &= \begin{bmatrix} \boldsymbol{\theta}^{k-1} - \mathbf{z}^k (r^k)^{-1} \tilde{e}^k \\ (r^k)^{-1} \tilde{e}^k \end{bmatrix} \end{aligned} \tag{23}$$

where $\tilde{e}^k = y^k - \hat{y}^k \in \mathbb{R}$ is the model error.

Sparcification. The recursive computation of $\mathbf{Q}^k$ increases quadratically with the number of data, and computation of $\boldsymbol{\theta}^k$ become costly over time. Therefore, a sparsification for KRLS is vital to reduce processing time and memory [19]. Sparcification techniques involve creation of a data dictionary with only a subset of the most relevant stored to update the matrix $\mathbf{K}^k$ and the parameter vector $\boldsymbol{\theta}^k$. Because this paper uses KRLS in local models, one must define a local dictionary $\mathcal{D}_i^k = [\mathbf{d}_{i1}^k, \ldots, \mathbf{d}_{in_i}^k]^T \in \mathbb{R}^{n_i \times m}$ for each rule, and only the dictionary associated with the most compatible rule, $\hat{v}_i^k | i = \operatorname{argmax}_j \{\rho_j^k\}$ is modified at k.

This process specifies a kernel matrix $\mathbf{K}_i^k$ and a parameters vector θ_i^k for each fuzzy rule, which must be updated under the same conditions.

This paper uses the novelty criterion (NC) to control the insertion of elements in each local dictionary [12,18]. Novelty criterion first computes the distance of $\mathbf{x}^k$ to all elements $\mathbf{d}_{ij}^k \in \mathcal{D}_i^k$ of the dictionary of rule $\mathcal{R}_i^k$:

$$\mathrm{dis}_{\mathbf{x}} = \min_{\forall \mathbf{d}_{ij}^k \in \mathcal{D}_i^k} \left\| \mathbf{x}^k - \mathbf{d}_{ij}^k \right\|. \tag{24}$$

If $\mathrm{dis}_{\mathbf{x}}$ is smaller than a threshold δ, then $\mathbf{x}^k$ is not added into the dictionary. Otherwise, the algorithm computes the approximated modeling error, $\mathscr{E}_i^k$ for the rule $\mathbf{v}_i^k$. If the error with $\mathbf{x}^k$ is lower than the error obtained without $\mathbf{x}^k$, then the input data is added to the dictionary. Otherwise, it is discarded. The value of δ is set as $\delta = 0.1\nu_{ij}^k$ [18]. The use of NC-based sparcification technique reduces time and space complexities of the ePL-KRLS+ model from $\mathcal{O}(k^2)$ to $\mathcal{O}(R^2)$, where R is the number of fuzzy rules of the model at step k.

Adaptive Tuning of the Kernel Parameters. The Gaussian kernel (3) is commonly used in many kernel-based methods. Gaussian kernels require parameter ν, called the kernel size. Choosing an appropriate value of ν is complex, and impacts the model approximation ability: if ν is too large, then the process reduces to linear regression; if ν is too small, then all the data look distinct, resulting in over-fitting. In this paper, ν is adjusted using a recursive version of the Levenberg-Marquardt algorithm [15]. Let $\nu_i^k = [\nu_{i1}^k, \ldots, \nu_{in_i}^k]$ be the vector of kernel parameters to be optimized, where each ν_{ij}^k is associated with an element $\mathbf{d}_{ij}^k$ in the local dictionary $\mathcal{D}_i^k = [\mathbf{d}_{i1}^k, \ldots, \mathbf{d}_{in_i}^k]$. The most suitable values for ν_i^k can be found by minimizing the local error function:

$$\tilde{e}_i^k = \Gamma_i^N(\mathbf{x}^k)\,(y^k - \hat{y}_i^k) \tag{25}$$

The vector of parameters ν_i^k is updated as follows:

$$\nu_i^k = \nu_i^{k-1} + \mathbf{P}_i^k \boldsymbol{\nabla}_i^k \tilde{e}_i^k \tag{26}$$

where $\mathbf{P}_i^k \in \mathbb{R}^{n_i}$ is found using

$$\mathbf{P}_i^k = \left[\mathbf{P}_i^{k-1} - \frac{\mathbf{P}_i^{k-1} \boldsymbol{\nabla}_i^k [\boldsymbol{\nabla}_i^k]^T \mathbf{P}_i^{k-1}}{1 + [\boldsymbol{\nabla}_i^k]^T \mathbf{P}_i^{k-1} \boldsymbol{\nabla}_i^k} \right]^T \tag{27}$$

and $\boldsymbol{\nabla}_i^k \in \mathbb{R}^{n_i}$ is the vector of the derivatives of $\tilde{e}_i^k$ with respect to $\nu_{i1}^k, \ldots, \nu_{in_i}^k$:

$$\boldsymbol{\nabla}_i^k = -\left[\frac{\partial \tilde{e}_i^k}{\partial \nu_{i1}^{k-1}}, \ldots, \frac{\partial \tilde{e}_i^k}{\partial \nu_{in_i}^{k-1}} \right]^T$$

$$= \Gamma_i^N(\mathbf{x}^k) \begin{bmatrix} \theta_{i1}^{k-1} \frac{\|\mathbf{x}^k - \mathbf{d}_{i1}^k\|^2}{(\nu_{i1}^{k-1})^3} \kappa(\mathbf{x}^k, \mathbf{d}_{i1}^k) \\ \vdots \\ \theta_{in_i}^{k-1} \frac{\|\mathbf{x}^k - \mathbf{d}_{in_i}^k\|^2}{(\nu_{in_i}^{k-1})^3} \kappa(\mathbf{x}^k, \mathbf{d}_{in_i}^k) \end{bmatrix} \tag{28}$$

with $\mathbf{P}_i^1 = \Omega\mathbf{I}, 0 < \Omega < 1000$ and $\nu_{i1}^1 = 0.5$. Algorithm 1 summarizes ePL-KRLS+.

3 Computational Experiments

This section evaluates the performance of ePL-KRLS for stock market price forecasting. Data comprises the daily S&P 500 from 4 January 2010 to 29 December 2017 with a total of 2,013 observations. This section summarizes performance evaluation and forecasting results.

3.1 Methodology

ePL-KRLS+ performance is compared with traditional forecasting algorithms such as ARIMA, ANFIS and MLP neural network, and with state-of-the-art evolving fuzzy modeling approaches, namely eTS, xTS, eTS+, ePL, ePL+, eTS-LS-SVM and eTS-KRLS+ [3,4,10,11,14,20]. Performance evaluation of the algorithms considers two criteria: i) forecasting error and; ii) complexity. Traditional forecasting error measures, the root mean squared error (RMSE) and non-dimensional index error (NDEI) are considered. They are computed as follows:

$$\text{RMSE} = \sqrt{\frac{1}{T}\sum_{k=1}^{T}(y^k - \hat{y}^k)^2} \tag{29}$$

$$\text{NDEI} = \frac{\text{RMSE}}{\mathbf{std}([y^1, \dots, y^T])} \tag{30}$$

where $\hat{y}^k$ is the k-th forecasted value, y^k the k-th actual value and T is the sample size.

Complexity is assessed using average memory storage (bytes), and total processing time (seconds). Considering the evolving fuzzy approaches, the number of rules/neurons at the end of the simulations is also recorded. The data for all simulations were normalized in the [0,1] interval.

Forecasts are one step ahead. Each evaluation considers $[x^k, x^{k-1}]$ as inputs because correlation for S&P 500 time series decreases to non-significant values for higher-order lags. The holdout technique is used for data partitioning: 25% of samples for training, and the remaining 75% for testing.

The parameters are defined as follows: the MLP neural network has one hidden layer with seven neurons trained with backpropagation algorithm. The activation functions are hyperbolic tangents. Initialization of weights is performed randomly at values around 10^{-2}. The stopping criterion was defined as 500 epochs. The ANFIS model was granularized with three fuzzy sets for each input variable, 9 fuzzy rules and the grid partition as the data space partitioning method. The stopping criterion was defined as 50 epochs. For the ARIMA(p, d, q), the AR order is $p = 2$, the degree of integration is $d = 1$ and the MA order is $q = 1$. Parameters of evolving fuzzy approaches were obtained from related

Algorithm 1 – The ePL-KRLS+ learning algorithm

Input: data samples, $\mathbf{x}^k = [x_1^k, \ldots, x_n^k]$, $k = 1, 2, \ldots$

Output: model output

Initialize the parameters α, β, τ, γ and λ

1. Read new data $\mathbf{x}^k$

2. Compute the compatibility index ρ_i^k using (7)

3. Compute the arousal index a_i^k using (6)

4. **If** $a_i^k \geq \tau$, $\forall i \in \{1, \ldots, R\}$ **then:**

5. $\mathbf{x}^k$ is a new rule center; set $R = R + 1$

6.   Initialize the local dictionary $\mathcal{D}_R^k$ and the consequent parameters θ_R^k

7. **Else:**

8.   Update the most compatible rule center $\hat{\mathbf{v}}_i^k$ using (8)

9.   Update the kernel size (26), the consequent parameters (23) and the rule radius (13)

10.   Add $\mathbf{x}^k$ to local dictionary if it was not consistent with $\mathcal{D}_i^k$

11. Compute ρ_{ij}^k for $i, j = 1, \ldots, R$, $i \neq j$ using (9)

12. **If** $\rho_{ij}^k \geq \gamma$ **then:**

13.   Merge rules $\mathbf{v}_i^k$, $\mathbf{v}_j^k$ and set $R = R - 1$

14. Compute the utility measure $\mathcal{U}_i^k$ using (10)

15. **If** $\mathcal{U}_i^k < \epsilon$ **then:**

16.   Remove rule $\mathbf{v}_i^k$ and set $R = R - 1$

17. Compute the normalized firing degree $\Gamma_i^N(\mathbf{x}^k)$ using (5)

18. Compute the local output $\hat{y}_i^k$ using (2) and the global output: $\hat{y}^k$ using (4)

works. The initial setup of ePL-KRLS+ considers $\alpha = 0.01$, the $\beta = 0.20$, $\tau = 0.80$, $\gamma = 0.20$, $\lambda = 10^{-4}$ and $\nu_0 = 0.5$. Notice that models parameters where selected based on simulations considering the RMSE values.

3.2 Results and Discussion

Table 1 shows the forecasting performance in terms of RMSE, NDEI, and computational costs. Results are based on the test set, and the best results are highlighted in bold. The classic ARIMA shows the worst RMSE and NDEI values, and MLP and ANFIS have poor performance as well. Amongst the evolving techniques, eTS-LS-SVM, eTS-KRLS and ePL-KRLS outperform all the remaining competitors. eTS-LS-SVM and eTS-KRLS show a slightly better performance than ePL-KRLS+. Comparison between ePL+ and ePL-KRLS+ shows that using KRLS marginally improves accuracy.

Figure 2 shows the actual and the forecasted values for 2017. Interestingly, the ePL-KRLS+ gives a good fit of the stock market index, indicating the potential of the method. Further, regarding the computational complexity of the models,

Table 1. Performance evaluation for S&P 500 index forecasting.

Models	RMSE	NDEI	Rules/neurons	Time (s)	Memory (b)
ARIMA	0.017	0.082	–	3.67	14338
MLP	0.013	0.060	7	7.01	22394
ANFIS	0.012	0.055	9	3.92	12983
eTS	0.008	0.037	9	0.34	871
xTS	0.010	0.046	6	0.43	789
eTS+	0.009	0.040	**2**	**0.18**	**312**
ePL	0.009	0.041	3	0.73	485
ePL+	0.009	0.040	**2**	0.88	519
eTS-LS-SVM	**0.005**	**0.024**	**2**	1.77	2295
eTS-KRLS	**0.005**	0.026	**2**	1.23	1729
ePL-KRLS+	0.007	0.032	**2**	0.95	1394

eTS, eTS+, ePL, and ePL+ outperform the remaining algorithm. The need to store past input data makes batch ARIMA, MLP, and ANFIS potentially costly over time. Moreover, retraining is needed whenever new samples are input. Algorithms eTS-LS-SVM, eTS-KRLS and ePL-KRLS+ show higher temporal and spatial complexity than the remaining evolving fuzzy models. This is because of the initial construction of the rule-based structure, and the inclusion of new elements in the dictionaries when learning the consequents. However, they still require lower computational cost than ARIMA, MLP, and ANFIS. The results are similar when comparing the number of rules/neurons, which is another way to measure complexity.

Fig. 2. Actual S&P 500 and one-step-ahead ePL-KRLS+ forecast.

4 Conclusion

This paper has addressed an enhanced fuzzy rule-based evolving participatory learning with kernel recursive least squares algorithm for stock market prices forecasting. The algorithm is a novel recursive, incremental approach to update the structure and parameters of a functional fuzzy rule-based forecasting model. The algorithm combines concepts of evolving systems, participatory learning, and kernel methods to produce higher forecast accuracy, robustness, stability, with acceptable computational cost, and has higher autonomy than its predecessors. Computational results concerning one-step-ahead forecasting using S&P 500 stock market index show that the proposed algorithm achieves higher accuracy than traditional benchmarks such as ARIMA, MLP, ANFIS, and state-of-the-art evolving fuzzy models, with an acceptable computational cost. Future investigation will consider ways to automatically set participatory learning clustering procedure parameters and applications to distinct financial data sets.

Acknowledgments. The authors thank the Brazilian Ministry of Education (CAPES), and the Brazilian National Council for Scientific and Technological Development (CNPq) for a fellowship, and grant 305906/2014-3.

References

1. Adebiyi, A.A., Adewumi, A.O., Ayo, C.K.: Comparison of arima and artificial neural network models for stock price prediction. J. Appl. Math. 1–7 (2014)
2. Agrawal, J., Chourasia, V., Mittra, A.: State-of-the-art in stock prediction techniques. Int. J. Adv. Res. Electr. Electron. Instrum. Eng. **2**(4), 1360–1366 (2013)
3. Angelov, P., Filev, D.: An approach to online identification of Takagi-Sugeno fuzzy models. Trans. Syst. Man Cybern. Part B (Cybern.) **34**(1), 484–498 (2004)
4. Angelov, P., Filev, D.P., Kasabov, N.: Evolving Intelligent Systems: Methodology and Applications. Wiley, Hoboken (2010)
5. Bacchetta, P., Mertens, E., Van Wincoop, E.: Predictability in financial markets: what do survey expectations tell us? J. Int. Money Finan. **28**(3), 406–426 (2009)
6. Bollerslev, T., Marrone, J., Xu, L., Zhou, H.: Stock return predictability and variance risk premia: statistical inference and international evidence. J. Financ. Quant. Anal. **49**(3), 633–661 (2014)
7. Engel, Y., Mannor, S., Meir, R.: The kernel recursive least-squares algorithm. Trans. Sig. Process. **52**(8), 2275–2285 (2004)
8. Fama, E.F.: Efficient capital markets: a review of theory and empirical work. J. Finan. **25**(2), 383–417 (1970)
9. Kim, Y., Enke, D.: Using neural networks to forecast volatility for an asset allocation strategy based on the target volatility. Proced. Comput. Sci. **95**, 281–286 (2016)
10. Komijani, M., Lucas, C., Araabi, B.N., Kalhor, A.: Introducing evolving Takagi-Sugeno method based on local least squares support vector machine. Evolv. Syst. **3**(2), 81–93 (2012)
11. Lima, E., Hell, M., Ballini, R., Gomide, F.: Evolving fuzzy modeling using participatory learning. Evol. Intell. Syst.: Methodol. Appl. 67–86 (2010)
12. Liu, W., Principe, J.C., Haykin, S.: Kernel Adaptive Filtering: A Comprehensive Introduction. Wiley, Hoboken (2011)
13. Lughofer, E.: Evolving Fuzzy Systems: Methodologies, Advances Concepts and Applications. Springer, Heidelberg (2011). https://doi.org/10.1007/978-3-642-18087-3

14. Maciel, L., Gomide, F., Ballini, R.: Enhanced evolving participatory learning fuzzy modeling: an application for asset returns volatility forecasting. Evol. Syst. **5**(2), 75–88 (2013)
15. Ngia, L.S.H., Sjoberg, J., Viberg, M.: Adaptive neural nets filter using a recursive Levenberg-Marquardt search direction. In: 32th IEEE Conference on Signals, Systems and Computers. pp. 697–701 (1998)
16. Phan, D.H.B., Sharma, S.S., Narayan, P.K.: Stock return forecasting: some new evidence. Int. Rev. Financ. Anal. **40**, 38–51 (2015)
17. Rather, A.M., Agarwal, A., Sastry, V.: Recurrent neural network and a hybrid model for prediction of stock returns. Expert Syst. Appl. **42**(6), 3234–3241 (2015)
18. Richard, C., Bermudez, J.C.M., Honeine, P.: Online prediction of time series data with kernels. Trans. Sig. Process. **57**(3), 1058–1067 (2009)
19. Scholkopf, B., Smola, A.J.: Learning with Kernels: Regularization, Optimization, and Beyond. MIT Press, Cambridge (2001)
20. Shafieezadeh-Abadeh, S., Kalhor, A.: Evolving takagi-sugeno model based on online gustafson-kessel algorithm and kernel recursive least square method. Evol. Syst. **7**(1), 1–14 (2016)
21. Silva, L.R.S.d.: Aprendizagem participativa em agrupamento nebuloso de dados. mestrado. Universidade Estadual de Campinas (2003). http://libdigi.unicamp.br/document/?code=vtls000296353. Accessed 27 Mar 2017
22. Vieira, R.G., Gomide, F., Ballini, R.: Kernel evolving participatory fuzzy modeling for time series forecasting (Manuscript submitted for publication at the IEEE World Congress on Computational Intelligence)
23. Yager, R.R.: A model of participatory learning. Trans. Syst. Man, Cybern. **20**(5), 1229–1234 (1990)

A Fuzzy Approach Towards Parking Space Occupancy Detection Using Low-Quality Magnetic Sensors

Renato Lopes Moura and Peter Sussner[(⊠)]

Universidade Estadual de Campinas, Campinas, SP, Brazil
{ra163050,sussner}@ime.unicamp.br

Abstract. The detection of vehicles in parking spaces is an important problem for the administration of large-sized parking lots. Economic reasons suggest the use of low cost and low quality magnetic sensors for this purpose. The traditional approach consists in applying thresholds to the signals for the x, y and z axes. Passing these threshold values indicates that a vehicle is located in the corresponding parking space. The literature also includes a straightforward extension of this threshold approach using fuzzy logic. The fuzzy approach described in this paper differs from the aforementioned approaches as well as other ones in the literature since it incorporates additional expert knowledge into a fuzzy rule-based decision system.

Keywords: Fuzzy rule base · Parking space occupancy detection
Anisotropic magneto-resistive sensor · Digital signal analysis

1 Introduction

One of the technologies that smart cities should be equipped with is smart parking in-ground vehicle detection using magnetic or infrared sensors. According to internet sources [1], the benefits include:

1. Live, full information on each and every parking space;
2. Guides drivers to available spaces, improving traffic flow and reducing pollution;
3. Instant information on overstays for infringement enforcement;
4. Facilitates simple, ticketless, barrier free payment systems;
5. Comprehensive information enables profitable future planning;
6. Best possible use of available space.

The smart parking problem consists basically in deciding whether a given parking space can be considered empty or occupied. More precisely, the approach presented in this paper yields three different responses:

© Springer International Publishing AG, part of Springer Nature 2018
G. A. Barreto and R. Coelho (Eds.): NAFIPS 2018, CCIS 831, pp. 374–384, 2018.
https://doi.org/10.1007/978-3-319-95312-0_32

- **E**: The parking spot is empty;
- **T**: A car is transiting into or out of the parking space;
- **O**: The parking spot is occupied by a vehicle.

Hence, we are dealing with a classification problem that can be solved using a number of different methodologies including Bayesian inference, artificial neural networks, and fuzzy logic. Note that statistical methods such as Bayesian inference [3] require some prior knowledge about a probability distribution and that a conventional neural network acts as a black box, i.e., a learning algorithm is responsible for tuning the network's parameters [4] but the result is difficult to interpret for the human practitioner. Neuro-fuzzy approaches [5] combine the main advantages of fuzzy systems and neural networks, namely interpretability and learning capability. On the one hand, such an approach is is not viable for the problem that is discussed in this paper since we only have a few training samples at hand. On the other hand, G. Yoshizawa and E. Ferdinando of the company Cogneti-Tec [2] helped us to relate parts of the signals generated by the magnetic sensors to the events **E**, **T**, and **O** above.

Therefore, we chose to develop a fuzzy rule-based approach whose final response is determined by a crisp decision rule. Our novel approach towards the vehicle detection problem has the following properties:

1. Computationally inexpensive compared to statistical methods and neural networks;
2. Allows to connect the AMR sensor to a microcontroller with low processing power;
3. Capable of handling uncertainties;
4. Capable of incorporating advanced expert knowledge described by linguistic expressions;
5. Methodology is interpretable by human practitioners.

Section 2 provides a few more details on the problem of vehicle detection through magnetic sensors. Section 3 discusses two approaches used to handle this problem, namely the widely used threshold approach and the existing fuzzy approach of Jian et al. [6]. Section 4 introduces our new fuzzy approach together with its main advantages. Section 5 exhibits the results produced by our methodology in applications to some test signals generated by the anisotropic magneto-resistive sensor. We finish with some concluding remarks.

2 Description of the Problem

Recent advances of research and industry and the low cost of anisotropic magneto-resistive (AMR) sensors make them attractive for several types of applications related to magnetic field disturbance. One of the most well-known applications is vehicle detection, since vehicle presence causes a noticeable variation of the Earth's magnetic field due to the existence of ferrous material in their chassis.

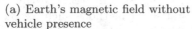

(a) Earth's magnetic field without   (b) Earth's magnetic field with ve-
vehicle presence   hicle presence

Fig. 1. Illustration of a variation in the earth's magnetic field.

Figure 1 illustrates these observations. The vectors representing the Earth's magnetic field are visualized without and with vehicle presence.

Since we are dealing with vectors, the sensor outputs the magnetic readings in three axis (x, y, and z). Evidently, AMR sensors produced by different manufacturers have different precisions and exhibit signals in non-standardized scales. Moreover, these sensors tend to be significantly affected by temperature variations throughout the day. These factors significantly increase the difficulty of the automated parking space occupancy detection problem, especially when using an approach that is merely based on evaluating if the signal emitted by an AMR sensor surpasses a certain threshold.

3   Current Approaches

3.1   Threshold Approach

Some authors [7,8] suggest the use of approaches that are based on the following simple idea: If the signal emitted by a sensor surpasses a pre-determined threshold, then it appears reasonable to assume that a vehicle is present,

However, this approach fails to consider natural variations in the signal, even without any other types of interference such as the ones caused by cars. Figure 2 illustrates the variations of the magnetic signal caused by temperature and sunlight throughout the day.

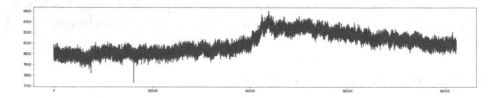

Fig. 2. Variations of a magnetic signal throughout a day.

In this scenario, the algorithm is subject to produce false alarms or should be dynamically adjusted according to the time of the day as well as the temperature and weather conditions.

3.2   The Fuzzy Approach of Jian et al.

As an improvement, Jian et al. proposed a fuzzy approach [6] that replaces
the crisp thresholds by fuzzy sets. To begin with, the means of the first n signal
values for each axis are calculated. Formally, if $i = 1, \ldots, n$ denote the first n data
points and if x_i, y_i, and z_i denote respectively the signal values corresponding
to the x, y, and z axes, then the means $\bar{x}$, $\bar{y}$, and $\bar{z}$ are given by

$$\bar{x} = \frac{\sum_{i=1}^{n} x_i}{n}, \ \bar{y} = \frac{\sum_{i=1}^{n} y_i}{n}, \ \bar{z} = \frac{\sum_{i=1}^{n} z_i}{n}. \tag{1}$$

Then Jian et al. proceed by calculating the absolute values of the differences
between the current reading and this mean. Formally, we have $\tilde{x}_i = |x_i - \bar{x}|$,
$\tilde{y}_i = |y_i - \bar{y}|$, and $\tilde{z}_i = |z_i - \bar{z}|$. These values serve as inputs for a Mamdani-
Assilian- style fuzzy inference system [9]. Figure 3 shows the Gaussian member-
ship functions that were used to model the fuzzy sets *low*, *medium* and *high* in
the antecedent part of the fuzzy rule base. The same fuzzy sets were used for all
axes.

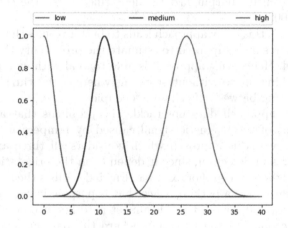

Fig. 3. Membership functions of the fuzzy sets *low*, *medium* and *high*.

The consequent part of the fuzzy rule base describes the probability of parking space occupancy in terms of the fuzzy sets *very low*, *low*, *moderate*, *high*, and
very high whose membership functions are depicted in Fig. 4 below.

The rule base suggested by Jian et al. includes the following rules:

1. **If** $\tilde{x}_i$ is *low* and $\tilde{y}_i$ is *low* and $\tilde{z}_i$ is *low* then the **probability of occupancy**
   is *very low*
2. **If** $\tilde{x}_i$ is *low* and $\tilde{y}_i$ is *low* and $\tilde{z}_i$ is *medium* then the **probability of occupancy** is *low*
3. **If** $\tilde{x}_i$ is *low* and $\tilde{y}_i$ is *medium* and $\tilde{z}_i$ is *medium* then the **probability of
   occupancy** is *medium*

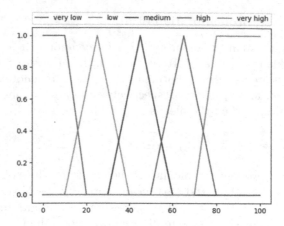

Fig. 4. Membership functions that are meant to describe the probability of parking spot occupancy.

Finally, the resulting output fuzzy is defuzzified using the centroid method in the range 0–100%.

According to [6], this fuzzy approach leads to better results than the threshold method since the result is supposed to estimate the probability that the parking space is occupied. Hence, this approach is able to deal with cars passing slowly through the parking space without stopping whereas the threshold approach would be alternating between "free" and "occupied".

However, this approach does not tackle the problems that are due to the natural variations of the magnetic signal caused by temperature and weather conditions. Moreover, this approach which considers all three axes separately can lead to failures in detection, since it depends on the orientation of the axes when the sensor was installed. For example, Fig. 5 describes the scenario where a car enters backwards into a parking space that is perpendicular to the sidewalk. Note that, in the y and z axes, there is almost no difference between the signal values that correspond to the situations where the parking space is free and where the parking space is occupied.

4   Our Fuzzy Approach

This paper is concerned with the problem of parking spot occupancy detection using AMR sensors, that is a central point to intelligent traffic management in big parking sites such as shopping malls and airports. Note that detecting vacant parking spaces on the basis of digital signals bears some similarities with the noise detection phase of an image denoising method. Several authors have suggested the use of a fuzzy rule base followed by a crisp decision rule for impulse noise detection in digital images [11]. In this case, the result of the image noise detector at a pixel location x only depends on the values in an $N \times N$ window that is centered at x.

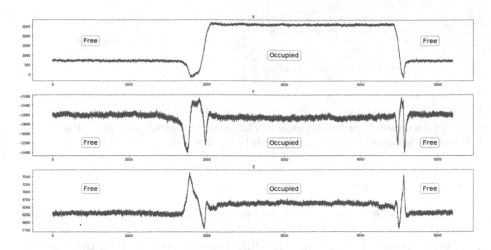

Fig. 5. Scenario where the fuzzy approach of Jian et al. is likely to fail.

Here, we are are considering a sliding window of size $N = 2T + 1$ which is centered at the position that is under consideration. Hence, the center of the window as well as the previous T points and the next T points are taken into account. Since only current and past values but no future values may be part of the decision process, our method will lead to a delay of T time units. Apart from the values in the window that is centered at the current location x, we also record the smallest mean in a window that is located before a window whose values have a variance that is considered to be big. To this end, we first used a crisp decision rule: The variance of the values in a window is considered to be big if its degree of membership in the fuzzy set *big* is larger than or equal to its degree of membership in the fuzzy set *small* depicted in Fig. 7.

Note that the variance in the window centered at location i is given by the equation

$$\frac{\sum_{k=-T}^{T} (v_{i+k})^2}{N}.$$

In the following, we evaluate the following quantities:

1. The **variance** in the current window;
2. The **difference** between the mean in the current window and the smallest mean in a window before the occurence of a big variance.

The difference between means can be better visualized in the Fig. 6:

As stated in [7,8], the vector magnitude deviation from the earth's magnetic field is a reliable method for detecting vehicle presence since we are not concerned with the vehicle's direction of movement. The vector magnitude at location i is given by

$$v_i = \sqrt{x_i^2 + y_i^2 + z_i^2}$$

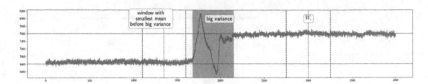

Fig. 6. Consider the location x. The mean of the values in $W_i = \{i + k \mid k = -T, -T + 1, \ldots, T\}$ corresponds to the upper line. The smallest mean in a window that occurs before a big variance in values corresponds to the lower line.

Let us suggest the following rules:

1. If the **variance** is *big*, then a car is *in transition* into or out of the parking spot.
2. If the **variance** is *small* and the **difference** is *positive*, then the **parking spot** is *occupied*.
3. If the **variance** is *small* and the **difference** is *not positive*, then **parking spot** is *free*.

As mentioned before, the fuzzy sets *small* and *big* that describe the variance of the Euclidean norm of the signal vector are depicted in Fig. 7. More precisely, the membership functions of the fuzzy sets *small* and *big* are determined by the following equations:

$$\mu_{small}(x) = \begin{cases} 1 & \text{if } x \leq 8000 \\ \frac{20000 - x}{12000} & \text{if } 8000 < x \leq 20000 \\ 0 & \text{otherwise} \end{cases} \quad \mu_{big}(x) = \begin{cases} 0 & \text{if } x \leq 8000 \\ \frac{x - 8000}{12000} & \text{if } 8000 < x \leq 20000 \\ 1 & \text{otherwise} \end{cases}$$

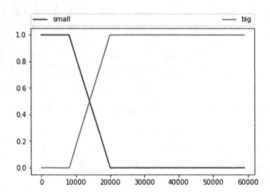

Fig. 7. Membership functions *small* and *big* for the variance of the values in a window.

Similarly, the membership functions of *not positive* and *positive* are shown in Fig. 8 below and are given by the following equations:

$$\mu_{notpositive}(x) = \begin{cases} 1 & \text{if } x \le 20 \\ \frac{50-x}{30} & \text{if } 20 < x \le 50 \\ 0 & \text{c.c.} \end{cases} \qquad \mu_{positive}(x) = \begin{cases} 0 & \text{if } x \le 20 \\ \frac{x-20}{30} & \text{if } 20 < x \le 50 \\ 1 & \text{c.c.} \end{cases}$$

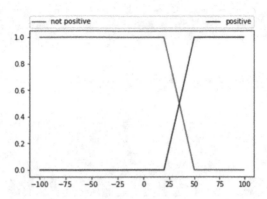

Fig. 8. Membership functions of the variable difference.

The membership functions above were determined manually on the basis of the three training scenarios that are visualized in Fig. 9. In this paper, we considered real-valued inputs corresponding to the variance within the current window of size N and a difference between the mean of the values in the current window and the smallest mean in a window before a big variance occurs. The rules were activated using Zadeh's compositional rule of inference. As an output, we receive degrees of membership in the fuzzy sets *in transition*, *occupied*, and *free*. The maximum of these three values was taken in order to produce a crisp decision.

5   Experimental Results

The subsequent experiments were performed using a sensor that has a sampling frequency of 40 Hz and three different vehicles with different metal compositions of their chassis, inducing different disturbances of the earth's magnetic field as verified on [10]. In the following, we will refer to these vehicles as Vehicles 1, 2 and 3. We adopted the following color scheme:

green: Our method indicates that the parking space is free;
blue: Our method indicates that a car is passing through;
red: Our method indicates that the parking space is occupied.

The membership functions and the size of the moving window were manually adjusted in the training phase in which we merely considered the simplest scenario:

- the monitored parking space that is perpendicular to the sidewalk.
- A single vehicle enters the parking space from the front and remains in the parking space for a while before leaving.

Specifically, we used the training set consisting of the three signals that are depicted in Fig. 9.

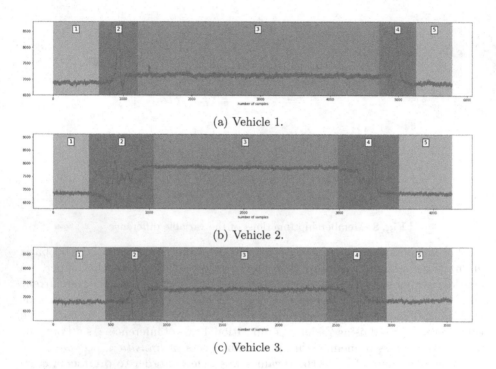

(a) Vehicle 1.

(b) Vehicle 2.

(c) Vehicle 3.

Fig. 9. Magnitudes of the signals that were used as a training set. (Color figure online)

Using the knowledge of our experts from Cogneti-Tec [2], the parameters of the method were adjusted to correctly identify when the parking space is empty (Time Intervals 1 and 5), when the vehicle is entering of leaving the parking space (Intervals 2 and 4) and when the parking space is occupied (Interval 3). As a result of this process, we obtained the membership functions presented in Sect. 4 and 501 as a suitable window size. Note that, for this application, it suffices to determine parameters that allow the correct identification of the three main situation within a few seconds.

In the testing phase, we applied our method to a more complex scenario where the sensor is parallel to the sidewalk and the order of events is as follows:

1. Initially there is no vehicle around (interval 1)
2. Vehicle 1 passes through the parking spot that is equipped with the sensor and parks in the parking space in front of the considered parking space with the sensor (intervals 2 and 3)

3. Vehicle 2 parks in the parking spot with the sensor, behind Vehicle 1 (intervals 4 and 5)
4. Vehicle 3 parks in the parking spot behind the one that has the sensor (on the interval 5)
5. Vehicle 2 leaves (interval 6)
6. The parking spot is free (interval 7)

Our method was able to detect the car passing through the parking space under consideration and that the car did not remain in the observed parking space. Moreover, we were able to detect the arrival of the second vehicle, that it parked in the observed space and that it left (Fig. 10).

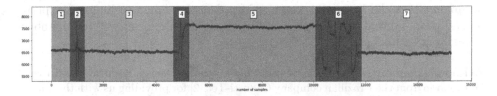

Fig. 10. Results for test case 1.

This scenario was repeated several times while changing the order of the vehicles. Nevertheless, our method managed to correctly identify all the events as can be seen in Fig. 11.

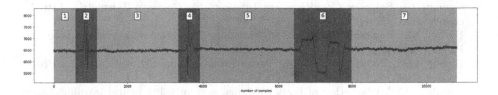

Fig. 11. Results for test case 2.

In the last scenario, the sensor is parallel to the sidewalk and there are already two cars parked: one in front of the parking spot monitored by the sensor and one behind. Then, a third vehicle arrives and is parallel parked in this spot.

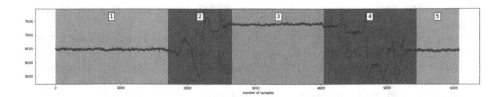

Fig. 12. Results for test case 3.

We can see that, in this case, the driver of the vehicle takes more time to maneuver which is clearly represented by the larger "transition" period seen in Intervals 2 and 4. Nevertheless, the method correctly identified all events (Fig. 12).

6   Concluding Remarks

This paper introduces a simple, but effective approach towards detecting vehicles in parking spaces on the basis of signals produced by AMR sensors. We argued that our approach is relatively unaffected by common pitfalls such as signal variations due to differences in temperature and vehicle chassis compositions. Our methodology yields degrees of certainty regarding the parking spot occupancy that can be used to reach a crisp decision. Some preliminary experimental results reveal that the proposed methodology performs well in a number of different scenarios.

Acknowledgements. The authors would like to thank George Yoshizawa and Érick Ferdinando from the Brazilian company Cogneti-Tec [2] for providing us with this interesting problem including the data used in the experiments. This work was supported in part by CNPq under grant numbers 134611/2017-9 and 313145/2017-2.

References

1. Benefits of In-Ground Vehicle Detection Sensors (n.d.). https://www.smartparking.com/technologies/in-ground-vehicle-detection-sensors
2. Cogneti-Tec Soluções Cognitivas em Internet das Coisas e Telemetria. http://cogneti-tec.com.br/
3. Barber, D.: Bayesian Reasoning and Machine Learning. Cambridge University Press, Cambridge (2012)
4. Bishop, C.M.: Neural Networks for Pattern Recognition. Oxford University Press, Oxford (1995)
5. Abraham, A.: Adaptation of fuzzy inference system using neural learning, fuzzy system engineering: theory and practice. In: Nedjah, N., et al. (eds.) Studies in Fuzziness and Soft Computing, pp. 53–83. Springer, Germany (2005). https://doi.org/10.1007/11339366_3. ISBN 3-540-25322-X
6. Jian, Z., Hongbing, C., Jie, S., Haitao, L.: Data fusion for magnetic sensor based on fuzzy logic theory. In: 2011 Fourth International Conference on Intelligent Computation Technology and Automation, vol. 1, pp. 87–92 (2011)
7. Honeywell: Application Note 218 Vehicle Detection Using AMR Sensors
8. Caruso, M.J., Withanawasam, L.S.: Vehicle Detection and Compass Applications using AMR Magnetic Sensors, Honeywell, SSEC, 12001 State Highway 55, Plymouth, MN USA 55441
9. Mamdani, E.H., Assilian, S.: An experiment in linguistic synthesis with a fuzzy logic controller. Int. J. Man Mach. Stud. **7**(1), 1–13 (1975)
10. Markevicius, V., Navikas, D., Daubaras, A., Cepenas, M., Zilys, M., Andriukaitis, D.: Vehicle influence on the earth's magnetic field changes. Elektronika Ir Elektrotechnika **20**(4), 43–48 (2014). ISSN 1392-1212
11. Schuster, T., Sussner, P.: An adaptive image filter based on the fuzzy transform for impulse noise reduction. Soft. Comput. **21**(13), 3659–3672 (2017)

Adaptive Fuzzy Learning Vector Quantization (AFLVQ) for Time Series Classification

Renan Fonteles Albuquerque[(⊠)], Paulo D. L. de Oliveira[(⊠)],
and Arthur P. de S. Braga[(⊠)]

Federal University of Ceará, Fortaleza, Brazil
{rfonteles,daving,arthurp}@dee.ufc.br

Abstract. Over the past decade, a variety of research fields studied high-dimensional temporal data for pattern recognition, signal processing, fault detection and other purposes. Time series data mining has been constantly explored in the literature, and recent researches show that there are important issues yet to be addressed in the field. Currently, neural network based algorithms have been frequently adopted for solving classification problems. However, these techniques generally do not take advantage from expert knowledge about the processed data. In contrast, Fuzzy-based techniques use expert knowledge for performing data mining classification but they lack on adaptive behaviour. In this context, Hybrid Intelligent Systems (HIS) have been designed based on the concept of combining the adaptive characteristic of neural networks with the informative knowledge from fuzzy logic. Based on HIS, we introduce a novel approach for Learning Vector Quantization (LVQ) called Adaptive Fuzzy LVQ (AFLVQ) which consists in combining a Fuzzy-LVQ neural network with adaptive characteristics. In this paper, we conducted experiments with a time series classification problem known as Human Activity Recognition (HAR), using signals from a tri-axial accelerometer and gyroscope. We performed multiple experiments with different LVQ-based algorithms in order to evaluate the introduced method. We performed simulations for comparing three approaches of LVQ neural network: Kohonen's LVQ, Adaptive LVQ and the proposed AFLVQ. From the results, we conclude that the proposed hybrid Adaptive-Fuzzy-LVQ algorithm outperforms several other methods in terms of classification accuracy and smoothness in learning convergence.

Keywords: Pattern recognition · Time series classification
Artificial neural networks · Fuzzy logic · Hybrid Intelligent Systems
Neuro-fuzzy · Learning Vector Quantization

1 Introduction

Machine learning (ML) is a research field from Artificial Intelligence (AI) which studies techniques for building systems capable of learning automatically by

© Springer International Publishing AG, part of Springer Nature 2018
G. A. Barreto and R. Coelho (Eds.): NAFIPS 2018, CCIS 831, pp. 385–397, 2018.
https://doi.org/10.1007/978-3-319-95312-0_33

experience. Within this area, there is a specific field dedicated to working with temporal data or time series. Time series data mining techniques have been constantly explored in the literature in the past decade and it is still an important topic frequently addressed by researches nowadays. Numerous works have contributed in multiple advances in time series data mining techniques [2,11,12,15,17,21]. Applications such as classification, clustering, anomaly detection and forecast are examples of common data mining tasks applied in time series.

Time series classification has been subject of various researches that explored temporal data collected from sensors, such as accelerometers, electrocardiogram (ECG) and electroencephalogram (EEG) as sources for multi-class identification. Examples of applications are: Human Activity Recognition (HAR) [15], ECG-based [2,5,17,21] and EEG-based [11] classification.

As time series data may have complex characteristics, it is necessary to apply sophisticated solutions that can handle the nonlinear operations. Among the data mining techniques, the Artificial Neural Network (ANN) is an interesting computational intelligent approach for addressing the problem due to its adaptive and generalization capabilities. Classical neural network based algorithms such as Support Vector Machine (SVM) [17], Multi-Layer Perceptron (MLP) [15], Learning Vector Quantization (LVQ) [5,12] have been employed in time series classification problems. Furthermore, deep learning based neural networks such as Deep MLP [2] and Convolution Neural Network (CNN) [21] have been increasingly explored in related studies.

Another category of intelligent classifiers is based on Hybrid Neural Systems (HNS) [19]. HNS are systems which combine artificial neural networks with multiple intelligent methods in a single model for solving a specific problem. This approach aims to extract specific advantages from different techniques in order to build a more robust system. A type of hybrid neural system is Fuzzy Neural Network. In this approach, concepts of Fuzzy Sets and Artificial Neural Networks are combined in a unique method.

Hybrid Fuzzy-LVQ neural networks have been widely explored in the literature. For instance, in [5], the authors presented a new model for data classification using LVQ-based neural network combined with type-2 fuzzy logic. In this work, a fuzzy inference system was employed to determine which network's prototype is the nearest to an input vector. This new method was implemented and tested with two data sets for comparing its effectiveness against the original LVQ algorithm and a type-1 Fuzzy-LVQ. In a different approach, [8] employed a FLVQ-based algorithm with wavelet transformation for classifying abnormalities in images of inner surface of the eye. In this work, the authors compared the FLVQ method with other two methods: Levenger-Marquardt (LM) and Adaptive Neuro-Fuzzy Inference System (ANFIS).

In [4], a study is conducted on the application of the FLVQ model proposed by [18] for probability distribution identification. In a more practical way, a classification algorithm based on Generalized Fuzzy-LVQ method was designed and implemented in a FPGA [1,10]. Furthermore, in [20] three different variation of LVQ algorithms are introduced and compared: Fuzzy-soft LVQ, batch LVQ and Fuzzy-LVQ. As a motivation, few works have explored Fuzzy-LVQ strategies for dealing with high-dimensional temporal data. Therefore, we believe hybrid Fuzzy-LVQ classifiers have the potential to be explored more deeply.

In this paper, we present a hybrid neuro-fuzzy algorithm that combines adaptive LVQ-ANN proposed by [3] and the Fuzzy-LVQ introduced in [7]. In our study, we employ the proposed technique for classifying human activities through tri-axial accelerometer time series patterns. The next sections of this paper are organized as follows: in Sect. 2 we introduce the main concepts of Learning Vector Quantization theory. In Sect. 3 we present the fundamentals of the proposed method (AFLVQ). In Sect. 4, we describe the methodology of this work, including a brief description of the data set used and the performed experiment. In Sect. 5 we present the results of the simulations and in Sect. 6 we conclude the article by summarizing the results and suggesting future works.

2 Learning Vector Quantization (LVQ)

Learning Vector Quantization (LVQ) is a prototype-based supervised classification algorithm which adopts a learning strategy based on similarity measures (distance functions) and winner-take-all approach. LVQ is a neural network based method proposed by Kohonen [13]. Kohonen's LVQ is a supervised competitive learning algorithm substantiated in concepts of vector distance as a similarity measure.

Its architecture is composed by a layered feedforward network which has a competitive layer where the neurons compete among them based on a distance metric, or a similarity measure, between training instances and prototypes. Generally, Euclidean Distance is chosen as the distance metric for LVQ implementation. This method aims to divide the data space into distinct regions and defining a vector prototype (or neuron) for each region. This process is also known as Vector Quantization.

2.1 Kohonen's LVQ1

The learning method in LVQ consists in using the input vector as guidance for organizing the prototypes in specific regions that defines a class. Firstly, a set of prototypes is initialized and for each prototype is assigned a class. Each class must be represented by at least one prototype, a class can have multiple prototypes, and one prototype only represents a unique class. Then, during the learning process, each instance from the training set is compared with all network's prototypes, using a similarity measure. LVQ-based algorithms are classified as competitive learning due to the selection of the closest prototype within the set of P prototypes:

$$w = arg\ \min_{i=1}^{P}\ d(\boldsymbol{x}_j, \boldsymbol{p}_i) \tag{1}$$

where w is the index of the winner prototype (the closest prototype of an specific instance x_j). The distance is measured by a distance function. The Euclidean distance, or L_2-norm, is generally used to calculate this distance.

$$d(x_j, p_i) = \|x_j - p_i\|^2 = \sqrt{\sum_{k=1}^{n} (x_{jk} - p_{ik})^2} \tag{2}$$

where n is the dimension of the instance x_j, which is the same for p_i. If the class of an instance is equal to the class of the closest prototype (winner proto-type), this prototype is moved towards the instance, otherwise it moves away. Consider t as the iteration counter of the training algorithm. The learning rule for Kohonen's LVQ1 algorithm is given by:

$$p_w(t+1) = \begin{cases} p_w(t) + \alpha(t)[x_j - p_w(t)] & \text{if } C(p_w) = C(x_j); \\ p_w(t) - \alpha(t)[x_j - p_w(t)] & \text{if } C(p_w) \neq C(x_j). \end{cases} \tag{3}$$

For all prototypes $p_i(t)$ where $i \neq w$, the prototypes remains the same. In our experiments, we adopted a linearly decreasing learning rate $\alpha(t) = \alpha(0)(1 - \frac{t}{N})$, where $\alpha(0)$ is the initial learning rate and N is the maximum number of training iterations.

2.2 Kohonen's LVQ2

Kohonen introduced in 1988 the LVQ2 algorithm, another variation similar to the original LVQ [14]. However, the learning process is based on two prototypes p_{1st} and p_{2nd} that are the first and second nearest prototypes to an instance x_j, respectively. One of them must belong to the correct class and the other to a incorrect class. Furthermore, these prototypes must fall into a zone defined around the mid plane between them. For an instance x_j and the two nearest prototypes p_{1st} and p_{2nd}, let d_{1st} and d_{2nd} be the distances of x_j to p_{1st} and p_{2nd}, respectively. Then x_j will fall into a window of a width w if Eq. 4 is satisfied. It is recommended to adopt the width w between 0.2 and 0.3 [13]. The prototype LVQ2 learning rule is given by the Eq. 5.

$$min\left(\frac{d_{1st}}{d_{2nd}}, \frac{d_{2nd}}{d_{1st}}\right) > s, \text{ where } s = \frac{1-w}{1+w} \tag{4}$$

$$\begin{aligned} p_{1st}(t+1) &= p_{1st}(t) - \alpha(t)[x_j - p_{1st}(t)] \\ p_{2nd}(t+1) &= p_{2nd}(t) + \alpha(t)[x_j - p_{2nd}(t)] \end{aligned} \tag{5}$$

2.3 Quantization Error (Q_E)

In prototype-based algorithms, the prototypes can be considered quantization vectors, as they represent a specific region in the input data [16]. For evaluating the vector quantization in a prototype-based algorithm, a Quantization Error

(Q_E) can be used. This error metric is based on the average of the distances between prototypes and the instances of the data.

$$Q_E = \frac{1}{N} \sum_{j=1}^{N} \|x_j - p_w\|^2 \qquad (6)$$

where N is the number of instances, x_j is the jth instance, and p_w is a prototype that represents the class of x_j. Given a set of prototypes $\mathcal{P} = \{p_1, p_2, \ldots, p_k\}$, w represents the index of the closest prototype to the instance x_j. The value w can be calculated by the Eq. 1.

3 Adaptive Fuzzy Learning Vector Quantization (AFLVQ)

3.1 Fundamentals on AFLVQ

The Adaptive-Fuzzy-LVQ model is inspired by two concepts:

Adaptability
Adaptive LVQ-ANN is a specific variation of LVQ-based methods which has the capability of adjusting their architecture to improve network performance during the training process. In general, the adaptive characteristics implies the ability of making changes in the network's structure by including or removing prototypes (codebooks or neurons). In previous work [3], a study was conducted on a proposed adaptive LVQ algorithm, applied to human activity recognition using data collected from a tri-axial accelerometer. In [3], the Kohonen's LVQ algorithm was modified to include an adaptive step at the end of each epoch during the network training. The adaptive process consisted of two stages:

- **Prototype inclusion:** The inclusion of new prototypes is based on Kohonen's Self-Organizing Map [13] applied on misclassified samples. The number of prototypes to be included is calculated based on the quantity of misclassified instances of a specific class. Hence, the greater the presence of misclassified instances of a class C_i, the greater the number of new prototypes $(p_{new} \rightarrow C_i)$ that will be included to represent this class.
- **Prototype removal:** The removal of prototypes is determined by a score calculated for each prototype. For a prototype k, its score can be calculated by $score_k = A_k - B_k$, where A_k and B_k represent how many times this prototype has been a winner and correctly classified and misclassified, respectively. Removal of a prototype will be done whenever this score is lower than a removal threshold (ψ). Low scores indicate that a prototype frequently classify incorrectly instances or do not contribute significantly to the classification performance.

The neural network growth is restricted by a variable called Budget. Therefore, the number of prototypes will not overstep the pre-defined architecture size. Further implementation details about this method can be found in [3].

Fuzzy

In the proposed AFLVQ method, the fuzzy part is based on the Fuzzy-LVQ introduced by Chung [7]. Its algorithm consists in optimizing a fuzzy objective function by minimizing the network output error, calculated by the difference of the class membership of the target and actual values, and minimizing the distances between training patterns and competing neurons. In their works, Chung and Lee [7] define the following objective function:

$$Q^m(U, \mathcal{V}) = \sum_{j=1}^{N} \sum_{i=1}^{P} [(t_{ji})^m - (\mu_{ji})^m] d(\boldsymbol{x}_j, \boldsymbol{p}_i) \tag{7}$$

subject to the following constraints: $\sum_{i=1}^{c} \mu_{ji} = 1; \forall j$ and $\mu_{ji} \in [0, 1]; \forall j, i$. The term $d(x_j, p_i)$ represents the distance between the ith prototype and the jth instance (See Eq. 1). The fuzziness parameter m define weights for the membership functions for each prototype in a manner that the greater the value of m, the smoother is the learning process. The target class membership value of neuron i for input pattern j is represented by $t_{ji} \in \{0, 1\}$. Hence, the FLVQ learning rule and the membership updating rule will be:

$$\boldsymbol{p}_i(t+1) = \boldsymbol{p}_i(t) + \alpha(t)[(t_{ji})^m - (\mu_{ji})^m][\boldsymbol{x}_j - \boldsymbol{p}_i(t)]; \forall i \tag{8}$$

$$\mu_{ji} = \left[\sum_{\ell=1}^{P} \left(\frac{d(\boldsymbol{x}_j, \boldsymbol{p}_i)}{d(\boldsymbol{x}_j, \boldsymbol{p}_\ell)} \right)^{\frac{1}{m-1}} \right]^{-1} \tag{9}$$

Note that the previous equations are only valid when the number of prototypes P is equal to the number of instances N. For LVQ architectures with multiple prototypes per class, we introduce a competitive step in the training process. In Fig. 1 the FLVQ network is described.

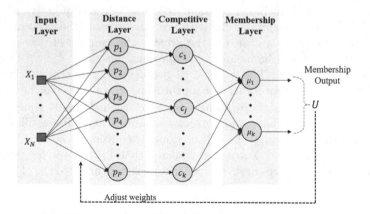

Fig. 1. Fuzzy-LVQ architecture

In the figure, we have an input layer that receives the instances from the dataset. The distance layer calculates the distance between each prototype to the presented instance. Then, in the competitive layer (called MIN layer by Chung [7]), only the closest prototype from each class is chosen to undergo the fuzzy competition. Therefore, the membership computations and parametric vector modification will be applied at the maximum number of classes in the problem or k prototypes.

3.2 AFLVQ Algorithm

The presented Adaptive-Fuzzy-LVQ algorithm can be divided in two stages:

- **Training:** During the training stage, the instances from the training dataset is presented to the neural network, and the prototypes are adjusted based on Chung's Fuzzy-LVQ algorithm [7].
- **Adaptation:** After completing an epoch, the resulted LVQ-network is evaluated in order to verify the need for adaptation.If there is a need for adaptation, the adaptive method from our previous work [3] removes or includes prototypes, according to criteria that aim to improve the network performance.

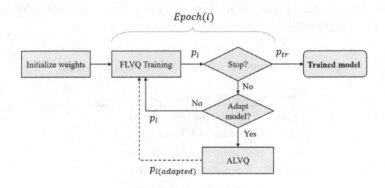

Fig. 2. Flow-chart of training an Adaptive-Fuzzy-LVQ model

Figure 2 presents a flowchart describing the process of training an Adaptive-Fuzzy-LVQ model. First, the prototypes' weights are initialized. We used the Kohonen Self-Organizing Map for initializing the prototypes. Afterwards, the FLVQ is executed during a whole epoch i. In the end of each epoch, the algorithm verifies the need for adaptation and, depending on this decision, the network is adapted ($p_{i(adapted)}$) or not (p_i). Then, the cycle restarts in the next epoch. When the stop criteria are satisfied, the algorithm returns the trained model.

4 Methodology

4.1 Dataset

The dataset used in our experiments was selected from the UCI Machine Learning Repository [9]. It is an Activity Recognition database, built from the recordings of 30 volunteers executing multiple activities while carrying a waist-mounted smartphone. This dataset was introduced by Anguita [6] and it contains the collection of data from embedded inertial sensors: a tri-axial accelerometer and a gyroscope. In this time series classification problem, the aim is to recognize activities or actions performed by humans based on the information retrieved from body-worn motion sensors. The selected activities classes are: *standing, sitting, laying down, walking, walking downstairs* and *walking upstairs*. The dataset is composed by 10299 samples which of them 7352 compose the training set and 2947 the test set, representing a division of 70% and 30% for training and test data, respectively.

Features

In order to properly represent the input data, multiple features were extracted from the sensors' raw data. As it is a time series classification problem, the features from each instance were calculated based on multiple observations in an ordered sequence, usually called sub-series or data window. Examples of extracted features are described in Table 1. These features were selected and extracted by Anguita [6].

Table 1. Example of extracted features from signals

Function	Description
mean	Mean value
std	Standard deviation
max	Largest values in array
min	Smallest value in array
sma	Signal magnitude area
correlation	Correlation coefficient
energy	Average sum of the squares

The total features extracted was 561. Thus, for an instance $x \in \mathbb{R}^D$, its dimension is given by $D = 561$.

4.2 Experiments

Experiments were designed to verify the algorithm's performance based on the number of prototypes (N_c). The occurrence of adaptation is strongly dependent on the amount of prototypes in the network. Hence, by analyzing this aspect, we

Table 2. Experiment's attributes and values

Attributes	Epochs	P	Algorithms	α_0	w	m
Values	100	$\{N_c, 5N_c, 10N_c\}$	LVQ1, LVQ2, FLVQ, ALVQ1, ALVQ2, AFLVQ	0.09	0.2	1.4

can demonstrate the influence of adaptation in the training performance. The attributes of the conducted experiments are described in Table 2.

The number of prototypes (P) depends on the number of classes (N_c). Therefore, since there are $N_c = 6$ classes in the dataset, the experimented networks will have 6, 30 and 60 prototypes, respectively. We executed the experiments with different variations of LVQ: LVQ1 and LVQ2 from Kohonen [13], FLVQ from Chung [7], ALVQ1 and ALVQ2 from our previous work [3] and AFLVQ, which is the approach presented in this paper. Note that w and m are specific parameter for LVQ2 and FLVQ learning rules.

5   Results

The classification performances of LVQ-based algorithms with networks composed by 6, 30 and 60 prototypes are presented in Table 3. For networks with fewer prototypes, the adaptation is generally nonexistent. Take, for example, the case where there is only one prototype representing each class, thus $P = N_c = 6$. In this case, it is unlikely that a prototype will be removed. As each prototype is representing a class by itself, eventually all prototypes will be chosen as a winner during the training process. Hence, in this case, the non-adaptive LVQ algorithm is equivalent to its adaptive version. As we can see in Table 3, for $P = 6$ the algorithms LVQ1, LVQ2 and FLVQ are equivalent to ALVQ1, ALVQ2 and AFLVQ, respectively.

As the number of prototypes grows, they reduce their chance to be chosen, and after one epoch, many prototypes may be removed for not being chosen at least once. Adaptation is more effective for greater number of prototypes. As we can see in the cases where $P = 30$ and $P = 60$, the training accuracy in adaptive methods increases. However, in some cases there is a cost of reducing its generalization. Hence, it is necessary to properly select the number of prototypes (P) in order to avoid overfitting. Note also that, in all three scenarios presented in Table 3, LVQ2 and ALVQ2 had the same results. In other words, ALVQ2 has not been adapted during training. This can also be seen in Fig. 3.a, where the chart of LVQ2 (Orange) and ALVQ2 (Green dots) are overlapping.

In Fig. 3 we present the evolution of two error measures throughout the epochs: classification error (C_E) and quantization error (Q_E). Classification error can be calculated by: $C_E = N_{miss}/N$, where C_E is the classification error, N_{miss} is the number of misclassified instances and N is the total of instances. Quantization error is described in Sect. 2.3. In Fig. 3.(a) we can notice that FLVQ and AFLVQ are methods which converge smoothly, with minor oscillations when

Table 3. Experiment results for $P = \{N_c, 5N_c, 10N_c\}$

P	Set	LVQ1	LVQ2	FLVQ	ALVQ1	ALVQ2	AFLVQ
N_c	Training	83.62%	98.83%	86.48%	83.62%	98.83%	86.48%
$P = 6$	Test	82.76%	95.45%	86.19%	82.76%	95.45%	86.19%
$5N_c$	Training	90.61%	99.67%	90.44%	90.61%	99.67%	91.04%
$P = 30$	Test	84.63%	93.96%	87.61%	86.73%	93.96%	87.58%
$10N_c$	Training	94.08%	99.62%	93.27%	94.27%	99.62%	94.07%
$P = 60$	Test	87.75%	91.99%	89.18%	88.12%	91.99%	89.65%

comparing to LVQ1 and ALVQ1. Regarding the proposed AFLVQ, it outperforms all other algorithms, except LVQ2 and ALVQ2, which have demonstrated to be significantly superior than the other algorithms, for this specific dataset. ALVQ2 and LVQ2 also presented the fastest convergence by reaching low error values in less than 20 epochs.

In Fig. 3.(b), taking as an example the best model (LVQ2 and ALVQ2), we can observe that initially the $Q_E = 15.78$. After training, the error increased to $Q_E = 30.45$, instead of reducing. Strange as it may seem, low Q_E does not necessarily mean a well trained model, neither high Q_E means a poorly trained model. However, there are limits for acceptable Q_E values that may change accordingly to the disposition of the data. It is extremely important to evaluate the relationship between quantization error and classification error in order to properly choose a classification model.

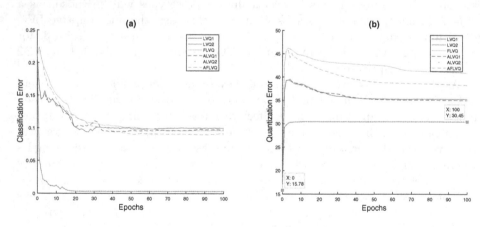

Fig. 3. Classification Error (C_E) and Quantization Error (Q_E) evolution throughout the epochs (Color figure online)

The dataset employed in our experiments has a specific characteristic where pairs of classes share very similar instances. For example, the class *walking* and *walking downstairs* present similar patterns. This can be evidenced by examining the confusion matrix obtained from the best trained model. In Table 4, we observe that most of the misclassifications were caused by mistaking class 4 for class 5, or the other way around. This characteristic existing in this dataset is well suited for LVQ2 learning rules, which explains the remarkable results obtained through this algorithm.

Table 4. Confusion matrix for the best test result (LVQ2 and ALVQ2 with $N_c = 6$)

	C_1	C_2	C_3	C_4	C_5	C_6	**Recall**
C_1	478	7	11	0	0	0	96.37%
C_2	18	450	3	0	0	0	95.54%
C_3	5	18	397	0	0	0	94.52%
C_4	0	2	0	448	40	1	91.24%
C_5	0	0	0	29	503	0	94.55%
C_6	0	0	0	0	0	537	100.00%
Precision	95.41%	94.34%	96.59%	93.92%	92.63%	99.81%	**95.45%**

6 Conclusion

In this paper, we presented a novel Adaptive-Fuzzy-LVQ method applied in human activity classification from high-dimensional signals of motion sensors. We conducted experiments in order to evaluate different variations of the LVQ algorithm to compare with the proposed method. It is possible to conclude from the results that employing AFLVQ provides considerable improvements in accuracy on the classification of time series, comparing to other LVQ-based algorithms. In general, AFLVQ outperformed all variations, except for LVQ2 and ALVQ2.

Comparing LVQ1 and FLVQ, we observed that both are very similar in overall accuracy. However, the learning process in LVQ1 seems to be more unstable while FLVQ presents a smoother evolution over training epochs and tends to converge to better results. The learning rate weighted by the membership values of each prototype justifies the FLVQ to be smoother as the adjustment rate will be relative to the distance of a prototype to a specific instance. From the obtained results, we can also conclude that LVQ-based algorithms are effective for performing classification of high-dimensional time series, since in most experiments, they demonstrated convergence to high training accuracy. Regarding generalization, all algorithms achieved test accuracy between 82.76% and 95.45%, which is satisfactory, considering the problem complexity.

Concerning future works, we intend to combine Kohonen's LVQ2 with Fuzzy Logic to evaluate its performance in the problem addressed in this paper. Based on the experiment's results, we understand that fuzzy can improve LVQ-based algorithms by smoothing the training convergence, avoiding major oscillation that may result in poor classification performance. Once LVQ2 and ALVQ2 have presented the best classification performances, we can improve their learning rule by including fuzzy aspects. Furthermore, we intend to explore the changes in performance by varying the fuzziness parameter m, as well as the learning rate α. Properly tuning these parameters is important, as they can significantly influence the training results. Finally, we plan to work with Type-2 Fuzzy sets for evaluating its performance in dealing with uncertainties present in input data.

Acknowledgements. We would like to express our gratitude to the Coordination for the Improvement of Higher Education Personnel (CAPES) for the financial support.

References

1. Afif, I.N., Wardhana, Y., Jatmiko, W.: Implementation of adaptive fuzzy neuro generalized learning vector quantization (AFNGLVQ) on field programmable gate array (FPGA) for real world application. In: 2015 International Conference on Advanced Computer Science and Information Systems (ICACSIS), pp. 65–71. IEEE (2015)
2. Al Rahhal, M.M., Bazi, Y., AlHichri, H., Alajlan, N., Melgani, F., Yager, R.R.: Deep learning approach for active classification of electrocardiogram signals. Inf. Sci. **345**, 340–354 (2016)
3. Albuquerque, R.F., Braga, A.P.d.S., Torrico, B.C., Reis, L.L.N.d.: Classificação de dinâmicas de sistemas utilizando redes neurais LVQ adaptativas. In: Conferência Brasileira de Dinâmica, Controle e Aplicações - DINCON (2017)
4. Alfa, G.D., Kurniasari, D., Usman, M., et al.: Neural network fuzzy learning vector quantization (FLVQ) to identify probability distributions. Int. J. Comput. Sci. Netw. Secur. (IJCSNS) **16**(10), 16 (2016)
5. Amezcua, J., Melin, P., Castillo, O.: New Classification Method Based on Modular Neural Networks with the LVQ Algorithm and Type-2 Fuzzy Logic. Springer, Heidelberg (2018). https://doi.org/10.1007/978-3-319-73773-7
6. Anguita, D., Ghio, A., Oneto, L., Parra, X., Reyes-Ortiz, J.L.: A public domain dataset for human activity recognition using smartphones. In: ESANN (2013)
7. Chung, F.L., Lee, T.: Fuzzy learning vector quantization. In: Proceedings of 1993 International Joint Conference on Neural Networks, IJCNN 1993, Nagoya, vol. 3, pp. 2739–2743. IEEE (1993)
8. Damayanti, A.: Fuzzy learning vector quantization, neural network and fuzzy systems for classification fundus eye images with wavelet transformation. In: 2017 2nd International Conferences on Information Technology, Information Systems and Electrical Engineering (ICITISEE), pp. 331–336, November 2017
9. Dheeru, D., Karra Taniskidou, E.: UCI machine learning repository (2017). http://archive.ics.uci.edu/ml
10. Fajar, M., Jatmiko, W., Agus, I.M., et al.: FNGLVQ FPGA design for sleep stages classification based on electrocardiogram signal. In: 2012 IEEE International Conference on Systems, Man, and Cybernetics (SMC), pp. 2711–2716. IEEE (2012)
11. Hajinoroozi, M., Mao, Z., Jung, T.P., Lin, C.T., Huang, Y.: EEG-based prediction of driver's cognitive performance by deep convolutional neural network. Sig. Process. Image Commun. **47**, 549–555 (2016)
12. Jain, B.J., Schultz, D.: Asymmetric learning vector quantization for efficient nearest neighbor classification in dynamic time warping spaces. Pattern Recogn. **76**, 349–366 (2018)
13. Kohonen, T.: The self-organizing map. Proc. IEEE **78**(9), 1464–1480 (1990)

14. Kohonen, T., Barna, G., Chrisley, R.: Statistical pattern recognition with neural networks: benchmarking studies. In: IEEE International Conference on Neural Networks, vol. 1, pp. 61–68 (1988)
15. Nakano, K., Chakraborty, B.: Effect of dynamic feature for human activity recognition using smartphone sensors. In: 2017 IEEE 8th International Conference on Awareness Science and Technology (iCAST), pp. 539–543, November 2017
16. Peres, S.M., Rocha, T., Biscaro, H.H., Madeo, R.C.B., Boscarioli, C.: Tutorial sobre fuzzy-c-means e fuzzy learning vector quantization: abordagens híbridas para tarefas de agrupamento e classificação. Revista de Informática Teórica e Aplicada 19(1), 120–163 (2012)
17. Rajesh, K.N., Dhuli, R.: Classification of ECG heartbeats using nonlinear decomposition methods and support vector machine. Comput. Biol. Med. 87, 271–284 (2017)
18. Sakuraba, Y., Nakamoto, T., Moriizumi, T.: New method of learning vector quantization using fuzzy theory. Syst. Comput. Jpn. 22(13), 93–103 (1991)
19. Wermter, S.: Hybrid Neural Systems. Springer, Heidelberg (2000)
20. Wu, K.L., Yang, M.S.: A fuzzy-soft learning vector quantization. Neurocomputing 55(3–4), 681–697 (2003)
21. Xia, Y., Wulan, N., Wang, K., Zhang, H.: Detecting atrial fibrillation by deep convolutional neural networks. Comput. Biol. Med. 93, 84–92 (2018)

A Fuzzy C-means-based Approach for Selecting Reference Points in Minimal Learning Machines

José A. V. Florêncio$^{(\boxtimes)}$, Madson L. D. Dias, Ajalmar R. da Rocha Neto$^{(\boxtimes)}$, and Amauri H. de Souza Júnior

Department of Teleinformatics, Federal Institute of Ceará, Treze de Maio Avenue, Fortaleza, Ceará 2081, Brazil
{jose.florencio,madson.dias}@ppgcc.ifce.edu.br,
{ajalmar,amauriholanda}@ifce.edu.br
http://ppgcc.ifce.edu.br/

Abstract. This paper introduces a new approach to select reference points of minimal learning machines (MLM) for classification tasks. The proposal is based on the Fuzzy C-means algorithm and consists of selecting data samples from regions where no overlapping between classes exists. Such an idea has been empirically shown capable of achieving simpler decision boundaries in comparison to the standard MLM, and thus less susceptible to overfitting. Experiments were performed using UCI data sets. The proposal was able to both reduce the number of reference points and achieve competitive performance when compared to conventional approaches for selecting reference points.

Keywords: Machine learning · Minimal learning machines
Fuzzy-C means

1 Introduction

The Minimal Learning Machine (MLM, [1]) is a supervised learning algorithm that has recently been applied to a diverse range of problems, such as fault detection [2], ranking of documents [3], and robot navigation [4].

The basic operation of MLM consists in a linear mapping between the geometric configurations of points in the input space and the respective points in the output space. The geometric configuration is captured by two distance matrices (input and output), computed between the training/learning points and a subset of it whose elements are called reference points (RPs). The learning step in the MLM consists of fitting a linear regression model between these two distance matrices. In the test phase, given an input, the MLM predicts its output by first computing distances in the input space and then using the learned regression

---

The authors would like to thank the IFCE for supporting their research.

© Springer International Publishing AG, part of Springer Nature 2018
G. A. Barreto and R. Coelho (Eds.): NAFIPS 2018, CCIS 831, pp. 398–407, 2018.
https://doi.org/10.1007/978-3-319-95312-0_34

model to predict distances in the output space. Those distances are then used to provide an estimate to the output.

The determination of the RPs, including its quantity, is fundamental to the quality of the surface boundary generated by the MLM model. In this regard, the original formulation of the MLM training algorithm establishes a random reference points choice, leaving just the number of points definition determined by the user. The random reference points choice ignores the data set disposition.

Clustering algorithms, such fuzzy C-means [5], are unsupervised machine learning methods that aim to separate objects into groups, based on the characteristics that these objects have. The basic idea is to bring objects with similar characteristics together into one group. The similarity of these objects is defined according to pre-established criteria.

The use of Fuzzy C-Means together with classification methods occurs mainly for the reduction of datasets with a large number of samples. In addition, some works use this algorithm and its variations in several types of applications, such as the classification of epilepsy risk [6], feature selection [7] and image segmentation [8].

In this work, it is proposed to use the algorithm Fuzzy C-Means for the selection of reference points of Minimum Learning Machines applied to problems of patterns recognition. Simulation with real-world datasets was performed to validate the proposal. The proposal was able to significantly reduce the number of reference points, as well as maintain its capacity of generalization equivalent or superior when compared to the approaches proposed in the original article [1].

The remainder of the paper is organized as follows. Section 2 briefly describes the MLM. Section 3 emphasizes the need for novel methods for selecting RPs. Section 4 introduces the Fuzzy C-Means MLM. Section 5 reports the empirical assessment of the proposal and the conclusions are outlined in Sect. 6.

2   Minimal Learning Machine

The Minimal Learning Machine is a supervised method whose training step consists of fitting a multiresponse linear regression model between distances computed from the input and output spaces. Output prediction for new incoming inputs is achieved by estimating distances in the output spaces using the underlying linear model followed by a search/optimization procedure in the space of possible outputs.

Basic Formulation. Let us define the learning problem as the problem of approximating a smooth continuous *target function* $f : \mathcal{X} \to \mathcal{Y}$ from the data $\mathcal{D} = \{(\mathbf{x}_n, \mathbf{y}_n = f(\mathbf{x}_n))\}_{n=1}^{N}$, where $\mathbf{x}_n \in \mathcal{X}$ and $\mathbf{y}_n \in \mathcal{Y}$. We call $\mathcal{X}$ and $\mathcal{Y}$ the input and output spaces, respectively. Henceforth, we assume $\mathcal{X} = \mathbb{R}^D$ and $\mathcal{Y} = \mathbb{R}^S$.

The MLM aims to approximate the target function f through the use of surrogate functions $\delta_k : \mathcal{Y} \to \mathbb{R}_+$ and $d_k : \mathcal{X} \to \mathbb{R}_+$. The surrogate functions are distance functions taken from fixed points $\mathcal{R} = \{(\mathbf{m}_k, \mathbf{t}_k = f(\mathbf{m}_k)) \in \mathcal{D}\}_{k=1}^{K}$, also called reference points. More precisely, we have $d_k(\mathbf{x}) = d(\mathbf{x}, \mathbf{m}_k)$ and $\delta_k(\mathbf{y}) = \delta(\mathbf{y}, \mathbf{t}_k)$, with both $d(\cdot, \cdot)$ and $\delta(\cdot, \cdot)$ given by the Euclidean distance function. In addition, we refer to the set $\{\mathbf{m}_k\}_{k=1}^{K}$ as input reference points, and $\{\mathbf{t}_k\}_{k=1}^{K}$ as the corresponding output reference points.

We assume the existence of a mapping between the spaces induced by the distance functions δ and d. Formally, we have $g_k : \prod_{j=1}^{K} d_j(\mathcal{X}) \to \delta_k(\mathcal{Y})$, or equivalently $g_k : \mathbb{R}_+^K \to \mathbb{R}_+$. Considering now the data $\mathcal{D}$, we collect the distances taken in the input space between the data points and the input reference points in a matrix $\mathbf{D} \in \mathbb{R}_+^{N \times K}$. Similarly, take the pointwise distance matrix in the output space between the N data points (outputs $\mathbf{y}_n$) and the output reference points to be represented by $\mathbf{\Delta} \in \mathbb{R}_+^{N \times K}$. Using the data, we are interested in finding the mapping g_k using the model $\Delta_{n,k} = g_k(\mathbf{D}_{n,\cdot}) + \epsilon_n$ for all $n = 1, \ldots, N$. The term ϵ_n represents the residuals whereas $\mathbf{D}_{n,\cdot}$ denotes the n-th row of the matrix $\mathbf{D}$; by the same token, $\Delta_{n,k}$ stands for the element in the n-th row and k-th column of $\mathbf{\Delta}$.

The MLM assumes that the mappings g_k can be sufficiently well approximated by linear models. In doing so, we have that distances in the output space can be approximated by a linear combination of distances in the input space, i.e., $\Delta_{n,k} = \mathbf{D}_{n,\cdot} \mathbf{b}_k + \epsilon_n$, where $\mathbf{b}_k \in \mathbb{R}^K$ represents the coefficients of the linear mapping g_k. Putting all the mappings together for all data points, we represent the so-called *distance regression* model of the MLM in a matrix form given by

$$\mathbf{\Delta} = \mathbf{D}\mathbf{B} + \varepsilon, \tag{1}$$

where the matrix $\mathbf{B} \in \mathbb{R}^{K \times K}$ comprises the K vectors of coefficients $\mathbf{b}_k$ in its columns.

Given that, the MLM computes a function $h_{\mathbf{B}}(\mathbf{x}) : \mathcal{X} \to \mathcal{Y}$ given by:

$$h_{\mathbf{B}}(\mathbf{x}) = \arg\min_{\mathbf{y}} \sum_{k=1}^{K} \left[\delta_k^2(\mathbf{y}) - \left(\sum_{i=1}^{K} d_i(\mathbf{x}) B_{i,k} \right)^2 \right]^2, \tag{2}$$

where $\delta_k(\mathbf{y}) = \|\mathbf{y} - \mathbf{t}_k\|$ represents the Euclidean distance between $\mathbf{y}$ and the k-th output reference point $\mathbf{t}_k$; similarly, $d_i(\mathbf{x}) = \|\mathbf{x} - \mathbf{m}_i\|$ denotes the Euclidean distance between $\mathbf{x}$ and the i-th input reference point $\mathbf{m}_i$; K denotes the number of reference points.

Learning Algorithm. The learning algorithm of the Minimal Learning Machine simply requires the (i) selection of the set reference points $\{(\mathbf{m}_k, \mathbf{t}_k)\}$; and (ii) determination of the parameters $\mathbf{B}$. With regard to the selection of reference points, in the original proposal, the MLM assigns the reference points randomly from the available data points for learning. This paper focus on alternatives to such random assignment.

Since the reference points are taken from the data, we have that $K \leq N$. The number of reference points K controls the model capacity, thus it can be used to avoid overfitting. Under the normal conditions where the number of selected reference points is smaller than the number of training points (i.e., $K < N$), the matrix $\mathbf{B}$ can be approximated by the usual least squares estimate

$$\hat{\mathbf{B}} = (\mathbf{D}^T\mathbf{D})^{-1}\mathbf{D}^T\boldsymbol{\Delta}, \tag{3}$$

where $\mathbf{D}$ and $\boldsymbol{\Delta}$ are the pairwise distance matrices between the data and the reference points in the input and output space respectively.

Out-of-Sample Prediction. Predicting the outputs for new input data mainly refers to solving the minimization problem embedded in Eq. (2). For an out-of-sample input point $\mathbf{x}$ whose distances from the K input reference points $\{\mathbf{m}_k\}_{k=1}^{K}$ are computed, i.e., $d_1(\mathbf{x})\ldots d_K(\mathbf{x})$, we then estimate the distances between its unknown output $\mathbf{y}$ and the output reference points using the linear model between distances, that is

$$\hat{\delta}_k(\mathbf{y}) = \sum_{i=1}^{K} d_i(\mathbf{x})\hat{B}_{i,k}, \qquad \forall k = 1,\ldots,K. \tag{4}$$

Together the estimates $\hat{\delta}_1(\mathbf{y})\ldots\hat{\delta}_K(\mathbf{y})$ can be used to locate $\mathbf{y}$ in the $\mathcal{Y}$-space. The location of $\mathbf{y}$ can be estimated from the minimizer given in Eq. (2) and rewritten here to emphasize the dependence of $\mathbf{y}$:

$$\hat{\mathbf{y}} = \arg\min_{\mathbf{y}} \sum_{k=1}^{K} \left((\mathbf{y} - \mathbf{t}_k)^T(\mathbf{y} - \mathbf{t}_k) - \hat{\delta}_k^2(\mathbf{y}) \right)^2. \tag{5}$$

It is worth mentioning that $\hat{\delta}_k(\mathbf{y})$ is not a function of $\mathbf{y}$ but rather a point estimate of the actual distance function $\delta_k(\mathbf{y}) = \|\mathbf{y} - \mathbf{t}_k\|_2$. Thus, from an optimization perspective, $\hat{\delta}_k(\mathbf{y})$ must be treated as a constant.

For the classification case, where outputs $\mathbf{y}_n$ are represented using the 1-of-S encoding scheme[1]. It was showed in [9] that under the assumption that the classes are balanced, the optimal solution to Eq. (5) is given by

$$h_{\mathbf{B}}(\mathbf{x}) = \hat{\mathbf{y}} = \mathbf{t}_{k^*}, \tag{6}$$

where $k^* = \arg\min_k \hat{\delta}_k(\mathbf{y})$. It means that output predictions for new incoming data can be carried out by simply selecting the output of the nearest reference point in the output space, estimated using the linear model $\hat{\mathbf{B}}$. This method was named Nearest Neighbor MLM (NN-MLM).

3   Empirical Analysis of RP Selection

Similarly to the support vector machines (SVM, [10]) and the relevance vector machines (RVM, [11]), the out-of-sample phase of the MLM requires the determination of a subset of points taken from the training set. However, unlike the

---

[1] A S-level qualitative variable is represented by a vector of S binary variables or bits, only one of which is *on* at a time. Thus, the j-th component of an output vector $\mathbf{y}$ is set to 1 if it belongs to class j and 0 otherwise.

SVM and RVM, the choice of such points is not a by-product of the training step. In the original MLM proposal, the choice of such points is random, leaving just the number of points determined by the user. However, such an approach is not a warranty of model effectiveness. In other words, a model generated from a random reference points selection can provide bad decision boundaries that can be overfitting or underfitting. In Fig. 1 we show examples of decision boundaries generated by underfitting, overfitting and appropriate-fitting models.

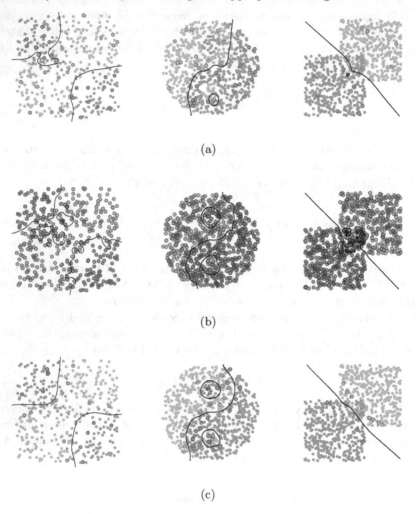

Fig. 1. Examples of decision boundaries generated by (a) underfitting, (b) overfitting and (c) appropriate-fitting models.

As can be seen in Fig. 1(a), using a small value for K and making the choice at random can lead to an unrepresentative subset, disregarding possibly important regions. Conversely, with the use of all points in the data set as reference points, a rather complex decision surface is created, as can be seen in Fig. 1(b). The ideal choice is to make reference points not in "confusing" regions, in other words, regions of data overlap. In Fig. 1(c) the ideal choice of a subset of reference points is presented. It is interesting to note that the number of points in Fig. 1(a) and (c) is the same.

4   Proposal: Fuzzy C-means MLM (FCM-MLM)

Our proposal, called fuzzy C-means MLM (FCM-MLM), uses the fuzzy C-means [5] algorithm as the main tool for select reference points (RPs) in minimal learning machines. As in the original proposal, FCM-MLM requires only a single hyper-parameter K, which in our case denotes the maximum number of RPs. The number of RPs obtained by the FCM-MLM will always be less than or equal to the parameter K (i.e. $|\mathcal{R}| \leq K$). This is possible due to the execution of an extra step for removing points that are in heterogeneous regions—regions that contain data of different classes. This is carried out by eliminating subsets of data (derived from fuzzy C-means) that contain patterns of different classes.

Briefly, FCM-MLM comprises three steps. In the first step, the fuzzy C-means algorithm runs under all the data set. After that, the next step is to perform the removal of prototypes resulting from the execution of fuzzy C-means that group patterns from different classes. Finally, the set of homogeneous prototypes is selected as RPs. The main idea of the proposal is to ensure that reference points are well distributed over the input space and, jointly, to ensure that such points are not located in class-overlapping regions. Algorithm 1 presents the pseudocode of the FCM-MLM method.

Algorithm 1. FCM-MLM

Input: Initial RPs number (K), data set inputs ($\mathcal{X}$) and outputs ($\mathcal{Y}$)
Output: Regression model ($\hat{\mathbf{B}}$), set of PRs inputs ($\mathcal{R}$) and outputs ($\mathcal{T}$)
 1: Apply the fuzzy C-means algorithm in the whole data set

$$\{c_k\}_{k=1}^{K}, \{\mu_{nk}\}_{n=1,k=1}^{N,K} \leftarrow C\text{–MEANS}(\mathcal{X}, K)$$

   where $\mu_{nk} \in [0,1]$ are the cluster membership values.
 2: Create K subsets with label of the data

$$\mathcal{Y}^k \leftarrow \{y_n \in \mathcal{Y} \mid k = \arg\max_{1 \leq i \leq K} \mu_{ni}\}, \qquad 1 \leq k \leq K$$

 3: Create the RP set only with the closest patterns of homogeneous centroids

$$\mathcal{R} \leftarrow \bigcup_k \left\{ \arg\min_{x_n \in \mathcal{X}} \|x_n - c_k\| \right\}, \quad \forall k : |\mathcal{Y}^k| = 1$$

   The set $\mathcal{T}$ are given by the corresponding output of the elements in $\mathcal{R}$.
 4: Compute the distance matrices $\mathbf{D}_x$ using $\mathcal{X}$ and $\mathcal{R}$; and $\boldsymbol{\Delta}_y$ using $\mathcal{Y}$ and $\mathcal{T}$
 5: Compute $\hat{\mathbf{B}}$ using the Eq. (3)
 6: **return** $\mathcal{R}, \mathcal{T}, \hat{\mathbf{B}}$

---

5   Simulations and Discussion

For a qualitative analysis, we have applied FCM-MLM, RN-MLM and FL-MLM to solve an artificial problem. The problem, well-known Ripley (RIP) dataset problem consists of two classes where the data for each class have been generated by a mixture of two Gaussian distributions (Fig. 2).

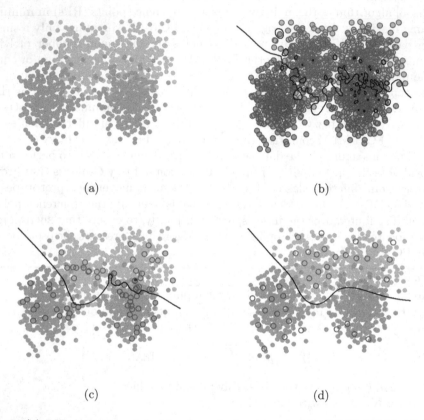

(a)   (b)

(c)   (d)

Fig. 2. (a) RIP data set and decision boundaries generated by (b) FL-MLM, (c) RN-MLM, and (d) FCM-MLM.

Based on the Fig. 2, we can infer that FCM-MLM produced better decision boundary when compared to the other algorithms. In the Fig. 2(c) and (d), one can see that the number of RPs for FCM-MLM is lower than the number of RPs for RN-MLM. Moreover, the decision boundary generated from the FCM-MLM is more smoothed than the other models.

Experiments with real-world benchmarking data sets were also carried out in this work. We used UCI data sets [12]; Heart (HEA), Haberman's Survival (HAB), Vertebral Column Pathologies (VCP), Breast Cancer Wisconsin (BCW), Statlog Australian Credit Approval (AUS), Pima Indians Diabetes

(PID) and Human Immunodeficiency Virus protease cleavage (HIV). In addition, three well-known artificial data sets were also used in our simulations, Two Moon (TMN), Ripley (RIP) and Banana (BNA). Some description, abbreviation, number of patterns (# *patterns*) and number of features (# *features*) about the aforementioned datasets are presented in the Table 1.

Table 1. General description for datasets used in this work.

Dataset	Abbreviation	# patterns	# train	# test	# features
Heart	HEA	270	216	54	13
Haberman's Survival	HAB	306	245	61	3
Vertebral Column	VCP	310	248	62	6
Breast Cancer W.	BCW	688	550	138	9
Australian Credit A.	AUS	690	552	138	14
Pima Indians Diabetes	PID	768	614	154	8
Two Moon	TMN	1001	801	200	2
Ripley	RIP	1250	1000	250	2
HIV-1 Protease Cleavage	HIV	3272	2617	655	8
Banana	BNA	5300	4240	1060	2

The performance of our proposal is compared to two variants of the MLM, regarding the selection of RPs. The first variant is the full MLM (FL-MLM), in which the set of reference points is equal to the training set (i.e., $K = N$). The second variant is the random MLM (RN-MLM), where we randomly select K reference points from the training data. It corresponds to the original proposal. A combination of the k-fold cross-validation and holdout methods was used in the experiments. The holdout method with a 80% training and 20% test division was used to estimate the performance metrics. In Table 2 we report the performance metrics of each RP selection method.

The adjustment of the parameter K for the FCM-MLM and the RN-MLM model was performed using grid search combined with 10-fold cross-validation. The RPs were selected in the range of 5–100% (with a step size of 5%) of the available training samples. The classification error was used to choose the best value of K. Each experiment was performed for 30 independent runs.

In order to verify the possible equivalence between the classifier accuracies, we perform a statistical hypothesis tests. Such tests aim to establish the limits beyond which two samples should no longer be considered to be taken from the same population, but as belonging to two different populations. That being said, we adopted a non-parametric test, named Friedman, which does not rely on any assumptions about the form of distribution that is taken to have generated the accuracy values. Besides, it can be used when comparing three or more classifiers [13]. Given the null and the alternative hypothesis, that all algorithms are equivalent or not, respectively; whether the test provided a significance level value less than or equal to the chosen significance level (in our case, 0.01), the test suggests that the observed data is inconsistent with the null hypothesis and, thus, the null hypothesis should be rejected. Information about superiority and inferiority can also be infered by Friedman statistical hypothesis test.

Table 2. Performance comparison – Accuracy (ACC) and reduction percentage in comparison with the training set (RED) – with the FCM-MLM, RN-MLM and FL-MLM; and results of statistical tests. The symbols ✓ and ✗ with respect to the Friedman statistical test means equivalence and no equivalence, respectively.

Data set	Metric	FCM-MLM		RN-MLM			FL-MLM	
HEA	ACC	71.36	6.03	70.00	6.29	✓	**72.16**±**6.29**	✓ ✗
	RED	**59.44**	16.50	33.12	18.10			
HAB	ACC	**72.13**	4.47	71.97	4.27	✓	68.09±4.98	✗ ✗
	RED	**88.01**	11.58	80.20	14.22			
VCP	ACC	**84.78**	4.48	82.58	4.39	✗	82.15±4.23	✗ ✓
	RED	**80.60**	5.99	56.51	26.86			
BCW	ACC	**97.00**	1.31	96.98	1.40	✓	96.96±1.27	✓ ✓
	RED	**86.91**	4.65	62.61	22.71			
AUS	ACC	69.13	3.06	**72.15**	3.05	✗	70.97±3.59	✓ ✓
	RED	**60.65**	5.13	55.59	22.22			
PID	ACC	73.44	2.86	**74.59**	2.58	✗	73.16±2.38	✓ ✗
	RED	**84.61**	8.11	75.92	16.10			
TMN	ACC	**99.87**	0.22	99.82	0.28	✓	99.87±0.22	✓ ✓
	RED	**63.63**	23.71	61.92	20.72			
RIP	ACC	**89.81**	1.88	89.75	1.77	✓	88.32±1.61	✗ ✗
	RED	**87.30**	11.01	76.64	18.83			
HIV	ACC	**86.68**	1.30	86.50	1.30	✓	85.99±1.14	✗ ✗
	RED	**96.73**	1.39	75.32	23.16			
BNA	ACC	88.16	0.90	**89.87**	0.81	✗	87.58±0.89	✓ ✗
	RED	78.19	23.94	**89.33**	2.54			

By analyzing the Table 2 one can conclude that the performances of the FCM-MLM were equivalent or even superior to those achieved by the RN-MLM and FL-MLM for each data sets evaluated. Moreover, one can also see that our proposal achieves sparse solutions, i.e., the FCM-MLM produces a reduced set of RPs.

6   Conclusions

In this paper, we propose an algorithm to select the reference points of the MLM for classification tasks based on the fuzzy C-means algorithm. Three strategies of

MLM reference point selection are evaluated. Our proposal called FCM-MLM is able to obtain the RP subset for MLMs. The experimental results indicate that the FCM-MLM represents a good alternative to the random selection, providing a competitive classifier while maintaining its simplicity.

References

1. de Souza Júnior, A.H., Corona, F., De Barreto, G.A., Miché, Y., Lendasse, A.: Minimal learning machine: a novel supervised distance-based approach for regression and classification. Neurocomputing **164**, 34–44 (2015)
2. Coelho, D.N., De Barreto, G.A., Medeiros, C.M., Santos, J.D.A.: Performance comparison of classifiers in the detection of short circuit incipient fault in a three-phase induction motor. In: 2014 IEEE Symposium on Computational Intelligence for Engineering Solutions, CIES 2014, Orlando, FL, USA, 9–12 December 2014, pp. 42–48. IEEE (2014)
3. Alencar, A.S.C, Caldas, W.L., Gomes, J.P., de Souza, A.H., Aguilar, P.A., Rodrigues, C., Wellington, F., de Castro, M.F., Andrade, R.M.C.: MLM-rank: a ranking algorithm based on the minimal learning machine. In: 2015 Brazilian Conference on Intelligent Systems, BRACIS 2015, Natal, Brazil, 4–7 November 2015, pp. 305–309. IEEE (2015)
4. Marinho, L.B., Almeida, J.S., Souza, J.W.M., de Albuquerque, V.H.C., Rebouças Filho, P.P.: A novel mobile robot localization approach based on topological maps using classification with reject option in omnidirectional images. Expert Syst. Appl. **72**, 1–17 (2017)
5. Bezdek, J.C.: Pattern Recognition with Fuzzy Objective Function Algorithms. Advanced Applications in Pattern Recognition, 2nd edn. Plenum Press, New York (1987)
6. Prabhakar, S.K., Rajaguru, H.: PCA and K-means clustering for classification of epilepsy risk levels from EEG signals—a comparitive study between them. In: 2015 International Conference on Intelligent Informatics and Biomedical Sciences (ICIIBMS), pp. 83–86, November 2015
7. Wang, D., Nie, F., Huang, H.: Unsupervised feature selection via unified trace ratio formulation and *K*-means clustering (TRACK). In: Calders, T., Esposito, F., Hüllermeier, E., Meo, R. (eds.) ECML PKDD 2014. LNCS (LNAI), vol. 8726, pp. 306–321. Springer, Heidelberg (2014). https://doi.org/10.1007/978-3-662-44845-8_20
8. Wang, L., Pan, C.: Robust level set image segmentation via a local correntropy-based k-means clustering. Pattern Recogn. **47**(5), 1917–1925 (2014)
9. Mesquita, D.P.P., Gomes, J.P.P., Souza Junior, A.H.: Ensemble of efficient minimal learning machines for classification and regression. Neural Process. Lett. **46**, 751–766 (2017)
10. Vapnik, V.N.: The Nature of Statistical Learning Theory. Springer, New York (1995). https://doi.org/10.1007/978-1-4757-2440-0
11. Tipping, M.E.: Sparse Bayesian learning and the relevance vector machine. J. Mach. Learn. Res. **1**, 211–244 (2001)
12. Lichman, M.: UCI machine learning repository (2013)
13. Demšar, J.: Statistical comparisons of classifiers over multiple data sets. J. Mach. Learn. Res. **7**, 1–30 (2006)

Prey-Predator Model Under Fuzzy Uncertanties

Chryslayne M. Pereira[1], Moiseis S. Cecconello[2(✉)], and Rodney C. Bassanezi[1]

[1] Universidade Estadual de Campinas - Unicamp, Campinas, Brazil
[2] Universidade Federal de Mato Grosso - UFMT, Cuiabá, Brazil
moiseis@ufmt.com

Abstract. This work studies the influence of fuzzy uncertainties on the asymptotic behavior of the solution of a prey-predator model. Here, initial conditions and parameters are interpreted as fuzzy variable. The population densities at a specific time are also interpreted as a fuzzy variable in which the possibility distribution function depends on the possibility distribution functions of the parameters. We provide closed formulas for expected values of some equilibrium points. We also compare the expected value of the fuzzy solution with the deterministic solution providing computational simulations in order see the difference between theses approaches.

Keywords: Fuzzy dynamical systems · Expected value
Fuzzy variables · Possibility theory · Fuzzy solutions

1 Introduction

It is not always possible to know exactly the initial number of individuals or the carrying capacity in a given environment in applied problems of population dynamics. In general, one gets information by means of linguistic statements such as *the initial condition is approximately* x_0 or *the carrying capacity is about* k_0. To the extent that the label *approximately* is imprecise, it can be modeled as a fuzzy set. Thus, linguistic statements like these can be seen as fuzzy restrictions on the values taken by the variable of interest [1].

Zadeh proposed a fuzzy restriction as a possibility distribution with its membership function playing the role of a possibility distribution function. In the context of population dynamics, let us suppose that the label *approximately* x_0 is modeled by a fuzzy set $\boldsymbol{x}_0$ with membership $\mu_{\boldsymbol{x}_0}(x)$. Thus, given a specific numerical value $x = u_0$, the value $\mu_{\boldsymbol{x}_0}(u_0)$ is the degree of possibility that the actual initial condition of the dynamical system assumes the value u_0 given the proposition *the initial condition is approximately* x_0. Thus, the membership function $\mu_{\boldsymbol{x}_0}(x)$ is the distribution of the possibility associated with the variable *initial condition*.

Once we do not have precise information about the actual value of initial condition or parameters we can not require a precise description of the state of

© Springer International Publishing AG, part of Springer Nature 2018
G. A. Barreto and R. Coelho (Eds.): NAFIPS 2018, CCIS 831, pp. 408–418, 2018.
https://doi.org/10.1007/978-3-319-95312-0_35

the population on a fixed time $t > 0$. It is reasonable to look for a description of the state of the system by means of a fuzzy restriction as *the state of the system at time τ is approximately u_0*. This, in turn, defines a possibility distribution on the values assumed by the state of the system.

Several approaches have been presented in order to consider fuzzy uncertainties on differential equations. Some authors use the H – derivative to obtain solutions with fuzzy uncertainties [2–11]. Others construct the fuzzy solutions of differential equations by means of a family of differential inclusions [12–15]. A third approach consists in applying the Zadeh extension principle on the initial conditions of deterministic solutions to obtain fuzzy solutions of differential equations [16–24].

However, when dealing with fuzzy uncertainties as fuzzy restrictions on the values taken by the variable of interest, or possibility distribution function, one faces the problem of describing how the possibility distribution function evolves over time. One gets a similar problem when looking at the initial condition as a random variable described by probability distribution functions [25–28]. Thus, here we follow a similar approach used in probability theory to deal with fuzzy uncertainties on initial conditions and parameters. As we will see, interpreting in this way, we end up with the approach of applying the Zadeh extension principle on the initial condition and parameters of deterministic solutions. Therefore, considering parameters and initial conditions as possibility distribution functions of fuzzy variables lead us to define fuzzy solutions by taking Zadeh extension of deterministic solutions which is similar to the third approach previously described.

A naive approach to handle deterministic differential equations with uncertainties, fuzzy or probabilistic, on parameters of the dynamical systems would be to obtain a representative value of these parameters by means of some statistical procedure [31]. These representative values, in turn, are inserted in the equation and the analysis is carried on. That is, by this approach, we deal with uncertainties prior to the analysis of the dynamical systems. Here we are going to think in another direction. First, we are interested in describing how the possibility distribution function evolves over time and, after that, we calculate a representative value of such a fuzzy variable. As we will see, these two approaches may lead to distinct results.

Thus, in order to measure the effects of the fuzzy uncertainties on the dynamics we wonder about the expected value of the values assumed by the state of the system. We do this by comparing the expected value with the deterministic solution defined by the expected value of the initial condition and parameters. We provide closed-form expressions for the expected value of the fuzzy variable described by the logistic equation and for the fuzzy variable that represents the maximum growth time.

The organization of this article is as follows: in Sect. 2 we discuss some basic concepts on fuzzy sets and fuzzy variables; in Sect. 3 we present the prey-predator model we are considering in this work; in Sect. 4 we discuss about the expected value of the fuzzy variable that describes the state of the system at time $t > 0$; in Sect. 5 we discuss about the expected value of the equilibrium points; in Sect. 6 we provide numerical simulations to illustrate our main results.

2   Some Basic Concepts

2.1   Fuzzy Sets

As it is well known, given a set U, a fuzzy subset of U is characterized by a function defined on U taking values on $[0, 1]$ [33]. This function is called *membership function*. Given $\alpha \in [0, 1]$, an α-cut or α-level of a fuzzy set is defined as the set of points of U where the membership function is greater than or equal to α. Precisely, if u is a fuzzy set of U with membership function $\mu_u : U \to [0, 1]$ then, for $0 < \alpha \leq 1$, the α-cut of u is a subset of U given by

$$[u]^\alpha = \{x \in U : \mu_u(x) \geq \alpha\}$$

and, for $\alpha = 0$,

$$[u]^0 = \overline{\{x \in U : \mu_u(x) > 0\}}$$

is the *support* of u [4].

Let us denote by $\mathcal{F}(U)$ the set of fuzzy subsets of $U \subset \mathbb{R}$, in which the α-cuts are non-empty, compact and (simply-) connected for every $\alpha \in [0, 1]$. We can measure the distance between two fuzzy sets in the following way: given two points $u, v \in \mathcal{F}(U)$, the distance between u and v is defined by

$$d_\infty(u, v) = \sup_{\alpha \in [0,1]} d_H([u]^\alpha, [v]^\alpha), \tag{1}$$

where d_H is the Hausdorff distance for compact sets. We also denote by $\chi_{\{A\}}$ the characteristic function of the set A.

In this work, we are interested in *fuzzy variables* taking values on $U = [0, +\infty)$, the set of non-negative real numbers. However, we describe the following concepts for a general set $U \subset \mathbb{R}^n$.

A fuzzy subset u of U, defined by a membership function $\mu_u : U \to [0, 1]$, induces, according to Zadeh [1], a *possibility distribution function* on the set of values of a variable of interest ξ. That is, if ξ is a fuzzy variable then $\mu_u(x)$ is the degree of possibility that ξ assumes the particular value x. In the context of possibility theory, $\mu_u(x) = 0$ means that it is impossible that the variable ξ assumes the value x. The quantity $\mu_u(x)$ represents the degree of possibility of the assignment $\xi = x$, where some values x being more possible than others. The closer the value $\mu_u(x)$ is to 1, the more possible it is that x is the actual value of the variable.

It is well known [1] that given a subset $A \subset U$, the *possibility measure* of A is defined by

$$\text{Pos}_\mu(A) = \sup_{x \in A} \mu_u(x),$$

and the *necessity measure* of A is defined by

$$\text{Nec}_\mu(A) = 1 - \text{Pos}_\mu(A^c),$$

where A^c stands for the complement set of A in U. The *credibility measure* of A, $A \subset U$, according to [32], is defined as

$$\mathrm{Cr}_\mu(A) = \frac{1}{2} \left(\mathrm{Pos}_\mu(A) + \mathrm{Nec}_\mu(A) \right).$$

We point out that $\mathrm{Pos}_\mu(A)$ is a measure of the possibility of the fuzzy variable ξ to assume values in A. Note that $\mathrm{Pos}_\mu(\emptyset) = 0$ and $\mathrm{Pos}_\mu(U) = 1$. On the other hand, $\mathrm{Nec}_\mu(A)$ can be seen as a measure of the fuzzy variable ξ not assume values in A^c. Thus, both numbers are measures for the question if an event A either occur or not [36]. Thus, the $\mathrm{Cr}_\mu(A)$ is the average of these two answers to the occurrence of an event A.

In order to get a representative value of a fuzzy variable ξ, it is important to define the concept of its expected value [1]. The *expected value* of ξ is defined as

$$E[\xi] = \int_0^\infty \mathrm{Cr}_\mu([r, +\infty)) dr - \int_{-\infty}^0 \mathrm{Cr}_\mu((-\infty, r)) dr,$$

provided that at least one of these integrals is finite [32].

Now, when U is the set of non-negative real numbers, if we define the quantities

$$\xi'_\alpha = \inf\{x : \mu_u(x) \geq \alpha\} \quad \text{and} \quad \xi''_\alpha = \sup\{x : \mu_u(x) \geq \alpha\},$$

for all $\alpha > 0$, then the previous formula becomes

$$E[\xi] = \frac{1}{2} \int_0^1 (\xi'_\alpha + \xi''_\alpha) \, d\alpha, \tag{2}$$

provided that ξ'_α and ξ''_α are finite [32]. We emphasize that the definition of $\mathrm{Pos}(A)$, $\mathrm{Nec}(A)$ and $\mathrm{Cr}(A)$ depends on the possibility distribution function μ_u of the fuzzy variable ξ.

2.2 Transformations of Fuzzy Variables

Consider now a continuous function g defined on some subset of the real numbers. If ξ is a fuzzy variable then so is $\eta = g(\xi)$, and there is a natural way to define a possibility distribution function $\mu_{g(u)}(x)$ to $g(\xi)$ from the distribution $\mu_u(x)$ of ξ as it follows: given $A \subset U$, then $g(\xi)$ assumes values on A if and only if ξ assumes values on $g^{-1}(A)$. Thus, by definition, the possibility of $g(\xi)$ to assume values in A is the same as the possibility of ξ to assume values on $g^{-1}(A)$. Thus, following [1], it turns out that

$$\mathrm{Pos}_\eta(A) = \sup_{y \in A} \mu_{\hat{g}(u)}(y) = \sup_{x \in g^{-1}(A)} \mu_u(x) = \mathrm{Pos}_\xi \left(g^{-1}(A) \right),$$

and, as a consequence, we obtain the possibility distribution function of the fuzzy variable $\eta = g(\xi)$ by taking

$$\mu_{\hat{g}(u)}(y) = \sup_{x \in g^{-1}(y)} \mu_u(x). \tag{3}$$

We remark that expression (3) is the Zadeh extension of g as given in [33]. This is the reason why we denote the possibility distribution function of $g(\xi)$ by $\mu_{\hat{g}(u)}(y)$. We also remark that this approach is similar to that one followed in the context of transformation of random variables in probability theory (see, for instance, [25]).

According to [34], if g is monotone (increasing or decreasing) then the expected value of the fuzzy variable $\eta = g(\xi)$ can be computed by

$$E\left[g\left(\xi\right)\right] = \frac{1}{2} \int_0^1 \left(g(\xi'_\alpha) + g(\xi''_\alpha)\right) d\alpha. \tag{4}$$

This formula will be useful in the following sections.

2.3   Several Fuzzy Variables

One faces the problem of uncertainties on several variables in population dynamics and other applications. Before proceeding to the next section, let us present the main ideas on this subject.

Let ξ_1 and ξ_2 be fuzzy variables with possibility distribution functions μ_{u_1} and μ_{u_2}, respectively, both defined on U. Following [1], these variables define a fuzzy variable, namely $\eta = (\xi_1, \xi_2)$, on $U \times U$, in which its joint possibility distribution function $\mu_u : U \times U \to [0, 1]$ is given by

$$\mu_u(x, y) = \min\{\mu_{u_1}(x), \mu_{u_2}(y)\}. \tag{5}$$

We are assuming that the variables ξ_1 and ξ_2 are unrelated, or *non-interactive*, in the sense that a specific value of ξ_1 gives no information about the possible values that ξ_2 can assume.

3   A Prey-Predator Model

We are considering the prey-predator model given by the system of differential equations

$$\begin{cases} \dfrac{dx}{dt} = a_1 x - b_1 x^2 - c_1 xy, & x(0) = x_o > 0, \\[4mm] \dfrac{dy}{dt} = a_2 y - b_2 y^2 + c_2 xy, & y(0) = y_o > 0, \end{cases} \tag{6}$$

in which the parameters are all non negative except possibly a_2. Let $\varphi_t(x_o, y_o, p)$ be the solution of Eq. (6) at (x_o, y_o) and a vector of parameters p. As is well known, the application $\varphi_t : \mathbb{R}^2 \to \mathbb{R}^2$ is the flow acting on the phase space $\mathbb{R}^2$. Thus, for every initial condition $(x_o, y_o) \in \mathbb{R}^2$ we have a deterministic solution $\varphi_t(x_o, y_o, p) = (x(t, x_o, y_o, p), y(t, x_o, y_o, p))$.

Equation (6) has Jacobian matrix at $(\bar{x}, \bar{y})$ given by

$$J(\bar{x}, \bar{y}) = \begin{pmatrix} a_1 - 2b_1\bar{x} - c_1\bar{y} & -c_1 \\ c_2 & a_2 - 2b_2\bar{y} + c_2\bar{x} \end{pmatrix} \tag{7}$$

and thus it turns out that:

(a) The equilibrium point $q_1 = (0,0)$ is unstable;
(b) The equilibrium point $q_2 = (a_1/b_1, 0)$ is unstable;
(c) The equilibrium point $q_3 = (0, a_2/b_2)$ is unstable provided that $c_1 a_2 < a_1 b_2$;
(d) The equilibrium point

$$q_4 = \left(\frac{a_1 b_2 - a_2 c_1}{b_1 b_2 + c_1 c_2}, \frac{a_2 b_1 + a_1 c_2}{b_1 b_2 + c_1 c_2} \right)$$

is unstable provided that $a_1 b_2 < a_2 c_1$.

The following analyzes the behavior of the solution $x(t)$ and $y(t)$ under fuzzy uncertainties.

4 Fuzzy Uncertainties on the Model

Due to the lack of complete information or error of measurements, more often than not, one needs to deal with imprecision on the parameters. A naive approach to deal with uncertainties in models like the previous one defined by Eq. (6), it could be to compute the average values of the parameters and then analyzing the dynamics by means of its deterministic solution using these average values for the parameters.

Thus, let us assume that parameters and initial conditions are under restriction given by fuzzy label as *approximately*, for instance. That is, we are assuming that these variables satisfy a statement like *the variable ξ is approximately ξ_o*. Thus, according to Zadeh, the membership function of the fuzzy label *approximately* is the possibility distribution function of ξ. In works like [24] and [16], in case of having fuzzy uncertainties on the initial conditions, the authors define the *fuzzy solution* of Eq. (6) as the Zadeh's extension of the deterministic flow $\varphi_t : \mathbb{R}^2 \to \mathbb{R}^2$. In case of having fuzzy uncertainties on other parameters of Eq. (6), we consider those parameters as initial conditions of a differential equation with zero derivative and proceed as before, taking the Zadeh's extension on the initial condition of the deterministic flow.

Here, however, we are going to take another direction. Since we are assuming that x_o, y_o and some parameter, or vector of parameters, p in Eq. (6) are fuzzy variables, these quantities $x(t, x_o, y_o, p)$ and $y(t, x_o, y_o, p)$, for a fixed $t > 0$, are fuzzy variables as well. Following the recipe described in the previous sections, for a fixed $t > 0$, we can obtain the possibility distribution function of $\varphi_t(x_o, y_o, p)$ by means of the Zadeh's extension on the parameters x_o, y_o and p of the functions $x(t, x_o, y_o, p)$ and $y(t, x_o, y_o, p)$. That is, the number of pray and predators at a fixed time $t \geq 0$ are fuzzy variables whose possibility distribution are $\hat{x}(t, \boldsymbol{x}_o, \boldsymbol{y}_o, \boldsymbol{p})$ and $\hat{y}(t, \boldsymbol{x}_o, \boldsymbol{y}_o, \boldsymbol{p})$.

Although these two approaches seem different, the relationship between them is as it follows [17].

Theorem 1 ([17]). *Let the applications $\hat{\pi}_x$ and $\hat{\pi}_y$ be the Zadeh's extensions of the orthogonal projections $\pi_x : \mathbb{R}^2 \to \mathbb{R}$ and $\pi_y : \mathbb{R}^2 \to \mathbb{R}$ on the x and y axis, respectively. Then it follows that:*

$$\hat{x} = \hat{\pi}_x \circ \hat{\varphi}_t \qquad \hat{y} = \hat{\pi}_y \circ \hat{\varphi}_t.$$

4.1   Fuzzy Uncertainties on Equilibrium Points

We consider fuzzy uncertainties on the parameters so that the equilibrium points are also fuzzy variables whose the possibility distribution function are the Zadeh's extension of the expressions that define such equilibrium points. In [17] the authors have proved if an equilibrium point is asymptotically stable then $\hat{x}$ and $\hat{y}$ converge to the Zadeh's extension of the x and y coordinates of the expressions that define such equilibrium point. In other words, we have that $\hat{q}$ is the Zadeh's extension of an equilibrium point then

$$\hat{x} = \hat{\pi}_x \circ \hat{\varphi}_t \to \hat{\pi}_x \circ \hat{q}$$

$$\hat{y} = \hat{\pi}_y \circ \hat{\varphi}_t \to \hat{\pi}_y \circ \hat{q}$$

Theorem 2. *Suppose that the fuzzy set c_1 is possibility distribution function of the fuzzy variables c_1 and furthermore suppose that*

$$\eta'_\alpha = \inf\{x : \mu_{c_1}(x) \geq \alpha\}, \qquad \eta''_\alpha = \sup\{x : \mu_{c_1}(x) \geq \alpha\}.$$

Then it turns out that:

(a) The α-cuts of the fuzzy set $\hat{\pi}_x \circ \hat{q}$ are

$$\left[\frac{a_1 b_2 - a_2 \eta''_\alpha}{b_1 b_2 + c_2 \eta''_\alpha}, \frac{a_1 b_2 - a_2 \eta'_\alpha}{b_1 b_2 + c_2 \eta'_\alpha} \right].$$

(b) The α-cuts of the fuzzy set $\hat{\pi}_y \circ \hat{q}$ are

$$\left[\frac{a_2 b_1 + a_1 c_2}{b_1 b_2 + c_2 \eta''_\alpha}, \frac{a_2 b_1 + a_1 c_2}{b_1 b_2 + c_2 \eta'_\alpha} \right].$$

Proof. Since π_x, π_y and each coordinate of the equilibrium point q_4 are continuous functions for $c_1 > 0$ then we have for a continuous function f that $[\hat{f}(\boldsymbol{u})]^\alpha = f([\boldsymbol{u}]^\alpha)$. In both cases, the x and y coordinates are decreasing functions with respect to c_1 and this proves the statement.

Next we look at the expected values of fuzzy solutions and equilibrium points.

5   Expected Values of Fuzzy Solutions and Equilibrium Points

We are interested, in this section, in the behavior of the expected values of the fuzzy solution defined by considering x_o, y_o and c_1 as fuzzy variables whose the possibility distribution functions are $\boldsymbol{x_o}, \boldsymbol{y_o}, \boldsymbol{c_1}$, respectively. To this end, we have the following theorem.

Theorem 3. *Let $q_x(c_1)$ and $q_y(c_1)$ the fuzzy variables defined by the x and y coordinates of q_4 respectively. If c_1 is a fuzzy variable with triangular possibility distribution function $\mu_{c_1}(x) = (c - \varepsilon/c/c + \varepsilon)$ then we have that:*

$$E[q_x] = -\frac{a_2}{c_2} + \frac{b_2 A}{2c_2^2 \varepsilon} \ln\left(\frac{B + c_2\varepsilon}{B - c_2\varepsilon}\right)$$

$$E[q_y] = \frac{A}{2c_2\varepsilon} \ln\left(\frac{B + c_2\varepsilon}{B - c_2\varepsilon}\right)$$

in which $A = a_2b_1 + a_1c_2$ and $B = b_1b_2 + c_2c$.

Proof. To prove the first statement we must observe that the α-cuts of $\mu_{c_1}(x)$ are the intervals $[\eta'_\alpha, \eta''_\alpha]$ where

$$\eta'_\alpha = c - (1 - \alpha)\varepsilon \quad \text{and} \quad \eta''_\alpha = c + (1 - \alpha)\varepsilon.$$

Since the expected value of the fuzzy variable $q_x c$ is given by

$$E[q_x(c_1)] = \frac{1}{2} \int_0^1 [q_x(\eta'_\alpha) + q_x(\eta''_\alpha)] \, d\alpha,$$

integrating we obtain the desired result.

On the other hand, the second statement can be prove similarly taking into account the expression that defines $q_y(c_1)$.

We have also the following theorem.

Theorem 4. *Suppose that the fuzzy sets $\boldsymbol{x_o}, \boldsymbol{y_o}$ and $\boldsymbol{c_1}$ are possibility distribution functions of the fuzzy variables x_o, y_o and c_1, respectively, and let q_x and q_y be the fuzzy variables defined by the x and y coordinates of q_4. If the equilibrium point $q_4(c_1)$ is asymptotically stable for all $c_1 \in [\boldsymbol{c_1}]^0$ then we have that:*

(a) The expected value of the fuzzy variable $x(t)$ converges to the expected value of q_x. That is, $E[x(t)] \rightarrow E[q_x]$ as $t \rightarrow \infty$.

(b) The expected value of the fuzzy variable $y(t)$ converges to the expected value of q_y. That is, $E[y(t)] \rightarrow E[q_y]$ as $t \rightarrow \infty$.

Proof. Since $q_4(c_1)$ is asymptotically stable for all $c_1 \in [\boldsymbol{c_1}]^0$, according to [35] (Corollary 14, p. 12), the family of function indexed by t, $x(t) : K \rightarrow \mathbb{R}$, $K = [\boldsymbol{x_o}]^0 \times [\boldsymbol{y_o}]^0 \times [\boldsymbol{c_1}]^0$, converges uniformly to $f : K \rightarrow \mathbb{R}$, defined by $f(x_o, y_o, c_1) = q_x$, as $t \rightarrow \infty$. That is, given $\varepsilon > 0$, there is a $T > 0$ such that for all $t > T$ we have $|x(t, x_o, y_o, c_1) - q_x(c_1)| < \varepsilon$ for all $(x_o, y_o, c_1) \in K$. Thus,

$$|E\left[x(t, x_o, y_o, c_1)\right] - E\left[q_x(c_1)\right]| = \left|\frac{1}{2}\int_0^1 \left[x(t, h'_\alpha) - q_x(\zeta''_\alpha) + x(t, h''_\alpha) - q_x(\zeta'_\alpha)\right] d\alpha\right|$$

$$\leq \frac{1}{2}\int_0^1 |x(t, h'_\alpha) - q_x(\zeta''_\alpha)|\, d\alpha + \frac{1}{2}\int_0^1 |x(t, h''_\alpha) - q_x(\zeta'_\alpha)|\, d\alpha$$

$$< \varepsilon$$

in which $h'_\alpha = (\xi'_\alpha, \eta'_\alpha, \zeta'_\alpha)$ and $h''_\alpha = (\xi''_\alpha, \eta''_\alpha, \zeta''_\alpha)$. This inequality proves the first statement.

We can prove the second statement analogously.

Once that $q_x(E(c_1))$ is not necessarily equal to $E(q_x(c_1))$ then we can conclude from last theorem that the expected value a the fuzzy solution are not necessarily equal a deterministic solution, at least not near an equilibrium point. Thus, although we are not able to find a closed-formula for the expected value of fuzzy solutions of Eq. (6) from last statement we can conclude that the two approaches of dealing with uncertainties discussed here provide different numerical values.

6   Worked Example

In order to illustrate the results obtained in previous sections, let us consider x_o, y_o and c_1 as fuzzy variables given by the triangular fuzzy possibility distribution functions $\mu_{x_o}(x) = (5/6/7)$, $\mu_{y_o}(x) = (0.01/0.51/1.01)$ and $\mu_{c_1}(x) = (0.0150/0.0250/0.0400)$. The others parameters of Eq. (6) are: $a_1 = 0.1, b_1 = 0.01, a_2 = -0.02, b_2 = 0.01$ and $c_2 = 0.005$.

The expected value of the fuzzy variable c_1 is 0.0250 and so the equilibrium point $q_4 = (6.6667, 1.3333)$ is asymptotically stable. However, since c_1 is a fuzzy variable thus q_4 is also a fuzzy variable and, as predicted by Theorem 3, the

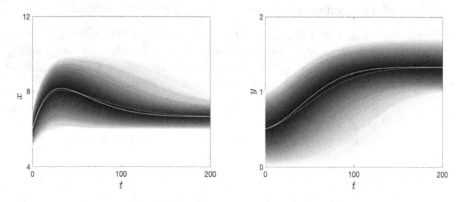

Fig. 1. Projections of the fuzzy solutions of Eq. (6). The white curve represents the deterministic solution calculated using the expected values of the fuzzy variables x_o, y_o and c_1. The red curve represents represents the expected value of the fuzzy projections. (Color figure online)

expected value of this fuzzy variable is $E(q_4) = (6.7119, 1.3560)$. By Theorem 4, the expected value of the projections of the fuzzy solution of Eq. (6), the red curves in Fig. 1, converges to $E(q_4)$ as the time evolves.

7 Conclusion

In this work we have interpreted the initial condition and parameters of a prey-predator model as fuzzy variables in which the possibility distribution function is given by a membership function of fuzzy sets. These fuzzy sets represent a label acting as a restriction on the values taken by the variables of interest. As we have shown, this approach leads to different results than the standard approach in which the uncertainties are handled apart from the dynamical system. Finally, we would like to point out that if we see parameters and initial conditions as possibility distributions functions of fuzzy variables then Zadeh's extension of deterministic solutions is the natural way to define fuzzy solutions for autonomous differential equations.

References

1. Zadeh, L.A.: Fuzzy sets as a basis for a theory of possibility. Fuzzy Sets Syst. **1**, 3–28 (1978)
2. Nieto, J.J., Rodriguez-Lopez, R.: Analysis of a logistic differential model with uncertainty. Int. J. Dyn. Syst. Differ. Equ. **1**(3), 164–176 (2008)
3. Bede, B., Rudas, I.J., Bencsik, A.L.: First order linear fuzzy differential equations under generalized differentiability. Inf. Sci. **177**(7), 1648–1662 (2007)
4. Diamond, P., Kloeden, P.: Metric Spaces of Fuzzy Sets: Theory and Applications. World Scientific, Singapore (1994)
5. Diamond, P.: Brief note on the variation of constants formula for fuzzy differential equations. Fuzzy Sets Syst. **129**(1), 65–71 (2002)
6. Kandel, A., Byatt, W.J.: Fuzzy differential equations. In: Proceedings of the International Conference Cybernetics and Society, Tokyo, pp. 1213–1216 (1978)
7. Malinowski, M.T.: On random fuzzy differential equations. Fuzzy Sets Syst. **160**(21), 3152–3165 (2009)
8. Song, S., Cheng, W., Lee, E.S.: Asymptotic equilibrium and stability of fuzzy differential equations. Comput. Math. Appl. **49**(7–8), 1267–1277 (2005)
9. Kaleva, O.: The cauchy problem for fuzzy differential equations. Fuzzy Sets Syst. **35**(3), 389–396 (1990)
10. Kaleva, O.: Fuzzy differential equations. Fuzzy Sets Syst. **24**(3), 301–318 (1987)
11. Seikkala, S.: On the fuzzy initial value problem. Fuzzy Sets Syst. **24**(3), 319–330 (1987)
12. Diamond, P.: Time-dependent differential inclusions, cocycle attractors and fuzzy differential equations. IEEE Trans. Fuzzy Syst. **7**(6), 734–740 (1999)
13. Rzezuchowski, T., Wasowski, J.: Differential equations with fuzzy parameters via differential inclusions. J. Math. Anal. Appl. **255**(1), 177–194 (2001)
14. Vorobiev, D., Seikkala, S.: Towards the theory of fuzzy differential equations. Fuzzy Sets Syst. **125**(2), 231–237 (2002)

15. Hullermeier, E.: Numerical methods for fuzzy initial value problems. Int. J. Uncertain. Fuzziness Knowl.-Based Syst. **7**(5), 439–461 (1999)
16. Cecconello, M.S., Bassanezi, R.C., Brandao, A.V., Leite, J.: Periodic orbits for fuzzy flows. Fuzzy Sets Syst. **230**, 21–38 (2013)
17. Cecconello, M.S., Leite, J., Bassanezi, R.C., Silva, J.D.M.: About projections of solutions for fuzzy differential equations. J. Appl. Math. **2013**, 1–9 (2013)
18. Khastan, A., Nieto, J.J., Rodrigues-Lopez, R.: Variation of constant formula for first order fuzzy differential equations. Fuzzy Sets Syst. **177**(1), 20–33 (2011)
19. Nieto, J.J., Otero-Espinar, M.V., Rodrigues-Lopez, R.: Dynamics of the fuzzy logistic family. Discret. Contin. Dyn. Syst.: Ser. B **14**(2), 699–717 (2010)
20. Perng, J.W.: Limit-cycle analysis of dynamic fuzzy control systems. Soft. Comput. **17**(9), 1553–1561 (2013)
21. Li, J., Zhao, A., Yan, J.: The Cauchy problem of fuzzy differential equations under generalized differentiability. Fuzzy Sets Syst. **200**(1), 1–24 (2012)
22. Buckley, J.J., Feuring, T.: Fuzzy differential equations. Fuzzy Sets Syst. **110**(1), 43–54 (2000)
23. Oberguggenberger, M., Pittschmann, S.: Differential equations with fuzzy parameters. Math. Comput. Model. Dyn. Syst. **5**(3), 181–202 (1999)
24. Mizukoshi, M.T., Barros, L.C., Chalco-Cano, Y., Roman-Flores, B.H., Bassanezi, R.C.: Fuzzy differential equation and the extension principle. Inf. Sci. **177**(17), 3627–3635 (2007)
25. Lasota, A., Mackay, M.C.: Chaos, Fractals and Noise: Stochastics Aspects of Dynamics. Springer, New York (1994). https://doi.org/10.1007/978-1-4612-4286-4
26. Casaban, M.C., Cortes, J.C., Romero, J.V., Rosello, M.D.: Probabilistic solution of random SI-type epidemiological models using the random variable transformation technique. Commun. Nonlinear Sci. Numer. Simul. **24**, 86–97 (2015)
27. Casaban, M.C., Cortes, J.C., Navarro-Quiles, A., Romero, J.V., Rosello, M.D., Villanueva, R.J.: A comprehensive probabilistic solution of random SIS-type epidemiological models using the random variable transformation technique. Commun. Nonlinear Sci. Numer. Simul. **32**, 199–210 (2016)
28. Dorini, F.A., Cecconello, M.S., Dorini, L.B.: On the logistic equation subject to uncertainties in the environmental carrying capacity and initial population density. Commun. Nonlinear Sci. Numer. Simul. **33**, 160–173 (2016)
29. Kozlowski, J.: Optimal allocation of resources to growth and reproduction: implications for age and size at maturity. Trends Ecol. Evol. **7**(1), 15–19 (1992)
30. Samuelson, P.A., Nordhaus, W.D.: Microeconomics, 17th edn. McGraw-Hill, New York City (2001)
31. Huggins, R.M., Yip, P.S.F., Lau, E.H.Y.: A note on the estimation of the initial number of susceptible individuals in the general epidemic model. Stat. Probab. Lett. **67**(4), 321–330 (2004)
32. Liu, B., Liu, Y.K.: Expected value of fuzzy variable and fuzzy expected value models. IEEE Trans. Fuzzy Syst. **10**(4), 445–450 (2002)
33. Zadeh, L.A.: Fuzzy sets. Inf. Control **8**, 338–353 (1965)
34. Hong, D.H.: Note on the expected value of a function of a fuzzy variable. J. Appl. Math Inform. **27**(4), 773–778 (2009)
35. Mizukoshi, M.T., Cecconello, M.S.: Dynamics and bifurcations of fuzzy nonlinear dynamical systems. Fuzzy Sets Syst. **319**, 81–92 (2017)
36. Dubois, D., Prade, H.: Possibility theory and its applications: where do we stand? In: Kacprzyk, J., Pedrycz, W. (eds.) Springer Handbook of Computational Intelligence, pp. 31–60. Springer, Heidelberg (2015). https://doi.org/10.1007/978-3-662-43505-2_3

Modeling and Simulation of Methane Dispersion in the Dam of Santo Antonio – Rondônia/Brazil

Geraldo L. Diniz[1](✉) and Evanizio M. Menezes Jr.[2]

[1] Federal University at Mato Grosso, Cuiabá, MT 78060-900, Brazil
geraldo@ufmt.br
[2] Federal University at Amazonas, Campus Humaitá, Manaus, AM 69080-900, Brazil
evaniziojr@ufam.edu.br

Abstract. We present in this paper a mathematical model of the dispersion of atmospheric methane that is proposed for the surface of lake in the region of the Santo Antônio Hydroelectric Dam in the state of Rondônia of Brazil. The model was elaborated from a general diffusion-advection-reaction for methane in which the diffusion coefficient was evaluated with techniques of fuzzy-logic-based. The numerical approximation was obtained with the use of the finite element method (FEM) for the spatial approximations and the Crank-Nicolson method for the temporal approximations. The approach provided scenarios for the directional fields of methane fluxes for different time periods and the results suggest a relation to the location in the reservoir, with flooded biomass, and with advective components for the dispersion of the gas.

Keywords: Fuzzy FEM · Environmental contamination
Greenhouse gases · Fuzzy biomathematical · Ecological models

1 Introduction

The dynamics of tropical ecosystems has received significant attention considering the effects that global warming can have on the Amazon, as well as the need to understand the effects of land cover change, regional bio-geochemical cycles and the role of tropical terrestrial ecosystems in the carbon balance [4]. Activities based on natural resource explorations can significantly alter landscapes and ecosystems, such as energy flows between natural systems. The climatic characteristics of a region can be altered by local development, such as changes in the hydrological cycle in a hydroelectric project.

The best hydro-power alternatives available are found in the Amazon region, where 51% of the total hydroelectric potential is concentrated in Brazil, and up to the year 2000, only 5% of the regional hydroelectric potential was in operation. The region to be flooded in a hydroelectric reservoir has as a characteristic the

Supported by CAPES–Brazil.

© Springer International Publishing AG, part of Springer Nature 2018
G. A. Barreto and R. Coelho (Eds.): NAFIPS 2018, CCIS 831, pp. 419–430, 2018.
https://doi.org/10.1007/978-3-319-95312-0_36

destruction of part of the native forest, resulting in an increase in the temperature of the air, causing changes in its capacity to retain water vapor.

In addition to the potential for electricity generation, the Madeira River historically is a natural navigation route that goes back to a prehistoric times, extended from the first Portuguese flags that ventured through the region and which today represents an important regional integration route, in the transportation of people and cargo. Therefore, it should be considered that the hydroelectric potential of Jirau (3,300 MW) and Santo Antônio (3,150 MW) could be combined with other hydroelectric projects and a combined waterway system that will allow South American integration, due to its proximity to Bolivia and Peru, opening spaces for energy infrastructure projects and transport between the three countries, boosting regional development.

Thus, it is evaluated that a hydroelectric project can produce the alteration of regional climatic characteristics. The emission of greenhouse gases in the Amazon can generate imbalance in natural ecosystems, altering hydrological cycles and, among other factors, increase the absorption of the so-called atmospheric window.

Methane gas in the atmosphere is related to the greenhouse effect, contributing with about 20% of the observed effect, besides being one of the main sinks of the radical hydroxyl (OH), thus influencing the oxidizing capacity of the atmosphere. The determinant conditions for the production and release of this gas into the atmosphere depend on many factors, such as decomposing organic matter, temperature, pH, among others [4]. The ability to monitor environmental factors that may generate imbalance for the region becomes a challenge, considering the dimensions involved and the possible logistics required for implementation.

Almost research on methane in the atmosphere uses observation towers whose measurements are made only on the vertical axis and assuming a homogeneous distribution over the surface in the tower region [9,22]. While the researches using fuzzy models, for the most part, are models applied to reactors for control mechanisms [17–19].

Considering this context, the implementation of methodologies that can contribute to an understanding of the dynamics of greenhouse gases for the region becomes relevant. In this sense, we propose this study in which the mathematical model and numerical codes are presented for computational simulation of scenario. In addition, due to the uncertainties of parameters that appear in the model, it was essential to use techniques fuzzy-logic-based [1].

2   Methane Gas in Tropical Reservoirs

Despite the increase in measurements, natural emissions of greenhouse gases in aquatic environments in the Amazon are still poorly understood [3]. The emission of methane gas has grown over the years, in a proportion that arouses concern in the scientific community. Data on the evolution of methane in the atmosphere can be seen in Fig. 1(a). And in the last 20 years, hydroelectric reservoirs have been identified as important sources of greenhouse gases. The release of methane to the atmosphere occurs due to the action of methanobacteria *Archaea*, responsible for the decomposition of organic matter to obtain energy [8,21].

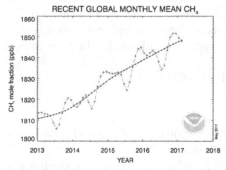

(a) Methane emissions, adapted from [13].   (b) Flooding of the S. Antônio reservoir in February 2012.

Fig. 1. Methane emission and flooding S. Antônio dam.

We can observe, in Fig. 1(b), organic matter being flooded in February 2012 in the reservoir of the Santo Antônio hydroelectric plant. In aquatic environments subject to seasonal flooding, the emission of methane by boiling can be an important source for the atmosphere, and can correspond to up to 90% of the total emission of this gas [10].

Methane is, after carbon dioxide, the most impacting greenhouse gas, with an infrared radiation absorption band between 7 and 8 µm of the so-called atmospheric window. In this region of the spectrum, gases with absorption capacity may have some relation with the radiation balance in the Earth-atmosphere system, contributing to its imbalance.

Although we have natural sinks, we have observed a surplus of methane gas emitted in relation to that removed annually, causing an increase in atmospheric concentration [6,7]. Per molecule, methane is a greenhouse gas 20 times more effective than carbon dioxide.

The broader analysis of emissions of 5000 lakes consists of only 2% represent tropical lakes with the majority of carbon dioxide emission studies, neglecting the effect of CH_4 emission. Considering the context above and taking into account the dimensions of the reservoirs and the variations in water levels throughout the year, samples taken at specific places in the reservoirs do not have the power to measure the flow of greenhouse gases such as methane effectively.

The emission of methane into the atmosphere depends on several interacting factors according to [12], such as climate, texture of surrounding soils, fauna and flora, land use and also geochemical processes reflect the difference in emissions between lakes and rivers; the nutrients, where a larger amount of decomposed organic matter, fixed by photosynthesis, is recycled as carbon dioxide in oxides and methane under anoxic conditions.

High primary production may, during the summer, reduce the emission of carbon dioxide into the atmosphere by acting as a collector; according to [2] some of the largest methane fluxes were obtained in eutrophic environments. High temperature has the potential to increase biological processes, including decomposition of organic matter by bacteria, which allows the release of methane (CH_4) and carbon dioxide (CO_2) into the atmosphere. The atmospheric concentration of CH_4 is controlled by the reaction with hydroxyl radicals in the troposphere via reaction

$$CH_4 + OH \rightarrow H_20 + CH_3$$

This reaction is largely responsible for the water vapor in the atmosphere; the measurements of CH_4 fluxes can vary with high frequency and are strongly influenced by external factors such as climatic conditions. Winds and rains mixed with water can intensify greenhouse gas emissions [20].

Thus, the use of mathematical modeling and an assembly of concepts and fuzzy systems is justified for a better understanding of the phenomena described here because they have demonstrated great potential for studies of the modeling in Ecology [1].

3   Mathematical Model

The modeling of substances such as methane presents the challenge of obtaining relevant knowledge of the behavior of this gas in the environment and the use of this knowledge to make the necessary simplifications in the implementation of the model. Difficulties in obtaining methane measurements in Amazonian regions are still a challenge, either because of the dimensions involved or because of the processes by which the gas is produced and emitted to the surface.

The equation used in this paper is well-known as diffusion-advection equation, which models phenomena related to the dispersion of pollutants in the aquatic systems and is associated with analyzes in mathematical ecology and general situations [15]. Taking into account the models proposed by [5,11,14], an equation adapted for the situation will be presented here. Denote by $c(x, y, t)$ the concentration of methane at a given position of the domain Ω at a time $t \in (0, T]$, the proposed problem can be presented in generic form by

$$\frac{\partial c(x, y, t)}{\partial t} = -\text{diffusion} - \text{transport} - \text{decay} + \text{source},$$

according to [16].

Thus, Eq. 1 represents the mathematical model that describes the phenomenon.

$$\frac{\partial c}{\partial t} = -\text{div}(\alpha \nabla C) - \text{div}(\vec{V} c) - \sigma c + f(x, y, t), \quad \forall t \in (0, T] \text{ and } (x, y) \in \Omega \ (1)$$

In Eq. (1) the term $\text{div}(\alpha \nabla c)$ is representing the diffusion to the environment, that is the natural scattering due to molecular movements or movements related to turbulence, depends on the gas itself in this case, position, time, and temperature; the term $\mathrm{div}(\overrightarrow{V}c)$ models the advective transport, related to external agents, in this case the transport of methane following the direction given by the vector of the dominant wind currents, in this approach we are considering $\mathrm{div}(\overrightarrow{V}) = 0$, as a conservative system; the term σc refers to phenomena related to changes suffered by the gas when reacting with the environment over time, and the function $f(x, y, t)$ represents the source gas term.

The part of the reservoir that will be considered for the computational simulations is depicted in Fig. 2.

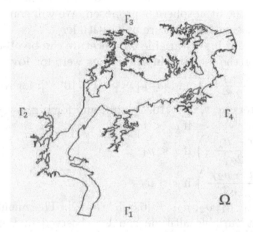

Fig. 2. The part of the S. Antônio reservoir considered.

Using the representation of Fig. 4, we must now establish the boundary conditions for the problem, which will considered be of the Robin type, in part of the boundary and of the Neumann type in another part of the border, as follows.

$$- \alpha \left. \frac{\partial c}{\partial \eta} \right|_{\Gamma_3 \cup \Gamma_4} = \beta c, \tag{2}$$

$$- \alpha \left. \frac{\partial c}{\partial \eta} \right|_{\Gamma_1 \cup \Gamma_2} = 0, \tag{3}$$

where η is the external unit vector, and the initial condition that will be considered, for the time $t = 0$ is given by

$$c(x, y, 0) = c_0(x, y) \tag{4}$$

This condition represent the initial distribution of methane on the domain considered, and the Eqs. (1) to (4) that model this phenomenon are denominated strong or classical formulation. It is important to emphasize that the diffusion is variable in relation to the altitude of the polluting source and with the temperature, expected behavior due to the chemical reactions that can occur in the atmosphere and the variation of air density, for example.

Thus, in this approach to the problem, we need to consider the uncertainty aspect of the diffusion coefficient (α). Hence, the need to use fuzzy-logic-based to treatment of this uncertainty.

4   Fuzzy Approach to Diffusion

According to [11,14], effective diffusion has a dependence relation with transport, so a model based on two inputs, altitude and temperature related to bands of the vertical layer of the atmosphere is proposed. We will consider that diffusion increases with increasing temperature and altitude.

The membership for input variable temperature can be observed in Fig. 3(a), where the choice for the membership functions were for **low** a gaussian type 1 by $f(x) = \exp\left(-\dfrac{(x-\mu)^2}{2\sigma^2}\right)$ and $[\sigma,\,\mu] = [8,\,4\times 10^{-16}]$; for **medium** a number fuzzy with parameters $[a,\,b,\,c] = [10,\,28,\,46]$, and for **high** a gaussian type 2 by

$$f(x) = \begin{cases} 1 & \text{if } \mu_1 < x < \mu_2 \\[2mm] \exp\left(-\dfrac{(x-\mu_1)^2}{2\sigma_1^2}\right) & \text{if } x < \mu_1 \\[2mm] \exp\left(-\dfrac{(x-\mu_2)^2}{2\sigma_2^2}\right) & \text{if } x > \mu_2 \end{cases}$$

with parameters $[\sigma_1,\,\mu_1,\,\sigma_2,\,\mu_2] = [6.5,\,52,\,8,\,62]$. The membership functions for the other input variable altitude can be observed in Fig. 3(b), where the choice for the membership functions were for **low** a gaussian type 1 with parameters $[\sigma,\,\mu] = [125,\,1.5\times 10^{-14}]$; for **medium** a number fuzzy with parameters $[a,\,b,\,c] = [200,\,500,\,800]$, and for **high** a gaussian type 2 with parameters $[\sigma_1,\,\mu_1,\,\sigma_2,\,\mu_2] = [122,\,1000,\,272,\,2080]$.

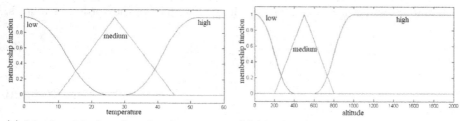

(a) Membership functions for the temperature.   (b) Membership functions for the altitude.

Fig. 3. Membership functions for input variables.

The membership functions for the output variable diffusion coefficient can be observed in Fig. 4(a), where the choice for the membership functions were for **low** a gaussian type 2 with parameters $[\sigma_1,\,\mu_1,\,\sigma_2,\,\mu_2] =$

$[5 \times 10^{-7}, 0, 0.0006, 0.0086]$; **low medium** was fuzzy number with parameters $[a, b, c] = [0.0015, 0.003, 0.0045]$, **medium** was fuzzy number with parameters $[a, b, c] = [0.0035, 0.005, 0.0065]$, **high medium** was fuzzy number and parameters $[a, b, c] = [0.0055, 0.007, 0.0085]$, and **high** was a gaussian type 2 with parameters $[\sigma_1, \mu_1, \sigma_2, \mu_2] = [0.0003, 0.009, 0.002, 0.011]$.

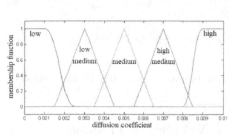

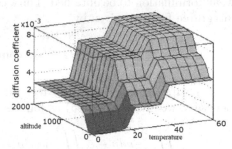

(a) Membership functions for the diffusion coefficient (α).

(b) Graphic for the output results by inference and defuzzified output.

Fig. 4. The membership functions for the output variable and the graphic result.

Based on these variables and considering some suggestions of expertise, we propose the following base rule for the problem.

- If temperature is **low** 'and' altitude is **low**, 'then' the diffusion coefficient is **low**.
- If temperature is **low** 'and' altitude is **medium**, 'then' the diffusion coefficient is **low medium**.
- If temperature is **low** 'and' altitude is **high**, 'then' the diffusion coefficient is **low medium**.
- If temperature is **medium** 'and' altitude is **low**, 'then' the diffusion coefficient is **low medium**.
- If temperature is **medium** 'and' altitude is **medium**, 'then' the diffusion coefficient is **medium**.
- If temperature is **medium** 'and' altitude is **high**, 'then' the diffusion coefficient is **high medium**.
- If temperature is **high** 'and' altitude is **low**, 'then' the diffusion coefficient is **medium**.
- If temperature is **high** 'and' altitude is **medium**, 'then' the diffusion coefficient is **high medium**.
- If temperature is **high** 'and' altitude is **high**, 'then' the diffusion coefficient is **high**.

For the rule base above we use a Mamdani-type fuzzy inference system as the inference operator, and the centroid method was employed for defuzzification method (see Fig. 4(b)).

5   Weak Formulation of the Problem

To obtain the numerical approximation for the solution of the strong formulation it is necessary to establish the problem in its weak formulation, which consists of using variational principles defined in convenient metric spaces, whose formulation allows the existence and uniqueness of solution to the problem in its weak formulation to be obtained. Thus, considering the space of square Lebesgue integrable function given by

$$\mathcal{L}^2(\Omega) = \left\{ \nu : \Omega \to \mathbb{R}, \text{ com } \iint_\Omega [\nu(x,y)]^2 d\mu < \infty \right\},$$

and multiplying Eq. (1) by any test function ν and integrating in the sense of Lebesgue we arrive at Eq. (5) as follows.

$$\iint_\Omega \frac{\partial c}{\partial t} \nu d\mu - \alpha \iint_\Omega \Delta c \nu d\mu + \iint_\Omega \nabla(\vec{V}c)\nu d\mu + \iint_\Omega \sigma c \nu d\mu$$
$$= \iint_\Omega f(x,y,t)\nu d\mu. \tag{5}$$

Now, applying Green's first identity to the diffusive term, we arrive at the weak formulation of the problem given by Eq. (6).

$$\iint_\Omega \frac{\partial c}{\partial t} \nu d\mu - \alpha \iint_\Omega \nabla c \cdot \nabla \nu d\mu + \iint_\Omega \nabla(\vec{V}c)\nu d\mu + \iint_\Omega \sigma c \nu d\mu 0$$
$$= \beta \int_{\partial\Gamma} c \nu d\gamma + \iint_\Omega f(x,y,t)\nu d\mu. \tag{6}$$

6   Discretization of the Problem

The finite elements method (FEM) applied to this problem requires for the computational simulation, a discretization of the problem for the spatial variables. The chosen option was triangular elements and linear base functions, better known as the standard Galerkin method. Using the inner product notation given by

$$\iint_\Omega f(u)g(u)d\mu = (f,g)_\Omega \text{ and } \int_\Gamma f(u)g(u)d\gamma = \langle f,g\rangle_\Gamma,$$

the weak formulation becomes Eq. (7).

$$\left(\frac{\partial c}{\partial t},\nu\right)_\Omega + \alpha\,(\nabla c,\nabla \nu)_\Omega + V_x \left(\frac{\partial c}{\partial x},\nu\right)_\Omega + V_y \left(\frac{\partial c}{\partial y},\nu\right)_\Omega + (\sigma c,\nu)_\Omega$$
$$= \beta\,\langle c,\nu\rangle_\Gamma + (f,\nu)_\Omega \tag{7}$$

where V_x and V_y represent the components given by the projections of the velocity of wind in each coordinates axes.

7   Results

The results used routines found in standard computational environments, whose graphic interface allows facilities in the analysis of the treated phenomena, besides the use of fuzzy concepts and characteristics. The region of the reservoir of the hydroelectric dam of Santo Antônio that was considered is closer to the dam of the reservoir, where we have environmental impact due to the flooding of nearby regions, thus being a place with potential for high emissions. In order to obtain the simulations for region 1 of the reservoir in which we adopt a mesh with 257.835 nodes generated through software Gmesh®. We used the parameters in Table 1, with 2000 iterations, and [0, 48] hours for interval of time, for which we consider that the diffusion coefficient is constant for an altitude range of 200 m of the surface and average temperature of 25 °C, that resulted by centroid method for defuzzification the coefficient $\alpha = 0.007 \, \mathrm{km^2/h}$.

Table 1. Value of parameters used in simulation

Parameter	Symbol	Value	Unit
Velocity of wind	V	3.6	km/h
Direction of wind	θ	$-3\pi/4$	Radians
Charge of methane	met	0.018	$\mu\mathrm{mol/km^2/h}$
Permeability in the border	β	0.01	km/h
Decay coefficient	σ	1.2×10^{-5}	$\mathrm{km^2/h}$
Diffusion coefficient	α	0.007	$\mathrm{km^2/h}$

Figure 5(a) shows the discretization for the part of region of study, named by Ω domain.

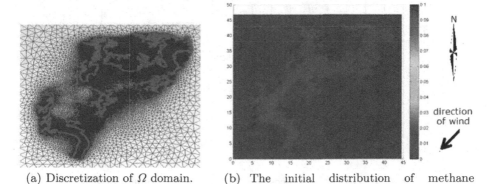

(a) Discretization of Ω domain.

(b) The initial distribution of methane ($\mu\mathrm{mol/km^2}$) on the domain Ω.

Fig. 5. The discretized domain and the initial condition for the simulation.

Based on these parameters we were able to run the simulation shown in Fig. 5(b) until Fig. 6(b), where the process of dispersion and increase of methane concentration is evident.

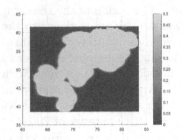

(a) The distribution of methane (μmol/km^2) on the Ω in $t = 12$ hours.

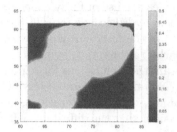

(b) The distribution of methane (μmol/km^2) on the Ω in $t = 48$ hours.

Fig. 6. The distribution of methane on the domain for 2 instances of time.

Figure 7 shows the evolution process for 2 nodes in the domain over time, for the period of time considered $[0, 48]$, where the first node (206103) has coordinates $(x, y) = (21.3373; 21.8410)$ and second node (224471) has coordinates $(x, y) = (19.1414; 35.2565)$.

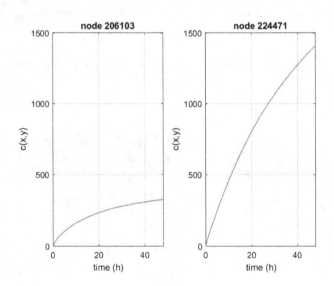

Fig. 7. The evolution process for 2 nodes of the domain in 48 h.

8   Conclusion

The various scenarios portray the dispersion of methane gas in the atmosphere, in a two-dimensional domain over the Santo Antônio hydroelectric reservoir, a complex phenomenon due to difficulties encountered in the advective dynamics, such as chemical reactions and natural processes, anthropogenic actions, in particular on the Amazon region. Thus, the scenario presented in the simulation, considering the dark waters in peripheral regions and amount of flooded biomass was shown to be consistent with theoretical studies on the gas emission process, presenting a gradient profile for the central region of the river.

The diffusive coefficient was estimated taking into account fuzzy concepts and procedures for factors such as temperature and altitude, where it is expected with the increase of altitude and temperature, allowing for greater diffusion, which is obtained through the proposed rules bases.

Finally, we believe that this paper brings relevant contributions to the study of methane dispersion in the Santo Antônio dam region, which has no study similar to this type until now.

References

1. Barros, L.C., Bassanezi, R.C., Lodwick, W.A.: A First Course in Fuzzy Logic, Fuzzy Dynamical Systems, and Biomathematics. Studies in Fuzziness and Soft Computing, 1st edn. Springer, Berlin (2017). https://doi.org/10.1007/978-3-662-53324-6
2. Barros, N., Cole, J., Travink, L., Prairie, Y., Bastiviken, D., Huszar, V., Giorgio, P.D., Roland, F.: Carbon emission from hydroelectric reservoirs linked to reservoir age and latitude. Nat. Geosci. **4**, 593–596 (2011)
3. Bastviken, D.: Organic Compounds-Methane. Elsevier, Amsterdan (2009)
4. Cicerone, R., Oremland, R.: Biogeochemical aspects of atmospheric methane. Glob. Biogeochem. Cycles **2**, 299–327 (1988)
5. Edelstein-Keshet, L.: Mathematical Models in Biology, Classics in Applied Mathematics, vol. 46. SIAM, Philadelphia (2005)
6. Hein, R., Crutzen, P., Heimann, M.: An inverse modeling approach to investigate the global atmospheric methane cycle. Glob. Biogeochem. Cycles **11**(1), 43–76 (1997)
7. Houweling, S., Kaminski, T., Dentener, F., Lelieveld, J., Heimann, M.: Inverse modeling of methane sources and sinks using the adjoint of a global transport model. J. Geophys. Res. **104**, 137–160 (1999)
8. Keppler, F., Hamilton, J., Brab, M., Röckmann, T.: Methane emissions from terrestrial plants under aerobic conditions. Nature **439**, 187–191 (2006)
9. Lamb, B., Cambaliza, M., Davis, K., Edburg, S., Ferrara, T., Floerchinger, C., Heimburger, A.M., Herndon, S., Lauvaux, T., Lavoie, T., Lyon, D., Miles, N., Prasad, K., Richardson, S., Roscioli, J., Salmon, O., Shepson, P., Stirm, B., Whetstone, J.: Direct and indirect measurements and modeling of methane emissions in Indianapolis Indiana. Environ. Sci. Technol. **50**(16), 8910–8917 (2016)
10. Marani, L., Alvala, P.C.: Methane emissions from lakes and floodplains in Pantanal Brazil. Atmos. Environ. **41**(8), 1627–1633 (2007)

11. Marchuk, G.I.: Mathematical Models in Environmental Problems. Studies in Mathematics and its Applications, vol. 16. North-Holland, Amsterdan (1986)
12. Meybeck, M.: Riverine transport of atmospheric carbon sources, global typology and budget. Water Air Soil Pollut. **70**, 443–463 (1993)
13. NOOA: Methane measurements. https://esrl.noaa.gov/gmd/ccgg/trends_ch4/ (2017). Accessed 26 May 2017
14. Okubo, A.: Diffusion and Ecological Problems: Mathematical Models. Springer, Berlin (1980). https://doi.org/10.1007/BF02851862
15. Poletti, E.C.C., Meyer, J.F.C.A.: Dispersion of pollutants in reservoir system: mathematical modeling via fuzzy logic and numerical approximation (in Portuguese). Biomatemática **19**, 57–68 (2009)
16. Prestes, M.F.B., Meyer, J.F.C.A., Poletti, E.C.C.: Dispersion of polluting matter in aquatic mean: mathematical model, numerical approximation and computational simulation - Salto Grande reservoir, Americana/SP (in Portuguese). Biomatemática **23**, 43–56 (2013)
17. Robles, A., Latrille, E., Ruano, M., Steyer, J.: A fuzzy-logic-based controller for methane production in anaerobic fixed-film reactors. Environ. Technol. **38**(1), 42–52 (2017)
18. Ruan, J., Chen, X., Huang, M., Zhang, T.: Application of fuzzy neural networks for modeling of biodegradation and biogas production in a full-scale internal circulation anaerobic reactor. J. Environ. Sci. Health Part A Tox Hazard Subst. Environ. Eng. **52**(1), 7–14 (2017)
19. Turkdogan-Aydinol, F., Yetilmezsoy, K.: A fuzzy-logic-based model to predict biogas and methane production rates in a pilot-scale mesophilic UASB reactor treating molasses wastewater. J. Hazard. Mater. **182**(1–3), 460–470 (2010)
20. UNESCO: The UNESCO/IHA Measurement specification guidance for evaluating the GHG status of man-made freshwater reservoirs, 1st. edn. UNESCO, New York (2009)
21. Wuebbles, D., Hayhoe, K.: Atmospheric methane and global change. Earth-Sci. Rev. **57**, 177–210 (2002)
22. Xu, X., Yuan, F., Hanson, P., Wullschleger, S.D., Thornton, P., Riley, W., Song, X., Graham, D., Song, C., Tian, H.: Four decades of modeling methane cycling in terrestrial ecosystems. Biogeosciences **13**, 3735–3755 (2016)

Solution to Convex Variational Problems with Fuzzy Initial Condition Using Zadeh's Extension

Michael M. Diniz[1]([✉]), Luciana T. Gomes[2], and Rodney C. Bassanezi[3]

[1] Instituto Federal de São Paulo, São José dos Campos, SP, Brazil
michael.diniz@ifsp.edu.br
[2] Departamento de Física, Química e Matemática,
Universidade Federal de São Carlos, Sorocaba, SP, Brazil
[3] IMECC, Universidade Estadual de Campinas, Campinas, SP, Brazil

Abstract. This paper investigates conditions to solve a fuzzy variational problem using Zadeh's extension. The fuzzy problem is obtained by extending a classical one in the initial condition. The solution to the problem is a fuzzy bunch of functions (fuzzy set of functions) obtained by extending the classical solution in the initial condition. For convex functionals the resulting functional value is a fuzzy number that is proved to be the smallest element in a partial order.

Keywords: Fuzzy variational problem · Zadeh's extension
Partial order

1 Introduction

Conditions for existence of solutions to variational problems have been extensively explored in the literature (see [7] for instance), but in the fuzzy context few results are available. As an example, recently, [4] studied fuzzy variational problems with fuzzy boundary conditions and the Euler-Lagrange conditions considering differentiability and integrability of fuzzy-number-valued functions.

The variational problem is commonly stated as

$$\min_{x \in C^{(1)}[t_0, t_f]} J(x) = \int_{t_0}^{t_f} F(t, x, \dot{x}) \, dt \tag{1}$$

$$x(t_0) = x_0 \qquad x(t_f) = x_f$$

where F is a real function, x_0 and x_f are real numbers and $C^{(1)}[t_0, t_f]$ is the space of continuously differentiable functions from the closed interval $[t_0, t_f]$ to $\mathbb{R}$.

The fuzzy version explored by [4] defines minimization using a partial order. Also, the author regards x, $\dot{x}$ and F as fuzzy-number-valued functions with t as a real variable. Calculation is on α-cuts, generalizing the classical methods, but taking into account the fuzzy specifics.

© Springer International Publishing AG, part of Springer Nature 2018
G. A. Barreto and R. Coelho (Eds.): NAFIPS 2018, CCIS 831, pp. 431–438, 2018.
https://doi.org/10.1007/978-3-319-95312-0_37

The present paper is based on Zadeh's extension principle: each solution to problem (1) where the initial condition is in the support of a fuzzy set of initial conditions $\widehat{x}_0$ will be used to build a fuzzy subset of functions. The derivative and integral of the classical functions will also be extended and, assuming some conditions, the extended functional will take values in fuzzy numbers. We will prove that the fuzzy subset of functions minimizes the extension of convex functionals J in x in the same partial order as in [4].

2 Preliminaries

Definitions, notation and results used in this paper are presented in what follows.

Let $\hat{u}$ be a fuzzy subset of $\mathbb{X}$ and denote by $\mu_{\hat{u}} : \mathbb{X} \to [0,1]$ the *membership function* of $\hat{u}$. The *support* of $\hat{u}$ is the classical set $\operatorname{supp} \hat{u} = \{x \in \mathbb{X} : \mu_{\hat{u}}(x) > 0\}$. The α-*levels* are defined by $[\hat{u}]^\alpha = \{x \in \mathbb{X} \,|\, \mu_{\hat{u}}(x) \geq \alpha\}$ for all $\alpha \in (0,1]$ and $[\hat{u}]^0 = \overline{\operatorname{supp} \hat{u}}$.

Definition 1 *(See* [1]*). A fuzzy subset $\hat{u}$ is a* fuzzy number *if $\mu_{\hat{u}} : \mathbb{R} \to [0,1]$ and*

1. *all α-levels of $\hat{u}$ are nonempty, with $0 \leq \alpha \leq 1$;*
2. *all α-levels of $\hat{u}$ are closed intervals of $\mathbb{R}$;*
3. *$\operatorname{supp} \hat{u} = \{x \in \mathbb{R} : \mu_{\hat{u}}(x) > 0\}$ is bounded.*

The symbol $\mathcal{F}(\mathbb{R})$ stands for the set of fuzzy numbers. The α-levels of a fuzzy number $\hat{u}$ are closed intervals denoted by $[\hat{u}]^\alpha = [u_L^\alpha, u_R^\alpha]$.

Definition 2. *Let $\hat{u}$ and $\hat{v}$ be two fuzzy numbers. The partial order relation $\preceq_F$ is defined as: $\hat{u} \preceq_F \hat{v}$ if $u_L^\alpha \leq v_L^\alpha$ and $u_R^\alpha \leq v_R^\alpha$ for all $\alpha \in [0,1]$.*

If $\hat{u} \neq \hat{v}$ and Definition 2 holds we write $\hat{u} \prec_F \hat{v}$. Usually, the order of Definition 2 is known as fuzzy-max order.

Definition 3. *Let P be a set partially ordered by $\preceq$. Then*

- *x is the smallest element of P if $x \preceq y$, for all $y \in P$.*
- *x is the minimal element of P if for all $y \in P$, $y \preceq x$ implies $y = x$.*

Definition 4. *Let $\mathbb{X}$ be a metric space, $\hat{J} : \mathbb{X} \to \mathcal{F}(\mathbb{R})$ and $\hat{x}^* \in \mathbb{X}$. The element $\hat{x}^*$ is a* local minimum *of $\hat{J}$ on $\mathbb{X}$ if $\hat{J}(\hat{x}^*) \preceq_F \hat{J}(\hat{x})$, for all $\hat{x}$ in a neighborhood of $\hat{x}^*$ in $\mathbb{X}$. If $\hat{J}(\hat{x}^*) \preceq_F \hat{J}(\hat{x})$, for all $\hat{x} \in \mathbb{X}$, $\hat{x}^*$ is a* global minimum *of $\hat{J}$ on $\mathbb{X}$.*

Note that $\hat{J}(\hat{x}^*)$ is the smallest element of the image $Im(\hat{J})$ of $\mathbb{X}$ through $\hat{J}$ according Definition 3.

Definition 5 *(See* [1,8]*) (Zadeh's Extension Principle). Let $f : \mathbb{U} \to \mathbb{Z}$ be a function and $\hat{u}$ be a fuzzy subset of $\mathbb{U}$. Zadeh's extension of f is a function $\hat{f}$ that maps $\hat{u}$ to the fuzzy subset $\hat{f}(\hat{u})$ of $\mathbb{Z}$ with membership function*

$$\mu_{\hat{f}(\hat{u})}(z) = \begin{cases} \sup\limits_{x = f^{-1}(z)} \mu_{\hat{u}}(x) & \text{if } f^{-1}(z) \neq \emptyset \\ 0 & \text{if } f^{-1}(z) = \emptyset \end{cases}$$

where $f^{-1}(z) = \{x : f(x) = z\}$ is the preimage of z.

Theorem 1 *(See* [5]*). Let* $\mathbb{U}$ *and* $\mathbb{V}$ *be two Hausdorff spaces and* $f : \mathbb{U} \to \mathbb{V}$ *be a function. If* f *is continuous, then* $\widehat{f} : \mathcal{F}_K(\mathbb{U}) \to \mathcal{F}_K(\mathbb{V})$ *is well-defined and*

$$[\widehat{f}(\hat{u})]^\alpha = f([\hat{u}]^\alpha) \tag{2}$$

for all $\alpha \in [0,1]$, *where* $\mathcal{F}_K(\mathbb{U})$ *denotes the family of all fuzzy subsets of* $\mathbb{U}$ *with compact and non-empty* α-*levels.*

Definition 6. *Let* $f : [a,b] \subset \mathbb{R} \to \mathbb{R}$ *and* $g : [a,b] \subset \mathbb{R} \to \mathbb{R}$. *The inequality* $f(\cdot) \leq g(\cdot)$ *is said to hold if* $f(t) \leq g(t)$, *for all* $t \in [a,b]$. *If* $f(\cdot) \leq g(\cdot)$ *and* $f(t^*) < g(t^*)$ *for some* $t^* \in [a,b]$ *then we write* $f(\cdot) < g(\cdot)$.

It is well-known that not all pairs of functions are comparable.

Definition 7. *A set of real functions* X *parameterized by* λ *is increasing in* λ *if* $\lambda_1 < \lambda_2$ *implies* $f_{\lambda_1}(\cdot) \leq f_{\lambda_2}(\cdot)$ *and it is decreasing in* λ *if* $\lambda_1 < \lambda_2$ *implies* $f_{\lambda_1}(\cdot) \geq f_{\lambda_2}(\cdot)$ *for* $\lambda_1, \lambda_2 \in I$, $f_{\lambda_1}(\cdot), f_{\lambda_2}(\cdot) \in X$ *and* I *is a real interval.*

Definition 8. *Let* X *and* Y *be two sets of continuous functions that map* $[a,b]$ *into* $\mathbb{R}$. *The distance* d_{cf} *between* X *and* Y *is*

$$d_{cf}(X,Y) = d_H(X,Y) = \max\{\sup_{x \in X} \inf_{y \in Y} d_f(x,y), \sup_{y \in Y} \inf_{x \in X} d_f(y,x)\}.$$

where d_H is Pompeiu-Hausdorff distance for sets of functions and d_f is a distance for functions.

Definition 9. *Let* $\mathbb{X}$ *be a function space from* $[a,b]$ *into* $\mathbb{R}$, $\widehat{f} \in \mathcal{F}(\mathbb{X})$ *and* $\widehat{g} \in \mathcal{F}(\mathbb{X})$. *The distance* $\widetilde{d}_{cf}$ *between* $\widehat{f}$ *and* $\widehat{g}$ *is*

$$\widetilde{d}_{cf}(\widehat{f}, \widehat{g}) = \sup_{\alpha \in [0,1]} d_{cf}([\widehat{f}(\cdot)]^\alpha, [\widehat{g}(\cdot)]^\alpha).$$

Definition 10 *(See* [2,3]*). Let* x_λ *be a real function, continuously parameterized by* $\lambda \in [\lambda_L, \lambda_R] \subset \mathbb{R}$ *and differentiable for each fixed* $\lambda \in [\lambda_L, \lambda_R]$. *We define the derivative of the extension of* $x_\lambda(t)$ *in* λ *as*

$$\frac{d\widehat{x}_{\widehat{\lambda}}(t)}{dt} = \frac{\widehat{dx_{\widehat{\lambda}}(t)}}{dt}.$$

where $\lambda \in \mathcal{F}([\lambda_L, \lambda_R])$ *and* $\dfrac{\widehat{dx_{\widehat{\lambda}}(t)}}{dt}$ *is Zadeh's extension of* $\dfrac{dx_\lambda(t)}{dt}$ *in* λ.

Similarly, we can define the integral of functions with fuzzy parameters as follows.

Definition 11 *(See* [2,3]*). Let* x_λ *be a real function, continuously parameterized by* $\lambda \in [\lambda_L, \lambda_R] \subset \mathbb{R}$ *and integrable for each fixed* $\lambda \in [\lambda_L, \lambda_R]$. *We define the integral of the extension of* $x_\lambda(t)$ *in* λ *as*

$$\int_a^b \widehat{x}_{\widehat{\lambda}}(t)\, dt = \widehat{\int_a^b x_{\widehat{\lambda}}(t)\, dt}$$

where $\lambda \in \mathcal{F}([\lambda_L, \lambda_R])$ *and* $\widehat{\int_a^b x_{\widehat{\lambda}}(t)dt}$ *is Zadeh's extension of* $\int_a^b x_\lambda(t)\, dt$ *in* λ.

Theorem 2 *(See [7]). Let D be a domain in $\mathbb{R}^2$ and for given a_1, b_1, set*

$$\mathcal{D} = \{x \in C^{(1)}[a, b] : x(a) = a_1, x(b) = b_1; (x(t), x'(t)) \in D\}.$$

If $f(t, x, z)$ is [strongly] convex on $[a, b] \times D$, then

$$F(y) = \int_a^b f(t, x(t), x'(t))dt$$

is [strictly] convex on $\mathcal{D}$. Hence each $y \in \mathcal{D}$ for which

$$\frac{d}{dt} f_z[x(t)] = f_x[x(t)]$$

on (a, b), minimizes F on $\mathcal{D}$ [uniquely].

We define Zadeh's extension of (1) in the initial condition as a minimization of the extension in the initial condition of the classical functional in (1) according to the fuzzy partial order in Definition 2. A solution to this fuzzy variational problem is a fuzzy set of functions (fuzzy bunch of functions) that solves this fuzzy minimization problem. The initial condition is a fuzzy number and the final condition is crisp.

Remark 1. It is well-known that integral is a continuous functional. If F is continuous in t, x and $\dot{x}$, and x and $\dot{x}$ depends continuously on x_0, then the mapping that takes x_0 into $J(x) = J(x(\cdot, x_0))$ is continuous. Hence if $\widehat{x}_0$ is a fuzzy number, the extension $\widehat{J}(\widehat{x}(\cdot, \widehat{x}_0))$ is also a fuzzy number [5,6].

3   Results

In what follows, the symbol $(X_F, \widetilde{d}_{cf})$ stands for the set of all fuzzy subsets of functions $\widehat{x}(t, \widehat{x}_0)$ having a continuous mapping from $[\widehat{x}_0]^0$ to $[\widehat{x}(\cdot, \widehat{x}_0)]^0$ and a continuous mapping from $[\widehat{x}_0]^0$ to $[\dot{\widehat{x}}(\cdot, \widehat{x}_0)]^0$ with the metric $\widetilde{d}_{cf}$ of Definition 9.

Theorem 3. *Let*

$$\min_{x \in C^1([t_0, t_f])} J(x) = \int_{t_0}^{t_f} F(t, x, \dot{x}) \, dt \tag{3}$$
$$x(t_0) = x_0 \quad x(t_f) = x_f$$

such that F is continuous and strongly convex and the solution $x^*(\cdot, x_0)$ is continuous and monotonic in $x_0 \in [x_0^L, x_0^R]$. If the initial condition $x_0 = \widehat{x}_0 \in \mathcal{F}(\mathbb{R})$ and $\widehat{J}$ is obtained via Zadeh's extension of J then $\widehat{J}(\widehat{x}^*) \preceq_F \widehat{J}(\widehat{x})$ for all $\widehat{x} \in (X_F, \widetilde{d}_{cf})$.

Proof. Remark 1 assures that $\widehat{J}(\widehat{x}^*(\cdot, \widehat{x}_0))$ is a fuzzy number and that $\widehat{J}(\widehat{x}(\cdot, \widehat{x}_0))$ is also a fuzzy number if $\widehat{x} \in (X_F, \widetilde{d}_{cf})$. Now we will show that

$$\widehat{J}(\widehat{x}^*(\cdot, \widehat{x}_0))_L^\alpha \le \widehat{J}(\widehat{x}(\cdot, \widehat{x}_0))_L^\alpha \tag{4}$$

and

$$\widehat{J}(\widehat{x}^*(\cdot, \widehat{x}_0))_R^\alpha \le \widehat{J}(\widehat{x}(\cdot, \widehat{x}_0))_R^\alpha \tag{5}$$

in order to prove that $\widehat{J}(\widehat{x}^*(\cdot, \widehat{x}_0)) \preceq_F J(\widehat{x}(\cdot, \widehat{x}_0))$ for all $\widehat{x} \in (X_F, \widetilde{d}_{cf})$. From Theorem 1,

$$\widehat{J}(\widehat{x}^*(\cdot, \widehat{x}_0))_L^\alpha = \min[\widehat{J}(\widehat{x}^*(\cdot, \widehat{x}_0))]^\alpha = \min J([\widehat{x}^*(\cdot, \widehat{x}_0)]^\alpha). \tag{6}$$

And for any $\widehat{x}(t, \widehat{x}_0) \in (X_F, \widetilde{d}_{cf})$,

$$\widehat{J}(\widehat{x}(\cdot, \widehat{x}_0))_L^\alpha = \min J([\widehat{x}(\cdot, \widehat{x}_0)]^\alpha). \tag{7}$$

Suppose that the fuzzy bunches of functions $x^*(\cdot, x_0^*)$ and $x(\cdot, \overline{x}_0)$ minimize $J([\widehat{x}^*(\cdot, \widehat{x}_0)]^\alpha)$ and $J([\widehat{x}(\cdot, \widehat{x}_0)]^\alpha)$. Then $\overline{x}_0 \in [\widehat{x}_0]^\alpha$ resulting in $x^*(\cdot, \overline{x}_0) \in [\widehat{x}^*(\cdot, \widehat{x}_0)]^\alpha$. As a consequence,

$$J(x^*(\cdot, x_0^*)) \le J(x^*(\cdot, \overline{x}_0)). \tag{8}$$

Since F is convex, $x^*(t, \overline{x}_0)$ is the (only) global solution to the classical case with initial condition $\overline{x}_0$ (see Theorem 2), meaning

$$J(x^*(\cdot, \overline{x}_0)) \le J(x(\cdot, \overline{x}_0))$$

for all $x(\cdot, \overline{x}_0) \in C^{(1)}[t_0, t_f]$. Hence

$$J(x^*(\cdot, x_0^*)) \le J(x^*(\cdot, \overline{x}_0)) \le J(x(\cdot, \overline{x}_0)) = \widehat{J}(\widehat{x}(\cdot, \widehat{x}_0))_L^\alpha.$$

This proves (4).

In order to prove (5), we first write

$$\widehat{J}(\widehat{x}^*(\cdot, \widehat{x}_0))_R^\alpha = \max J([\widehat{x}^*(\cdot, \widehat{x}_0)]^\alpha) \quad \text{and} \quad \widehat{J}(\widehat{x}(\cdot, \widehat{x}_0))_R^\alpha = \max J([\widehat{x}(\cdot, \widehat{x}_0)]^\alpha).$$

Let $x^*(\cdot, x_0^*)$ e $x(\cdot, \overline{x}_0)$ maximize $J([\widehat{x}^*(\cdot, \widehat{x}_0)]^\alpha)$ and $J([x(\cdot, \widehat{x}_0)]^\alpha)$. Then $x^*(\cdot, x_0^*)$ is the global solution to the classical case of minimizing J with initial condition x_0^*, meaning

$$J(x^*(\cdot, x_0^*)) \le J(x(\cdot, x_0^*)) \tag{9}$$

for all $x(\cdot, x_0^*) \in C^{(1)}[t_0, t_f]$. Since $x_0^* \in [\widehat{x}_0]^\alpha$, we have $x(\cdot, x_0^*) \in [\widehat{x}(\cdot, \widehat{x}_0)]^\alpha$ and

$$J(x^*(\cdot, x_0^*)) \le J(x(\cdot, x_0^*)) \le J(x(\cdot, \overline{x}_0)) = \max J([\widehat{x}(\cdot, \widehat{x}_0)]^\alpha) = \widehat{J}(\widehat{x}(\cdot, \widehat{x}_0))_R^\alpha$$

Then

$$\widehat{J}(\widehat{x}^*(\cdot, \widehat{x}_0))_R^\alpha \le \widehat{J}(\widehat{x}(\cdot, \widehat{x}_0))_R^\alpha$$

which proves (5).

This result is illustrated in the next section.

4   Example

Example 1. Consider the following fuzzy variational problem

$$\min_{x \in C^1[0,2]} J(x) = \int_0^2 [x^2(t) + 2x(t)\dot{x}(t) + \dot{x}^2(t)]dt$$

$$\widehat{x}(0) = \widehat{x}_0 \quad x(2) = -3$$

where $\widehat{x}_0 = (-1; 0; 1)$, i.e., $\widehat{x}_0$ is a triangular fuzzy number with $[-1, 1]$ as support and 0 as core.

Using Euler-Lagrange equation we have the following necessary condition for optimality:

$$2x^*(t) - 2\ddot{x}^*(t) = 0$$

with boundary conditions $x(0) = x_0$ and $x(2) = -3$. The solution to this boundary value problem is

$$x^*(t, x_0) = \frac{x_0 e^{-2} + 3}{e^{-2} - e^2}e^t - \frac{3 + e^2 x_0}{e^{-2} - e^2}e^{-t} \tag{10}$$

Its derivative in t is

$$\frac{dx^*(t, x_0)}{dt} = \frac{x_0 e^{-2} + 3}{e^{-2} - e^2}e^t + \frac{3 + e^2 x_0}{e^{-2} - e^2}e^{-t}.$$

The solution and its derivative depend continuously on the initial condition x_0 for $x_0 \in [-1, 1] = [\widehat{x}_0]^0$. Moreover, F is continuous and convex with respect to all variables, therefore, we conclude by Theorem 3 that $\widehat{x}^*(\cdot, \widehat{x}_0)$ is the solution of the fuzzy variational problem in X_F.

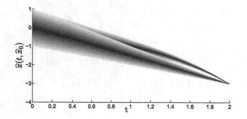

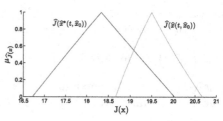

Fig. 1. Fuzzy optimal solution $\widehat{x}^*(\cdot, \widehat{x}_0)$ and feasible fuzzy solution $\widehat{x}(\cdot, \widehat{x}_0)$ for $\widehat{x}_0 = (-1; 0; 1)$.

Fig. 2. Optimal fuzzy functional $\widehat{J}(\widehat{x}^*)$ and perturbation $\widehat{J}(\widehat{x})$.

First Comparison. Let's compare the optimal solution given by Eq. (10) with a feasible perturbation given by Eq. (11),

$$x(t, x_0) = -\left(\frac{3 + x_0}{2}\right)t + x_0. \tag{11}$$

Both $\widehat{x}^*$ and $\widehat{x}$ satisfy the boundary conditions $\widehat{x}(0) = (-1; 0; 1)$ and $\widehat{x}\left(\dfrac{\pi}{2}\right) = 1$. Figure 1 shows the graphics of both functions. Based on Fig. 2 it is possible to note that $\widehat{J}(\widehat{x}^*(\cdot, \widehat{x}_0)$ is smaller than $\widehat{J}(\widehat{x}(\cdot, \widehat{x}_0)$, as expected by Theorem 3.

Second Comparison. Let's compare the optimal solution given by Eq. (10) with a feasible perturbation given by Eq. (12),

$$x(t, x_0) = \left[\left(\frac{x_0 e^{-2} + 3}{e^{-2} - e^2}\right) e^t - \left(\frac{x_0 e^2 + 3}{e^{-2} - e^2}\right) e^{-t}\right] \left(\frac{sen(10\pi t)}{10} + 1\right). \quad (12)$$

This feasible solution depends on x_0 and its derivative is given by Eq. (13) (Figs. 3 and 4),

$$\frac{dx(t, x_0)}{dt} = \left(\frac{x_0 e^{-2} + 3}{e^{-2} - e^2}\right) e^t + \left(\frac{x_0 e^2 + 3}{e^{-2} - e^2}\right) e^{-t}. \quad (13)$$

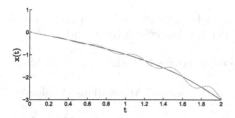

Fig. 3. Optimal solution $x^*(\cdot, x_0)$ and feasible solution $x(\cdot, x_0)$ for $x_0 = 0$.

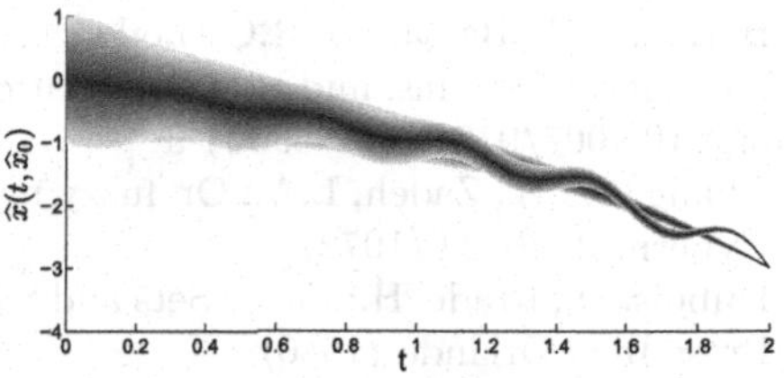

Fig. 4. Fuzzy optimal solution $\widehat{x}^*(\cdot, x_0)$ and feasible fuzzy solution $\widehat{x}(\cdot, x_0)$ for $\widehat{x}_0 = (-1; 0; 1)$.

Figure 5 illustrates the fact that $\widehat{J}(\widehat{x}^*(\cdot, \widehat{x}_0)) \prec \widehat{J}(\widehat{x}(\cdot, \widehat{x}_0))$ as was expected from Theorem 3.

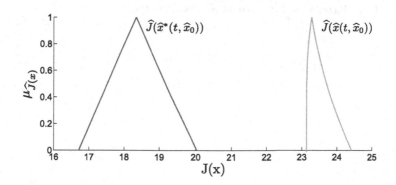

Fig. 5. Membership functions of $\widehat{J}(\widehat{x}^*(\cdot, \widehat{x}_0))$ and $\widehat{J}(\widehat{x}(\cdot, \widehat{x}_0))$, illustrating that $\widehat{J}(\widehat{x}^*(\cdot, \widehat{x}_0)) \prec \widehat{J}(\widehat{x}(\cdot, \widehat{x}_0))$.

5   Conclusion

We proposed a fuzzy version with fuzzy initial condition from a classical variational problem with fixed boundary conditions. To this end, we used Zadeh's extension of the solutions in the initial condition and, consequently, the functional was also obtained via extension. The solution was regarded as the resulting fuzzy bunch of functions that minimizes the fuzzy functional in a partial order. Convexity assured uniqueness of the classical global solution and provided means to extending it to a fuzzy bunch of functions. This fuzzy bunch of functions was proved to minimize the fuzzy-number-valued functional in a partial order. A quadratic variational problem illustrated the result. The minimum fuzzy value of the functional, compared to other possible values, provided visual illustration.

References

1. Barros, L.C., Bassanezi, R.C., Lodwick, W.A.: A First Course in Fuzzy Logic, Fuzzy Dynamical Systems, and Biomathematics. Springer, Heidelberg (2016). https://doi.org/10.1007/978-3-662-53324-6
2. Chang, S.S.L., Zadeh, L.A.: On fuzzy mapping and control. IEEE Trans. Syst. Man Cybern. **2**, 30–34 (1972)
3. Dubois, D., Prade, H.: Fuzzy Sets and Systems: Theory and Applications. Academic Press Inc., Orlando (1980)
4. Farhadinia, B.: Necessary optimality conditions for fuzzy variational problems. Inf. Sci. **181**, 1348–1357 (2012)
5. Gomes, L.T., Barros, L.C., Bede, B.: Fuzzy Differential Equations in Various Approaches. Springer, Berlin (2015). https://doi.org/10.1007/978-3-319-22575-3
6. Nguyen, H.T.: A note on the extension principle for fuzzy sets. J. Math. Anal. Appl. **64**, 369–380 (1978)
7. Troutman, J.L.: Variational Calculus and Optimal Control: Optimization with Elementary Convexity. Springer, New York (2012). https://doi.org/10.1007/978-1-4612-0737-5
8. Zadeh, L.A.: Fuzzy sets. Inf. Control **8**, 338–353 (1965)

Necessary Optimality Conditions for Interval Optimization Problems with Inequality Constraints Using Constrained Interval Arithmetic

Gino G. Maqui-Huamán[1]([✉]) [iD], Geraldo Silva[1] [iD], and Ulcilea Leal[2] [iD]

[1] Department of Applied Mathematics, Institute of Biosciences,
Humanities and Exact Sciences, São Paulo State University (UNESP),
São José do Rio Preto, SP 15054-000, Brazil
ginomaqui@gmail.com, gsilva@ibilce.unesp.br
[2] Federal University of Triângulo Mineiro (UFTM), Iturama, MG 38280-000, Brazil
ulcilea.leal@uftm.edu.br

Abstract. This article is devoted to obtaining necessary optimality conditions for optimization problems with interval-valued objective and interval inequality constraints. These objective and constraint functions are obtained from continuous functions by using constrained interval arithmetic. We give a concept of derivative for this class of interval-valued functions and we find necessary conditions based on Karush-Kunh-Tucker theorem in their interval version. We present an example to illustrate our results.

Keywords: Interval optimization problem
Constrained interval arithmetic

1 Introduction

In mathematical programming models, from a practical point of view, it is usually difficult to determine the coefficients of an objective function as a real number since, most often, these possess inherent uncertainty and/or inaccuracy [3]. Given that this is the usual state, we consider interval mathematical programming as one approach to tackle uncertainty and inaccuracies in the objective function coefficients [4,5,8].

Constrained interval arithmetic (CIA) was introduced in [6]. The idea is to consider an interval through a constrained parametric representation and operate via this parametric representation. CIA is the complete implementation of united extension introduced in [10] and so the interval-valued functions obtained via CIA preserve the properties of the crisp function.

In this article we consider interval-valued functions obtained from crisp functions applying CIA. We introduce a concept of differentiability for this class of

Supported by CAPES.

© Springer International Publishing AG, part of Springer Nature 2018
G. A. Barreto and R. Coelho (Eds.): NAFIPS 2018, CCIS 831, pp. 439–449, 2018.
https://doi.org/10.1007/978-3-319-95312-0_38

functions which is equivalent to the concept of differentiability previously introduced in [1,2]. This concept is based in the constrained parametric representation of the images of each element of domain which is an interval.

Based in the constrained parametric representation of the image of each element of the domain we also introduce the concept constraint qualification for the interval optimization problem. We show that the newly introduced concept of constraint qualification for the interval optimization problem is important to obtain the Karush-Kuhn-Tucker theorem in their interval version using the constrained interval arithmetic.

2   Linear Representation of an Interval and Constrained Interval Arithmetic

Given a bounded and closed interval A, we denote the extreme points of A by $\underline{a}$ and $\bar{a}$, where $\underline{a} \leq \bar{a}$, i.e., $A = [\underline{a}, \bar{a}]$. We denote by $\mathbb{I}$ the family of all bounded and closed intervals,

$$\mathbb{I} = \{[\underline{a}, \bar{a}] \ : \ \underline{a} \leq \bar{a}, \ \underline{a}, \bar{a} \in \mathbb{R}\}.$$

Lodwick [6,7] was primarily concerned about constrained interval arithmetic (CIA). This new arithmetic is derived directly from the united extension of [10]. To this end, an interval is redefined into an equivalent form as the real-valued function of one variable and two coefficients or parameters over the compact domain $[0, 1]$.

Definition 1. *An interval $A = [\underline{a}, \bar{a}]$ is the real single-valued function $A^I(\lambda_a)$*

$$A^I(\lambda_a) = (1 - \lambda_a)\underline{a} + \lambda_a\bar{a}, \tag{1}$$
$$= w_a\lambda_a + \underline{a}, \ 0 \leq \lambda_a \leq 1.$$

Here $w_a = \bar{a} - \underline{a} \geq 0$ is the width of the interval.

Strictly speaking, in (1), since the numbers $\underline{a}$ and $\bar{a}$ (consequently w_a) are known (inputs or data), they are *coefficients*, whereas λ_a is varying, although constrained between 0 and 1. Hence the name "constrained interval arithmetic". This means that $A^I(\lambda_a)$ is a single-valued real function with two coefficients. Moreover, we write λ_a to denote the parameter associated to the interval A.

To simplify the notation we will write $\lambda, \lambda_1, \lambda_2, ...$ to denote the parameters associate to each interval. So the constrained parametric representation of an interval A will be (see [1])

$$A = [\underline{a}, \bar{a}] = \{a(\lambda) = w_a\lambda + \underline{a} \ : \ \lambda \in [0, 1]\} \tag{2}$$

The algebraic operations for CIA are defined as follows. Let $A = \{a(\lambda_1) : \lambda_1 \in [0,1]\}$ and $B = \{b(\lambda_2) : \lambda_2 \in [0,1]\}$ two intervals, then

$$A * B = C$$
$$= [\underline{c}, \bar{c}]$$
$$= \{a(\lambda_1) * b(\lambda_2) : \lambda_1, \lambda_2 \in [0,1]\}$$
$$= \{c : c = a(\lambda_1) * b(\lambda_2), \lambda_1, \lambda_2 \in [0,1]\}$$

where $\underline{c} = \min\{c\}$, $\bar{c} = \max\{c\}$, $0 \leq \lambda_1 \leq 1, 0 \leq \lambda_2 \leq 1$ (3)

and $* \in \{+, -, \times, \div\}$.

It is clear from (3) that constrained interval arithmetic is a constrained global optimization problem.

Remark 1. From CIA [6] we know that, for dependent operations, we consider the same constrained parametric representation for the same intervals involved in the algebraic operations, i.e., $A * A = \{a(\lambda) * a(\lambda) : \lambda \in [0,1]\}$, where $* \in \{+, -, \times, \div\}$.

CIA is the complete implementation of the united extension, and possesses an algebra which has desired properties. For instance $A - A = [0,0] = \{0\}$, $A \div A = [1,1] = \{1\}$ when $0 \notin A$ and possess a distributive law $A \times (B + C) = A \times B + A \times C$. For more details of properties of the constrained parametric representation of an interval and CIA see [6,7].

Several partial order relations on $\mathbb{I}$ have been introduced in the literature. For instance, the usual order relation is $\preceq_{LU}$ defined by (see [4,5,9,11,12]).

$$A \preceq_{LU} B \text{ iff } \underline{a} \leq \underline{b} \text{ and } \bar{a} \leq \bar{b}.$$

Now considering the constrained parametric representation of an interval we have the following order relations on $\mathbb{I}$.

Definition 2. *For $A, B \in \mathbb{I}$ we write*

(i) $A \overset{\preceq}{=} B$ *iff* $a(\lambda) \leq b(\lambda), \forall \lambda \in [0,1]$;
(ii) $A \preceq B$ *iff* $A \overset{\preceq}{=} B$ *and* $A \neq B$; *equivalently*
   $A \preceq B$ *iff* $a(\lambda) \leq b(\lambda), \forall \lambda \in [0,1]$ *and there exists* $\lambda_0 \in [0,1]$ *such that* $a(\lambda_0) < b(\lambda_0)$;
(iii) $A \prec B$ *iff* $a(\lambda) < b(\lambda), \forall \lambda \in [0,1]$.

The idea of the previous definition of order is to compare level by level.

3 Interval Valued Function and Differentiability

In this section we uses interval-valued functions $F : \mathbb{R} \to \mathbb{I}$. Functions which are generated from a real-valued function considering the parameters as intervals. For this, we denote by $\mathbb{I}^k$ the product space, i.e.,

$$\mathbb{I}^k = \underbrace{\mathbb{I} \times \mathbb{I} \times ... \times \mathbb{I}}_{k \text{ times}}.$$

We also denote by $\mathcal{C}^k$ a k-uple of k-intervals. That is $\mathcal{C}^k \in \mathbb{I}^k$, where

$$\mathcal{C}^k = (C_1, ..., C_k), \quad C_j = [\underline{c_j}, \overline{c_j}], \quad j = 1, ..., k.$$

Since each interval C_j has a constrained parametric representation $c_j(\lambda_j)$ we can write the constrained parametric representation of $\mathcal{C}^k$ by

$$\mathcal{C}^k = \Big\{ c(\lambda) \; : \; c(\lambda) = (c_1(\lambda_1), ..., c_k(\lambda_k)), \quad c_j(\lambda_j) = (\overline{c_j} - \underline{c_j})\lambda_j + \underline{c_j}, \\ \lambda = (\lambda_1, ..., \lambda_k), \; 0 \le \lambda_j \le 1, \; j = 1, ..., k \Big\}.$$

Let us consider the function $f : \mathbb{R} \times \mathbb{R}^k \to \mathbb{R}$. For each $c = (c_1, ..., c_k) \in \mathbb{R}^k$, which are parameters involved with function f, we can write $f_c : \mathbb{R} \to \mathbb{R}$. For instance, f_c can represent the objective function of an optimization problem which has k-parameters (k-coefficient) $c_1, ..., c_k$, with $c_j \in \mathbb{R}$. In the present article we are going to consider constrained interval arithmetic to obtain $F_{\mathcal{C}^k}$ from f_c.

Definition 3 ([1]). *Let $f : \mathbb{R} \times \mathbb{R}^k \to \mathbb{R}$ be a function and let $c = (c_1, ..., c_k) \in \mathbb{R}^k$ be parameters involved with f. For each k-uple of intervals $\mathcal{C}^k$, we define a constrained parametric representation of $F_{\mathcal{C}^k}(x)$ by*

$$F_{\mathcal{C}^k}(x) = \left\{ f_{c(\lambda)}(x) \; : \; f_{c(\lambda)} : \mathbb{R} \to \mathbb{R}, \; c(\lambda) \in \mathcal{C}^k \right\}. \tag{4}$$

Proposition 1. *Let $f : \mathbb{R} \times \mathbb{R}^k \to \mathbb{R}$ be a continuous function in the second argument $c \in \mathbb{R}^k$. Then the interval-valued functions $F_{\mathcal{C}^k} : \mathbb{R} \to \mathbb{I}$ given by expression (4) is well defined and*

$$F_{\mathcal{C}^k}(x) = \left[\min_{\lambda \in [0,1]^k} f_{c(\lambda)}(x), \; \max_{\lambda \in [0,1]^k} f_{c(\lambda)}(x) \right], \tag{5}$$

for all $x \in \mathbb{R}$.

Proof. Since f is a continuous function in the second argument and $c(\lambda) = (c_1(\lambda_1), ..., c_k(\lambda_k))$, with $\lambda = (\lambda_1, ..., \lambda_k) \in [0,1]^k$, is continuous in λ then for each x fixed $f_{c(\lambda)}(x)$ is continuous in λ. So we have that $\min_{c(\lambda) \in \mathcal{C}^k} f_{c(\lambda)}(x)$ and $\max_{c(\lambda) \in \mathcal{C}^k} f_{c(\lambda)}(x)$ exist and

$$\min_{\lambda \in [0,1]^k} f_{c(\lambda)}(x) = \min_{c(\lambda) \in \mathcal{C}^k} f_{c(\lambda)}(x) \quad \text{and} \quad \max_{\lambda \in [0,1]^k} f_{c(\lambda)}(x) = \max_{c(\lambda) \in \mathcal{C}^k} f_{c(\lambda)}(x).$$

Thus we obtain (5). $\qquad \square$

Example 1. Consider the interval-valued function $F_{\mathcal{C}^1} : \mathbb{R} \to \mathbb{I}$ defined by

$$F_{\mathcal{C}^1}(x) = [1, 3]x^2 - 2x.$$

Clearly $F_{\mathcal{C}^1}$ is obtained from $f_c(x) = cx^2 - 2x$ by applying (4). In fact, in this case $\mathcal{C}^1 = [1, 3]$, $c(\lambda) = 2\lambda + 1$ and the constrained parametric representation of $F_{\mathcal{C}^1}(x)$ is given by

$$F_{\mathcal{C}^1}(x) = \left\{ f_{c(\lambda)}(x) \; : \; \lambda \in [0,1] \right\} = \left\{ (2\lambda + 1)x^2 - 2x \; : \; \lambda \in [0,1] \right\}.$$

Since $f_{c(\lambda)}(x)$ is linear in λ, from (5), we have

$$F_{\mathcal{C}^1}(x) = [x^2 - 2x, 3x^2 - 2x] = [1,3]x^2 - 2x.$$

Next we will give a concept of derivative for an interval-valued function. This concept is based on the differentiability of each element of the constrained parametric representation.

Definition 4. *Let $X \subset \mathbb{R}$ be an open set and let $F_{\mathcal{C}^k} : X \to \mathbb{I}$ be an interval-valued function. Suppose that $f_{c(\lambda)}$ is differentiable at x_0 for each $\lambda \in [0,1]^k$. Then we define the derivative of $F_{\mathcal{C}^k}$ at x_0, denoted by $F'_{\mathcal{C}^k}(x_0)$, by the constrained parametric representation*

$$F'_{\mathcal{C}^k}(x_0) = \left\{ f'_{c(\lambda)}(x_0) \; : \; c(\lambda) \in \mathcal{C}^k, \, \lambda \in [0,1]^k \right\}.$$

We say that $F_{\mathcal{C}^k}$ is differentiable at $x_0 \in X$ iff $F'_{\mathcal{C}^k}(x_0) \in \mathbb{I}$.

Proposition 2. *Let $X \subset \mathbb{R}$ be an open set and let $F_{\mathcal{C}^k} : X \to \mathbb{I}$ be an interval-valued function. Suppose that $f_{c(\lambda)}$ is differentiable at x_0 for each $\lambda \in [0,1]^k$ and $f'_{c(\lambda)}(x_0)$ is continuous at λ. Then $F_{\mathcal{C}^k}$ is differentiable and*

$$F'_{\mathcal{C}^k}(x_0) = \left[\min_{\lambda \in [0,1]^k} f'_{c(\lambda)}(x_0) \, , \, \max_{\lambda \in [0,1]^k} f'_{c(\lambda)}(x_0) \right]. \tag{6}$$

Proof. Since $f'_{c(\lambda)}(x_0)$ is continuous as a function of λ then $\min_{\lambda \in [0,1]^k} f'_{c(\lambda)}(x_0)$ and $\max_{\lambda \in [0,1]^k} f'_{c(\lambda)}(x_0)$ exist and (6) holds.

4 Interval Optimization Problems with Inequality Constraints

In this section we consider the following (scalar) interval optimization problem with interval inequality constraints

$$\begin{aligned} \textbf{(CIO)} \quad \min \quad & F_{\mathcal{C}^k}(x) \\ \text{subject to} \quad & G_{i,\mathcal{C}^{l_i}}(x) \overset{\preceq}{=} 0, \; i = 1, 2, ..., m \\ & x \in X \end{aligned}$$

where $F_{\mathcal{C}^k}, G_{i,\mathcal{C}^{l_i}} : X \to \mathbb{I}$ are interval-valued functions, every $G_{i,\mathcal{C}^{l_i}}$ is a constraint of the problem **(CIO)** and X is a nonempty open subset of $\mathbb{R}$.

A way to interpret a solution for problem (CIO) is to use the partial order relations given in Definition 2, the constrained parametric representation of an interval-valued function (4) and we may follow a similar solution concept to the Pareto optimal solution. For this we denote by $N_\delta(x^*)$ the δ-neighborhood of x^*.

Definition 5. *Let* $x^* \in X$.

(i) *It is said to be a (local) strict minimum for* $F_{\mathcal{C}^k}$ *iff there does not exist another* $x \in X$, $x \neq x^*$, $(x \in X \cap N_\delta(x^*))$ *such that* $F_{\mathcal{C}^k}(x) \stackrel{\preceq}{=} F_{\mathcal{C}^k}(x^*)$.

(ii) *It is said to be a (local) minimum for* $F_{\mathcal{C}^k}$ *iff there does not exist another* $x \in X$, $x \neq x^*$, $(x \in X \cap N_\delta(x^*))$ *such that* $F_{\mathcal{C}^k}(x) \preceq F_{\mathcal{C}^k}(x^*)$.

(iii) *It is said to be a (local) weak minimum for* $F_{\mathcal{C}^k}$ *iff there does not exist another* $x \in X$, $x \neq x^*$, $(x \in X \cap N_\delta(x^*))$ *such that* $F_{\mathcal{C}^k}(x) \prec F_{\mathcal{C}^k}(x^*)$.

Lemma 1. *If* $x^* \in X$ *is a strict minimum, then* x^* *is a minimum, and consequently* x^* *is a weak minimum.*

Proof. The proof follows immediately from Definition 5.

Example 2. Let $F_{\mathcal{C}^1} : \mathbb{R} \to \mathbb{I}$ be defined, as in the Example 1, by

$$F_{\mathcal{C}^1}(x) = [1,3]x^2 - 2x.$$

In this case, the constrained parametric representation of $F_{\mathcal{C}^1}$ is

$$F_{\mathcal{C}^1}(x) = \{f_{c(\lambda)}(x) = (2\lambda + 1)x^2 - 2x \ : \ \lambda \in [0,1]\}.$$

Then $x^* = 1$ is a strict minimum for $F_{\mathcal{C}^1}$. In fact, if there exists another $x \in \mathbb{R}$, $x \neq 1$, such that $F_{\mathcal{C}^1}(x) \stackrel{\preceq}{=} F_{\mathcal{C}^1}(1)$ then

$$f_{c(\lambda)}(x) \leq f_{c(\lambda)}(1), \ \ \forall \lambda \in [0,1];$$

equivalently, for all $\lambda \in [0,1]$,

$$(2\lambda + 1)x^2 - 2x \leq (2\lambda + 1) - 2 \Leftrightarrow (2\lambda + 1)(x - 1)(x + 1) \leq 2(x - 1).$$

If $x > 1$ we have, for all $\lambda \in [0,1]$,

$$(2\lambda + 1)(x + 1) \leq 2 \Leftrightarrow x \leq \frac{2}{2\lambda + 1} - 1 \leq 1,$$

which is absurd. In the same way, if $x < 1$ we have, for all $\lambda \in [0,1]$,

$$(2\lambda + 1)(x + 1) \geq 2 \Leftrightarrow x \geq \frac{2}{2\lambda + 1} - 1,$$

so $x \geq 1$ which is absurd.

Below, we will consider the conditions that must be satisfied so that a certain feasible point of the problem (**CIO**) be optimal. Such conditions, commonly known as first order conditions, involve the first order interval derivative. We also present constraint interval versions of well known optimization results.

We denote by

$$M = \{x \in X : G_{i,\mathcal{C}^{l_i}}(x) \stackrel{\preceq}{=} 0, \ i = 1, 2, ..., m\}$$

the feasible solution set of problem (**CIO**).

For simplicity, we define $I = \{1, 2, .., m\}$ and for every feasible point $x \in M$ the set of index of the active constraints:

$$I(x) = \left\{ i \in I : 0 \in G_{i,\mathcal{C}^{l_i}}(x) \right\}.$$

As in the previous sections, we use the constrained parametric representation given by (5) for all interval expressions.

Remark 2. We associate to the (**CIO**) problem its equivalent constrained parametric representation given by

$$\begin{aligned}
(\mathbf{CCIO}) \quad \min \quad & f_{c(\lambda^k)}(x) \\
\text{subject to} \quad & g_{i,c(\lambda^{l_i})}(x) \leq 0, \quad i = 1, 2, ..., m \\
& x \in X,
\end{aligned}$$

where $\lambda^k \in [0, 1]^k$ is a vector with k components where each component is related to its respective component of the interval k-uple $\mathcal{C}^k$, $\lambda^{l_i} \in [0, 1]^l$, l_i means that there are l new parameters in the constraint i. $f_{c(\lambda^k)}, g_{i,c(\lambda^{l_i})} : \mathbb{R} \to \mathbb{R}$ are assumed to be differentiable functions.

From the Remark 1 it is clear that the coordinates of the parameters $c(\lambda^k)$ and $c(\lambda^{l_i})$, $i = 1, 2, ..., m$ will be interdependent, as in the next example.

Example 3. Consider the problem

$$\begin{aligned}
(\mathbf{CIO1}) \quad \min \quad & [0, 3]x^2 + [-1, 2]x + [1, 4] \\
\text{subject to} \quad & [0, 3]x + [-2, 0] \stackrel{\preceq}{=} 0 \\
& [-1, 2]x \stackrel{\preceq}{=} 0.
\end{aligned}$$

Here

- $F_{\mathcal{C}^3}(x) = [0, 3]x^2 + [-1, 2]x + [1, 4];$
- $G_{1,\mathcal{C}^{1_1}}(x) = [0, 3]x + [-2, 0];$
- $G_{2,\mathcal{C}^{0_2}}(x) = [-1, 2]x.$

The interval $[0, 3]$ of the cost function and constraint 1 is interdependent, as well interval $[-1, 2]$ in the cost function and constraint 2. It follows from (4) and Remark (1), that:

- $f_{c(\lambda^3)}(x) = (0 + 3\lambda_1)x^2 + (-1 + 3\lambda_2)x + (1 + 3\lambda_3);$
- $g_{1,c(\lambda^{1_1})}(x) = (0 + 3\lambda_1)x + (-2 + 2\lambda_4);$
- $g_{2,c(\lambda^{0_2})}(x) = (-1 + 3\lambda_2)x.$

with $(\lambda_1, \lambda_2, \lambda_3, \lambda_4) \in [0, 1]^4$, consequently the constrained equivalent associated problem is

$$\begin{aligned}
(\mathbf{CCIO1}) \quad \min \quad & (0 + 3\lambda_1)x^2 + (-1 + 3\lambda_2)x + (1 + 3\lambda_3) \\
\text{subject to} \quad & (0 + 3\lambda_1)x + (-2 + 2\lambda_4) \leq 0, \\
& (-1 + 3\lambda_2)x \leq 0,
\end{aligned}$$

with $(\lambda_1, \lambda_2, \lambda_3, \lambda_4) \in [0, 1]^4$.

5   Necessary Conditions of Interval Optimality

In this section, we will present first order interval conditions so that a feasible point of (**CIO**) problem is optimal. For this, we will use the (**CCIO**) equivalent problem.

Next, we present a geometrical characterization of local optimality for our problem (**CIO**).

Proposition 3. *Let* $F_{\mathcal{C}^k}, G_{i,\mathcal{C}^{l_i}} : X \to \mathbb{I}$, $i \in I$ *be differentiable interval-valued functions. If* $x^* \in M$ *is a local weak minimum of* $F_{\mathcal{C}^k}$ *over* M, *then the system*

$$F'_{\mathcal{C}^k}(x^*) \cdot d \prec 0$$
$$G'_{i,\mathcal{C}^{l_i}}(x^*) \cdot d \prec 0, \ i \in I(x^*), \tag{7}$$

has no solution $d \in \mathbb{R}$.

Proof. Let $x^* \in M$ a local weak minimum of (**CIO**). Suppose by contradiction that, exist a direction $d \in \mathbb{R}$ that resolves the system (7). This implies that it also solves the equivalent constrained parametric representation system below

$$f'_{c(\lambda^k)}(x^*) \cdot d < 0, \ \lambda^k \in [0,1]^k$$
$$g'_{i,c(\lambda^{l_i})}(x^*) \cdot d < 0, \ i \in I(x^*) \ \lambda^{l_i} \in [0,1]^{l_i}. \tag{8}$$

These contradicts the local optimality of x^* for every λ in the problem (**CCIO**), consequently, exist $\overline{x} \in M$, such that,

$$f_{c(\lambda^k)}(\overline{x}) < f_{c(\lambda^k)}(x^*), \ \forall \lambda^k \in [0,1]^k.$$

Minimizing and maximizing $f_{c(\lambda^k)}(\overline{x})$ and $f_{c(\lambda^k)}(x^*)$ in λ^k,

$$F_{\mathcal{C}^k}(\overline{x}) < F_{\mathcal{C}^k}(x^*)$$

therefore $\overline{x} \in M$ contradicts the minimality of x^*.

Another important result in optimization is the Fritz John theorem that we present their interval version next.

Theorem 1. *Let* $x^* \in M$ *be a local weak minimum of (**CIO**) and* $F_{\mathcal{C}^k}, G_{i,\mathcal{C}^{l_i}} :$ $X \to \mathbb{I}$ *be differentiable interval-valued functions with their constraint parametric representations continuous in* λ^k *and* λ^{l_i}, $i \in I$, *respectively. Then, there exist scalars* $\delta_0, \delta_i \in \mathbb{R}$, $i \in I$, *not all simultaneously zero, such that*

$$0 \in \delta_0 \cdot F'_{\mathcal{C}^k}(x^*) + \sum_{i \in I} \delta_i \cdot G'_{i,\mathcal{C}^{l_i}}(x^*); \tag{9}$$

$$\delta_0, \delta_i \geq 0, \ i \in I; \tag{10}$$

$$0 \in \delta_i \cdot G_{i,\mathcal{C}^{l_i}}(x^*), \ i \in I. \tag{11}$$

Proof. Let $x^* \in M$, be a weak local minimum of **(CIO)** problem, considering Remark 2, x^* is also solution of **(CCIO)** for every $(\lambda^k, \lambda^{l_1}, \lambda^{l_2}, ..., \lambda^{l_m}) \in [0, 1]^{k+l_1+l_2+...+l_m}$. From the differentiability of $f_{c(\lambda^k)}$ and $g_{c(\lambda^{l_i})}$, $i \in I$, we have that there exist δ_0, δ_i, $i \in I$ not all nulls, such that

$$\delta_0 \cdot f'_{c(\lambda^k)}(x^*) + \sum_{i \in I} \delta_i \cdot g'_{i,c(\lambda^{l_i})}(x^*) = 0$$

$$\delta_0, \delta_i \geq 0, \ i \in I$$

$$\delta_i \cdot g_{i,c(\lambda^{l_i})}(x^*) = 0, \ i \in I.$$

From, continuity of $f'_{c(\lambda^k)}(x^*)$ at λ^k, and continuity of $g'_{i,c(\lambda^{l_i})}(x^*)$ at λ^{l_i}, $\forall i \in I$ we obtain,

$$0 \in \ \delta_0 \cdot [\min_{\lambda^k \in [0,1]^k} f'_{c(\lambda^k)}(x^*), \max_{\lambda^k \in [0,1]^k} f'_{c(\lambda^k)}(x^*)]$$
$$+ \sum_{i \in I} \delta_i \cdot [\min_{\lambda^k \in [0,1]^k} g'_{i,c(\lambda^{l_i})}(x^*), \max_{\lambda^k \in [0,1]^k} g'_{i,c(\lambda^{l_i})}(x^*)];$$

$$\delta_0, \delta_i \geq 0, \ i \in I;$$

$$0 \in \delta_i \cdot [\min_{\lambda^k \in [0,1]^k} g_{i,c(\lambda^{l_i})}(x^*), \max_{\lambda^k \in [0,1]^k} g_{i,c(\lambda^{l_i})}(x^*)], \ i \in I.$$

From the above it follows immediately the desired result.

In the last proof, it is important to note that the multipliers are the same for all $\lambda = (\lambda^k, \lambda^{l_1}, \lambda^{l_2}, ..., \lambda^{l_m}) \in [0, 1]^{k+l_1+l_2+...+l_m}$, consequently they are independent of the parameter λ.

As it is well known in classical optimization, Karush-Kuhn-Tucker conditions, which provides nonzero multiplier associated with the cost function, are obtained by imposing some constraints qualification. Next we present these type results in an interval version involving constraint interval arithmetic.

We say that the set $\{V_i\}_{i \in I}$ of interval elements is *independent* if for every $\lambda \in [0, 1]$ the set of vectors $\{v_i(\lambda)\}_{i \in I}$ is linearly independent.

Definition 6. *We say that the* **(CIO)** *problem satisfies the constraint qualification at x^* if the set $\left\{ G'_{i,\mathcal{C}^{l_i}}(x^*) \right\}_{i \in I(x^*)}$ is independent.*

Theorem 2. *Let $x^* \in M$ be a weak minimum for* **(CIO)**, *let $F_{\mathcal{C}^k}, G_{i,\mathcal{C}^{l_i}} : X \to \mathbb{I}$ be differentiable interval-valued functions with their constraint parametric representations continuous in λ^k and λ^{l_i}, $i \in I$, respectively, and suppose that* **(CIO)**

problem satisfies the constraint qualification in x^*. Then there exist $\mu_i \in \mathbb{R}, i \in I$, such that

$$0 \in F'_{\mathcal{C}^k}(x^*) + \sum_{i \in I} \mu_i \cdot G'_{i,\mathcal{C}^{l_i}}(x^*); \tag{12}$$

$$\mu_i \geq 0, \ i \in I; \tag{13}$$

$$0 \in \mu_i \cdot G_{i,\mathcal{C}^{l_i}}(x^*), \ i \in I. \tag{14}$$

Proof. From the last theorem, there exist multipliers $\delta_0, \delta_i, \ (i \in I)$ verifying the Eqs. (9), (10) and (11). If $\delta_0 = 0$, by Eq. (9) we obtain $\sum_{i \in I} \delta_i \cdot g'_{i,c(\lambda^{l_i})}(x^*) = 0$ with $\delta_i \geq 0$ and not all zero, which contradicts the constraint qualification. Now just define

$$\mu_i = \frac{\delta_i}{\delta_0}, \ i \in I,$$

and analogously to the last proof we obtain the desired result.

A feasible point $x^* \in M$ for (**CIO**) problem is called a *Karush-Kuhn-Tucker point* if there exists $\mu_i \in \mathbb{R}, \ i \in I$ verifying the Eqs. (12), (13) and (14).

Example 4. Consider the following (scalar) interval optimization problem

$$\begin{aligned}
(\textbf{CIO2}) \quad \min \quad & [1,3]x^2 - 2x \\
\text{subject to} \quad & x^2 - [\tfrac{1}{2}, \tfrac{3}{2}] \overset{\preceq}{=} 0, \\
& -x \overset{\preceq}{=} 0
\end{aligned}$$

Let x^* be an optimal point of our (**CIO2**) problem, from KKT conditions (Theorem 2) we can see that $x^* \in [\frac{1}{3}, \sqrt{\frac{1}{2}}]$. Consequently, for example, if $x^* = \frac{1}{3}$ it saturates the constraint $G_{1,\mathcal{C}^{1_1}}$ $(I(x^*) = \{1\})$. Since $\{G'_{i,\mathcal{C}^{l_i}}(\frac{1}{3})\}$ is independent the (**CIO2**) problem satisfies the constraint qualification (Definition 6), and the interval KKT conditions guarantee the existence of $\mu_1, \mu_2 \in \mathbb{R}$ such that

$$0 \in F'_{\mathcal{C}^k}(x^*) + \sum_{i \in I} \mu_i \cdot G'_{i,\mathcal{C}^{l_i}}(x^*);$$

$$\mu_i \geq 0, \ i \in I;$$

$$0 \in \mu_i \cdot G_{i,\mathcal{C}^{l_i}}(x^*), \ i \in I.$$

As x^* does not saturate $i = 2$, we have $\mu_2 = 0$.

Since $F'_{\mathcal{C}^k}(\frac{1}{3}) = [-\frac{4}{3}, 0]$, $G'_{1,\mathcal{C}^{1_1}}(\frac{1}{3}) = \frac{2}{3}$ and $G'_{2,\mathcal{C}^{0_2}}(\frac{1}{3}) = -1$, it is obtained $\mu_1 = 0$.

6   Conclusions

In this paper we have considered optimization problems with constraints where the parameters (coefficient) of the objective function and the constraints are intervals and used the constrained interval arithmetic recently introduced by Lodwick [6]. And, for the constrained problem, we provided both interval Fritz-John and interval Karush-Kuhn-Tucker necessary conditions of optimality. For the Karush-Kuhn-Tucker theorem in their interval version, we use the concept of constraint qualification in the interval version.

References

1. Bhurjee, A.K., Panda, G.: Efficient solution of interval optimization problem. Math. Methods Oper. Res. **76**(3), 273–288 (2012). https://doi.org/10.1007/s00186-012-0399-0
2. Bhurjee, A., Panda, G.: Nonlinear fractional programing problem with inexact parameter. J. Appl. Math. Inf. **31**(5–6), 853–867 (2013). https://doi.org/10.14317/jami.2013.853
3. Costa, T., Bouwmeester, H., Lodwick, W., Lavor, C.: Calculating the possible conformations arising from uncertainty in the molecular distance geometry problem using constraint interval analysis. Inf. Sci. **415–416**, 41–52 (2017). https://doi.org/10.1016/j.ins.2017.06.015
4. Inuiguchi, M., Kume, Y.: Goal programming problems with interval coefficients and target intervals. Eur. J. Oper. Res. **52**(3), 345–360 (1991). https://doi.org/10.1016/0377-2217(91)90169-v
5. Ishibuchi, H., Tanaka, H.: Multiobjective programming in optimization of the interval objective function. Eur. J. Oper. Res. **48**(2), 219–225 (1990). https://doi.org/10.1016/0377-2217(90)90375-1
6. Lodwick, W.A.: Constrained Interval Arithmetic. Citeseer (1999). http://www-math.cudenver.edu/ccm/reports/index.shtml
7. Lodwick, W.A.: Interval and fuzzy analysis: a unified approach. In: Advances in Imaging and Electron Physics, pp. 75–192. Elsevier (2007). https://doi.org/10.1016/s1076-5670(07)48002-8
8. Lodwick, W.A.: An overview of flexibility and generalized uncertainty in optimization. Comput. Appl. Math. **31**(3), 569–589 (2012). https://doi.org/10.1590/s1807-03022012000300008
9. Moore, R.E.: Interval Analysis. 1966. Prince-Hall, Englewood Cliffs (1969)
10. Strother, W.L.: Continuity for multi-valued functions and some applications to topology. Ph.D. thesis, Tulane University of Louisiana (1952)
11. Wu, H.C.: The Karush-Kuhn-Tucker optimality conditions in an optimization problem with interval-valued objective function. Eur. J. Oper. Res. **176**(1), 46–59 (2007). https://doi.org/10.1016/j.ejor.2005.09.007
12. Wu, H.C.: The Karush-Kuhn-Tucker optimality conditions in multiobjective programming problems with interval-valued objective functions. Eur. J. Oper. Res. **196**(1), 49–60 (2009). https://doi.org/10.1016/j.ejor.2008.03.012

Order Relations, Convexities, and Jensen's Integral Inequalities in Interval and Fuzzy Spaces

Tiago Mendonça da Costa[1]([⊠])(iD), Yurilev Chalco-Cano[2](iD),
Laécio Carvalho de Barros[3], and Geraldo Nunes Silva[4](iD)

[1] Instituto de Ciências Exatas e Naturais,
Universidade Federal do Pará, Belém, Pará, Brazil
grafunjo@yahoo.com.br
[2] Instituto de Alta Investigación, Universidad de Tarapacá,
Casilla 7D, Arica, Chile
ychalco@uta.cl
[3] Instituto de Matemática, Estatística e Computação Científica,
Universidade Estadual de Campinas, Campinas, São Paulo, Brazil
laeciocb@ime.unicamp.br
[4] Departamento de Matemática Aplicada, Universidade Estadual Paulista,
São José do Rio Preto, São Paulo, Brazil
gsilva@ibilce.unesp.br

Abstract. This study presents new interval and fuzzy versions of the Jensen's integral inequality, which extend the classical Jensen's integral inequality for real-valued functions, using Aumann and Kaleva integrals. The inequalities for interval-valued functions are interpreted through the preference order relations given by Ishibuchi and Tanaka, which are useful for dealing with interval optimization problems. The order relations adopted in the space of fuzzy intervals are extensions of those considered the interval spaces.

Keywords: Jensen's integral inequality · Interval-valued functions Fuzzy-interval-valued functions

1 Introduction

Since the Jensen's integral inequality is applicable in several fields of mathematics, many versions of this inequality have been developed in different mathematical spaces. For instance, in [6] are presented interval and fuzzy versions of the Jensen's integral inequality, where the interval version is given by means of the Aumman integral and of the order relation "⊆" given by Moore [12], and the

Supported by (PNPD/CAPES/UFPA), FONDECYT 1151154, (CNPq) grant 306546/2017-5, and -CEPID-CEMEAI through São Paulo Research Foundation, FAPESP grant 13/07375-0, respectively.

© Springer International Publishing AG, part of Springer Nature 2018
G. A. Barreto and R. Coelho (Eds.): NAFIPS 2018, CCIS 831, pp. 450–463, 2018.
https://doi.org/10.1007/978-3-319-95312-0_39

fuzzy version is given by means of the Kaleva integral and of the order relation "$\subseteq_\mathcal{F}$" which is an extension of "$\subseteq$". On the other hand, it is well-known that "$\subseteq$" compares only nested intervals. Consequently, the Jensen's integral inequalities versions given in [6] are applicable only for the class of nested intervals and for the class of nested fuzzy intervals. Moreover, the Jensen's integral inequality versions given in [6] does not extend the classical Jensen's integral inequality for real-valued functions.

The main objective of this study is to present interval and fuzzy versions of Jensen's integral inequality using order relations that allow us to compare non-nested intervals as well as non-nested fuzzy intervals, which complement the versions given in [6] and which extend the classical Jensen's integral inequality for real-valued functions. In order to achieve this goal, we use the order relations given by Ishibuchi and Tanaka [10], which have the desired property, and which can represent the decision makers' preference between intervals. The order relations in the space of fuzzy intervals are constructed level-by-level through the order relations considered in interval spaces.

This study is organized as follows: Sect. 2 presents some preliminary results on interval spaces. Section 3 recalls the order relations given in [10] and presents some definitions of convexity for interval-valued functions that will be useful to provide the new interval versions of Jensen's integral inequality in Sect. 4. In Sect. 5 is presented preliminary results in the space of fuzzy intervals, which are used in Sect. 6, where the order relations are defined level-wise through the order relations given in interval spaces, and the concepts of convexity for fuzzy-interval-valued functions are presented. Section 6 also provides new fuzzy versions of the Jensen's integral inequality. Section 7 presents our final considerations.

2   Preliminaries: Interval Space

Consider the space $\mathcal{K}_C$ of all closed and bounded intervals of real numbers, that is, $\mathcal{K}_C = \{[\underline{a}, \overline{a}] \mid \underline{a}, \overline{a} \in \mathbb{R}$ and $\underline{a} \leqslant \overline{a}\}$, which is called the interval space. Given $A = [\underline{a}, \overline{a}], B = [\underline{b}, \overline{b}] \in \mathcal{K}_C$ and $\lambda \in \mathbb{R}$, the interval arithmetic operations are defined by

$$A + B = [\underline{a}, \overline{a}] + [\underline{b}, \overline{b}] = [\underline{a} + \underline{b}, \overline{a} + \overline{b}] \text{ and } \lambda \cdot [\underline{a}, \overline{a}] = \begin{cases} [\lambda\underline{a}, \lambda\overline{a}] & \text{if } \lambda \geqslant 0, \\ [\lambda\overline{a}, \lambda\underline{a}] & \text{if } \lambda < 0. \end{cases} \quad (1)$$

Moreover, $A = B$ if and only if $\underline{a} = \underline{b}$ and $\overline{a} = \overline{b}$. In this presentation S represents a non-empty subset of $\mathbb{R}$.

A map $F : S \to \mathcal{K}_C$ such that $F(t) = [\underline{f}(t), \overline{f}(t)]$ for all $t \in S$, where $\underline{f}, \overline{f} : S \to \mathbb{R}$ are real-valued functions with $\underline{f}(t) \leqslant \overline{f}(t)$ for all $t \in S$, it is called an **interval-valued function**. The functions $\underline{f}$ and $\overline{f}$ are called the lower and upper functions of F, respectively.

Given the interval-valued functions $F, G : S \to \mathcal{K}_C$ and $\lambda \in \mathbb{R}$, the notation used for the algebraic operations between interval-valued functions is given by $(F + G)(t) := F(t) + G(t)$ and $(\lambda \cdot F)(t) := \lambda \cdot F(t)$.

Remark 1 ([6]). Let $F : [a, b] \to \mathcal{K}_C$ be an interval-valued function, with $F(t) = [\underline{f}(t), \overline{f}(t)]$, where $\underline{f}, \overline{f} : [a, b] \to \mathbb{R}$. Given a real-valued function $g : S \to \mathbb{R}$ such that $g(S) \subseteq [a, b]$, then $(F \circ g)$ means the interval-valued function $(F \circ g) : S \to \mathcal{K}_C$ that associates each $t \in S$ to the interval $(F \circ g)(t) = [(\underline{f} \circ g)(t), (\overline{f} \circ g)(t)] = [\underline{f}(g(t)), \overline{f}(g(t))] \in \mathcal{K}_C$.

It is known that the space $(\mathcal{K}_C, d_H)$, where $d_H(A, B) = \max\{|\underline{a} - \underline{b}|, |\overline{a} - \overline{b}|\}$ is the Pompeiu-Hausdorff distance between $A = [\underline{a}, \overline{a}], B = [\underline{b}, \overline{b}] \in \mathcal{K}_C$, it is a complete and separable metric space (see [1,8]). Let S^d be the set of all accumulation points in $S \subseteq \mathbb{R}$.

Definition 1 *(see [1,8]). Let $F : S \to \mathcal{K}_C$ be an interval-valued function. Then $L \in \mathcal{K}_C$ is called a limit of F at $t_0 \in S^d$ if for every $\epsilon > 0$ there exists $\delta(\epsilon, t_0) = \delta > 0$ such that $d_H(F(t), L) < \epsilon$ for every $t \in S$ with $0 < |t - t_0| < \delta$. This is denoted by $\lim_{t \to t_0} F(t) = L$.*

Theorem 1 *(see [1,8]). Let $F : S \to \mathcal{K}_C$ be such that $F(t) = [\underline{f}(t), \overline{f}(t)]$ for all $t \in S$. Given $t_0 \in S^d$, it follows that $\lim_{t \to t_0} F(t) = \left[\lim_{t \to t_0} \underline{f}(t), \lim_{t \to t_0} \overline{f}(t) \right].$*

Definition 2 *(see [1,8]). Let $F : S \to \mathcal{K}_C$ be an interval-valued function. Then F is said to be d_H-continuous at $t_0 \in S$ if for every $\epsilon > 0$ there exists $\delta(\epsilon, t_0) = \delta > 0$ such that $d_H(F(t), L) < \epsilon$ for every $t \in S$ with $|t - t_0| < \delta$.*

An interval-valued function $F : [a, b] \longrightarrow \mathcal{K}_C$ is said to be measurable if and only if $\{(t, x) : x \in F(t)\} \in \mathcal{A} \times \mathcal{B}$, where $\mathcal{A}$ denotes the $\sigma-$algebra composed of all Lebesgue-measurable subsets of $\mathbb{R}$ and $\mathcal{B}$ denotes the $\sigma-$algebra composed of all Borel-measurable subsets of $\mathbb{R}$. F is said to be integrably bounded on $[a, b]$ if and only if there exists a Lebesgue-integrable function $h : [a, b] \longrightarrow [0, +\infty)$ such that $|x| \leqslant h(t)$ for all x and t such that $x \in F(t)$.

Given an interval-valued function $F : [a, b] \longrightarrow \mathcal{K}_C$, the Aumann integral $((IA)-$ integral, for short) of F over $[a, b]$ is defined in [2] by $(IA) \int_a^b F(t)dt = \left\{ \int_a^b f(t)dt : f \in S(F) \right\}$, where $S(F) := \{f \in L^1([a, b]) : f(t) \in F(t)$ for almost every $t \in [a, b]\}$ and $L^1([a, b])$ is the space of all functions $f : [a, b] \longrightarrow \mathbb{R}$ that are Lebesgue-integrable over $[a, b]$. We say that the (IA)-integral of F over $[a, b]$ exists (or that F is (IA)-integrable over $[a, b]$) if $S(F) \neq \varnothing$.

Theorem 2 *(see [1,8,11]). If an interval-valued function $F : [a, b] \to \mathcal{K}_C$ is measurable and integrably bounded on $[a, b]$, then F is $(IA)-$ integral over $[a, b]$.*

Theorem 3 *(see [1,8,11]). Let $F : [a, b] \longrightarrow \mathcal{K}_C$ be such that $F(t) = [\underline{f}(t), \overline{f}(t)]$ for all $t \in [a, b]$. Then F is $(IA)-$integrable over $[a, b]$ if and only if $\underline{f}$ and $\overline{f}$ belongs to $L^1([a, b])$. Moreover, if F is $(IA)-$integrable over $[a, b]$, then $(IA) \int_a^b F(t)dt = \left[\int_a^b \underline{f}(t)dt, \int_a^b \overline{f}(t)dt \right].$*

3   Order Relations and Concepts of Convexity for Interval-Valued Functions

This section recalls some interesting interval order relations for comparing intervals, which represent the decision makers' preference between intervals. These order relations are defined by means of upper bound, lower bound, centre and radius of the intervals. Based on these order relations, concepts of convexity for interval-valued functions are presented.

3.1   Order Relations on $\mathcal{K}_C$

It is attributed to Moore the introduction of (2) and (3) (see [12]). This is why he is considered the pioneer of the study about interval relations order.

$$[\underline{a}, \overline{a}] <_M [\underline{b}, \overline{b}] \quad \text{if and only if} \quad \overline{a} < \underline{b}, \tag{2}$$

$$[\underline{a}, \overline{a}] \subseteq [\underline{b}, \overline{b}] \quad \text{if and only if} \quad \underline{b} \leqslant \underline{a} \text{ and } \overline{a} \leqslant \overline{b}. \tag{3}$$

It is easy to see that "$<_M$" is a partial order relation that extends the usual strict order relation "$<$" defined on $\mathbb{R}$. From (2), it follows that two intervals are comparable via "$<_M$" if and only if they are disjoint. On the other hand, although "$\subseteq$" is a partial order relation that does not extend the usual order relation "$\leqslant$" defined on $\mathbb{R}$, from (3) it is clear that two intervals may be comparable via "$\subseteq$" if they are not disjoint. To be more precise, two intervals are comparable via "$\subseteq$" if and only if one of these intervals is nested in the other.

In order to handle interval minimization and interval maximization problems, Ishibuchi and Tanaka [10] introduced the following partial order relations that allow to compare two intervals under different conditions from those given in (2) and (3). The order relations related with interval maximization problems are:

$$[\underline{a}, \overline{a}] \leq_{LR} [\underline{b}, \overline{b}] \Leftrightarrow \underline{a} \leqslant \underline{b} \text{ and } \overline{a} \leqslant \overline{b} \tag{4}$$

and

$$[\underline{a}, \overline{a}] <_{LR} [\underline{b}, \overline{b}] \Leftrightarrow [\underline{a}, \overline{a}] \leq_{LR} [\underline{b}, \overline{b}] \text{ and } [\underline{a}, \overline{a}] \neq [\underline{b}, \overline{b}]. \tag{5}$$

Other two order relations suggested by Ishibuchi and Tanaka [10] in order to deal with maximization problems in which the use of "$\leq_{LR}$" and/or "$<_{LR}$" are not suitable, are given as:

$$[\underline{a}, \overline{a}] \leq_{CW} [\underline{b}, \overline{b}] \Leftrightarrow \frac{\overline{a} + \underline{a}}{2} \leqslant \frac{\overline{b} + \underline{b}}{2} \text{ and } \frac{\overline{a} - \underline{a}}{2} \geqslant \frac{\overline{b} - \underline{b}}{2} \tag{6}$$

and

$$[\underline{a}, \overline{a}] <_{CW} [\underline{b}, \overline{b}] \Leftrightarrow [\underline{a}, \overline{a}] \leq_{CW} [\underline{b}, \overline{b}] \text{ and } [\underline{a}, \overline{a}] \neq [\underline{b}, \overline{b}]. \tag{7}$$

By defining the partial order relations "$\leq_{LC}$" and "$<_{LC}$" on $\mathcal{K}_C$ as

$$[\underline{a}, \overline{a}] \leq_{LC} [\underline{b}, \overline{b}] \Leftrightarrow \underline{a} \leqslant \underline{b} \text{ and } \frac{\overline{a} + \underline{a}}{2} \leqslant \frac{\overline{b} + \underline{b}}{2} \tag{8}$$

and

$$[\underline{a}, \overline{a}] <_{LC} [\underline{b}, \overline{b}] \Leftrightarrow [\underline{a}, \overline{a}] \leq_{LR} [\underline{b}, \overline{b}] \text{ and } [\underline{a}, \overline{a}] \neq [\underline{b}, \overline{b}], \tag{9}$$

respectively, Ishibuchi and Tanaka [10] proved that, for two given intervals $[\underline{a}, \overline{a}], [\underline{b}, \overline{b}] \in \mathcal{K}_C$, one has that

$$[\underline{a}, \overline{a}] \leq_{LC} [\underline{b}, \overline{b}] \quad \text{if and only if} \quad [\underline{a}, \overline{a}] \leq_{LR} [\underline{b}, \overline{b}] \text{ or } [\underline{a}, \overline{a}] \leq_{CW} [\underline{b}, \overline{b}] \tag{10}$$

and

$$[\underline{a}, \overline{a}] <_{LC} [\underline{b}, \overline{b}] \quad \text{if and only if} \quad [\underline{a}, \overline{a}] <_{LR} [\underline{b}, \overline{b}] \text{ or } [\underline{a}, \overline{a}] <_{CW} [\underline{b}, \overline{b}]. \tag{11}$$

Ishibuchi and Tanaka [10] also use (4) to handle interval minimization problems, and in order to deal with maximization problems in which the use of "$\leq_{LR}$" and/or "$<_{LR}$" are not suitable, they defined the following order relations:

$$[\underline{a}, \overline{a}] \leq_{CW^\cdot} [\underline{b}, \overline{b}] \Leftrightarrow \frac{\overline{a} + \underline{a}}{2} \leq \frac{\overline{b} + \underline{b}}{2} \text{ and } \frac{\overline{a} - \underline{a}}{2} \leq \frac{\overline{b} - \underline{b}}{2} \tag{12}$$

and

$$[\underline{a}, \overline{a}] <_{CW^\cdot} [\underline{b}, \overline{b}] \Leftrightarrow [\underline{a}, \overline{a}] \leq_{CW^\cdot} [\underline{b}, \overline{b}] \text{ and } [\underline{a}, \overline{a}] \neq [\underline{b}, \overline{b}]. \tag{13}$$

By defining the partial order relations "$\leq_{RC^\cdot}$" and "$<_{RC^\cdot}$" on $\mathcal{K}_C$ as

$$[\underline{a}, \overline{a}] \leq_{RC^\cdot} [\underline{b}, \overline{b}] \Leftrightarrow \overline{a} \leq \overline{b} \text{ and } \frac{\overline{a} + \underline{a}}{2} \leq \frac{\overline{b} + \underline{b}}{2} \tag{14}$$

and

$$[\underline{a}, \overline{a}] <_{RC^\cdot} [\underline{b}, \overline{b}] \Leftrightarrow [\underline{a}, \overline{a}] \leq_{RC^\cdot} [\underline{b}, \overline{b}] \text{ and } [\underline{a}, \overline{a}] \neq [\underline{b}, \overline{b}], \tag{15}$$

respectively, Ishibuchi and Tanaka [10] proved that, for two given intervals $[\underline{a}, \overline{a}], [\underline{b}, \overline{b}] \in \mathcal{K}_C$, one has that

$$[\underline{a}, \overline{a}] \leq_{RC^\cdot} [\underline{b}, \overline{b}] \quad \text{if and only if} \quad [\underline{a}, \overline{a}] \leq_{LR} [\underline{b}, \overline{b}] \text{ or } [\underline{a}, \overline{a}] \leq_{CW^\cdot} [\underline{b}, \overline{b}] \tag{16}$$

and

$$[\underline{a}, \overline{a}] <_{RC^\cdot} [\underline{b}, \overline{b}] \quad \text{if and only if} \quad [\underline{a}, \overline{a}] <_{LR} [\underline{b}, \overline{b}] \text{ or } [\underline{a}, \overline{a}] <_{CW^\cdot} [\underline{b}, \overline{b}]. \tag{17}$$

In terms of applications, it is the semantics of a problem that indicates which one of these order relations it is more suitable to deal with such a problem. This fact has motivated the elaboration of a diversity of mathematical concepts where each one of them is based on an specific order relation on $\mathcal{K}_C$. Here we recall some convexity concepts for interval-valued functions where each convexity concept is interpreted through a particular order relation on $\mathcal{K}_C$.

3.2 Convexity for Interval-Valued Functions

In the sequel, some concepts of convexity for interval-valued functions that are based on the partial order relations "$\subseteq$", "$\leq_{LR}$", "$\leq_{CW}$", and "$\leq_{CW^*}$." are presented. Moreover, necessary and sufficient conditions for convexity of interval-valued functions are provided.

Henceforth S denotes an interval in $\mathbb{R}$ and F denotes an interval-valued function $F : S \to \mathcal{K}_C$.

Definition 3 *(see [6]). F is said to be $\subseteq -convex$ if*

$$(1-\lambda)F(t_0)+\lambda F(t_1) \subseteq F((1-\lambda)t_0+\lambda t_1)) \text{ for any } t_0, t_1 \in S \text{ and } \lambda \in [0,1]. \quad (18)$$

Definition 4 *(see [5]). F is said to be $LR-convex$ if*

$$F((1-\lambda)t_0 + \lambda t_1) \leq_{LR} (1-\lambda)F(t_0) + \lambda F(t_1) \text{ for any } t_0, t_1 \in S \text{ and } \lambda \in [0,1]. \quad (19)$$

Definition 5. *F is said to be $CW-convex$ if*

$$F((1-\lambda)t_0 + \lambda t_1) \leq_{CW} (1-\lambda)F(t_0) + \lambda F(t_1) \text{ for any } t_0, t_1 \in S \text{ and } \lambda \in [0,1]. \quad (20)$$

Definition 6. *F is said to be $CW^*-convex$ if*

$$F((1-\lambda)t_0 + \lambda t_1) \leq_{CW^*} (1-\lambda)F(t_0) + \lambda F(t_1) \text{ for any } t_0, t_1 \in S \text{ and } \lambda \in [0,1]. \quad (21)$$

It is easy to see that both Definitions 4, 5 and 6 extend the classical concept of convexity for real-valued functions.

Proposition 1. *Given a convex set $S \subseteq \mathbb{R}$, let $F : S \to \mathcal{K}_C$ be such that $F(t) = [\underline{f}(t), \overline{f}(t)]$ for all $t \in S$. Thus,*

(i) F is $\subseteq -convex$ if and only if $\underline{f}$ is convex and $\overline{f}$ is concave;
(ii) F is $LR-convex$ if and only if $\underline{f}$ is convex and $\overline{f}$ is convex;
(iii) F is $CW-convex$ if and only if $(\overline{f} + \underline{f})$ is convex and $(\overline{f} - \underline{f})$ is concave;
(iv) F is $CW^*-convex$ if and only if $(\overline{f} + \underline{f})$ and $(\overline{f} - \underline{f})$ are convex.

Proof. This result follows directly from Definitions 3, 4, 5 and 6, respectively.

4 Some Jensen's Inequalities for Interval-Valued Functions

Recently, using the $\subseteq -$convexity, Costa [6] showed the following result.

Theorem 4 *(Jensen's Interval Inequality with the $\subseteq$ $-$convexity)* *([6]).*
Let $g : [0,1] \rightarrow (a,b)$ be a Lebesgue-integrable real-valued function. Given a
$\subseteq$ $-$convex interval-valued function $F : [a,b] \rightarrow \mathcal{K}_C$ by $F(t) = [\underline{f}(t), \overline{f}(t)]$,
where $\underline{f}, \overline{f} : [a,b] \rightarrow \mathbb{R}$ are real functions such that $(\underline{f} \circ g)$ and $(\overline{f} \circ g)$ are
Lebesgue-integrable over $[0,1]$, it follows that

$$(IA) \int_0^1 F(g(t))dt \subseteq F \left(\int_0^1 g(t)dt \right).$$

That is, the interval $(IA) \int_0^1 F(g(t))dt$ is nested in the interval $F \left(\int_0^1 g(t)dt \right)$.

Next we present our interval versions of Jensen's integral inequality using the
concepts of $LR-$convexity, $CW-$convexity, and of CW^*-convexity. These versions are interpreted by means of order relations that allow us to compare
non-nested intervals, and consequently, they complement that one given in Theorem 4. Moreover, these inequalities extend the classical Jensen's integral inequality for real-valued functions.

Theorem 5 *(Jensen's Interval Inequality with the $LR-$convexity).* *Let*
$g : [0,1] \rightarrow (a,b)$ be a Lebesgue-integrable real-valued function. Given a
$LR-$convex interval-valued function $F : [a,b] \rightarrow \mathcal{K}_C$ by $F(t) = [\underline{f}(t), \overline{f}(t)]$,
where $\underline{f}, \overline{f} : [a,b] \rightarrow \mathbb{R}$ are real functions such that $(\underline{f} \circ g)$ and $(\overline{f} \circ g)$ are
Lebesgue-integrable over $[0,1]$, it follows that

$$F \left(\int_0^1 g(t)dt \right) \leq_{LR} (IA) \int_0^1 F(g(t))dt. \tag{22}$$

Proof. Since F is $LR-$convex, from item (ii) in Proposition 1, it follows that $\underline{f}$
and $\overline{f}$ are convex. Thus, from Jensen inequality for convex real-valued functions,
it follows that

$$\underline{f} \left(\int_0^1 g(t)dt \right) \leq \int_0^1 \underline{f}(g(t))dt \quad and \quad \overline{f} \left(\int_0^1 g(t)dt \right) \leq \int_0^1 \overline{f}(g(t))dt. \tag{23}$$

On the other hand, from Remark 1 and Theorem 3, it follows that

$$F \left(\int_0^1 g(t)dt \right) = \left[\underline{f} \left(\int_0^1 g(t)dt \right), \overline{f} \left(\int_0^1 g(t)dt \right) \right] \tag{24}$$

and

$$(IA) \int_0^1 F(g(t))dt = \left[\int_0^1 \underline{f}(g(t))dt, \int_0^1 \overline{f}(g(t))dt \right]. \tag{25}$$

Therefore, from (23), (24), (25), and from definition of $\leq_{LR}$, it follows that (22)
holds. □

Theorem 6 *(Jensen's Interval Inequality with the $CW-$convexity).*
Let $g : [0,1] \rightarrow (a,b)$ be a Lebesgue-integrable real-valued function. Given a

$CW-convex$ interval-valued function $F : [a, b] \rightarrow \mathcal{K}_C$ by $F(t) = [\underline{f}(t), \overline{f}(t)]$, where $\underline{f}, \overline{f} : [a, b] \rightarrow \mathbb{R}$ are real functions such that $(\underline{f} \circ g)$ and $(\overline{f} \circ g)$ are Lebesgue-integrable over $[0, 1]$, it follows that

$$F \left(\int_0^1 g(t)dt \right) \leq_{CW} (IA) \int_0^1 F(g(t))dt. \tag{26}$$

Proof. Since F is $LR-convex$, from item (ii) in Proposition 1, it follows that $(\overline{f} + \underline{f})$ is convex and $(\overline{f} - \underline{f})$ is concave. Thus, from Jensen inequality for convex and concave real-valued functions, it follows that $(\overline{f} + \underline{f}) \left(\int_0^1 g(t)dt \right) \leq \int_0^1 (\overline{f} + \underline{f})(g(t))dt$ and $\int_0^1 (\overline{f} - \underline{f})(g(t))dt \leq (\overline{f} - \underline{f}) \left(\int_0^1 g(t)dt \right)$. Consequently,

$$\frac{\overline{f} \left(\int_0^1 g(t)dt \right) + \underline{f} \left(\int_0^1 g(t)dt \right)}{2} \leq \frac{\int_0^1 \overline{f}(g(t))dt + \int_0^1 \underline{f}(g(t))dt}{2} \tag{27}$$

and

$$\frac{\overline{f} \left(\int_0^1 g(t)dt \right) - \underline{f} \left(\int_0^1 g(t)dt \right)}{2} \geq \frac{\int_0^1 \overline{f}(g(t))dt - \int_0^1 \underline{f}(g(t))dt}{2}. \tag{28}$$

On the other hand, from Remark 1 and Theorem 3, it follows that

$$F \left(\int_0^1 g(t)dt \right) = \left[\underline{f} \left(\int_0^1 g(t)dt \right), \overline{f} \left(\int_0^1 g(t)dt \right) \right] \tag{29}$$

and

$$(IA) \int_0^1 F(g(t))dt = \left[\int_0^1 \underline{f}(g(t))dt, \int_0^1 \overline{f}(g(t))dt \right]. \tag{30}$$

Therefore, from (28), (29), (27), and from definition of $\leq_{CW}$, it follows that (26) holds. $\square$

Theorem 7 (Jensen's Interval Inequality with the $CW^*-convexity$). Let $g : [0, 1] \rightarrow (a, b)$ be a Lebesgue-integrable real-valued function. Given a $CW^*-convex$ interval-valued function $F : [a, b] \rightarrow \mathcal{K}_C$ by $F(t) = [\underline{f}(t), \overline{f}(t)]$, where $\underline{f}, \overline{f} : [a, b] \rightarrow \mathbb{R}$ are real functions such that $(\underline{f} \circ g)$ and $(\overline{f} \circ g)$ are Lebesgue-integrable over $[0, 1]$, it follows that

$$F \left(\int_0^1 g(t)dt \right) \leq_{CW^*} (IA) \int_0^1 F(g(t))dt. \tag{31}$$

Proof. This result is obtained using similar argumentation to that used in the prove of Theorem 6.

5   Preliminaries: Space of Fuzzy Intervals

This section recalls concepts and results from fuzzy literature used in this presentation.

Definition 7 *(see [3,8]). A fuzzy subset A of $\mathbb{R}$ is characterized by a function $\tilde{u} : \mathbb{R} \to [0,1]$ called the membership function of A. In general, in order to simplify the notation, a fuzzy subset A of $\mathbb{R}$ is presented as being its membership function $\tilde{u}$. That is, a fuzzy subset A of $\mathbb{R}$ is a function $\tilde{u} : \mathbb{R} \to [0,1]$. In this study this representation is adopted. Moreover, the family of all fuzzy subset of $\mathbb{R}$ is denoted by $\mathcal{F}(\mathbb{R})$.*

   A fuzzy subset $\tilde{u}$ of $\mathbb{R}$ is called a real fuzzy interval if it has the following properties:

 (i) *$\tilde{u}$ is normal, i.e., there exists $\bar{x} \in \mathbb{R}$ such that $\tilde{u}(\bar{x}) = 1$;*
 (ii) *$\tilde{u}$ is fuzzy convex, i.e., $\min\{\tilde{u}(x_1), \tilde{u}(x_2)\} \leqslant \tilde{u}(\lambda x_1 + (1-\lambda)x_2)$ for all $x_1, x_2 \in \mathbb{R}$ and for all $\lambda \in [0,1]$;*
 (iii) *$\tilde{u}$ is upper semicontinuous on $\mathbb{R}$, i.e., given $\bar{x} \in \mathbb{R}$, for every $\epsilon > 0$ there exists $\delta > 0$ such that $\tilde{u}(x) - \tilde{u}(\bar{x}) < \epsilon$ for all $x \in \mathbb{R}$ with $|x - \bar{x}| < \delta$;*
 (iv) *$\tilde{u}$ is compactly supported, i.e., $cl\{x \in \mathbb{R} : 0 < \tilde{u}(x)\}$ is compact, where $cl(A)$ denotes the closure of a classical set A.*

The family of all real fuzzy intervals is denoted by $\mathcal{F}_C(\mathbb{R})$.

Definition 8 *(see [3,8]). Given $\tilde{u} \in \mathcal{F}_C(\mathbb{R})$, the level sets of $\tilde{u}$ are given by $[\tilde{u}]^\alpha = \{x \in \mathbb{R} : \alpha \leqslant \tilde{u}(x)\}$ for all $\alpha \in (0,1]$ and by $[\tilde{u}]^0 = cl\{x \in \mathbb{R} : 0 < \tilde{u}(x)\}$. These sets are called the $\alpha-$level sets of $\tilde{u}$ for all $\alpha \in [0,1]$.*

Theorem 8 *([8,13]). $\tilde{u} \in \mathcal{F}_C(\mathbb{R})$ if and only if $[\tilde{u}]^\alpha \subset \mathbb{R}$ is a nonempty, bounded, and closed interval for each $\alpha \in [0,1]$.*

Theorem 9 *([3,9]). Let u be a fuzzy interval and let $[\tilde{u}]^\alpha = [\underline{u}^\alpha, \overline{u}^\alpha]$. Then the functions $\underline{u}, \overline{u} : [0,1] \to \mathbb{R}$, defining the endpoints of the $\alpha-$level sets, satisfy the following conditions:*

 (i) *$\underline{u}(\alpha) = \underline{u}^\alpha \in \mathbb{R}$ is bounded, non-decreasing, left-continuous function on $(0,1]$ and it is right-continuous at 0.*
 (ii) *$\overline{u}(\alpha) = \overline{u}^\alpha \in \mathbb{R}$ is bounded, non-increasing, left-continuous function on $(0,1]$ and it is right-continuous at 0.*
 (iii) *$\underline{u}(1) \leqslant \overline{u}(1)$.*

Theorem 10 *([3,9]). Consider the functions $\underline{u}, \overline{u} : [0,1] \longrightarrow \mathbb{R}$ satisfying the following conditions:*

 (i) *$\underline{u}(\alpha) = \underline{u}^\alpha \in \mathbb{R}$ is bounded, non-decreasing, left-continuous function on $(0,1]$ and it is right-continuous at 0.*
 (ii) *$\overline{u}(\alpha) = \overline{u}^\alpha \in \mathbb{R}$ is bounded, non-increasing, left-continuous function on $(0,1]$ and it is right-continuous at 0.*
 (iii) *$\underline{u}(1) \leqslant \overline{u}(1)$.*

Then there is a fuzzy interval $\tilde{u} \in \mathcal{F}_C(\mathbb{R})$ such that $[\tilde{u}]^\alpha = [\underline{u}^\alpha, \overline{u}^\alpha]$ for every $\alpha \in [0,1]$.

Remark 2 ([6]). Let $\tilde{F} : [a,b] \to \mathcal{F}_C(\mathbb{R})$ be whose α–levels are $F_\alpha : [a,b] \to \mathcal{K}_C$ such that $F_\alpha(t) = \left[\underline{f}^\alpha(t), \overline{f}^\alpha(t) \right]$ with $\underline{f}^\alpha, \overline{f}^\alpha : [a,b] \to \mathbb{R}$ for all $\alpha \in [0,1]$. From Theorem 9, it follows that

(I) $\underline{f}(\alpha,t) = \underline{f}^\alpha(t) \in \mathbb{R}$ is bounded, non-decreasing, left-continuous function on $(0,1]$ and it is right-continuous at 0 with respect to α for all $t \in [a,b]$.

(II) $\overline{f}(\alpha,t) = \overline{f}^\alpha(t) \in \mathbb{R}$ is bounded, non-increasing, left-continuous function on $(0,1]$ and it is right-continuous at 0 with respect to α for all $t \in [a,b]$.

(III) $\underline{f}(1,t) \leqslant \overline{f}(1,t)$ for all $t \in [a,b]$.

In particular, given a real-valued function $g : S \to \mathbb{R}$ such that $g(S) \subseteq [a,b]$, we can define the fuzzy-interval-valued function $(\tilde{F} \circ g) : S \to \mathcal{F}_C(\mathbb{R})$, which associates to each $t \in S$ the value $\left(\tilde{F} \circ g \right)(t) \in \mathcal{F}(\mathbb{R})$ such that $\left[\left(\tilde{F} \circ g \right)(t) \right]^\alpha = (F_\alpha \circ g)(t) = \left[\underline{f}^\alpha(g(t)), \overline{f}^\alpha(g(t)) \right] \in \mathcal{K}_C$ for all $t \in S$ and for all $\alpha \in [0,1]$. $\left(\tilde{F} \circ g \right)$ is well defined because $\underline{f}(\alpha,t)$ and $\overline{f}(\alpha,t)$ satisfy $I - III$ for all $t \in [a,b]$. In particular, it follows that $\underline{f}(\alpha,g(t))$ and $\overline{f}(\alpha,g(t))$ satisfy $I - III$ for all $t \in S$, and from Theorem 10, it follows that $\left(\tilde{F} \circ g \right)(t) \in \mathcal{F}_C(\mathbb{R})$.

It is well-known that, given $\lambda \in \mathbb{R}$ and $\tilde{u}, \tilde{v} \in \mathcal{F}_C(\mathbb{R})$, then the multiplication by scalar $\lambda \odot \tilde{u}$ and the multiplication $\tilde{u} \oplus \tilde{v}$ are characterized level-wise, respectively, by $[\lambda \odot \tilde{u}]^\alpha = \lambda \cdot [\tilde{u}]^\alpha$ and $[\tilde{u} \oplus \tilde{v}]^\alpha = [\tilde{u}]^\alpha + [\tilde{v}]^\alpha$ for all $\alpha \in [0,1]$. Another well-known fact is that, $\tilde{u} = \tilde{v}$ if and only if $[\tilde{u}]^\alpha = [\tilde{v}]^\alpha$ for all $\alpha \in [0,1]$.

A fuzzy-interval-valued map $\tilde{F} : S \to \mathcal{F}_C(\mathbb{R})$ is called a **fuzzy-interval-valued function**. Given a fuzzy-interval-valued function $\tilde{F} : S \to \mathcal{F}_C(\mathbb{R})$, the interval-valued function $F_\alpha : S \to \mathcal{K}_C$ given by $F_\alpha(t) = [\tilde{F}(t)]^\alpha$ for all $t \in S$ is called the α–**level of** $\tilde{F}$ for all $\alpha \in [0,1]$.

Theorem 11 *(see* [8,14]*). The space $\mathcal{F}_C(\mathbb{R})$ equipped with the supremum metric, i.e., $d_\infty(\tilde{u}, \tilde{v}) = \sup\limits_{\alpha \in [0,1]} d_H([\tilde{u}]^\alpha, [\tilde{v}]^\alpha)$, it is a complete metric space.*

Definition 9 *(see* [3,8]*). A fuzzy-interval-valued function $\tilde{F} : S \to \mathcal{F}_C(\mathbb{R})$ is said to be continuous at $t_0 \in S$ if for any $\epsilon > 0$, there exists $\delta(\epsilon, t_0) = \delta > 0$ such that $d_\infty \left(\tilde{F}(t), \tilde{F}(t_0) \right) < \epsilon$ for all $t \in S$ with $|t - t_0| < \delta$.*

Remark 3 ([6]). A fuzzy-interval-valued function $\tilde{F} : [a,b] \to \mathcal{F}_C(\mathbb{R})$ is said to be integrably bounded if there exists a Lebesgue-integrable function $h : [a,b] \to [0,+\infty)$ such that $|x| \leqslant h(t)$ for all x and t such that $x \in F_0(t)$. A fuzzy-interval-valued function $\tilde{F} : [a,b] \to \mathcal{F}_C(\mathbb{R})$ is said to be strongly measurable [11] if and only if its α–levels $F_\alpha : [a,b] \to \mathcal{K}_C$ are measurable for all $\alpha \in [0,1]$. This concept is equivalent to the concept of measurability given in [14] and such equivalence can be obtained through Theorem III-2 and Theorem III-30 given in [4].

Definition 10 *([11]). Let $\tilde{F} : [a,b] \to \mathcal{F}_C(\mathbb{R})$ be a fuzzy-interval-valued function. The integral of $\tilde{F}$ over $[a,b]$, denoted by $(FA)\int_a^b \tilde{F}(t)dt$, it is defined levelwise by*

$$\left[(FA)\int_a^b \tilde{F}(t)dt \right]^\alpha = (IA)\int_a^b F_\alpha(t)dt = \left\{ \int_a^b f(t)dt : f \in S(F_\alpha) \right\}$$

for all $\alpha \in [0,1]$. $\tilde{F}$ is $(FA)-$integrable over $[a,b]$ if $(FA)\int_a^b \tilde{F}(t)dt \in \mathcal{F}_C(\mathbb{R})$.

Theorem 12 *([14]). Given a fuzzy-interval-valued function $\tilde{F} : [a,b] \to \mathcal{F}_C(\mathbb{R})$, if $\tilde{F}$ is strongly measurable and integrably bounded, then $\tilde{F}$ is $FA-$integrable over $[a,b]$.*

Given the fuzzy-interval-valued function $\tilde{F} : [a,b] \to \mathcal{F}_C(\mathbb{R})$, whose $\alpha-$levels are given by $F_\alpha : [a,b] \to \mathcal{K}_C$ for all $\alpha \in [0,1]$, if F_α is d_H-continuous for all $\alpha \in [0,1]$, then from Definition 10 and from Theorem 3, it is follows that $\tilde{F}$ is $FA-$integrable over $[a,b]$.

Theorem 13 *(see [8,11]). Let $\tilde{F} : [a,b] \to \mathcal{F}_C(\mathbb{R})$ be the fuzzy-interval-valued function, whose $\alpha-$levels $F_\alpha : [a,b] \to \mathcal{K}_C$ are given by $F_\alpha(t) = \left[\underline{f}_\alpha(t), \overline{f}_\alpha(t) \right]$ for all $t \in [a,b]$ and for all $\alpha \in [0,1]$. Then $\tilde{F}$ is integrable over $[a,b]$ if and only if $\underline{f}_\alpha, \overline{f}_\alpha \in S(F_\alpha)$ for all $\alpha \in [0,1]$. Moreover, if $\tilde{F}$ is $FA-$integrable over $[a,b]$, then*

$$\left[(FA)\int_a^b \tilde{F}(t)dt \right]^\alpha = (IA)\int_a^b F_\alpha(t)dt = \left[\int_a^b \underline{f}_\alpha(t)dt, \int_a^b \overline{f}_\alpha(t)dt \right] \text{ for all } \alpha \in [0,1].$$

6 Jensen's Integral Inequalities for Fuzzy-Interval-Valued Functions

This section presents new versions of Jensen's integral inequality for fuzzy-interval-valued functions. To this end, order relations on $\mathcal{F}_C(\mathbb{R})$ and concepts of convexity for fuzzy-interval-valued functions are defined as extensions from those mentioned in Sect. 3.

Henceforth S denotes an interval in $\mathbb{R}$ and $\tilde{F}$ denotes a fuzzy-interval-valued function $\tilde{F} : S \to \mathcal{F}_C(\mathbb{R})$.

6.1 Order Relations, and Concepts of Convexity for Fuzzy-Interval-Valued Functions

Given $\tilde{u}, \tilde{v} \in \mathcal{F}_C(\mathbb{R})$, let $\subseteq_\mathcal{F}$, $\leq_{LR_\mathcal{F}}$, $\leq_{CW_\mathcal{F}}$, $\leq_{CW_\mathcal{F}^*}$ be the relations given on $\mathcal{F}_C(\mathbb{R})$, respectively, by

$$\tilde{u} \subseteq_\mathcal{F} \tilde{v} \Leftrightarrow [\tilde{u}]^\alpha \subseteq [\tilde{v}]^\alpha \ \forall \alpha \in [0,1], \quad \tilde{u} \leq_{LR_\mathcal{F}} \tilde{v} \Leftrightarrow [\tilde{u}]^\alpha \leq_{LR} [\tilde{v}]^\alpha \ \forall \alpha \in [0,1],$$

$\tilde{u} \leq_{CW_{\mathcal{F}}} \tilde{v} \Leftrightarrow [\tilde{u}]^\alpha \leq_{CW} [\tilde{v}]^\alpha \ \forall \alpha \in [0,1],$ and $\tilde{u} \leq_{CW^{*}_{\mathcal{F}}} \tilde{v} \Leftrightarrow [\tilde{u}]^\alpha \leq_{CW^{*}} [\tilde{v}]^\alpha \ \forall \alpha \in [0,1].$

Since "$\subseteq$", "$\leq_{LR}$", "$\leq_{CW}$", and "$\leq_{CW^{*}}$" are partial order relations on $\mathcal{K}_C$, it follows that $\subseteq_{\mathcal{F}}$, $\leq_{LR_{\mathcal{F}}}$, $\leq_{CW_{\mathcal{F}}}$, $\leq_{CW^{*}_{\mathcal{F}}}$ are partial order relations on $\mathcal{F}_C(\mathbb{R})$.

Definition 11 *(see [6]).* $\tilde{F}$ *is said to be* $\subseteq_{\mathcal{F}}$ *−convex if*

$$(1 - \lambda)\tilde{F}(t_0) + \lambda\tilde{F}(t_1) \subseteq_{\mathcal{F}} \tilde{F}((1 - \lambda)t_0 + \lambda t_1)) \text{ for any } t_0, t_1 \in S \text{ and } \lambda \in [0,1]. \tag{32}$$

Definition 12 *(see [5]).* $\tilde{F}$ *is said to be* $LR_{\mathcal{F}}$*−convex if*

$$\tilde{F}((1 - \lambda)t_0 + \lambda t_1) \leq_{LR_{\mathcal{F}}} (1 - \lambda)\tilde{F}(t_0) + \lambda\tilde{F}(t_1) \text{ for any } t_0, t_1 \in S \text{ and } \lambda \in [0,1]. \tag{33}$$

Definition 13. $\tilde{F}$ *is said to be* $CW_{\mathcal{F}}$*−convex if*

$$\tilde{F}((1 - \lambda)t_0 + \lambda t_1) \leq_{CW_{\mathcal{F}}} (1 - \lambda)\tilde{F}(t_0) + \lambda\tilde{F}(t_1) \text{ for any } t_0, t_1 \in S \text{ and } \lambda \in [0,1]. \tag{34}$$

Definition 14. $\tilde{F}$ *is said to be* $CW^{*}_{\mathcal{F}}$*−convex if*

$$\tilde{F}((1 - \lambda)t_0 + \lambda t_1) \leq_{CW^{*}_{\mathcal{F}}} (1 - \lambda)\tilde{F}(t_0) + \lambda\tilde{F}(t_1) \text{ for any } t_0, t_1 \in S \text{ and } \lambda \in [0,1]. \tag{35}$$

Proposition 2. *Let* $\tilde{F} : S \to \mathcal{F}_C(\mathbb{R})$ *be such that* $F_\alpha : S \to \mathcal{K}_C$ *is its* α*−level for all* $\alpha \in [0,1]$. *Thus,*

(i) $\tilde{F}$ is $\subseteq_{\mathcal{F}}$ −convex if and only if F_α is $\subseteq$ −convex for all $\alpha \in [0,1]$.
(ii) $\tilde{F}$ is $LR_{\mathcal{F}}$−convex if and only if F_α is LR−convex for all $\alpha \in [0,1]$.
(iii) $\tilde{F}$ is $CW_{\mathcal{F}}$−convex if and only if F_α is CW−convex for all $\alpha \in [0,1]$.
(iv) $\tilde{F}$ is $CW^{*}_{\mathcal{F}}$−convex if and only if F_α is CW^{*}−convex for all $\alpha \in [0,1]$.

Proof. This result follows directly from the definitions of "$\subseteq_{\mathcal{F}}$", "$\leq_{LR_{\mathcal{F}}}$", "$\leq_{CW_{\mathcal{F}}}$", "$\leq_{CW^{*}_{\mathcal{F}}}$", and from Definitions 11, 12, 13 and 14, respectively.

6.2 Some Jensen's Inequalities for Fuzzy-Interval-Valued Function

Recently, using the $\subseteq_{\mathcal{F}}$ −convexity, Costa [6] showed the following result.

Theorem 14 *(Fuzzy Jensen's Inequality with the* $\subseteq_{\mathcal{F}}$ *−convexity)* *([6]).* *Let* $g : [0,1] \to (a,b)$ *be a Lesbesgue-integral function. Given a* $\subseteq_{\mathcal{F}}$ *−convex fuzzy-interval-valued function* $\tilde{F} : [a,b] \to \mathcal{F}_C(\mathbb{R})$ *whose* α*−levels* $F_\alpha : [a,b] \to \mathcal{K}_C$ *are given by* $F_\alpha(t) = \left[\underline{f}_\alpha(g(t)), \overline{f}_\alpha(g(t))\right]$, *where* $\underline{f}_\alpha, \overline{f}_\alpha : [a,b] \to \mathbb{R}$ *and* $(\underline{f}_\alpha \circ g)$ *and* $(\overline{f}_\alpha \circ g)$ *are Lesbesgue-integrable over* $[0,1]$ *for all* $\alpha \in [0,1]$, *then*

$$(FA) \int_0^1 \tilde{F}(g(t))dt \subseteq_{\mathcal{F}} \tilde{F}\left(\int_0^1 g(t)dt\right). \tag{36}$$

We now provide our fuzzy versions of Jensen's integral inequality based on the $LR_{\mathcal{F}}$−convexity, $CW_{\mathcal{F}}$−convexity, and on the $CW_{\mathcal{F}}^*$−convexity.

Theorem 15 *(Fuzzy Jensen's Inequality with the $LR_{\mathcal{F}}$−convexity).* *Let*
$g : [0,1] \to (a,b)$ be a Lesbesgue-integral function. Given a $LR_{\mathcal{F}}$−convex fuzzy-interval-valued function $\tilde{F} : [a,b] \to \mathcal{F}_C(\mathbb{R})$ whose α−levels $F_\alpha : [a,b] \to \mathcal{K}_C$ are given by $F_\alpha(t) = \left[\underline{f}_\alpha(g(t)), \overline{f}_\alpha(g(t)) \right]$, where $\underline{f}_\alpha, \overline{f}_\alpha : [a,b] \to \mathbb{R}$ and $(\underline{f}_\alpha \circ g)$ and $(\overline{f}_\alpha \circ g)$ are Lebesgue-integrable over $[0,1]$ for all $\alpha \in [0,1]$, then

$$\tilde{F}\left(\int_0^1 g(t)dt \right) \leq_{LR_{\mathcal{F}}} (FA) \int_0^1 \tilde{F}(g(t))dt. \tag{37}$$

Proof. *From definition of $\leq_{LR_{\mathcal{F}}}$ and definition of (FA)−integral, and from Remark 2, it follows that (37) holds if and only if*

$$\tilde{F}_\alpha\left(\int_0^1 g(t)dt \right) \leq_{LR} (IA) \int_0^1 \tilde{F}_\alpha(g(t))dt \quad \text{for all } \alpha \in [0,1]. \tag{38}$$

Since $\tilde{F}$ is $LR_{\mathcal{F}}$−convex, then from item (ii) in Proposition 2, it follows that F_α is LR−convex for all $\alpha \in [0,1]$. Then, applying Theorem 5 for each $\alpha \in [0,1]$, it follows that (38) holds. Therefore, (37) also holds. □

Using similar argumentation to that used in the proof of Theorem 15, one can obtain easily a proof of the following two results.

Theorem 16 *(Fuzzy Jensen's Inequality with the $CW_{\mathcal{F}}$− convexity).* *Let*
$g : [0,1] \to (a,b)$ be a Lesbesgue-integral function. Given a $CW_{\mathcal{F}}$−convex fuzzy-interval-valued function $\tilde{F} : [a,b] \to \mathcal{F}_C(\mathbb{R})$ whose α−levels $F_\alpha : [a,b] \to \mathcal{K}_C$ are given by $F_\alpha(t) = \left[\underline{f}_\alpha(g(t)), \overline{f}_\alpha(g(t)) \right]$, where $\underline{f}_\alpha, \overline{f}_\alpha : [a,b] \to \mathbb{R}$ and $(\underline{f}_\alpha \circ g)$ and $(\overline{f}_\alpha \circ g)$ are Lebesgue-integrable over $[0,1]$ for all $\alpha \in [0,1]$, then

$$\tilde{F}\left(\int_0^1 g(t)dt \right) \leq_{CW_{\mathcal{F}}} (FA) \int_0^1 \tilde{F}(g(t))dt. \tag{39}$$

Theorem 17 *(Fuzzy Jensen's Inequality with the $CW_{\mathcal{F}}^*$−convexity).* *Let*
$g : [0,1] \to (a,b)$ be a Lesbesgue-integral function. Given a $CW_{\mathcal{F}}^*$−convex fuzzy-interval-valued function $\tilde{F} : [a,b] \to \mathcal{F}_C(\mathbb{R})$ whose α−levels $F_\alpha : [a,b] \to \mathcal{K}_C$ are given by $F_\alpha(t) = \left[\underline{f}_\alpha(g(t)), \overline{f}_\alpha(g(t)) \right]$, where $\underline{f}_\alpha, \overline{f}_\alpha : [a,b] \to \mathbb{R}$ and $(\overline{f}_\alpha \circ g)$ and $(\overline{f}_\alpha \circ g)$ are Lebesgue-integrable over $[0,1]$ for all $\alpha \in [0,1]$, then

$$\tilde{F}\left(\int_0^1 g(t)dt \right) \leq_{CW_{\mathcal{F}}^*} (FA) \int_0^1 \tilde{F}(g(t))dt. \tag{40}$$

7   Conclusion

In this presentation we introduce Theorems 5, 6, 7, 15, 16 and 17 that provide new interval and fuzzy versions of Jensen's integral inequality. Different from the inequalities given in [6], these Jensen's inequalities are extensions of the classical Jensen's integral inequality for real-valued functions, and these inequalities are interpreted by means of order relations that allow us to compare non-nested intervals as well as to compare non-nested fuzzy intervals. From analytic viewpoint, these inequalities allow one to obtain interval numeric estimations for a class of integrals of interval-valued function and for a class of integrals of fuzzy-interval-valued functions (also interpreted as fuzzy interval expected values as is done by de Barros et al. in [7] and by Puri and Ralescu in [14]) through the values that the interval integrands and fuzzy integrands assume at real numbers, respectively. Our next step is to try to extend these inequalities for fuzzy functions, that is, for applications of type $\tilde{\tilde{F}} : \mathcal{F}_C(\mathbb{R}) \to \mathcal{F}_C(\mathbb{R})$.

References

1. Aubin, J.P., Cellina, A.: Differential Inclusions: Set-Valued Maps and Viability Theory. Grundlehren der mathematischen Wissenschaften. Springer, Heidelberg (1984). https://doi.org/10.1007/978-3-642-69512-4
2. Aumann, R.J.: Integrals of set-valued functions. J. Math. Anal. Appl. **12**(1), 1–12 (1965)
3. Bede, B.: Mathematics of Fuzzy Sets and Fuzzy Logic. Studies in Fuzziness and Soft Computing, vol. 295. Springer, Heidelberg (2013). https://doi.org/10.1007/978-3-642-35221-8
4. Castaing, C., Valadier, M.: Convex Analysis and Measurable Multifunctions. Lecture Notes in Mathematics, vol. 580. Springer, Berlin (1977). https://doi.org/10.1007/BFb0087685
5. Chalco-Cano, Y., Lodwick, W.A., Rufian-Lizana, A.: Optimality conditions of type KKT for optimization problem with interval-valued objective function via generalized derivative. Fuzzy Optim. Decis. Mak. **12**, 305–322 (2013)
6. Costa, T.M.: Jensen's inequality type integral for fuzzy-interval-valued functions. Fuzzy Sets Syst. **327**, 31–47 (2017)
7. de Barros, L.C., Bassanezi, R.C., Lodwick, W.A.: A First Course in Fuzzy Logic, Fuzzy Dynamical Systems, and Biomathematics: Theory and Applications. Studies in Fuzziness and Soft Computing, vol. 347. Springer, Heidelberg (2017). https://doi.org/10.1007/978-3-662-53324-6
8. Diamond, P., Kloeden, P.E.: Metric Spaces of Fuzzy Sets: Theory and Applications. World Scientific, Singapore (1994)
9. Goetschel, R., Voxman, W.: Elementary fuzzy calculus. Fuzzy Sets Syst. **18**(1), 31–43 (1986)
10. Ishibuchi, H., Tanaka, H.: Multiobjective programming in optimization of the interval objective function. Eur. J. Oper. Res. **48**, 219–225 (1990)
11. Kaleva, O.: Fuzzy numbers fuzzy differential equations. Fuzzy Sets Syst. **24**(3), 301–317 (1987)
12. Moore, R.E.: Interval Analysis. Prentice-Hall, Englewood Cliffs (1966)
13. Negoita, C.V., Ralescu, D.A.: Applications of Fuzzy Sets to Systems Analysis. Wiley, New York (1975)
14. Puri, M.L., Ralescu, D.A.: Fuzzy random variables. J. Math. Anal. Appl. **114**, 409–422 (1986)

Fuzzy Initial Value Problem: A Short Survey

Marina Tuyako Mizukoshi[✉]

Instituto de Matemática e Estatística, Universidade Federal de Goiás,
Campus II - Samambaia, Goiânia, GO 74690-900, Brazil
tuyako@ufg.br
http://www.ime.ufg.br

Abstract. This article provides a survey of the available literature on
Fuzzy Initial Value Problem (FIVP) and various different interpretations. The fuzzy differential equations can be studied using the derivative concept or without it. The Malthusian population model with fuzzy
initial condition is used to illustrate the different approaches, namely,
Hukuhara derivative, gh-differentiability, $\pi-$derivative and Zadeh's
extension applied to derivative operator using the differentiability and
differential inclusion theory, Zadeh's extension principle applied in deterministic solution without derivative concept.

Keywords: FIVP · Differential inclusion · Fuzzy differentiability
Stability · Hukuhara derivative

1 Introduction

The modeling of various phenomenon is frequently made by using deterministic
differential systems

$$x'(t) = f(t, x(t)); \quad x(0) = x_0, \tag{1}$$

where $x(t), x_0 \in \mathbb{R}^n$ and $f : \mathbb{R} \times \mathbb{R}^n \to \mathbb{R}^n$ is a function that satisfies some
existence conditions.

In practice, exact knowledge of the initial condition or parameters of (1)
may be unavailable, or difficult to obtain. Generally, their values are imprecise
because they are either approximately known, or result from observations prone
to error. As Rouvray [44], "all scientific pronouncements have some inherent
uncertainty about them and cannot be assumed to be strictly valid". A way
to address imprecise initial conditions is to rewrite (1) as Fuzzy Initial Value
Problem (FIVP)

$$x'(t) = f(t, x(t)); \quad X(0) = X_0,$$

where $X_0 \in \mathcal{F}(\mathbb{R})$ and $f : [0, T] \times \mathcal{F}(\mathbb{R}^n) \to \mathcal{F}(\mathbb{R}^n)$ with $x(t) \in \mathcal{F}(\mathbb{R}^n)$ or
$f : [0, T] \times \mathbb{R}^n \to \mathcal{F}(\mathbb{R}^n)$ for $x(t) \in \mathbb{R}^n$.

© Springer International Publishing AG, part of Springer Nature 2018
G. A. Barreto and R. Coelho (Eds.): NAFIPS 2018, CCIS 831, pp. 464–476, 2018.
https://doi.org/10.1007/978-3-319-95312-0_40

Uncertainty was formally admitted into sciences about four centuries ago and since then the modeling has been dominated by stochastic methods. In the 1930's, differential inclusions theory was introduced by the Polish and French mathematicians Zaremba [47] and Marchaud [31] as a generalization of differential equations by considering the uncertainty in direction the vector velocity. They studied the so-called paratingent and contingent equations, respectively. In 1962, Wazewski [48] proved that the solutions can be understood in the Caratheodory sense of absolute continuity satisfying the differential inclusion almost everywhere for all time.

Fuzzy set and possibility theory are notions that have introduced recently. Interval analysis and fuzzy set theory emerged in 1959 and 1965, respectively with Moore [34] and Zadeh [46]. Subsequently in [29] was proposed the study of the interval theory using Constraint Interval Arithmetic (CIA). According with Lodwick and Dubois [30] interval analysis is not only useful but necessary to the understanding of fuzzy interval analysis especially in the context of linear systems.

The initial value problem is discussed in [41] and a comprehensive overview of the computational aspects are given in [12]. Differential inclusion in the framework of fuzzy set theory was first discussed by Baidosov [4] as follows:

$$x'(t) \in F(t, x(t)),$$

where the right side is a fuzzy multivalued function. Alternatively, Aubin [1] assumes taht the right side of differential inclusion is a fuzzy set. Hullermeier [18] has suggests to solve the FIVP looking at it as a family of differential inclusions.

The term fuzzy differential equation was introduced in 1978 by Kandel and Byatt [22] and an extended version of this short note was published two years later [23]. The concept of differentiability and integrability for fuzzy multivalued were introduced by Puri and Ralescu [39]. The Cauchy problem for first-order fuzzy differential equation was investigated by Kaleva [21], Seikkala [45], Ouyang and Wu [37] using a extension of Hukuhara derivative.

Fuzzy differential equations can be studied from a point of view discrete [2, 27, 28, 42, 43] or continuous [7, 15, 21, 22, 45]. In the continuous case we have two different approaches: in the first one without the derivative concept by differential inclusion theory [13, 14] or Zadeh's extension principle [8, 33]; in the second case, the differentiability is considered, Hukuhara derivative [21], $\pi-$derivative [9], gh-differentiability [6], extension in derivative operator [17]. The spaces of all closed and bounded interval of $\mathbb{R}$ are not linear spaces and therefore the subtraction is not well defined. As a consequence, alternative formulations for subtraction have been suggested and so on there are different definitions for the differentiability in a fuzzy differential equation. The concept of fuzzy derivative, leads the connection between interval and fuzzy theories. By considering the generalized derivative some authors obtained both the solution to linear interval systems and to fuzzy differential equations. Another point of view is the dynamic systems obtained by means of a Mamdani type fuzzy rule-based system [19, 38].

Oberguggenberger and Pittshmann [20,36] studied differential equations system with fuzzy parameter by applying the Zadeh's extension in equations and solution operators. Finally, fuzzy periodic solutions were studied in [10,11,14,35].

We consider the Fuzzy Initial Value Problem, in particular one dimensional Malthusian model to concretize our review of the different approaches to solution and stability of fuzzy differential equations.

2   Basic Concepts

First of all, we provide some notation and recall known results.

We denote by $\mathcal{K}^n$ the family of all the nonempty compact subsets of $\mathbb{R}^n$. For $A, B \in \mathcal{K}^n$ and $\lambda \in \mathbb{R}$ the operations of addition and scalar multiplication are defined by

$$rA + B = \{a + b|\ a\ \in A, b \in B\} \qquad \lambda A = \{\lambda a \mid a \in A\}.$$

Let X be metric space. A fuzzy subset U of X is given by a mapping $\mu_U : X \to [0,1]$ such that the set of ordered pairs $(x, \mu_U(x)), x \in X$ indicates the degree of each x in U. The degrees 0 and 1 represent, respectively, the non-belonging and the maximum belonging of x to fuzzy subset U. To simplify the notation we indicate the membership function μ_U by U.

Let U be a fuzzy set in $\mathbb{R}^n$, the n-dimensional Euclidian space, we define $[U]^\alpha = \{x \in \mathbb{R}^n / U(x) \geq \alpha\}$ the α-level of U, with $0 < \alpha \leq 1$. For $\alpha = 0$ we have $[U]^0 = \text{supp}(U) = \{x \in \mathbb{R}^n \mid U(x) > 0\}$, the support of U.

A fuzzy set U is called compact if $[U]^\alpha \in \mathcal{K}^n$, $\forall \alpha \in [0,1]$. We will denote by $\mathcal{F}(\mathbb{R}^n)$ the space of all the compact fuzzy sets whose $\alpha-$level are compact and connected set in $\mathbb{R}^n$.

The operations of addition and scalar multiplication on $\mathcal{F}(\mathbb{R}^n)$ for all $\alpha-$levels are defined by

$$[U + V]^\alpha = [U]^\alpha + [V]^\alpha; \quad \text{and} \quad [\lambda U]^\alpha = \lambda[U]^\alpha, \ \forall \alpha \in [0,1]. \tag{2}$$

The metric on $\mathcal{F}(U)$ is given by

$$d_\infty(U, V) = \sup_{0 \leq \alpha \leq 1} d_H([U]^\alpha, [V]^\alpha),$$

where d_H is the usual Pompieu-Hausdorff metric defined for compact subsets of $\mathbb{R}^n$. This metric turns the space $(\mathcal{F}(\mathbb{R}), d_\infty)$ into a complete metric space [40].

Zadeh [46] proposed the so called extension principle [5], which became an important tool in fuzzy set theory. The idea is that each function, $f : X \to Y$, induces a corresponding function $\widehat{f} : \mathcal{F}(X) \to \mathcal{F}(Y)$ (i.e., $\widehat{f}$ is a function mapping fuzzy sets in X to fuzzy sets in Y) defined for each fuzzy set U in X by

$$\widehat{f}(U)(y) = \begin{cases} \sup\limits_{u \in f^{-1}(y)} U(u), \text{ if } f^{-1}(y) \neq \emptyset \\ \qquad\qquad 0, \text{ if } f^{-1}(y) = \emptyset. \end{cases} \tag{3}$$

The function $\widehat{f}$ is said to be obtained from f by the extension principle.

An important result of extension principle is the characterization of the levels of the image of a fuzzy set through $\widehat{f}$, where f is a continuous function.

Theorem 1 [2]. *If $f : \mathbb{R}^n \longrightarrow \mathbb{R}^n$ is continuous, then the Zadeh's extension $\widehat{f} : \mathcal{F}(\mathbb{R}^n) \longrightarrow \mathcal{F}(\mathbb{R}^n)$ is well-defined and*

$$\left[\widehat{f}(U)\right]^\alpha = f([U]^\alpha), \forall \alpha \in [0, 1]. \tag{4}$$

Relation (4) continues to be valid if $f : W \to \mathbb{R}^n$, and W is an open subset in $\mathbb{R}^n$. Moreover, according to Román-Flores et al. [42] it was shown that $\widehat{f}$ is a continuous function with respect to Pompieu-Hausdorff metric extended to $\mathcal{F}(\mathbb{R}^n)$.

Let us consider the following differential inclusion,

$$\begin{cases} x'(t) \in F(t, x(t)) \\ x(t_0) = x_0 \in X_0 \end{cases} \tag{5}$$

where $F : [t_0, T] \times \mathbb{R}^n \to \mathcal{K}^n$ is a set-valued function and $X_0 \in \mathcal{K}^n$.

A function $x(t, x_0)$ with the initial condition $x_0 \in X_0$ is a solution of (5) in interval $[t_0, T]$ if it is absolutely continuous and satisfies (5) for all $t \in [t_0, T]$, (for more details, see [1]). The attainable set in time $t \in [t_0, T]$, associated with problem (5), is the subset of $\mathbb{R}^n$ given by

$$\mathcal{A}_t(X_0) = \{x(t, x_0) \ / \ x(\cdot, x_0) \text{ is solution of (5) with } x_0 \in X_0\}.$$

The set-valued function F allows modeling of certain types of uncertainty [26] and because for each pair $(t, x) \in [t_0, T] \times \mathbb{R}^n$, the derivative may not be known precisely, but known to be an element of the set $F(t, x)$.

The following presents a survey of Fuzzy Initial Value Problems (FIVP) considering differential inclusion, extension principle and fuzzy differentiability.

3   FIVP and Some Interpretations

Consider the FIVP

$$\begin{cases} X'(t) = f(t, X(t)) \\ X(0) = X_0. \end{cases} \tag{6}$$

To study the stability of solutions of (6) we need to understand what is a solution of (6).

We consider the deterministic Malthusian problem

$$\begin{cases} x'(t) = -ax(t) \\ x(0) = x_0. \end{cases}, \tag{7}$$

where $0 < a \in \mathbb{R}$ and $x : [0, T] \times \mathbb{R} \to \mathbb{R}$ to illustrate the different approach to fuzzy differential equations.

3.1   FIVP with Differentiability

In this section we consider the problem 6 where is a fuzzy function f that indicates a fuzzy direction and the trajectory for FIVP are or different deterministic solutions with a membership degree to each of them or a function that assigns to each instant $t \in [0, T]$ a fuzzy subset.

First Approach: The first interpretation about (6) appears in 1987 with Seikkala and Kaleva using the Hukuhara derivative. In this interpretation, $\hat{f} : [0, T] \times \mathcal{F}(\mathbb{R}^n) \rightarrow \mathcal{F}(\mathbb{R}^n), X_0 \in \mathcal{F}(\mathbb{R}^n)$ and the solution is a fuzzy-set-valued function $X : [0, T] \rightarrow \mathcal{F}(\mathbb{R}^n)$. Next, we present the concept established by Seikkala [45] for $\mathcal{F}(\mathbb{R})$, which was used by author to rewrite problem (6) from the one dimensional case into a bidimensional system of ordinary differential equations.

Definition 1. *Let $I = [0, T], T \in \mathbb{R}_+$ be a real interval. The application $X : I \longrightarrow \mathcal{F}(\mathbb{R})$ is called a fuzzy process. We denoted $[X(t)]^\alpha = [x_1^\alpha(t), x_2^\alpha(t)], t \in I, \alpha \in [0, 1]$. The derivative $X'(t)$ of a fuzzy process X is defined by $[X'(t)]^\alpha = [(x_1^\alpha(t))', (x_2^\alpha(t))'], 0 < \alpha \leq 1$ provided that the equation defines a fuzzy number $X'(t) \in \mathcal{F}(\mathbb{R})$.*

Let $[X(t)]^\alpha = [x_1^\alpha(t), x_2^\alpha(t)]$ and $f : [0, T] \times \mathcal{F}(\mathbb{R}) \rightarrow \mathcal{F}(\mathbb{R})$ is a continuous mapping. Applying the Zadeh's extension principle to f, (6) in the one dimensional case can be rewritten as:

$$\begin{cases} (x_1^\alpha)'(t) = f_1(x_1^\alpha, x_2^\alpha), x_1^\alpha(0) = x_{01}^\alpha \\ (x_2^\alpha)'(t) = f_2(x_1^\alpha, x_2^\alpha), x_2^\alpha(0) = x_{02}^\alpha, \end{cases} \quad (8)$$

for $t \in [0, T)$ e $\alpha \in [0, 1]$, where

$$\begin{cases} f_1(x_1^\alpha, x_2^\alpha) = min\{f(x)/x \in [X]^\alpha\} \\ f_2(x_1^\alpha, x_2^\alpha) = max\{f(x)/x \in [X]^\alpha\}. \end{cases}$$

Note that the fuzzy problem has been reduced to an initial value problem in $\mathbb{R}^2$.

Example 1. [3]: Suppose the size a population occurs in accordance the law of Malthusian growth. Then, FIVP associated to (7) is given by:

$$\begin{cases} X'(t) = -aX(t) \\ X(0) = X_0, \end{cases} \quad (9)$$

where $X(t), X_0 \in \mathcal{F}(\mathbb{R})$, $0 < a \in \mathbb{R}$ and the $\alpha-$levels of $X(t)$ are given by $[X]^\alpha = [x_1^\alpha, x_2^\alpha]$.

Using the Hukuhara derivative it follows that: $[X'(t)]^\alpha = [(x_1^\alpha)'(t), (x_2^\alpha)'(t)], \alpha \in [0, 1], t \in [0, T]$. By Zadeh's extension principle we have that $f(t, X(t)) = -aX(t)$ such that the $\alpha-$levels are:

$[f(t, X(t))]^\alpha = [min\{-ax_1^\alpha(t), -ax_2^\alpha(t)\}, max\{-ax_1^\alpha(t), -ax_2^\alpha(t)\}], \forall \alpha \in [0, 1], t \in [0, T]$.

Thus for $a > 0$

$$\begin{cases} (x_1^\alpha)'(t) = -ax_2^\alpha(t) \\ (x_2^\alpha)'(t) = -ax_1^\alpha(t) \\ x_1^\alpha(0) = x_{01}^\alpha, x_2^\alpha(0) = x_{02}^\alpha \ , \ \alpha \in [0,1]. \end{cases} \tag{10}$$

The solutions of (10) are given by

$$\begin{cases} x_1^\alpha(t) = \ \dfrac{1}{2}\left(x_{01}^\alpha - x_{02}^\alpha\right)e^{at} + \left(\dfrac{1}{2}x_{01}^\alpha + x_{02}^\alpha\right)e^{-at} \\ x_2^\alpha(t) = +\dfrac{1}{2}\left(-x_{01}^\alpha + x_{02}^\alpha\right)e^{at} + \dfrac{1}{2}\left(x_{01}^\alpha + x_{02}^\alpha\right)e^{-at} \end{cases} \tag{11}$$

Therefore, the solution $X(t)$ of (9) has $\alpha-$levels given by (11) for $a \geq 0$. Then, $diam(x_1^\alpha(t), x_2^\alpha(t)) = |x_{01}^\alpha - x_{02}^\alpha|e^{at}, \forall \alpha \in [0,1], t > 0$.

This means that, the solution is more fuzzy when t is increasing, implying that we do not have the stability condition that we had in deterministic theory.

Then, we consider another type of differentiability in (6) because the Hukuhara derivative leads to solutions with increasing support. A complete review about as type of differentiability of fuzzy multivalued as compared solutions of fuzzy differential equations can be found in [16].

Second Approach: Strongly generalized differentiability was defined by considering the lateral Hukuhara derivative (four cases) and a generalization is given in [6], which is called weakly generalized differentiable. The advantage these definitions is that if g is differentiable on (a,b), then $f : (a,b) \to \mathcal{F}(\mathbb{R})$ such that $f(x) = c \odot g(x), \forall x \in (a,b)$ is the strongly generalized differentiable on (a,b) and $f'(x) = c \odot g'(x)$. In this context the fuzzy differential equation has no unique solution but we can choose among the solutions to find one solution with increasing or decreasing support. This feature allow us to choose singular points where the solutions change monotonicity, such points are called switch points.

From Theorem 4.2.4 in [17], the problem (6) on some interval $[t_0, t_0 + k]$ with $X(t), X(t_0) \in \mathcal{F}(\mathbb{R})$ is the union of the following two ordinary differential equations:

$$\begin{cases} (x_1^\alpha(t))' = f_1^\alpha(t, x_1^\alpha, x_2^\alpha) \\ (x_2^\alpha(t))' = f_2^\alpha(t, x_1^\alpha, x_2^\alpha) \\ x_1^\alpha(t_0) = x_{01}^\alpha, x_2^\alpha(t_0) = x_{02}^\alpha \end{cases} \tag{12}$$

$$\begin{cases} (x_1^\alpha(t))' = f_2^\alpha(t, x_1^\alpha, x_2^\alpha) \\ (x_2^\alpha(t))' = f_1^\alpha(t, x_1^\alpha, x_2^\alpha) \\ x_1^\alpha(t_0) = x_{01}^\alpha, x_2^\alpha(t_0) = x_{02}^\alpha, \end{cases} \tag{13}$$

where were considered the following $\alpha-$levels $[X(t_0)]^\alpha = [x_{01}^\alpha, x_{02}^\alpha], [X(t)]^\alpha = [x_1^\alpha, x_2^\alpha], [f(t, X(t))]^\alpha = [f_1^\alpha(t, x_1^\alpha, x_2^\alpha), f_2^\alpha(t, x_1^\alpha, x_2^\alpha)], \alpha \in [0,1]$.

Third Approach: Chalco-Cano et al. [9] studied fuzzy differential equations using the $\pi-$derivative. The spaces of all closed and bounded interval of $\mathbb{R}$ by using the Radstrøm Embedding Theorem guarantee the existence of a real normed linear space [9]. The π-derivative for fuzzy interval valued functions is a generalization of π-derivative for set-valued mappings. The mapping $X : [a, b] \to \mathcal{F}(\mathbb{R})$ is called a fuzzy function and $[X(t)]^\alpha = X^\alpha(t) = [f^\alpha(t), g^{\alpha(t)}]$, $t \in [a, b]$, $0 \le \alpha \le 1$. In this context, we have in (6) that $f : [0, T] \times \mathcal{F}(\mathbb{R} \to \mathcal{F}(\mathbb{R}$ is a continuos function and $X_0 \in \mathcal{F}(\mathbb{R})$. A solution in this case is a fuzzy function $x : [0, T] \to \mathcal{F}(\mathbb{R}$ which satisfies the FIVP for each $t \in [0, T]$. Then, if $[X(t)]^\alpha = [x_1^\alpha(t), x_2^\alpha(t)], [X(t_0)]^\alpha = [x_{01}^\alpha(t), x_{02}^\alpha(t)], [f(t, X(t))]^\alpha = [f_1^\alpha(t, x_1^\alpha(t), x_2^\alpha(t)), f_2^\alpha(t, x_1^\alpha(t), x_2^\alpha(t))]$ such that $\pi([X(t)]^\alpha) = (x_1^\alpha(t), x_2^\alpha(t) - x_1^\alpha(t) = \delta^\alpha(t)), \pi([X(t_0)]^\alpha) = (x_{01}^\alpha(t), x_{02}^\alpha(t) - x_{01}^\alpha(t) = \delta_0^\alpha(t))$ and $\pi([f(t, X(t))]^\alpha) = (g_1^\alpha(t, x_1^\alpha(t), \delta^\alpha(t)), g_2^\alpha(t, x_1^\alpha(t), \delta^\alpha(t)))$. Thus (6) is equivalent to solving

$$\begin{array}{l} (x_1^\alpha)'(t) = g_1^\alpha(t, x_1^\alpha(t), \delta^\alpha(t)) \\ (x_2^\alpha)'(t) = g_2^\alpha(t, x_1^\alpha(t), \delta^\alpha(t)) \end{array} \tag{14}$$

whose solution is $[x_1^\alpha(t), x_1^\alpha(t) + \delta^\alpha(t)]$, if $\delta^\alpha(t) > 0$ and $[x_1^\alpha(t) + \delta^\alpha(t), x_1^\alpha(t)]$, if $\delta^\alpha(t) < 0$.

Fourth Approach: Actually, Gomes and Barros [16,17] studied (6) by considering that f is a function that indicates the direction of stable variable X in a time t using extended derivative operator $\widehat{D}$. This approach is not equivalent to any another considered because as $\widehat{D}$ is a fuzzy derivative, then we can to consider as Hukuhara derivative as strongly generalized derivative. Besides, provided some conditions it is possible to obtain the same solution via differential inclusion theory and Zadeh's extension of the deterministic solution. In this context, the FIVP (6) becomes

$$\begin{cases} \widehat{D}X(t) = f(t, X(t)) \\ X(0) = X_0, X_0, \widehat{D}X(t) \in \mathcal{F}(\mathbb{R}^n). \end{cases} \tag{15}$$

Gomes et al. [16] proved that if f is continuous and has an unique solution in deterministic autonomous IVP then the solution obtained via extension of deterministic solution and $\widehat{D}$-derivative is the same. Besides, in some conditions for f and X_0 the FIVP has at least two solutions.

Example 2. Considering the problem (9) for $a > 0$ in context of the strong generalized differentiability we have that (12) is the same that (10) in according with Hukuhara differentiable solution. Now, for (13) we have

$$\begin{cases} (x_1^\alpha(t))' = -ax_1^\alpha(t) \\ (x_2^\alpha(t))' = -ax_2^\alpha(t) \\ x_1^\alpha(t_0) = x_{01}^\alpha, x_2^\alpha(t_0) = x_{02}^\alpha, \end{cases} \tag{16}$$

whose solution is $x_1^\alpha(t) = x_{01}^\alpha e^{-at}, x_2^\alpha(t) = x_{02}^\alpha e^{-at}$ such that $\lim\limits_{t \to +\infty} x_1^\alpha(t) = \lim\limits_{t \to +\infty} x_2^\alpha(t) = 0$.

Now, by considering the π−derivative, the solution for FIVP (9) with the initial condition being a fuzzy number whose α−levels are given by $[x_{01}^{\alpha}, x_{01}^{\alpha}]$ is

$$
\begin{aligned}
(x_1^{\alpha})'(t) &= -ax_1^{\alpha}(t) \\
(\delta^{\alpha})'(t) &= -a\delta^{\alpha}(t)) \\
x_{10}^{\alpha}(t_0) &= x_{01}^{\alpha}, \delta^{\alpha}(t) = x_{02}^{\alpha} - x_{01}^{\alpha}
\end{aligned}
\tag{17}
$$

with solution $[X(t)]^{\alpha} = [x_{01}^{\alpha}e^{-at}, x_{01}^{\alpha}e^{-at}]$.

Finally, the solution of (10) via the interpretation of (6) considering the extended derivative operator is

$[\hat{D}X(.)]^{\alpha} = D[X(.)]^{\alpha} = \{Dx(.) : x(t) = x_0e^{-at}, x_0 \in [X_0]^{\alpha}\} = \{ax(.) : X(t) = x_0e^{-at}, x_0 \in [X_0]^{\alpha}\} = [aX(.)]^{\alpha}$.

Therefore, in all this cases the diameter of solutions are decreasing.

3.2   FIVP Without Differentiability

In this section we study FIVP without the derivative concept. We consider uncertainties in modelling(coefficient and/or initial condition) via IVP with fuzzy initial condition. The preference less or more about the trajectories are considered by the value of its membership degrees. For each $\alpha-$ level of fuzzy initial condition we have a family of trajectories with the same membership degree.

First Approach: The interpretation for (6) is considering a family of differential inclusions. Hüllermeier [18] proposal that (6), where $f : [0, T] \times \mathbb{R}^n \longrightarrow \mathcal{F}(\mathbb{R}^n)$ is a fuzzy multivalued and $X_0 \in \mathcal{F}(\mathbb{R}^n)$, can be rewritten as a family of differential inclusions

$$
\begin{cases}
x'(t) \in [f(t, x(t))]^{\alpha} \\
x(0) \in \quad\quad [X_0]^{\alpha},
\end{cases}
\tag{18}
$$

where $[f(t, x(t))]^{\alpha}$ e $[X_0]^{\alpha}$ are the α−levels of the fuzzy subsets $f(t, x(t))$ and X_0, respectively and $x : I \longrightarrow \mathbb{R}^n$. For each $\alpha \in [0, 1]$, we say that $x : [t_0, T] \longrightarrow \mathbb{R}^n$, is an α-solution of (6) if it is a solution of (5). We will denote by $\mathcal{A}_t([X_0]^{\alpha}) := \mathcal{A}_t^{\alpha}$, $t_0 \leq t \leq T$, the attainable set of the α-solutions, that is,

$$
\mathcal{A}_t^{\alpha} = \mathcal{A}_t([X_0]^{\alpha}) = \{x(t, x_0) \ / \ x(., x_0) \text{ is solution of (5) with } x_0 \in [X_0]^{\alpha}\}.
$$

According Gomes [17], the solution is a fuzzy bunch of functions $X(t) \in \mathcal{F}(AC([0, t]; \mathbb{R}^n))$ whose elements satisfy (18) a.e. in $[0, t]$. But, here we don't have the fuzzy derivative concept in equations of (6).

Diamond, in [13], uses the Representation Theorem to prove that $\mathcal{A}_t^{\alpha}$ are the α-levels of a fuzzy set $\mathcal{A}_t(X_0)$ in $\mathbb{R}^n$ for all $t_0 \leq t \leq T$. The fuzzy set $\mathcal{A}_t(X_0)$ will be said to be the attainable set of problem (6). In [14] the author studied Lyapunov stability and periodicity of the fuzzy solution set for both the time-dependent and autonomous case.

Second Approach: Oberguggenberger and Pittschmann [36] studied (6) when the coefficients and initial conditions are fuzzy subsets. The authors define the equation, restriction and solution for

$$
\begin{cases}
x'(t) = f(x) \\
x(0) = x_0, \ x_0 \in \mathbb{R}^n,
\end{cases}
\tag{19}
$$

and apply the Zadeh's extension principle in operator to obtain a solution for (6). In addition, they establish in formal way the concepts for fuzzy solution and fuzzy componentwise solution. Also, in this interpretation we don't have the fuzzy derivative concept and $f : [0,T] \times \mathbb{R}^n \to \mathcal{F}(\mathbb{R}^n)$ was obtained from a continuous function $g : [0,T] \times X \to \mathbb{R}^n$ by applying Zadeh's extension principle and $X_0 \in \mathcal{F}(\mathbb{R}^n)$.

Buckley et al. [8] in similar way to idea of Oberguggenberger and Pittschmann also obtained solutions for (6) by fuzzifying the deterministic solution using the Zadeh's extension principle.

Mizukoshi et al. [33] was prove that the solution obtained via family of differential inclusions for (6) and the solution via Zadeh's extension principle of deterministic solution is equivalent in some conditions. In the context of Mizukoshi et al., if the coefficients and/or initial condition are fuzzy subsets then $\hat{X}(t, X_0)$ is a solution for (6) by applying the Zadeh's extension principle in deterministic solution. The point of view in [32,33] was that if the solutions satisfy the concept of flow, then we can to stablish results equilibrium and stability (for more details see [32]).

Let $\varphi_t(x_0)$ be the solution (unique) of (19) for each x_0 in time t, defined on its maximal interval of existence $I(x_0)$. For each $t \in I(x_0)$, the family of mappings $\varphi_t : X \longrightarrow X$ defined by $\varphi_t(x_0) = \varphi(t, x_0)$ such that $\varphi_0 = I$, where I is the identity mapping on X and $\varphi_{t+s} = \varphi_t \circ \varphi_s, t, s \in \mathbb{R}^+$, where "$\circ$" is the composition operation.

The mapping $\hat{\varphi}_t : \mathcal{F}(X) \to \mathcal{F}(X)$ obtained by applying the Zadeh's extension principle on the initial condition in $\varphi_t : X \to X$ is a fuzzy flow for (6).

Recall that $\bar{X} \in \mathcal{F}(X)$ is a fuzzy equilibrium for the FIVP if $\hat{\varphi}_t(\bar{X}) = \bar{X}, \forall t \geq 0$ or equivalentely, $\hat{\varphi}_t(\bar{X})]^\alpha = [\bar{X}]^\alpha, \forall \alpha \in [0,1]$.

Let $\bar{X}$ be an equilibrium point of the FIVP (6). We say that,

(a) $\bar{X}$ is stable if and only if for all $\varepsilon > 0$, there is a $\delta > 0$, such that, for every $X \in \mathcal{F}(\mathbb{R}^n)$ with $D(X, \bar{X}) < \delta$, then $D(\hat{\varphi}_t(X), \bar{X}) < \varepsilon, \forall t \geq 0$.

(b) $\bar{X}$ is asymptotically stable if it is stable and, in addition, $\exists r > 0$ such that $\lim_{t \to +\infty} D(\hat{\varphi}_t(X), \bar{X}) = 0$, for all X satisfying $D(X, \bar{X}) < r$.

Example 3. Consider the fuzzy Malthusian problem

$$\begin{cases} x'(t) = ax(t) \\ x(0) = X_0, \end{cases} \tag{20}$$

where $X_0 \in \mathcal{F}(\mathbb{R}), a, x(t) \in \mathbb{R}$.

In according Hullermeier's interpretation, the FIVP (9) is the following family of differential inclusions:

$$\begin{cases} X'(t) \in [-aX(t)]^\alpha \\ X(0) \in [X_0(t)]^\alpha \end{cases} \tag{21}$$

where $X : [0,T] \times \mathbb{R} \to \mathcal{F}(\mathbb{R})$ and X_0 is a a fuzzy number.

The α-levels are given by $[X]^\alpha = [x_1^\alpha, x_2^\alpha]$ and $[X(0)]^\alpha = [x_{01}^\alpha, x_{02}^\alpha]$, in such way (21) is given as

$$\begin{cases} [X'(t)]^\alpha \in [-ax_1^\alpha, -ax_2^\alpha] \\ ([X(0)])^\alpha \in [x_{01}^\alpha, x_{02}^\alpha] \end{cases}, 0 \le \alpha \le 1. \tag{22}$$

Then, for (22) we have that the solution set is the attainable set give by level sets:

$S([X(0)]^\alpha, \tau) = \{x(.) : x(t) \in [x_{01}^\alpha e^{-at}, x_{02}^\alpha e^{-at}], 0 \le t \le \tau\}$ and
$A([X(0)]^\alpha, t) = [x_{01}^\alpha e^{-at}, x_{02}^\alpha e^{-at}]$.

Note that for $a > 0$, the diameter is decreasing.

According Mizukoshi et al. to obtain the solution for (20), firstly we consider the deterministic solution of (7), $x(t) = x_0 e^{at}$. If the initial condition is a fuzzy number whose α-levels are $[X(0)]^\alpha = [x_{01}^\alpha, x_{02}^\alpha]$, then the solution for (7) with fuzzy initial condition is given by $[X(t)]^\alpha = [X_0 e^{at}]^\alpha = [x_{01}^\alpha e^{at}, x_{02}^\alpha e^{at}]$. By definition we have that the α-levels of the flow of (20) with fuzzy initial condition are given by $[\hat{\varphi}_t(X_0)]^\alpha = \varphi_t([X_0]^\alpha) = [x_{01}^\alpha e^{-at}, x_{02}^\alpha e^{-at}]$.

Then, the fuzzy equilibrium is obtained from equality $\varphi_t(X_0) = X_0$, i. e., $\varphi_t([X_0]^\alpha) = [X_0]^\alpha$. From $[x_{01}^\alpha e^{-at}, x_{02}^\alpha e^{-at}] = [x_{01}^\alpha, x_{02}^\alpha], \forall t \ge 0$, if $x_{01}^\alpha = x_{02}^\alpha = 0$. That is $\chi_{\{0\}}$ is an equilibrium for (20). Moreover, in [32] is proved that $\chi_{\{0\}}$ is asymptotically stable.

Therefore, in this approaches the diameter of solutions are decreasing.

The FIVP was discussed only in the continuous case, but Barros et al. [2] studied a discrete fuzzy dynamical system

$$x_{n+1} = \widehat{f}(x_n), \tag{23}$$

where $\widehat{f} : \mathcal{F}(\mathbb{R})^n \to \mathcal{F}(\mathbb{R})^n$ and $x_0 \in \mathcal{F}(\mathbb{R})^n$. The authors used the semigroup properties, i.e, that the solution of (23) is a flow and obtained results about stability of fixed points using the Pompieu-Hausdorff distance. In [25] the authors studied nonautonomous difference inclusions as difference inclusion cocycles, which generalize set-valued semigroups establishing the existence of a difference cocycle attractor.

Other authors have studied (6) with periodic behavior using a different approach. Cecconello et al. [10] establish a result of stability conditions by using Zadeh's extension principle. Nieto et al. [24] studied the existence of solutions of a class of first-order linear fuzzy differential equations using generalized differentiability. Diamond [14] presented results about periodicity of the fuzzy solution set for the time-dependent and autonomous case via fuzzy differential inclusion.

4   Conclusion

The FIVP has two approaches: (1) The first one analysis the FIVP without the derivative fuzzy concept that are fuzzy differential inclusions or by applying the Zadeh's extension principle in deterministic solution; (2) The second one applies the concept of derivative for fuzzy functions, but the Hukuhara derivative lead us to solutions with increasing support; $\pi-$derivative we have solutions with decreasing diameter; gh-differentiability the uniqueness is not present and we obtain one solution with decreasing support while the others have increasing support. The solutions for the extended operator are all found in other approaches but they are not easy to obtain analytically. The FIVP for gh-differentiability and $\pi-$derivative were established only for unidimensional case. The concepts of equilibrium point and classification are also open issues for the extended operator.

References

1. Aubin, J.P.: Fuzzy differential inclusions. Probl. Control Inf. Sci. Theory **19**(1), 55–67 (1990)
2. Bassanezi, R.C., de Barros, L.C., Tonelli, P.A.: Attractors and asymptotic stability for fuzzy dynamical systems. Fuzzy Sets Syst. **113**, 473–483 (2000)
3. Barros, L.C., Bassanezi, R.C., Lodwick, A.W.: A First Course in Fuzzy Logic, Fuzzy Dynamical Systems, and Biomathematics: Theory and Applications. Studies in Fuzziness and Soft Computing. Springer, New York (2017). https://doi.org/10. 1007/978-3-662-53324-6
4. Baidosov, V.A.: Fuzzy differential inclusions. Prikl. Matem. Mekhan. **54**(1), 12–17 (1990)
5. Barros, L.C., Bassanezi, R.C., Tonelli, P.A.: On the continuity of Zadeh's extension. In: Proceedings of the 7th IFSA World Congress, Praga, pp. 3–8 (1997)
6. Bede, B., Gal, S.G.: Generalizations of the differentiability of fuzzy number valued functions with applications to fuzzy differential equation. Fuzzy Set Syst. **151**, 581–599 (2005)
7. Bhaskar, T.G., Lakshmikantham, V., Devi, V.: Revisiting fuzzy differential equations. Nonlinear Anal. **58**, 351–358 (2004)
8. Buckley, J.M., Eslami, E., Feuring, T.: Fuzzy Mathematics in Economics and Enginnering. Springer, New York (2002). https://doi.org/10.1007/978-3-7908-1795-9
9. Chalco-Cano, Y., Román-Flores, H., Jiménez-Gamero, M.D.: Fuzzy differential equations with $\pi-$derivative. In: Proceedings of the Joint 2009 International Fuzzy Systems Association World Congress and 2009 European Society of Fuzzy Logic and Technology Conference, Lisbon, pp. 20–24 (2009)
10. Cecconello, M.S., Bassanezi, R.C., Brandão, A.V., Leite, J.: Periodic orbits for fuzzy flows. Fuzzy Sets Syst. **230**, 21–38 (2013)
11. Chen, M., Li, D., Xue, X.: Periodic problems of first order uncertain dynamical systems. Fuzzy Sets Syst. **162**, 67–78 (2011)
12. Corliss, G.F.: Survey of interval algorithms for ordinary differential equations. Appl. Math. Comput. **31**, 112–120 (1989)
13. Diamond, P.: Time-dependent differential inclusions, cocycle attractors and fuzzy differential equations. IEEE Trans. Fuzzy Syst. **7**, 734–740 (1999)
14. Diamond, P.: Stability and periodicity in fuzzy differential equations. IEEE Trans. Fuzzy Syst. **8**, 583–590 (2000)
15. Diamond, P.: Brief note on the variation of constants formula for fuzzy differential equations. Fuzzy Sets Syst. **129**, 65–71 (2002)

16. Gomes, L.T., de Barros, L.C., Bede, B.: Fuzzy differential equations. Fuzzy Differential Equations in Various Approaches. SM, pp. 69–113. Springer, Cham (2015). https://doi.org/10.1007/978-3-319-22575-3_4

17. Gomes, L.T.: On fuzzy differential equations. Ph.D. thesis, IMECC, UNICAMP (2014)

18. Hullermeier, E.: An approach to modeling and simulation of uncertain dynamical systems. Int. J. Uncertain. Fuzziness Knowl.-Bases Syst. 5, 117–137 (1997)

19. de Silva, J.D.M., Leite, J., Bassanezi, R.C., Cecconello, M.S.: Stationary points-I: one-dimensional p-fuzzy dynamical systems. J. Appl. Math. 2013, 1–11 (2013)

20. Jafelice, R.S.M., de Barros, L.C., Bassanezi, R.C., Gomide, F.: Methodology to determine the evolution of asymptomatic HIV population using fuzzy set theory. Int. J. Uncertain. Fuzziness Knowl.-Based Syst. 13(1), 39–58 (2005)

21. Kaleva, O.: Fuzzy differential equations. Fuzzy Sets Syst. 24, 301–317 (1987)

22. Kandel, A., Byatt, W.J.: Fuzzy differential equations. In: Proceedings of the International Conference on Cybernetics and Society, Tokyo, pp. 1213–1216 (1978)

23. Kandel, A., Byatt, W.J.: Fuzzy processes. Fuzzy Sets Syst. 4(2), 117–152 (1980)

24. Khastan, A., Nieto, J.J., Rodríguez-López, R.: Periodic boundary value problems for the first-order linear differential equations with uncertainty under generalized differentiability. Inf. Sci. 222, 544–558 (2013)

25. Kloeden, P.E., Schmalfuss, B.: Asymptotic behaviour of nonautonomous difference inclusions. Syst. Control Lett. 33, 275–280 (1998)

26. Krivan, V., Colombo, G.: A non-stochastic approach for modelling uncertainty in population dinamics. Bull. Math. Biol. 60, 721–751 (1998)

27. Kupka, J.: On devaney chaotic induced fuzzy and set-valued dynamical systems. Fuzzy Sets Syst. 177, 34–44 (2011)

28. Lakshmikantham, V., Vatsala, A.S.: Basic theory of fuzzy difference equations. J. Differ. Equ. Appl. 8, 957–968 (2002)

29. Lodwick, W.A.: Constrained interval arithmetic. Technical report, University Of Colorado at Denver, Denver (1999)

30. Lodwick, W.A., Dubois, D.: Interval linear systems as a necessary step in fuzzy linear systems. Fuzzy Sets Syst. 281, 227–251 (2015)

31. Marchaud, A.: Sur les champs de demicônes et les équations differentielles du premier ordre. Bull. Soc. Math. France 62, 1–38 (1934)

32. Mizukoshi, M.T., de Barros, L.C., Bassanezi, R.C.: Stability of fuzzy dynamics systems. Int. J. Uncertain. Fuzziness Knowl.-Based Syst. 17, 69–83 (2009)

33. Mizukoshi, M.T., Barros, L.C., Chalco-Cano, Y., Roman-Flores, H., Bassanezi, R.C.: Fuzzy differential equation and the extension principle. Inf. Sci. 177, 3627–3635 (2007)

34. Moore, R.E.: Automatic error analysis in digital computation, Technical Report LMSD-48421, Lockheed Missile and Space Division, Sunnyvale, CA (1959). http://interval.lousiana.edu/Moores_early_papers/bibliography.html

35. Nieto, J.J., Rodríguez-López, R.: Bounded solutions for fuzzy differential and integral equations. Chaos, Solitons Fractals 27, 1376–1386 (2006)

36. Oberguggenberger, M., Pittschmann, S.: Differential equations with fuzzy parameters. Math. Comput. Model. Dyn. Syst. 5, 181–202 (1999)

37. Ouyang, H., Wu, Y.: On fuzzy differential equations. Fuzzy Sets Syst. 32, 321–325 (1989)

38. Peixoto, M.S., Barros, L.C., Bassanezi, R.C.: Predator-prey fuzzy model. Ecol. Model. 214, 39–44 (2008)

39. Puri, M.L., Ralescu, D.A.: Differentials for fuzzy functions. J. Math. Anal. Appl. 91, 552–558 (1983)

40. Puri, M.L., Ralescu, D.A.: Fuzzy random variables. J. Math. Anal. Appl. **114**, 409–422 (1986)
41. Rihm, R.: Interval methods for initial value problems in ODEs. In: Topics in Validated Computations, pp. 173–206 (1994)
42. Román-Flores, H., Barros, L.C., Bassanezi, R.C.: A note on Zadeh's extension. Fuzzy Sets Syst. **117**, 327–331 (2001)
43. Román-Flores, H.: A note on transitivity in set-valued discrete systems. Chaos, Solitons Fractals **17**(1), 99–104 (2003)
44. Rouvray, D.H.: The treatment of uncertainty in the sciences. Endevour **21**(4), 154–158 (1997)
45. Seikkala, S.: On the fuzzy initial value problem. Fuzzy Sets Syst. **24**, 309–330 (1987)
46. Zadeh, L.A.: Fuzzy sets. Inf. Control **8**, 338–353 (1965)
47. Zaremba, S.K.: Sur les équations au paratingent. Bull. Sci. Math. **60**, 139–160 (1936)
48. Wasewski, T.: Sur une genéralisation de la notion des solutions d'une équation au contingent. Bull. Acad. Pol. Sci. Ser. Math. Astronom. Phys. **10**(1), 1–15 (1962)

Zadeh's Extension of the Solution of the Euler-Lagrange Equation for a Quadratic Functional Type

Jônathas D. S. Oliveira[1(✉)], Luciana T. Gomes[2], and Rodney C. Bassanezi[3]

[1] IMECC, Universidade Estadual de Campinas, Campinas, SP, Brazil
jonathas.math.oliveira@gmail.com
[2] Departamento de Física Química e Matemática,
Universidade Federal de São Carlos, Sorocaba, SP, Brazil
lucianagomes.math@gmail.com
[3] IMECC, Universidade Estadual de Campinas, Campinas, SP, Brazil
rodney@ime.unicamp.br

Abstract. This article aims to provide some necessary conditions so that the solution of the Euler-Lagrange equation, arising from the necessary conditions of a variational calculation problem is increasing with respect to the initial condition. This result is important for the study of the variational calculus problems with fuzzy initial condition and arbitrary final condition, since it establishes conditions to apply Zadeh's extension with respect to the initial condition for a quadratic functional type.

Keywords: Zadeh's extension · Variational calculus
Euler-Lagrange equation

1 Introduction

Variational problems are a type of optimization problem that are characterized by being defined in spaces of functions (infinite dimension), and whose functional objective is represented by an integral operator. Formal study and how to obtain necessary and sufficient conditions for this type of problem is known as variational calculus.

The problem of variational calculus is an optimization problem. Bellman and Zadeh [2] studied optimization in fuzzy context where the fuzzy set of feasible solutions to structure decision making strategies in fuzzy environments.

The fuzzy set theory was introduced by Zadeh [15] and initially was applied for control theory. However, it was diffusing for other areas and consequently for calculus of variation. In [7], for example, the author presents a version for the fuzzy variational problem with optimality conditions established via Euler-Lagrange equations and by using the concept of differentiability and integrability of fuzzy functions that can be parameterized by left and right functions of their level sets.

© Springer International Publishing AG, part of Springer Nature 2018
G. A. Barreto and R. Coelho (Eds.): NAFIPS 2018, CCIS 831, pp. 477–488, 2018.
https://doi.org/10.1007/978-3-319-95312-0_41

Differently from the approach in [7] we are interested in studying the extremes of functional for variational problem using the same point of view of Diniz [5]. Diniz proved that under some conditions the minimum of a variational calculation problem with initial fuzzy condition and fixed final condition is the Zadeh's extension of the classical solution in relation to the initial condition with a partial order relation. Diniz demonstrated that for the fuzzy variational calculus problem with initial fuzzy condition and fixed boundary conditions it is always possible to apply Zadeh's extension to solve it provided that the solutions of the Euler-Lagrange equations never intersect.

We are interested in verifying that with the conditions established in Diniz and some additional hypotheses the same result is valid for variational problems with fuzzy initial condition and arbitrary final condition. However, the hypothesis of increasing monotonicity of the solution with respect to the initial condition is not always true for this type of problem. Then, we added conditions to enable the application of Zadeh's extension and to obtain the solution of variational problem.

We used a type of quadratic functional to obtain a candidate for minimum that is monotonic decreasing in relation to the initial condition. The choice of function seems to be the best in our construction because the minimum of this functional is the solution of second order differential equation. The study can also be done with linear functional and other types of quadratic functional.

2   Preliminaries

In this section we present basic concepts and results, referring to fuzzy set theory and variational calculus necessary to obtain the proposed Theorem 2.

2.1   Variational Calculus

The definitions and results presented regarding variational calculus were extracted from the references [6,8,11,12,14].

The simplest form of a variational problem is finding a continuously differentiable function $x : [a, b] \rightarrow \mathbb{R}$ that minimizes

$$J(x(t)) := \int_a^b F(t, x(t), \dot{x}(t))dt, \tag{1}$$

where $[a, b]$ is a fixed interval and $F : [a, b] \times \mathbb{R} \times \mathbb{R} \mapsto \mathbb{R}$ is a given function.

Moreover, the functional minimization is often subject to some restrictions. In our case, we will work with problems with given initial condition and arbitrary final condition, that is, $x(a) = x_a$ and $x(b)$ is arbitrary.

We say that $x^*(t)$ is an extreme of the functional if it is a maximum or a minimum of (1).

In order to try to find a solution to problem (1), Theorem 1 establishes the necessary conditions for existence of an extreme. These conditions consist of the resolution of the Euler-Lagrange equation, together with the initial condition and the transversality condition according to Theorem 1.

Theorem 1. *Let J be a functional of the form of Eq. (1) where $x(a) = x_a$ and $x(b)$ is arbitrary and F has continuous first and second partial derivatives with respect to x, $\dot{x}$ and t. If J has an extremum at x^*, then x^* satisfies the **Euler-Lagrange equation**:*

$$\frac{\partial F}{\partial x}(t, x^*(t), \dot{x}^*(t)) - \frac{d}{dt}\left(\frac{\partial F}{\partial \dot{x}}(t, x^*t), \dot{x}^*(t))\right) = 0, \tag{2}$$

with the initial condition $x(a) = x_a$ and the transversality condition

$$\frac{\partial F}{\partial \dot{x}}(b, x(b), \dot{x}^*(b)) = 0. \tag{3}$$

Diniz [5] shows that the solution of the variational calculus problem with fixed final condition is monotonous with respect to the initial condition, which means that when the classical solution is extended via Zadeh's extension in relation to initial condition the trajectories never intersect and it is possible to apply Zadeh's extension principle. The following example shows that the result is not assured when the final condition is arbitrary.

Example 1. Consider the following variational calculus problem

$$\min \int_0^1 \left[x(t)^2 - 3x(t)\dot{x}(t) + \dot{x}(t)^2\right] dt,$$

where $x(0) = x_0$ and $x(1)$ is arbitrary.

We will use the Euler-Lagrange equation to pursue a candidate to the minimum of the problem and show that the solution is not monotone with respect to the initial condition, that is, for two different values of x_0 in $[x_{0i}, x_{0f}]$, trajectories intersect.

So,

$$\frac{\partial F}{\partial x}(t, x^*(t), \dot{x}^*(t)) - \frac{d}{dt}\left[\frac{\partial F}{\partial \dot{x}}(t, x^*(t), \dot{x}^*(t))\right] = 0.$$

$$\ddot{x}^*(t) - x^*(t) = 0$$

Therefore, a necessary condition for $x^*(t)$ to be the solution of the variational problem is that it satisfies the second-order differential equation $\ddot{x}^*(t) - x(t) = 0$ [4] , whose general solution is given by $x^*(t) = c_1 e^t + c_2 e^{-t}$, where c_1 and c_2 are constants. To find the constants we use the initial condition $x(0) = x_0$ and the transversality conditions

$$\frac{\partial F}{\partial \dot{x}}(t, x^*(1), \dot{x}^*(1)) = 0$$

$$2\dot{x}^*(1) - 3x^*(1) = 0$$

$$-ec_1 + 5e^{-1}c_2 = 0.$$

Therefore, c_1 and c_2 must satisfy the system

$$\begin{cases} c_1 + c_2 = x_0 \\ -ec_1 + 5e^{-1}c_2 = 0, \end{cases} \tag{4}$$

whose solution is $c_1 = \frac{5e^{-1}x_0}{5e^{-1}-e}$ and $c_2 = x_0 - \frac{5e^{-1}x_0}{5e^{-1}-e}$ Therefore, the solution, dependent on the initial condition, is given by

$$x^*(t) = \left(\frac{5e^{-1}x_0}{5e^{-1}-e}\right)e^t + \left(x_0 - \frac{5e^{-1}x_0}{5e^{-1}-e}\right)e^{-t}.$$

The Fig. 1 illustrates the solution graph for $x_0 = 1$ and $x_0 = 2$ and shows that the solution is not monotonic with respect to the initial condition.

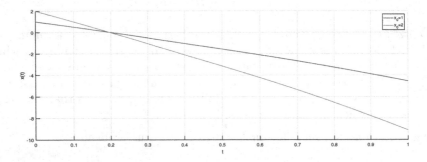

Fig. 1. Solution of the Euler-Lagrange equation for $x_0 = 1$ and $x_0 = 2$

2.2 Fuzzy Sets

The concepts and results described here on fuzzy set theory, as well as more details, can be found in [3,10,13,15].

Definition 1 (Fuzzy set). *Let U be a (classic universal) set. A fuzzy subset $\widehat{u}$ of U is defined by a function $\varphi_{\widehat{u}}$, called the **membership** function (of $\widehat{u}$)*

$$\varphi_{\widehat{u}} : U \to [0, 1].$$

The value of $\varphi_{\widehat{u}}(x) \in [0,1]$ indicates the degree to which the element x of U belongs to the fuzzy set $\widehat{u}$.

Definition 2 (support). *The classic subset of U*

$$supp\,\widehat{u} = \{x \in U : \varphi_{\widehat{u}}(x) > 0\}.$$

*is called the **support** of $\widehat{u}$.*

The **support** of $\widehat{u}$ and has a fundamental role in the interrelation between classical and fuzzy set theory. Interestingly, unlike fuzzy subsets, a support is a crisp set.

Definition 3 (α-cut). *Let $\widehat{u}$ be a fuzzy subset of U and $\alpha \in [0,1]$. The α-cut of the subset $\widehat{u}$ is the set $[\widehat{u}]^\alpha$ of U defined by*

$$[\widehat{u}]^\alpha = \{x \in U : \varphi_{\widehat{u}}(x) \geq \alpha\}, 0 < \alpha \leq 1$$

and $[\widehat{u}]^0 = \widehat{supp\,\widehat{u}}$.

Definition 4 (Zadeh's Extension Principle). *Let f be a function such that $f : X \mapsto Z$ and let $\widehat{u}$ be a fuzzy subset of X. Zadeh's extension of f is the function $\widehat{f}$ which applied to $\widehat{u}$ gives us the fuzzy subset $\widehat{f}(\widehat{u})$ of Z with the membership function given by*

$$\varphi_{\widehat{f}(\widehat{u})}(z) = \begin{cases} \sup_{f^{-1}(z)} \varphi_{\widehat{u}}(x) & \text{if } f^{-1}(z) \neq \emptyset \\ 0 & \text{if } f^{-1}(z) = \emptyset, \end{cases}$$

where $f^{-1}(z) = \{x; f(x) = z\}$ is the preimage of z.

Note that if f is bijective, then the membership function of $f(\widehat{u})$ is given as follows

$$\varphi_{\widehat{f}(\widehat{u})}(z) = \sup_{\{x:f(x)=z\}} \varphi_{\widehat{u}}(x) = \varphi_{\widehat{u}}(f^{-1}(z))$$

where f^{-1} is the inverse function of f.

Definition 5 (Fuzzy Number). *A fuzzy subset $\widehat{u}$ is called a **fuzzy number** when the universal set on which $\varphi_{\widehat{u}}$ is defined is the set of all real numbers $\mathbb{R}$ and satisfies the following conditions:*

(i) *all the α-levels of $\widehat{u}$ are not empty for $0 \leq \alpha \leq 1$;*
(ii) *all the α-levels of $\widehat{u}$ are closed intervals of $\mathbb{R}$;*
(iii) *$supp\,\widehat{u} = \{x \in \mathbb{R} : \varphi_{\widehat{u}}(x) > 0\}$ is bounded.*

Definition 6 (Triangular Fuzzy Number). *A fuzzy number $\widehat{u}$ is said to be **triangular** if its membership function is given by*

$$\varphi_{\widehat{u}}(x) = \begin{cases} 0 & \text{if } x \leq a \\ \frac{x-a}{p-a} & \text{if } a < x \leq p \\ \frac{x-b}{p-b} & \text{if } p < x \leq b \\ 0 & \text{if } x \geq b, \end{cases}$$

where a, p, b are given numbers. Let us denote a triangular fuzzy number by $(a; p; b)$.

Definition 7. *Let $\widehat{v}$ and $\widehat{u}$ fuzzy numbers. We say that $[\widehat{u}]^\alpha \preceq [\widehat{v}]^\alpha$ ($[\widehat{u}]^\alpha \prec [\widehat{v}]^\alpha$), if $\widehat{u}_L^\alpha \leq \widehat{u}_L^\alpha$ ($\widehat{u}_L^\alpha < \widehat{u}_L^\alpha$) and $\widehat{u}_R^\alpha \leq \widehat{u}_R^\alpha$ ($\widehat{u}_R^\alpha < \widehat{u}_R^\alpha$), $\forall \alpha \in [0,1]$, where $[\widehat{u}]^\alpha = [\widehat{u}_L^\alpha, \widehat{u}_R^\alpha]$ and $[\widehat{v}]^\alpha = [\widehat{v}_L^\alpha, \widehat{v}_R^\alpha]$.*

Definition 8 establishes a relation of partial order between fuzzy numbers, and is this relation of order that we are using to optmize a fuzzy functional.

Definition 8. *Let $\widehat{v}$ and $\widehat{u}$ fuzzy numbers. We say that $\widehat{u} \preceq \widehat{v}$ ($\widehat{u} \prec \widehat{v}$), if only if, respectively, $[\widehat{u}]^\alpha \preceq [\widehat{v}]^\alpha$ ($[\widehat{u}]^\alpha \prec [\widehat{v}]^\alpha$), $\forall \alpha \in [0,1]$.*

3   Results

Consider the following variational calculus problem

$$\min_{x \in C^1([0,t_f])} J(x) = \int_0^{t_f} \left[ax(t)^2 + bx(t)\dot{x}(t) + c\dot{x}(t)^2 \right]$$

(5)

$$x(0) = x_0 \text{ and } x(t_f) \text{ is arbitrary.}$$

In this section we derive conditions on the candidate of the solution of the problem (5) so that it is increasing in x_0

Applying the Euler-Lagrange equation for this problem we obtain

$$\frac{\partial F}{\partial x}(t, x^*(t), \dot{x}^*(t)) - \frac{d}{dt}\left[\frac{\partial F}{\partial \dot{x}}(t, x^*(t), \dot{x}^*(t)) \right] = 0$$

$$2ax^*(t) + b\dot{x}^*(t) - \frac{d}{dt}[bx^*(t) + 2c\dot{x}^*(t)] = 0$$

$$c\ddot{x}^*(t) - ax^*(t) = 0.$$

Then a necessary condition for $x^*(t)$ to be a solution of the variational problem (5) is that the function is solution of differential equation $c\ddot{x}^*(t) - ax^*(t) = 0$, with initial condition $x(0) = x_0$ and boundary condition

$$\frac{\partial F}{\partial \dot{x}}(t, x^*(t_f), \dot{x}^*(t_f)) = 0.$$

(6)

The differential equation can be solved analytically since it is a linear second-order differential equation with constant coefficients. To solve it we will divide in four cases.

1st case: a and c have the same sign. In this case the characteristic equation of the differential equation $c\ddot{x}^*(t) - ax^*(t) = 0$ has two distinct real roots and the general solution is given by

$$x^*(t) = c_1 e^{\sqrt{\frac{a}{c}}t} + c_2 e^{-\sqrt{\frac{a}{c}}t}.$$

(7)

By using the initial condition $x(0) = x_0$ and the Eq. (6), we have

$$c_1 = x_0 - \frac{k_1 x_0}{k_1 - k_2} \text{ and } c_2 = \frac{k_1 x_0}{k_1 - k_2}$$

Therefore the solution of Euler-Lagrange equation, dependent on the initial solution is given by

$$x^*(t, x_0) = \left(x_0 - \frac{k_1 x_0}{k_1 - k_2} \right) e^{\sqrt{\frac{a}{c}}t} + \left(\frac{k_1 x_0}{k_1 - k_2} \right) e^{-\sqrt{\frac{a}{c}}t}.$$

(8)

where $k_1 = be^{\sqrt{\frac{a}{c}}t_f} + 2\sqrt{ace}^{\sqrt{\frac{a}{c}}t_f}$ and $k_2 = be^{-\sqrt{\frac{a}{c}}t_f} - 2\sqrt{ace}^{-\sqrt{\frac{a}{c}}t_f}$. Now, we will establish conditions for (8) to be monotone increasing in relation to x_0. Calculating the partial derivative with respect to x_0

$$\frac{\partial x^*}{\partial x_0}(t, x_0) = \left(1 - \frac{k_1}{k_1 - k_2} \right) e^{\sqrt{\frac{a}{c}}t} + \left(\frac{k_1}{k_1 - k_2} \right) e^{-\sqrt{\frac{a}{c}}t},$$

as $e^{\sqrt{\frac{a}{c}}t}$ and $e^{-\sqrt{\frac{a}{c}}t}$ are always positive $\forall\ t \in [0, t_f]$, we will need to analyze the signal of $\left(1 - \frac{k_1}{k_1 - k_2}\right)$ and $\frac{k_1}{k_1 - k_2}$, dividing in two cases.

– If $b = 0$ in (8, so

$$k_1 = 2\sqrt{ace}^{\sqrt{\frac{a}{c}}t_f}\ \text{e}\ k_2 = -2\sqrt{ace}^{-\sqrt{\frac{a}{c}}t_f}.$$

Therefore

$$\frac{k_1}{k_1 - k_2} = \frac{2\sqrt{ace}^{\sqrt{\frac{a}{c}}t_f}}{2\sqrt{ace}^{\sqrt{\frac{a}{c}}t_f} + 2\sqrt{ace}^{-\sqrt{\frac{a}{c}}t_f}} \Rightarrow 0 < \frac{k_1}{k_1 - k_2} \leq 1$$

so $\frac{\partial x^*}{\partial x_0}(t, x_0)$ is always positive. Thus (t, x_0) is always increasing with respect to the condition initial

– If $b \neq 0$, so

$$\frac{k_1}{k_1 - k_2} = \frac{(b + 2\sqrt{ac})\,e^{\sqrt{\frac{a}{c}}t_f}}{(b + 2\sqrt{ac})\,e^{\sqrt{\frac{a}{c}}t_f} + (2\sqrt{ac} - b)\,e^{-\sqrt{\frac{a}{c}}t_f}}.$$

Then, a sufficient condition for $0 < \frac{k_1}{k_1 - k_2} \leq 1$ is that $2\sqrt{ac} - b > 0 \Rightarrow b < 2\sqrt{ac}$ and $b + 2\sqrt{ac} > 0 \Rightarrow b > -2\sqrt{ac}$, this is, $b < |2\sqrt{ac}|$ and, In this case, $\frac{\partial x^*}{\partial x_0}(t, x_0)$ is always positive and $x^*(t, x_0)$ is always increasing with respect to the initial condition. Note that if $b > 0$ then $b + 2\sqrt{ac} > 0$ and $x^*(t, x_0)$ is always increasing with respect to the initial condition if $b < 2\sqrt{ac}$.

2nd case: $a = 0$ and $c \neq 0$. If $a = 0$ and $c \neq 0$, the Euler-Lagrange equation is reduced to $c\ddot{x}(t) = 0$, Here the general solution is given by

$$x(t) = c_1 + c_2 t. \tag{9}$$

By using the initial and the boundary condition, we have

$$c_1 = x_0\ \text{e}\ c_2 = \frac{-bx_0}{bt_f + 2c}$$

Then the solution of the Euler-Lagrange equation is

$$x^*(t, x_0) = x_0 + \left(\frac{-bx_0}{bt_f + 2c}\right) t$$

In this case, $\frac{\partial x^*}{\partial x_0}(t, x_0) = 1 - \frac{bt}{bt_f + 2c}$, and a condition sufficient for $\frac{\partial x^*}{\partial x_0}(t x_0) \geq$ and if $0 \leq \frac{bt}{bt_f + 2c} < 1$, for all t, and thus $x^*(t, x_0)$ will be monotonous with respect to the initial condition.

We analyze under what conditions $0 \leq \frac{bt}{bt_f + 2c} < 1$. Note that if b and c has the same sign then the inequality is always satisfied because for $t_f > t$, $\forall\ t \in [0, t_f)$, we have

$$1 = \frac{bt_f}{bt_f} > \frac{bt}{bt_f} > \frac{bt}{bt_f + 2c} \geq 0.$$

3rd case: $a \neq 0$ and $c = 0$. In this case, the Euler-Lagrange equation provides that the optimal solution $x^*(t)$ must satisfy $ax^*(t) = 0$, which implies that $x^{(}t) = 0$. If $x_0 = 0$, then the null function is the unique solution of the variational problem, otherwise the variational problem has no solution. In both situations we have nothing to analyze.

4th case: $a \cdot c < 0$. If $a \cdot c < 0$, the characteristic equation associated to $c\ddot{x}^*(t) - ax^*(t) = 0$ has two complex conjugate roots, namely $r_1 = \sqrt{-\frac{a}{c}}i$ and $r_2 = -\sqrt{-\frac{a}{c}}i$. Then the general solution is given by

$$x^*(t) = c_1 \cos\left(\sqrt{-\frac{a}{c}}t\right) + c_2 \sin\left(\sqrt{-\frac{a}{c}}t\right). \tag{10}$$

By using the initial $x(0) = x_0$ and the boundary condition we obtain

$$c_1 = x_0 \text{ and } c_2 = \frac{x_0\left[2c\sqrt{-\frac{a}{c}}\sin\left(\sqrt{-\frac{a}{c}}t_f\right) - b\cos\left(\sqrt{-\frac{a}{c}}t_f\right)\right]}{b\sin\left(\sqrt{-\frac{a}{c}}t_f\right) + 2c\sqrt{-\frac{a}{c}}\cos\left(\sqrt{-\frac{a}{c}}t_f\right)}.$$

To analysis the general case of c_2 signal is hard then we study when $b = 0$. If $b = 0$, then

$$c_2 = \frac{x_0 \sin\left(\sqrt{-\frac{a}{c}}t_f\right)}{\cos\left(\sqrt{-\frac{a}{c}}t_f\right)} = x_0 \tan\left(\sqrt{-\frac{a}{c}}t_f\right).$$

Therefore, the general solution is given by

$$x^*(t, x_0) = x_0 \cos\left(\sqrt{-\frac{a}{c}}t\right) + x_0 \tan\left(\sqrt{-\frac{a}{c}}t_f\right) \sin\left(\sqrt{-\frac{a}{c}}t\right). \tag{11}$$

Now, we establish conditions for $x^*(t, x_0)$ to be monotonically increasing with respect to the initial condition

$$\frac{\partial x^*}{\partial x_0}(t, x_0) = \cos\left(\sqrt{-\frac{a}{c}}t\right) + \tan\left(\sqrt{-\frac{a}{c}}t_f\right) \sin\left(\sqrt{-\frac{a}{c}}t\right).$$

A sufficient condition for $\frac{\partial x^*}{\partial x_0}(t, x_0) > 0 \ \forall \ t \in [0, t_f]$ is that

$$0 \leq t < \sqrt{-\frac{c}{a}\frac{\pi}{2}},$$

and in this case $x^*(t, x_0)$ is monotonically increasing with respect to the initial condition x_0.

The following Theorem provides conditions to obtain a monotonic increasing solution for (5).

Theorem 2. *Consider the following variational problem*

$$\min_{x \in C^1([0,t_f])} J(x) = \int_0^{t_f} \left[ax(t)^2 + bx(t)\dot{x}(t) + c\dot{x}(t)^2\right] dt \tag{12}$$

$$x(0) = x_0 \quad x(t_f) = arbitrary.$$

(i) If $ac > 0$ and $b = 0$ or

(ii) If $ac > 0$, $b \neq 0$ and $b < |2\sqrt{ac}|$ or

(iii) If $a = 0$, $c \neq 0$ and $0 \leq \frac{bt}{bt_f + 2c} < 1$, $\forall\, t \in [0, t_f]$ or

(iv) If a and c are non-zero, have different signs, $b = 0$ and $0 \leq t < \sqrt{-\frac{c}{a}\frac{\pi}{2}}$ $\forall$ $t \in [0, t_f]$

Then, if it exists, the extremum $x^*(t, x_0)$ of (12) is monotonically increasing with respect to the initial condition.

3.1   Example

Example 2. In this example, we will solve the Euler-Lagrange equation of the following minimization problem

$$\min_{x \in C^1([0,2])} \widehat{J}(x) = \int_0^2 \left[x(t)^2 + \dot{x}(t)^2 \right] dt$$

$$x(0) = \widehat{x_0} \quad x(t_f) = arbitrary. \tag{13}$$

where $\widehat{x_0}$ is the triangular fuzzy number $(1; 1.5; 2)$ and the fuzzy functional is obtained by applying the Zadeh's extension in composition of the following functions:

$$x_0 \xrightarrow{\phi_1} x_{x_0}(t) \xrightarrow{\phi_2} F(t, x, \dot{x}) \xrightarrow{\phi_3} \int_{t_0}^{t_f} F(t, x, \dot{x})dt = J(x), \tag{14}$$

where

$$\phi_1 : [x_0^L, x_0^R] \subset \mathbb{R} \mapsto C^1([a, b], \mathbb{R}), \tag{15}$$

$$\phi_2 : C^1([a, b], \mathbb{R}) \mapsto C([a, b], \mathbb{R}), \tag{16}$$

$$\phi_3 : C([a, b], \mathbb{R}) \mapsto \mathbb{R}, \tag{17}$$

Thus by applying the Zadeh's extension in each of the functions in (14), we obtain:

$$\widehat{x_0} \xrightarrow{\widehat{\phi_1}} \widehat{x}_{\widehat{x_0}}(t) \xrightarrow{\widehat{\phi_2}} \widehat{F}(t, \widehat{x}, \widehat{\dot{x}}) \xrightarrow{\widehat{\phi_3}} \widehat{\int_{t_0}^{t_f} F(t, \widehat{x}, \widehat{\dot{x}})dt} = \widehat{J}(\widehat{x_0}). \tag{18}$$

By the using the Nguyen's Theorem ([1, 3, 9]) and the continuity of the functional with respect to the initial condition we can show that $\widehat{J}(\widehat{x_0})$ is a fuzzy number.

First, note that $J(x)$ given in (13) is a quadratic functional with $a = c = 1$ and $b = 0$ in (5). Therefore, by item i) of Theorem 2, the solution to Problem (13), if there is, is monotonically increasing with respect to the initial condition.

Therefore, the necessary condition of optimality for the classical problem, is the solution of the Euler-Lagrange equation

$$F_x(t, x(t), \dot{x}(t)) - \frac{d}{dt}(F_{\dot{x}}(t, x(t), \dot{x}(t)) = 0 \tag{19}$$

$$2x(t) - \frac{d}{dt}(2\dot{x}(t)) = 0 \tag{20}$$

$$x(t) - \ddot{x}(t) = 0. \tag{21}$$

Thus a necessary condition for $x(t)$ to be a solution to Problem (13) is that this function is the solution of the second-order differential equation $x(t) - \ddot{x}(t) = 0$, this is, $x^* = c_1 e^t + c_2 e^{-t}$. To find the values of constants c_1 and c_2, we use the initial condition $x(0) = x_0$ in

$$c_1 + c_2 = x_0,$$

together with the boundary conditions

$$F_{\dot{x}}(t_f, x(t_f), \dot{x}(t_f)) = 0$$
$$c_1 e^2 - c_2 e^{-2} = 0.$$

Then $c_1 = \frac{x_0}{1+e^4}$ and $c_2 = \frac{x_0 e^4}{1+e^4}$ and follows that the extremum of the classical problem dependent on the initial condition is given by

$$x^*(t) = \frac{x_0}{1+e^4}\left(e^t + e^{4-t}\right). \tag{22}$$

Figures 2 and 3 illustrate, respectively, the Zadeh's extension of solution (2), considering x_0 the triangular fuzzy number $\widehat{x}_0 = (1; 1.5; 2)$, $\widehat{J}(\widehat{x}_0)$ and evolution of the value of the functional with respect to $x_0 \in [1, 2]$.

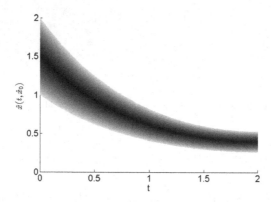

Fig. 2. Zadeh's extension of solution (2) with $\widehat{x}_0 = (1; 1.5; 2)$.

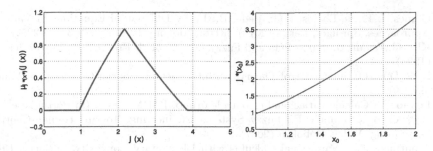

Fig. 3. $\widehat{J}(\widetilde{x_0})$ and evolution of the value of the functional with respect to x_0.

In [5] it is shown that the Zadeh's extension, with respect to initial condition is the minimum solution for the variational calculus problem with fixed final condition using a partial order relation. Using this same relation of order, the same result is valid when the final condition is free.

4   Conclusions

This paper establishes conditions for the solution of the Euler-Lagrange equation, of a certain type of quadratic functional, be increasing with respect to the initial condition. In this case, the trajectories do not intersect, allowing the application of Zadeh's extension. Theorem 2 shows that these conditions are obtained via the parameters of this quadratic functional.

References

1. Bede, B.: Mathematics of Fuzzy Sets and Fuzzy Logic. Springer, Heidelberg (2013). https://doi.org/10.1007/978-3-642-35221-8
2. Bellman, R.E., Zadeh, L.A.: Decision-making in a fuzzy environment. Manag. Sci. **17**(4), B-141 (1970)
3. De Barros, L.C., Bassanezi, R., Lodwick, W.: A First Course in Fuzzy Logic, Fuzzy Dynamical Systems, and Biomathematics. Springer, Berlin (2016). https://doi.org/10.1007/978-3-662-53324-6
4. Boyce, W.E., DiPrima, R.C., Haines, C.W.: Elementary Differential Equations and Boundary Value Problems, vol. 9. Wiley, New York (1969)
5. Diniz, M.M.: Otimização de funções, funcionais e controle fuzzy. Tese de doutorado, Instituto de Matemática, Estatística e Computação Científica (IMECC), Universidade Estadual de Campinas (UNICAMP) (2016). (in Portuguese)
6. Elsgolts, L.E., Yankovsky, G.: Differential Equations and the Calculus of Variations. Mir, Moscow (1977)
7. Farhadinia, B.: Necessary optimality conditions for fuzzy variational problems. Inf. Sci. **181**(7), 1348–1357 (2011)
8. Gelfand, I.M., Fomin, S.V.: Calculus of variations. Courier Corporation (1963). Revised English edition translated and edited by Richard A. Silverman

9. Gomes, L.T., de Barros, L.C., Bede, B.: Fuzzy Differential Equations in Various Approaches. Springer, Berlin (2015). https://doi.org/10.1007/978-3-319-22575-3
10. Klir, G., Yuan, B.: Fuzzy Sets and Fuzzy Logic, vol. 4. Prentice Hall, New Jersey (1995)
11. Kirk, D.E.: Optimal Control Theory: An Introduction. Courier Corporation, Mineola (2012)
12. Leitao, A.: Cálculo variacional e controle ótimo. IMPA (2001). (in Portuguese)
13. Pedrycz, W., Gomide, F.: Fuzzy Systems Engineering: Toward Human-Centric Computing. Wiley, Hoboken (2007)
14. Troutman, J.L.: Variational Calculus with Elementary Convexity. Springer, Heidelberg (2012). https://doi.org/10.1007/978-1-4684-0158-5
15. Zadeh, L.: Fuzzy sets. Inf. Control **8**(3), 338–353 (1965)

Estimating the Xenobiotics Mixtures Toxicity on Aquatic Organisms: The Use of α-level of the Fuzzy Number

Magda S. Peixoto[1], Claudio M. Jonsson[2], Lourival C. Paraiba[2],
Laécio C. Barros[3($\boxtimes$)], and Weldon A. Lodwick[4]

[1] Universidade Federal de São Carlos, Sorocaba, São Paulo 18052-780, Brazil
magda@ufscar.br
[2] Embrapa Meio Ambiente, Jaguariúna, São Paulo 13820-000, Brazil
{claudio.jonsson,lourival.paraiba}@embrapa.br
[3] Universidade Estadual de Campinas, Campinas, São Paulo 13083-859, Brazil
laeciocb@ime.unicamp.br
[4] University of Colorado, Denver, CO 80217-3364, USA
Weldon.Lodwick@ucdenver.edu

Abstract. Agricultural practices that use various xenobiotics can contaminate surface water and groundwater with xenobiotics mixtures concentrations which cause serious risks to water quality and to the health of aquatic organisms that inhabit them. Xenobiotics in water when present as mixtures can exacerbate or reduce the toxic effects in aquatic organisms, when compared to the toxic effects of each individual component concentrations of the xenobiotics mixture. The objective of this study is to develop a mathematical method using α-level of the fuzzy numbers with less accounts and simpler calculations to sort ecotoxicological effects in aquatic organisms of xenobiotics mixtures concentrations occurring in water, classifying them into antagonistic, additive or synergistic and also establishing the magnitude of the effects of concentrations of mixtures. The proposed method in this paper using fuzzy numbers can be suggested in protocols established by regulatory agencies to classify ecotoxicological effects of xenobiotics mixtures in water.

Keywords: Mixtures · Fuzzy numbers · α-level · Ecotoxicological

1 Introduction

Agricultural practices that use various xenobiotics can contaminate surface water [3,22] and groundwater [5,6,14] with xenobiotics mixtures concentrations that can cause serious risks to water quality and to the health of aquatic organisms that inhabit them [1,9,18]. Xenobiotics when present in water as mixtures can

M. S. Peixoto—Financial support by Coordination for the Improvement of Higher Education Personnel (CAPES).

© Springer International Publishing AG, part of Springer Nature 2018
G. A. Barreto and R. Coelho (Eds.): NAFIPS 2018, CCIS 831, pp. 489–499, 2018.
https://doi.org/10.1007/978-3-319-95312-0_42

exacerbate or reduce the toxic effects on aquatic organisms, when compared to the toxic effects of each individual component concentrations of the xenobiotics mixture. For instance, the combination of the insecticides quinalphos and phenthoate showed synergistic toxicity to tilapia *Oreochromis mossambicus* [7]. The association of the fungicides piraclostrobin and epiconazole increased the toxicity to microalgae in 13.6 times when compared to the toxicity of individuals compounds [20]. According to Nair et al. [16], the insecticide combination malathion-endosulfan showed a "more than additive" effect to juveniles of rohu fishes (*Labeo rohita*). Qu et al. [21] proposed an ecological risk assessment of pesticide residues for wetland ecosystems and the risks of eight pesticides in Taihu Lake wetland were assessed, as single substances and in mixtures.

The simultaneous presence of substances in the aquatic compartment can also be derived from the commercial formulations that contain more than one active ingredient, or from the mixture of products in the spray tanks [19,23]. In this context, the use of more than one active ingredient is seen as an advantage due to the reduced cost and the reduced spraying of the recommended dose. Also because the increase of the number of pest species to be chemically controlled.

When two xenobiotics enter concurrently in a biological system there is a need to characterize the toxic effect of the combination in relation to the toxic effect of each compound individually [12]. Some methods allow the classification of such chemical interactions. In this classification, additivity can be generalized for two compounds that act independently on the same target and their effects are additive. Synergism is defined as an interaction among compounds producing a higher effect (more than additive effect) when compared with the individual effect of each compound. Conversely, antagonist compounds would reduce the effect [8].

The toxicity of a compound can be expressed by the value of the median effective concentration (LC_{50}), or concentration that affects 50% of individuals in a population in a given time interval. Therefore, the smaller this value, the more toxic the compound [15]. Thus, by knowing the LC_{50} values for the individual compounds and LC_{50} values for the compounds in the mixture (with their confidence intervals), one can classify the magnitude of the mixtures effect when compared to the individual component concentrations of mixture. Also it is possible to establish confidence intervals for the magnitude of the effect [12].

A fuzzy set has been defined as a collection of objects with membership values between 0 (complete exclusion) and 1 (complete membership). The membership values express the degrees to each object with respect to the properties or distinctive features to the collection. Recently, a fuzzy model have been applied in the field of mixture toxicity prediction to solve the limitations of existing prediction models of mixture toxicity [24].

Then, the objective of this study was to develop a mathematical method using α-level of the fuzzy numbers with less accounts and simpler calculations to sort ecotoxicological effects in aquatic organisms of xenobiotics mixtures concentrations occurring in water. The method allows classify the mixture into antagonistic, additive or synergistic and also establish the magnitude of the effects of concentrations of mixtures. The legislation establishing limits of chemicals in water bodies in Brazil [4] does not report such limits for chemicals mixtures. Thus, the importance in detecting a synergistic action, that results in a potentiation of the effect, contributes to the establishment of public policies in order to improve the water quality standards.

2   Additive Toxicity

Toxicity was defined by the median effective concentration LC_{50}, that is, the concentration calculated to produce 50% of effect and 95% confidence intervals according to the procedures of [11,12]. The procedures for determining the additive index is based on the toxic unit concept in which each component in the mixture contributes to toxicity.

Definition 1. *The contributions of two components of chemical mixtures are summed accordingly*

$$(A_m/A_i) + (B_m/B_i) = S,$$

where A and B are chemicals, A_i and B_i are toxicities (LC_{50}) of the individual chemicals, A_m and B_m are toxicities (LC_{50}) of the mixtured chemicals and S is the sum of biological activity [12].

Definition 2. *The additive index is defined by*

$$AI = \begin{cases} (1/S) - 1.0 & if\ S \le 1.0 \\ (-S) + 1.0 & if\ S > 1.0 \end{cases} \tag{1}$$

The range for additive index is derived by selecting values of 95% confidence interval yielding the greatest derivation from the additive index. The lower limits of the individual toxicants – A_i and B_i – and the upper limits of the mixtures – A_m and B_m – are substituted for LC_{50} to determine the lower limit of the index. Analogously, the upper limits of the mixture – A_m and B_m – are substituted into the formula to determine the upper limit of the index.

If the range overlaped zero, then toxicity of the chemicals in combination is considered additive [12].

We suggest [11,12] for a detailed study of the procedures for classifying the mixture into antagonistic, additive or synergistic.

Remark: The additive toxicity of n chemicals in a mixture is assessed by adding the contributions of additional chemicals according to the formula

$$(A_m^1/A_i^1) + (A_m^2/A_i^2) + (A_m^3/A_i^3) + \cdots + (A_m^n/A_i^n) = S$$

3   Fuzzy Numbers

Next, we develop brief reviews of the concept of fuzzy numbers, and we detail the method suggested in this paper.

Fuzzy Sets and Fuzzy Logic have become one of the emerging areas in contemporary technologies of information processing. Fuzzy Sets Theory was first developed by [25] in the mid-1960s to represent uncertain and imprecise knowledge. It provides an approximate but effective means of describing the behavior of the system that is too complex, ill defined, or not easily analyzed mathematically.

Definition 3. *Let U be a classical non-empty set. A fuzzy subset F of U is described by a function,*

$$F : U \to [0, 1],$$

called membership function of fuzzy set F [25].

The value $F(x) \in [0, 1]$ indicates the membership degree of the element x of U in fuzzy set F, with $F(x) = 1$ and $F(x) = 0$ designating, the belonginness and not-belongingness of x in F, respectively. Note that the membership function of empty, $\emptyset$, and universe, U, sets are, respectively, $\emptyset(x) = 0$ and $U(x) = 1$ for all $X \in U$ [13].

Linguistic variables (or fuzzy) are variables whose values are fuzzy sets [17].

The set of all elements that belong to a fuzzy set A with at least α degree is called α-level of A and denoted by $[A]^{\alpha}$.

Definition 4. *Let A be a fuzzy subset of X and $\alpha \in [0, 1]$. The α-level of A is the subset of X defined by*

$$[A]^{\alpha} = \{x \in X / A(x) \geq \alpha\}$$

for $0 < \alpha \leq 1$.

So, the set $[A]^{\alpha}$ consists of those elements of the universe X whose membership degree is larger than α [10].

A very special class of fuzzy sets is the so-called "fuzzy numbers". This is due to the fundamental role that they play in fuzzy modeling. In this sense, the majority of the fuzzy sets belongs to the fuzzy numbers class [13].

Definition 5. *A fuzzy subset A in $\mathbb{R}$ is called fuzzy number when:*

1. all α-levels of A are non-empty with $0 \leq \alpha \leq 1$, that is, A must be normal;
2. all α-levels of A are closed intervals of $\mathbb{R}$;
3. the support of A, that is, $supp A = \{x \in \mathbb{R} / A(x) > 0\}$.

Definition 6. *Let us represent the α-levels of the fuzzy numbers A by*

$$[A]^{\alpha} = [a_1^{\alpha}, a_2^{\alpha}].$$

A fuzzy subset F of real numbers is called triangular if its membership function is a triangular function. This function is specified by three parameters, $F(x : a, b, c)$, such as:

$$F(x : a, b, c) = \begin{cases} 0 & \text{if } x < a \\ \frac{x-a}{b-a} & \text{if } a \leq x < b \\ \frac{c-x}{c-b} & \text{if } b \leq x < c \\ 0 & \text{if } x \geq c \end{cases}$$

where a, b, c are given numbers.

The α-levels of triangular fuzzy numbers have the following simplified from:

$$[a_1^\alpha, a_2^\alpha] = [(b - a)\alpha + a, (b - c)\alpha + c], \tag{2}$$

fo ll $\alpha \in [0, 1]$.

The great advantage of fuzzy numbers is that it is possible to compute with them. Thus, we can define arithmetic operations on fuzzy numbers.

Definition 7. *Let A and B be fuzzy numbers and λ a real number.*

1. *the addition of A and B produces a third fuzzy number $A + B$, whose membership function is given by:*

$$\psi_{A+B}(x) = \begin{cases} sup_{\phi(z)}min[\psi_A(x), \psi_B(x)] & if\ \phi(z) \neq 0 \\ 0 & if\ \phi(z) = 0 \end{cases}$$

where $\phi(z) = \{(x, y) : x + y = z\}$.

2. *the subtraction of two fuzzy numbers A and B produces a third fuzzy number $A - B$, whose membership function is given by:*

$$\psi_{A-B}(x) = \begin{cases} sup_{\phi(z)}min[\psi_A(x), \psi_B(x)] & if\ \phi(z) \neq 0 \\ 0 & if\ \phi(z) = 0 \end{cases}$$

where $\phi(z) = \{(x, y) : x - y = z\}$.

3. *the multiplication of λ by fuzzy number A produces a third fuzzy number λA, whose membership function is given by:*

$$\psi_{\lambda A}(x) = \begin{cases} sup_{\{x:\lambda x = z\}}min[\psi_A(x)] & if\ \lambda \neq 0 \\ \chi_{\{0\}}(x) & if\ \lambda = 0 \end{cases}$$

where $\chi_{\{0\}}$ is the characteristic function of $\{0\}$.

4. *the multiplication of A and B produces a third fuzzy number $A.B$, whose membership function is given by:*

$$\psi_{A.B}(x) = \begin{cases} sup_{\phi(z)}min[\psi_A(x), \psi_B(x)] & if\ \phi(z) \neq 0 \\ 0 & if\ \phi(z) = 0 \end{cases}$$

where $\phi(z) = \{(x, y) : x.y = z\}$.

5. *the division of A and B, if $0 \notin supp(B)$, produces a third fuzzy number A/B, whose membership function is given by:*

$$\psi_{A/B}(x) = \begin{cases} sup_{\phi(z)}min[\psi_A(x), \psi_B(x)] & if\ \phi(z) \neq 0 \\ 0 & if\ \phi(z) = 0 \end{cases}$$

where $\phi(z) = \{(x, y) : x/y = z\}$.

From concept of α-level we have a "practical method" to obtain the results of each arithmetic operation between fuzzy numbers, because the arithmetic operations with fuzzy numbers are closely linked to the interval mathematics [2].

Theorem 1. *Let A and B be fuzzy numbers with α-levels $[A]^\alpha = [a_1^\alpha, a_2^\alpha]$ and $[B]^\alpha = [b_1^\alpha, b_2^\alpha]$, respectively; and λ a real number. Then, we have the following properties:*

1. *the addition between A and B is a fuzzy number $A + B$, whose α-levels are given by*
$$[A + B]^\alpha = [A]^\alpha + [B]^\alpha = [a_1^\alpha + b_1^\alpha, a_2^\alpha + b_2^\alpha].$$

2. *the subtraction between A and B is a fuzzy number $A - B$ whose α-levels are given by*
$$[A - B]^\alpha = [A]^\alpha - [B]^\alpha = [a_1^\alpha - b_2^\alpha, a_2^\alpha - b_1^\alpha].$$

3. *the multiplication of a real number λ by the fuzzy number A produces is a fuzzy number λA, whose α-levels are given by*
$$[\lambda A]^\alpha = \lambda[A]^\alpha = [\lambda a_1^\alpha, \lambda a_2^\alpha] \quad if \ \lambda \geq 0$$

or

$$[\lambda A]^\alpha = \lambda[A]^\alpha = [\lambda a_2^\alpha, \lambda a_1^\alpha] \quad if \ \lambda < 0.$$

4. *the multiplication of a fuzzy number A by a fuzzy number B is a fuzzy number $A.B$, whose α-levels are*
$$[A.B]^\alpha = [A]^\alpha.[B]^\alpha = [\min P^\alpha, \max P^\alpha],$$

where $P^\alpha = \{a_1^\alpha b_1^\alpha, a_1^\alpha b_2^\alpha, a_2^\alpha b_1^\alpha, a_2^\alpha b_2^\alpha\}$.
5. *the division of a fuzzy number A by a fuzzy number B, if $0 \notin supp B$, is a fuzzy number A/B, whose α-levels are given by*
$$\left[\frac{A}{B}\right]^\alpha = \frac{[A]^\alpha}{[B]^\alpha} = [a_1^\alpha, a_2^\alpha].\left[\frac{1}{b_2^\alpha}, \frac{1}{b_1^\alpha}\right].$$

Proof: See [2].
Thus, it is enough to consider the interval arithmetic operations.

4   Additive Toxicity Using α-level of the Fuzzy Numbers

In this section, we have developed a mathematical method using α-level of the fuzzy numbers to sort ecotoxicological effects in aquatic organisms of xenobiotics mixtures concentrations occurring in water, classifying them into antagonistic, additive or synergistic and also establishing the magnitude of the effects of concentrations of mixtures.

We have proposed to use α-level of the fuzzy numbers to classify the mixture prepared by the adaptation of classic method (1). By these means, we intend to simplify the calculus of the additive index.

For this method, we consider the values of the LC_{50} individually and in combination of the chemicals A and B and the $100(1 - \alpha)\%$ confidence interval individually and in combination of each chemicals as being the α-level.

Definition 8. *Let* $[A]^\alpha$ *be the* α-*level of the fuzzy number* (LC_{50}) *the xenobiotic* A *individually,* $[A_m]^\alpha$ *the* α-*level of the fuzzy number* (LC_{50}) *the xenobiotic* A *in combination,* $[B]^\alpha$ *the* α-*level of the fuzzy number* (LC_{50}) *the xenobiotic* B *individually,* $[B_m]^\alpha$ *the* α-*level of the fuzzy number* (LC_{50}) *the xenobiotic* B *in combination, then we define the interval sum of biological activity, denoted* IS, *by*

$$\frac{[A_m]^\alpha}{[A]^\alpha} + \frac{[B_m]^\alpha}{[B]^\alpha} = IS. \tag{3}$$

And considering $1 = [1, 1]$, we have

Definition 9. *The fuzzy additive index,* FAI, *is defined by*

$$FAI = \begin{cases} ([1,1]/IS) - [1,1] & if \; SM \le 1.0 \\ IS(-[1,1]) + [1,1] & if \; SM > 1.0 \end{cases} \tag{4}$$

where SM *is the arithmetic mean between the lower limit and the upper one of* IS.

We can see (4) as the adaptation of classic method (1).

5   Results

In this section we use the method of the Sect. 4 to classify the toxicity of mixtures. Fish are exposed simultaneously to more than one contaminant because some chemicals are applied as combinations to increase efficacy or reduce costs [12].

In Table 1 the columns 2, 3 and 4 are available in [12]. Note that the column 4 is the values obtained by classic model (Sect. 2). In column 5, we have the corresponding fuzzy number obtained through the mathematical model proposed in Sect. 4 with $\alpha = 0.05$.

It has been highlighted that is possible to choose any α, but it was chosen $\alpha = 0.05$ in order to compare the results by the classic method [11] and the results obtained by our method because the classical method considers 95% confidence intervals.

Considering case 1 of the Table 1:

Let A and B be two xenobiotics such that:

- LC_{50} of xenobiotic A individually is equal to 0.0312 $\mu g/L$;
- LC_{50} of xenobiotic B individually is equal to 0.049 mg/L;
- LC_{50} of xenobiotic A in combination is equal to 0.03 $\mu g/L$;
- LC_{50} of xenobiotic B in combination is equal to 0.03 mg/L;
- 95% confidence interval of xenobiotic A individually is equal to [0.0266, 0.0366];
- 95% confidence interval of xenobiotic B individually is equal to [0.0279, 0.0633];
- 95% confidence interval of xenobiotic A in combination is equal to [0.0272, 0.0331];
- 95% confidence interval of xenobiotic B in combination is equal to [0.0272, 0.0331].

Table 1. Toxicity and additive indices for xenobiotics, pairs of xenobiotics combinations against rainbow trout in soft water at $12\,^\circ$C.

Xenobiotics	LC_{50} and interval 95% confidence interval individually	LC_{50} and interval 95% confidence interval in combination	additive index and range	fuzzy additive index
Antimycin	0.0312	0.03		
$\mu g/L$	[0.0266, 0.0366]	[0.0272, 0.0331]	-0.574	[-1.4307, -0.1729]
Dibrom	0.049	0.03	[-1.43, -0.173]	Antagonism
mg/L	[0.0279, 0.0633]	[0.0272, 0.0331]	Antagonism	
TFM	1.81	1.16		
lampricide, mg/L	[1.53, 2.14]	[0.998, 1.35]	-0.326	[-0.8083, 0.0287]
Bayer 73	0.0346	0.0237	[-0.808, 0.0295]	Additive
lampricide, mg/L	[0.0204, 0.0275]	[0.0204, 0.0275]	Additive	
Malathion	70	3.44		
$\mu g/L$	[59.2, 82.7]	[2.92, 4.06]	7.20	[5.0851, 10.0106]
Delnav	47.2	3.44	[5.09, 10.0]	Synergism
$\mu g/L$	[42.4, 52.6]	[2.92, 4.06]	Synergism	

According to Sect. 2, the classic method to calculate the additive index and the range is

$$S = \frac{0.03}{0.0312} + \frac{0.03}{0.049} = 1.574.$$

Since $S > 1$, then

$$-1.574 + 1.0 = -0.574.$$

Thus, the additive index for A and B is equal to -0.574.
Next, we calculate the range for A and B:

- the lower limit of range is equal to $(-S) + 1.0 = -2.431 + 1.0 = -1.43$, since $S = \frac{0.0331}{0.0266} + \frac{0.0331}{0.0279} = 2.43 > 1$.
 And
- the upper limit of range is equal to $(-S) + 1.0 = -1.173 + 1.0 = -0.173$, since $S = \frac{0.0272}{0.0366} + \frac{0.0272}{0.0633} = 1.173 > 1$.

Hence, the range of mixture is equal to $[-1.43, -0.173]$, i.e., $0 \notin [-1.43, -0.173]$ and $[-1.43, -0.173] \subset \mathbb{R}_-^*$. Therefore, the toxicity of A mixed with B is antagonistic because the fuzzy additive index belongs to negative values.

Now, the fuzzy method to calculate the fuzzy additive index is

- for the xenobiotic A, we defined the fuzzy number LC_{50} of A individually is equal to $A = (x : 0.0264, 0.0312, 0.0369)$, that is, $[A]^{0.05} = [0.0266, 0.0366]$;

- for the xenobiotic B, we defined the fuzzy number LC_{50} of B individually is equal to $B = (x : 0.0268, 0.048, 0.0641)$, that is, $[B]^{0.05} = [0.0279, 0.0633]$;
- for the xenobiotic A, we defined the fuzzy number LC_{50} of A_m in combination is equal to $A_m = (x : 0.0271, 0.03, 0.0333)$, that is, $[A_m]^{0.05} = [0.0272, 0.0331]$;
- for the xenobiotic B, we defined the fuzzy number LC_{50} of B_m in combination is equal to $B_m = (x : 0.0271, 0.03, 0.0333)$, that is, $[B_m]^{0.05} = [0.0272, 0.0331]$;

that is,

$$IS = \frac{[0.0272, 0.0331]}{[0.0266, 0.0366]} + \frac{[0.0272, 0.0331]}{[0.0279, 0.0633]} = [1.1729, 2.4307].$$

Since $SM > 1$, then

$$FAI = [1.1729, 2.4307](-[1, 1]) + [1, 1] = [-1.4307, -0.1729].$$

Therefore, the fuzzy additive index is equal to $[-1.4307, -0.1729]$. Observe that, we get the same range of the combination between A and B of the classic method (1). In this way, the toxicity of A mixed with B is antagonism.

Nevertheless, the quantity of the mathematical calculations is smaller than the classic method and the formula (4) is easiest solution than the classic formula (1).

Analogously, we determine the values of the Table 1 for cases 2 and 3.

6   Conclusions

In this study we develop a method using fuzzy numbers to sort ecotoxicological effects in aquatic organisms of xenobiotics mixtures concentrations occurring in water, classifying them into antagonistic, additive or synergistic and also establishing the magnitude of the effects of concentrations of mixtures.

It can be observed that the values obtained by the fuzzy model are very close to the values found in the literature but there are less accounts and simpler calculations. It has been used the formulas (1) of Sect. 2 three times to determine the values of the additive index, the lower limit and the upper one of the confidence interval. And it has been used the formulas (4) just once to determine the fuzzy additive index of Sect. 4.

Furthermore, it can be applied our method for any confidence interval. Simply take the α-level as desired. It has been highlighted that is possible to choose any α, but it was chosen $\alpha = 0.05$ in order to compare the results by the classic method [11] and the results obtained by our method.

The method developed using fuzzy numbers can be suggested in protocols established by regulatory agencies to classify ecotoxicological effects of xenobiotics mixtures in water. In this way, public policies would be implemented to ensure the health of the aquatic environment.

Acknowledgments. The first author acknowledges the Coordination for the Improvement of Higher Education Personnel (CAPES) and the São Paulo Research Foundation

(FAPESP), projects numbers 2016/04299-9, the second author acknowledges the São Paulo Research Foundation (FAPESP), project numbers 2010/06294-8, and the fourth author acknowledges the National Council for Scientific and Technological Development (CNPq), project numbers 306546/2017-5, for the financial support.

References

1. Andreu-Sanchez, O., Paraíba, L.C., Jonsson, C.M., Carrasco, J.M.: Acute toxicity and bioconcentration of fungicide tebuconazole in zebrafish (Danio rerio). Environ. Toxicol. (Print) **27**, 109–116 (2012)
2. Barros, L.C., Bassanezi, R.C., Lodwick, W.A.: A First Course in Fuzzy Logic, Fuzzy Dynamical Systems, and Biomathematics: Theory and Applications. Springer, Germany (2016). https://doi.org/10.1007/978-3-662-53324-6
3. Blanchoud, H., Moreau-Guigon, E., Farrugia, F., Chevreuil, M., Mouchel, J.M.: Contribution by urban and agricultural pesticide uses to water contamination at the scale of the Marne watershed. Sci. Total Environ. **375**, 168–179 (2007)
4. Brasil, Resolução CONAMA No. 357, de 17 de março de 2005, Diario Oficial da União, Brasilia (2005)
5. Caldas, S.S., Demoliner, A., Costa, F.P., D'Oca, M.G.M., Primel, E.G.: Pesticide residue determination in groundwater using solid-phase extraction and high-performance liquid chromatography with diode array detector and liquid chromatography-tandem mass spectrometry. J. Braz. Chem. Soc. **21**(4), 642–650 (2010)
6. Carbo, L., Souza, V., Dores, E.F.G.C., Ribeiro, M.L.: Determination of pesticides multiresidues in shallow groundwater in a cotton-growing region of Mato Grosso. J. Braz. Chem. Soc. **19**(6), 1111–1117 (2008)
7. Durairaj, S., Selvarajan, V.R.: Synergistic action of organophosphorous insecticides on fish Oreochromis mossambicus. J. Exp. Biol. **16**(1), 51–53 (1995)
8. Jonsson, C.M., Aoyma, H.: In vitro effect of agriculture pollutants and their joint action on Pseudokirchneriella subcapitata acid phosphatase. Chemosphere **69**, 849–855 (2007)
9. Jonsson, C.M., Paraiba, L.C., Mendoza, M.T., Sabater, C., Carrasco, J.M.: Bioconcentration of the insecticide pyridaphenthion by the green algae Chlorella saccharophila. Chemosphere **143**, 321–325 (2007)
10. Klir, G.J., Yuan, B.: Fuzzy Sets and Fuzzy Logic: Theory and Applications. Prentice Hall, Upper Saddle River (1995)
11. Litchfield, J.T., Wilcoxon, F.: J. Pharmacol. Exp. Ther. **96**, 99–113 (1949)
12. Marking, L.L.: Method for assessing additive toxicity of chemical mixtures. In: Mayer, F.L., Hamelink, J.L. (eds.) Aquatic Toxicology and Hazard Evaluation, pp. 99–108. American Society for Testing and Materials (1985)
13. Massad, E., Oretga, N.R.S., Barros, L.C., Struchiner, C.J.: Fuzzy Logic in Action: Applications in Epidemiology and Beyond. Springer, Heidelberg (2008). https://doi.org/10.1007/978-3-540-69094-8
14. Mott, T.: Groundwater and rainwater contamination by pesticides in an agricultural region of Mato Grosso state in central Brazil. Ciência Saúde Coletiva **17**(6), 1557–1568 (2012)
15. Murty, A.S.: Toxicity tests and test methodology. In: Murty, A.S. (ed.) Toxicity of Pesticides to Fish, vol. 1, pp. 117–154. CRC Press, Boca Raton (1986)

16. Nair, V.P., Nair, J.R.A., Mercy, T.V.A.: The individual and combined acute lethal toxicity of selected biocides on the juveniles of rohu (Labeo rohita) under tropical conditions. O Mundo da Saúde (Online) **54**(3), 267–274 (2007)

17. Pedrycs, W., Gomide, F.: An Introduction to Fuzzy Sets: Analysis and Design. Massachusets Institute of Technology (1998)

18. Prestes, E.B., Jonsson, C.M., Castro, V.L.S., Paraiba, C.C.M.: Avaliação da toxicidade crônica de piraclostrobin, epoxiconazol e sua mistura em Daphnia similis. Ecotoxicol. Environ. Contam. **8**, 113–117 (2013)

19. Prestes, E.B., Jonsson, C.M., Castro, V.L.S.: Avaliação da toxicidade aguda de piraclostrobin, epoxiconazol e sua mistura em Daphnia similis. Pesticidas **22**, 43–50 (2012)

20. Prestes, E.B., Jonsson, C.M., Castro, V.L.S.: Avaliação da toxicidade de piraclostrobin, epoxiconazol e sua mistura em alga pseudokirchneriella subcapitata. Pesticidas **21**, 39–46 (2011)

21. Qu, C.S., Chen, W., Bi, J., Huang, L., Li, F.Y.: Ecological risk assessment of pesticide residues in Taihu Lake wetland, China. Ecol. Modell. **222**(2), 287–292 (2011)

22. Rissato, S.R., Galhiane, M.S., Ximenes, V.F., Andrade, R.M., Talamoni, J.L., Libânio, M., Almeida, M.V., Apon, B.M., Cavalari, A.A.: Organochlorine pesticides and polychlorinated biphenyls in soil and water samples in the Northeastern part of São Paulo State Brazil. Chemosphere **65**(11), 1949–1958 (2006)

23. Tesolin, G.A.S., Marson, M.M., Jonsson, C.M., Nogueira, A.J.A., Franco, D.A.S., Almeida, S.D.B., Matallo, M.B., Moura, M.A.M.: Avaliação da toxicidade de herbicidas usados em cana-de-açúcar para o Paulistinha (Danio rerio). O Mundo da Saúde (Online) **38**, 86 (2014)

24. Wang, Z., Chen, J., Huang, L., Wang, Y., Cai, X., Qiao, X., Dong, Y.: Integrated fuzzy concentration addition-independent action (IFCA-IA) model outperforms two-stage prediction (TSP) for predicting mixture toxicity. Chemosphere **74**, 735–740 (2009)

25. Zadeh, L.A.: Fuzzy sets. Inf. Control **8**, 338–353 (1965)

An Approach for Solving Interval Optimization Problems

Fabiola Roxana Villanueva$^{(\boxtimes)}$ and Valeriano Antunes de Oliveira$^{(\boxtimes)}$

Department of Applied Mathematics, Institute of Biosciences, Letters and Exact
Sciences, UNESP - São Paulo State University, São José do Rio Preto, SP, Brazil
olita_villanueva@hotmail.com, antunes@ibilce.unesp.br

Abstract. This work considers an optimization problem where the
objective function possesses interval uncertainty in the coefficients. In
this sense, first, an order relation will be defined for the interval space
and, from this, it will be defined a solution concept for the interval problem in question. Subsequently, it will be shown that an interval problem
is equivalent to a bi-objective problem. Finally, it will be established the
necessary conditions of Fritz John and Karush-Kuhn-Tucker types for
the interval-valued optimization problem.

Keywords: Multi-objective optimization problems
Classical necessary condition of Fritz John · Karush-Kuhn-Tucker
Interval optimization problems · LU order relation · LU solutions
Necessary optimality conditions for the LU-solution

1 Introduction

The importance of the study of interval analysis from a theoretical point of
view, as well as its applications, is well known (Aubin and Cellina [1], Aubin
and Frankowska [2]). Many advances in interval analysis have been motivated
by mathematical programming and control theory (Aubin and Franskowska [3]).
In general, problems have coefficients as fixed and deterministic values. However, there are many situations where this assumption is not valid, the problems
involve some kind of uncertainty. Thus, methods to deal with uncertainty are
needed. We assume an interval type uncertainty in order to solve these problems.

The work is organized as follows. The next section is devoted to give some
preliminaries. The third section presents the interval optimization problem, the
lower and upper, LU in short, definition of order relation, then, we present the
LU-solution, the strictly weakly LU-solution, the weakly LU-solution, and their
local definitions, respectively, and after this, we show that to determine the
LU-solutions of the interval problem is equivalent to determining the solution
of a classical multi-objective optimization problem. In Sect. 4, we give the necessary conditions of Fritz John and Karush-Kuhn-Tucker types. Finally, some
conclusions are given.

Supported by CAPES.

© Springer International Publishing AG, part of Springer Nature 2018
G. A. Barreto and R. Coelho (Eds.): NAFIPS 2018, CCIS 831, pp. 500–507, 2018.
https://doi.org/10.1007/978-3-319-95312-0_43

2   Preliminaries

In this section we follow the classical nomenclature of Sawaragi et al. [4].

Let $\mathbb{R}^p$ be the p-dimensional Euclidean space. Let $x = (x_1, \ldots, x_p) \in \mathbb{R}^p$ and $y = (y_1, \ldots, y_p) \in \mathbb{R}^p$,

(1) $x \geqq y$ means $x_i \geq y_i$ for all $i = 1, \ldots, p$;
(2) $x \geq y$ means $x \geqq y$ and $x \neq y$, i.e., $x_i \geq y_i$ for all $i = 1, \ldots, p$ and we have at least one strict inequality;
(3) $x > y$ means $x_i > y_i$ for all $i = 1, \ldots, p$.

Let us denote $\mathbb{R}^p_+$, the nonnegative orthant of $\mathbb{R}^p$, i.e.,

$$\mathbb{R}^p_+ := \{y \in \mathbb{R}^p : y_i \geq 0 \text{ for } i = 1, \ldots, p\}.$$

Let us denote $\mathring{\mathbb{R}}^p_+$, the positive orthant of $\mathbb{R}^p$, i.e.,

$$\mathring{\mathbb{R}}^p_+ := \{y \in \mathbb{R}^p : y_i > 0 \text{ for } i = 1, \ldots, p\}.$$

The Multi-objective Optimization problem. Let's consider:

$$(P) \begin{cases} \text{minimize } f(x) = (f_1(x), f_2(x), \ldots, f_p(x)) \\ \text{subject to } x \in X \subset \mathbb{R}^n. \end{cases}$$

Definition 1 (Pareto Optimal Solution). *A point $\overline{x} \in X$ is said to be a (globally) Pareto optimal solution (or efficient or non-dominated, or non-inferior solution) to the problem (P) if there exists no $x \in X$ such that $f(x) \leq f(\overline{x})$.*

In some cases a slightly weaker solution concept than Pareto optimality is often used. It is called *weak Pareto optimality*, which corresponds to the case in which we consider the positive orthant

$$\mathring{\mathbb{R}}^p_+ = \{y \in \mathbb{R}^p : y > 0\}.$$

Definition 2 (Weak Pareto Optimal Solution). *A point $\overline{x} \in X$ is said to be a weak Pareto optimal solution to the problem (P) if there is no $x \in X$ such that $f(x) < f(\overline{x})$.*

The Multi-objective Programming Problem. Let's consider:

$$(P') \begin{cases} \text{minimize } f(x) = (f_1(x), \ldots, f_p(x)) \\ \text{subject to } x \in X = \{x \mid g(x) = (g_1(x), \ldots, g_m(x)) \leqq 0\}. \end{cases}$$

That is, X is supposed to be specified by m inequality constraints. Here, all the functions f_i $(i = 1, \ldots, p)$ and g_j $(j = 1, \ldots, m)$ are assumed to be continuously differentiable.

Definition 3 (Kuhn-Tucker's Constraint Qualification). *Let us consider the problem* (P'), $\overline{x} \in X$ *is said to satisfy the Kunh-Tucker constraint qualification if, for any* $h \in \mathbb{R}^n$ *such that* $\langle \nabla g_j(\overline{x}), h \rangle \leq 0$ *and for any* $j \in J(\overline{x}) = \{j \mid g_j(\overline{x}) = 0\}$, *there exist* $\overline{t} > 0$, *a vector-valued function* θ *on* $[0, \overline{t}]$ *differentiable at* $t = 0$, *and a real number* $\alpha > 0$, *such that*

$$\theta(0) = \overline{x}, \quad g(\theta(t)) \leqq 0 \quad \text{for any } t \in [0, \overline{t}], \quad \dot{\theta}(0) = \alpha h.$$

Next, we have the necessary conditions for multi-objective programming problems for weak Pareto optimality:

Theorem 1 *[The Kuhn-Tucker Condition]. Suppose that at the problem* (P'), $\overline{x} \in X$ *satisfies the Kuhn-Tucker constraint qualification. Then, a necessary condition for* $\overline{x}$ *to be a weak Pareto optimal solution to* (P') *is that there exist* $\overline{\lambda} \in \mathbb{R}^p$ *and* $\overline{\mu} \in \mathbb{R}^m$ *such that*

(i) $\langle \overline{\lambda}, \nabla f(\overline{x}) \rangle + \langle \overline{\mu}, \nabla g(\overline{x}) \rangle = 0$,
(ii) $\langle \overline{\mu}, g(\overline{x}) \rangle = 0$,
(iii) $\overline{\lambda} \geq 0, \overline{\mu} \geq 0$,

where $\langle \overline{\lambda}, \nabla f(\overline{x}) \rangle$ *and* $\langle \overline{\mu}, \nabla g(\overline{x}) \rangle$ *stand for* $\sum_{i=1}^{p} \overline{\lambda}_i \nabla f_i(\overline{x})$ *and* $\sum_{j=1}^{m} \overline{\mu}_j \nabla g_j(\overline{x})$, *respectively.*

3   The Interval Optimization Problem

The optimization problem that will be considered has the feasible set

$$\mathcal{X} = \{x \in \mathbb{R}^n \mid g_i(x) \leq 0, i = 1, \dots, m\},$$

where $g : \mathbb{R}^n \to \mathbb{R}^m$ is a real vector-valued function, and $F : \mathbb{R}^n \to \mathbb{I}(\mathbb{R})$ is the interval objective function given as $F(x) = [f_1(x), f_2(x)]$, with $f_i : \mathbb{R}^n \to \mathbb{R}$, $i = 1, 2$, and $f_1(x) \leq f_2(x) \, \forall x \in X$. In this way, we have the following interval-valued problem:

$$(IP) \begin{cases} \text{minimize } F(x) = [f_1(x), f_2(x)] \\ \text{subject to } g_i(x) \leq 0, \quad i = 1, \dots, m. \end{cases}$$

We are going to use the differentiability concept of an interval function, by just considering the differentiability of the extreme functions of the interval function.

Definition 4. *Let* $F :\,]a, b[\to \mathbb{I}(\mathbb{R})$ *be an interval function and* $x_0 \in\,]a, b[$. *We say that* F *is differentiable extremal, E-differentiable in short, at* x_0 *if and only if the extreme functions of real value* f_1 *and* f_2 *are differentiable at* x_0.

Next, will be defined the order relation concept proposed by Kulish and Miranker [5]:

Definition 5. *Let $A = [a_1, a_2]$ and $B = [b_1, b_2]$ be two intervals. The order relation lower and upper, LU in short, is defined by*

1. *$A \leq_{LU} B$ if and only if $a_1 \leq b_1$ and $a_2 \leq b_2$.*
2. *$A \leq_{LU} B$ if and only if $A \leq_{LU} B$ and $A \neq B$, that is, $a_1 < b_1$ and $a_2 \leq b_2$ or $a_1 \leq b_1$ and $a_2 < b_2$ or $a_1 < b_1$ and $a_2 < b_2$.*
3. *$A <_{LU} B$ if and only if $a_1 < b_1$ and $a_2 < b_2$.*

The concept of the LU-solution for the optimization problem above is given by:

Definition 6. *Let $\bar{x}$ be a feasible solution of the problem (IP), i.e., $\bar{x} \in \mathcal{X}$. Then,*

(i) *$\bar{x}$ is a **LU-solution of the problem (IP)** if there does not exist $x \in \mathcal{X}$, $x \neq \bar{x}$, such that $F(x) \leq_{LU} F(\bar{x})$.*
    *Furthermore, $\bar{x}$ is a **local LU-solution of the problem (IP)** if there does not exist $x \in N_\epsilon(\bar{x}) \cap \mathcal{X}$, $x \neq \bar{x}$, such that $F(x) \leq_{LU} F(\bar{x})$, where $N_\epsilon(\bar{x})$ is an $\epsilon-$neighborhood of $\bar{x}$.*
(ii) *$\bar{x}$ is a **strictly weakly LU-solution of the problem (IP)** if there does not exist $x \in \mathcal{X}$ such that $F(x) \leq_{LU} F(\bar{x})$.*
    *Furthermore, $\bar{x}$ is a **local strictly weakly LU-solution of the problem (IP)** if there does not exist $x \in N_\epsilon(\bar{x}) \cap \mathcal{X}$ such that $F(x) \leq_{LU} F(\bar{x})$.*
(iii) *$\bar{x}$ is a **weakly LU-solution of the problem (IP)** if there does not exist $x \in \mathcal{X}$ such that $F(x) <_{LU} F(\bar{x})$.*
    *Furthermore, $\bar{x}$ is a **local weakly LU-solution of the problem (IP)** if there does not exist $x \in N_\epsilon(\bar{x}) \cap \mathcal{X}$ such that $F(x) <_{LU} F(\bar{x})$.*

The set of the LU-solutions is given by:

$$\overline{\mathcal{X}}_{LU} = \left\{ \bar{x} \in \mathbb{R}^n \,|\, \bar{x} = \arg\min_{x \in \mathcal{X}} F(x) \right\}.$$

Remark 1. *1. The following relations are immediate:*

$$\boxed{LU\text{-solution} \Rightarrow strictly\ weakly\ LU\text{-solution} \Rightarrow weakly\ LU\text{-solution.}}$$

2. The theory that will be developed for the weakly LU-solutions is applicable to the LU-solutions and to the strictly weakly LU-solutions as well.

The method to obtain the weakly LU-solutions is equivalent to the resolution of a classical bi-objective optimization problem, derived from the interval problem in question. This is seen in the next Theorem:

Theorem 2. *If $\bar{x}$ is a weakly LU-solution of the problem (IP), then $\bar{x}$ is a weak Pareto optimal solution of the following classical bi-objective problem:*

$$(BP)_{LU} \begin{cases} minimize\ (f_1(x), f_2(x)) \\ subject\ to\ g_i(x) \leq 0,\ i = 1, \ldots, m. \end{cases}$$

Conversely, if $\bar{x}$ is a weak Pareto optimal solution of $(BP)_{LU}$, then $\bar{x}$ is a weakly LU-solution of the problem (IP).

Proof. Let $\bar{x}$ be a weakly LU-solution of the problem (IP), that is, there does not exist $x \in \mathcal{X}$ such that

$$
\begin{aligned}
&F(x) <_{LU} F(\bar{x}) \\
\Leftrightarrow\ &[f_1(x), f_2(x)] <_{LU} [f_1(\bar{x}), f_2(\bar{x})] \\
\Leftrightarrow\ &f_1(x) < f_1(\bar{x}) \text{ and } f_2(x) < f_2(\bar{x}). \tag{1}
\end{aligned}
$$

Suppose that $\bar{x}$ does not solve the classical problem $(BP)_{LU}$. Then there exist a solution $x \in \mathcal{X}$ such that

$$
\begin{aligned}
&(f_1(x), f_2(x)) < (f_1(\bar{x}), f_2(\bar{x})) \\
\Leftrightarrow\ &f_1(x) < f_1(\bar{x}) \text{ and } f_2(x) < f_2(\bar{x}),
\end{aligned}
$$

which contradicts the weakly LU-optimality of $\bar{x}$ (see (1)). Conversely, if $\bar{x} \in \mathcal{X}$ is a weak Pareto optimal solution of the problem $(BP)_{LU}$, then there does not exist $x \in \mathcal{X}$ such that

$$
\begin{aligned}
&(f_1(x), f_2(x)) < (f_1(\bar{x}), f_2(\bar{x})) \\
\Leftrightarrow\ &f_1(x) < f_1(\bar{x}) \text{ and } f_2(x) < f_2(\bar{x}) \\
\Leftrightarrow\ &[f_1(x), f_2(x)] <_{LU} [f_1(\bar{x}), f_2(\bar{x})] \\
\Leftrightarrow\ &F(x) <_{LU} F(\bar{x}),
\end{aligned}
$$

which is the definition of weakly LU-optimality for $\bar{x}$.

4   Necessary Optimality Conditions for the LU-Solution

The next two theorems present the necessary conditions of Fritz John and Karush-Kuhn-Tucker type for the weakly LU-solution using the continuous differentiability of f_1 and f_2 of the interval objective function $F(x) = [f_1(x),\ f_2(x)]$ and also the continuous differentiability of the restriction function g in order to apply classical theorems that require just the continuous differentiability of the involved functions.

Theorem 3 (Necessary condition of Fritz John). *Let's suppose that the interval-valued function $F : \mathbb{R}^n \to \mathbb{I}(\mathbb{R})$ is continuously E-differentiable and $g : \mathbb{R}^n \to \mathbb{R}^m$ is continuously differentiable. If $\bar{x} \in \mathcal{X}$ is a local weakly LU-solution of the problem (IP), then there exist multipliers $0 \leq (\lambda, \mu) \in \mathbb{R}^{2+m}$ with $(\lambda, \mu) \neq (0, 0)$, such that*

$$
\mu_i g_i(\bar{x}) = 0, \quad i = 1, \ldots, m;
$$

$$
\lambda_1 \nabla f_1(\bar{x}) + \lambda_2 \nabla f_2(\bar{x}) + \sum_{i=1}^{m} \mu_i \nabla g_i(\bar{x}) = 0.
$$

Proof. If $\overline{x}$ is a local weakly LU-solution of the problem (IP), then, by Theorem 2, $\overline{x}$ is a local weak Pareto optimal solution of the problem $(BP)_{LU}$. Applying the necessary conditions of Fritz John at $\overline{x}$ (Theorem 2.1) of Maciel, Santos and Sottosanto [8], there exist multipliers $0 \leqq (\lambda, \mu) \in \mathbb{R}^{2+m}$ with $(\lambda, \mu) \neq (0, 0)$, such that

$$\mu_i g_i(\overline{x}) = 0, \quad i = 1, \ldots, m;$$

$$\lambda_1 \nabla f_1(\overline{x}) + \lambda_2 \nabla f_2(\overline{x}) + \sum_{i=1}^{m} \mu_i \nabla g_i(\overline{x}) = 0.$$

To ensure the positivity of λ, some regularity condition must be presupposed (see Luenberger [6], Miettinen [7], Maciel et al. [8]), and these regularity conditions are called *constraints qualification*. We are going to consider the Kuhn-Tucker constraint qualification of Sawaragi et al. [4] presented at the preliminaries. The following theorem will present the necessary conditions of Karush-Kuhn-Tucker for a local weakly LU-solution of the interval-valued optimization problem.

Theorem 4 (Necessary condition of Karush-Kuhn-Tucker). *Let's suppose that the interval-valued function $F : \mathbb{R}^n \to \mathbb{I}(\mathbb{R})$ is continuously E-differentiable and $g : \mathbb{R}^n \to \mathbb{R}^m$ is continuously differentiable. If $\overline{x} \in \mathcal{X}$ satisfies the Kuhn-Tucker constraint qualification and is a local weakly LU-solution of the problem (IP), then there exist (Lagrange) multipliers $0 \leqq (\lambda, \mu) \in \mathbb{R}^{2+m}$ with $\lambda_1 + \lambda_2 = 1$, such that*

$$\mu_i g_i(\overline{x}) = 0, \quad i = 1, \ldots, m; \tag{2}$$

$$\lambda_1 \nabla f_1(\overline{x}) + \lambda_2 \nabla f_2(\overline{x}) + \sum_{i=1}^{m} \mu_i \nabla g_i(\overline{x}) = 0. \tag{3}$$

Proof. If $\overline{x}$ is a local weakly LU-solution of the problem (IP), then, by Theorem 2, $\overline{x}$ is a local weak Pareto optimal solution of the problem $(BP)_{LU}$. Applying the Theorem 1 at $\overline{x}$, there exist (Lagrange) multipliers $0 \leq (\lambda, \mu) \in \mathbb{R}^{2+m}$ such that (2) and (3) hold and also satisfying the property that $\lambda_1 + \lambda_2 = 1$. In fact, if (λ_1, λ_2) and $(\mu_1, \ldots, \mu_m)$ serve as sets of multipliers then, for all $c > 0$, $(\widetilde{\lambda_1}, \widetilde{\lambda_2}) := (c\lambda_1, c\lambda_2)$ and $(\widetilde{\mu_1}, \ldots, \widetilde{\mu_m}) := (c\mu_1, \ldots, c\mu_m)$ also serve as sets of multipliers.

If $\widetilde{\lambda_1} = c\lambda_1 \neq 0$ and $\widetilde{\lambda_2} = c\lambda_2 \neq 0$ we can always choose $c := \frac{1}{\lambda_1 + \lambda_2}$ such that

$$\widetilde{\lambda_1} + \widetilde{\lambda_2} = c\lambda_1 + c\lambda_2 = c(\lambda_1 + \lambda_2) = \frac{1}{\lambda_1 + \lambda_2}(\lambda_1 + \lambda_2) = 1.$$

If $\widetilde{\mu_1} = c\mu_1 \neq 0, \ldots, \widetilde{\mu_m} = c\mu_m \neq 0$ we have that

$$\mu_i g_i(\overline{x}) = 0, \quad \text{for all } i = 1, \ldots, m$$

$$\Leftrightarrow c\mu_i g_i(\overline{x}) = 0, \quad \text{for all } i = 1, \ldots, m \text{ and for all } c > 0$$

$$\Leftrightarrow \widetilde{\mu_i} g_i(\overline{x}) = 0, \quad \text{for all } i = 1, \ldots, m$$

and

$$\lambda_1 \nabla f_1(\overline{x}) + \lambda_2 \nabla f_2(\overline{x}) + \sum_{i=1}^{m} \mu_i \nabla g_i(\overline{x}) = 0$$

$$\Leftrightarrow c \left(\lambda_1 \nabla f_1(\overline{x}) + \lambda_2 \nabla f_2(\overline{x}) + \sum_{i=1}^{m} \mu_i \nabla g_i(\overline{x}) \right) = 0, \qquad \text{for all } c > 0$$

$$\Leftrightarrow c\lambda_1 \nabla f_1(\overline{x}) + c\lambda_2 \nabla f_2(\overline{x}) + \sum_{i=1}^{m} c\mu_i \nabla g_i(\overline{x}) = 0$$

$$\Leftrightarrow \widetilde{\lambda_1} \nabla f_1(\overline{x}) + \widetilde{\lambda_2} \nabla f_2(\overline{x}) + \sum_{i=1}^{m} \widetilde{\mu_i} \nabla g_i(\overline{x}) = 0.$$

Therefore, we have that there exist (Lagrange) multipliers $0 \leqq (\widetilde{\lambda}, \widetilde{\mu}) \in \mathbb{R}^{2+m}$ with $\widetilde{\lambda_1} + \widetilde{\lambda_2} = 1$, such that

$$\widetilde{\mu_i} g_i(\overline{x}) = 0, \quad i = 1, \ldots, m;$$

$$\widetilde{\lambda_1} \nabla f_1(\overline{x}) + \widetilde{\lambda_2} \nabla f_2(\overline{x}) + \sum_{i=1}^{m} \widetilde{\mu_i} \nabla g_i(\overline{x}) = 0.$$

5   Conclusion

We considered an interval optimization problem and the LU order relation which defines the concept of solution for the problem. With this in hand, we presented a method to determine the LU-solutions. Then using the continuous differentiability of the involved functions and classical necessary conditions, we gave the necessary conditions of the Fritz John and Karush-Kuhn-Tucker types for the interval optimization problem. The next step is to consider an interval optimal control problem and give necessary conditions of the Maximum Principle type. Similar results can be obtained using other order relations such as the UC order relation (see Ishibuchi and Tanaka [10]) and LS order relation (see Chalco-Cano et al. [11]) by just changing the objective function of the bi-objective problem $(f_1(x), f_2(x))$ for $(f_2(x), f^C(x))$ and $(f_1(x), f^S(x))$, respectively, where $f^C(x) = \frac{f_1(x) + f_2(x)}{2}$ and $f^S(x) = f_2(x) - f_1(x)$.

Acknowledgements. The authors are thankful with the UNESP, CAPES, CBSF and NAFIPS for let them to present and develop this work.

References

1. Aubin, J., Cellina, A.: Differential Inclusions. Springer, New York (1984). https://doi.org/10.1007/978-3-642-69512-4
2. Aubin, J., Franskowska, H.: Set-Valued Analysis. Birkhäuser, Boston (1990)

3. Aubin, J., Franskowska, H.: Introduction: set-valued analysis in control theory. In: Set-Valued Analysis, vol. 8 (2000)
4. Sawaragi, Y., Nakayama, H., Tanino, T.: Theory of multiobjective optimization. Mathematics in Science and Engineering, vol. 176. Richard Bellman, University of Southern California (1985)
5. Kulish, U.W., Miranker, W.L.: Computer Arithmetic in Theory and in Practice. Academic Press, Cambridge (1981)
6. Luenberger, D.G.: Linear and Nonlinear Programming, 2nd edn. Addison-Wesley Publishing Company, Nova York (1986)
7. Miettinen, K.M.: Nonlinear Multiobjective Optimization. Kluwer Academic Publishers, Boston (1999)
8. Maciel, M.C., Santos, S.A., Sottosanto, G.N.: Regularity conditions in differentiable vector optimization revisited. J. Optim. Theory Appl. **142**, 385–398 (2009)
9. Oliveira, V.A., Silva, G.N.: On sufficient optimality conditions for multiobjective control problems. J. Glob. Optim. **64**, 721–744 (2016)
10. Ishibuchi, H., Tanaka, H.: Multi-objective programming in optimization of the interval objective function. Eur. J. Oper. Res. **48**, 219–225 (1990)
11. Chalco-Cano, Y., Lodwick, W.A., Rufián-Lizana, A.: Optimality conditions of type KKT for optimizarion problem with interval-valued objective function via generalized derivative. Fuzzy Optim. Decis. Making **12**, 305–322 (2013)

A Brief Review of a Method for Bounds on Polynomial Ranges over Simplexes

Ralph Baker Kearfott[(⊠)] and Dun Liu

University of Louisiana at Lafayette, Lafayette, LA, USA
{rbk,dxl8898}@louisiana.edu

Abstract. We are concerned with tools to find bounds on the range of certain polynomial functions of n variables. Although our motivation and history of the tools are from crisp global optimization, bounding the range of such functions is also important in fuzzy logic implementations. We review and provide a new perspective on one such tool. We have been examining problems naturally posed in terms of barycentric coordinates, that is, over simplexes. There is a long history of using Bernstein expansions to bound ranges of polynomials over simplexes, particularly within the computer graphics community for 1-, 2-, and 3-dimensional problems, with some literature on higher-dimensional generalizations, and some work on use in global optimization. We revisit this work, identifying efficient implementation and practical application contexts, to bound ranges of polynomials over simplexes in dimensions, 2, 3, and higher.

1 Introduction

Finding bounds on the range of a function

$$f(x) = f(x_0, \ldots, x_n) : \mathcal{D} \subseteq \mathbb{R}^{n+1} \to \mathbb{R}$$

of $n+1$ variables $x_0, \ldots, x_n$ over some region $\mathcal{D}$ is an important problem occurring in many forms and contexts. Indeed, evaluating the value of a function over fuzzy inputs involves finding its range over the appropriate non-fuzzy ("crisp") sets (namely, alpha-cuts); such evaluations are central to computing fuzzy outputs based on fuzzy inputs; see [6] for an excellent explanation of this. Finding the exact range (or good numerical approximation thereof), or equivalently, the global optimization problem, is known, for general continuous f and general compact region $\mathcal{D}$, to be NP-hard, and is not NP-hard only for certain special classes of f and $\mathcal{D}$. However, non-sharp bounds, if not too wide, can be a useful basic tool, even in global optimization algorithms. The literature is replete with discussion of the relationship between bounding the range of a function and fuzzy computations, as pointed out, say, in the overviews [8,11], in [24], or the more recent work [22]. (A plethora of references is omitted.) One survey on fuzzy optimization, including its relationship to crisp optimization techniques, is [18].

© Springer International Publishing AG, part of Springer Nature 2018
G. A. Barreto and R. Coelho (Eds.): NAFIPS 2018, CCIS 831, pp. 508–518, 2018.
https://doi.org/10.1007/978-3-319-95312-0_44

Various approaches to easily computing usable bounds on ranges, many depending on particular contexts and applications, have been proposed and implemented. A common context is where the region $\mathcal{D}$ is defined by lower and upper bounds $\underline{x}_i$ and $\overline{x}_i$ on each of the variables x_i, that is, $\underline{x}_i \leq x_i \leq \overline{x}_i$, $0 \leq i \leq n$. We speak of such regions as a *box*, or *interval vector*, and write

$$x = (x_0, \ldots, x_n) = ([\underline{x}_0, \overline{x}_0], \ldots, [\underline{x}_n, \overline{x}_n]).$$

In this context, simple (or "naive") interval arithmetic, introduced and explained in numerous texts and reviews, such as our relatively recent work [10], is a possibility. However, such bounds are not guaranteed to be sufficiently close to the actual range to be useful; experts in interval arithmetic have extensively studied techniques to obtain the best possible bounds with the minimal amount of work.

1.1 Simplexes

In various applications, range bounds are required, not over a box, but over a *simplex* $\mathcal{S}$. In particular, to within an affine transformation, an n-dimensional simplex in $\mathbb{R}^{n+1}$ is characterized by:

$$\mathcal{S}_c = \left\{ (x_0, \ldots, x_n), \quad x_i \geq 0, \quad \sum_{i=0}^{n} x_i = 1 \right\}, \tag{1}$$

Thus, a simplex can be viewed as a box $x \in \mathbb{R}^{n+1}$ subject to the additional (commonly occurring) constraint $\sum_{i=0}^{n} x_i = 1$. Alternatively, a simplex can be viewed as a volume in $\mathbb{R}^n$ bounded by affine constraints:

$$\mathcal{S} = \left\{ \sum_{i=0}^{n} x_i P_i, \quad x_i \geq 0, \quad P_i \in \mathbb{R}^q, \quad q \geq n, \quad \sum_{i=0}^{n} x_i = 1 \right\}. \tag{2}$$

The P_i in (2) are called the *vertexes* of $\mathcal{S}$, and the x_i are called the *barycentric coordinates* of points in $\mathcal{S}$. (In (1), the vertexes are the coordinate vectors in $\mathbb{R}^{n+1}$.) Simplexes are often specified in terms of their vertexes:

$$\mathcal{S} = \langle P_0, \ldots, P_q \rangle \quad \text{(and usually } q = n \text{)}. \tag{3}$$

When $n = 1$, a simplex corresponds to a line segment, when $n = 2$, a simplex corresponds to a triangle in $\mathbb{R}^3$, while simplexes with $n = 3$ correspond to tetrahedra.

Remark 1. In branch and bound algorithms for global optimization, the simplex $\mathcal{S}$ is partitioned or subdivided in various ways. That is,

Definition 1. *A subdivision of a simplex $\mathcal{S}$ is a set of simplexes $\{\mathcal{S}_1, \ldots, \mathcal{S}_m\}$ such that*

$$\mathcal{S} = \bigcup_{i=1}^{m} \mathcal{S}_i, \quad \mathcal{S}_i \cap \mathcal{S}_j \text{ lies on the boundary of } \mathcal{S}_i \text{ and } \mathcal{S}_j \text{ for each } i, j. \tag{4}$$

A common subdivision in this context is *bisection* into two simplexes, $\mathcal{S}_1$ and $\mathcal{S}_2$, by replacing P_i by $\frac{1}{2}(P_i + P_j)$ in $\mathcal{S}_1$ and replacing P_j by $\frac{1}{2}(P_i + P_j)$ in $\mathcal{S}_2$, for some suitably chosen i and j. If $\mathcal{S}$ is identified with $\mathcal{S}_c$ and $f(x_0, \ldots x_n)$ is defined on $\mathcal{S}_c$, we can identify each of $\mathcal{S}_1$ and $\mathcal{S}_2$ with $\mathcal{S}$, but then the domain of f (or coefficients of f if f is a polynomial in the barycentric coordinates) needs to be rescaled to the barycentric coordinates over $\mathcal{S}_1$ and over $\mathcal{S}_2$.

1.2   Bernstein Polynomials

Univariate Bernstein Polynomials. If f is a polynomial, or in some cases a rational function, bounds on the ranges can be computed with *Bernstein polynomials*, both over boxes and simplexes. Bernstein polynomials have been analyzed by approximation theorists for over a century. The $d + 1$ Bernstein polynomials of degree d form a basis for degree d polynomials over the interval $[0, 1]$, and are given by

$$B_i^{(d)}(t) = \binom{d}{i} t^i (1 - t)^{d-i}, \quad 0 \leq i \leq d, \tag{5}$$

Following the notation in [2, p. 7], the properties of Bernstein polynomials that make them useful are:

$$\left. \begin{array}{ll} \sum_{i=0}^{d} B_i^{(d)}(t) \equiv 1 & \text{(partition of unity).} \\ B_i^{(d)}(t) \geq 0 \quad \text{for } t \in [0, 1] & \text{(non-negativity).} \\ B_i^{(d)}(t) = (1 - t)B_i^{(d-1)}(t) + tB_{i-1}^{(d-1)}(t) & \text{(recursion).} \end{array} \right\} \tag{6}$$

The general form (5) combined with the partition of unity property makes Bernstein polynomials suitable for representation of functions defined on simplexes. Observe, for $n = 1$, if $x_0 = 1 - t$ and $x_1 = t$, $B_i^{(1)} = x_i$. For a general function f defined on $[0, 1]$, the Bernstein approximation to f by a degree d polynomial is

$$f(t) \approx f_{B^{(d)}}(t) = \sum_{i=0}^{d} f\left(\frac{i}{d}\right) B_i^{(d)}(t). \tag{7}$$

For continuous f, the Bernstein approximation converges relatively slowly to f as the degree n increases, but the f_B are very smooth (without the overshoot in high degree polynomial interpolation and regression, or even in splines), and the convergence is uniform. Furthermore:

1. Because of the partition of unity property (6), $f_{B^{(d)}}(t)$ is a weighted average of the $f(t_i) = f(i/d)$, and therefore lies between the smallest and largest values of f at these $n + 1$ equally spaced sample points t_i.
2. Suppose f is a homogeneous degree d polynomial of two variables:

$$f(x_0, x_1) = \sum_{i=0}^{d} a_i x_0^i x_1^{d-i} \quad \text{with the condition } x_0 + x_1 = 1.$$

Then

$$f(x_0, x_1) = \sum_{i=0}^{d} \frac{a_i}{\binom{d}{i}} B_i^{(d)}(t), \quad \text{where } x_0 = t, \ x_1 = 1 - t, \ \text{ and} \left.\vphantom{\sum_{i=0}^{d}}\right\}$$

$$\min_{\substack{x_0 \in [0,1], \\ x_0 + x_1 = 1}} f(x_0, x_1) = \min_{0 \le i \le d} \frac{a_i}{\binom{d}{i}}, \quad \max_{\substack{x_0 \in [0,1], \\ x_0 + x_1 = 1}} f(x_0, x_1) = \max_{0 \le i \le d} \frac{a_i}{\binom{d}{i}}. \tag{8}$$

Property 2 concerns a homogeneous polynomial defined on a one-dimensional simplex ($n = 1$), and shows a quick way of computing exact bounds on the range of that polynomial. For example, a multi-dimensional analog of Property 2 can be used directly in algorithms for quadratic programming problems.

Property 1 can be used in conjunction with (7) for vector valued functions $\vec{f}(t)$, in particular for $\vec{f}(t) \in \mathbb{R}^2$ or $\mathbb{R}^3$. In that case, the resulting curve $\vec{f}(t)$, $0 \le t \le 1$ is called a *Bézier curve*, and the values $\vec{f}(i/d)$, usually given as discrete points in $\mathbb{R}^2$ or $\mathbb{R}^3$ rather than with reference to an underlying function, are called the *control points*. In 1959, Paul de Casteljau at Citroën (and independently, Pierre Bézier at Renault) at developed an ingenious algorithm, based on the above properties of Bernstein polynomials, to evaluate Bézier curves, for computer aided geometric design. The de Casteljau algorithm is now ubiquitous throughout the computer science literature and common in implementations. Furthermore, the de Casteljau algorithm is found in the literature on global optimization. Information about Bézier curves is available in numerous papers and course notes; a somewhat recent review is [3].

In addition to Bézier curves, the de Casteljau algorithm has been studied in the context of bounding ranges of functions. Thus, the plethora of literature on the de Casteljau algorithm within the computer aided geometric design literature is available to designers of global optimization algorithms.

Multivariate Bernstein Polynomials. Computer-aided geometric design researchers, as well as global optimization experts and others, have examined generalizations of the definition (5), properties (6) and the approximation (7) to $n > 1$. Tensor products of the $B_i(d)$ are used, with generalizations to both boxes and simplexes. Here, we focus on such generalizations to homogeneous polynomials and n-simplexes in $\mathbb{R}^{n+1}$.

In much of the literature, the multi-dimensional forms are written down with the aid of multi-indexes. Loosely following the notation in [14] for the simplicial extension, we have

Definition 2. *A multi-index $\vec{i}$ is simply an $(n + 1)$-vector of indexes:*

$$\vec{i} = (i_0, \ldots, i_n), \quad |\vec{i}| = \sum_{j=0}^{n} i_j, \quad \text{and} \quad x^{\vec{i}} = \prod_{j=0}^{n} x_j^{i_j}, \quad \binom{d}{\vec{i}} = \frac{d!}{\prod_{j=0}^{n}(i_j!)}.$$

The multi-dimensional simplicial Bernstein functions are defined on the canonical n-simplex[1] $\mathcal{S}_c \in \mathbb{R}^{n+1}$ of (1). With the notation in Definition 2, the n-dimensional *simplicial Bernstein basis functions* corresponding to (5) are

$$B_{\vec{i}}^{(d,n)}(\vec{x}) = \binom{d}{\vec{i}} x^{\vec{i}} \quad \text{for } \vec{x} = (x_0, \dots x_n) \in \mathcal{S}_c \text{ and } |\vec{i}| = d, \tag{9}$$

where $\mathcal{S}_c$ is the canonical simplex (1). Observe that (9) corresponds to (5) for $n = 1$, $x_0 = 1 - t$, $x_1 = t$.

Definition 3. *A homogeneous degree d polynomial of $n+1$ variables is a polynomial of the form $f(x_0, \dots x_n) = \sum\limits_{|\vec{i}|=d} a_{\vec{i}} x^{\vec{i}}$, that is, a polynomial of $n + 1$ variables each of whose non-zero terms has total degree d.*

Remark 2. Corresponding to (8), if f is a homogeneous polynomial of $n + 1$ variables of the form in Definition 3 defined on $\mathcal{S}_c$, and $\vec{x} = (x_0, \dots, x_n)$, then

$$\left. \begin{aligned} f(\vec{x}) &= \sum_{|\vec{i}|=d} \frac{a_{\vec{i}}}{\binom{d}{\vec{i}}} B_{\vec{i}}^{(d,n)}(t), \quad \text{where } \vec{x} \in \mathcal{S}_c, \text{ and} \\ \min_{\vec{x} \in \mathcal{S}_c} f(\vec{x}) &= \min_{|\vec{i}|=d} \frac{a_{\vec{i}}}{\binom{d}{\vec{i}}}, \quad \max_{\vec{x} \in \mathcal{S}_c} f(\vec{x}) = \max_{|\vec{i}|=d} \frac{a_{\vec{i}}}{\binom{d}{\vec{i}}}. \end{aligned} \right\} \tag{10}$$

In fact, arbitrary (non-homogeneous) polynomials can be represented in terms of the Bernstein basis; this is necessary for many important problems when using Bernstein techniques in global optimization. The conversion process, for various n has been presented and studied in the literature; for example, an algorithm for the $n = 2$ case is given in [23]. We analyze the conversion for an important case in global optimization in Sect. 2 below.

An advantage of the Bernstein polynomial representation is that the de Casteljau algorithm can simultaneously compute coefficients and bounds over each element $\mathcal{S}_i$ of a subdivision (4) of $\mathcal{S}$, with respect to the local barycentric coordinates for $\mathcal{S}_i$, by taking combinations of the coefficients over $\mathcal{S}$. In particular:

Remark 3. if $\mathcal{S}_1$ and $\mathcal{S}_2$ are formed from bisection and the coefficients are known for $\mathcal{S}$, then the coefficients (and hence bounds on f, via (10)) can be computed in $\binom{d+n}{n+1}$ add-and-shift operations; see [14, Lemma 3.2].

For example, for quadratic programming problems ($d = 2$) with the constraints $\sum_{i=0}^{n} x_i = 1$, $x_i \geq 0$, $0 \leq i \leq n$, this requires $n + 2$ adds and shifts, and may require less if many of the coefficients are non-zero.

Remark 4. Points to make: The multidimensional de Casteljau algorithm is most clearly implemented using recursion within a programming language, a technique that can be very inefficient in certain environments. On the other hand, implementation without recursion leads to confusing indexing schemes. Identification of important problems representing special cases where the algorithm can be simplified may thus be of use.

---

[1] Not to be confused with the *standard simplex* in $\mathbb{R}^n$ of the literature, defined in terms of (2).

1.3 Alternatives and Previous Work

Garloff et al. (see [19–21] and earlier works) Leroy (see [7]), Nataraj et al. (see [13,15,16], etc.) and others have studied Bernstein polynomials in the context of global optimization, in particular, in the conjunction with interval arithmetic to supply mathematical guarantees on the results. Muñoz and Narkawicz (see [12]) consider efficient representation of Bernstein polynomials for symbolic computations to be incorporated in global optimization algorithms.

In [5], we analyze the interplay between interval arithmetic and the constraints defining a simplex, to obtain formulas that are superior to naive interval arithmetic.

Here, we identify applications for which the simplicial Bernstein form is natural and likely to be competitive. We look at sharpness of bounds on the range, and we count the number of operations. Exhaustive comparison of implementation efficiencies on appropriate problems will be in future work.

2 Quadratic Programming Problems

The quadratic programming problems naturally suited to benefiting from Bernstein representations can be written as

$$\left.\begin{aligned}
\text{minimize } f(\vec{x}) &= \sum_{i=0}^{n} \sum_{j=0}^{n} a_{i,j} x_i x_j + \sum_{i=0}^{n} b_i x_i \\
\text{subject to } \sum_{i=0}^{n} x_i &= 1, \quad x_i \geq 0, \quad i = 0, \ldots, n, \\
h_j(\vec{x}) &= \sum_{i=0}^{n} h_{i,j} x_i = r_j, \quad j = 1, \ldots, m.
\end{aligned}\right\} \quad (11)$$

The non-negativity conditions and the condition $\sum_{i=0}^{n} x_i = 1$ are common to many practical problems, and define the optimization to be over the unit simplex $\mathcal{S}$. However, the linear terms $\sum_{i=0}^{n} b_i x_i$ and $\sum_{i=0}^{n} h_{i,j} x_i$ in the objective function as well as in the m additional equality constraints are usually present, and cannot be easily eliminated. In the absence of these additional constraints, (11) would be of the form in (10) with $d = 2$, and no conversion would be required. However, for this $d = 2$ case, rewriting each x_k, $0 \leq k \leq n$ in terms of the basis functions $B_{\vec{i}}^{(2,n)}$ is relatively simple, so the objective function f and the constraints h_j are written in terms of barycentric coordinates over the canonical simplex $\mathcal{S}$. We have

$$x_k = 1 - \sum_{\substack{j=0 \\ j \neq k}}^{n} x_j, \quad \text{so} \quad x_k \left(1 - \sum_{\substack{j=0 \\ j \neq k}}^{n} x_j \right) = x_k - \sum_{\substack{j=0 \\ j \neq k}}^{n} x_k x_j = x_k^2. \quad (12)$$

Adding the sum $\sum_{\substack{j=0 \\ j \neq k}}^{n} x_k x_j$ to both sides thus gives

$$x_k = \sum_{j=0}^{n} x_k x_j = \sum_{j=0}^{n} \frac{B_{\vec{i}_{j,k}}^{(2,n)}}{\binom{d}{\vec{i}_k}} = 2 \sum_{\substack{j=0 \\ j \neq k}}^{n} B_{\vec{i}_{j,k}}^{(2,n)} + B_{\vec{i}_{k,k}}^{(2,n)}, \quad \text{where} \quad (13)$$

$\vec{i}_{j,k}$ is the multi-index with $n + 1$ entries whose j-th and k-th entries are 1 and all of whose other entries are 0, and

$$B^{(2,n)}_{\vec{i}_{j,k}} = \begin{cases} 2x_k x_j, \ j \neq k, \\ x_k^2, \ j = k. \end{cases} \tag{14}$$

Replacing linear terms in (11) using (13) and collecting terms gives the linear programming problem completely in terms of barycentric coordinates:

$$\left. \begin{array}{l} \text{minimize} \ \ f(\vec{x}) = \displaystyle\sum_{i=0}^{n} \sum_{j=0}^{n} \alpha_{i,j} B^{(2,n)}_{\vec{i}_{i,j}} \\[2ex] \text{subject to} \ \displaystyle\sum_{i=0}^{n} x_i = 1, \ \ x_i \geq 0, \ \ i = 0, \ldots, n, \\[2ex] h_j(x) = \displaystyle\sum_{i=0}^{n} \sum_{k=0}^{n} \gamma_{j,i,k} B^{(2,n)}_{\vec{i}_{i,k}} = r_j, \ \ j = 1, \ldots, m. \end{array} \right\} \tag{15}$$

The conversion process need be done only once, before beginning the branch and bound algorithm. Once the conversion is done, the de Casteljau algorithm may be applied separately to f and the h_j in (15), to obtain a barycentric representation of the quadratic program over each element of a subdivision of $\mathcal{S}$. As those elements of the subdivision are further subdivided, the process can be repeated.

Remark 5. Higher degree polynomials can be treated with this method, but the total number of adds and shifts for a bisection $\binom{d+n}{d+1}$ grows rapidly with the degree d, and implementation is not as simple. This may cause computations with high d to be impractical, although parallelization, and indeed, computations similar to fast Fourier transforms can possibly be used; see [1].

3   Examples

The following examples are applications that have appeared in the literature or in private correspondence.

Example 1. This example, namely, the Markowitz model of stock portfolio optimization, originates with [9]. It consists precisely of (11), with $b_i = 0$, $0 \leq i \leq n$, and $m = 1$. The objective f represents risk to be minimized, the $a_{i,j}$ are the correlations between holdings, and the constant r in the constraint c represents the required rate of return.

We will illustrate the use of Bernstein expansions in the de Casteljau algorithm, as in [14], to facilitate computations associated with a subdivision process. In Example 1. f and the h_j have separate Bernstein expansions, and their ranges subject to the barycentric condition $\sum_{i=0}^{N} x_i = 1$ can be computed in parallel (e.g. in a vector computation). For brevity, we illustrate the technique with f alone. The objective function of Example 1 is exactly the objective function f in

(2). We consider the case $d = 2$, $n = 2$. Then $\vec{x} = (x_0, \, x_1, x_2)$, and the objective function has 6 coefficients $a_{0,0}$, $a_{0,1}$, $a_{0,2}$, $a_{1,1}$, $a_{1,2}$, and $a_{2,2}$. In our illustration, set the coefficients the following:

$$a_{0,0} = 2, \quad a_{0,1} = 6, \quad a_{0,2} = 8, \quad a_{1,1} = 4, \quad a_{1,2} = 12, \quad a_{2,2} = 8,$$

so the objective function is

$$f\left(\vec{x}\right) = 2x_0^2 + 6x_0x_1 + 8x_0x_2 + 4x_1^2 + 12x_1x_2 + 8x_2^2. \tag{16}$$

To apply the de Casteljau algorithm, the homogeneous polynomial (16) needs to be transformed into Bernstein form according to Remark 2. The corresponding Bernstein coefficients after conversion are

$$\alpha_{0,0} = 2, \alpha_{0,1} = 3, \alpha_{0,2} = 4, \alpha_{1,1} = 4, \alpha_{1,2} = 6, \alpha_{2,2} = 8.$$

The objective function in Bernstein form is given by

$$f\left(\vec{x}\right) = 2B_{\vec{i}_{0,0}}^{(2,2)} + 3B_{\vec{i}_{0,1}}^{(2,2)} + 4B_{\vec{i}_{0,2}}^{(2,2)} + 4B_{\vec{i}_{1,1}}^{(2,2)} + 6B_{\vec{i}_{1,2}}^{(2,2)} + 8B_{\vec{i}_{2,2}}^{(2,2)}. \tag{17}$$

We introduce additional notation, similar to that in [14], to denote Bernstein coefficients associated with subdivisions of the original simplex, as follows.

$$\alpha_{0,0} = c_{\vec{i}_{0,0}} = c_{2,0,0} = 2; \quad \alpha_{0,1} = c_{\vec{i}_{0,1}} = c_{1,1,0} = 3; \quad \alpha_{0,2} = c_{\vec{i}_{0,2}} = c_{1,0,1} = 4$$
$$\alpha_{1,1} = c_{\vec{i}_{1,1}} = c_{0,2,0} = 4; \quad \alpha_{1,2} = c_{\vec{i}_{1,2}} = c_{0,1,1} = 6; \quad \alpha_{2,2} = c_{\vec{i}_{2,2}} = c_{0,0,2} = 8.$$

The indexes for these coefficients correspond to scaled barycentric coordinates on edges of the simplex, as illustrated in Fig. 1; they are related to the values of the polynomial at those points through (9). According to Remark 2, the objective function is bounded within $[2, 8]$.

We now illustrate how these coefficients are reassigned and combined to obtain the coefficients for the barycentric representation over the two sub-simplexes $\mathcal{S}_1$ and $\mathcal{S}_2$ obtained by bisecting the edge connecting P_1 and P_2. The corresponding computations may be done efficiently with the Edge DeCasteljau algorithm in [14]. In particular, running EdgeDecasteljau($b_{2,2}[\mathcal{S}]$, $\frac{1}{2}; \dots$) gives the coefficients in Fig. 2(b). The de Casteljau algorithm computes these coefficients through a simple averaging process of adjacent coefficients. One reads off the Bernstein coefficients of the polynomials in barycentric form for $\mathcal{S}_1$ and $\mathcal{S}_2$ directly from this diagram, as illustrated in Fig. 3 According to Remark 2, f is bounded over $\mathcal{S}_1$ by $[2, 8]$, and f is bounded over $\mathcal{S}_2$ by $[2, 6]$. If we continued the process within a branch and bound algorithm, $\mathcal{S}_1$ and / or $\mathcal{S}_2$ can replace $\mathcal{S}$, for the process to be repeated. Note that, since f is quadratic, the ranges $[2, 8]$ and $[2.6]$ are exact ranges over $\mathcal{S}_1$ amd $\mathcal{S}_2$, respectively.

Example 2. The condition $\sum_{i=0}^{n} x_i^2 = 1$ is equivalent to requiring the optimum be on the unit $(n+1)$-sphere, a much-studied condition with many applications. As one of many examples of this, the work [17] deals with minimization of a homogeneous polynomial on a sphere. If each x_i only occurs to even powers in the objective and other constraints, a change of variables $\tilde{x}_i = x_i^2$ transforms the problem to optimization over the unit simplex.

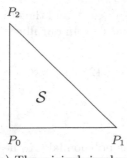

(a) The original simplex.

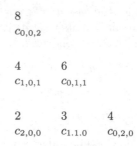

(b) Corresponding coefficient labeling.

Fig. 1. Simplex coefficients for Example 1.

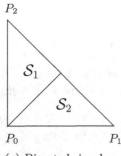

(a) Bisected simplex.

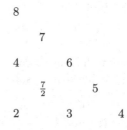

(b) de Casteljau output.

Fig. 2. Bisecting the simplex.

8
$c_{0,0,2}$

4   7
$c_{1,0,1}$ $c_{0,1,1}$

2       $\frac{7}{2}$ 6
$c_{2,0,0}$ $c_{1.1.0}$ $c_{0,2,0}$

(a) Coefficients for $\mathcal{S}_1$.

6
$c_{0,0,2}$

$\frac{7}{2}$ 5
$c_{1,0,1}$ $c_{0,1,1}$

2   3   4
$c_{2,0,0}$ $c_{1.1.0}$ $c_{0,2,0}$

(b) Coefficients for $\mathcal{S}_2$.

Fig. 3. Bernstein coefficients for $\mathcal{S}_1$ and $\mathcal{S}_2$.

Example 3. Minimization over the ℓ_1 sphere is minimization subject to the condition $\sum_{i=0}^{n} |x_i| = 1$. Minimization over the first orthant of the ℓ_1-sphere can thus be viewed as minimization over the standard simplex S. If, for example, there is some symmetry across orthants, techniques for optimization over a simplex (and in particular, the Bernstein representation) can be used. This would happen, for example, if most of the individual variables only occur with even powers, or if there are a few variables that just occur with odd powers. However, associated branch and bound algorithms are likely to be practical only for relatively small n.

4 Summary

We have reviewed simplexes and the simplicial Bernstein form for multivariate polynomials in the context of global optimization and, indirectly, computing α-cuts of a fuzzy logic output. We have identified contexts commonly occurring in applications in which the simplicial Bernstein form and computations with it simplify and are likely to prove practical and competitive, and have presented some preliminary examples. Actual comparisons and computations, including within a fuzzy logic context, will appear in future work.

References

1. Bezerra, L.: Efficient computation of Bézier curves from their Bernstein-Fourier representation. Appl. Math. Comput. **220**, 235–238 (2013)
2. Böhm, W., Farin, G., Kahmann, J.: A survey of curve and surface methods in CAGD. Comput. Aided Geom. Des. **1**(1), 1–60 (1984)
3. Farouki, R.T.: The Bernstein polynomial basis: a centennial retrospective. Comput. Aided Geom. Des. **29**(6), 379–419 (2012)
4. Hu, C., Kearfott, R.B., de Korvin, A. (eds.): Knowledge Processing with Interval and Soft Computing. Advanced Information and Knowledge Processing. Springer, New York (2008). https://doi.org/10.1007/978-1-84800-326-2
5. Karhbet, S., Kearfott, R.B.: Range bounds of functions over simplices, for branch and bound algorithms. Reliab. Comput. **25**, 53–73 (2017). Special volume containing refereed papers from SCAN 2016, guest editors Vladik Kreinovich and Warwick Tucker
6. Kreinovich, V.: Relations between interval and soft computing, pp. 75–97. In: Hu, C. et al. [4] (2008)
7. Leroy, R.: Convergence under subdivision and complexity of polynomial minimization in the simplicial Bernstein basis. Reliab. Comput. **17**(1), 11–21 (2012)
8. Lodwick, W.A., Jamison, K.D.: Special issue: interfaces between fuzzy set theory and interval analysis. Fuzzy Sets Syst. **135**(1), 1–3 (2003). Interfaces between fuzzy set theory and interval analysis
9. Markowitz, H.: Portfolio selection. J. Finance **7**(1), 77–91 (1952)
10. Moore, R.E., Kearfott, R.B., Cloud, M.J.: Introduction to Interval Analysis. SIAM, Philadelphia (2009)
11. Moore, R., Lodwick, W.: Interval analysis and fuzzy set theory. Fuzzy Sets Syst. **135**(1), 5–9 (2003). Interfaces between fuzzy set theory and interval analysis
12. Muñoz, C., Narkawicz, A.: Formalization of a representation of Bernstein polynomials and applications to global optimization. J. Autom. Reason. **51**(2), 151–196 (2013)
13. Nataraj, P.S.V., Arounassalame, M.: An interval newton method based on the Bernstein form for bounding the zeros of polynomial systems. Reliab. Comput. **15**(2), 109–119 (2011)

14. Peters, J.: Evaluation and approximate evaluation of the multivariate Bernstein-Bézier form on a regularly partitioned simplex. ACM Trans. Math. Softw. **20**(4), 460–480 (1994)

15. Ray, S., Nataraj, P.S.V.: A new strategy for selecting subdivision point in the Bernstein approach to polynomial optimization. Reliab. Comput. **14**(1), 117–137 (2010)

16. Ray, S., Nataraj, P.S.V.: A matrix method for efficient computation of Bernstein coefficients. Reliab. Comput. **17**(1), 40–71 (2012)

17. So, A.M.-C.: Deterministic approximation algorithms for sphere constrained homogeneous polynomial optimization problems. Math. Program. **129**(2), 357–382 (2011)

18. Tang, J.F., Wang, D.W., Fung, R.Y.K., Yung, K.-L.: Understanding of fuzzy optimization: theories and methods. J. Syst. Sci. Complex. **17**(1), 117 (2004)

19. Titi, J., Garloff, J.: Fast determination of the tensorial and simplicial Bernstein forms of multivariate polynomials and rational functions. Reliab. Comput. **25**, 24–37 (2017)

20. Titi, J., Garloff, J.: Matrix methods for the simplicial Bernstein representation and for the evaluation of multivariate polynomials. Appl. Math. Comput. **315**, 246–258 (2017)

21. Titi, J., Hamadneh, T., Garloff, J.: Convergence of the simplicial rational Bernstein form. In: Le Thi, H.A., Pham Dinh, T., Nguyen, N.T. (eds.) Modelling, Computation and Optimization in Information Systems and Management Sciences. AISC, vol. 359, pp. 433–441. Springer, Cham (2015). https://doi.org/10.1007/978-3-319-18161-5_37

22. Ullah, A., Li, J., Hussain, A., Shen, Y.: Genetic optimization of fuzzy membership functions for cloud resource provisioning. In: 2016 IEEE Symposium Series on Computational Intelligence (SSCI), pp. 1–8, December 2016

23. Waggenspack, W.N., Anderson, D.C.: Converting standard bivariate polynomials to Bernstein form over arbitrary triangular regions. Comput. Aided Des. **18**(10), 529–532 (1986)

24. Walster, G.W., Kreinovich, V.: Computational complexity of optimization and crude range testing: a new approach motivated by fuzzy optimization. Fuzzy Sets Syst. **135**(1), 179–208 (2003). Interfaces between fuzzy set theory and interval analysis

Moore: Interval Arithmetic in C++20

W. F. Mascarenhas[✉]

University of São Paulo, São Paulo, Brazil
walter.mascarenhas@gmail.com

1 Introduction

This article presents the Moore library for interval arithmetic in C++20. It gives examples of how the library can be used, and explains the basic principles underlying its design. It also describes how the library differs from the several other good libraries already available [1–12]. The Moore library is not compliant with the recent IEEE standards for interval arithmetic [13,14], and it will never be, but it would fair to rank in the top five in terms of compliance among the libraries in [1–12], the first and only truly compliant being [5], followed by [9], which is almost compliant. Of course, the library has limitations, and some of them are addressed in the last section, but only by playing a bit with it you will be able to tell whether its pluses offset its minuses.

The library was written mainly for myself and my students, to be used in our research about interval arithmetic and scientific computing in general. It is also meant to be used by other people, and its open source code and manual are available upon request to me. It is distributed under the Mozilla 2.0 license.

The Moore library will be useful for people looking for better performance or more precise types of endpoints for their intervals. To emphasize this point, Sect. 7 presents experiments showing that it is competitive with well known libraries, and it is significantly faster than some of them. The library will be most helpful for people using single or double precision arithmetic for most of their computation, with sporadic use of higher precision to handle critical particular cases. In this scenario the Moore library offers tools which are not available "out of the box" in other libraries, if available at all.

I assume that you are familiar with interval arithmetic, and understands me when I say that the library satisfies all the usual containment requirements of interval arithmetic. I also assume that you have some experience with templates, but you do not need to be familiar with the feature of C++20 which distinguishes most the Moore library from the others: *Concepts* [15], which are described in Sect. 3.

In the rest of the article I present the library, starting from the basic operations and moving to more advanced features, and present extensions of the library for linear algebra and automatic differentiation.

2 Hello Interval World

The Moore library can be used by people with varying degrees of expertise. Non experts can simply follow what is outlined in the code below:

© Springer International Publishing AG, part of Springer Nature 2018
G. A. Barreto and R. Coelho (Eds.): NAFIPS 2018, CCIS 831, pp. 519–529, 2018.
https://doi.org/10.1007/978-3-319-95312-0_45

```
#include "moore/config/minimal.h"
...
using namespace Ime::Moore;
...
UpRounding r;
TInterval<> x(2.0, 3.0);
TInterval<> y("[-1/3, 2/3]");

for(int i = 0; i < 10; ++i)
{
  y = (sin(x) - (y/x + 5.0) * y) * 0.05;
  cout << y << endl;
}
```

With the Moore library you construct intervals by providing their endpoints as numbers or strings, and then use them in arithmetical expressions as if they were numbers. The library also provides trigonometric and hyperbolic functions, their inverses, exponentials and logarithms, and convenient ways to read and write intervals to streams.

The file /moore/config/minimal.h contains the required declarations for using the library with double endpoints. The line

```
UpRounding r;
```

is mandatory. It sets the rounding mode to upwards, and the rounding mode is restored when r is destroyed. This is like one of the options in the boost library [1], but the Moore library uses only one rounding policy. In fact, giving fewer options instead of more is my usual choice. I only care about concrete use cases motivated by my own research, instead of all possible uses of interval arithmetic. I prefer to provide a better library for a few users rather than trying to please a larger audience which I will never reach.

Intervals are represented by the class template TInterval<E>, which is parameterized by a single type E. The letter E stands for *endpoint*, and both endpoints of the same interval are of the same type E, but intervals of different types may have different types of endpoints, and we can operate with them, as illustrated below. The default value for E is double, so that TInterval<> represents the plain vanilla intervals with endpoints of type double available in other libraries.

The library does not contain class hierarchies, virtual methods or policy classes. On the one hand, you can only choose the type of the endpoints defining the intervals of the form $[a, b]$ with $-\infty \leq a \leq b \leq +\infty$, or the empty interval. On the other hand, I do believe that it goes beyond what is offered by other libraries in its support of generic endpoints, intervals and operations. The library works with several types of endpoints "out of the box," that is, it provides tested code in which several types of endpoints can be combined, as in the example below. It also implements other kinds of convex subsets of the real line. For instance, it has classes to represent intervals of the form (a,b], [a,b) or (a,b), in which the "openness" of the endpoints can be decided at compile or runtime, and these half open intervals are used to implement tight arithmetic operations.

The code below illustrates the use of intervals with four types of endpoints:

```
TInterval<>            x(5,6);
TInterval<float>       y(-1,2);
TInterval<__float128> z("[-inf,4]");
TInterval<RealEnd<256>> w("[-1/3,2/3]");

auto h = x | y | 0.3;                    // the convex hull of x,y and 3
auto i = x & y & z & w;                  // the intersection of x,y,z and w
auto j = sin(z * x/cos(y * z)) - exp(w);
```

- The interval x has endpoints of type double.
- y has endpoints of type float.
- The endpoints of z have quadruple precision.
- w has endpoints of type RealEnd<256>, which are floating point numbers with $N = 256$ bits of mantissa, and you can choose other values for N.
- The compiler deduces that h is an interval with endpoints of type double, which is the appropriate type for storing the convex hull of x, y and 0.3.
- It also deduces that RealEnd<256> is the appropriate type of endpoints for the interval representing the intersection of x, y, z and w, and this is the endpoint type for j.

I ask you not to underestimate the code above. It is difficult to develop the infrastructure required to handle intervals with endpoints of different types in expressions as natural as the ones in that code. In fact, there are numerous issues involved in dealing with intervals with generic endpoints, and simply writing generic code with this purpose is not enough. The code must be tested, and my experience shows that it may compile for some types of endpoints and may not compile for others.

3   Concepts

The Moore Library differs significantly from the previous C++ interval arithmetic libraries due to its use of *Concepts*, a feature which will be part of the C++20 standard [15]. Concepts improve the diagnostic of errors in the compilation of templated C++ code, and they can be motivated by the following example. Suppose we write the code below to compute the length of intervals of types Interval provided by several libraries.

```
template <typename Interval>        // Code in a header file somewhere.
double length(Interval const& i) {  // Interval is meant to be a type
   return sup(i) - inf(i);          // provided by an existing interval
}                                   // arithmetic library.
```

This code works as long as the functions inf and sup are provided, either by the original library for the type Interval or by an adapter. However, it will not take long for someone to code something like the snippet below and get indecipherable error messages about infs, sups and strings.

```
void unlucky()        // code in a source file unrelated to intervals.
{
  std::string str("I know nothing about intervals!!!");
  std::cout << length(str) << std::endl;
}
```

When reading the error messages about infs and sups of strings in the compilation of the unlucky function, the programmer may not be aware of the chain of inclusions leading to the header file containing the declaration of the function length for intervals, and the length function for strings will be declared in yet another header file. It will be difficult to relate the error messages to the code which is apparently being compiled, and unexperienced programmers will get lost. Even people experienced with templates will tell you how frustrating these error messages can be, and this is indeed a problem with templates.

We could solve this problem by telling the compiler what an interval is. Knowing that strings are not intervals, it would not consider the function template length below as an option for strings, and there would be no meaningless error messages about infs and sups of strings.

```
template <Interval I>       // Telling the compiler that I must be an
double length(I const& i) { // interval for this function template to
  return sup(i) - inf(i);   // be considered.
}
```

In essence, this is what a concept is: a way to tell the compiler whether a type should be be considered in the instantiation of templates. In the Moore library concepts are used, for example, to tell whether a type represents an interval (the Interval concept) or an endpoint (the End concept), or when there exist an exact conversion from endpoints of type T to endpoints of type E (the Exact<T,E> concept.) We then can code as follows and the compiler will instantiate the appropriate templates. In the end, concepts allow us to operate naturally with intervals and endpoints of different types.

```
template <Interval I>                 // sum of intervals of the same type
I operator+(I const&, I const&)

template <Interval I, Interval J>     // sum of intervals when there
requires Exact<EndOf<J>, EndOf<I>>()  // is an exact conversion from
I operator+(I const&, J const&)       // J to I.

template <Interval I, Interval J>     // sum of intervals when there
requires Exact<EndOf<I>, EndOf<J>>()  // is an exact conversion from
J operator+(I const&, J const&  )     // I to J.

template <Interval I, End E>          // sum of an intervals and an
requires Exact<E, EndOf<I> >()        // endpoint when there is an
I operator+(I const&, E const&)       // exact conversion form E to I.
```

The code above also presents an alternative way to enforce concepts: the requires clauses. These clauses make sure that the operator+ will be considered only when there is an obviously consistent type for the output.

Overall, the motivation for concepts is clear and intuitive. Their problems lie in the details and the crucial question: How should we tell the compiler what an interval or and endpoint is (or any concept, really)? I do not know the best answer to this question, and neither does the rest of the C++ community. This is why concepts are taking so long to become part of the C++ standard.

This ignorance should not prevent us from using concepts. They are a great tool, and we can do a lot with what is already available. With time, as concepts and our experience with them evolve, we will improve our code. For now the Moore library tells the compiler in an ad-hoc way what intervals and endpoints are. It basically lists explicitly which types qualify for a concept, and avoids the more elaborate schemes to declare concepts which are already available, for two reasons: First, their current implementation has bugs (it does not handle recursion properly, for instance.) Second, it is difficult to list precisely and concisely all the requirements which would characterize intervals and endpoints. I would not be able to do it even if the current implementation of concepts were perfect.

The last questions are then: do concepts work for interval arithmetic? Are they worth the trouble? I would not have written this article if my answer to these questions were not an enthusiastic "yes!!", and I invite you to try out the library and verify whether you share my enthusiasm.

4   Input and Output

Flexible and precise input and output are essential for an interval arithmetic libary. The Moore library accepts as input interval literals and streams as follows

```
try {
  TInterval<> x("[]");       // the empty interval
  x = "[-inf, 1]";           // -inf = minus infinity
  x = "[2.0e-20, 1/3]";      // rational numbers are ok
  x = "[-2.345, 0x23Ap+4]";  // hexadecimal floats too
  std::cin >> x;             // reading from an input stream
} catch(...){}
```

As the code above indicates, the library throws an exception when the string literal meant to represent an interval is invalid. Strings in hexadecimal notation are handled exactly, and by using them for both input and output you can persist intervals without rounding errors. In the other formats the resulting interval is usually the tightest representable interval containing the input, the only exception being contrived rational numbers for which it would take an enormous amount of memory or time to compute this tight enclosure. In these rare cases you may get a memory allocation exception or need to wait forever.

Properly formatted output is important to visualize the results of interval computations, and the library implements an extension of the usual printf format to specify how intervals are written to streams. This extension is needed in order to align numbers properly in columns when printing vectors and matrices. For example, the code below creates a 3×3 matrix of intervals (a box matrix) and writes it to the standard output. The output is formatted according to the string

"11.2E3W26", which extends the argument "+11.2E" passed to printf to write floating point numbers in scientific notation (E), showing the plus sign (+), with 11 characters per number and 2 digits after the decimal point. We add the suffix "3W26" to the format to ensure that exponents are printed with 3 digits and each interval is 26 characters wide. Without this extension the output would not be as well as organized at it is below.

```
using I = TInterval<>;
text_format() = "+11.2E3W26";
TBoxMatrix<> a( { { I(0x1p-1021,0x1p+100), I()                },
                 { I("[-inf,0]"),          I("[0,inf]") },
                 { I(-12343,0),            I(50,10000)  } } );
std::cout << a << std::endl;
```

This is the output:

```
[ +4.45E-308, +1.27E+030] [                         ]
[       -INF, +0.00E+000] [ +0.00E+000,        +INF]
[ -1.24E+004, +0.00E+000] [ +5.00E+001, +1.00E+004]
```

5   Linear Algebra

Besides plain intervals, the library provides vectors of intervals, called boxes, and matrices with interval entries (box matrices) The arithmetic operations involving vectors and matrices are implemented using expression templates and one can write code as the one below, which handles the three dimensional vectors x and y and the 3×3 matrix a in a natural way.

```
using I = TInterval<>;

TBox<> x( {I(1,3), I(2,4), I(1,5)} );
TBox<> y( {I(1,2), I(2,3), I(2,3)} );

TBoxMatrix<> a( { { I(1,1), I(0,1), I(3,5) },
                 { I(2,1), I(2,2), I(4,7) },
                 { I(2,1), I(2,2), I(3,5) } });

TBox<> z = a * x + 2 * y + x;
TBox<> w = tr(a) * y + dot(y,z) * x; // tr(a) = transposed(a)
```

6   Automatic Differentiation

The Moore library is part of a larger collection of tools for scientific computing, called *Ime library*. As part of the work of my student Fernando Medeiros, the Ime library provides classes for automatic differentiation, and I now describe how these automatic differentiation tools by Fernando and myself are integrated with the Moore library. First, we use a function template to declare the function which we want to differentiate.

```
template <typename T>
T example(T const& x) {
  return exp( sqrt(exp(x)/ 3) + x) * (2 * x) - 10;
}
```

Once we have declared `example`, it is easy to compute its derivative using interval arguments. For instance, the function `newton_step` below performs one step of Newton's method for solving the equation $f(x) = 0$. In this code the type `ADT<I>` represents the usual pair of function value and derivative used in forward automatic differentiation schemes.

```
template <Interval I>
void newton_step(I& x, ADT<I> (*f)(I const& i)) {
  auto fd = adt(x, example); // evaluating f and its derivative
  x &= x - fd.f / fd.d;      // x = (x - f(x)/f'(x)) intersected with x
}

void calling_newton() {
  TInterval<> x(1,2);
  newton_step(x, example);
}
```

The library Ime also provides automatic differentiation for functions of several variables, like in the example below in which we print the enclosure of the function value and gradient of the given multivariate function.

```
template <typename T, int N>
T multivariate_example(StaticVector<T,N> const& x) {
  return exp( sqrt(exp(x[0] + x[1] / 3) + x[2]) * (2 * x[3])) / x[4];
}

void print_function_value_and_gradient() {
  using I = TInterval<>;
  text_format() = "+10.4E";
  StaticVector<I,5> x( {I(1,2), I(-2,3), I(3,4), I(-1,1), I(1,2)} );
  std::cout << adtnf(x, multivariate_example);
}
```

This is the output:

```
  f = [+2.730E-05,+1.832E+04]
g[0] = [-3.509E+05,+3.509E+05]
g[1] = [-1.170E+05,+1.170E+05]
g[2] = [-1.748E+04,+1.748E+04]
g[3] = [+5.724E-05,+3.596E+05]
g[4] = [-1.832E+04,-1.365E-05]
```

7   Experiments

This section presents the results of experiments comparing the Moore library with three other interval arithmetic libraries: boost interval [1], Filib [3] and

Table 1. Normalized times for the Lebesgue function

Moore	Filib	boost	P1788
1	3.8	1.1	268.5

libieeep1788 [9]. In summary, the experiments show that, for arithmetic operations, the Moore library is slightly faster than the boost library, it is significantly faster than the libieeep1788 library, and it is faster than the Filib library. However, in double precision the elementary functions (sin, cos, etc.) in Filib are significantly faster than the Moore library, which is in turn significantly faster than the boost library and the libieeep1788 library.

Besides the difference in speed, there is a difference in the accuracy of the elementary functions. When using IEEE754 double precision, due to the way in which argument reduction is performed, the boost and Filib libraries can lead to errors of order 10^{-8} in situations in which the Moore library and the libieeep1788 library lead to errors of the order 10^{-16}. In fact, in extreme cases these other libraries can produce intervals of length 2 when the sharpest answer would be an interval of length of order 10^{-16}.

The Moore library was implemented to be used in my research, and the experiments reflect this. I present timings related to my current research about the stability of barycentric interpolation [16–18]. In this research I look for parameters $w_0, \ldots w_n$ which minimize the maximum of the *Lebesgue function*

$$\mathcal{L}(\mathbf{w}; \mathbf{x}, t) := \sum_{k=0}^{n} \left| \frac{w_k}{t - x_k} \right| \Big/ \left| \sum_{k=0}^{n} \frac{w_k}{t - x_k} \right| \tag{1}$$

among all $t \in [-1, 1]$, for a given vector $\mathbf{x}$ of nodes, and I use interval arithmetic to find such minimizers and validate them.

The first experiment timed the evaluation of the Lebesgue function for 257 Chebyshev nodes of the second kind [16], with interval weights, at a million points t. I obtained the normalized times in Table 1 (the time for the Moore library was taken as the unit.) This table indicates that for the arithmetic operations involved in the evaluation of the Lebesgue function (1) the Moore library is more efficient that the boost, Filib and libieeep1788 libraries. The difference is slight between Moore and boost (10%), more relevant between Moore and Filib (about 300%) and very significant between Moore and libieeeP1788 (about 25000%).

In the second experiment, myself and my former student Tiago Montanher considered the computation of the roots of functions which use only arithmetic operations, like the Lebesgue function in Eq. (1) and its derivatives with respect to its parameters. The data for this experiment was generated with an interval implementation of Newton's method which can use any one of the four libraries mentioned above. We compared the times for the solution of random polynomial equations, with the polynomials and their derivatives evaluated by Horner's method. We obtained the times in Fig. 1, which corroborate the data in Table 1.

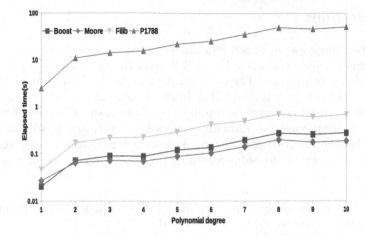

Fig. 1. Times for Newton's method with polynomials, in log scale.

Table 2. Time for 10^6 evaluations of the elementary functions with random intervals.

Function	Moore	Filib	boost	P1788
sin	0.552	0.032	1.444	9.320
cos	0.156	0.032	1.560	10.172
tan	0.124	0.020	0.756	2.476
atan	0.408	0.036	10.424	10.656
exp	0.308	0.164	4.532	4.644
asin	0.356	0.088	16.572	16.156
acos	0.368	0.088	16.724	16.748
log	0.272	0.044	5.404	5.272

The first two experiments show that the Moore library is competitive for arithmetic operations, but they tell only part of the history about the relative efficiency of the four libraries considered. In order to have a more balanced comparison, in the third and last experiment I compared the times that the four libraries mentioned above take to evaluate of the elementary functions (sin, cos, exp, etc.) using the IEEE 754 double precision arithmetic. The results of this experiment are summarized in Table 2 below, which shows that the Filib library is much faster than the Moore library in this scenario, and the Moore library is much faster than the other two libraries.

I emphasize that I tried to be fair with all libraries and, to the best of my knowledge, I used the faster options for each library. For instance, I used the boost library on its unprotected mode, which does not change rounding modes in order to evaluate arithmetical expressions. The code was compiled with gcc 6.2.0 with flag -O3 and NDEBUG defined (the flag -frounding-math should also be used when compiling the Moore library.)

8   Limitations

The Moore library was designed and implemented using a novel feature of the C++ language called concepts [15], and it pays the price for using the bleeding edge of this technology. The main limitations in the library are due to the current state of concepts in C++. For instance, only the latest versions of the gcc compiler support concepts, and today the library cannot be used with other compilers. Concepts are not formally part of C++ yet, and it will take a few years for them to reach their final form and become part of the C++ standard.

Additionally, several decisions regarding the library were made in order to get around bugs in gcc's implementations of concepts and in the supporting libraries, and in order to reduce the compilation time. The code would certainly be cleaner if we did not care about these practical issues, but without the compromises we took using the library would be more painful.

Another limitation is the need to guard the code by constructing an object of type `UpRounding`. In other words, the code must look like this

```
UpRounding r;
code using the Moore library
```

A similar requirement is made by the most efficient rounding policy for the boost library, but that library allows users to choose other policies for rounding, although the resulting code is less efficient. Things are different with the Moore library: as the buyers of Henry Ford's cars in the 1920s, its users can choose any rounding mode as they want, so long as it is upwards. Users wanting to mix code from the Moore library with code requiring rounding to nearest will need to resort to kludges like this one:

```
{
UpRounding r;
do some interval operations
}
back to rounding to nearest
{
UpRounding r;
do more interval operations
}
```

References

1. Brönnimann, H., Melquiond, G., Pion, S.: The design of the boost interval arithmetic library. Theoret. Comput. Sci. **351**(1), 111–118 (2006)
2. Hofschuster, W., Krämer, W.: C-XSC 2.0 – A C++ library for extended scientific computing. In: Alt, R., Frommer, A., Kearfott, R.B., Luther, W. (eds.) Numerical Software with Result Verification. LNCS, vol. 2991, pp. 15–35. Springer, Heidelberg (2004). https://doi.org/10.1007/978-3-540-24738-8_2
3. Lerch, M., Tischler, G., von Gudenberg, J.W.: FILIB++, a fast interval library supporting containment computations. ACM Trans. Math. Softw. **32**(2), 299–324 (2006)

4. Goualard, F.: Gaol: not just another interval library. http://www.sourceforge.net/projects/gaol/. Accessed 21 Sept 2016
5. Heimlich, O.: GNU octave interval package. http://octave.sourceforge.net/interval/. Accessed 21 Sept 2016
6. Nadezhin, D., Zhilin, S.: JInterval library: principles, development, and perspectives. Reliable Comput. **19**, 229–247 (2014)
7. Lambov, B.: Interval arithmetic using SSE-2. In: Hertling, P., Hoffmann, C.M., Luther, W., Revol, N. (eds.) Reliable Implementation of Real Number Algorithms: Theory and Practice. LNCS, vol. 5045, pp. 102–113. Springer, Heidelberg (2008). https://doi.org/10.1007/978-3-540-85521-7_6
8. Rouillier, F., Revol, N.: Motivations for an arbitrary precision interval arithmetic and the MPFI library. Reliable Comput. **11**, 275–290 (2005)
9. Nehmeier, M.: libieeep1788: A C++ implementation of the IEEE interval standard P1788. In: 2014 IEEE Conference on Norbert Wiener in the 21st Century (21CW), pp. 1–6 (2014)
10. Knueppel, O.: PROFIL/BIAS - a fast interval library. Computing. **53**(3–4), 277–287 (1994). http://www.ti3.tu-harburg.de/knueppel/profil/. Accessed 21 Sept 2016
11. Rump, M.: INTLAB - INTerval LABoratory. In: Csendes, T. (ed.) Developments in Reliable Computing, pp. 77–104. Kluwer Academic Publishers, Dordrecht (1999)
12. C++ Interval Arithmetic Programming Reference, MC68175/D, Sun Microsystems (1996). http://docs.sun.com/app/docs/doc/819-3696-10. Accessed 21 Sept 2016
13. 1788-2015 IEEE Standard for Interval Arithmetic, IEEE Std. (2015). https://standards.ieee.org/findstds/standard/1788-2015.html. Accessed 21 Sept 2016
14. P1788.1-Standard for Interval Arithmetic (simplified), IEEE Std. (2015). https://standards.ieee.org/develop/project/1788.1.html. Accessed 21 Sept 2016
15. Wikipedia: Introduction to C++ concepts. https://en.wikipedia.org/wiki/Concepts_(C%2B%2B). Accessed 21 Sept 2016
16. Mascarenhas, W.F.: The stability of barycentric interpolation at the chebyshev points of the second kind. Numer. Math. **128**, 265–300 (2014)
17. Mascarenhas, W.F., de Camargo, A.P.: The effects of rounding errors in the nodes on barycentric interpolation. Numer. Math. **135**, 113–141 (2017)
18. de Camargo, A.P., Mascarenhas, W.F.: The stability of extended floater-hormann interpolants. Numer. Math. **136**, 287–313 (2017)

Towards Foundations of Fuzzy Utility: Taking Fuzziness into Account Naturally Leads to Intuitionistic Fuzzy Degrees

Christian Servin[1] and Vladik Kreinovich[2(✉)]

[1] Computer Science and Information Technology Department,
El Paso Community College, 919 Hunter, El Paso, TX 79915, USA
cservin@gmail.com
[2] Department of Computer Science, University of Texas at El Paso,
El Paso, TX 79968, USA
vladik@utep.edu

Abstract. The traditional utility-based decision making theory assumes that for every two alternatives, the user is either absolutely sure that the first alternative is better, or that the second alternative is better, or that the two alternatives are absolutely equivalent. In practice, when faced with alternatives of similar value, people are often not fully sure which of these alternatives is better. To describe different possible degrees of confidence, it is reasonable to use fuzzy logic techniques. In this paper, we show that, somewhat surprisingly, a reasonable fuzzy modification of the traditional utility elicitation procedure naturally leads to intuitionistic fuzzy degrees.

1 Formulation of the Problem

Need to Help People Make Decisions. In many practical situations, we need to make a decision, i.e., we need to select an alternative which is, for us, better than all other possible alternatives.

If the set of alternatives is small, we can easily make such a decision: indeed, we can easily compare each alternative with every other one, and, based on these comparisons, decide which one is better. However, when the number of alternatives becomes large, we have trouble making decisions. Even in simple situations, when we are looking for cereal in a supermarket, there are usually so many selections that we just ignore most of them and go with a familiar one – instead of the optimal one.

The situation is even more complicated if we are trying to make a decision not on behalf of ourselves, but rather on behalf of a company or a community. In this case, even comparing two alternatives is not easy: it requires taking into account interests of different people involved, so the decision making process becomes even more complicated.

© Springer International Publishing AG, part of Springer Nature 2018
G. A. Barreto and R. Coelho (Eds.): NAFIPS 2018, CCIS 831, pp. 530–537, 2018.
https://doi.org/10.1007/978-3-319-95312-0_46

Traditional Approach to Decision Making: The Notion of Utility. The traditional approach to decision making was originally motivated by the idea of money.

We all know what money is, but when money was invented, it was a revolutionary idea that made economic exchange much easier. Indeed, before money was invented, people exchanged goods by barter: chicken for a shirt, jewelry for boots, etc. Thus, to make a proper decision, every person needed to be able to compare every two items with each other: how many chickens is this person willing to exchange for a shirt, how many boots for a golden earing, etc. For n goods, we have $\dfrac{n \cdot (n-1)}{2} \approx \dfrac{n^2}{2}$ possible pairs. So, each person had to have in mind a table of $n^2/2$ numbers.

With money as a universally accepted means of exchange, all the person needs to do is to decide, for each of n items, how much he or she is willing to pay for 1 unit. So, to successfully make decisions, it is sufficient to know n numbers – the values of each of n items. Then, even when we want to barter, we can easily decide how many chickens are worth a shirt: it is sufficient to divide the price of a shirt by the price of a chicken.

A similar idea can be used to compare different alternatives. All we need is to have a numerical scale, i.e., a 1-parametric family of "standard" alternatives whose quality increases with the increase in the value of the parameter. This can be the money amount. Alternatively, this can be the probability p of a lottery in which we get something very good: the larger the probability, the more preferable the lottery.

Then, instead of comparing every alternative with every other alternative, we simply compare every alternative with alternatives on the selected scale, and thus, for each alternative, we find the numerical value of the standard alternative which is equivalent to a given one. This numerical value is known as the *utility* $u(a)$ of a given alternative a; see, e.g., [3,4,6,8,11].

In terms of utility, an alternative a is better than the alternative a' if and only the utility $u(a)$ of the alternative a is larger than the utility $u(a')$ of the alternative a'. Thus, once we have found the utility $u(a)$ of each alternative, then it is easy to predict which alternative the person will select: he/she will select the alternative for which the utility $u(a)$ is the largest possible.

How to Actually Find the Utility. From the algorithmic viewpoint, the fastest way to find the utility of a given alternative a based on binary comparisons is to use bisection. Usually, we have an a prior lower bound and an a priori upper bound for the desired utility $u(a) : \underline{u} \leq u(a) \leq \overline{u}$. In other words, we know that the desired utility $u(a)$ is somewhere in the interval $[\underline{u}, \overline{u}]$. In this procedure, we will narrow down this interval.

Once an interval is given, we can compute its midpoint $\widetilde{u} = \dfrac{\underline{u} + \overline{u}}{2}$ and compare a with the corresponding standard alternative $s(\widetilde{u})$.

If a is exactly equivalent to the resulting standard alternative, this means that we have found the exact value of the utility $u(a)$: it is equal to $\tilde{u}$. However, such exact equivalences are rare; in most cases, we will find out that:

- either a is better than $s(\tilde{u})$; we will denote it by $s(\tilde{u}) < a$; or
- the standard alternative is better: $a < s(\tilde{u})$.

In the first case, the preference $s(\tilde{u}) < a$ means that $\tilde{u} < u(a)$. Thus, we know that $u(a) \in [\tilde{u}, \overline{u}]$. In other words, we have a new interval containing the desired utility. We can obtain this new interval if we replace the previous lower bound $\underline{u}$ with the new lower bound $\tilde{u}$.

In the second case, the preference $a < s(\tilde{u})$ means that $u(a) < \tilde{u}$. Thus, we know that $u(a) \in [\underline{u}, \tilde{u}]$. In other words, we have a new interval containing the desired utility. We can obtain this new interval if we replace the previous upper bound $\overline{u}$ with the new upper bound $\tilde{u}$.

In both cases, the width of the interval is decreased by a factor of 2. Then, we can repeat this procedure, and in k steps, we get $u(a)$ with accuracy 2^{-k}. For example, in 7 steps, we get an accuracy of 1%.

Need to take Fuzziness into Account. The above procedure works well if a person is absolutely sure about his/her preferences. In practice, we are often not 100% sure about our preferences, especially when we compare alternatives of nearby value.

It is reasonable to describe this uncertainty in fuzzy terms. For example, if we use money as a standard scale, then for each alternative a, instead of having a single amounts of money equivalent to this item, we may have different amounts with different degree of certainty. In other words, instead of the above crisp model, in which a person has an exact utility value $u(a)$ for each alternative a, we know have a fuzzy model in which for each person and for each alternative a, we have a membership function $\mu_a(u)$ that describes, for each possible value u, to what extend this value u is equivalent to the alternative a; see, e.g., [2,5, 7,9,10,12].

How to Elicit Fuzzy Utility: A Reasonable Idea. We know how to elicit crisp utility $u(a)$ of a given alternative a: we need to compare the alternative a with different values u_0 of the scale. In the case of fuzzy utility, it is reasonable to apply the same procedure. The only difference is that now, since the utility value $u(a)$ is fuzzy, this comparison will not lead to a crisp "yes"-"no" answer; instead, we will get a fuzzy answer – the degree to which it is possible that a is better than u_0 (and, if needed, the degree to which it is possible that a is worse than u_0).

Remaining Open Problems and What we do in this Paper. In the crisp case, we can determine the utility value $u(a)$ from the results of the user's comparisons.

To deal with the more realistic fuzzy case, we need to be able to extract the fuzzy utility from the fuzzy answers to different comparisons. This is the question that we deal with in this paper.

Interestingly, it turns out that in this context, intuitionistic fuzzy degrees (see, e.g., [1]) naturally appear – in other words, instead of a single degree of confidence in each corresponding statement, we now get *two* degrees:

- the degree to which this statement is true, and
- the degree to which this statement is false,

and, in contrast to the traditional fuzzy logic, these degrees do not add up to 1.

2 Analysis of the Problem

What Happens if we Compare the Alternative a with a Fixed Value u_0 on the Utility Scale? As we have mentioned earlier, while in the crisp case, each alternative a is equivalent to a single number $u(a)$ on the utility scale, in general, the utility of an alternative is characterized not by a single number, but rather with a membership function $\mu_a(u)$. This function describes, for each value u from the utility scale, to what extent the alternative a is equivalent to u.

What will happen is we compare the alternative a to a value u_0 on the utility scale? In the crisp case, since the changes that a is exactly equivalent to a_0 are slim, we have either $a < u_0$ or $u_0 < a$. So, we can ask whether a is better than u_0, or we can ask whether u_0 is better than a – whatever question we ask, we get the exact same information.

Let us first consider the question of whether a is better than u_0, i.e., whether $u_0 < a$. How can we extend this to the fuzzy case? To perform this extension, it is convenient to take into account that while from the purely mathematical viewpoint, $<$ is a relation – and in mathematics, relations usually treated differently than functions – from the computational viewpoint, $<$ is simply a function. Just like $+$ is a function that takes two numbers and returns a number which is their sum, the relation $<$ is a function that takes two numbers and returns a boolean value: true or false.

Since $<$ can be naturally treated as function, the question of how to extend this to fuzzy becomes a particular case of a more general question of how to extend functions to fuzzy – and this extension is well known, it is described by Zadeh's extension principle. Let us recall how this principle is usually derived.

Zadeh's Extension Principle and How it is Usually Derived. Suppose that we have a function $y = f(x_1, \ldots, x_n)$ of n real-valued variables, and we have fuzzy information about the values $x_1, \ldots, x_n$, i.e., we know membership functions $\mu_1(x_1), \ldots, \mu_n(x_n)$ that describes our knowledge about the inputs $x_1, \ldots, x_n$. Based on this information, what do we know about $y = f(x_1, \ldots, x_n)$?

Intuitively, Y is a possible value of the variable y if there exists values $X_1, \ldots, X_n$ for which X_1 is a possible value of x_1 and ... and X_n is a possible value of x_n and $Y = f(X_1, \ldots, X_n)$. We know the degrees, $\mu_i(X_i)$ to which each real number X_i is a possible values of the input x_i. To combine these degrees into our degree of confidence in a composite and-statement, we can use an "and"-operation (t-norm), the simplest of which is $\min(a, b)$. Thus, for each tuple $(X_1, \ldots, x_n)$ for which $Y = f(X_1, \ldots, X_n)$, our degrees of confidence is the above and-statement is $\min(\mu_1(X_1), \ldots, \mu_n(X_n))$.

The existential quantifier "there exists" is, in effect, an "or": it means that either this property is true for one tuple, or for another tuple, etc. Thus, to find the degree to which the value Y is possible, we need to apply an "or"-operation (t-conorm) to the degrees of confidence of the corresponding and-statements. The simplest "or"-operation is $\max(a, b)$. Thus, we arrive at the following formula for the degree $\mu(Y)$ to which Y is a possible value of the variable y:

$$\mu(Y) = \max\{\min(\mu_1(X_1), \ldots, \mu_n(X_n)) : f(X_1, \ldots, X_n) = Y\}.$$

This formula – first proposed by L. Zadeh himself – is known as *Zadeh's extension principle.*

Let us Apply Zadeh's Extension Principle to our Problem: Resulting Formulas. In our case, we have a Boolean-valued function $f(x_1, x_2) = (x_1 < x_2)$ of $n = 2$ real-valued variables. When we compare an alternative a with fuzzy utility $\mu_a(u)$ with a crisp value u_0, Zadeh's extension principle takes the following form:

- for the value $y =$"true", the degree $\mu_+(a < u_0)$ that the statement $a < u_0$ is true is equal to

$$\mu_+(a < u_0) = \max(\mu_a(u) : u < u_0);$$

- for the value $y =$"false", the degree $\mu_-(a < u_0)$ that the statement $a < u_0$ is false is equal to

$$\mu_-(a < u_0) = \max(\mu_a(u) : u \geq u_0).$$

Let us Analyze the Resulting Formulas. Intuitively, since in fuzzy logic negation is represented by the function $1 - a$ (in the sense that our degree of believe that A is false is estimated as 1 minus degree that A is true), we should expect that $\mu_+(a < u_0) + \mu_-(a < u_0) = 1$. Let us show, however, that this is not the case.

Indeed, let us consider a typical case when $\mu_a(u)$ is a fuzzy number, i.e., when for some value U:

- the function $\mu_a(u)$ increases to 1 when $u \leq U$, and
- this function decreases from 1 when $u \geq U$.

When $u_0 < U$, then the function $\mu_a(u)$ is increasing for all $u < u_0$ and thus, $\mu_+(a < u_0) = \mu_a(u_0)$. On the other hand, since $u_0 < U$ and for $u = U$, we have $\mu_a(U) = 1$, we get $\mu_-(a < u_0) = 1$. Thus,

$$\mu_+(a < u_0) + \mu_-(a < u_0) = 1 + \mu_a(u) \neq 1,$$

unless, of course, we consider absolutely impossible values u for which $\mu_a(u) = 0$.

Similarly, when $u_0 \geq U$, then the function $\mu_a(u)$ is decreasing for all $u > u_0$ and thus, $\mu_-(a < u_0) = \mu_a(u_0)$. On the other hand, since $u_0 \geq U$ and for $u = U$, we have $\mu_a(U) = 1$, we get $\mu_+(a < u_0) = 1$. Thus, in this case too, we have

$$\mu_+(a < u_0) + \mu_-(a < u_0) = 1 + \mu_a(u) \neq 1,$$

unless, of course, we consider absolutely impossible values u for which $\mu_a(u) = 0$.

So, we get Intuitionistic Fuzzy Degrees. In the traditional fuzzy logic, the sum of degrees to which each statement is true and to which this same statement is false is always equal to 1. This means that when we compare alternatives, we get beyond the traditional fuzzy logic.

How can we describe where we are? This is not the only case when the degrees of confidence in a statement and in its negation doe not add up to 1. To describe such cases, K. Atanassov came up with an idea of *intuitionistic fuzzy logic* (see, e.g., [1]), in which, for each statement, we have *two* degrees:

- the degree to which this statement is true, and
- the degree to which this statement is false,

and these degrees do not necessarily add to 1. Our analysis this leads us to a conclusion that the result of comparing two alternatives is an intuitionistic fuzzy degree.

3   Discussion

What we got is Somewhat Different from Intuitionistic Fuzzy Logic. There is a minor difference between what we observe when comparing two alternative and the traditional intuitionistic fuzzy logic is that:

- in the intuitionistic fuzzy logic, the sum of positive and negative degrees is always smaller than or equal to 1, while
- in our case, the sum is always greater than or equal to 1.

However, such (minor) generalization of intuitionistic fuzzy logic has been proposed in the past.

There is also a way to reconcile the results of comparing alternatives with the traditional intuitionistic fuzzy logic. Indeed, in general, Zadeh's extension principle, we compute the degree to which y is a *possible* value. In particular,

$\mu_+(a < u_0)$ is the degree to which is possible that $a < u_0$, and $\mu_-(a < u_0)$ is a degree to which it is possible that $a \geq u_0$. Instead, we can consider degrees $n_+(a < u_0)$ and $n_-(a < u_0)$ to which it is *necessary* that $a < u_0$ and that $a \geq u_0$ – defined, as usual, as 1 minus the degree to which the opposite statement is possible. Then, we get

$$n_+(a < u_0) = 1 - \mu_-(a < u_0)$$

and

$$n_-(< u_0) = 1 - \mu_+(a < u_0).$$

From the fact that $\mu_+(a < u_0) + \mu_-(a < u_0) \geq 1$, we can now conclude that

$$n_+(a < u_0) + n_-(a < u_0) = 2 - (\mu_+(a < u_0) + \mu_-(a < u_0)) \leq 1.$$

Thus, the degrees of necessity are consistent with the traditional intuitionistic fuzzy logic.

We can Still Reconstruct the Original Membership Function from the Results of Expert Elicitation. We assume that the expert's preferences are described by a membership function $\mu_a(u_0)$. As we have mentioned, as a result of expert elicitation, we do not get this function, we get instead a more complex construct, in which for each possible value u_0, we get two degrees $\mu_+(a < u_0)$ and $\mu_-(a < u_0)$.

We should mention, however, that from this construct, we can uniquely reconstruct the original membership function. Indeed, as have shown:

– when $u_0 \leq U$, then we have $\mu_+(a < u_0) = \mu_a(u_0)$ and $\mu_-(a < u_0) = 1$; and
– when $u_0 \geq U$, then we have $\mu_-(a < u_0) = \mu_a(u_0)$ and $\mu_+(a < u_0) = 1$.

In both cases, we thus have

$$\mu_a(u_0) = \min(\mu_+(a < u_0), \mu_-(a < u_0)).$$

Acknowledgments. This work was supported in part by the Perkins Grant administered by the Grants Management Office of the El Paso Community College and by the US National Science Foundation grant HRD-1242122.

References

1. Atanassov, K.T.: Intuitionistic Fuzzy Sets: Theory and Applications. Springer, Heidelberg (1999). https://doi.org/10.1007/978-3-7908-1870-3
2. Belohlavek, R., Dauben, J.W., Klir, G.J.: Fuzzy Logic and Mathematics: A Historical Perspective. Oxford University Press, New York (2017)
3. Fishburn, P.C.: Utility Theory for Decision Making. Wiley, New York (1969)
4. Fishburn, P.C.: Nonlinear Preference and Utility Theory. The John Hopkins Press, Baltimore (1988)

5. Klir, G., Yuan, B.: Fuzzy Sets and Fuzzy Logic. Prentice Hall, Upper Saddle River (1995)
6. Luce, R.D., Raiffa, R.: Games and Decisions: Introduction and Critical Survey. Dover, New York (1989)
7. Mendel, J.M.: Uncertain Rule-Based Fuzzy Systems: Introduction and New Directions. Springer, Cham (2017). https://doi.org/10.1007/978-3-319-51370-6
8. Nguyen, H.T., Kosheleva, O., Kreinovich, V.: Decision making beyond Arrow's 'impossibility theorem', with the analysis of effects of collusion and mutual attraction. Int. J. Intell. Syst. **24**(1), 27–47 (2009)
9. Nguyen, H.T., Walker, E.A.: A First Course in Fuzzy Logic. Chapman and Hall/CRC, Boca Raton, Florida (2006)
10. Novák, V., Porfiliova, I., Močkoř, J.: Mathematical Principles of Fuzzy Logic. Kluwer, Boston (1999)
11. Raiffa, H.: Decision Analysis. Addison-Wesley, Reading (1970)
12. Zadeh, L.A.: Fuzzy sets. Inf. Control **8**, 338–353 (1965)

Solving Transhipment Problems with Fuzzy Delivery Costs and Fuzzy Constraints

Juan Carlos Figueroa-García[✉], Jhoan Sebastian Tenjo-García, and Camilo Alejandro Bustos-Tellez

Universidad Distrital Francisco José de Caldas, Bogotá, Colombia
jcfigueroag@udistrital.edu.co,
{jstenjog,caabustost}@correo.udistrital.edu.co

Abstract. This paper shows a method for solving transhipment problems where its delivery costs and constraints are defined using information coming from experts. Then, we use fuzzy numbers to represent delivery costs and constraints, and an iterative algorithm based on the cumulative membership function of a fuzzy set to find an overall solution among fuzzy delivery times and constraints.

Keywords: Fuzzy transhipment · Fuzzy numbers
Cumulative membership function

1 Introduction and Motivation

Transhipment problems are important problems in logistics and production planning since they involve not only customers and suppliers but transhipment points with different conditions/characteristics (see Pinedo [19], Johnson and Montgomery [15], and Sipper and Bulfin [27]). The classical transportation problem only considers crisp transportation costs, so it is often solved using classical optimization techniques (except in large-scale cases where metaheuristics help).

In several transhipment problems we have transportation times/costs and requirements that cannot be defined as constants, so we need an alternative source to define them. Statistical analysis help to obtain such information in most of cases (see Heyman and Sobel [12]), but in some cases there is not enough statistical information to get them. This way, we suggest to ask the experts of the system to get reliable information, so their perceptions about transportation times and requirements of customers can be summarized as fuzzy sets.

This paper focuses on a transhipment problem with fuzzy transportation times and fuzzy demands. Based on Linear Programming (LP) models proposed by Bazaraa et al. [1], we extend those results to a fuzzy environment using the method proposed by Figueroa-García [6], and Figueroa-García and López-Bello [8,9].

© Springer International Publishing AG, part of Springer Nature 2018
G. A. Barreto and R. Coelho (Eds.): NAFIPS 2018, CCIS 831, pp. 538–550, 2018.
https://doi.org/10.1007/978-3-319-95312-0_47

This paper is divided into five principal sections. Section 1 shows the introduction and motivation of the work. In Sect. 2, the classical transhipment model is presented; Sect. 3 presents some basics on fuzzy sets/numbers. In Sect. 4, the Fuzzy Transhipment Problem (FTP) and the proposed method for solving it are presented. Section 5 shows an application example, and finally some concluding remarks are presented in Sect. 6.

2   The Crisp Transhipment Problem

A transhipment problem is similar to a classical transportation problem in the sense that an amount of some resource is required to be sent from a set of plants to a customer. The transhipment problem addresses a situation where a transportation mean (car, aircraft, train, ship, etc.) is required to cover a specified route no matter its capacity, and some mandatory routes are required to be covered (this is quite common in practice where not all connections between destinations are easy to be covered).

Thus, there is a specific time associated to cover a route, and some destinations are available only by transhipment i.e. some routes cannot be directly reached, so there is a need for transhipment through some available routes. In addition, every destination requires a specific amount of transportation means.

The mathematical form of a transhipment problem (minimizing the total transportation time) is as follows:

$$z = \text{Min} \sum_i \sum_j t_{ij} x_{ij} \tag{1}$$

$$s.t.$$

$$\sum_j x_{ij} \geqslant b_i \ \forall \ i \in \mathbb{N}_m \tag{2}$$

$$\sum_k x_{ki} \geqslant b_i \ \forall \ i \in \mathbb{N}_m \tag{3}$$

$$x_{ij} \in \mathbb{Z}^+ \ \forall \ \{i, j\} \in \mathbb{N}_m \tag{4}$$

Index sets:

$i, j, k \in \mathbb{N}_m$ is the set of m routes

Parameters:

$t_{ij} \in \mathbb{R}$ is the delivering time that a transportation mean coming off from the i_{th} route takes to cover the j_{th} route
$b_i \in \mathbb{Z}$ is the amount of transportation means needed in the i_{th} route

Decision variables:

$x_{ij} \in \mathbb{Z}$ is the amount of transportation means coming off from the route i_{th} to be assigned to the j_{th} route

$x_{ki} \in \mathbb{Z}$ is the amount of transportation means coming off from the route k_{th} to be assigned to the i_{th} route

Equation (2) ensures to every route $i \in \mathbb{N}_m$ to be covered by a needed amount of b_i transportation means and all of them are to be assigned to the j_{th} route. Equation (3) ensures to cover all b_i transportation means from a set of k routes. In general, what we want is to satisfy the required transportation means b_i from the k routes to then dispatch all of them to the j routes, at a minimal transportation time.

On the other hand, the transportation objective is to minimize the total time used in transportation/transhipment of all b_i via different routes. This time (including transhipment) is computed using t_{ij} which is the time used by a transportation mean coming off the i_{th} route to take the j_{th}. The total elapsed time depends on some other aspects such as setup times at every origin/destination per route plus transportation times t_{ij}. This way, the main idea of this transhipment problem is to minimize the total elapsed time z as defined in Eq. (1).

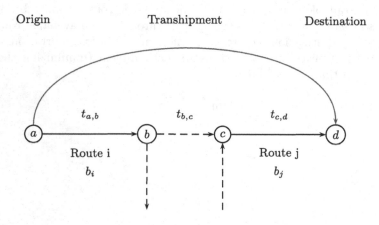

Fig. 1. Transhipment problem (Color figure online)

Figure 1 shows how the transhipment problem works. Suppose that you want to come from origin a to destination d (blue line), but there is no a direct available connection between both places. The continuous black line indicates two mandatory routes i which covers $a \rightarrow b$ and route j which covers $c \rightarrow d$ using an amount of means b_i and b_j. The dashed line between $b \rightarrow c$ indicates a non-mandatory (but possible) route which makes the connection between $a \rightarrow d$ possible, so we consider this non-mandatory route as a transhipment route. The vertical dashed line getting from b indicates that all b_i means must be dispatched to other routes, and the vertical dashed line coming to c indicates that other routes can come to c to supply needed transportation means b_j.

3   Fuzzy Sets/Numbers

A fuzzy set is denoted by emphasized capital letters $\tilde{A}$ with a membership function $\mu_{\tilde{A}}(x)$ over a universal set $x \in X$. $\mu_{\tilde{A}}(x)$ measures the membership of a value $x \in X$ regarding the concept/word/label A. $\mathcal{P}(X)$ is the class of all crisp sets, $\mathcal{F}(X)$ is the class of all fuzzy sets, $\mathcal{F}(\mathbb{R})$ is the class of all real-valued fuzzy sets, and $\mathcal{F}_1(\mathbb{R})$ is the class of all fuzzy numbers. Thus, a fuzzy set A is a set of ordered pairs of an element x and its membership degree, $\mu_A(x)$, i.e.,

$$\tilde{A} = \{(x, \mu_{\tilde{A}}(x)) \mid x \in X\}. \tag{5}$$

Then, a fuzzy number is:

Definition 1. *Let $\tilde{A} : \mathbb{R} \to [0,1]$ be a fuzzy subset of the reals. Then $\tilde{A} \in \mathcal{F}_1(\mathbb{R})$ is a Fuzzy Number (FN) iff there exists a closed interval $[x_l, x_r] \neq \emptyset$ with a membership function $\mu_{\tilde{A}}(x)$ such that:*

$$\mu_{\tilde{A}}(x) = \begin{cases} c(x) & \text{for } x \in [c_l, c_r], \\ l(x) & \text{for } x \in [-\infty, x_l], \\ r(x) & \text{for } x \in [x_r, \infty], \end{cases} \tag{6}$$

where $c(x) = 1$ for $x \in [c_l, c_r]$, $l : (-\infty, x_l) \to [0,1]$ is monotonic non-decreasing, continuous from the right, i.e. $l(x) = 0$ for $x < x_l$; $l : (x_r, \infty) \to [0,1]$ is monotonic non-increasing, continuous from the left, i.e. $r(x) = 0$ for $x > x_r$.

The α-cut of a set $\tilde{A} \in \mathcal{F}_1(\mathbb{R})$ namely $^{\alpha}\tilde{A}$ is defined as follows:

$$^{\alpha}\tilde{A} = \{x \mid \mu_{\tilde{A}}(x) \geqslant \alpha\} \ \forall \ x \in X, \tag{7}$$

$$^{\alpha}\tilde{A} = \left[\inf_x {}^{\alpha}\mu_{\tilde{A}}(x), \ \sup_x {}^{\alpha}\mu_{\tilde{A}}(x) \right] = [\check{A}_{\alpha}, \hat{A}_{\alpha}]. \tag{8}$$

The cumulative function $F(x)$ of a probability distribution $f(x)$ is:

$$F(x) = \int_{-\infty}^{x} f(t) \, dt, \tag{9}$$

where $x \in \mathbb{R}$. Its fuzzy version is as follows (see Figueroa-García and López-Bello [8,9], Figueroa-García [6], Pulido-López et al. [21]).

Definition 2 (Cumulative Membership Function). *Let $\tilde{A} \in \mathcal{F}(\mathbb{R})$ be a fuzzy set and $X \subseteq \mathbb{R}$, then the cumulative membership function (CMF) of $\tilde{A}$, $\psi_{\tilde{A}}(x)$ is defined as:*

$$\psi_{\tilde{A}}(x) = Ps_{\tilde{A}}(X \leqslant x), \tag{10}$$

Equation (10) shows the possibility of the set $X \leqslant x$ i.e. $Ps(X \leqslant x)$. In probability theory $F(\infty) = 1$ while in fuzzy theory $1 < \psi(\infty) < \Lambda$ where Λ is the cardinality (or total area) of $\tilde{A}$:

$$\Lambda_{\tilde{A}} = \int_{-\infty}^{\infty} \mu_{\tilde{A}}(t) \, dt. \tag{11}$$

To normalize $\psi_{\tilde{A}}(x)$ can be divide it by $\Lambda_{\tilde{A}}$:

$$\psi_{\tilde{A}}(x) = \frac{1}{\Lambda_{\tilde{A}}} \int_{-\infty}^{x} \mu_{\tilde{A}}(t)\, dt. \tag{12}$$

Figure 2 presents the CMF $\psi_{\tilde{A}}$ of a triangular fuzzy number:

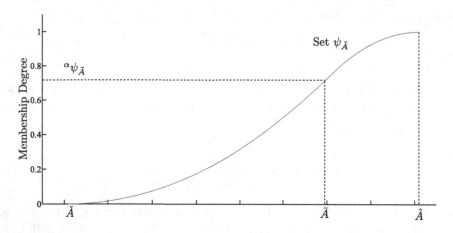

Fig. 2. Cumulative membership function $\psi_{\tilde{A}}$ of a triangular fuzzy set

4 The Proposed Fuzzy Transhipment Optimization Method

The mathematical programming form of the transhipment problem with fuzzy transportation times, fuzzy constraints, and positive integer amount of transportation means is as follows:

$$\tilde{z} = \text{Min} \sum_{i} \sum_{j} \tilde{t}_{ij} x_{ij} \tag{13}$$

$$s.t.$$

$$\sum_{j} x_{ij} \gtrsim \tilde{b}_i \; \forall \, i \in \mathbb{N}_m \tag{14}$$

$$\sum_{k} x_{ki} \gtrsim \tilde{b}_i \; \forall \, i \in \mathbb{N}_m \tag{15}$$

$$x_{ij} \in \mathbb{Z}^+ \; \forall \, \{i, j\} \in \mathbb{N}_m \tag{16}$$

Index sets:

$i, j, k \in \mathbb{N}_m$ is the set of m routes

Parameters:

$\tilde{t}_{ij} \in \mathcal{F}_1$ is the uncertain delivering time that a transportation mean coming off from the i_{th} route takes to cover the j_{th} route

$\tilde{b}_i \in \mathcal{F}_1$ is the uncertain amount of transportation means needed in the i_{th} route

Decision variables:

$x_{ij} \in \mathbb{Z}$ is the amount of transportation means coming off from the route i_{th} to be assigned to the j_{th} route

$x_{ki} \in \mathbb{Z}$ is the amount of transportation means coming off from the route k_{th} to be assigned to the i_{th} route

We focus on a transhipment problem where its transportation times see (13) (total transportation time) and the amount of required transportation means are uncertain (see Eq. (14) for the outgoing means from the i_{th} route and Eq. (15) for the incoming means to the i_{th} route), so they are defined by the experts of the system as fuzzy numbers (see Definition 1). Basically, this model considers uncertainty over transportation times since there are several sources that can affect them such as climate, immigration times, road conditions, etc. and uncertain amount of transportation means since the exact amount of goods to be sent are uncertain itself (customers demands, clients requirements, etc.)

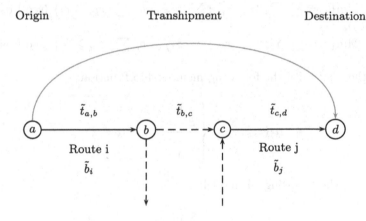

Fig. 3. Fuzzy transhipment problem

So that, in the fuzzy transhipment problem (see Fig. 3) we consider $\tilde{t}_{ij}$ as a fuzzy number (see Eq. (6)), and every requirement $\tilde{b}_i$ of Eqs. (14) and (15) as linear fuzzy sets since customers have no deterministic requirements due to market conditions, uncertain demands, etc.:

$$\mu_{\tilde{b}_i}(f(\mathbf{x}), \check{b}_i, \hat{b}_i) = \begin{cases} 1, & f(\mathbf{x}) \geqslant \hat{b}_i \\ \dfrac{f(\mathbf{x}) - \check{b}_i}{\hat{b}_i - \check{b}_i}, & \check{b}_i \leqslant f(\mathbf{x}) \leqslant \hat{b}_i \\ 0, & f(\mathbf{x}) \leqslant \check{b}_i \end{cases} \qquad (17)$$

where $\check{b}_i \in \mathbb{Z}$ is the lower bound of b_i, and $\hat{b}_i \in \mathbb{Z}$ is the upper bound of b_i.

In the Zimmermann soft constraints model $\mu_{\tilde{b}_i}(f(\mathbf{x}))$ is a linear fuzzy set that represents a soft $\geqslant$ namely $\gtrsim$ where $f(\mathbf{x})$ is the left side of every constraint i.e. $\sum_j x_{ij}$ in Eq. (14) and $\sum_k x_{ki}$ in Eq. (15) seen as the universe of discourse of $\tilde{b}$ and for which every $f(\mathbf{x})$ returns a membership degree. This way, we use an iterative version of the Zimmermann soft constraints model (extensively applied in fuzzy optimization with linear fuzzy constraints) to solve this FTP.

4.1   The Proposed Method

Based on fuzzy optimization concepts introduced by Rommelfanger [25], Ramík [23], Inuiguchi and Ramík [13], Fiedler et al. [5], Ramík [22], Inuiguchi et al. [14], we apply the proposal of Figueroa-García [6] and Figueroa-García and López-Bello [8,9] in order to minimize $\sum_i \sum_j \tilde{t}_{ij} x_{ij}$.

1- Iterative method:

- Set $\alpha \in [0,1]$,
- Compute $\psi_{\tilde{t}_{ij}}$ and $^{\alpha}\psi_{\tilde{t}_{ij}} \ \forall \, (i,j)$,

2- Soft constraints method:

- Set $\check{z} = \text{Min}\{z = \sum_i \sum_j {}^{\alpha}\psi_{\tilde{t}_{ij}} x_{ij} : \sum_j x_{ij} \geqslant \check{b}_i, \sum_k x_{ki} \geqslant \check{b}_i\}$ (see Eqs. (14), (15), (16)),
- Set $\hat{z} = \text{Min}\{z = \sum_i \sum_j {}^{\alpha}\psi_{\tilde{t}_{ij}} x_{ij} : \sum_j x_{ij} \geqslant \hat{b}_i, \sum_k x_{ki} \geqslant \hat{b}_i\}$ (see Eqs. (14), (15), (16)),
- Define the set $\tilde{z}$ with the following membership function:

$$\mu_{\tilde{z}}(z, \check{z}, \hat{z}) = \begin{cases} 1, & z \leqslant \check{z} \\ \dfrac{\hat{z} - z}{\hat{z} - \check{z}}, & \check{z} \leqslant z \leqslant \hat{z} \\ 0, & z \geqslant \hat{z} \end{cases} \tag{18}$$

- Thus, solve the following LP model:

$$\text{Max } \{\lambda\},$$
$$s.t.$$

$$\sum_i \sum_j {}^{\alpha}\psi_{\tilde{t}_{ij}} x_{ij} + \lambda(\hat{z} - \check{z}) = \hat{z}, \tag{19}$$

$$\sum_j x_{ij} - \lambda(\hat{b}_i - \check{b}_i) \geqslant \check{b}_i \ \forall \, i \in \mathbb{N}_m \tag{20}$$

$$\sum_k x_{ki} - \lambda(\hat{b}_i - \check{b}_i) \geqslant \check{b}_i \ \forall \, i \in \mathbb{N}_m \tag{21}$$

$$x_{ij} \geqslant 0 \ \forall \, \{i,j\} \in \mathbb{N}_m \tag{22}$$

3- Convergence:

- If $\lambda^* = \alpha$ then stop and return λ^* as the overall satisfaction degree of $\tilde{z}, \tilde{t}_{ij}$, and $\tilde{b}_i$; if $\lambda^* \neq \alpha$ then go to Step 1 and set $\lambda^* = \alpha$.

The set $\tilde{z}$ is defined over a universe of discourse $z = \sum_i \sum_j t_{ij} x_{ij}$ and it is intended to represent *minimum transportation cost* so its highest membership degree is given by $\check{z}$ and its lowest membership degree is given by $\hat{z}$. We point out that no matter the crisp objective z is, the Zimmermann method looks for maximizing the overall satisfaction degree of all soft constraints (via **x**) and the goal z, so it is always required to maximize λ.

4.2   Other Approaches

Several approaches to the fuzzy transportation problem have been proposed by Sakawa et al. [26], Chanas et al. [3], and Lee and Kim [16] which are based on fuzzy constraints only. The problem $\text{Min}\{z = \tilde{c}x : \tilde{A}x \gtrsim \tilde{b}, x \geqslant 0\}$ has been treated by Herrera and Verdegay [11], Peidro et al. [18], Najafi et al. [17] and Pishvaee and Khalaf [20] who defuzzified $\tilde{A}$ using the Yager index (see Yager [28]); Donga and Wan [17] who defuzzified $\tilde{A}$ using the mean-value of a fuzzy set proposed by Dubois and Prade [4], and Rena et al. [24] who defuzzified $\tilde{A}$ using the Heilpern's definition of the average value of a fuzzy set (see Heilpern [10]).

5   Application Example

We fuzzified the Tanker Scheduling problem proposed by Bazaraa et al. (see [1]) composed by 4 mandatory routes over 6 origins/destinations. We use triangular fuzzy transportation times $T = (\check{t}, \bar{t}, \hat{t})$ and requirements as defined in Eq. (17). Mandatory routes b_i and their requirements $\check{b}_i, \hat{b}_i$ are shown next.

R_1 : Dhahran $\rightarrow$ New York $\check{b}_1 = 8$ $\hat{b}_1 = 12$
R_2 : Marseilles $\rightarrow$ Istanbul $\check{b}_2 = 4$ $\hat{b}_2 = 7$
R_3 : Naples $\rightarrow$ Mumbai $\check{b}_3 = 3$ $\hat{b}_3 = 6$
R_4 : New York $\rightarrow$ Marseilles $\check{b}_4 = 1$ $\hat{b}_4 = 6$

Transportation times are triangular fuzzy sets $T(\check{t}, \bar{t}, \hat{t})$ as shown in Table 1.

5.1   The Proposed Solution

We solve the problem using the method shown in Sect. 4.1. First, we initially set $\alpha = 0.2$ to then compute $\psi_{\tilde{t}_{ij}}$ and $^\alpha\psi_{\tilde{t}_{ij}} \, \forall \, (i,j)$ as shown in (12) and Fig. 2. After 2 iterations we found an optimal $\lambda^* = 0.4396$ and the following solution:

$$x_{14} = 10, \; x_{21} = x_{23} = x_{31} = 5, \; x_{42} = 4; \; x_{22} = 2.$$

For instance $x_{14} = 10$ means that 10 ships should be sent from New York to Marseilles, $x_{31} = 5$ means that six ships came off from the route Naples $\rightarrow$

Table 1. Fuzzy transportation times

Orig/Dest	Naples	Marseilles	Istanbul	New York	Dhahran	Mumbai
Naples	—	(0.5,1,2)	(1.5,2,3.5)	(12,14,17)	(4,7,10)	(6,7,9)
Marseilles	(0.5,1,2.5)	—	(1,3,4.5)	(11,13,16)	(5.5,8,9)	(7,8,12)
Istanbul	(1.8,2,4)	(1.5,3,5)	—	(14,15,17)	(3.5,5,7)	(4,5,6)
New York	(11,14,15.8)	(9.5,13,15)	(14.5,15,17.5)	—	(14,17,18)	(17,20,24)
Dhahran	(5,7,8.5)	(6,8,10)	(4.5,5,7)	(15,17,18.5)	—	(1.5,3,5)
Mumbai	(5.5,7,9.5)	(7,8,11)	(3,5,8)	(18,20,23.5)	(1.5,3,5.5)	—

Mumbai and should be sent from New York → Marseilles which implies to cover the route Mumbai → New York. $x_{22} = 2$ means that two ships should be sent from Marseilles → Istanbul and then be sent from Dhahran → New York, so they must cover the route Istanbul → Dhahran, etc.

Note that $\lambda^* \leqslant 0.5$ because $\mathbf{x} \in \mathbb{Z}^+$, so there is a chance of having some excesses in some constraints.

Table 2. Optimal fuzzy transportation times and requirements

Orig/Dest	Naples	Marseilles	Istanbul	New York	Dhahran	Mumbai	b_i
Naples	—	1.09	2.21	14.11	6.82	7.17	9.76
Marseilles	1.21	—	2.76	13.11	7.47	8.66	5.32
Istanbul	2.43	3.02	—	15.17	5.02	4.94	4.32
New York	13.52	12.41	15.45	—	16.3	20.04	3.20
Dhahran	6.76	7.88	5.33	16.76	—	3.02	—
Mumbai	7.14	8.41	5.11	20.22	3.14	—	—

Table 2 shows defuzzified transportation times and requirements (note that the obtained optimal $\lambda^* = 0.4396$ does not change with the initial $\alpha = \{0.2, 0.8\}$). Figure 4 shows the set $\tilde{z}$ of optimal transhipment time for a final optimal degree $\lambda^* = 0.4386$.

We point out that $\lambda^* = 0.4396$ is an overall satisfaction degree of $\tilde{z}$, $\tilde{t}_{ij}$ and $\tilde{b}_i$. Also note that the algorithm finds the same overall satisfaction degree no matter the value of $\alpha \in [0, 1]$ we use to initialize it. Figure 5 shows the values of λ^* per iteration for 2 different starting values $\alpha = \{0.2, 0.8\}$ where both points reach $\lambda^* = 0.4396$.

5.2   Pre-defuzzified Solution

As we referred before, some other approaches use centroids and/or Yager indexes of t_{ij} to later solve a soft constraints model.

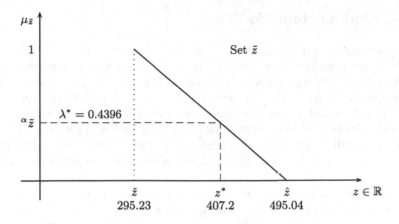

Fig. 4. Optimal solution of the problem.

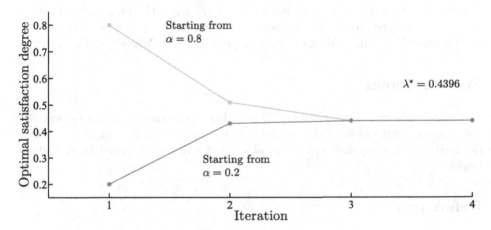

Fig. 5. Convergence of the proposed method for $\alpha = \{0.2, 0.8\}$

(i) First, we defuzzify $\tilde{t}_{ij}$ using the Yager index for triangular fuzzy sets e.g. $I(t) = 0.5(\bar{t} + 0.5(\check{t} + \hat{t}))$ as proposed by Herrera and Verdegay [11]. The obtained optimal $\lambda^* = 0.4395$ results into $\check{z} = 299.1, \hat{z} = 501.2, z^* = 412.37$ which is more expensive than the obtained results.

(ii) Now, if we defuzzify $\tilde{t}_{ij}$ using the centroid of a triangular fuzzy set e.g. $C(t) = (\check{t} + \bar{t} + \hat{t})/3$, then the obtained optimal $\lambda^* = 0.4391$ results into $\check{z} = 299.46, \hat{z} = 501.6, z^* = 412.84$ which is still more expensive.

(iii) Finally, if we defuzzify $\tilde{t}_{ij}$ using the mode of a triangular fuzzy set e.g. $M(t) = \bar{t}$, then the obtained optimal $\lambda^* = 0.3353$ results into $\check{z} = 298, \hat{z} = 500, z^* = 432.27$ which again is more expensive than our proposal.

6   Concluding Remarks

We proposed a fuzzified version of the crisp transhipment problem whose solution is given by the algorithm proposed by Figueroa-García [6] and Figueroa-García and López-Bello [8,9] with satisfactory results. The main goal of the algorithm is to obtain an optimal overall satisfaction degree ($\lambda^* = 0.4396$ in the application example), and the defuzzified solution of the parameters of the problem.

The algorithm reaches an overall satisfaction degree λ^* based on the cumulative membership function $\psi_{\tilde{A}}$ of a fuzzy set (which can be computed for non-convex and non-linear membership functions), and it operates as a defuzzification degree for $\tilde{z}$, $\tilde{t}_{ij}$ and $\tilde{b}_i$. The optimal solution provides the amount of ships per route, and single values of t_{ij} and b_i, useful to handle uncertainty in order to provide expected shipping dates/costs to customers.

The pre-defuzzified solution obtains an average-based solution of the problem by reducing its complexity using the Yager index. This leads to have an optimal membership regarding the fuzzy constraints only, while the proposed algorithm satisfies both fuzzy goal and constraints at a maximum degree, as intended by the Bellman-Zadeh fuzzy decision making principle (see Bellman and Zadeh [2]).

Further Topics

Transportation problems involving Type-2 fuzzy sets (see Figueroa-García [7]) are a natural extension of this model and the applied algorithm. Also a more detailed description/analysis of the algorithm will be discussed in a journal paper.

References

1. Bazaraa, M.S., Jarvis, J.J., Sherali, H.D.: Linear Programming and Networks Flow. Wiley, Hoboken (2009)
2. Bellman, R.E., Zadeh, L.A.: Decision-making in a fuzzy environment. Manag. Sci. **17**(1), 141–164 (1970)
3. Chanas, S., Delgado, M., Verdegay, J., Vila, M.: Interval and fuzzy extensions of classical transportation problems. Transp. Plan. Technol. **17**, 202–218 (1993). https://doi.org/10.1080/03081069308717511
4. Dubois, D., Prade, H.: The mean value of a fuzzy number. Fuzzy Sets Syst. **24**(3) (1987). https://doi.org/10.1016/0165-0114(87)90028-5
5. Fiedler, M., Nedoma, J., Ramík, J., Rohn, J., Zimmermann, K.: Linear Optimization Problems with Inexact Data. Springer, Heidelberg (2006). https://doi.org/10.1007/0-387-32698-7
6. Figueroa-García, J.C.: Mixed production planning under fuzzy uncertainty: a cumulative membership function approach. In: Workshop on Engineering Applications (WEA), vol. 1, pp. 1–6. IEEE (2012). https://doi.org/10.1109/WEA.2012.6220081

7. Figueroa-García, J.C., Hernández, G.: A transportation model with interval type-2 fuzzy demands and supplies. In: Huang, D.-S., Jiang, C., Bevilacqua, V., Figueroa, J.C. (eds.) ICIC 2012. LNCS, vol. 7389, pp. 610–617. Springer, Heidelberg (2012). https://doi.org/10.1007/978-3-642-31588-6_78

8. Figueroa-García, J.C., López-Bello, C.A.: Linear programming with fuzzy joint parameters: a cumulative membership function approach. In: 2008 Annual Meeting of the IEEE North American Fuzzy Information Processing Society (NAFIPS), pp. 1–6. IEEE (2008). https://doi.org/10.1109/NAFIPS.2008.4531293

9. Figueroa-García, J.C., López-Bello, C.A.: Pseudo-optimal solutions of FLP problems by using the cumulative membership function. In: Annual Meeting of the North American Fuzzy Information Processing Society (NAFIPS), vol. 28, pp. 1–6. IEEE (2009). https://doi.org/10.1109/NAFIPS.2009.5156464

10. Heilpern, S.: The expected valued of a fuzzy number. Fuzzy Sets Syst. **47**(1), 81–86 (1992). https://doi.org/10.1016/0165-0114(92)90062-9

11. Herrera, F., Verdegay, J.: Three models of fuzzy integer linear programming. Eur. J. Oper. Res. **83**(2), 581–593 (1995)

12. Heyman, D.P., Sobel, M.J.: Stochastic Models in Operations Research, Vol. II: Stochastic Optimization. Dover Publishers, New York (2003)

13. Inuiguchi, M., Ramík, J.: Possibilistic linear programming: a brief review of fuzzy mathematical programming and a comparison with stochastic programming in portfolio selection problem. Fuzzy Sets Syst. **111**, 3–28 (2000). https://doi.org/10.1016/S0165-0114(98)00449-7

14. Inuiguchi, M., Ramik, J., Tanino, T., Vlach, M.: Satisficing solutions and duality in interval and fuzzy linear programming. Fuzzy Sets Syst. **135**(1), 151–177 (2003). https://doi.org/10.1016/S0165-0114(02)00253-1

15. Johnson, L.A., Montgomery, D.C.: Operations Research in Production Planning, Scheduling and Inventory Control. Wiley, Hoboken (1974)

16. Lee, Y., Kim, S.: Production-distribution planning in supply chain considering capacity constraints. Comput. Ind. Eng. **43**, 169–190 (2002)

17. Najafi, H.S., Edalatpanah, S., Dutta, H.: A nonlinear model for fully fuzzy linear programming with fully unrestricted variables and parameters. Alex. Eng. J. **55**(1), 2589–2595 (2016). https://doi.org/10.1016/j.aej.2016.04.039

18. Peidro, D., Mula, J., Poler, R., Verdegay, J.: Fuzzy optimization for supply chain planning under supply, demand and process uncertainties. Fuzzy Sets Syst. **160**, 2640–2657 (2009)

19. Pinedo, M.L.: Scheduling: Theory, Algorithms, and Systems, 5th edn. Springer, Heidelberg (2016)

20. Pishvaee, M.S., Khalaf, M.F.: Novel robust fuzzy mathematical programming methods. Appl. Math. Model. **40**(1), 407–418 (2016). https://doi.org/10.1016/j.apm.2015.04.054

21. Pulido-López, D.G., García, M., Figueroa-García, J.C.: Fuzzy uncertainty in random variable generation: a cumulative membership function approach. Commun. Comput. Inf. Sci. **742**(1), 398–407 (2017). https://doi.org/10.1007/978-3-319-66963-2_36

22. Řamík, R.: Inequality relation between fuzzy numbers and its use in fuzzy optimization. Fuzzy Sets Syst. **16**, 123–138 (1985). https://doi.org/10.1016/S0165-0114(85)80013-0

23. Ramík, J.: Optimal solutions in optimization problem with objective function depending on fuzzy parameters. Fuzzy Sets Syst. **158**(17), 1873–1881 (2007). https://doi.org/10.1016/j.fss.2007.04.003

24. Rena, A., Wang, Y., Xue, X.: Interactive programming approach for solving the fully fuzzy bilevel linear programming problem. Knowl.-Based Syst. **99**(1), 103–111 (2016). https://doi.org/10.1016/j.knosys.2016.01.044
25. Rommelfanger, H.: A general concept for solving linear multicriteria programming problems with crisp, fuzzy or stochastic values. Fuzzy Sets Syst. **158**(17), 1892–1904 (2007). https://doi.org/10.1016/j.fss.2007.04.005
26. Sakawa, M., Nishizaki, I., Uemura, Y.: Fuzzy programming and profit and cost allocation for a production and transportation problem. Eur. J. Oper. Res. **131**, 1–15 (2001). https://doi.org/10.1016/S0377-2217(00)00104-1
27. Sipper, D., Bulfin, R.: Production: Planning, Control and Integration. McGraw-Hill College, New York City (1997)
28. Yager, R.: A procedure for ordering fuzzy subsets of the unit interval. Inf. Sci. **24**(1), 143–161 (1981)

How to Gauge Repair Risk?

Francisco Zapata[1] and Vladik Kreinovich[2]([✉])

[1] Department of Industrial, Manufacturing, and Systems Engineering,
University of Texas at El Paso, El Paso, TX 79968, USA
`fazg74@gmail.com`
[2] Department of Computer Science, University of Texas at El Paso,
El Paso, TX 79968, USA
`vladik@utep.edu`

Abstract. At present, there exist several automatic tools that, given a software, find locations of possible defects. A general tool does not take into account a specificity of a given program. As a result, while many defects discovered by this tool can be truly harmful, many uncovered alleged defects are, for this particular software, reasonably (or even fully) harmless. A natural reaction is to repair all the alleged defects, but the problem is that every time we correct a program, we risk introducing new faults. From this viewpoint, it is desirable to be able to gauge the repair risk. This will help use decide which part of the repaired code is most likely to fail and thus, needs the most testing, and even whether repairing a probably harmless defect is worth an effort at all – if as a result, we increase the probability of a program malfunction. In this paper, we analyze how repair risk can be gauged.

1 Formulation of the Problem

Traditional Approach to Software Testing. The main objective of software is to compute the desired results for all possible inputs. From this viewpoint, a reasonable way to test the software is:

- to run it on several inputs for which we know the desired answer, and
- to compare the results produced by this software with the desired values.

This was indeed the original approach to software testing.

It Turned out that Experts can Detect Some Software Defects Without Running the Program. Once it turns out that on some inputs, the program is not producing the desired result, the next step is to find – and correct – the defect that leads to the wrong answer.

After going through this procedure many times, programmers started seeing common patterns in the original defect locations. For example, a reasonably typical mistake is forgetting to initiate the value of the variable. In this case, we may get different results depending on what happens to be the initial value stored in the part of the computer memory which is allocated for this variable.

© Springer International Publishing AG, part of Springer Nature 2018
G. A. Barreto and R. Coelho (Eds.): NAFIPS 2018, CCIS 831, pp. 551–558, 2018.
https://doi.org/10.1007/978-3-319-95312-0_48

This defect is even more dangerous if the variable is a *pointer*, i.e., crudely speaking, if it stores not the actual value of the corresponding object, but rather the memory address at which the actual value is stored. In this case, if we do not initialize the pointer, not only can we access the wrong value, but we may also end up with a non-existing address or an address outside the memory segment in which your program is allowed to operate – at which point the program stops, since it either does not know what value to pick or is not allowed to pick up the corresponding value (this is known as *segmentation fault*).

In many programming language that do not automatically check the array indices, another typical defect is asking for a value $a[i]$ of an array a for an index i which is outside the array's range. In this case, the compiler obediently finds the corresponding space in the memory, not realizing that it is beyond the place of the original array – this can overwrite important information; this is known as *buffer overrun*.

Static Analysis Tools. Once programmers realized that there are certain patterns of code typical for software defects, they started to come up with automatic tools for detecting such patterns and thus, warning the user of possible defects of different potential severity.

At present, there are many such tools – Coverity [1], Fortify, Lint, etc. – and most of these tools are efficiently used in practice; see, e.g., [3].

Some "Defects" Found by Static Analysis Tools Do not Harm the Program's Functionality. For the purpose of this paper, it is important to mention that not all "defects" uncovered by a static analysis tool are actually hurting the program.

For example, some programs have *extra variables*, i.e., variables which are never used. This happens if a programmer originally planned to use the variable, started coding with it, then changed her mind but forgot to delete all the occurrences of this variable. Static analysis tools mark it as a possible detect, since in some situations, it is indeed an indication that some important value is never used. However, in many other cases, it may be syntactically clumsy, but does not cause any problem for the program.

Another defect that may not necessarily be harmful is the *logically dead* code, when a branch in a branching code is never visited. For example, if as part of the computations, we compute a square root of some quantity, it makes sense to make sure that this quantity is non-negative. When this quantity appears as result of long computations, it may happen that, due to rounding errors, a small non-negative value becomes small negative. In this case, it makes sense, if the value is negative, to replace this with 0. However, if we write a code this way, but we only use it to compute the square root of an input which is always non-negative (e.g., of the weight), then the branch corresponding to a negative value is never used. In some cases, this may be a real defect, indicating that we may have missed something that would lead to the possibility of this condition. However, in cases like described above, this "defect" is mostly harmless.

Yes another example of a possible defect is indentation. In some programming languages like Python, indentation is used to indicate the end of the condition or the end of the loop. However, in most other programming languages, indentation is ignored by the compiler, it simply helps people better understand each other's code. A static analysis tool will indicate the discrepancy between the indentation and the actual end of the condition or of a loop as a defect – and it indeed may be a defect. However, in many cases, it is just a sloppiness of a programmer that does not affect the program's execution.

Correcting Non-harmful Defects May Cause Real Problems. Once a static analysis tool marks a piece of code as containing a possible defect, a natural reaction is to repair this part of the code.

The problem is that, every time you change even a few lines of software, this may introduce additional faults – and this time, serious ones. The only way to avoid this problem is to thoroughly test the changed software. However, an extensive testing – that would, in principle, reveal all new faults – is very expensive. As a result, many of these changes have to be performed without complete testing, thus introducing many possible points of failure at every place where the code was changed.

We Need to Gauge Repair Risk. To make the repair effort cost-efficient, it would be useful to know which defect repair have the highest risk of causing a problem after the fix. This way, we can focus our testing effort on these defects, and save money by performing only limited testing of low-risk repairs.

And if an alleged defect is usually harmless but its repair may cause trouble, maybe a better strategy would be to keep this alleged defect in place. This is specially true for legacy software, software that was developed before static analysis tools became ubiquitous. If we apply such a tool to this software, we may find lots of alleged defects, but since the program has been running successfully for many years, it is highly probable that most of these alleged defects are actually harmless.

What We do in This Paper. In this paper, we describe how repair risk can be gauged.

In our analysis, we use two different approaches: a probabilistic approach and a fuzzy-based approach. Interestingly, both approaches lead to the same expression for the repair risk, which makes us confident that this is indeed the correct expression for the repair risk.

2   Analysis of the Problem: Which Factors Determine the Risk

First Factor: How Big are the Changes. Every time we change a line of code, we increase a risk. The more lines of code we change, the more we increase the risk.

Thus, one of the factors affecting the risk is the number L of lines of code that has been changed.

Second Factor: How Frequently are the Changed Lines Used. Simple errors, when a piece of code always produced wrong results, are usually mostly filtered out by simple testing.

As a result, a faulty piece of code usually leads to correct results, but sometimes, for some combination of inputs, produces an erroneous value.

If we run this piece of code once, the chances that we accidentally hit the wrong inputs are small, so most probably, this will not lead to any serious problem. However, if this piece of code appears inside a loop, then for each program run, this piece of code runs many times with different inputs. As a result, it becomes more and more possible that in one of these inputs, we will get a wrong result – and thus, that the overall software will fail.

Thus, the second factor that we need to take into account is the number of iterations I that this particular piece of code is repeated in the program.

For example, if this piece of code is inside a for-loop that repeats 1000 times, then $I = 1000$. If this piece of code is inside a double for-loop – i.e., a for-loop for which each of its 1000 iterations is itself a for-loop with 1000 iterations (as often happens with matrix operations), we get $I = 1000 \times 1000 = 10^6$.

What We Want. We want to be able to gauge the repair risk based on these two parameters: L and U.

Two Types of Software Errors. As we have mentioned, there are, in effect, two types of software errors:

- rarer *fatal* error that practically always lead to a wrong result or, more generally, to a program malfunction; and
- more frequent *subtle* error which are usually harmless, but can cause trouble for a certain (reasonably rare) combination of inputs.

In our analysis, we need to take into account both types of software errors.

3   How to Gauge Repair Risk: Probabilistic Approach

Taking Fatal Errors into Account. Let p_f denote the probability that a line of code contains a fatal error. Then, the probability that a line of code *does not* contain a fatal error is equal to $1 - p_f$.

Software errors in different lines are reasonably independent. Thus, the probability that an L-line new piece of code does not contain a fatal error can be computed as a product of L probabilities corresponding to each of the lines, i.e., as $(1 - p_f)^L$.

Taking Subtle Errors into Account. Let p_s denote the probability that one run of a line will lead to a fault. So, the probability that a line performs correctly during one run is equal to $1 - p_s$.

Faults on different lines are, as we have mentioned, reasonably independent. Also, inputs corresponding to different iterations are reasonably independent. When we run an L-line piece of new code I times, this means that we perform a running-of-one-line process $I \cdot L$ times. Thus, the probability that all lines will run correctly on all iterations is equal to the product of $I \cdot L$ individual probabilities, i.e., to the value $(1 - p_s)^{I \cdot L}$.

Taking Both Errors into Account. Fatal and subtle errors are reasonably independent; e.g., as we have mentioned, discovering a fatal error does not prevent the software from having subtle error.

We know that the probability that fatal errors will not affect the result is equal to $(1 - p_f)^L$. We also know that the probability that subtle errors will not affect the result is equal to $(1 - p_s)^{I \cdot L}$. Thus, due to independence, the probability that the new piece of code will perform correctly, i.e., that neither of the two types of errors will surface, is equal to the product of these two probabilities, i.e., to the value

$$P = (1 - p_f)^L \cdot (1 - p_s)^{I \cdot L}. \tag{1}$$

Resulting Criteria for Repair Risk. Ideally, we would like to know the probability of the program's fault. However, this requires that we know two parameters p_f and p_s, which may be difficult to get.

In the first approximation, it would be sufficient to simply *order* different repaired piece of code by risk – so that, in realistic situations with limited resources, we should concentrate all the testing on the pieces with the highest repair risk – and among probably harmless alleged defects, only repair those whose repair risk is the lowest.

From the viewpoint of such comparison, comparing the probabilities is equivalent to comparing their logarithms

$$\log(P) = L \cdot \log(1 - p_f) + I \cdot L \cdot \log(1 - p_s).$$

This is, in turn, equivalent to comparing the ratios

$$\frac{\log(P)}{\log(1 - p_s)} = I \cdot L + c \cdot L = L \cdot (I + c),$$

where we denoted

$$c \stackrel{\text{def}}{=} \frac{\log(1 - p_f)}{\log(1 - p_s)}.$$

So, we arrive at the following conclusion.

Probabilistic Case: Conclusion. To gauge the risk of repairing an alleged defect, we need to know:

– the number of lines L changed in the process of this repair, and
– the number of times I that this piece of code is repeated during one run of the software.

The relative repair risk is represented by the product

$$L \cdot (I + c), \tag{2}$$

for some constant c.

Comment. Note that, in contrast to the expression for probability, which required two parameters, this expression requires only one parameter – and one parameter is easier to experimentally determine than two.

4   How to Gauge Repair Risk: Fuzzy Approach

Need to Go Beyond the Traditional Probabilistic Approach. To follow through with the probabilistic approach, we needed to make an assumption that faults corresponding to different lines and/or different iterations are completely independent. While in the first approximation, this assumption may sound reasonable, it is clear that in reality, this assumption is only approximately true: programmers know that a fault in one line often causes faults in the neighboring lines as well.

This can happen if the same mistake appears in different lines due to the same programmer's misunderstanding, or due to the fact that the second line may be obtained from the first one by editing – and so, an undetected error in the first line is simply copied into the second one.

Ideally, in addition to probabilities of one line being correct, we should also consider:

– a separate probability of two lines being correct – which is, in general, different from the square of the first probability,
– a separate probability that three lines are being correct, etc.

However, as we have mentioned earlier, even obtained two probabilities is difficult. Obtaining many others – corresponding to different numbers of lines and different numbers of iterations – would be practically impossible. What can we do?

Solution: Fuzzy Approach. Lotfi Zadeh faced a similar problem when he decided to analyze expert knowledge. Expert knowledge contains many imprecise ("fuzzy") rules that uses imprecise words from natural language like "small".

For each such word, and for each value x of the corresponding quantity, we can ask the expert to gauge to what extent the given value satisfies the given property: e.g, to what extent the value x is small. We can call the resulting estimate the degree of belief, the degree of confidence, we can call it a subjective probability – the name does not change anything.

The problem appears if we take into account that the condition of an expert rule contains usually not just one simple statement like "x is small", but an "and"-combination of several such statements. For example, a typical expert rule for driving a car would say something like "if we are going fast *and* the car in front decelerates a little bit, *and* the road is reasonably slippery, then we need to break gently".

To utilize this rule, we need to find the subjective probability (degree of confidence) that for a given velocity v, for a given distance d to the car in front, etc. the corresponding "and"-condition is satisfied.

How can we find this condition? Ideally, we should elicit this subjective probability from the expert for each possible combination of the inputs $(v, d, \ldots)$ However, for a large number of parameters, the number of such combinations becomes astronomical, and there is no way to ask an expertthe resulting millions and billions of questions.

What Zadeh proposed – and what is one of the main ideas behind what he called *fuzzy logic* (see, e.g., [2,4,5,7–9]) is that, since we cannot elicit all degree of belief in "and"-statement $A \& B$ from the experts, we thus need to come up with an algorithm $f_\&(a, b)$ that would:

- given degree of belief a in the statement A and b in the statement B,
- return an estimate $f_\&(a, b)$ for the expert's degree of confidence in the "and"-statement $A \& B$.

This algorithm should satisfy some reasonable properties. For example, since $A \& B$ means the same as $B \& A$, it is reasonable to require that $f_\&(a, b) = f_\&(b, a)$, i.e., in mathematical terms, that the operation $f_\&(a, b)$ is *commutative*.

Similarly, since $A \& (B \& C)$ means the same as $(A \& B) \& C$, it is reasonable to require that $f_\&(a, f_\&(b, c)) = f_\&(f_\&(a, b), c)$, i.e., that the operation $f_\&(a, b)$ is *associative*. An "and"-operation $f_\&(a, b)$ that satisfies these and other similar properties is known as a *t-norm*.

There are many possible t-norms. One of them is the product $f_\&(a, b) = a \cdot b$, that corresponds to the case when all the events are independent. However, there are many other t-norms – that correspond to possible dependence.

Let us Apply this Approach to our Problem. In this approach, we no longer assume independence. To compute the subjective probability (degree of confidence) in an "and"–combination of different events, instead of a product, we can use an appropriate t-norm $f_\&(a, b)$. Thus, instead of the formula (1), we get a more complex formula

$$P = f_\&(1 - p_f, \ldots, 1 - p_f \ (L \text{ times}), 1 - p_s, \ldots, 1 - p_s \ (I \cdot L \text{ times})). \qquad (3)$$

It is known (see, e.g., [6]) that every t-norm can be approximated, with arbitrary accuracy, by t-norms of the type $f_\&(a, b) = h^{-1}(h(a) \cdot h(b))$, for some strictly increasing function $h(x)$, where $h^{-1}(x)$ denotes an inverse function, for which $h^{-1}(h(x)) = x$. So, for all practical purposes, we can safely assume that our t-norm is exactly of this type.

For such t-norms, $f_\&(a, b, \ldots, c) = h^{-1}(h(a) \cdot h(b) \cdot \ldots \cdot h(c))$. Thus, the formula (3) takes the form

$$P = h^{-1}\left((h(1 - p_f))^L \cdot (h(1 - p_s))^{I \cdot L}\right). \qquad (4)$$

Comparing such values is equivalent comparing the values

$$h(P) = (h(1 - p_f))^L \cdot (h(1 - p_s))^{I \cdot L},$$

or, equivalently, the value

$$\log(h(P)) = L \cdot \log(h(1 - p_f)) + I \cdot L \cdot \log(h(1 - p_s)),$$

or the value

$$\frac{\log(h(P))}{\log(h(1 - p_s))} = I \cdot L + c \cdot L = L \cdot (I + c),$$

where

$$c \stackrel{\text{def}}{=} \frac{\log(h(1 - p_f))}{\log(h(1 - p_s))}.$$

Conclusion. The fact that in this more general not-necessarily-independent case, we get the same expression $L \cdot (I + c)$ for repair risk makes us confident that this is indeed the correct expression.

Acknowledgments. This work was supported in part by the US National Science Foundation grant HRD-1242122.

References

1. Almossawi, A., Lim, K., Sinha, T.: Analysis tool evaluation: coverity prevent. Final report, Carnegie Mellon University, Pittsubrgh, Pannsylvania (2006). http://www.cs.cmu.edu/~aldrich/courses/654-sp09/tools/cure-coverity-06.pdf
2. Belohlavek, R., Dauben, J.W., Klir, G.J.: Fuzzy Logic and Mathematics: A Historical Perspective. Oxford University Press, New York (2017)
3. Emanuelsson, P., Nilsson, U.: A comparative study of industrial static analysis tools. Electron. Notes Theor. Comput. Sci. **217**, 5–21 (2008)
4. Klir, G., Yuan, B.: Fuzzy Sets and Fuzzy Logic. Prentice Hall, Upper Saddle River (1995)
5. Mendel, J.M.: Uncertain Rule-Based Fuzzy Systems. Springer, Cham (2017). https://doi.org/10.1007/978-3-319-51370-6
6. Nguyen, H.T., Kreinovich, V., Wojciechowski, P.: Strict Archimedean t-norms and t-conorms as universal approximators. Int. J. Approx. Reason. **18**(3–4), 239–249 (1998)
7. Nguyen, H.T., Walker, E.A.: A First Course in Fuzzy Logic. Chapman and Hall/CRC, Boca Raton (2006)
8. Novák, V., Perfilieva, I., Močkoř, J.: Mathematical Principles of Fuzzy Logic. Kluwer, Boston, Dordrecht (1999)
9. Zadeh, L.A.: Fuzzy sets. Inf. Control **8**, 338–353 (1965)

Interval Type II Fuzzy Rough Set Rule Based Expert System to Diagnose Chronic Kidney Disease

Mona Abdolkarimzadeh[1], M. H. Fazel Zarandi[1(✉)], and O. Castillo[2]

[1] Department of Industrial Engineering,
Amirkabir University of Technology, Tehran, Iran
{mabdolkarimzadeh, zarandi}@aut.ac.ir
[2] Tijuana Institute Technology, Tijuana, Mexico
ocastillo@tectijuana.mx

Abstract. Chronic kidney disease is a worldwide public health problem with an increasing incidence and prevalence, poor outcomes, and high cost. Diagnosis of Chronic Kidney Disease has always been a challenge for physicians. This paper presents an effective method for diagnosis of Chronic Kidney Disease based on interval Type-II fuzzy. This proposed system includes three steps: pre-processing (feature selection), Type-II fuzzy classification, and system evaluation. Fuzzy Rough QuickReduct algorithm feature selection is used as the preprocessing step in order to exclude irrelevant features and to improve classification performance and efficiency in generating the classification model. Rough set theory is a very useful tool for describing and modeling vagueness in ill-defined environments. In the type-II fuzzy classification step, an "indirect approach" is used for II fuzzy system modeling by implementing the Sugeno index for determining the number of rules in the fuzzy clustering approach. In the proposed system, the process of diagnosis faces vagueness and uncertainty in the final decision. The results that were obtained show that interval Type-II fuzzy has the ability to diagnose Chronic Kidney Disease with an average accuracy of 90%.

Keywords: Chronic kidney disease · Interval type-II fuzzy · Rough set
Diagnosis · Feature selection

1 Introduction

1.1 Chronic Kidney Disease

Chronic kidney disease includes conditions that damage your kidneys and decrease their ability to keep you healthy by doing the jobs listed. If kidney disease gets worse, wastes can build to high levels in your blood and make you feel sick. Also, kidney disease increases your risk of having heart and blood vessel disease. These problems may happen slowly over a long period of time. When kidney disease progresses, it may eventually lead to kidney failure, which requires dialysis or a kidney transplant to maintain life. The number of persons with kidney failure who are treated with dialysis

© Springer International Publishing AG, part of Springer Nature 2018
G. A. Barreto and R. Coelho (Eds.): NAFIPS 2018, CCIS 831, pp. 559–568, 2018.
https://doi.org/10.1007/978-3-319-95312-0_49

and transplantation is projected to increase from 340 000 in 1999 to 651 000 [1]. Unfortunately, chronic kidney disease is underdiagnosed and undertreated, resulting in lost opportunities for prevention [2, 3] in part because of a lack of agreement on a definition and classification of stages in the progression of chronic kidney disease [4] and a lack of uniform application of simple tests for detection and evaluation. Chronic kidney disease affects approximately 11% of the U.S. adult population (20 million people from 1988 to 1994). The prevalence of earlier stages of disease (10.8%) is more than 100 times greater than the prevalence of kidney failure (0.1%). Adverse outcomes of chronic kidney disease, including loss of kidney function and development of kidney failure can often be prevented or delayed through early detection and treatment.

1.2 Fuzzy Logic System

The theory of Fuzzy logic was introduced by Prof. Zadeh. In this theory an element belongs to a set according to the membership function values. Theory of FSs is an expansion of the traditional sets theory in which an element either is or is not a set member [5]. The fuzzy logic systems (FLSs) are well known for their ability to model linguistics and system uncertainties. Due to this ability, FLSs have been successfully used for many real world applications, including modeling and controlling [6–8].

1.3 Interval Type-II Fuzzy

Type II fuzzy sets have grades of membership that are themselves fuzzy. A type II membership grade can be any subset in [0, 1]. When the secondary memberships are either zero or one, we call them interval type II sets [9]. As Type II fuzzy logic is better suited for modeling linguistic terms [10] in this study, we use the Type II FLS and introduce a type II fuzzy system for diagnosing Chronic Kidney disease. A type II fuzzy set denoted as $\tilde{A}$, is characterized by a type-II membership function $\mu_{\tilde{A}}(x, u): U \times I \rightarrow I$ where $x \in U$ and $u \in J_x \subseteq [0, 1]$ i.e.

$$\tilde{A} = \{((x, u), \mu_{\tilde{A}}(x, u)) | \forall x \in X, \forall u \in J_x \subseteq [0, 1]\} \tag{1}$$

Where $0 \leq \mu_{\tilde{A}}(x, u) \leq 1$. $\tilde{A}$ can also be expressed as:

$$\tilde{A} = \int_{x \in X} \int_{u \in J_x} \mu_{\tilde{A}}(x, u)/(x, u), J_x \subseteq [0, 1] \tag{2}$$

The upper membership function (UMF) and lower membership function (LMF) of $\tilde{A}$ are two type 1 membership function that bound the FOU. The UMF of $\tilde{A}$ is the upper bound of the FOU($\tilde{A}$) and denoted $\overline{\mu_{\tilde{A}}}(x) \ \forall x \in X$, and the LMF is the lower bound of the FOU($\tilde{A}$) and denoted $\underline{\mu_{\tilde{A}}}(x) \ \forall x \in X$.

$$\overline{\mu_{\tilde{A}}}(x) = \overline{FOU}(\tilde{A}), \forall x \in X \tag{3}$$

$$\underline{\mu_{\tilde{A}}}(x) = \underline{FOU}(\tilde{A}), \forall x \in X \tag{4}$$

Figure 1 shows the bounds of type-II membership function for Gaussian MF. A structure of a type-II fuzzy logic system shows in Fig. 2.

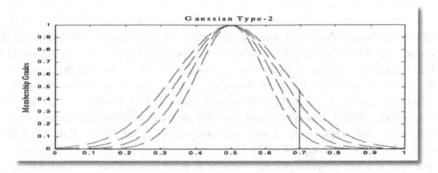

Fig. 1. The type-II membership function [10]

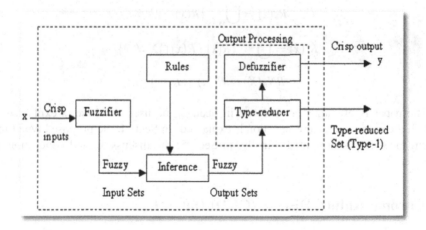

Fig. 2. A structure of a type-II fuzzy logic system [10]

Figure 2 shows the structure of an IT2 FLS. IT2 FLS contain the four mentioned major components (rules, fuzzifier, inference engine, and output processor) but the only difference between T1 and T2 structures is in the output processing part. In type-I

FLSs, output processing consists of a defuzzifier which transforms the fuzzy output of the system into a crisp value. But, output processing component in an IT2 FLS has two parts: Type reducer and defuzzifier. So before defuzzifying the output, it should be transformed from type-II to type-1. After type reduction, the output becomes a type-I FS and then we can implement various dufuzzification methods to obtain the crisp output [10]. Due to this ability, I2FLSs have been successfully used for many real world applications, including modeling and controlling [11–13].

1.4 Rough Set Theory

These two denominators (fuzzy and rough) have been successfully used in various uncertainty information processing systems. The RST, attributed by prof. Pawlak, is based on the research in the logical properties of information systems, and the uncertainty in information systems which are expressed by a boundary region [14]. RST has been generalized in many ways to tackle various problems. In particular, in 1990, Dubois and Prade [15] combined concepts of vagueness expressed by membership degrees in fuzzy sets [16] and indiscernibility in RST to obtain fuzzy rough set theory (FRST). FRST has been used e.g., for feature selection, instance selection, classification, and regression. There are many application areas that have been addressed by FRST, see e.g. [17–21].

For the sake of simplicity we assume that R is an equivalence relation. Let X is a subset of U. R-lower approximation of X ($R_*(x)$) and R-upper approximation of X ($R^*(x)$) and R-boundary region of X ($RN_R(X)$) are as follows:

$$R_*(x) = \bigcup_{x \in U} \{R(x) \subseteq X\} \tag{5}$$

$$R^*(x) = \bigcup_{x \in U} \{R(x) : R(x) \cap X \neq \emptyset\} \tag{6}$$

$$RN_R(X) = R^*(x) - R_*(x) \tag{7}$$

The paper is organized as follows: in Sect. 2, the used database is explained. In Sect. 3, the proposed feature selection is explained. In Sect. 4, the proposed type fuzzy system modeling is presented. Finally, in Sect. 5, the discussion and conclusion are presented.

2 Chronic Kidney Disease (CKD) Dataset

In this study, the Chronic Kidney database gathered from the Chamran Hospital in Tehran, Iran [22]. This data set contains 600 samples, 2 classes and fifteen features for each sample. These classes are assigned to the values that named as patient and healthy. The attributes of Chronic Kidney dataset are given in Table 1.

Table 1. The attributes of chronic kidney disease dataset

The number of attribute	The name of attribute	The values of attribute
1	Sex	Male – Female
2	Age*	2–100
3	Blood pressure max*	6–22
4	Blood pressure min	5–13
5	FBS	41–600
6	Bacteria*	Yes, No
7	Blood urea*	5–138
8	Serum creatinine*	0.5–12
9	Na - sodium*	120–150
10	K - potassium	2–8
11	Hemoglobin*	4–21
12	Rbc - red blood cells	2–7
13	Wbc - white blood cells*	0.4–40
14	Diabetes	Yes, No
15	Anemia	Yes, No

3 Feature Selection

The number of features in the raw dataset can be enormously large. This enormity may cause serious problems to many data mining systems. Feature selection is one of the oldest existing methods that deal with these problems. A method is used to compute reducts for fuzzy rough sets, where only the minimal elements in the discernibility matrix are considered. First, relative discernibility relations of conditional attribute are defined and relative discernibility relations are used to characterize minimal elements in the discernibility matrix. Then, an algorithm to compute the minimal elements is developed. Finally, novel algorithms to find proper reducts with the minimal elements are designed [23]. In general, there are two methods for choosing a feature by Rough sets: Measure the dependencies between features and Detection Matrix Method. In the first method, the degree of dependence between the features is calculated by the Eq. 8.

$$\gamma(c, d) = \frac{|POS_c(d)|}{U} \tag{8}$$

$$POS_c(d) = U_{X \in U/IND(d)} \underline{C}(X) \tag{9}$$

Which in Eq. 9, C is a set of conditional properties, and $POS_c(d)$ denotes a set of samples that are obtained in the positive region resulting from the division of samples into equivalence classes and finally a set the features that have the most dependency are introduced as optional features.

This method was used and the most important variables between the possible candidates were selected. Based on the results of this feature selection method, the number of features was reduced to 8, which show by star in Table 1, and we used these features in our proposed system.

4 Type - II Fuzzy System Modeling

4.1 Determining the Number of Rules

In a fuzzy clustering algorithm, we should use a cluster validity index to determine the most suitable number of clusters. In this study, we used the validity index proposed by Fukuyama and Sugeno [24]. This validity index can find the number of clusters as the minimum of its function with respect to c. This index is defined as:

$$FS(c) = \sum_{i=1}^{c} \sum_{j=1}^{n} \mu_{ij}^{m} ||x_j - a_i||^2 - \sum_{i=1}^{c} \sum_{j=1}^{n} \mu_{ij}^{m} ||a_i - \bar{a}||^2 = J_m(\mu, a) + K_m(\mu, a) \quad (10)$$

Where $\bar{a} = \sum_{i=1}^{c} a_i/c$. $J_m(\mu, a)$ is the FCM objective function which measures the compactness and $K_m(\mu, a)$ measures the separation. This cluster validity index is implemented to determine the most suitable number of clusters or rules. The best number of clusters based on this cluster validity index is obtained in five clusters. So, the system contains five rules.

4.2 The Proposed Type - II Fuzzy Model

In the, we obtain fuzzy model with five rules, eight inputs and one output. The inputs are age, blood pressure (max), bacteria, urea, creatinine, Na, hemoglobin and wbc. The output of our rule-base is an interval type II fuzzy set that must be type reduced and then defuzzify. We used centroid type reduction and defuzzifier. The proposed system used the mamdani fuzzy inference method. Figures 4, 5 and 6 show the memberships functions of samples of features. In the proposed model, Gaussian membership function was used. The numbers of rules consist five.these rules are as follow:

Rule 1:   IF (Age isr in1cluster1) AND (blood pressure (max) isr in2cluster1) AND (bacteria isr in3cluster1) AND (urea isr in4cluster1) AND (creatinine isr in5cluster1) AND (Na isr in6cluster1) AND (hemoglobin isr in7cluster1) AND (wbc isr in8cluster1) THEN (out isr cluster1).

Rule 2:   IF (Age isr in1cluster2) AND (blood pressure (max) isr in2cluster2) AND (bacteria isr in3cluster2) AND (urea isr in4cluster2) AND (creatinine isr in5cluster2) AND (Na isr in6cluster2) AND (hemoglobin isr in7cluster2) AND (wbc isr in8cluster2) THEN (out isr cluster2).

Rule 3:   IF (Age isr in1cluster3) AND (blood pressure (max) isr in2cluster3) AND (bacteria isr in3cluster3) AND (urea isr in4cluster3) AND (creatinine isr in5cluster3) AND (Na isr in6cluster3) AND (hemoglobin isr in7cluster3) AND (wbc isr in8cluster3) THEN (out isr cluster3).

Rule 4:   IF (Age isr in1cluster4) AND (blood pressure (max) isr in2cluster4) AND (bacteria isr in3cluster4) AND (urea isr in4cluster4) AND (creatinine isr in5cluster4) AND (Na isr in6cluster4) AND (hemoglobin isr in7cluster4) AND (wbc isr in8cluster4) THEN (out isr cluster4).

Rule 5:   IF (Age isr in1cluster5) AND (blood pressure (max) isr in2cluster5) AND (bacteria isr in3cluster5) AND (urea isr in4cluster5) AND (creatinine isr in5cluster5) AND (Na isr in6cluster5) AND (hemoglobin isr in7cluster5) AND (wbc isr in8cluster5) THEN (out isr cluster5).

Figure 3 represents the type-II fuzzy rules of the proposed system.

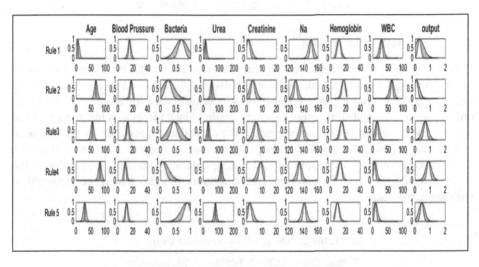

Fig. 3. Type-II-fuzzy rule-based

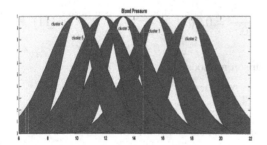

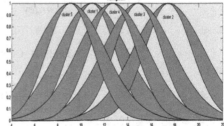

Fig. 4. Membership function of blood pressure   **Fig. 5.** Membership function of hemoglobin

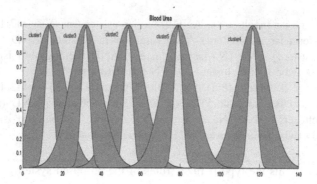

Fig. 6. Membership function of blood urea

4.3   Performance Evaluation

In this study, we used classification accuracy as criteria for evaluating the performance of the proposed system. For this purpose, we divided the CKD data set to training data and testing data. Training data consists of 480 sample data for modeling and developing the system and 120 sample data as testing data for evaluating the proposed system. By using confusion matrix method, the classification accuracy of the proposed system for diagnosis of chronic kidney disease was obtained about 90% (Eq. 11). Table 2 represents the test results of 120 testing data. As you can see in Table 3, the accuracy of the proposed method is greater than the method used in the previous article with the same data.

Table 2. The result of confusion matrix

Testing data	Class - healthy	Class - disease
Class - healthy	47	6
Class - disease	5	62

$$accuracy = \frac{47 + 62}{120} = 0.90 \tag{11}$$

Table 3. Comparison methods

Methods	Accuracy
Type - I fuzzy [6]	80%
Proposed method	90%

5   Conclusion

This paper represents an Interval type-II fuzzy rule-based expert system as an assistance system for diagnosing chronic kidneys function disease. This system uses the results of the prescribed measurement of chronic kidney as input data and by entering the input data, the output of the system will be a crisp value. In this study, we focused on identifying the rules and the parameters of the type-II fuzzy system. We used an Interval type-II fuzzy classification based on Sugeno index and FCM algorithm for determining the number of clusters and values of parameters. The classification accuracy of the proposed system for diagnosis of chronic kidney disease was obtained about 90%.

References

1. United States Renal Data System: Excerpts from the 2000 U.S. renal data system annual data report: atlas of end stage renal disease in the United States. Am. J. Kidney Dis. **36**, S1–S279 (2000)
2. McClellan, W.M., Knight, D.F., Karp, H., Brown, W.W.: Early detection and treatment of renal disease in hospitalized diabetic and hypertensive patients: important differences between practice and published guidelines. Am. J. Kidney Dis. **29**, 368–375 (1997). PMID: 9041212
3. Coresh, J., Wei, G.L., McQuillan, G., Brancati, F.L., Levey, A.S., Jones, C., et al.: Prevalence of high blood pressure and elevated serum creatinine level in the United States: findings from the third National Health and Nutrition Examination Survey (1988–1994). Arch. Intern. Med. **161**, 1207–1216 (2001). PMID: 11343443
4. Hsu, C.Y., Chertow, G.M.: Chronic renal confusion: insufficiency, failure, dysfunction, or disease. Am. J. Kidney Dis. **36**, 415–418 (2000). PMID: 10922323
5. Zadeh, L.A.: Fuzzy logic = computing with words. IEEE Trans. Fuzzy Syst. **4**, 103–111 (1996)
6. Fazel Zarandi, M.H., Abdolkarimzadeh, M.: Fuzzy rule based expert system to diagnose chronic kidney disease. In: Melin, P., Castillo, O., Kacprzyk, J., Reformat, M., Melek, W. (eds.) NAFIPS 2017. AISC, vol. 648, pp. 323–328. Springer, Cham (2018). https://doi.org/10.1007/978-3-319-67137-6_37
7. Abdolkarimzadeh, L., Azadpour, M., Fazel Zarandi, M.H.: Two hybrid expert system for diagnosis air quality index (AQI). In: Melin, P., Castillo, O., Kacprzyk, J., Reformat, M., Melek, W. (eds.) NAFIPS 2017. AISC, vol. 648, pp. 315–322. Springer, Cham (2018). https://doi.org/10.1007/978-3-319-67137-6_36
8. Fazel Zarandi, M.H., Seifi, A., Ershadi, M.M., Esmaeeli, H.: An expert system based on fuzzy bayesian network for heart disease diagnosis. In: Melin, P., Castillo, O., Kacprzyk, J., Reformat, M., Melek, W. (eds.) NAFIPS 2017. AISC, vol. 648, pp. 191–201. Springer, Cham (2018). https://doi.org/10.1007/978-3-319-67137-6_21
9. Liang, Q., Mendel, J.M.: Interval type-2 fuzzy logic systems: theory and design. IEEE Trans. Fuzzy Syst. **8**(5), 535–550 (2000)
10. Mendel, J.M., John, R.I.: Type-2 fuzzy sets made simple. IEEE Trans. Fuzzy Syst. **10**(April), 117–127 (2002)
11. Husseini, Z.M., Fazel Zarandi, M.H.: Type-2 fuzzy approach in multi attribute group decision making problem. In: Melin, P., Castillo, O., Kacprzyk, J., Reformat, M., Melek, W. (eds.) NAFIPS 2017. AISC, vol. 648, pp. 73–81. Springer, Cham (2018). https://doi.org/10.1007/978-3-319-67137-6_8
12. Fazel Zarandi, M.H., Seifi, A., Esmaeeli, H., Sotudian, Sh.: A type-2 fuzzy hybrid expert system for commercial burglary. In: Melin, P., Castillo, O., Kacprzyk, J., Reformat, M., Melek, W. (eds.) NAFIPS 2017. AISC, vol. 648, pp. 41–51. Springer, Cham (2018). https://doi.org/10.1007/978-3-319-67137-6_5

13. Sadat Asl, A.A., Fazel Zarandi, M.H.: A type-2 fuzzy expert system for diagnosis of leukemia. In: Melin, P., Castillo, O., Kacprzyk, J., Reformat, M., Melek, W. (eds.) NAFIPS 2017. AISC, vol. 648, pp. 52–60. Springer, Cham (2018). https://doi.org/10.1007/978-3-319-67137-6_6

14. Pawlak, Z.: Rough sets. Int. J. Comp. Sci. **11**, 341–356 (1982)

15. Dubois, D., Prade, H.: Rough fuzzy sets and fuzzy rough sets. Int. J. Gen Syst **17**, 91–209 (1990)

16. Zadeh, L.A.: Fuzzy sets. Inf. Control **8**, 338–353 (1965)

17. Huang, B., Zhuang, Y., Li, H., Wei, D.: A dominance intuitionistic fuzzy-rough set approach and its applications. Appl. Math. Model. **37**, 7128–7141 (2013)

18. Yu, X.D.: A new patterns recognition method based on fuzzy rough sets. Appl. Mech. Mater. **380–384**, 3795–3798 (2013)

19. Bhatt, R.B., Gopal, M.: FRCT: fuzzy-rough classification trees. Pattern Anal. Appl. **11**, 73–88 (2008)

20. Leung, Y., Fischer, M.M., Wu, W.-Z., Mi, J.-S.: A rough set approach for the discovery of classification rules in interval-valued information systems. Int. J. Approx. Reason. **47**, 233–246 (2008)

21. Zarandi, F., Hossein, M., Kazemi, A.: Application of rough set theory in data mining for decision support systems (DSSs). J. Optim. Ind. Eng. 25–34 (2010)

22. Chamran hospital in iran. http://www.chamranhospital.ir

23. Hu, Q., Yu, D., Guo, M.: Fuzzy preference based rough sets. Inf. Sci. **180**(10), 2003–2022 (2010)

24. Fukuyama, Y., Sugeno, M.: A new method of choosing the number of clusters for the fuzzy c-means method. In: Proceeding of Fifth Fuzzy Systems Symposium, pp. 247–250 (1989)

Parameter Optimization for Membership Functions of Type-2 Fuzzy Controllers for Autonomous Mobile Robots Using the Firefly Algorithm

Marylu L. Lagunes, Oscar Castillo$^{(\boxtimes)}$, Fevrier Valdez, Jose Soria, and Patricia Melin

Tijuana Institute of Technology, Tijuana, BC, Mexico
ocastillo@tectijuana.mx

Abstract. This paper describes the comparison of dynamic adjustment parameters in the firefly algorithm using type-1 and type-2 fuzzy logic for the optimization of a fuzzy controller. The adjustment is performed to improve the behavior of the method. Fuzzy systems use fuzzy sets by defining membership functions, which indicate how much an element belongs to the fuzzy set. Type-2 fuzzy logic assigns degrees of belonging that are fuzzy and this can be viewed as an extension of type-1 fuzzy logic. The Firefly algorithm has 3 main parameters Beta, Gamma and Alpha with a range of 0 to 1 each, which need to the dynamically adjusted to improve the performance of the algorithm.

Keywords: Type-1 fuzzy logic · Type-2 fuzzy logic · Fuzzy logic
Firefly algorithm · Dynamic adjustment

1 Introduction

This paper focuses on the performance of the firefly algorithm, [1–5] by making one of its main parameters to be dynamic, with this aiming to obtain better results in terms of error minimization when optimizing a fuzzy controller of an autonomous mobile robot [6–9]. Applying a fuzzy dynamic adjustment, a simple fuzzy system of one input and one output was developed to make the adjustment to the alpha parameter, which in the firefly algorithm represents the randomization values of fireflies movement in the search of mimimizing the objective function. This paper is organized as follows. Section 2 describes Type-1 and Type-2 Fuzzy Logic, Sect. 3 presents in detail the Firefly Algorithm, Sect. 4 describes the Model of Fuzzy Dynamic Adjustment, Sect. 5 shows the Fuzzy Controller Autonomous Mobile Robot, Sect. 6 describes the Experimentation, Sect. 7 presents in detail the results obtained in the experiments and finally Sect. 8 describes the conclusions.

2 Type-1 and Type-2 Fuzzy Logic

Fuzzy logic [10–12], emerged based on the principle of incompatibility between accuracy and complexity, since normal computational procedures are so exact that they are incompatible with the complexity of human reasoning, and for this reason fuzzy

© Springer International Publishing AG, part of Springer Nature 2018
G. A. Barreto and R. Coelho (Eds.): NAFIPS 2018, CCIS 831, pp. 569–579, 2018.
https://doi.org/10.1007/978-3-319-95312-0_50

logic uses linguistic variables that are words like those humans use and not exact numbers as they use machines.

Type-2 fuzzy logic [13–15] is an extension of the classical fuzzy logic that gives values to the linguistic variables and improves the inference in the type-1 fuzzy sets. In Figs. 1 and 2, type-1 and type-2 the triangular membership functions are illustrated, respectivelly.

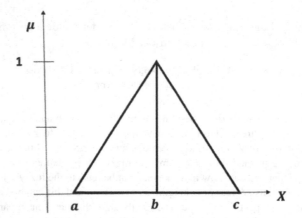

Fig. 1. Type-1 triangular membership functions.

$$
triangular(u; a, b, c) = \begin{cases} 0, & u \leq a \\ \frac{u-a}{b-c}, & a \leq x \leq b \\ \frac{c-x}{c-b}, & b \leq x \leq c \\ 0, & c \leq x \end{cases} \tag{1}
$$

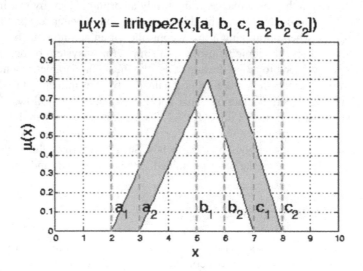

Fig. 2. Type-2 triangular membership functions.

$$\tilde{\mu}(x) = \left[\underline{\mu}(X), \overline{\mu}(x)\right] = \text{itritype2}(x, [a_1, b_1, c_1, a_2, a_2, a_2]) \tag{2}$$

where $a_1 < a_2, b_1 < b_2, c_1 < c_2$

$$\mu_1(x) = \max\left(\min\left(\frac{x - a_1}{b_1 - a_1}, \frac{c_1 - x}{c_1 - b_1}\right), 0\right)$$

$$\mu_2(x) = \max\left(\min\left(\frac{x - a_2}{b_2 - a_2}, \frac{c_2 - x}{c_2 - b_2}\right), 0\right)$$

3 Firefly Algorithm

The Firefly algorithm was developed by Yang [16] in 2008, and its inspiration is based on the behavior of the flickering fireflies. The algorithm has 3 important rules.

1. All the fireflies are unisex so that a firefly can be attracted by any other.
2. The attractiveness is in conformity with the brightness and it is minimized when its distance is greater, the less bright firefly will move towards the brightest, if there is not a brighter firefly they will move at random.
3. The brightness of a firefly is given by its function fitness.

Attractiveness Equation
The equation for defining attraction is:

$$\beta = \beta_0 e^{-\gamma r^2} \tag{3}$$

β_0 = Attractive initial in $r = 0 \in [0, 1]$.
$e \approx 2{,}71828$ Natural logarithm.
r = Distance between each of two fireflies.
γ = Determines the variation of attractiveness by increasing the distance between the fireflies $\in [0, 1]$.

Movement Equation
The equation for defining the movement is:

$$x_i^{t+1} = x_i^t + \beta_0 e^{-\gamma r_{ij}^2}\left(x_j^t - x_i^t\right) + \alpha_t \varepsilon_i^t \tag{4}$$

Where:

x_i^{t+1} = Next position.
x_i^t = Actual position.
$\beta_0 e^{-\gamma r_{ij}^2}\left(x_j^t - x_i^t\right)$ = The attraction.
α_t = The randomization with α being the parameter of randomness $\alpha \in [0, 1]$.

ε_i^t = Vector of random numbers extracted from a Gaussian distribution.

Distance Equation
The equation for defining the distance is:

$$r_{ij} = \sqrt{\sum_{k=1}^{d} \left(x_{i,k} - x_{j,k} \right)^2}, \tag{5}$$

Where:

r_{ij} = Euclidian distance between two fireflies i and j.
$x_{i,k}$ = Is the kth component of the spatial coordinate.
x_i = The i-th firefly.
d = Number of dimensions.

4   Model of Fuzzy Dynamic Adjustment

As it was already mentioned previously, a fuzzy system was developed to make the dynamic adjustment to the alpha parameter, the fuzzy system has an input that represents the iterations and an output that is the value that takes alpha for each iteration, in the following Figs. 3 and 4 input and output are observed of fuzzy models where each has a range [0 1] and as linguistic variable low, medium and high Figs. 5 and 6.
Model Type-1 and Type-2 Fuzzy Systems
The equation for defining the iteration variable is:

$$Iteration = \frac{Current\ Iteration}{Maximum\ of\ Iterations} \tag{6}$$

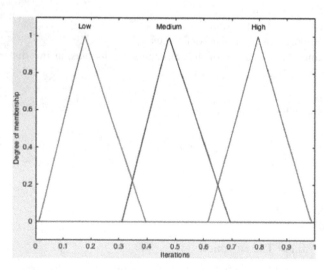

Fig. 3. Type-1 Fuzzy system input iterations with 3 triangular membership functions.

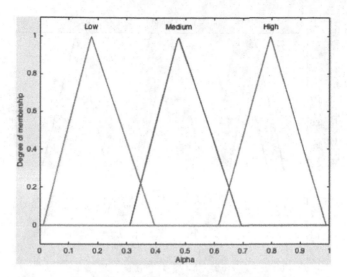

Fig. 4. Type-1 fuzzy systems output alpha

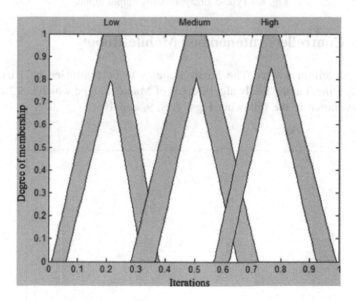

Fig. 5. Type-2 fuzzy systems input iterations.

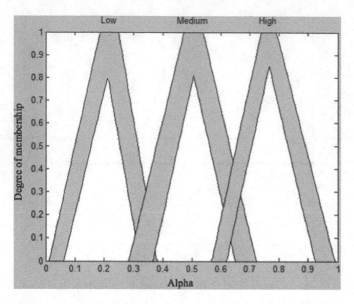

Fig. 6. Type-2 fuzzy systems output alpha.

5 Fuzzy Controller Autonomous Mobile Robot

The firefly algorithm was co The Fuzzy systems to be optimized [17] using fuzzy dynamic adjustment in the firefly algorithm, is of Mamdani type which has 2 inputs and 2 outputs as shown in the following Figs. 7, 8, 9, and 10.

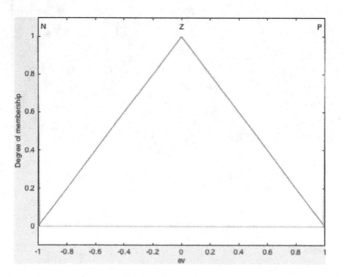

Fig. 7. Type-1 fuzzy system input 1 linear error (e_v).

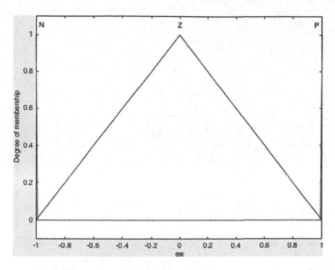

Fig. 8. Type-1 fuzzy systems input 2 angular error (e_w).

The First input is the linear error (e_v) and represents the error in linear velocity, with 3 triangular membership functions with their respective linguistic variables Negative, Zero and Positive, in a range of -1 to 1.

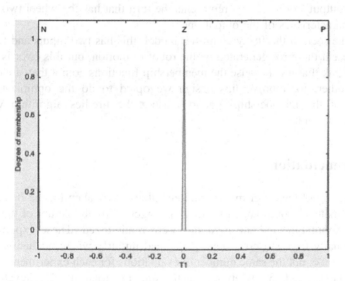

Fig. 9. Type-1 fuzzy systems output 1 torque 1 (t_1).

The Second input of the angular error (e_w) represents the error in angular velocity and is designed with the same characteristics as the first input.

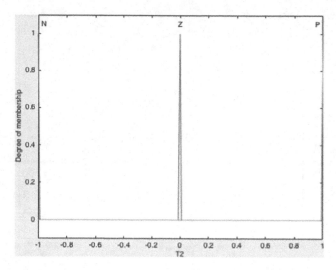

Fig. 10. Type-1 fuzzy systems output 2 torque 2 (t_2).

First output is called the Torque 1 (t_1). This output represents the rotation of the wheel one that the autonomous robot can give. It consists of 3 triangular membership functions with Negative, Zero and positive as linguistic variables respectively, with a range of −1 to 1.

Second output Torque 2 (t_2) represents the turn that has the wheel two and meets the same characteristics of the output one.

As can be seen in the fuzzy controller model, this has two inputs and two outputs which represent the error generated by the robot in motion, but this error is very high, one might think that it is because the membership functions do not they are overlapped with each other, to improve this design we opted to do the optimization of the parameters of the membership functions using the fireflies algorithm with fuzzy dynamic adjustment.

6 Experimentation

At the beginning of the experiments, random values were taken for the Beta, alpha and gamma parameters with a range of 0 to 1 as suggested by the author of the algorithm, just as the population and the iterations were randomized. These experiments performed manually trial and error were performed unfairly when evaluating the fitness function since it is not the same number of evaluations for each experiment for example a population of 35 and 550 iterations which results in a total of 19,250 evaluations and a population of 25 fireflies and 100 iterations give a total of 2500 evaluations to the functions fitness, took that information obtained to given an idea of how many evaluations could be used for the next experimentation, experiments ware performed with 30,000 evaluation per experiment, although they seem Many evaluations the optimization problem is more complicated than a benchmark function therefore it asks for more evaluation time. We show in the following Tables 2 and 3 the results obtained from the fuzzy controller optimization, with dynamic adjustment of type-1 and type-2.

Table 1. Firefly algorithm parameters

Experiment	Firefly	Iterations	Beta	Gamma	Alpha	Error
1	25	100	1	0.1	0.5	0.26
2	30	500	1	0.1	0.6	0.075920923
3	30	550	1	0.1	0.7	0.082229181
4	30	1000	1	0.1	0.8	0.0739153
5	35	550	0.3	0.1	0.3	0.2818
6	35	550	0.4	0.1	0.7	**0.004817189**
7	40	300	0.3	0.1	0.8	0.073915
8	45	600	0.9	0.1	0.1	0.3279
9	45	750	0.4	0.1	0.2	0.110954
10	50	680	0.6	0.1	0.9	0.0759
11	50	800	0.5	0.1	0.6	0.1952

Table 2. Dynamic adjustment type-1 firefly algorithm parameters

Experiment	Firefly	Iterations	Beta	Gamma	Alpha	Error
1	12	2500	0.3	0.2	D	0.01572
2	15	2000	0.3	0.2	D	0.3739
3	20	1500	0.2	0.2	D	0.01562
4	25	1200	0.1	0.2	D	0.0077152
5	30	1000	0.3	0.2	D	0.24964
6	40	750	0.5	0.1	D	0.07927
7	50	600	0.7	0.1	D	0.039156
8	50	600	0.8	0.1	D	0.053435

This Table 1 describes the values the Beta, alpha and gamma parameters the Firefly Algorithm.

7 Simulation Results

In this work the dynamic adjustment to the alpha parameter of the firefly algorithm was proposed, to improve the performance of the method in the optimization of the parameters of the membership functions, of an autonomous mobile controller, using type-1 and type-2 fuzzy logic.

As we can see the results of the experiment doing the dynamic adjustment by iteration are better than changing the value manually by experiment. The results show that type-2 fuzzy logic helps more the behavior of the method in this particular case of

Table 3. Dynamic adjustment type-2 firefly algorithm parameters

Experiment	Firefly	Iterations	Beta	Gamma	Alpha	Error
1	12	2500	0.3	0.2	D	0.075921
2	15	2000	0.3	0.2	D	0.073915
3	20	1500	0.2	0.2	D	0.077152
4	25	1200	0.1	0.2	D	0.077152
5	30	1000	0.3	0.2	D	0.0048172
6	40	750	0.5	0.1	D	**0.00065927**
7	50	600	0.7	0.1	D	0.073915
8	50	600	0.8	0.2	D	0.013943

decreasing the error which produces the simulation of the robot using the mean square error (mse). As can be seen in the previous Table 2, the results of the dynamic adjustment with type-1 fuzzy logic.

8   Conclusions

Fuzzy logic is a tool that is serving as an interface between the reasoning of the human being and the actions of the machines, in this regard fuzzy logic can have linguistic variables that are used in fuzzy control systems.

In this work, the dynamic adjustment of the alpha parameter of the firefly algorithm was performed to improve the performance of the algorithm when optimizing the fuzzy controller of a mobile autonomous fuzzy controller. Test experiments were carried out to take a base of evaluations to the objective function, and with these data, experiments were carried out for optimization. We envision as future work applying the proposed method to the optimization of fuzzy systems in other applications.

References

1. Lagunes, M.L., Castillo, O., Soria, J.: Methodology for the optimization of a fuzzy controller using a bio-inspired algorithm. In: Melin, P., Castillo, O., Kacprzyk, J., Reformat, M., Melek, W. (eds.) NAFIPS 2017. AISC, vol. 648, pp. 131–137. Springer, Cham (2018). https://doi.org/10.1007/978-3-319-67137-6_14
2. Yang, X.-S.: Nature-Inspired Metaheuristic Algorithms. Luniver Press, Bristol (2010)
3. Yang, X.-S., He, X.: Firefly algorithm: recent advances and applications. Int. J. Swarm Intell. **1**(1), 36–50 (2013)
4. Soto, C., Valdez, F., Castillo, O.: A review of dynamic parameter adaptation methods for the firefly algorithm. In: Melin, P., Castillo, O., Kacprzyk, J. (eds.) Nature-Inspired Design of Hybrid Intelligent Systems. SCI, vol. 667, pp. 285–295. Springer, Cham (2017). https://doi.org/10.1007/978-3-319-47054-2_19
5. Solano-Aragón, C., Castillo, O.: Optimization of benchmark mathematical functions using the firefly algorithm. In: Castillo, O., Melin, P., Pedrycz, W., Kacprzyk, J. (eds.) Recent Advances on Hybrid Approaches for Designing Intelligent Systems. SCI, vol. 547, pp. 177–189. Springer, Cham (2014). https://doi.org/10.1007/978-3-319-05170-3_13

6. Amador-Angulo, L., Castillo, O.: Comparative analysis of designing differents types of membership functions using bee colony optimization in the stabilization of fuzzy controllers. In: Melin, P., Castillo, O., Kacprzyk, J. (eds.) Nature-Inspired Design of Hybrid Intelligent Systems. SCI, vol. 667, pp. 551–571. Springer, Cham (2017). https://doi.org/10.1007/978-3-319-47054-2_36

7. Caraveo, C., Valdez, F., Castillo, O.: Optimization mathematical functions for multiple variables using the algorithm of self-defense of the plants. In: Melin, P., Castillo, O., Kacprzyk, J. (eds.) Nature-Inspired Design of Hybrid Intelligent Systems. SCI, vol. 667, pp. 631–640. Springer, Cham (2017). https://doi.org/10.1007/978-3-319-47054-2_41

8. Castillo, O., Martinez Marroquin, R., Melin, P., Valdez, F., Soria, J.: Comparative study of bio-inspired algorithms applied to the optimization of type-1 and type-2 fuzzy controllers for an autonomous mobile robot. Inf. Sci. **192**, 19–38 (2012)

9. Castillo, O., Neyoy, H., Soria, J., Melin, P., Valdez, F.: A new approach for dynamic fuzzy logic parameter tuning in ant colony optimization and its application in fuzzy control of a mobile robot. Appl. Soft Comput. **28**, 150–159 (2015)

10. Zadeh, L.A.: Fuzzy sets. Inf. Control **8**, 338–353 (1965)

11. Zadeh, L.A.: Fuzzy logic. Computer **21**(4), 83–93 (1988)

12. Zadeh, L.A.: On Fuzzy Algorithms, pp. 127–147 (1996)

13. Karnik, N.N., Mendel, J.M.: Operations on type-2 fuzzy sets. Fuzzy Sets Syst. **122**, 327–348 (2001)

14. Karnik, N.N., Mendel, J.M., Liang, Q.: Type-2 fuzzy logic systems. IEEE Trans. Fuzzy Syst. **7**(6), 643–658 (1999)

15. Liang, Q., Mendel, J.M.: Interval type-2 fuzzy logic systems: theory and design. IEEE Trans. Fuzzy Syst. **8**(5), 535–550 (2000)

16. Yang, X.-S.: Firefly algorithm. Nature-Inspired Metaheuristic Algorithms, pp. 79–90. Luniver Press, Bristol (2008)

17. Astudillo, L., Melin, P., Castillo, O.: Optimization of a fuzzy tracking controller for an autonomous mobile robot under perturbed torques by means of a chemical optimization paradigm. In: Castillo, O., Melin, P., Kacprzyk, J. (eds.) Recent Advances on Hybrid Intelligent Systems. SCI, vol. 547, pp. 3–20. Springer, Berlin (2013). https://doi.org/10.1007/978-3-642-33021-6_1

Differential Evolution Algorithm Using a Dynamic Crossover Parameter with Fuzzy Logic Applied for the CEC 2015 Benchmark Functions

Patricia Ochoa, Oscar Castillo[✉], and José Soria

Division of Graduate Studies, Tijuana Institute of Technology, Tijuana, Mexico
Ochoa.martha@hotmail.com, ocastillo@tectijuana.mx,
jsoria57@gmail.com

Abstract. The study of metaheuristics has become an important area for research, these metaheuristics contain parameters and the literature provides us with a range of values in which the algorithm can have good results. For this paper we propose to use the Differential Evolution algorithm combined with fuzzy logic to enable having dynamic crossover parameter, and to complement this work we include the diversity variable based on Euclidean distance, which will help us to know if the individuals of the population are separated or near in the search space in other words is the exploration and the exploitation in the search space, and for this article we work with two types of Simple Multimodal and Hybrid functions belonging to set of CEC 2015 benchmark functions.

Keywords: Diversity · Crossover · Dynamic parameter adaptation
Fuzzy logic

1 Introduction

The Differential Evolution (DE) algorithm was originally proposed by Storn and Price in 1994, and the algorithm has several variants, but in this work we use the original algorithm. We in particular use the Differential Evolution algorithm and make dynamic one of its parameters, in this case crossover with the help of fuzzy logic and then apply it to set of CEC 2015 benchmark functions.

The use of benchmark functions to experiment with metaheuristics is an element of a study, there is currently a set of benchmark functions that have a broader level of complexity, and in this paper we decided to work with certain functions of the set of CEC 2015 functions benchmark some outstanding works of this competence are mentioned below: a self-optimization approach for L-SHADE incorporated with eigenvector-based crossover and successful-parent-selecting framework on CEC 2015 benchmark Set [6] (rank 1), a differential evolution algorithm with success-based parameter adaptation for CEC 2015 learning-based optimization [3] (rank 2), testing MVMO on learning-based real-parameter single objective benchmark optimization problems [13] (rank 3) and a neuro-dynamic differential evolution algorithm and solving CEC 2015 competition problems [14] (rank 3).

© Springer International Publishing AG, part of Springer Nature 2018
G. A. Barreto and R. Coelho (Eds.): NAFIPS 2018, CCIS 831, pp. 580–591, 2018.
https://doi.org/10.1007/978-3-319-95312-0_51

On the other hand we can find works where fuzzy logic is combined with meta-heuristics converting parameters that are fixed during the execution of the algorithm and with the help of fuzzy logic can be made dynamic, and this combination is what we are working on this paper [20–22]. Here are some works related to this type of combination: a new fuzzy bee colony optimization with dynamic adaptation of parameters using interval type-2 fuzzy logic for tuning fuzzy controllers [2], imperialist competitive algorithm with dynamic parameter adaptation using fuzzy logic applied to the optimization of mathematical functions [4], comparison between ant colony and genetic algorithms for fuzzy system optimization [7], fuzzy dynamic adaptation of parameters in the water cycle algorithm. in nature-inspired design of hybrid intelligent systems [9], differential evolution using fuzzy logic and a comparative study with other metaheuristics [10], an adaptive fuzzy control based on harmony search and its application to optimization [11], a fuzzy hierarchical operator in the grey wolf optimizer algorithm [12], fuzzy system optimization using a hierarchical genetic algorithm applied to pattern recognition [15], optimization of benchmark mathematical functions using the firefly algorithm with dynamic parameters [16] and Evolutionary method combining particle swarm optimization and genetic algorithms using fuzzy logic for decision making [17].

We have as inspiration to our work some other works where the diversity variable was used to improve the performance of the respective algorithm, and to mention some we have: Optimal design of fuzzy classification systems using PSO with dynamic parameter adaptation through fuzzy logic [8], Statistical Analysis of Type-1 and Interval Type-2 Fuzzy Logic in dynamic parameter adaptation of the BCO [1] and Optimization of fuzzy controller design using a new bee colony algorithm with fuzzy dynamic parameter adaptation [5].

The paper is organized in the following form: Sect. 2 describes the Differential Evolution algorithm. Section 3 describes the methodology using the fuzzy logic approach. Section 4 presents the experimentation with the set of CEC 2015 Benchmark functions. Section 5 finally offers some Conclusions.

2 Differential Evolution Algorithm

The use of the Differential Evolution algorithm is very wide at present time, for this article we use the original version of algorithm which has the following mathematical structure [10]:

Population structure

$$P_{x,g} = \left(x_{i,g}\right), \; i = 0, 1, \; \dots, \; Np, g = 0, 1, \; \dots, \; g_{max} \tag{1}$$

$$x_{i,g} = \left(x_{j,i,g}\right), \; j = 0, 1, \; \dots, \; D - 1 \tag{2}$$

$$P_{v,g} = \left(v_{i,g}\right), \; i = 0, 1, \; \dots, \; Np - 1, \; g = 0, 1, \; \dots, \; g_{max} \tag{3}$$

$$v_{i,g} = (v_{j,I,g}), j = 0, 1, \ldots, D - 1 \tag{4}$$

$$P_{v,g} = (u_{i,g}), \ i = 0, 1, \ldots, Np - 1, \ g = 0, 1, \ldots, g_{max} \tag{5}$$

$$u_{i,g} = (u_{j,I,g}), j = 0, 1, \ldots, D - 1 \tag{6}$$

Initialization

$$x_{j,i,0} = rand_j(0, 1) \cdot (b_{j,U} - b_{j,L}) + b_{j,L} \tag{7}$$

Mutation

$$v_{i,g} = x_{r0,g} + F(x_{r1,g} - x_{r2,g}) \tag{8}$$

Crossover

$$u_{i,g} = (u_{j,i,g}) = \begin{cases} v_{j,i,g} & if\,(rand_j(0,1) \le Cr\ or\ j = j_{rand}) \\ x_{j,i,g} & otherwise \end{cases} \tag{9}$$

Selection

$$x_{i,g+1} = \begin{cases} u_{i,g} & if\,f(u_{i,g}) \le f(x_{i,g}), \\ x_{i,g} & otherwise. \end{cases} \tag{10}$$

3 Methodology

The methodology applied in this paper is the use of the Differential Evolution algorithm combined with fuzzy logic, and this combination will help make the algorithm have dynamic parameters, in this case is the crossover parameter.

Figure 1 represents the structure of the Differential Evolution algorithm to which we have added the fuzzy system which gives as the output the crossover parameter, which means that dynamically the crossover parameter will change during the execution of the algorithm.

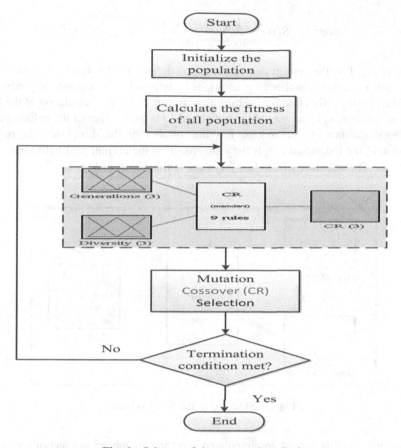

Fig. 1. Scheme of the proposed method

The fuzzy system is constructed with two inputs which are the generations and are calculated in Eq. 11 and the diversity is the second input, which is calculated with Eq. 12

$$\text{Generations} = \frac{\text{Current Generations}}{\text{Maximum of Generations}} \tag{11}$$

$$\text{Diversity } (S(t)) = \frac{1}{n_S} \sum_{i=1}^{n_S} \sqrt{\sum_{j=1}^{n_x} (x_{ij}(t) - \overline{x}_j(t))^2} \qquad (12)$$

Where Eq. 1, is the current generation and is defined by the number of generations elapsed and maximum number of generations is defined by the number of generations established for the DE to find the best solution. In Eq. 2, S is the population of the DE; t is the current time, n_s is the size of the individuals, i is the number of the individuals, n_x is the total number of dimensions, j is the number of the dimension, x_{ij} is the j dimension of the individual i, $\overline{x}_j$ is the j dimension of the current best individual of the individuals. The structure of the fuzzy system is illustrated in Fig. 2.

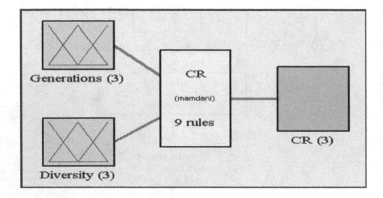

Fig. 2. Structure of the fuzzy system

The inputs and the output are granulated into three membership functions which are triangular. Figure 3 represents the membership functions and the parameters of each are described below:

Generations:

- M. F. 1 = Low [−0.5 0 0.5]
- M. F. 2 = Medium [0 0.5 1]
- M. F. 3 = High [0.5 1 1.5]

Diversity:

- M. F. 1 = Low [−0.5 0 0.5]
- M. F. 2 = Medium [0 0.5 1]
- M. F. 3 = High [0.5 1 1.5]

Mutation F parameter:

- M. F. 1 = Low [−0.5 0 0.5]
- M. F. 2 = Medium [0 0.5 1]
- M. F. 3 = High [0.5 1 1.5]

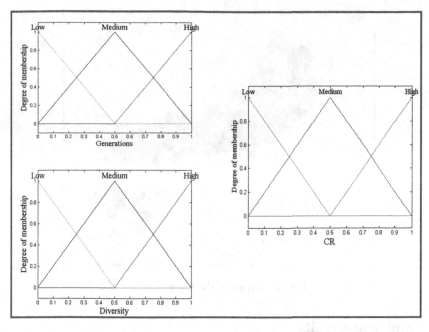

Fig. 3. Representation of the membership functions

Figure 4 outlines the rules of the fuzzy system and Fig. 5 shows the surface of the interval-type 2 fuzzy logic system.

1. If (Generations is Bajo) and (Diversity is Bajo) then (CR is Low)
2. If (Generations is Bajo) and (Diversity is Medio) then (CR is Medium)
3. If (Generations is Bajo) and (Diversity is Alto) then (CR is High)
4. If (Generations is Medio) and (Diversity is Bajo) then (CR is Low)
5. If (Generations is Medio) and (Diversity is Medio) then (CR is Medium)
6. If (Generations is Medio) and (Diversity is Alto) then (CR is High)
7. If (Generations is Alto) and (Diversity is Bajo) then (CR is Low)
8. If (Generations is Alto) and (Diversity is Medio) then (CR is Medium)
9. If (Generations is Alto) and (Diversity is Alto) then (CR is High)

Fig. 4. Rules for the fuzzy system

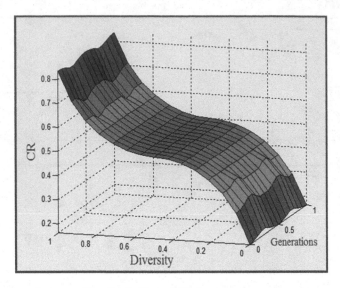

Fig. 5. Surface of the fuzzy system

4 Simulation Results

In this paper we used 6 functions of the set of CEC 2015 benchmark functions, which are described in Table 1. We performed experiments with the original algorithm and later using the proposed method, where a fuzzy system will help us to dynamically change the CR parameter. Previously we have already worked with the F parameter (mutation) to realize the dynamics and the work now for the CR parameter is a complementary part of our work and thus to be able to construct a more complete fuzzy system and to apply it to more complex systems.

Table 1. Summary of the CEC'15 learning-based benchmark suite

	No.	Functions	$Fi^* = Fi(x^*)$
Simple multimodal functions	3 (f1)	Shifted and rotated Ackley's function	300
	4 (f2)	Shifted and rotated Rastrigin's function	400
	5 (f3)	Shifted and rotated Schwefel's function	500
Hybrid functions	6 (f4)	Hybrid function 1 ($N = 3$)	600
	7 (f5)	Hybrid function 2 ($N = 4$)	700
	8 (f6)	Hybrid Function 3($N = 5$)	800

The parameters used for these experiments are shown in Table 2 where NP is the size of the population, D is the dimension of each individual, F is the mutation, CR is the crossover and GEN are the generations. For the case of the experiments using the original algorithm CR is assigned a value of 0.6 and subsequently for the following experiments CR is dynamic by using the fuzzy system. Experiments were only carried out with a number of dimensions of 10 and 30 with the same guidelines of the CEC 2015 competition.

Table 2. Parameters of the experiments

Parameters
NP = 250
D = 10 and 30
F = 0.6
CR = 0.5 and dynamic
GEN = 100000 and 300000

The results obtained with the original algorithm are shown in Table 3 for number of dimensions of 10, the table contains the best and the worst result, the mean and the standard deviation.

Table 3. Results for dimensions D = 10

Differential Evolution algorithm for D = 10			
	Best	Worst	Mean	Std
f1	3.21E+02	3.21E+02	3.21E+02	1.45E−01
f2	4.80E+02	5.64E+02	5.28E+02	2.01E+01
f3	2.42E+03	3.50E+03	2.97E+03	2.41E+02
f4	1.52E+05	1.37E+08	2.92E+07	2.78E+07
f5	7.14E+02	8.01E+02	7.53E+02	2.18E+01
f6	7.48E+04	1.39E+07	4.39E+06	4.13E+06

Table 4 shows the results for number of dimensions of 30 and contains the best and the worst result, the mean and the standard deviation.

Table 4. Results for dimensions D = 30

Differential Evolution algorithm for D = 30			
	Best	Worst	Mean	Std
f1	3.21E+02	3.21E+02	3.21E+02	7.28E−02
f2	9.31E+02	1.13E+03	1.05E+03	4.21E+01
f3	8.62E+03	1.09E+04	9.93E+03	4.54E+02
f4	7.26E+07	1.09E+09	4.44E+08	2.51E+08
f5	1.14E+03	4.05E+03	2.07E+03	5.89E+02
f6	1.33E+07	4.65E+08	1.43E+08	5.89E+02

Tables 5 and 6 represent the results obtained with a dynamic CR parameter for 10 and 30 dimensions respectively and contains the best and the worst result, the mean and the standard deviation.

Table 5. Results with CR dynamic for D = 10

D. E. with CR dynamic for D = 10				
	Best	Worst	Mean	Std
f1	5.15E−01	2.12E+01	1.62E+01	8.83E+00
f2	8.23E+00	1.55E+02	1.01E+02	5.36E+01
f3	1.76E+01	2.89E+03	2.16E+03	8.63E+02
f4	1.69E+04	1.12E+08	2.22E+07	2.68E+07
f5	4.65E−01	1.12E+02	4.45E+01	3.25E+01
f6	3.86E+04	1.55E+07	4.38E+06	3.73E+06

Table 6. Results with CR dynamic for D = 10

D. E. with CR dynamic for D = 30				
	Best	Worst	Mean	Std
f1	2.18E−01	2.15E+01	2.07E+01	3.86E+00
f2	1.00E+02	7.44E+02	6.43E+02	1.13E+02
f3	2.36E−01	1.02E+04	6.70E+03	4.48E+03
f4	5.04E+01	7.58E+08	2.41E+08	2.45E+08
f5	3.70E−01	1.75E+03	5.61E+02	6.36E+02
f6	9.63E+01	3.67E+08	9.88E+07	1.05E+08

Table 7 represents a comparison between the original method and the proposed method where CR is dynamic, and a comparison of the best results obtained for the number or dimensions of 10 and 30 is made.

Table 7. Comparison between DE and DE with dynamic CR

Comparison between DE and DE with dynamic CR				
D = 10			D = 30	
	Original	Proposed	Original	Proposed
f1	3.21E+02	5.15E−01	5.15E−01	2.18E−01
f2	4.80E+02	8.23E+00	9.41E+01	1.00E+02
f3	2.42E+03	1.76E+01	1.65E+03	2.36E−01
f4	1.52E+05	1.69E+04	5.94E+05	5.04E+01
f5	7.14E+02	4.65E−01	1.67E+01	3.70E−01
f6	7.48E+04	3.86E+04	3.86E+04	9.63E+01

Figures 6 and 7 show graphically the comparison of the best results between the original algorithm and the proposed method, and Fig. 6 shows the results for dimensions of 10 and Fig. 7 shows the results for dimensions of 30.

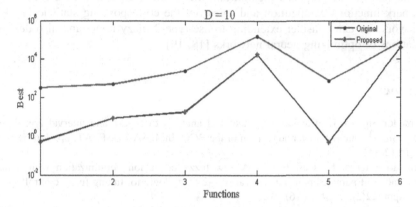

Fig. 6. Graphic to D = 10

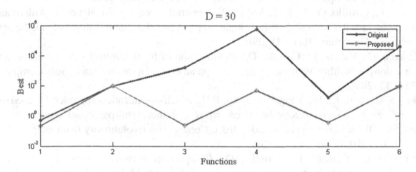

Fig. 7. Graphic to D = 30

We can notice graphically that the proposed method has better results than the original algorithm for both comparisons, the separation between the two methods is clearly shown.

5   Conclusions

The study carried out with the experimentation of at least 6 functions of CEC 2015 gives us an idea of the behavior of the CR parameter in the algorithm, with the obtained results we can notice that we are on the right track although we cannot yet affirm that the fuzzy system used is the optimal one since we need more experimentation and a statistical test to be able to demonstrate the improvement of the algorithm.

We can rescue the following points the use of the diversity variable helps the algorithm to obtain better results. We can also state that the use of the diversity variable helps the algorithm to obtain better results until now, we lack more experimentation, the goal is to use the whole set of the CEC 2015 benchmark functions and thus to make a comparison of our proposed method and the winning algorithm of the CEC 2015 benchmark functions competition and carry out the corresponding statistical test. As future work, we can consider extending to use type-2 fuzzy logic, and also deal with applications of optimizing neural networks [18, 19].

References

1. Amador-Angulo, L., Castillo, O.: Statistical analysis of type-1 and interval type-2 fuzzy logic in dynamic parameter adaptation of the BCO. In: IFSA-EUSFLAT, pp. 776–783, June 2015 (2015)
2. Amador-Angulo, L., Castillo, O.: A new fuzzy bee colony optimization with dynamic adaptation of parameters using interval type-2 fuzzy logic for tuning fuzzy controllers. Soft Comput. **22**(2), 1–24 (2016)
3. Awad, N., Ali, M.Z., Reynolds, R.G.: A differential evolution algorithm with success-based parameter adaptation for CEC 2015 learning-based optimization. In: 2015 IEEE Congress on Evolutionary Computation (CEC), pp. 1098–1105. IEEE, May 2015
4. Bernal, E., Castillo, O., Soria, J., Valdez, F.: Imperialist competitive algorithm with dynamic parameter adaptation using fuzzy logic applied to the optimization of mathematical functions. Algorithms **10**(1), 18 (2017)
5. Caraveo, C., Valdez, F., Castillo, O.: Optimization of fuzzy controller design using a new bee colony algorithm with fuzzy dynamic parameter adaptation. Appl. Soft Comput. **43**, 131–142 (2016)
6. Guo, S.M., Tsai, J.S.H., Yang, C.C., Hsu, P.H.: A self-optimization approach for L-SHADE incorporated with eigenvector-based crossover and successful-parent-selecting framework on CEC 2015 benchmark set. In: 2015 IEEE Congress on Evolutionary Computation (CEC), pp. 1003–1010. IEEE, May 2015
7. Martinez, C., Castillo, O., Montiel, O.: Comparison between ant colony and genetic algorithms for fuzzy system optimization. In: Castillo, O., Melin, P., Kacprzyk, J., Pedrycz, W. (Eds.) Soft computing for hybrid intelligent systems, pp. 71–86 (2008)
8. Melin, P., Olivas, F., Castillo, O., Valdez, F., Soria, J., Valdez, M.: Optimal design of fuzzy classification systems using PSO with dynamic parameter adaptation through fuzzy logic. Expert Syst. Appl. **40**(8), 3196–3206 (2013)
9. Méndez, E., Castillo, O., Soria, J., Sadollah, A.: Fuzzy dynamic adaptation of parameters in the water cycle algorithm. In: Melin, P., Castillo, O., Kacprzyk, J. (eds.) Nature-Inspired Design of Hybrid Intelligent Systems. SCI, vol. 667, pp. 297–311. Springer, Cham (2017). https://doi.org/10.1007/978-3-319-47054-2_20
10. Ochoa, P., Castillo, O., Soria, J.: Differential evolution using fuzzy logic and a comparative study with other metaheuristics. In: Melin, P., Castillo, O., Kacprzyk, J. (eds.) Nature-Inspired Design of Hybrid Intelligent Systems. SCI, vol. 667, pp. 257–268. Springer, Cham (2017). https://doi.org/10.1007/978-3-319-47054-2_17
11. Peraza, C., Valdez, F., Castillo, O.: An adaptive fuzzy control based on harmony search and its application to optimization. In: Melin, P., Castillo, O., Kacprzyk, J. (eds.) Nature-Inspired Design of Hybrid Intelligent Systems. SCI, vol. 667, pp. 269–283. Springer, Cham (2017). https://doi.org/10.1007/978-3-319-47054-2_18

12. Rodríguez, L., Castillo, O., Soria, J., Melin, P., Valdez, F., Gonzalez, C.I., Soto, J.: A fuzzy hierarchical operator in the grey wolf optimizer algorithm. Appl. Soft Comput. **57**, 315–328 (2017)
13. Rueda, J.L., Erlich, I.: Testing MVMO on learning-based real-parameter single objective benchmark optimization problems. In: 2015 IEEE Congress on Evolutionary Computation (CEC), pp. 1025–1032. IEEE, May 2015
14. Sallam, K.M., Sarker, R.A., Essam, D.L., Elsayed, S.M.: Neurodynamic differential evolution algorithm and solving CEC 2015 competition problems. In: 2015 IEEE Congress on Evolutionary Computation (CEC), pp. 1033–1040. IEEE, May 2015
15. Sánchez, D., Melin, P., Castillo, O.: Fuzzy system optimization using a hierarchical genetic algorithm applied to pattern recognition. In: Filev, D., et al. (eds.) Intelligent Systems 2014. AISC, vol. 323, pp. 713–720. Springer, Cham (2015). https://doi.org/10.1007/978-3-319-11310-4_62
16. Solano-Aragón, C., Castillo, O.: Optimization of benchmark mathematical functions using the firefly algorithm with dynamic parameters. In: Castillo, O., Melin, P. (eds.) Fuzzy Logic Augmentation of Nature-Inspired Optimization Metaheuristics. SCI, vol. 574, pp. 81–89. Springer, Cham (2015). https://doi.org/10.1007/978-3-319-10960-2_5
17. Valdez, F., Melin, P., Castillo, O.: Evolutionary method combining particle swarm optimization and genetic algorithms using fuzzy logic for decision making. In: IEEE International Conference on Fuzzy Systems, FUZZ-IEEE 2009, August 2009, pp. 2114–2119. IEEE (2009)
18. Valdez, F., Melin, P., Castillo, O.: An improved evolutionary method with fuzzy logic for combining Particle Swarm Optimization and Genetic Algorithms. Appl. Soft Comput. J. **11**(2), 2625–2632 (2011)
19. Valdez, F., Melin, P., Castillo, O.: Modular Neural Networks architecture optimization with a new nature inspired method using a fuzzy combination of Particle Swarm Optimization and Genetic Algorithms. Inf. Sci. J. **270**, 143–153 (2014)
20. Valdez, F., Melin, P., Castillo, O.: A survey on nature-inspired optimization algorithms with fuzzy logic for dynamic parameter adaptation. Expert Syst. Appl. J. **41**(14), 6459–6466 (2014)
21. Valdez, F., Melin, P., Castillo, O.: Toolbox for bio-inspired optimization of mathematical functions. Comp. Applic. Eng. Educ. **22**(1), 11–22 (2014)
22. Valdez, F., Melin, P., Castillo, O.: Comparative study of the use of fuzzy logic in improving particle swarm optimization variants for mathematical functions using co-evolution. Appl. Soft Comput. J. **52**, 1070–1083 (2017)

Corporate Control with a Fuzzy Network

A Knowledge Engineering Application

Gustavo Pérez Hoyos[(✉)]

Universidad Nacional de Colombia, Bogotá, Colombia
goosper@gmail.com

Abstract. This paper presents a Knowledge engineering application whereby a Fuzzy Network (FN) is used to build a computing model to reproduce corporate dynamics and to implement a Model Reference Adaptive Control (MRAC) strategy [2] for Corporate Control. This model is used as a *What If?* Environment to explore future consequences of actions planned within a strategic scenario context in terms of KPIs displayed in a Balanced ScoreCard (BSC) control board. Corporation's Strategy Map is required to plan the Knowledge Identification and Capture Activity (KICA) required to obtain the knowledge to be represented in the FN's Nodes Rule Bases. KICA produces linguistic variables as well as the qualitative relationships amongst them. A FN appears as a natural solution to model the knowledge distributed within the members participating in all analysis and decision making tasks along the organization. As an example an application done for a Utility Corporation is included.

1 Introduction

Since BSC officially appeared [1] it has become an important corporate control tool. However the feedback it provides through the KPIs takes a good time after an action has been taken. This occurs because the time constants involved in a corporate dynamics are rather long ranging from weeks to months depending of the strategic deep of decisions and the corresponding actions. This paper presents an enhancement of the BSC control strategy by means of a knowledge based computing model representing the corporate dynamics and implementing a Model Reference Adaptive Control (MRAC) strategy [3]. Since the Strategy Map [4] displays the inner causality in an organization's dynamic it is used to conduct the KICA required to identify the involved linguistic variables and the qualitative relationships amongst them. The model is constructed using a FN which stores the knowledge gathered through a KICA. This model is used to provide the **What If?** Environment to test different strategic scenarios and to identify the best actions to be taken on order to obtain desired KPI's values in a given time horizon. FN have already been used and reported [6, 7] as a tool to build Expert Systems using distributed knowledge.

© Springer International Publishing AG, part of Springer Nature 2018
G. A. Barreto and R. Coelho (Eds.): NAFIPS 2018, CCIS 831, pp. 592–599, 2018.
https://doi.org/10.1007/978-3-319-95312-0_52

2   BSC Control Structure

The current BSC control loop is displayed in Fig. 1. No matter how good the measurement system is the values entered for the KPIs in the BSC Board will reflect the consequences of the taken actions only after a good amount of time since the time constants involved in the corporate dynamics are some times weeks or even months long. The *What If?* Environment introduced here allows to test groups of actions, in fact, whole strategies, to examine the future KPI's values and so to determine the best strategy in terms of future KPI's values.

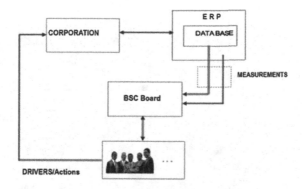

Fig. 1.   Current BSC control loop structure

3   Model Reference Adaptive Control Structure

The *What If?* Environment is obtained using the mentioned MRAC control strategy and this is showed in Fig. 2. The Corporate Model allows to test actions showing future KPI's values in the BSC Model Board.

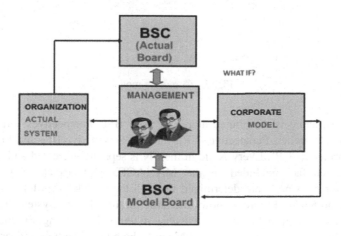

Fig. 2.   MRAC structure with *What If?* environment

Once the structure depicted in Fig. 2 is constructed all that is needed is a good Simulation Barnacle to start testing strategies and actions to be input to the model in order to observe over the BSC Model Board the resultant KPIs in the specified time horizon.

4   Corporate Model

Figure 3 shows a typical FN structure for the corporate model.

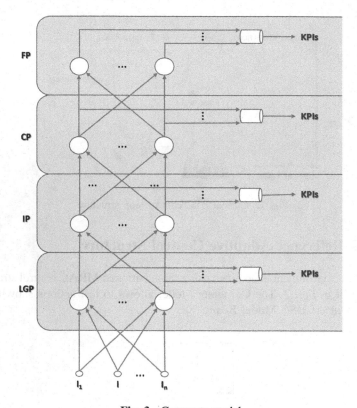

Fig. 3. Corporate model

The FN structure is distributed along the four BSC Perspectives: Financial Perspective (FP), Customer Perspective (CP), Internal Perspective (IP) and Learning and Grow Perspective (LGP). Every Node in the FN is separately created and tuned using the MRAC interface included in the MANAGEMENT Block (Fig. 2). Design parameters for each node are determined through the tuning procedure using criteria obtained in the KICA. Figure 4 shows the structure of a Fuzzy System (FS) in every node with the design parameters for every module. Depending on the particular dynamic associated to a company the FN could be either Feedforward or Recurrent.

The Rule Base for every Node will contain the fuzzy rules obtained after analyzing the results obtained with the KICA. The nodes producing the KIP's values are explicitly shown. All the fuzzy rules for every node contain the time as a linguistic variable in the Antecedent. Rule k is then explicitly written as (1). Although the Inputs go into nodes in the Learning and Growing Perspective in Fig. 3, inputs can actually go into any node in any perspective.

Rk: IF X1 is LX_1 AND ... AND Xm is LXm AND

Time is LT THEN Y1 is LY1 AND ... AND Yn is LYn   (1)

Where:
X1 ... Xm are the inputs to the node and Y1 ... Yn are the outputs,

$$LXi \ \varepsilon \ \{L, M, H\}, \ LYi \ \varepsilon \ \{L, M, H\}$$

$$LT \ \varepsilon \ \{VS, S, M, L, VL\}$$

With: L = Low, H = High, VS = Very Short, S = Short, M = Medium, L = Long, VL = Very Long.

Although only one layer is shown for the FN in each perspective, it is the particular dynamics for each corporation's value-creating processes which determines the FN's nodes structure for every perspective. The particular FN structure constructed for any particular organization will reflect the particular dynamics and the inner causality within the value-creating processes in that corporation.

5   The *What If?* Environment

The User Interface also contains the simulation environment so that managers and/or planners can perform the simulation tasks required to test any strategy. This environment keeps the record of all simulations to facilitate the supervision tasks required to identify the best strategies.

When a corporation has a comprehensive and well maintained Data Base with a few years of data, rules can be identified using a mining procedure with the help of Adaptive (Trainable) Fuzzy Systems [3]. This capability is built in the UNFUZZY tool [5] used for the developments reported in [6, 7].

Implementing this MRAC strategy with *What If?* Environment requires the committed participation of all members in the organization. KEA in particular requires the open and patient collaboration since many times it is necessary to go over some particular subjects in order to clearly identify the linguistic variables as well as the nature of relationships.

MRAC with *What If?* Environment was implemented in a local utility. It is currently used in the planning tasks and it has been taken as a pilot experience to scout new possibilities. The task force organized for this accomplishment is now working on model refinement to include risks, using the experience gathered in a recent work [7].

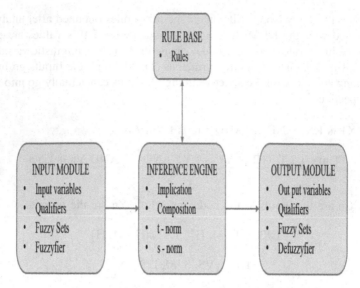

Fig. 4. Node's structure with design parameters

This means identifying the associated risks to every strategic goal and to actually acquire the heuristics associated to its assessment. This approach intends to reach the point where we can modify the fuzzy rules for the KPI nodes in the corporate model to include risk values. The BSC Boards, Corporate Board as well as Model Board, will also be modified to include the column corresponding to risk values.

6   Utility Company Application

The example presented in this paper is an application developed for a Utility Corporation. Figure 5 shows the strategy map for this utility, where the inputs are also indicated. As a result of KICA the input and output linguistic variables for every process in the map were identified as well as the proper fuzzy relationship relating each input-output pair in order to elaborate the fuzzy rules for every node. Identifying the proper fuzzy rules includes detecting the adequate set of Qualifying Linguistic Terms (QLT) as well as the Trend relationships, i.e. whether a variable appearing in the consequent of a particular fuzzy rule is growing or decaying when a variable in the antecedent is growing.

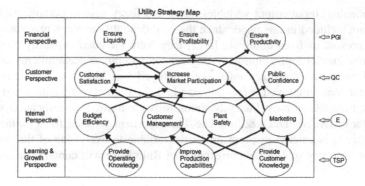

Fig. 5. Strategy map

(TSP) **Training and Selection Processes**

(E) **Empowerment**

(QC) **Quality Control**

(PGI) **Positioning & Growth Indexes**

7 Fuzzy Network Architecture

Figure 6 shows the final FN designed architecture. Five layers were required in order to implement the network using UNFUZZY [5], a software tool to implement and simulate FS and FN, developed at Universidad Nacional de Colombia.

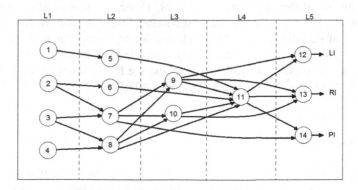

Fig. 6. FN architecture

Functionality, input/output variables, rules and fuzzy system parameters for every node are not included in this paper since the confidentiality agreement signed with the company prevent us from doing so. In [7] they were included explicitly because that was not an industrial application with a signed contract but an academic research work.

For all the nodes the fuzzy inference employed was Mandani (Minimum), membership functions employed were the standard L, Lambda (Triangle) and Gamma Functions, available in most Fuzzy Systems software tools. Defuzzification was done with Center of Gravity DeDuzzifier. The Time Horizon selected for a particular simulation exercise is entered as input linguistic variable to every one of the nodes. The antecedent for every rule in the corresponding Rule Base will contain the component *time is $\mathcal{L}T$*, where

$$\mathcal{L}T \in \{\text{Very Short, Short, Medium, Long, Very Long}\}$$

As it was stated before. Besides the inputs to the nodes in the following layer, The outputs of the nodes in every layer include the Key Performance Indexes (KPIs) that go to BSC simulation Board. Then Layer 1 outputs include the KPIs related to the Training and Selection Processes, Layer 2 outputs include the KPIs related to corporate Empowerment, Layer 3 outputs include KPIs related to Quality Control and Layer 4 outputs include KPIs related to Positioning and Growth.

8 Conclusions and Final Recommendations

- The Corporate Model constructed after the KICA using a Fuzzy Network is a good example of a working asset obtained through capitalizing organization knowledge.
- A Fuzzy Network is a versatile way to model corporate knowledge since this is distributed all over the people working in analysis and decision making activities within the organization's value-creating processes.
- A What If? Environment provides the BSC corporate control strategy a powerful tool to explore different strategic scenarios. Using different action scenarios and different time horizons to detect the best possible results in terms of KPIs requires a great deal of expertise achieved through consistently using this tool.
- Identifying the risks associated to every strategic goal as well as the associated heuristic to its assessment would allow to add the Risk column to the BSC Board, adding Risk Management to the MRAC STRATEGY.

References

1. Kaplan, R.S., Norton, D.P.: The Strategy Focused Organization, How Balanced Scorecard Companies Thrive in the New Business Environment. Harvard Business School Publishing Corporation, Brighton (2001)
2. Shastry, S., Bodson, M.: Adaptive Control: Stability, Convergence and Robustness. Dover Publications, Mineola (2011)

3. Wang, L.-X.: Adaptive Fuzzy Systems and Control. PTR Prentice Hall, Englewood Cliffs (1994)
4. Kaplan, R.S., Norton, D.P.: Strategy Maps, Converting Intangible Assets into Tangible Outcomes. Harvard Business School Publishing Corporation, Boston (2004)
5. Duarte, O., Pérez, G.: UNFUZZY: fuzzy logic system analysis, design simulation and implementation software. In: Proceedings of the 1999 EUSFLAT-ESTYLF Joint Conference, Mallorca (1999)
6. Perez, G.: A fuzzy logic based expert system for short term energy negotiations. In: Proceedings of the 18th Conference of the North American Fuzzy Information Processing Society, NAFIPS. IEEE (1999)
7. Perez, G.: Pipeline risk assessment using a fuzzy systems network. In: Proceedings of the IFSA (World Congress) - NAFIPS (Annual Meeting) 2013, IFSA-NAFIPS. IEEE (2013)
8. Perez, V.: FuzzyNet, A Software Tool to Implement Fuzzy Networks (to be published)

Author Index

Printed in the United States
By Bookmasters